DRUG DELIVERY:
FUNDAMENTALS & APPLICATIONS

SECOND EDITION

DRUG DELIVERY:
FUNDAMENTALS &
APPLICATIONS
SECOND EDITION

Edited by

Anya M. Hillery
Saint Louis University – Madrid Campus, Madrid, Spain

Kinam Park
Purdue University, West Lafayette, IN, USA

CRC Press
Taylor & Francis Group
Boca Raton London New York

CRC Press is an imprint of the
Taylor & Francis Group, an **informa** business

CRC Press
Taylor & Francis Group
6000 Broken Sound Parkway NW, Suite 300
Boca Raton, FL 33487-2742

© 2017 by Taylor & Francis Group, LLC
CRC Press is an imprint of Taylor & Francis Group, an Informa business

No claim to original U.S. Government works

Printed and bound in India by Replika Press Pvt. Ltd.

Printed on acid-free paper
Version Date: 20160603

International Standard Book Number-13: 978-1-4822-1771-1 (Paperback)

Visit the Taylor & Francis Web site at
http://www.taylorandfrancis.com

and the CRC Press Web site at
http://www.crcpress.com

Contents

SECTION I   Fundamental Issues

SECTION II   Parenteral Routes for Drug Delivery and Targeting

SECTION III   Nonparenteral Routes for Drug Delivery and Targeting

SECTION IV   Emerging Technologies

SECTION V   Toward Commercialization

Preface

Controlled drug delivery systems have evolved over the past six decades, from the sustained-release Spansule® technology of the 1950s to the highly sophisticated and targeted drug delivery systems of today. Numerous drug delivery systems (DDS) have been successfully developed for clinical applications over the years, and the demand for innovative technologies continues to grow, driving a variety of new developments in the field. This book describes the fundamental concepts and underlying scientific principles of drug delivery, current applications of drug delivery technologies, and potential future developments in the field. It is intended to serve both as a core textbook and as a valuable reference source for students, researchers, practitioners, and scientists in disciplines including the pharmaceutical and formulation sciences, chemical and biomedical engineering, materials science, medicine and oncology, the health sciences, and natural sciences.

In common with the first edition,* our aim is to provide a single, comprehensive, easy-to-read reference book that covers all aspects of controlled drug delivery. To this end, considerable attention has been paid to the overall layout and contents of the text. Chapter 1 opens with a historical introduction to the field of controlled drug delivery to provide relevant background details for the subsequent chapters.

Section I: Fundamental Issues serves as a comprehensive introduction to the fundamental concepts that underpin drug delivery and targeting. Chapter 2 describes the principles of controlled release, including the various mechanisms, types, and mathematical models of controlled release. Chapter 3 describes various technologies to enhance the water solubility of poorly soluble drugs, which has important implications for lead development in the drug discovery process, as well as for the formulation, bioavailability, and therapeutic efficacy of poorly soluble drugs. An important objective of this book is to provide a thorough understanding of the multitude of highly complex biological barriers to successful drug delivery and targeting that pertain in vivo. For this reason, an entire chapter (Chapter 4) is dedicated to providing a comprehensive overview of the characteristics and properties of the various types of epithelial interfaces in the body of relevance for drug delivery strategies; the factors that influence drug transport across these interfaces are also described.

Section II: Parenteral Routes for Drug Delivery and Targeting opens with a chapter on nanotechnology, the engineering and manufacturing of materials at the molecular scale, which offers the potential to revolutionize the drug delivery field. Chapter 5 focuses on the application of nanotechnology to drug delivery and targeting, and highlights several areas of opportunity. Various limitations of current drug delivery nanotechnologies are also described, in order to help guide future research; in particular, the anatomical, physiological, and pathological obstacles to the targeting concept are discussed. Chapter 6 describes a variety of long-acting injectables and implant platforms that are currently commercially available or at an advanced stage of development; this chapter also reinforces the general concepts and principles of controlled drug release introduced in Chapters 1 and 2.

Section III: Nonparenteral Routes for Drug Delivery and Targeting describes the major epithelial routes of drug delivery currently under investigation. In keeping with the objective to emphasize an understanding of the biological obstacles for successful drug delivery, each chapter of this section begins with a detailed consideration of the relevant anatomical and physiological barriers pertaining specifically to the route in question, as well as the implications therein to successful drug delivery and targeting via this route. The first epithelial route described is the oral route (Chapter 7), the most common and convenient of the existing administration methods for introducing drugs to the bloodstream. The oral route is discussed with respect to the various mechanisms of controlled release,

---

* Hillery, A.M., A.W. Lloyd, and J. Swarbrick. 2001. *Drug Delivery and Targeting: For Pharmacists and Pharmaceutical Scientists*. Boca Raton, FL: CRC Press.

regional targeting, strategies for improving bioavailability, and the use of vaccines. These same themes recur through the following chapters on the various other epithelial routes, many of which also serve as alternative portals of drug entry to the systemic circulation. The chapters in Section III deliberately follow a common format, in order to ease understanding and facilitate learning, and also to highlight the many similarities that exist between the various epithelial routes, as well as the unique attributes associated with each specific route.

Section IV: Emerging Technologies covers some of the new and exciting possibilities that are emerging as future directions in the field. Chapter 14 describes hydrogels and their applications to drug delivery, including as microfluidic chips, biosensors, and stimuli-sensitive DDS. A variety of sophisticated delivery approaches for overcoming the blood–brain barrier (BBB) are described in Chapter 15, as a means of delivering therapeutics to the central nervous system (CNS). Chapter 16 describes the most promising delivery vehicles emerging for gene therapy, including recent advances such as gene delivery systems that can target intracellular organelles. Chapter 17 provides a comprehensive account of vaccines, as well as the current and emerging vaccine delivery systems used for various routes of vaccination. The newly emerged field of theranostics, which holds great promise for personalized therapy, is described in Chapter 18, while Chapter 19 describes the leverage of techniques from the microelectronics industry to precisely fabricate DDS in the nanometer range and the application of such nanofabricated systems to drug delivery.

Section V: Toward Commercialization is an entirely new section for this edition, which reflects the onward success and progress of drug delivery in the 15 years since the publication of the first edition, as technology moves "from bench to bedside." Chapter 20 describes the more robust and successful methods currently used in drug discovery, design, and development, with particular emphasis on rationally integrating the drug discovery process with the requirements to optimize successful drug delivery, in order to optimize clinical success. The extensive regulatory development pathway for parenteral nanotechnologies is described in Chapter 21—for those working in the preclinical sector, it offers a comprehensive account of the regulatory hurdles that lie ahead. Chapter 22 provides a thorough analysis of the global drug delivery market and market forces, including the latest trends and developments. Chapter 23 presents an engaging account of the clinical translation of a liposomal product (ThermoDox®, a thermal-sensitive liposome for cancer therapy). It provides an illuminating insight, from the inventor's perspective, into the process—and difficulties—of guiding a DDS through initial funding, development, and preclinical and clinical trials.

In the conclusions of Chapter 24, we discuss some of the future directions for drug delivery and targeting, raise some of the challenges that need to be addressed, and propose some possible solutions and ways forward for research.

In keeping with our aim to produce an accessible, easily comprehensible book, we have endeavored to ensure that the text is clear, concise, and direct. Careful editing has ensured that the final text displays an overall continuity and integrated style. The book is characterized by the ample usage of carefully chosen figures, illustrations, and graphics. Many of the figures have been specially commissioned and are unique and original in the field. Collectively, the artwork greatly assists the clarity and visual appeal of the book, aids understanding, and facilitates our pedagogic, explanatory approach.

We welcome readers' suggestions, comments, and corrections on the text. Finally, we hope that you enjoy reading this book as much as we enjoyed editing it!

Anya M. Hillery
Kinam Park

Acknowledgments

Preparing the second edition of this book has been an exciting, challenging, and very enjoyable project. We are deeply grateful to many people for its successful completion. First and foremost, we thank our chapter contributors, listed on page xv et seq., for their time, effort, expertise, and excellent submissions.

We also thank R. Tyler Gabbard (Purdue University) for his excellent editorial assistance, Carol Cserneczky (Saint Louis University Madrid Campus, IT department) for his highly professional computer expertise, and Fernando Béjar (Saint Louis University Madrid Campus, Marketing and Communication) for his help in preparing some of the illustrations.

We acknowledge the support of our publishers and thank Jill Jurgensen and Barbara Norwitz at Taylor & Francis Group. We express our gratitude to Vijay Bose at SPi Global for the copy editing of this text. Also *mil gracias* to the design team at ICON Brandworks Madrid, Spain, for our stunning front cover.

We sincerely thank our co-editors for the first edition, Professor Andrew Lloyd (University of Brighton, UK) and Professor James Swarbrick (PhamaceuTech Inc.), for their interest and support for this edition. In particular, we have benefited greatly from Jim's encouragement, wise counsel, and good humor.

Finally, AMH thanks Mike, Danny, and Robbie Pinkney, for their brilliant support, encouragement, and patience during the preparation of this text.

Editors

Dr. Anya M. Hillery received her BSc in pharmacy from the School of Pharmacy, Trinity College Dublin, Ireland, in 1990 (awarded the Trinity College Gold Medal of Outstanding Achievement). She was awarded a scholarship by Syntex Research to carry out her PhD in pharmaceutics under the supervision of Professor Sandy Florence at the School of Pharmacy, Brunswick Square, University College London, United Kingdom. She continued with postdoctoral research studies at the Square upon being awarded a Maplethorpe Research Fellowship. Anya took up a lectureship position in Pharmaceutical Sciences at the Department of Pharmacy, University of Brighton, United Kingdom, in 1995, and became a senior lecturer in 1997. This was followed by a move in 1999 to Saint Louis University Madrid Campus, Spain, starting as a lecturer in health sciences, subsequently becoming Director of the Department of Science and Engineering, and then Vice Dean of the university in 2001. After some years at home to raise her young family, she has recently returned to full-time academia at SLU Madrid.

Professor Kinam Park received his PhD in pharmaceutics from the University of Wisconsin–Madison, Madison, Wisconsin in 1983. After his postdoctoral training in the Department of Chemical Engineering at the same university, he joined the faculty of the Department of Industrial and Physical Pharmacy, College of Pharmacy, Purdue University, West Lafayette in 1986. He was promoted to full professor of pharmaceutics in 1994. Since 1998, he has held a joint appointment in the Department of Biomedical Engineering and became Showalter Distinguished Professor of Biomedical Engineering in 2006. His research focuses on oral delivery, drug–device combination products, and long-term microparticle formulations. He is the founder of Akina, Inc. specializing in polymers for drug delivery. He is currently the editor in chief of the *Journal of Controlled Release*.

Contributors

Carmen Alvarez-Lorenzo
Department of Pharmacy and Pharmaceutical
  Technology
University of Santiago de Compostela
Santiago de Compostela, Spain

Mariam Badawi
Strathclyde Institute of Pharmacy &
  Biomedical Sciences
University of Strathclyde
Glasgow, Scotland, United Kingdom

Shyamanga Borooah
MRC Centre for Regenerative Medicine
University of Edinburgh
Edinburgh, Scotland, United Kingdom

J. Phillip Bowen
Center for Drug Design
Department of Pharmaceutical Sciences
College of Pharmacy
Mercer University
Atlanta, Georgia, USA

Terry L. Bowersock
Global Biological Research and Development
Zoetis, Inc.
Kalamazoo, Michigan, USA

David J. Brayden
School of Veterinary Medicine
University College Dublin
Dublin, Ireland

Marc B. Brown
MedPharm, Ltd.
Guildford, United Kingdom

Donna Cabral-Lilly
Taaneh, Inc.
Princeton, New Jersey, USA

Justin T. Clark
Department of Biomedical Engineering
Northwestern University
Evanston, Illinois, USA

Simon R. Corrie
Australian Institute for Bioengineering
  and Nanotechnology
The University of Queensland
St. Lucia, Queensland, Australia

Daan J.A. Crommelin
Department of Pharmaceutical Sciences
Utrecht University
Utrecht, the Netherlands

Tejal Desai
Department of Bioengineering and Therapeutic
  Sciences
University of California, San Francisco
San Francisco, California, USA

Baljean Dhillon
Centre for Clinical Brain Sciences
University of Edinburgh
and
Princess Alexandra Eye Pavilion
Edinburgh, Scotland, United Kingdom

Per Gisle Djupesland
OptiNose AS
Oslo, Norway

Alexander T. Florence
UCL School of Pharmacy
University College London
London, United Kingdom

Cade Fox
Department of Bioengineering and Therapeutic
  Sciences
University of California, San Francisco
San Francisco, California, USA

Kirsten Graeser
Roche Research and Early Development, TMo
Roche Innovation Center Basel
F. Hoffmann La Roche, Ltd.
Basel, Switzerland

Osman F. Güner
Center for Drug Design
Department of Pharmaceutical Sciences
College of Pharmacy
Mercer University
Atlanta, Georgia, USA

Don Hayes, Jr.
Departments of Pediatrics and Internal Medicine
The Ohio State University
Columbus, Ohio, USA

Anya M. Hillery
Department of Science and Engineering
Saint Louis University—Madrid Campus
Madrid, Spain

Allan S. Hoffman
Department of Bioengineering
University of Washington
Seattle, Washington, USA

Kohsaku Kawakami
Smart Biomaterials Group, Biomaterials Unit
International Center for Materials
    Nanoarchitectonics
National Institute for Materials Science
Tsukuba, Japan

Mark A.F. Kendall
Australian Institute for Bioengineering
    and Nanotechnology
The University of Queensland
St. Lucia, Queensland, Australia

Sung Wan Kim
Department of Pharmaceutics and
    Pharmaceutical Chemistry
University of Utah
Salt Lake City, Utah, USA

and

Department of Bioengineering
Hanyang University
Seoul, South Korea

Patrick F. Kiser
Department of Biomedical Engineering
and
Department of Obstetrics and Gynecology
Northwestern University
Evanston, Illinois, USA

Floriane Laurent
Department of Pharmacy and Pharmacology
University of Bath
Bath, United Kingdom

Kwang Suk Lim
Center for Controlled Chemical Delivery
Department of Pharmaceutics and
    Pharmaceutical Chemistry
University of Utah
Salt Lake City, Utah, USA

Zheng-Rong Lu
Department of Biomedical Engineering
Case Western Reserve University
Cleveland, Ohio, USA

Suman M. Mahan
Veterinary Medicine Research and
    Development
Zoetis, Inc.
Kalamazoo, Michigan, USA

Anthony S. Malamas
National Cancer Institute
National Institutes of Health
Bethesda, Maryland, USA

Heidi M. Mansour
College of Pharmacy
The University of Arizona
Tucson, Arizona, USA

James Matriano
DURECT Corporation
Cupertino, California, USA

Lawrence D. Mayer
Celator Pharmaceuticals, Inc.
Ewing, New Jersey, USA

and

Celator Pharmaceuticals Corporation
Vancouver, British Columbia, Canada

Roly Megaw
MRC Centre for Regenerative Medicine
University of Edinburgh
Edinburgh, Scotland, United Kingdom

Randall Mrsny
Department of Pharmacy and Pharmacology
University of Bath
Bath, United Kingdom

Priya Muralidharan
College of Pharmacy
The University of Arizona
Tucson, Arizona, USA

Paul B. Myrdal
College of Pharmacy
The University of Arizona
Tucson, Arizona, USA

David Needham
Department of Mechanical Engineering and
   Material Science
Duke University
Durham, North Carolina, USA

and

DNRF Niels Bohr Visiting Professor
Center for Single Particle Science and
   Engineering
University of Southern Denmark
Odense, Denmark

Kinam Park
Department of Pharmaceutics
and
Department of Biomedical Engineering
Purdue Universty
West Lafayette, Indiana, USA

Viralkumar F. Patel
Department of Pharmacy, Pharmacology, and
   Postgraduate Medicine
University of Hertfordshire
Hatfield, United Kingdom

Yvonne Perrie
Strathclyde Institute of Pharmacy and
   Biomedical Sciences
University of Strathclyde
Glasgow, Scotland, United Kingdom

Thomas Rades
Department of Pharmacy
Faculty of Health and Medical Sciences
University of Copenhagen
Copenhagen, Denmark

Louise Rosenmayr-Templeton
Tower Pharma Consulting
Vienna, Austria

Erica Schlesinger
Department of Bioengineering and Therapeutic
   Sciences
University of California, San Francisco
San Francisco, California, USA

Ronald A. Siegel
Department of Pharmaceutics
and
Department of Biomedical Engineering
University of Minnesota
Minneapolis, Minnesota, USA

Jonathan T. Su
Department of Biomedical Engineering
Northwestern University
Evanston, Illinois, USA

Clive G. Wilson
Strathclyde Institute of Pharmacy &
   Biomedical Sciences
University of Strathclyde
Glasgow, Scotland, United Kingdom

Jeremy C. Wright
DURECT Corporation
Cupertino, California, USA

Usir Younis
College of Pharmacy
The University of Arizona
Tucson, Arizona, USA

Haizhen A. Zhong
Department of Chemistry
University of Nebraska Omaha
Omaha, Nebraska, USA

1 Historical Introduction to the Field of Controlled Drug Delivery

Anya M. Hillery and Allan S. Hoffman

CONTENTS

1.1   INTRODUCTION

This chapter presents a historical overview of the field of controlled drug delivery, describing how it grew in the past 60 years from a very small field, to the immense size and importance it represents today for human and animal health. This chapter also highlights many of the people who were involved in the conception and design of the key controlled drug delivery systems (DDS), as well as details about the compositions of the materials used. We begin by considering some of the earliest drug delivery formulations, followed by a discussion of some of the key technologies in the history of controlled drug delivery. It should be noted at the outset that in the early days of controlled drug delivery, the term "controlled release" tended to refer specifically to zero-order drug release obtained via a rate-controlling membrane (RCM), whereas the terms "sustained release" and "extended release" referred to the prolonged drug release obtainable using other DDS such as the oral Spansules® and bioerodible implants. With the passage of time, however, the delineation of these definitions has blurred. Currently, all these terms are used interchangeably, and the term "sustained release" is widely used.

1.2   EARLY DRUG DELIVERY SYSTEMS

Conventional oral delivery systems release the drug immediately in the lumen of the gastrointestinal (GI) tract. The drug then dissolves in the GI fluids and permeates the gut wall to be absorbed into the systemic circulation via the underlying blood capillaries. There is no control over the release of the drug.

An early example of modifying drug release via the oral route was the use of enteric coatings. Tablets can be coated with the so-called enteric polymers, which are nonswelling and hydrophobic at the acidity of the stomach, but become ionized, and then dissolve and release the drug, once they enter the slightly alkaline pH of the intestinal region of the GI tract. Thus, drug release is delayed from the stomach to the small intestine. These "delayed release" coatings are useful to either (1) protect the stomach from drugs that can cause gastric irritation (e.g., aspirin) or (2) protect drugs that can be destroyed in the acidic gastric environment (e.g., some penicillins). Early coatings, introduced in the late 1800s, such as keratin and shellac suffered from storage instability and also, crucially, the pH at which they dissolved was too high for adequate dissolution in the small intestine, so that they were not very effective.

In 1951, cellulose acetate phthalate was introduced as an enteric-coating material (Malm et al. 1951). This polymeric cellulose derivative dissolved at a very weakly alkaline pH, such as found in the small intestine, making it highly suitable for enteric controlled-release applications. Many enteric-coating products followed, including the commercially very successful poly(methacrylates), now marketed as the Eudragit® L and Eudragit® S series by Evonik Industries. Figure 1.1 shows compositions of some enteric-coating polymers.

With respect to parenteral delivery, the development of controlled-release systems began in the 1930s, with the introduction of compressed pellets of hydrophobic compounds, which could provide sustained drug release over time, thereby allowing a reduction in the dosing frequency. Pellets consisting of compressed, finely powdered, estradiol particles were administered via subcutaneous (s.c.) implantation to animals, to cause rapid weight gain in the treated animals. Subsequently, other

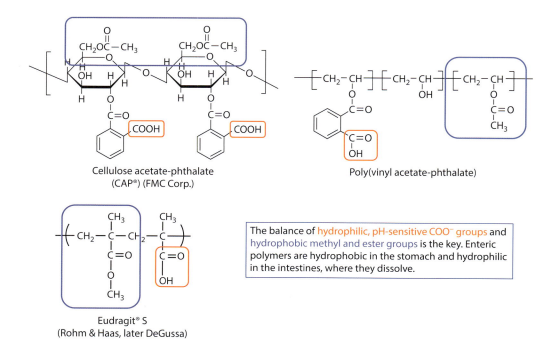

Cellulose acetate-phthalate
(CAP®) (FMC Corp.)

Poly(vinyl acetate-phthalate)

Eudragit® S
(Rohm & Haas, later DeGussa)

The balance of hydrophilic, pH-sensitive COO⁻ groups and hydrophobic methyl and ester groups is the key. Enteric polymers are hydrophobic in the stomach and hydrophilic in the intestines, where they dissolve.

FIGURE 1.1   Enteric-coating polymers.

pellet-type implants were developed using other steroidal hormones. The rate of sustained release of the hydrophobic drugs was determined by the relative hydrophobicity of the pellet (Chien 1982; Hoffman and Wright 2012).

1.3 THE SPANSULE® DELIVERY SYSTEM: THE FIRST CONTROLLED-RELEASE FORMULATION

Even though drug release could be delayed by using enteric coatings, these formulations still featured immediate release of the drug upon removal of the enteric coating. The next stage of technological development was the design of true controlled-release systems, designed to control the drug release rate throughout the lifetime of the formulation. The first of these was the Spansule oral DDS (Figure 1.2), introduced in 1952 by Smith, Kline & French (SKF) for the 12-hour delivery of dextroamphetamine sulfate (Dexedrine®). Each Spansule® capsule contains hundreds of tiny drug-loaded beads, coated with a variable layer of natural waxes, such as carnauba wax, beeswax, or glyceryl monostearate. On ingestion, the outer capsule rapidly disintegrates, liberating the drug-loaded beads. The waxy coating around the beads then gradually dissolves as they transit down the GI tract, to liberate the drug. The rate of drug release is controlled by the thickness and dissolution rate of the waxy coating. A single capsule contains subpopulations of beads with different coating thicknesses, to provide a sustained release of drug over time (Lee and Li 2010).

Subsequently, SKF introduced the cold remedy Contac® 600 (so called because each capsule contained 600 beads), which became the world's leading cold or allergy remedy after its launch in 1960. Each capsule contained four distinct populations of beads: a quarter with no coating, for

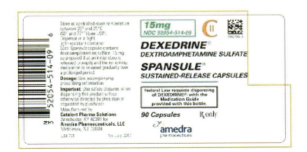

The Spansule® system: Sustained oral delivery

FIGURE 1.2   The Spansule system achieved "sustained" drug delivery kinetics over many hours.

immediate drug release; a quarter with a thin waxy coating, which dissolved after 3 hours; a quarter had a thicker coating so that drug release occurred after 6 hours; and the final quarter, with the thickest coating, dissolved after 9 hours. In total, the beads from a capsule provided cold/allergy relief over a sustained 12-hour period. Many advertisements at the time described the Contac® system as "tiny little time pills," which provided "12 hour cold or allergy relief," thereby introducing the general public to the concept of sustained release (Figure 1.2). Since then, many drugs have been reformulated in the Spansule® system, although the original waxy coatings have largely been replaced with more stable and reproducible, slowly dissolving, synthetic polymers.

1.4 CONTROLLED RELEASE USING A RATE-CONTROLLING MEMBRANE

1.4.1 DRUG DIFFUSION THROUGH A RATE-CONTROLLING MEMBRANE

Judah Folkman, an MD at Harvard University, was an early pioneer in the field of controlled drug delivery. He was circulating rabbit blood inside a Silastic® (silicone rubber [SR]) arteriovenous shunt, and when he exposed the tubing to hydrophobic anesthetic gases in the atmosphere surrounding the tubing, the rabbits went to sleep. He concluded that the gases were permeating across the SR tubing and absorbing into the blood. He proposed that sealed capsules of SR containing a drug could be implanted to act as a prolonged DDS (Folkman and Long 1964; Folkman et al. 1966; Hoffman 2008).

In this way, a *reservoir* of drug is contained within a RCM. The drug can diffuse out through the reservoir at a controlled rate. If certain conditions are filled, drug release remains constant, "zero order" with time. The principle of the RCM zero-order DDS depends on a RCM that does not vary in permeation properties over the period of use. The zero-order condition also assumes that no significant diffusional resistances will be introduced with time, such as the deposition of a thick layer of scar tissue due to a foreign body response. Then, if the drug concentration–driving force from inside to outside of the device is constant, the delivery rate will be constant over the period of use (Figure 1.3).

Another key pioneer in the origin of the controlled drug delivery field was Alejandro Zaffaroni, a superb pharmaceutical chemist and entrepreneur who had collaborated with Carl Djerassi at Syntex on the synthesis of the steroid levonorgestrel, which was used in the first contraceptive pill. Zaffaroni had been thinking about creating a company devoted to controlled drug delivery. When he heard about Judah Folkman's work, he went to visit him in Boston and Folkman agreed

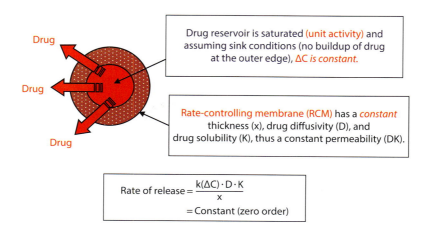

FIGURE 1.3 Membrane-controlled drug delivery systems. Zero-order delivery rate, controlled by a rate-controlling membrane.

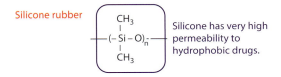

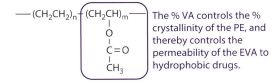

FIGURE 1.4 Silicone rubber and poly(ethylene-*co*-vinyl acetate) were the first polymers to be used as rate-controlling membrane barriers, for the controlled delivery of small hydrophobic drugs.

to become Chairman of the company's Scientific Advisory Board. In 1968, Zaffaroni founded the very first company dedicated to the development of controlled drug delivery materials and devices, which he called Alza, after the first two letters of each of his first and last names. One of the authors (ASH) was invited to become a consultant at Alza and was thus a witness to the origins and growth of the controlled DDS field and met most of the pioneers personally over the years.

The most common materials used as RCMs in the first devices were two polymers, SR and poly(ethylene-*co*-vinyl acetate) (EVA) (Figure 1.4). The EVA RCM is based on the copolymer of ethylene and vinyl acetate (VA). The VA disrupts the crystalline regions of the poly(ethylene) component, creating amorphous regions through which the drug can permeate (a drug cannot permeate through the crystalline region of a polymer). Thus, the higher the VA content of EVA, the higher the permeability of the drug through the EVA membrane. EVA RCMs may typically have as much as 40% VA.

A number of zero-order RCM DDS were developed in the 1970s and were approved for clinical use in the 1980s–1990s (Hoffman 2008). Typically, the drugs delivered were small and relatively hydrophobic, such as a variety of contraceptive steroids, as well as LHRH analogs (for treating prostate cancer) and pilocarpine (for treating glaucoma). Alza's first commercial product, the eye insert, Ocusert®, received FDA approval in 1974. The device released the antiglaucoma drug, pilocarpine, at a constant rate in the eye for 1 week, using an EVA RCM (Figure 1.5a). An EVA RCM was also used in Alza's intrauterine device (IUD), Progestasert®, approved in 1976, which provided zero-order controlled release of the contraceptive steroid progesterone, for over a month (Figure 1.5b).

In addition to the Alza Corp., others such as the Population Council, WHO, Upjohn/Pharmacia Pharmaceuticals, and Planned Parenthood were active in the development, approval process, and marketing of contraceptive drug devices. Norplant® is a controlled DDS birth control device that was developed by Sheldon Segal and Horatio Croxatto at the Population Council in New York in 1966 (Figure 1.5c). The original Norplant® consisted of a set of six small (2.4 mm × 34 mm) SR capsules, each filled with 36 mg of levonorgestrel (a progestin used in many birth control pills), for s.c. implantation in the upper arm. The implanted tubes had a 5-year duration of delivery, after which they had to be explanted surgically. Another set of six tubes could then be implanted if the patient desired it. Norplant® was approved for clinical trials in Chile in 1974 and finally approved from human use in Europe in the 1980s, followed by the United States in 1990. It was withdrawn by Wyeth Pharmaceuticals (who was marketing it in the United States) from the U.S. market in 2002 due to numerous "unwarranted" lawsuits. Production of the original Norplant® was discontinued globally in 2008.

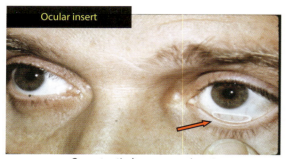

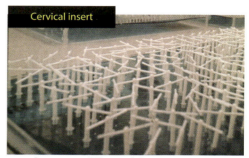

(a) *Ocusert* anti-glaucoma eye insert containing pilocarpine (RCM = EVA) (Alza)

(b) *Progestesert* IUD containing contraceptive steroid, progesterone (RCM = EVA) (Alza)

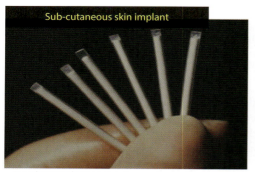

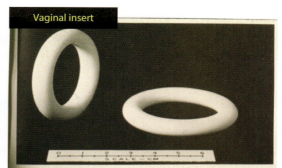

(c) *Norplant* implant containing contraceptive steroid, levonorgestrel (RCM = SR) (Population Council, Wyeth)

(d) *Vaginal ring* insert containing contraceptive steroid, medroxy-progesterone (RCM = SR) (WHO, Upjohn)

FIGURE 1.5 Examples of early drug delivery devices based on a drug reservoir and a rate-controlling membrane, to effect controlled release. (a) Ocusert®, an eye insert, (b) Progestasert®, an intrauterine device, (c) Norplant®, for subcutaneous implantation, and (d) an early prototype of a vaginal ring. SR, silicone rubber; EVA, poly[ethylene-*co*-vinyl acetate].

In 2006, Organon introduced a single-tube system, Implanon®, using EVA as the RCM. The implant provides controlled release of the contraceptive drug etonogestrel for up to 3 years. More recently, Valera Pharmaceuticals introduced Vantas®, a tubular s.c. implant made of Hydron®, a poly(hydroxyethylmethacrylate) (polyHEMA) hydrogel. The implant provides continuous delivery, for over a year, of the gonadotropin-releasing hormone (GnRH) analog, histrelin acetate, for the treatment of prostate cancer. Similar to Norplant® and Implanon®, the Vantas® implant is nondegradable and has to be surgically retrieved after use.

Vaginal rings were also designed as zero order, RCM, DDS. Gordon Duncan at Upjohn/ Pharmacia, with support of the WHO, developed an early example comprising a SR core, loaded with a contraceptive steroid and coated with SR (Figure 1.5d). Although it did not become commercialized, this ring laid the groundwork for the subsequent development of other vaginal rings, such as the Estring® and Femring®, which were approved in the late 1990s for the delivery of estradiol acetate, in the treatment of postmenopausal urogenital symptoms. NuvaRing®, developed at Merck, is made of EVA and has been used clinically to deliver estradiol for treating postmenopausal urogenital symptoms.

Zaffaroni was also interested in the potential of transdermal drug delivery. He patented the controlled delivery rate skin patch as a "Bandage for Administering Drugs" in 1971, shortly after he had founded Alza (Figure 1.6). The Alza skin patch is a reservoir system that incorporates two release mechanisms: an initial burst release of drug from the adhesive layer and zero-order release

United States Patent

[11] **3,598,122**

[72]	Inventor	**Alejandro Zaffaroni** Atherton, Calif.
[21]	Appl. No.	812,116
[22]	Filed	Apr. 1, 1969
[45]	Patented	Aug. 10, 1971
[73]	Assignee	**Alza Corporation**

3,249,109	5/1966	Maeth et al.	128/268
3,339,546	9/1967	Chen	128/156
3,444,858	5/1969	Russell	128/268 X
3,464,413	9/1969	Goldfarb et al.	128/268
3,518,340	6/1970	Raper	424/19
3,520,949	7/1970	Shepherd et al.	128/156 UX

Primary Examiner—Charles F. Rosenbaum
Attorney—Steven D. Goldby

[54] **BANDAGE FOR ADMINISTERING DRUGS**
15 Claims, 3 Drawing Figs.

[52] U.S. Cl. 128/268, 424/20, 424/28
[51] Int. Cl. A61f 7/02
[50] Field of Search .... 128/155- —156, 268, 296; 424/19—20, 28

[56] **References Cited**
UNITED STATES PATENTS

| 2,629,378 | 2/1953 | Barton | 128/268 |
| 3,053,255 | 9/1962 | Meyer | 128/268 |

ABSTRACT: Bandage for use in the continuous administration of systemically active drugs by absorption through the skin or oral mucosa comprising a backing member having on one surface thereof a reservoir containing a systemically active drug. The reservoir has a wall distant from the backing member and permeable to passage of the drug. A pressure-sensitive adhesive layer, also permeable to passage of the drug, is carried by the reservoir. The drug is in a form acceptable for absorption through the skin or the mucosa of the mouth.

FIGURE 1.6 Copy of the 1971 Alza patent for transdermal drug delivery using a skin patch ("bandage").

over an extended period (e.g., several days), facilitated by the RCM built into the patch and separating the drug reservoir from the skin surface. The skin patch technology is referred to as a transdermal therapeutic system (TTS). If the RCM does not change in properties during the contact time of the patch on the skin, the drug diffusion rate across the membrane and out of the patch will be constant. The delivery rate from the patch is designed to be much slower than the diffusion of the drug through the stratum corneum (the main resistance in the skin), thus rate control is determined by the patch and not the skin. This was referred to as "putting the major resistance to drug delivery into the device."

Alza developed other types of RCMs that contain micropores, e.g., a stretched polypropylene (PP) membrane (Celgard®), with micropores that may be prefilled with mineral oil or wax. This membrane is used in some controlled-release skin patches. Since PP is highly crystalline, the wax- or oil-filled pores represent the preferred diffusion pathway for the drug. As long as the pore volume and pore interconnections within the microporous RCM do not change with the time of the patch on the skin, the drug diffusion rate across the membrane will be constant. As earlier, if the overall drug delivery rate from the patch is designed to be much slower than that by diffusion through the stratum corneum, the patch will exhibit a zero-order drug delivery rate.

The first controlled delivery skin patch commercially available was Alza's product Transderm-Scop®, approved in 1979 for the transdermal delivery of scopolamine, a drug that alleviates the discomfort of motion sickness. It was developed with the idea that Alza could get funding from the U.S. space program for use of the patch in zero-gravity conditions. Many other TTS patches were subsequently developed by Alza and other companies, allowing once-a-day or even once-a-week dosing, with reduced side effects compared to the oral route (a further description of TTS is given in Chapter 9). It is important to note that some adhesives used to adhere the patch to the skin caused skin irritation.

1.4.2 OSMOTIC PRESSURE CONTROLLED RELEASE: WATER DIFFUSION THROUGH A RATE-CONTROLLING MEMBRANE

In the 1970s, an alternative method to achieving controlled release was developed, based on the principles of the osmotic pump. It utilizes a constant volume and constant concentration (saturated)

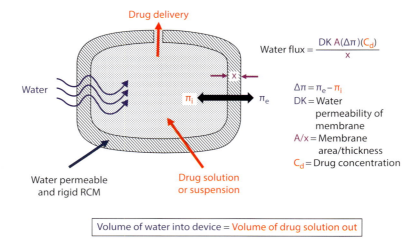

$$\text{Water flux} = \frac{DK\,A(\Delta\pi)(C_d)}{x}$$

$\Delta\pi = \pi_e - \pi_i$

DK = Water
            permeability of
            membrane

A/x = Membrane
            area/thickness

C_d = Drug concentration

Volume of water into device = Volume of drug solution out

FIGURE 1.7 The elementary osmotic pump. $\Delta\pi$ = the osmotic gradient, i.e., the difference between the osmotic pressure in the surrounding environment ($\Delta\pi_e$) and the osmotic pressure inside the device ($\Delta\pi_i$).

of a drug solution, or dispersion, inside a rigid, semipermeable membrane (the RCM). Water permeates through the RCM into the device, displacing an equal volume of drug solution out of the device, through a microscopic pore created in the membrane. The water permeates into the tablet due to an osmotic pressure difference between the osmotic pressure of water within the body fluids (e.g., the GI tract fluids) and the low osmotic pressure within the saturated drug condition inside. Figure 1.7 shows the elementary osmotic pump (EOP), developed by Felix Theeuwes and colleagues at Alza in 1975, for controlled-release oral drug delivery (Theeuwes 1975).

It is important to emphasize that, while these devices exhibit zero-order drug delivery, they operate on a completely different delivery mechanism from the diffusion-driven, RCM devices described earlier. For osmotic pressure control, the constant drug delivery rate is driven by a membrane-controlled, constant rate of *water* permeation *into* the device (in contrast to *drug* diffusion *out* of the device, as described earlier), which displaces an equal volume of a constant concentration of drug solution through the pore and out of the device. The rate of drug diffusion across the membrane is negligible.

Examples of such osmotic devices include

1. Many types of oral tablets: the oral EOP is shown in Figure 1.7, further examples are described in Chapter 7
2. The implanted Duros® titanium device, developed at Alza and now fabricated and marketed by Durect (see Chapter 6, Figure 6.4)
3. The programmable infusion pump, Alzet®, also developed at Alza and now fabricated and marketed by Durect, which is widely used in preclinical animal studies (see Chapter 6, Figure 6.3)

The exit pore may be formed in drug tablets by a laser beam; it is built into the Duros® implant and Alzet® pump devices. Cellulose acetate is used for the RCM membrane of many peroral drug tablets. A small amount (e.g., 10%) of poly(ethylene glycol) (PEG) of ≈3 kDa MW is added to stimulate the start-up of water permeation into the tablet and reduce somewhat the time lag for drug delivery from the device. In the case of the implanted Duros® device, the RCM is a polyurethane block copolymer, where some blocks are of PEG, to control the rate of water permeation into the implanted titanium cylinder. One drug solution used in the Duros® implant is LHRH, dissolved in a DMSO–water mixed solvent.

1.5   LONG-ACTING INJECTABLES AND IMPLANTS

The earlier *macroscopic* devices (transdermal patches, IUDs, ocular inserts, subdermal implants, etc.) of the 1960s and 1970s were then followed, from the mid- to late 1980s, by newer parenteral DDS in the *microscopic* size range. This period in particular saw important developments in the field of depot DDS: injected or implanted, drug-loaded, polymeric microparticles, wafers, and gels.

In 1974, Robert (Bob) Langer joined the lab of Judah Folkman as a postdoctoral fellow and studied the use of nondegradable polymeric matrix systems, for the sustained release of proteins. In a seminal article in *Nature* in 1976, they showed the sustained release of active proteins from various EVA-based matrices in the rabbit eye (Langer and Folkman 1976). Various proteins (soybean trypsin inhibitor, lysozyme, catalase, insulin, and tumor angiogenesis factor [TAF]), as well as other macromolecules (including heparin and DNA), were dispersed within polymeric matrices of HEMA, EVA copolymer, and PVA and implanted in rabbit corneas. The matrices were capable of releasing biochemically active drug molecules (as demonstrated, for example, by the neovascular response produced by a TAF-loaded implant) for periods exceeding 100 days. This was one of the earliest depot DDS and this pioneering work stimulated much research and interest in the drug delivery field.

The nondegradable depot DDS studied by Langer, Folkman, and others required surgical removal and also tended to be unsuitable for the delivery of hydrophilic drugs. Therefore, further research focused more on the use of degradable polymers, consisting of mixtures of drug/degradable polymer that were implanted or injected into the body and that could release drug for a sustained period of time. These implants and injections could provide "sustained release", rather than the zero order, controlled release of the RCM DDS described earlier.

A variety of options are now possible: (1) long-acting injections (LAIs) of liquid dispersions of solid microparticles, (2) LAIs comprising solutions that subsequently form gel-like masses upon injection, due to the temperature rise or solvent dilution occurring in vivo, and (3) s.c. implants of resorbable, polymer–drug solids, in the forms of wafers, discs, or other shapes.

The polymers used in these systems have most often been based on poly(esters), with the most well-known, and commonly used, degradable polymers in drug delivery being polyesters based on the copolymers of lactic acid and glycolic acid, i.e., poly(lactic-*co*-glycolic acid) (PLGA). Poly(glycolic acid) was prepared and patented by Edward Schmitt and Rocco Polistina at Davis & Geck of American Cyanamid Co. in 1967, for use as a degradable suture, which they named Dexon®. Ethicon, a J&J subsidiary, added lactic acid and prepared the PLGA degradable suture. Ethicon licensed Cyanamid's patent and began the manufacture of a PLGA degradable suture, Vicryl®, which continues today as a highly used, resorbable suture.

Following from their use as biodegradable sutures, PLGA polymers were developed in the 1980s at the Southern Research Institute (SRI) and the University of Alabama as spherical, drug-loaded, microparticle dispersions that were injected subcutaneously as depot DDS (Okada and Toguchi 1995; Anderson and Shive 2012). Lynda Sanders of Syntex also developed several degradable microparticle depot DDS formulations in collaboration with SRI and various companies in the mid- to late 1980s (Sanders et al. 1985).

One of the first companies to market a PLGA microparticle formulation was TAP Pharmaceuticals, a joint venture formed in 1977 between the Abbott Laboratories and the Japanese pharmaceutical company Takeda. They called it Lupron Depot®, a formulation of nafarelin (a GnRH agonist) loaded into PLGA solutions and formed into microparticles. The joint venture was dissolved in 2008 and most of the company merged with Takeda. Abbott ended up with the rights to the Lupron Depot®. PLGAs were subsequently also extensively studied as nanocarriers (NCs) in the 1990s and beyond, as described in Section 1.8.

Other degradable polymers used for depot DDS include poly(caprolactone) (PCL). Poly(ethylene oxide) (PEO) and its block copolymer with poly(ethylene terephthalate) was proposed as a degradable polymer for use in surgery in the late 1970s (Gilding and Reed 1979). Polyactive®, a block copolymer of poly(ethyleneglycol terephthalate)-*b*-poly(butylene terephthalate) and used for delivery of the anticancer drug, interferon, is based on the work of Jan Feijen at Twente University in the Netherlands (Jorgensen et al. 2006). A further group, the poly(hydroxyalkanoates) are linear poly(esters) formed in a biochemical fermentation process.

The poly(anhydrides) are a family of hydrolytically degradable polymers used in depot DDS that were conceived and synthesized in Bob Langer's laboratory at MIT (Rosen et al. 1983) and implanted in the brain by the surgeon Henry Brem. This work led to the commercial introduction in 1995 of Gliadel®, solid wafers, or discs of poly(anhydride), loaded with the cytotoxic drug, carmustine (*bis*-chloroethylnitrosourea [BCNU]), for the treatment of brain glioblastomas (see also Chapter 15, Section 15.4.2). It is historically interesting to note that Langer and Brem met as postdoctoral fellows in Judah Folkman's laboratory at Harvard Medical School.

Drug–polymer depot DDS formulations may also be injected into the body as a liquid solution of drug plus a "smart" polymer at room temperature, that phase separates to form a solid or a gel in vivo, as the polymer is warmed to body temperature and/or as the polymer solvent is diluted with tissue water (see also Chapter 6, Section 6.2.2). Solvents that have been used in the injectable solutions include *N*-methylpyrrolidone (NMP) and a solvent mixture of benzyl benzoate and benzyl alcohol (BB/BA). Three examples of the drug–polymer solutions that have been developed as depot DDS during the 1990s include the following:

1. Atrix's Atrigel®, an NMP solution of water-insoluble poly(lactic acid) (PLA) + drug. Richard Dunn of Atrix developed this injectable depot DDS in the 1990s; Atrix is now a partner of Pfizer.
2. Alza's Alzamer®, a benzyl benzoate solution of a water-insoluble poly(orthoester) containing a drug.
3. Macromed's ReGel®, an aqueous solution of a temperature-responsive triblock copolymer, PLGA–PEG–PLGA, containing a dispersion of paclitaxel as the drug. Kim, Jeong and collaborators at the University of Utah developed this novel, thermally responsive depot DDS (Jeong et al. 1997).

1.6 FURTHER DEVELOPMENTS IN ORAL CONTROLLED RELEASE

The 1950s and 1960s saw the introduction of a variety of semisynthetic, and synthetic, hydrophilic, gel-forming polymers such as hydroxypropylmethylcellulose (HPMC), hydroxypropyl cellulose, PEO, and Carbopol® by chemical companies such as Dow and Union Carbide. These polymers were "rated" by the FDA as generally recognized as safe (GRAS) materials, as a result of the Food Additives Amendment of 1958. One of the many applications of these polymers was in the field of oral sustained release, in the preparation of "swell and gel" hydrophilic matrix DDS.

In this type of DDS, a drug is dispersed in a hydrophilic polymer, such as HPMC, and compressed into a tablet. As the tablet becomes hydrated in the GI tract, the hydrophilic polymer chains hydrate, relax, and swell, thereby forming an outer, rubbery, viscous gel layer on the tablet surface (Wen et al. 2010). The gel layer slows both water penetration into, and drug diffusion out of, the tablet core. Sustained release of the drug is achieved as the API dissolves in the incoming fluids and then must diffuse out through the viscous, swollen, polymer chains (see also Chapter 7, Section 7.3.1.1). Nowadays, the majority of SR formulations for the oral route are based on "swell and gel" matrix tablets. Although easy to manufacture, drug release from this type of DDS is not zero order; instead, it typically follows first-order kinetics. The amount of drug released decreases with time because the drug diffusional path length through the matrix increases with time.

During the 1980s and 1990s, research focused on improving the delivery profile of the "swell and gel" matrix tablets, by modifying the geometry of the system in such a way as to effect a zero-order release rate. One successful design is called Geomatrix™, designed by Conte et al. at the University of Padua, Italy, and developed by Skye Pharmaceuticals (Conte et al. 1993). It is like a sandwich, with two outer layers of drug-impermeable, nonswelling polymer that surround a central, drug-loaded, dry polymer layer. The central dry layer gradually swells in vivo, becoming a hydrogel, which releases the drug through the outer, exposed, swollen "edges" of the tablet (see Chapter 7, Figure 7.3). The basis for achieving zero order with this design is understood by examining Fick's equation, as seen in Figure 1.3, and applying it at a moment during the gel swelling as it releases drug:

$$\text{Rate of release} = \left\{ k \cdot A \left(\frac{\Delta C}{x} \right) D \cdot K \right\} \tag{1.1}$$

where
  k is a constant
  A is the exposed area around the peripheral face of the swelling gel
  ΔC is the concentration driving force from the center of the swelling gel to the outer peripheral surface (where C is assumed to be a perfect sink concentration of 0)
  x is the gel length from the center to the outer edge of the device that the drug traverses across and out of the device at any moment
  D is the drug diffusion coefficient in the swelling gel
  K is the drug partition coefficient in the swelling gel

With reference to Equation 1.1, as the gel "swells and gels," a number of interrelated processes occur:

- "A" will increase, which will increase the delivery rate.
- $\Delta C/x$ will decrease overall, due to two processes (1) ΔC decreases because drug is being depleted from the gel and (2) the gel thickness, x, increases, as the gel swells. Both these effects will decrease $\Delta C/x$ and thus will decrease delivery rate.
- D and K of the drug should both increase with time, as water penetration and swelling increases, especially if the drug is partially polar.

If all these potential increases and decreases in delivery rate just balance out, then the delivery rate will remain constant during the main transit time in the GI tract. This Geomatrix™ DDS technology is currently used in a number of products that are available around the world.

A different mechanism of achieving zero-order controlled release for the oral route has already been described using osmotic pressure to control drug release (Section 1.4.2). Osmotic pressure forces the drug out, in a zero-order controlled fashion, through a tiny laser-drilled orifice in the tablet (Figure 1.7). Since the original EOP described by Theeuwes in 1975, a variety of different sophisticated systems have been developed, utilizing oral osmotic technology (see also Chapter 7, Section 7.3.3).

1.7 DRUGS ON SURFACES

During the 1960s–2000s, a number of different DDS were developed in which drugs were localized onto surfaces, including (1) the anticoagulant heparin, immobilized on blood-contacting surfaces; (2) drug–polymer matrices coated on drug-eluting stent (DES) surfaces; and (3) mucoadhesive-drug formulations that have enhanced residence times on mucosal surfaces.

1.7.1 HEPARINIZED SURFACES

Heparin was the first drug directly adsorbed or linked onto a biomaterial surface. In the late 1960s, it was physically adsorbed by ionic forces to a cationic surfactant (benzalkonium chloride), which was embedded into a graphite coating on the polymer surface (Gott et al. 1968) (Figure 1.8). The coating

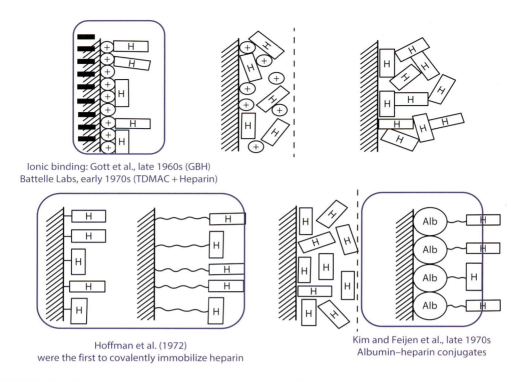

Ionic binding: Gott et al., late 1960s (GBH)
Battelle Labs, early 1970s (TDMAC + Heparin)

Hoffman et al. (1972)
were the first to covalently immobilize heparin

Kim and Feijen et al., late 1970s
Albumin–heparin conjugates

FIGURE 1.8 Heparin-coated polymers were one of the earliest polymeric drug delivery systems (1960s). GBH, graphite–benzalkoniumchloride–heparin; TDMAC, tridodecyl methyl ammonium chloride.

is called "GBH" for "graphite–benzalkoniumchloride–heparin." Heparin molecules are released by ion exchange with chloride ions from the blood plasma.

In the early 1970s, heparin was ionically linked to polymer surfaces by binding it to a cationic surfactant known as "TDMAC" (tridodecyl methyl ammonium chloride), a cationic surfactant that had been physically adsorbed by hydrophobic interactions of the dodecyl groups with surface molecules of the hydrophobic polymer (Figure 1.8). TDMAC was developed at the Battelle labs in Columbus, Ohio by Bob Leininger and Richard Falband, and marketed by Polysciences. Hoffman and colleagues in 1972 were among the first (if not the first) to *covalently* immobilize heparin on polymer surfaces (Hoffman et al. 1972). The heparin remained active, as indicated by its ability to bind antithrombin III and thrombin. They conjugated heparin to the OH groups of poly(HEMA) that had been radiation grafted onto the SR surfaces.

1.7.2 DRUG-ELUTING STENTS AND BALLOONS

Drug–polymer matrices have been coated onto stents and are known as drug-eluting stents, or DES (see also Chapter 6, Section 6.3.5). One of the earliest DES was the Cypher® stent of J&J, which was coated with a thin layer of a blend of poly(n-butyl methacrylate) and EVA, containing the smooth muscle cell antiproliferative drug, Sirolimus® (Suzuki et al. 2001). Taxus™, the DES of Boston Scientific and approved by FDA in 2004, uses a thermoplastic triblock elastomer poly(styrene-b-isobutylene-b-styrene) (SIBS). In this device, the drug paclitaxel is dispersed primarily as discrete nanoparticles (NPs) embedded in the SIBS matrix. Paclitaxel release involves the initial dissolution of drug particles from the paclitaxel/SIBS-coating surface, which exhibits an early burst release, followed by a sustained, slower, release of paclitaxel from the bulk of the coating (Ranade et al. 2004). This coating was developed and patented by Kennedy et al. at Akron University (Kennedy et al. 1990).

There is an active development and marketing of DES with struts that contain periodic "concavities" or roughened patches that are coated or filled with polymer–drug mixtures (e.g., PLA-drug) and provide sustained release of antiproliferative, cytostatic, and antithrombogenic drugs. These types of "patchy coatings" do not leave behind a residue of the polymer coating after the drug is gone.

Drug-eluting, bioresorbable stents (DEBS, also known as BDES) are also under testing; they are composed of a range of slow, bioresorbable polymers such as PLA, PCL, and their copolymers. Tyrosine polycarbonate is a bioresorbable polymer based on the amino acid tyrosine, developed by Joachim Kohn and collaborators (Kohn and Zeltinger 2005). In some cases, the totally bioresorbable stent may be coated with PLA–drug mixtures that release their drug cargoes within days and then resorb, while the stent body disappears much more slowly, allowing the blood vessel to heal from the injury sustained from the initial insertion of the stent into the blocked vessel (Abizaid and Ribamar-Costa 2010). Medtronic Corp. has recently developed a drug-coated balloon (DCB), a catheter containing a balloon that is expanded against a vessel blockage. The balloon is coated with the drug paclitaxel plus urea as an excipient. Called the In. Pact Admiral™, it received FDA approval in 2014 for treating lesions up to 18 cm in length.

1.7.3 MUCOADHESIVES

Mucoadhesive DDS are designed to adhere to mucosal surfaces, i.e., those epithelial interfaces with an overlying mucus layer, such as the GI tract, the nose, the lungs, the eye, etc. (as described in Section III of this book). A mucoadhesive DDS increases the residence time of a drug dosage form at the targeted mucosal surface, thereby allowing more time for the absorption process, as well as providing a greater concentration gradient to drive the process. To accomplish mucoadhesion, they are designed to interact strongly with the mucus layer, which is rich in secreted, highly hydrophilic glycoproteins. Mucoadhesive drug delivery polymers are similar to mucus in that they are highly hydrophilic, often negatively charged and highly H-bonding with the mucus layer. Poly(acrylic acid) has been a favorite mucoadhesive polymer, beginning with the seminal and pioneering work in the 1980s of Joseph Robinson and Kinam Park (Park and Robinson 1984; Park et al. 2011). Nicholas Peppas has proposed PEGylated methacrylate polymers as mucoadhesives (Serra et al. 2006).

1.8 NANOSCALE DDS

In the 1990s, the size of controlled-release DDS scaled down again so that the *micro* systems of the previous decade made way for technologies in the *nanometer* size range. Drug NCs are individual molecules or assemblies of many molecules, ranging in size from 1 or a few nm, up to several hundreds of nm. These NCs are conjugated to, complexed with, or encapsulate drug molecules.

Three basic technologies stimulated the growth of the field of nanoscale DDS:

1. PEGylation, which provided protection for biomolecular drugs and extended the circulation times of the nanoDDS
2. Active targeting to specific cells, using antibodies and ligands
3. The "enhanced permeability and retention" (EPR) effect, for passive targeting to tumor tissues

These technologies are described briefly here; a more detailed account is given in Chapter 5.

PEGylation. The most widely used synthetic polymer in nanoscale DDS is PEG. In 1969, Frank Davis of Rutgers University wanted to reduce the immunogenicity of the newly developed recombinant protein drugs that were available due to advances in biotechnology and molecular biology. He hypothesized that attaching a hydrophilic polymer might make the proteins less immunogenic. In the early 1970s, he discovered methoxy PEG (mPEG) in a company catalog; mPEG was a very convenient molecule with which to PEGylate a protein drug molecule. The first publication of a PEG–drug conjugate appeared in 1977, based on the PhD work of Davis's PhD student, Abraham Abuchowski (Abuchowski et al. 1977). Their coupling technology yielded PEG proteins that showed not only reduced immunogenicity, but also much longer circulation times than their non-PEGylated counterparts (Davis 2002).

PEG has been incorporated into a variety of nanoscale DDS as the outer "corona" of the NC, which enhances circulation times, reduces opsonization and removal by the RES in the liver, and may also enhance tumor uptake by the EPR effect. It is important to note, however, that evidence indicates that multiple injections of some PEGylated nanoscale DDS formulations can stimulate the formation of antibodies against PEG (Ishida et al. 2007).

Poly(carboxybetaine) (PCB) is an interesting "stealth-like" polymer that has recently been studied, and its performance compared to PEGylated NPs (Yang et al. 2014). In this study, PEGylated gold NPs were completely removed from the circulation in rats 50 hours after the first injection, or 24 hours after the second injection, which was attributed to IgM recognition, binding, and removal. In contrast, more than 50% of the PCB-coated gold NPs remained in circulation at both time points, demonstrating the potential of a PCB coating to enhance circulation time.

In addition to the work of Frank Davis and colleagues who developed PEG–drug conjugates, it is interesting to note that Helmut Ringsdorf had independently published a prescient article proposing drug–polymer conjugates as new therapeutic entities in 1975 (Ringsdorf 1975) (Figure 1.9).

Kopeček, Duncan, and colleagues in Prague and the United Kingdom designed, synthesized, and tested one of the earliest drug–polymer conjugates, which was based on the water-soluble polymer HPMA. The anticancer drug, doxorubicin, was conjugated to the polymer backbone by a tetrapeptide linkage that was a substrate for a lysosomal enzyme (cathepsin B), as shown in Figure 1.10. This allowed for the release of the drug intracellularly within lysosomes (Kopeček and Kopečková 2010).

Active targeting. The discovery of monoclonal antibodies (mAb) in 1975 by Niels Jerne, Georges Kohler, and Cesar Milstein (Köhler and Milstein 1975) led to their being awarded the 1984 Nobel Prize. mAb can be coupled to drug molecules, or NC DDS, and then used as targeting ligands to

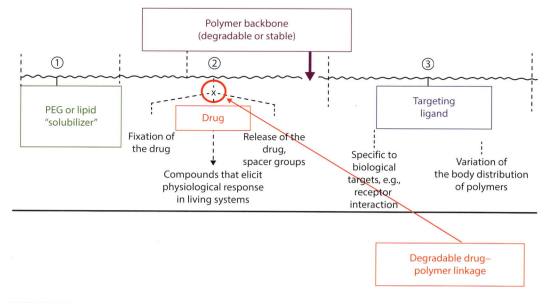

FIGURE 1.9 The components of a pharmacologically active polymer, as proposed by Helmut Ringsdorf in 1975. The polymer backbone can be degradable or stable; other components include (1) a poly(ethylene glycol) or lipid "solubilizer"; (2) the drug, which is attached to the polymer via a degradable linkage; and (3) a targeting ligand.

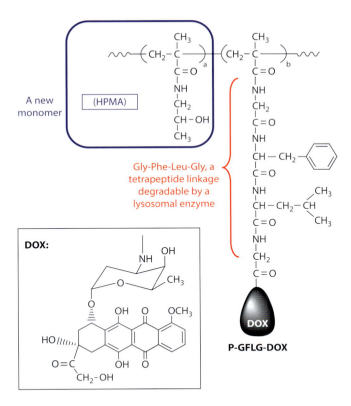

FIGURE 1.10   Drug–polymer conjugate. Poly(hydroxypropyl methacrylamide)–doxorubicin conjugate, with a tetrapeptide biodegradable linker (GFLG).

facilitate the uptake by specific cells. Erkki Ruoslahti and his PhD student Michael Pierschbacher discovered the integrin membrane receptor peptide ligand (RGD) and published it in PNAS in 1984. Many other peptide and small molecule ligands have since been used to target drugs, and NCs, to specific cells.

The EPR effect. From 1984, Hiroshi Maeda and colleagues published findings that showed tumoritropic accumulation of proteins and also their novel drug–polymer conjugate, SMANCS (Maeda et al. 1984; Matsumura and Maeda 1986; Maeda 2001). SMANCS is a conjugate of the polymer styrenemaleic acid (**SMA**) and the anticancer drug neocarzinostatin (**NCS**). They also coined the term "enhanced permeability and retention" (EPR) effect of tumor vasculature, which they described as an intrinsic pathological leakiness of the tumor blood supply, which coupled with poor lymphatic drainage, combine to facilitate the uptake and retention of SMANCS by tumor tissue but not by normal tissue (Figure 1.11).

The EPR effect is claimed to allow "passive targeting" to tumors, i.e., NCs of an appropriate size can passively enter the leaky tumor blood supply and thus accumulate in the tumor tissue. This is in contrast to the "active targeting" described earlier, which necessitates the addition of an appropriate targeting ligand.

Although these three technologies were all discovered around 1975–1984, the field of nanoscale drug delivery did not really begin to emerge until the 1990s. Since the 1990s, synthesis and testing of novel drug NCs continue to be the most active area of R&D in the drug delivery field. In addition, microfabrication techniques, leveraged from the microelectronics industry, are increasingly being used to manufacture NCs of extremely precise geometries and compositions (see also Chapter 19).

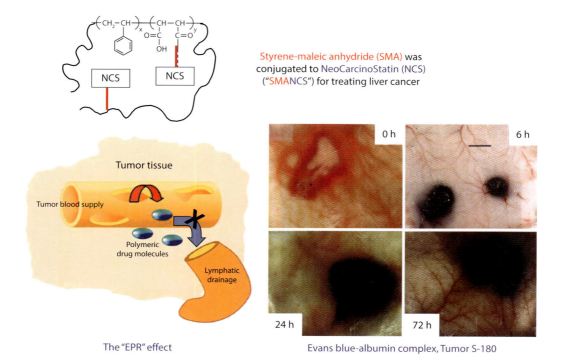

FIGURE 1.11 SMANCS drug–polymer conjugate. The "enhanced permeability and retention" effect for passive targeting to tumor tissues.

NCs come in a wide variety of compositions; they may also be partly, or totally, synthetic or natural. Figure 1.12 shows the basic molecular and NP ingredients of NCs. The synthetic polymeric NCs have a variety of structural compositions, such as homopolymers, random copolymers, block copolymers, and graft copolymers. They may also carry a net charge, e.g., polycations, such as poly(ethyleneimine) and poly(amidoamine), or poly(anions) such as poly(aspartic acid) and poly(propylacrylic acid). Some specific examples of NC are described next.

1.8.1 Liposomes

Since their discovery in 1965 and subsequent application to clinical therapy, liposomes have been a reliable dosage form in the clinic for almost 20 years (Bangham 1995; Torchilin 2005; Allen and Cullis 2013). Liposomes may be PEGylated either by conjugating PEG to either a phospholipid (a component of the liposome bilayer membrane) or a lipid that inserts into the lipid bilayer of a liposome (Figure 1.12).

One of the first liposomal DDS to be approved was the doxorubicin-loaded liposome named Doxil®. Barenholz has published a detailed history of the development of this liposomal DDS (Barenholz 2012). Other marketed liposomal DDS are described in Chapter 5.

1.8.2 Nanoparticle DDS

NPDDS may be drug–polymer mixtures or simply drug NPs with a surfactant stabilizer on their surfaces. Examples of polymeric NP carriers include the pioneering work of Patrick Couvreur on poly(alkylcyanoacrylate) NPs (Lambert et al. 2000). PLGA copolymers, developed as microparticle depot DDS a decade earlier, were also intensively investigated as NCs throughout the 1990s, and this work continues at the present time. PLGA NPs can also be PEGylated and

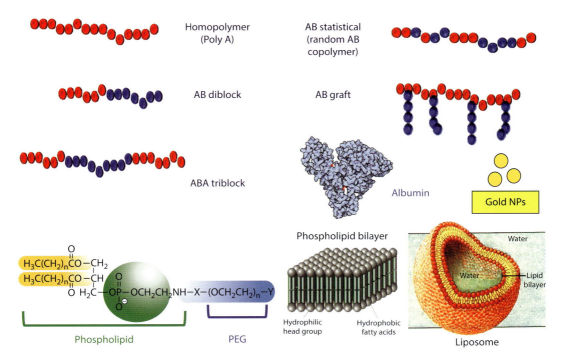

FIGURE 1.12 Composition of drug nanocarriers: polymers, phospholipids, PEGylated phospholipid, albumin and gold nanoparticles, and liposomes.

incorporate targeting-specific recognition ligands (e.g., Kreuter 2007; Danhier et al. 2012; Sealy 2014). Gold NPs have also been used to bind and deliver drugs. Natural polymeric NCs include proteins, such as albumin, which has been used to bind and deliver paclitaxel in the Abraxane®NP formulation (Celgene). NPs made from hemoglobin, collagen, and casein have also been extensively studied.

1.8.3 POLYPLEXES AND LIPOPLEXES

Cationic polymers have been used to complex negatively charged drugs, especially nucleic acid drugs, to form polyion complexes called "polyplexes." Important advances in the field were carried out by George and Catherine Wu (Wu and Wu 1987) who used poly(L-lysine) to complex DNA and form a polyplex, which they targeted to hepatocytes by conjugating a hepato-specific asialoglycoprotein to the carrier. Uptake of the soluble polyplex was achieved via receptor-mediated endocytosis, and they showed that the polyplex-delivered DNA was active within the targeted bacterial cells. Polycations used to form polyplexes include natural cationic polypeptides, synthetic polypeptides, and synthetic polymers. Polycations may also be cationic lipids, which form "lipoplexes" with nucleic acid drugs. Figure 1.13 lists the compositions of some of these polycations.

Patrick Stayton and Allan Hoffman at the University of Washington (Seattle, WA) have designed "smart" pH- and glutathione-responsive polymeric NCs that have the potential to be targeted to specific cells. The pH-responsive endosomolytic action of the NC means the DDS can escape from the endosomal compartment after cellular uptake. The drug, linked via a pendant disulfide bond to the polymer backbone, is subsequently released from the NC by glutathione reduction. If the active drug is a nucleic acid drug such as siRNA, the smart carrier will be negatively charged, forming polyplex intracellular DDS (Stayton et al. 2000).

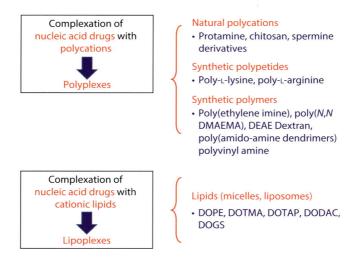

FIGURE 1.13 Polycations that form "polyplexes" and "lipoplexes."

1.8.4 POLYMERIC MICELLES

Kataoka, Okano, and Yokoyama pioneered the development of polymeric micelle DDS (Yokoyama et al. 1990; Nishiyama and Kataoka 2006). These micelles are formed by amphiphilic block copolymers, which contain both a hydrophilic polymer such as PEG and a hydrophobic polymer such as PLA or PLGA (Figure 1.14).

These copolymers self-associate at concentrations above the critical micelle concentration to form spherical micelles in aqueous solution. The hydrophobic blocks make up the micellar core, which can accommodate a poorly soluble drug; the outer shell is composed of the flexible tethered hydrophilic polymer strands. As can be seen in Figure 1.14, the introduction of functional groups on the micelle surface allows for cell-specific targeting. An interesting block copolymer micelle is

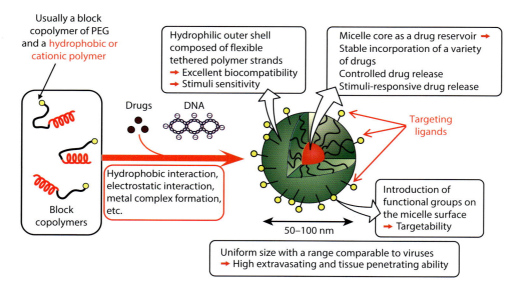

FIGURE 1.14 The PEGylated polymeric micelle delivery carrier.

that of PEG conjugated to a cationic block that complexes with a nucleic acid drug in the core of the PEGylated micelle (Harada and Kataoka 1995). A number of PEGylated polymeric micelles are in clinical trials at the present time. Several polymeric micelles have been studied with poly(HPMA), instead of PEG, as the "stealth" corona of the micelle (Talelli et al. 2010).

1.9 CONCLUSIONS

We have highlighted some of the key developments in the history of the field of controlled drug delivery. Attempts have been made to show how the early technologies and ideas paved the way for subsequent advances, and how the direction of drug delivery research was often shaped by the input, and the demands, from advances in other fields, such as biotechnology, molecular biology, polymer technology, and pharmacokinetics. The field of controlled drug delivery continues to develop and expand. The following chapters of this book provide much information on current and future directions in the discipline.

REFERENCES

Abizaid, A. and J. Ribamar-Costa. 2010. New drug-eluting stents: An overview on biodegradable and polymer-free next-generation stent systems. *Circulation* 3:384–393.

Abuchowski, A., J.R. McCoy, N.C. Palczuk et al. 1977. Effect of covalent attachment of polyethylene glycol on immunogenicity and circulating life of bovine liver catalase. *J Biol Chem* 252:3582–3586.

Allen, T.M. and P.R. Cullis. 2013. Liposomal drug delivery systems: From concept to clinical applications. *Adv Drug Deliv Rev* 65:36–48.

Anderson, J.M. and M.S. Shive. 2012. Biodegradation and biocompatibility of PLA and PLGA microspheres. *Adv Drug Deliv Rev* 64:72–82.

Bangham, A.D. 1995. Surrogate cells or trojan horses. The discovery of liposomes. *BioEssays* 17:1081–1088.

Barenholz, Y. 2012. Doxil®—The first FDA-approved nano-drug: Lessons learned. *J Control Release* 160:117–134.

Chien, Y.W. 1982. *Novel Drug Delivery Systems*. New York: Marcel Dekker, Inc.

Conte, U., L. Maggi, P. Colombo et al. 1993. Multi-layered hydrophilic matrices as constant release devices (Geomatrix™ Systems). *J Control Release* 26:39–47.

Danhier, F., E. Ansorena, J.M. Silva et al. 2012. PLGA-based nanoparticles: An overview of biomedical applications. *J Control Release* 161:505–522.

Davis, F.F. 2002. The origin of pegnology. *Adv Drug Deliv Rev* 54:457–458.

Folkman, J. and D.M. Long. 1964. The use of silicone rubber as a carrier for prolonged drug therapy. *J Surg Res* 4:139.

Folkman, J., D.M. Long, Jr., and R. Rosenbaum. 1966. Silicone rubber: A new diffusion property useful for general anesthesia. *Science* 154(3745):148–149.

Gilding, D.K. and A.M. Reed. 1979. Biodegradable polymers for use in surgery—Polyglycolic/poly(lactic acid) homo- and copolymers: 1. *Polymer* 20:1459–1464.

Gott, V.L., J.D. Whiffen, and S. Vallathan. 1968. GBH coatings on plastics and metals. *Ann NY Acad Sci* 146:21–29.

Harada, A. and K. Kataoka. 1995. Formation of polyion complex micelles in an aqueous milieu from a pair of oppositely charged block copolymers with poly(ethylene glycol) segments. *Macromolecules* 28:5294–5299.

Hoffman, A.S. 2008. The origins and evolution of "controlled" drug delivery systems. *J Control Release* 132:153–163.

Hoffman, A.S., G. Schmer, C. Harris et al. 1972. Covalent binding of biomolecules to radiation-grafted hydrogels on inert polymer surfaces. *Trans Am Soc Artific Intern Organ* 18:10.

Hoffman, A.S. and J.C. Wright. 2012. Historical overview of long acting injections and implants. In *Long Acting Injections and Implants*, eds. J. Wright and D. Burgess, pp. 11–24. New York: Springer.

Ishida, T., X.Y. Wang, T. Shimizu et al. 2007. PEGylated liposomes elicit an anti-PEG IgM response in a T cell-independent manner. *J Control Release* 122:349–355.

Jeong, B., Y.H. Bae, D.S. Lee et al. 1997. Biodegradable block copolymers as injectable drug-delivery systems. *Nature* 388:860–862.

Jorgensen, L., E.H. Moeller, M. van de Weert et al. 2006. Preparing and evaluating delivery systems for proteins. *Eur J Pharm Sci* 29:174–182.

Kennedy, J.P., G. Kaszas, J.E. Puskas et al. 1990. New thermoplastic elastomers. U.S. Patent 4,946,899.

Kim, S,W. and J. Feijen. 1985. Surface modification of polymers for improved blood compatibility, *CRC Crit Revs in Biocomp* 1:229–260.

Köhler, G. and C. Milstein. 1975. Continuous cultures of fused cells secreting antibody of predefined specificity. *Nature* 256:495–497.

Kohn, J. and J. Zeltinger. 2005. Degradable, drug-eluting stents: A new frontier for the treatment of coronary artery disease. *Expert Rev Med Dev* 2:667–661.

Kopeček, J. and P. Kopečková. 2010. HPMA copolymers: Origins, early developments, present, and future. *Adv Drug Deliv Rev* 62:122–149.

Kreuter, J. 2007. Nanoparticles—A historical perspective. *Int J Pharm* 331:1–10.

Lambert, G., E. Fattal, H. Pinto-Alphandary et al. 2000. Polyisobutylcyanoacrylate nanocapsules containing an aqueous core as a novel colloidal carrier for the delivery of oligonucleotides. *Pharm Res* 17:707–714.

Langer, R. and J. Folkman. 1976. Polymers for the sustained release of proteins and other macromolecules. *Nature* 263:797–800.

Lee, P.I. and J.-X. Li. 2010. Evolution of oral controlled release dosage forms. In *Oral Controlled Release Formulation Design and Drug Delivery*, eds. H. Wen and K. Park, pp. 21–31. Hoboken, NJ: Wiley.

Maeda, H. 2001. SMANCS and polymer-conjugated macromolecular drugs: Advantages in cancer chemotherapy. *Adv Drug Deliv Rev* 46:169–185.

Maeda, H., Y. Matsumoto, T. Konno et al. 1984. Tailor-making of protein drugs by polymer conjugation for tumor targeting: A brief review on SMANCS. *J Protein Chem* 3:181–193.

Malm, C.J., J. Emerson, and G.D. Hiait. 1951. Cellulose acetate phthalate as an enteric coating material. *J Am Pharm Assoc* 40(10):520–525.

Matsumura, Y. and H. Maeda. 1986. A new concept for macromolecular therapeutics in cancer chemotherapy: Mechanism of tumoritropic accumulation of proteins and the antitumor agent SMANCS. *Cancer Res* 46:6387–6392.

Nishiyama, N. and K. Kataoka. 2006. Current state, achievements, and future prospects of polymeric micelles as nanocarriers for drug and gene delivery. *Pharmacol Ther* 112:630–648.

Okada, H. and H. Toguchi. 1995. Biodegradable microspheres in drug delivery. *Crit Rev Ther Drug Carrier Syst* 12:1–99.

Park, K., I.C. Kwon, and K. Park. 2011. Oral protein delivery: Current status and future prospect. *React Funct Polym* 71:280–287.

Park, K. and J.R. Robinson. 1984. Bioadhesive polymers as platforms for oral-controlled drug delivery: Method to study bioadhesion. *Int J Pharm* 84:90154–90156.

Ranade, S.V., K.M. Miller, R.E. Richard et al. 2004. Physical characterization of controlled release of paclitaxel from the TAXUS™ Express²™ drug-eluting stent. *J Biomed Mater Res* 71A:625–634.

Ringsdorf, H. 1975. Structure and properties of pharmacologically active polymers. *J Polymer Sci Sympos* 51:135–153.

Rosen, H.B., J. Chang, G.E. Wnek et al. 1983. Bioerodible polyanhydrides for controlled drug delivery. *Biomaterials* 4:131.

Sanders, L.M., G.I. McRae, K.M. Vitale et al. 1985. Controlled delivery of an LHRH analogue from biodegradable injectable microspheres. *J Control Release* 2:187–195.

Sealy, C. 2014. Nanoparticles help the medicine go down; Breaking up helps nanoparticles delivery drugs; Nanoparticles make light of tumor detection; Nanoparticles stick like glue. *Nano Today* 9:1–5.

Serra, L., J. Dome, and N.A. Peppas. 2006. Design of poly(ethylene glycol)-tethered copolymers as novel mucoadhesive drug delivery systems. *Eur J Pharm Biopharm* 63:11–18.

Stayton, P.S., A.S. Hoffman, N. Murthy et al. 2000. Molecular engineering of proteins and polymers for targeting and intracellular delivery of therapeutics. *J Control Release* 65:203–220.

Suzuki, T., G. Kopia, and S. Hayashi, et al. 2001. Stent-based delivery of sirolimus reduces neointimal formation in a porcine coronary model. *Circulation* 104(10):1188–1193.

Talelli, M., C.J.F. Rijcken, C.F. van Nostrum et al. 2010. Micelles based on HPMA copolymers. *Adv Drug Deliv Rev* 62:231–239.

Theeuwes, F. 1975. Elementary osmotic pump. *J Pharm Sci* 64:1987–1991.

Torchilin, V.P. 2005. Recent advances with liposomes as pharmaceutical carriers. *Nat Rev Drug Discov* 4:145–160.

Wen, X., A. Nokhodchi, and A.R. Siahboomi. 2010. Oral extended release hydrophilic matrices: Formulation and design. In *Oral Controlled Release Formulation Design and Drug Delivery*, eds. H. Wen and K. Park, pp. 89–100. Hoboken, NJ: Wiley.

Wu, G.Y. and C.H. Wu. 1987. Receptor-mediated *in vitro* gene transformation by a soluble DNA carrier system. *J Biol Chem* 262:4429–4432.

Yang, W., S. Liu, T. Bai et al. 2014. Poly(carboxybetaine) nanomaterials enable long circulation and prevent polymer-specific antibody production. *Nano Today* 9:10–16.

Yokoyama, M., Y. Okano, K. Sakurai et al. 1990. Polymer micelles as novel drug carrier: Adriamycin-conjugated poly(ethylene glycol)–poly(aspartic acid) block copolymer. *J Control Release* 11:269–278.

Section I

Fundamental Issues

2 Principles of Controlled Release

Yvonne Perrie, Thomas Rades, and Kirsten Graeser

CONTENTS

2.1 INTRODUCTION

In the prevention and treatment of diseases using pharmacotherapy, the drug of choice is obviously a key factor. However, the site and duration of drug delivery, the rate of drug release, and the pharmacokinetic and side effect profiles of the drug all play pivotal roles in the success of a medical intervention. It is for these reasons that the development of suitable dosage forms or drug delivery systems (DDS; these terms will be used interchangeably in this chapter) is strictly required for any drug, in order to allow the safe, effective, and reliable administration of drugs to patients.

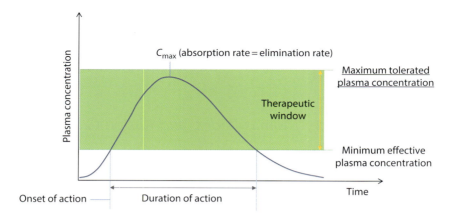

FIGURE 2.1 Typical plasma concentration–time profile, following oral administration. The therapeutic range is the concentration interval between the minimal effective concentration and the maximal tolerated concentration.

The rate and extent of absorption of a drug into the body, for example, from the gastrointestinal (GI) tract after oral administration, will determine its bioavailability. As described in Chapter 4, the bioavailability of drugs is influenced by physiological factors as well as by the physicochemical characteristics of the drug itself. However, factors associated with the route of administration and properties of the dosage form also have a major influence on the bioavailability of a drug.

An ideal DDS ensures that the active drug is available at the site of action for the appropriate duration of time, while ideally not displaying any physiological effect itself. The DDS should aim to provide drug concentrations at the site of action that are above the minimal effective concentration yet below the maximal tolerated plasma concentration (minimal toxic concentration) (Figure 2.1). Factors that control the drug concentration include the route of administration, the frequency of administration, the metabolism of the drug and its clearance rates, and importantly, the dosage form design itself. Thus, it is not surprising that many different dosage forms have been developed that affect the release and absorption of drugs. Therefore, in clinical practice, it is vital to consider both the drug treatment and the required release properties of the medicine. Since the majority of drugs are administered via the oral route, within this chapter, we mainly consider the options available in controlling drug release from oral solid dosage forms (SDFs), and only to a lesser extent from non-oral dosage forms, both in terms of rate and site of release. We consider how different release profiles can be achieved through the design of different release systems and use of appropriate excipients.

2.1.1 TYPES OF DRUG RELEASE

To understand the different types of drug release, in this chapter we categorize these into six basic groups: immediate release (IR), delayed release (DR), sustained release (SR), controlled release (CR), stimulus-sensitive release (SSR), and targeted release (TR). The pharmacopoeias, for example, the U.S. Pharmacopeia (USP), simply differentiate DDS into IR and modified-release (MR) systems; the latter can include DR, SR, and CR systems. Currently, SSR and TR systems are not defined in the pharmacopoeias, but certainly would also be categorized within the group of MR systems. We briefly describe the different types of drug release here:

- *Immediate release (IR, also called "fast release")*: In this type of drug release, the drug is released immediately after administration and is designed to give a fast onset of drug action (Figure 2.2a). This type of drug release is realized in a range of delivery systems including liquid and SDFs, and it is the most common type of drug release. Within the field

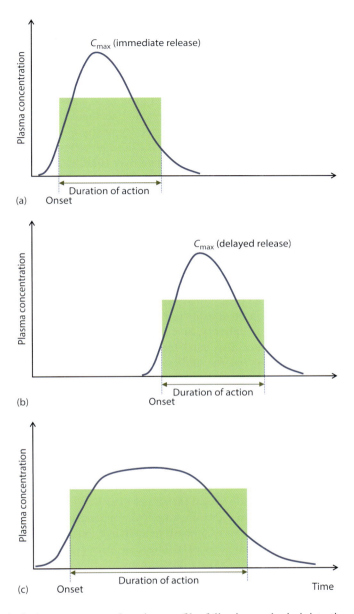

FIGURE 2.2 Typical plasma concentration–time profile, following oral administration, for (a) immediate release, (b) delayed release, and (c) sustained release dosage forms.

of oral drug delivery, typical dosage forms are "normal" disintegrating tablets, as well as chewable, effervescent, sublingual, and buccal tablets.

• *Modified release (MR)*: As stated earlier, this is a summarizing term, describing drug release that occurs sometime after administration (i.e., DR; Figure 2.2b), or for a sustained period (SR or CR; Figure 2.2c), or to a specific target in the body (TR). These systems tend to be used to improve therapeutic outcome of the drug intervention and/or to improve patient adherence. Modification of the time course and/or the site of drug release is more commonly achieved by SDFs, and oral SDFs can be designed to offer all MR formats.

- *Delayed release (DR)*: In this type of drug release, the drug is not released immediately after administration but some time after administration. Within the field of oral drug delivery, typical DR dosage forms are enteric-coated tablets (Figure 2.2b), which delay drug release until the small intestine.
- *Sustained release (SR, also called "prolonged release," or "extended release")*: In SR dosage forms, the drug is released over a prolonged period of time. The drug is thus released slower than in IR delivery systems, so that therapeutic plasma levels are maintained over a prolonged period of time (typically 8–12 hours) (Figure 2.2c). Sometimes, an initial fast release of a part of the dose gives rise to a fast onset of drug action, followed by a slower release of the drug over time, to maintain plasma levels. Such dosage forms would also be classified as SR systems. Within the field of oral drug delivery, typical dosage forms are polymer-coated pellets (reservoir systems) and matrix tablets.
- *Controlled release (CR)*: In this type of drug release, the drug is released at a predetermined rate for a prolonged period of time, so that effective drug plasma levels are maintained and controlled over an extended period of time. In contrast to SR dosage forms, CR delivery systems aim to control both the drug release from the dosage form and also the subsequent plasma levels; therefore, in principle at least, plasma levels are solely determined by the drug release kinetics from the DDS. The plasma profile from a CR DDS would then be similar to that achieved via an intravenous infusion. Within the field of oral drug delivery, an example is the oral osmotic system (OROS), but CR delivery systems are also used for other routes of administration, such as transdermal patches, ocular inserts, and implant devices, as discussed further in the following text.
- *Stimulus-sensitive release*: Drug release is a function of a specific biological/pathological signal (e.g., glucose concentration, to trigger the release of insulin), or an externally applied stimulus (e.g., temperature, or pH). Such systems are still in the early stages of development and are described further in Chapter 14.
- *Targeted release (TR)*: While all of the aforementioned types of drug release control the release of the drug from the dosage forms, or in the case of CR also the plasma levels, none of these dosage forms can influence the distribution of the drug in the body, for example, to specifically reach the site of drug action. However, with TR, it is attempted to deliver the drug to a specific target where the drug release is triggered, for example, through a time delay or as a result in a change in the environment surrounding the MR system (e.g., a change in pH). This can offer enhanced drug efficacy and reduced toxicity; however, TR of a drug does not ensure that the drug is retained at the site of action. Targeted drug delivery using a range of systems is the focus of Chapter 5 and will not be considered further here.

A word of caution: While we have just categorized and defined the different types of drug delivery, to highlight the different principles, it should be noted that for any given dosage form, a strict categorization may be far from simple. For example, in practice it is often difficult to clearly differentiate between SR and CR delivery systems. In fact, after a history of about 60 years of development in drug delivery, the terms CR and SR are often now used interchangeably. SR systems may also show initial IR properties. Furthermore, the release of DR systems, after the delay, may be of an IR or SR type. The development of any release system should not be guided by the desire to achieve a predetermined type of release, but by the principle to develop an optimal delivery system for the specific drug and treatment regime, to maximize patient adherence and therapeutic outcome.

2.1.2   Mechanisms of Controlling Drug Release

Drug release can be controlled via a variety of mechanisms. Chemical approaches utilize chemical degradation, or enzymatic degradation, to effect CR of drug molecules from a drug–polymer conjugate. Physical approaches focus on four main mechanisms:

2.1.2.1   Dissolution Control

The drug is associated with a polymeric carrier, which slowly dissolves, thereby liberating the drug. The polymeric carrier can be as follows:

1. A reservoir system (encapsulated dissolution system), whereby a drug core is surrounded by a polymeric membrane. The rate of drug release is determined by the thickness and dissolution rate of the membrane.
2. A matrix system, whereby the drug is distributed through a polymeric matrix. Dissolution of the matrix facilitates drug release.

2.1.2.2   Diffusion Control

The drug must diffuse through a polymeric carrier. Again, two main types of design system are used:

1. A reservoir system, whereby the drug is surrounded by a polymeric membrane, known as a rate-controlling membrane (RCM). The rate of drug release is dependent on the rate of diffusion through the RCM.
2. A matrix system (also known as a monolithic device), whereby the drug diffuses through a polymeric matrix.

2.1.2.3   Osmotic Pressure Control

Osmotic pressure induces the diffusion of water across a semipermeable membrane, which then drives drug release through an orifice of the DDS.

2.1.2.4   Ion-Exchange Control

Ion-exchange resins are water-insoluble polymeric materials that contain ionic groups. Charged drug molecules can associate with an ion-exchange resin via electrostatic interaction between oppositely charged groups. Drug release results from the exchange of bound drug ions with ions commonly available in body fluids (such as Na^+, K^+, or Cl^-).

However, in many cases, a combination of factors is involved, and this can be exploited to give a range of release profiles. Mechanisms of drug release will be discussed further in the following sections.

2.2   IMMEDIATE RELEASE DELIVERY SYSTEMS

As mentioned earlier, the primary role of IR dosage forms is to give a rapid onset of drug action and to achieve a high plasma concentration quickly. Most drugs exert their action through interaction with receptors in the body. To enter into the body, for example, from the GI tract and to interact with biological receptors, the drug must be in solution, and hence liquid dosage forms can offer advantages in this respect. If a drug is delivered in a SDF, disintegration and dissolution are required so that the drug is released into solution. Therefore, both solubility, and dissolution rate, control drug release from the delivery system. However, drug absorption across the epithelial barriers is also required; therefore, drug membrane permeability must also be considered. It should be noted though, that influencing permeability by means of drug delivery is much more difficult than influencing disintegration, dissolution, or solubility. The most common type of any solid dosage form (SDF) is the IR tablet. These tablets are therefore often referred to as plain or conventional tablets. If a fast onset of drug action is required from SDFs, key factors that can be considered are (1) improving drug dissolution, (2) improving drug solubility, and (3) improving dosage form disintegration.

2.2.1 IMPROVING DRUG DISSOLUTION RATES AND SOLUBILITY

During dissolution of solid drug particles in the GI tract, the molecules in the surface layer dissolve, creating a saturated solution of the drug at the surface of the particle, which in turn, through diffusion, leads to the formation of a drug concentration gradient around the particles: from saturation concentration (C_s) at the surface, to the equilibrium bulk concentration of the dissolved drug (C_t; initially this will be zero). This diffusion layer has to be crossed by further dissolving drug molecules, which then diffuse into the GI tract and eventually reach the epithelial membrane for absorption. The Noyes–Whitney equation is often used to describe the dissolution of particles and can be applied here to understand the dissolution process. It states that the rate of mass transport of drug molecules through the aforementioned diffusion layer (dm/dt) is directly proportional to the available surface area (A) and the concentration difference across the diffusion layer ($\Delta C = C_s - C_t$) and inversely proportional to the thickness of the diffusion layer (h). The term D is the diffusion coefficient of the dissolving molecules (in our case, the drug):

$$\frac{dm}{dt} = \frac{DA(C_s - C_t)}{h} \tag{2.1}$$

If the difference between the saturation concentration (C_s) and the concentration of drug dissolved (C_t) is large, i.e., if the solubility of the drug is high and the dose of the drug to be dissolved is low, Equation 2.1 may be simplified to

$$\frac{dC}{dt} = k\Delta C \tag{2.2}$$

From the Noyes–Whitney equation, we can see that the dissolution rate can be increased if the surface area available for dissolution is increased (i.e., if the size of the particles from which dissolution occurs is decreased) and if the saturation concentration (i.e., the solubility) of the drug is high.

For IR DDS, high drug solubility and a fast dissolution rate are of the utmost importance to achieve the desired release and thus the desired plasma concentration profile. However, if the solubility and dissolution rate of a given drug are low, there is a range of options to improve dissolution rate and solubility, such that an IR dosage form can be formulated. Mechanisms to improve the water solubility of solid oral IR dosage forms are discussed in detail in Chapter 3. To summarize here, some of the options include using salt forms of drugs, rather than using the free acid or base form; using micronized or even nanonized drug particles to prepare the SDF in order to improve the dissolution rate; and using amorphous forms of the drug rather than a crystalline form. Of these methods, currently the formulation of drugs as salts is a commonly used approach to improve solubility and dissolution rate.

To increase the solubility of a drug through salt formulation, the pKa values of the drug and the counterion need to be considered; as a "rule of thumb," for acidic drugs, the pKa of the counterions should be at least 2 pH values higher than that of the drug, and for basic drugs, the pKa of the counterion should be at least 2 pH units below that of the drug. In pharmaceutical development, a lot of effort is placed on finding the best salt form for a given drug, but it is fair to say that increasing dissolution rate and solubility are not the only concerns in this so-called salt screening process, but stability and processability also need to be considered. It should also be noted that not all drugs can be converted into a salt form, and even in cases where this is possible, it might not be the pursued option, due to stability or processing problems.

Another method to improve the solubility of a compound is to use a metastable polymorphic crystalline form or the amorphous form of the drug, instead of the stable crystalline polymorph. While using a metastable polymorph only in exceptional cases leads to a substantial increase in solubility, using the amorphous form instead of a crystalline form of the drug has gained considerable interest within the pharmaceutical industry in recent years as a suitable strategy to address solubility issues of poorly water-soluble compounds. The amorphous form does not have a crystal lattice structure, but the molecular arrangement is that of a "frozen liquid." The absence of a crystal lattice means that the molecules in an amorphous form are in a higher energetic state compared to the crystalline form and thus show an often very significant increase in solubility. However, the use of amorphous forms has to be treated with great caution as the amorphous form is thermodynamically unstable, and crystallization of the drug may occur during manufacturing, storage, or even administration.

2.2.2 IMPROVING DISINTEGRATION: FAST-DISPERSING PRODUCTS

From the Noyes–Whitney equation, we have seen that increasing the surface area available for dissolution will increase the dissolution rate. Thus, improving the disintegration time of a formulation has become a relatively common strategy, leading to the development of orally disintegrating tablets (ODTs) (also known as "fast melts") and orally dissolving/disintegrating films. Both systems offer the advantages of tablets (dose accuracy, ease of administration) combined with those of liquid dosage forms (easy to swallow and rapid drug absorption). In the case of ODTs, these are designed to disintegrate or even completely dissolve in a matter of seconds after contact with saliva, thereby removing the need to chew a tablet or swallow an intact tablet. FDA guidance on these formulations states that ODTs should have an in vitro disintegration time of approximately 30 seconds or less, determined using the USP disintegration test method, and that generally the ODT tablet weight should not exceed 500 mg. Current technologies used in commercially available ODTs can be divided into three main categories: (1) lyophilized tablets, (2) compressed tablets, and (3) others—including molded tablets, spray-dried powders, sugar floss, and films. Table 2.1 lists examples of various ODT technologies. In these systems, the ODT platforms offer bioequivalence to conventional oral SDFs.

TABLE 2.1
Examples of Orally Disintegrating Tablet Technology

Name	Proprietary Technology	Drug	Clinical Indication
Lyophilized tablets			
Zofran Zydis	Zydis	Ondansetron	Chemotherapy-associated nausea
Loperamide Lyoc	Lyoc	Loperamide	Diarrhea
Propulsid Quicksolv	Quicksolv	Cisapride	Oral prokinetic
Compressed tablets			
Lactimal ODT	Advatab	Lamotrigine	Epilepsy
Remeron Soltab	Orasolv/Durasolv	Mirtazapine	Depression
Prevacid Solutab	Flashtab	Lansoprazole	Heartburn, ulcers
Other			
Ralivia	Flashdose (cotton candy)	Tramadol HCl	Opioid analgesic
Benadryl FastMelt	Wow-Tab[a]	Diphenhydramine	Allergy
Zuplenz	PharmFilm (film)	Ondansetron	Chemotherapy-associated nausea

[a] Wow-Tab, without water tablet.

In the case of lyophilized systems, rapid disintegration can be promoted by forming the wafer-like, highly porous structure obtained using the freeze-drying process. In the case of the Zydis® system, the drug is dispersed in a matrix consisting of a water-soluble polymer (e.g., gelatin) to give the tablet structure and a sugar or sugar alcohol (typically mannitol). This highly porous structure promotes rapid penetration of water, dissolution of the matrix, and rapid disintegration.

The basis of compressed tablet ODTs is the use of superdisintegrants or effervescent agents, which improve the dispersion of the dosage form in the fluids of the GI tract. Superdisintegrants are commonly starch, cellulose, or poly(vinylpyrrolidone) (PVP) derivatives and examples include sodium starch glycolate, croscarmellose, and cross-linked PVP. As these are all cross-linked and water insoluble, when in contact with water, the polymers rapidly swell, resulting in fast disintegration.

Alternative approaches to enhancing disintegration time include the production of fibers from molten saccharides (sucrose, dextrose, or lactose) or polysaccharides. Flashdose® uses this technology, and the high solubility of the sugar components, combined with the porosity of the system, promotes rapid disintegration. In the case of Wow-Tab®, tablets are molded using highly soluble sugars that dissolve rapidly.

In addition to tablet formulations, thin-film technologies are now being used to prepare oral fast-dispersing systems. These films are several micrometers (e.g., 50–200 μm) in thickness and consist of hydrophilic polymers. The drug concentration is controlled through the concentration in the film and the film thickness. The films are prepared by liquid casting and evaporation of the solvent within ovens. The dried film is then cut into single unit doses. These systems offer rapid disintegration, but are limited in drug loading and require specific packaging (protection from moisture). There is also a range of buccal, sublingual, and nasal IR DDS available, and the reader is referred to Chapters 8 and 10 respectively, for further details on these systems.

2.3 DELAYED RELEASE ORAL DRUG DELIVERY SYSTEMS

There is a range of systems that aim to delay drug release until the delivery system reaches the small intestine or the colon. Ideally, upon reaching these sites, drug release should be rapid and therefore the drug concentration versus time profiles are often similar to that of an IR dosage form, but the time between administration of the DDS and drug release (and thus its appearance in the plasma) is delayed. These issues are introduced here and discussed further in Chapter 7 (Section 7.5).

2.3.1 DRUG RELEASE WITHIN THE INTESTINE

Enteric coatings can be added to SDFs to prevent the release of the drug before the delivery systems reach the small intestine. There are three main reasons for enteric coating and delaying drug release until the intestine is reached:

1. *Protection.* Either the drug has to be protected from the acidic environment of the stomach against degradation, or the stomach has to be protected from the drug that may lead to irritation when released in the stomach (i.e., to prevent gastric mucosal irritation). Examples of drugs that require protection from degradation in the stomach include omeprazole, pantoprazole, erythromycin, and penicillin. Examples of drugs that irritate the stomach include aspirin, ibuprofen, and naproxen.
2. *Local action.* The drug is supposed to act locally in the small intestine and a high drug concentration in this part of the GI tract is desired. Examples of drugs that act locally include mebendazole and piperazine.
3. *Enhanced absorption.* It may be beneficial to enterically coat the dosage form to achieve a high drug concentration in the small intestine as this is the primary site of absorption for many drugs. The small intestine, due to its large surface area and high blood perfusion rate, has a higher capacity for absorption than the stomach, and most drug absorption occurs

in the proximal jejunum. Drug absorption in the small intestine is primarily through diffusion (although facilitated transport, active transport, and endocytosis can also occur; see Chapter 4, Section 4.3) and usually follows first-order kinetics (if no active transport processes are involved). This also means that the amount of absorbed drug will be higher if the drug concentration at the absorptive site is higher. In diffusion-controlled absorption, the pKa of the drug will be a key factor dictating drug absorption as generally uncharged molecules are preferably absorbed (see Chapter 4, Box 4.1).

2.3.2 MECHANISMS OF ENTERIC COATING

Enteric coating relies on exploiting the pH of the intestine to promote drug release. The pH of the human stomach is usually around 1.5–2 in the fasted state (and can rise to 4–5 in the fed state). However, the pH of the small intestine is higher, usually between 6 (in the duodenum) and 6–7 (in the jejunum and ileum). Therefore, by coating an IR SDF (e.g., a tablet or pellets) with a polymer that is insoluble in the stomach pH but soluble at intestinal pH, pH-triggered release can be promoted as the SDF moves from the stomach into the small intestine. However, it is worth noting that the pH in the stomach is dependent on the fed or fasted state of the patient and that the residence time in the stomach can vary from less than 1 hour to over 6 hours or longer. Given this variability, it is important that the dissolution of the polymer in the stomach is as low as possible. pH-sensitive polymers used for enteric coatings can be grouped based on their chemical structures: (1) cellulose derivatives (such as cellulose acetatephthalate and hypromellose phthalate), (2) poly(vinyl) derivatives (such as poly(vinyl acetate phthalate)), and (3) poly(methacrylates) (e.g., Eudragit® S and L). Plasticizers may be added to the polymer to improve the flexibility of the film coating and avoid cracking of the film, which would result in rapid drug release. In the formation of the enteric coating, the SDF is spray coated with a solution or dispersion of the polymer using fluid bed coaters or drum coaters. Tablets, pellets, granules, powders, microparticles, and even capsules can all be enterically coated.

2.3.3 DRUG RELEASE WITHIN THE COLON

In contrast to the small intestine, the large intestine has a notably smaller surface area but can still act as a site of drug absorption, for example, for drugs that have not been completely absorbed in the small intestine. However, due to the reduced water content of the large intestine, diffusion of drugs from the contents of the large intestine to the mucosa can be hindered. Drug release systems that are able to target release in the colon can be useful for the treatment of local diseases such as colorectal cancer and Crohn's disease, but can also be used for systemic drug delivery; similarly to the small intestine, the pH conditions are more favorable than that of the stomach; in addition, the enzymatic activity in the colon is much lower than the small intestine. However, the microorganism concentration in the colon is high and may influence drug degradation, metabolism, and absorption. Examples of drugs that are intended for enteric delivery are shown in Table 2.2. Delaying drug release until the delivery system has reached the colon can be achieved by three main strategies (see also Chapter 7, Section 7.5.3):

1. *Enzymatic triggered release.* This strategy exploits the presence of microorganisms in high concentrations in the colon either to convert prodrugs into an active form or to enzymatically degrade a polymer coating to release the drug. A range of polymers can be used to promote drug release in the colon including natural and semisynthetic polysaccharides, such as amylose, chitosan, cyclodextrins, guar gum, pectin, inulin, and xylan. These polymers are not digested in the stomach or small intestine but can be degraded by the microflora in the colon. These polymers may be used either to form film coatings (e.g., pectin) or to build a matrix within which the drug is embedded. However, it should be noted that the microflora can be adversely affected by antibiotics and diet and therefore compromise the absorption from these release systems.

TABLE 2.2

Examples of Delayed Release Oral Drug Delivery Systems

Product	Drug	Clinical Indication
Small intestine		
Feosol	Iron	Iron deficiency
Deltasone	Prednisone	Inflammation
Vermox	Mebendazole	Antihelminthic
Large intestine		
Entocort	Budesonide	Morbus Crohn, colitis ulcerosa
Azulfidine	Sulfasalazine	Morbus Crohn, colitis ulcerosa

2. *pH controlled release.* Poly(methacrylate) derivatives that break down at the pH found in the colon (pH 7) can be used to promote pH-triggered release, as was described for release in the small intestine. A pH of 7 is reached at the distal part of the ileum, so drug release can occur here, and the drug will still pass into the colon.
3. *Time controlled release.* Using pH control alone is not without disadvantages; the upper part of the small intestine has a pH close to 7, and hence this approach can give poor control over the site of drug release. The process can be improved by using a temporal CR mechanism in addition to a pH-responsive coat. A double coating may be used, with an outer enteric coat (which stops drug release in the stomach) and an inner coat to slow down release, thus giving time for the system to move into the colon prior to release of the majority of the drug in the delivery system. The Eudracol® system uses this approach. Here, the drug is incorporated into pellets that are coated with a slow-release methacrylate polymer, followed by a Eudragit® FS 300 coat that dissolves at pH 7. These pellets are then filled into capsules or compressed to tablets.

2.4 SUSTAINED RELEASE DELIVERY SYSTEMS

We have already discussed how we can modify the onset of drug release with both rapid IR and DR being available options. However, the *duration* of drug action is also a key parameter to consider. Maintaining the drug concentration within a therapeutic window for an extended period of time is not easily achieved by increasing the dose, as this can result in plasma concentrations moving from the therapeutic window to the toxic range. It is possible to prolong the time that drug plasma concentrations are within the therapeutic range by repeat dosing with an IR formulation. However, this is less convenient for the patient and leads to fluctuations in the drug plasma concentration profiles (Figure 2.3). Furthermore, patients can forget to take a dose resulting in irregular dosing frequency and drug plasma concentrations moving outside the therapeutic window. Therefore, it is beneficial if dosing intervals can be prolonged to once a day rather than three or four times a day (Figure 2.3). This can improve patient adherence and prolongs the time the drug plasma concentration is within therapeutic range.

SR dosage forms can be defined as delivery systems that release the drug over a prolonged period. With an oral delivery system, maintaining constant plasma levels for an extended time interval requires that the rate of drug absorption from the GI tract should be equal to the elimination rate (both by excretion and metabolism) of the drug. However, these profiles are occasionally not achieved with SR formulations. This is due to the fact that drug release from the delivery system may not be the rate-limiting factor in drug absorption and factors including drug solubility, stability, permeability, and dose are all factors to consider. Therefore, not all drugs may be effectively formulated into SR oral dosage forms. Table 2.3 summarizes considerations of drug properties that make a drug suitable for formulation into a SR oral dosage form.

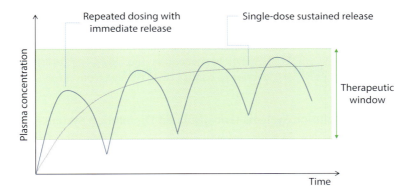

FIGURE 2.3 Comparison of plasma concentration–time profiles following (1) repeated dosing of an immediate-release dosage form and (2) a single dose of a sustained-release dosage form.

TABLE 2.3

Drug Properties to Consider When Formulating Sustained-Release Dosage Forms

Factor

Dose	Drugs with a high potency and thus low therapeutic dose are easier to formulate as a sustained release product. This ensures that the total dose required can be effectively incorporated within the SDF.
Therapeutic range	A large therapeutic range is advantageous. This can circumvent any possible toxicity concerns should the total amount of drug be released inappropriately (e.g., through failure of the sustained release coating).
Solubility	Low solubility drugs tend not to require sustained release, as their absorption is controlled by their dissolution profile. Through the Noyes–Whitney equation, we have seen that the dissolution rate of a drug is a function of its saturation solubility and the surface area of the drug particles. We can control dissolution rates by controlling particle sizes. Therefore, by modifying the particle size, we can reduce the dissolution rate and hence produce a sustained release rate.
Absorption	Generally, high absorption rates are useful, so that the drug *release rate* is the rate-limiting step for the appearance of the drug in the body.
Absorption window	The absorption window in the intestine should be large. If the drug is only absorbed in a certain part of the intestine, a large proportion of the dose may not be absorbed.
Elimination rates	If a drug has a short half-life, the dose required to maintain the therapeutic plasma concentrations can become too large to accommodate within the oral dosage form.

Should the drug candidate fit the requirements for formulation into a SR dosage form, there are a range of options that can be adopted, which can offer sustained drug release. The key physical controlling mechanisms of drug release from SR systems are dissolution and/or diffusion, from either matrix or reservoir constructs.

2.4.1 Dissolution-Controlled SR

The overall dissolution process for the drug particles of a SDF such as a tablet can be described as a two-step process. First, drug particles at the surface of the SDF dissolve and leave the surface. A boundary layer at the surface is formed, where a concentration gradient of the drug exists, from saturation concentration at the surface to the concentration in the bulk medium. Dissolved drug then diffuses further into the bulk medium. For poorly water-soluble drugs, the rate of dissolution of the drug is the release controlling step. Such drugs tend to have inherently SR properties, as their release is controlled by their dissolution profile. We have already discussed the Noyes–Whitney equation and its relevance for the dissolution of a drug (Equation 2.1). The parameters of the Noyes–Whitney equation can be modified to reduce the dissolution rate, in order to formulate a SR delivery system; for example, the particle size can be increased. An alternative approach is to use a slowly dissolving polymeric carrier to control drug release from the system, as described in the following text.

2.4.1.1 Reservoir Dissolution Systems (Encapsulated Dissolution Systems)

A reservoir of drug can be coated with a slowly dissolving polymeric film. Drug release can be controlled by controlling the thickness and the dissolution rate of the polymer film surrounding the drug reservoir. The delivery system is usually subdivided into smaller units, which are coated with coating layers of different thickness. The units are then loaded into capsules. This way, SR can be achieved, as the units with a thinner coating will dissolve faster and the units with thicker coating will dissolve at a later time. An example of this type of SR technology is the original Spansule® system (Smith, Kline & French Ltd.), described in Chapter 7 (Section 7.3).

2.4.1.2 Matrix Dissolution Systems

In matrix dissolution, a drug is homogeneously distributed throughout a polymeric matrix. As the polymer matrix slowly dissolves, drug molecules are released. Hydrophilic matrix tablets are a widely used SR dosage form for oral administration. However, although matrix *dissolution* can play an important role in drug release for these DDS, the predominant mechanism of drug release here is usually via drug *diffusion*—the drug diffuses through a viscous gel layer formed by the hydrophilic matrix, as described in Section 2.4.2.2.3.

2.4.2 DIFFUSION-CONTROLLED SR

In diffusion-controlled SR systems, drug molecules have to diffuse through either a polymer membrane (reservoir system), or a polymer matrix (matrix system), in order to be released. Two steps can be defined:

1. Liquid penetrates into the delivery system and dissolves the drug. This results in a concentration gradient of dissolved drug between the inside of the delivery system and the surrounding media.
2. Dissolved drug will then diffuse through the membrane (in case of reservoir systems), or through the matrix (in case of matrix systems) to be released. In reservoir systems, diffusion occurs through the membrane itself, or pores that form in the membrane. In matrix systems, diffusion often takes place through small pores that are formed upon contact of the matrix with the dissolution medium. Pore formers are typically water-soluble polymers, which are added to the matrix or reservoir.

2.4.2.1 Reservoir Diffusion Systems

In these systems, the drug is present in the core of the dosage form (the reservoir) and is surrounded by an inert polymer coating (also called a "film" or "membrane"). In a diffusion-controlled system, the coating should not dissolve; rather, the drug should continuously diffuse through the polymer coating (Figure 2.4), into the surrounding biological fluids (e.g., the GI tract, for oral DDS). For the process of drug diffusion out of the reservoir through the membrane, water must diffuse into the reservoir and dissolve the drug. The drug reservoir is on the inside of the coating, and due to the large amount of solid drug inside the system, the concentration of dissolved drug in the reservoir can be assumed to be constant (saturation concentration). The driving force for this type of DDS is drug diffusion out of the system, while maintaining saturation solubility within the system.

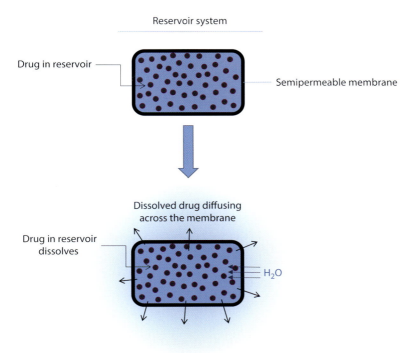

Reservoir system

Drug in reservoir

Semipermeable membrane

Dissolved drug diffusing
across the membrane

Drug in reservoir
dissolves

H_2O

FIGURE 2.4   Drug release from a reservoir system. Water diffuses into the dosage form across a semiperme-able membrane, dissolving drug within the reservoir. The drug within the reservoir dissolves and diffuses out of the dosage form.

The diffusion process is described by Fick's first law:

$$J = -D\left(\frac{dC}{dx}\right)$$
(2.3)

where
   J is the flux
   D is the diffusivity of the drug
   C is the drug concentration
   x is the area covered

It states that the amount of drug in solution that passes across a unit area (e.g., the membrane) is proportional to the concentration difference across this unit area. The passage of a drug across the area is termed the flux, J. Since we can assume a constant concentration of dissolved drug inside the reservoir systems and a negligible (i.e., also practically constant) concentration of the drug in the release medium (i.e., the GI tract), dC/dx can be assumed to be constant.

As the release of the drug occurs by diffusion of the drug molecules across the coating, the nature of this coating dictates the drug release profile, with both the thickness and the porosity being the controlling factors. A small quantity of a water-soluble polymer is sometimes added to the coating in order to form pores in the membrane, which aid in drug release.

As the surrounding film does not dissolve, the reservoir systems should maintain a constant size throughout the release process, and as long as a constant drug concentration gradient is maintained (i.e., undissolved drug remains in the reservoir), drug release can be considered constant, following a zero-order release rate:

$$M = kt \tag{2.4}$$

where
    M is the amount of drug released
    t is the time
    k is a constant

Thus, theoretically, zero-order release is possible. In practice, the surrounding film is usually saturated with drug, so that reservoir formulations usually result in an initial burst release of drug.

For non-pore-forming reservoir systems, the drug release rate can be described by the following equation:

$$\frac{M}{t} = CAK\frac{D}{l} \tag{2.5}$$

where
    C is the solubility of the drug
    A is the area
    l is the thickness of the membrane
    D is the diffusion coefficient of the drug in the membrane
    K is the partition coefficient of the drug between the membrane and the liquid

These factors can be adjusted to control the release rate, but can be regarded as constants for a given system.

For reservoir systems that form pores due to the incorporation of water-soluble polymers (e.g., hydroxypropyl methylcellulose), the release is described by

$$\frac{M}{t} = CAe\frac{D}{l}\tau \tag{2.6}$$

where
    the term e describes the porosity of the membrane
    τ is a measure of transport distance in the pores (also called the tortuosity)

For an oral DDS, if the polymer is not dissolved or degraded after transiting the GI tract, the insoluble coating may appear in the stool of the patient. This can cause the patient to assume the drug has not been released and absorbed (which may cause patient anxiety). In addition, if the polymer coating is faulty, "dose dumping" can occur. Given that the total amount of drug in an SR dosage form is usually two to three times higher than in an IR dosage form, this can give rise to toxicity issues. This risk can be reduced if pellets or granules are coated with an insoluble coating, which can then be either filled into hard gelatin capsules or compressed into tablets, e.g., Elantan LA®. Examples of polymers commonly used as coatings for oral reservoir systems are ethylcellulose and poly(ethylacrylate, methylmethacrylate, trimethylammoniummethacrylate chloride) copolymers (e.g., Eudragit® RS and RL).

2.4.2.2 Matrix Diffusion Systems

In matrix systems, the drug is dispersed in a lipid, or more often a polymer, that acts as a release matrix. This matrix can be insoluble, swellable, or erodible. The advantage of matrix systems compared to reservoir systems is that they are easy to produce using standard tableting technology.

There is no coating step required and dose dumping is unlikely to occur. However, a down side to these formulations is that it is difficult to achieve zero-order drug release.

In the following sections, the characteristics of different matrices will be described. It has to be noted that in the literature, several methods of classifying matrix systems can be found

1. On the basis of porosity of the matrix
   a. Homogeneous matrices (Figure 2.5a)
   b. Porous matrices (Figure 2.5b)
2. On the basis of the matrix former used
   a. Hydrophobic matrices (Figure 2.5a and b). Hydrophobic polymers used for matrix systems include poly(ethylene), poly(propylene), and ethylcellulose.
   b. Hydrophilic matrices (Figure 2.5c). Hydrophilic polymers used for matrix systems include methylcellulose, hydroxypropyl methylcellulose, polyethylene oxide, sodium alginate, xanthan gum, and poly(ethylacrylate, methylmethacrylate) copolymers such as Eudragit® NE 40D.
   c. Bioerodible matrices. Common polymers for erodible systems include poly(lactides) or copolymers of lactic and glycolic acid, which tend to be used for the parenteral route.

2.4.2.2.1 Homogeneous Matrix Systems

In homogenous polymer matrices (Figure 2.5a), initially the drug is homogeneously dispersed throughout the matrix. When this dosage form comes into contact with water, water diffuses into the matrix and the drug diffuses out. This results in the formation of a layer of drug-free matrix often referred to as the depletion zone. As more drug diffuses out, the layer of drug-free matrix becomes larger, and hence the remaining drug has to diffuse over a longer distance to be released, resulting in a decrease in the amount of drug released over time.

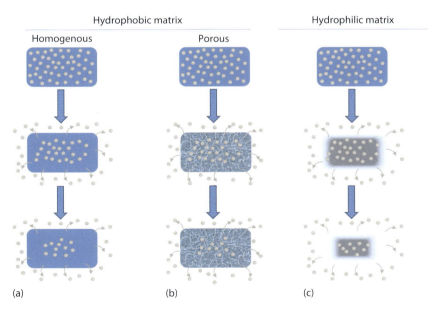

FIGURE 2.5 Drug release from matrix systems. (a) Homogenous polymer matrices with the drug initially homogeneously dispersed throughout the matrix. When this dosage form comes into contact with water, water diffuses into the matrix and the drug diffuses out. (b) Porous matrix systems where the drug leaves the matrix by dissolving in, and diffusing through, aqueous-filled pores of the matrix. (c) Bioerodible matrix systems where the drug release is controlled by the erosion/degradation of the matrix.

The release from a homogenous matrix can be described by the Higuchi equation:

$$Q = \frac{M}{A} = \sqrt{D2C_0C_st} \tag{2.7}$$

where
   $Q = M/A$ is the amount of drug released per surface area
   D is the diffusivity of the drug
   C_0 and C_s are the initial concentration of the drug in the matrix system and solubility concentration of the drug in the matrix at time t, respectively

The amount of drug released (M) does not follow zero-order kinetics but instead is a linear function of the square root of time, where k is a constant:

$$M = kt^{0.5} \tag{2.8}$$

2.4.2.2.2 Porous Matrix Systems

In porous matrix systems (Figure 2.5b), the drug leaves the matrix by dissolving in and diffusing through aqueous-filled pores of the matrix. Initially, drug located at the surface of the delivery system is dissolved, and therefore drug is released rapidly at the beginning. Once the drug molecules at the surface have diffused into the liquid (e.g., the GI tract), particles located deeper within the matrix will be dissolved and then diffuse through the pores out of the delivery system. As in the case of the homogenous matrix systems, as time proceeds, the drug molecules have a longer distance to travel, resulting in a tailing off of drug release with time. In these systems, the solubility of the drug as well as the shape and volume of the pores are important. These factors can be displayed in a different form of the Higuchi equation:

$$Q = \frac{M}{A} = \sqrt{D\frac{\varepsilon}{\tau}(2C_0 - \varepsilon C_s)C_st} \tag{2.9}$$

As in the case of drug release from a homogenous matrix, drug release in pore-forming matrix systems follows a linear function of the square root of time (Equation 2.8).

However, it should be mentioned that in order to apply the Higuchi equation, a few assumptions have to be made: the polymer should not swell or dissolve upon contact with the liquid and the diffusivity, D, must remain constant throughout the release process. If these prerequisites are not met, the Higuchi equation cannot be applied, as the geometry of the system changes constantly. As real systems usually do not fulfill all these requirements, a second equation has been introduced to describe drug release from a matrix:

$$\frac{Q}{Q_\infty} = kt^n \tag{2.10}$$

where
   Q is the amount of drug released
   Q_∞ is the amount of drug at infinite time
   n is the release exponent

The value of n describes the release kinetics of the system, and it is influenced by the geometry of the system. In the case of a thin-film matrix, n with a value of 0.5 denotes Fickian diffusion according to the Higuchi equation, whereas zero-order release is met if n equals 1.0.

2.4.2.2.3 Hydrophilic Swellable Matrix Systems

In hydrophilic matrix tablets for oral drug delivery, the drug is mixed with a hydrophilic gel-forming polymer such as hydroxypropyl methylcellulose (HPMC) and compressed into a tablet. On contact with water in vivo, the hydrophilic polymer chains swell and rapidly form a viscous gel matrix layer, on the tablet surface (see also Chapter 7, Figure 7.2). Diffusion is still the controlling mechanism for drug release: the drug has to diffuse through the gel layer to be released; although in many cases, as noted earlier (Section 2.4.1.2), polymer erosion and dissolution may also play a part in the release mechanism.

Generally, the polymer type and concentration are used to control drug release. Release rates are dependent on the movement kinetics of the polymer chains: the faster these relax, the faster the diffusion rate. Release rates can be slowed by increasing the viscosity of the gel network formed by

- Using polymers with high molecular weight
- Using polymer with a high degree of cross-linking

As such, mathematical modeling of drug release from these systems can be complex. The drug release mechanism may be defined by considering the magnitude of the drug release exponent. If $n = 0.5$, drug release is predominantly by Fickian diffusion; if $n = 1.0$, release is zero order, and if values of n lie between 0.5 and 1, release is thought to be a combination of the two. In addition to being easy to manufacture, these systems have the advantage that a burst release is unlikely to occur, as significant release of drug from the system can only take place once the matrix has swollen. Examples of hydrophilic matrix tablets for oral SR include Tegretol® XR (carbamazepine) and Glucophage® XR (metformin) (see also Chapter 7, Table 7.1).

2.4.2.2.4 Bioerodible Matrix Systems

Bioerodible matrix systems are SR systems where the drug release is not primarily governed by diffusion or dissolution, but is controlled by the erosion/degradation of the matrix (Figure 2.5c). They are usually also compressed into tablets. The drug is dispersed or dissolved throughout the tablet, and the erosion process can be described as a continuous liberation of polymer and thereby drug from the surface. As the matrix erodes, the tablet weight constantly decreases. We can also classify erodible or degrading matrices as hydrophilic matrix systems. Unlike the simple mathematical descriptions of release behavior for reservoir and matrix systems, the erosion process is more complex. The geometry of the system changes over time, and only very simplified equations have been proposed. However, for the first part of the erosion process, often release following zero-order kinetics is achieved.

The advantage of these systems is that they degrade completely and no "empty matrix" is excreted by the patient. The main disadvantage is the aforementioned difficulty in predicting or controlling the release kinetics of these systems. Additionally, care has to be taken, that the eroding or degrading polymers are nontoxic to the patient. As outlined earlier, common polymers for erodible systems include poly(lactides) or copolymers of lactic and glycolic acid, i.e., poly(lactide-co-glycolide) (PLGA), which tend to be used for the parenteral route. For example, Zoladex® subcutaneous implant comprises a PLGA matrix system for the delivery of the gonadorelin analog, goserelin. Lupron Depot® comprises a PLA/PLGA microsphere delivery system for the parenteral delivery of the GnRH analog, leuprolide, over a 1-, 3-, or 4-month period. The release rate is determined by the polymer composition and molecular weight. Further examples of bioerodible matrices for parenteral delivery are given in Chapter 6. For ophthalmic delivery, Ozurdex® is an intravitreal implant based on a slowly degrading PLGA matrix, which contains the anti-inflammatory corticosteroid dexamethasone for the treatment of macular edema and uveitis. The rod-shaped implant is injected directly into the vitreous cavity of the eye. The PLGA matrix slowly degrades in vivo, releasing dexamethasone over 6-month period.

TABLE 2.4

Examples of Gastro-Retentive Systems

Product	Drug	Clinical Indication
Proquin XR	Ciprofloxacin	Antibiotic
Zanocin	Ofloxazin	Antibiotic
Madopar HBS	Levodopa and benserazide hydrochloride	Parkinson's disease
Gaviscon effervescent floating liquid	Varies	Heartburn, antacid
Valrelease	Diazepam	Anxiety

2.4.3 GASTRO-RETENTIVE SUSTAINED RELEASE SYSTEMS

Promoting retention within the stomach or the upper small intestine can also prolong drug release. This can be achieved using a range of gastro-retentive systems including floating systems, high-density systems, and expandable systems. The residence time of the dosage form in the GI tract will depend on the size and the density of the system and the fed and fasted state of the patient. Examples of products available that utilize density differences include Madopar®. This formulation contains levodopa and benserazide hydrochloride as active drugs. The use of the gastro-retentive system is useful as levodopa has a short half-life and is absorbed in the upper parts of the small intestine, and therefore absorption is prolonged using this formulation. Table 2.4 shows a number of examples for gastro-retentive systems; they discussed in further detail in Chapter 7 (Section 7.5.1).

2.5 CONTROLLED RELEASE DELIVERY SYSTEMS

While SR dosage forms aim to prolong and extend drug release over a longer period of time, CR dosage forms aim to control the drug plasma concentration levels, by releasing the drug at a pre-determined rate. These systems aim to mimic drug infusions, which can give steady-state drug plasma concentrations by balancing the infusion kinetics with the elimination kinetics. While not all drugs benefit from CR delivery, drugs used for long-term treatment are especially suitable candidates for CR drug delivery, as the dosing frequency can be reduced and patient adherence improved. Unlike the previously discussed immediate, delayed, and SR formulations, CR is also used extensively for other routes of administration than the oral route, including transdermal, ophthalmic, vaginal, and parenteral delivery. The most common methods of achieving CR are via osmotic pressure, or using a rate-controlling membrane (RCM).

2.5.1 OSMOTIC PRESSURE–ACTIVATED CR SYSTEMS

Osmosis is the diffusion of molecules from regions of high concentration to regions of low concentration through a semipermeable membrane. This process is exploited in tablets formulated as oral osmotic systems (OROS). Using film coating technology, osmotic pump systems have been developed as CR systems. In these systems, the drug is incorporated within a reservoir and the tablet is coated with a semipermeable membrane that allows water to diffuse into the tablet core. However, unlike for SR reservoir systems, solubilized drug is not able to cross the membrane. When the tablet enters the GI tract, water diffuses across the semipermeable membrane and will dissolve the drug. As the drug dissolves, osmotic pressure builds within the tablet, forcing drug out through a hole drilled in the coating (Figure 2.6a). This is the simplest form of an OROS. Drug release from this system is controlled by the rate of water diffusion into the reservoir and also the rate at which the drug (in solution or suspension) can pass through the hole. The hole can be made by laser drilling, indentations in the tablet, or the use of pore formers in the coating. A drawback of these formulations is that the tablet must be exposed to a sufficient amount of fluid in order to build up an

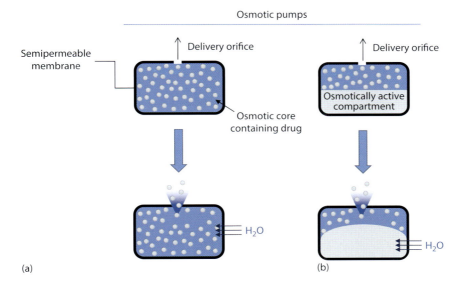

FIGURE 2.6 Oral osmotic systems. The drug is incorporated within a reservoir and the tablet is coated with a semipermeable membrane, which allows water to diffuse into the tablet core. Drug release is promoted by (a) the drug dissolving, building osmotic pressure within the tablet forcing the drug out through a hole drilled in the coating or (b) an osmotically active compartment that swells within the tablet forcing the drug out of the tablet via a drilled hole in the coating.

internal osmotic pressure. These systems are not generally used for poorly water-soluble drugs, as dissolution must occur rapidly to build up the required osmotic pressure for controlled drug release. To increase the osmotic pressure difference between the outside and inside of the delivery system, other water-soluble excipients can be added to the core.

The liquid influx into such an osmotic system (V/t) is dependent on the properties of the membrane (thickness h_m, permeability P_w, and surface area A_m) and the osmotic pressure difference between inside and outside of the system ($\pi_i - \pi_o$):

$$\frac{V}{t} = \frac{A_m P_w}{h_m}(\pi_i - \pi_o) \tag{2.11}$$

If π_i is constant, then a zero-order release rate is achieved.

This technology has been developed further by adding a drug-free layer called the "push layer," which is used to push the drug out and which is separated from the drug layer. The "push layer" usually contains osmotically active excipients and swellable polymers that expand upon contact with water (Figure 2.6b). With the exception of containing a push layer, these systems are otherwise prepared in an identical fashion to simple OROS, with an outer semipermeable membrane to allow water diffusion into the OROS (pull) and with a small hole for the drug to be pushed out (push). These systems are therefore often referred to as "push–pull" systems. The drug layer contains the drug but may additionally also contain an osmotically active excipient and suspending agents. This technology is used especially when the drug is poorly soluble and cannot build up an osmotic gradient on its own. Examples of products that use osmotic pressure technology for oral drug delivery are given in Table 2.5. The technology is also used for other delivery routes, including parenteral delivery, e.g., the Viadur® s.c. implant uses the Duros™ delivery technology based on osmotic pressure CR, to deliver the therapeutic peptide leuprolide acetate for 12 months, for the treatment of prostate cancer (see also Chapter 6).

TABLE 2.5

Example Products That Use Osmotic Pressure Controlled-Release Technology for Oral Delivery

Name	Drug	Clinical Indications
Adalat CC	Nifedipine	Hypertension and chronic stable angina
Procardia	Indomethacin	Analgesic (NSAID)
Cardura XL	Doxazosin Mesylate	Hypertension, benign prostatic hyperplasia
Concerta XL	Methylphenidate	Attention deficit hyperactivity disorder
Ditropan XL	Oxybutynin	Overactive bladder
DynaCirc CR	Isradipine	Hypertension
Glucotrol XL	Glipizide	Type 2 diabetes
Invega	Paliperidone	Schizophrenia; psychotic or manic symptoms

2.5.2 CONTROLLED RELEASE VIA A RATE-CONTROLLING MEMBRANE

In reservoir CR systems, the drug diffuses through a polymeric membrane at a controlled rate. In order to achieve constant drug release, a reservoir of the drug is encapsulated by a "rate-controlling membrane" (RCM; Figure 2.4; see also Chapter 1, Figure 1.3). These systems can use either nonporous or microporous membranes to achieve CR. As long as drug is present in the reservoir system, the release rate from the system will be constant and achieve a zero-order kinetic profile. Drug diffuses out of the dosage form, thereby reducing the concentration of dissolved drug within the system. Further drug from the reservoir dissolves, maintaining the concentration in the system. Consequently, with these systems, not all the dose is released from the system; there must be excess drug, in order to maintain a saturated drug concentration to drive zero-order release.

It is obvious that these systems show a large similarity with the SR reservoir systems described in Section 2.4.2.1. The method of release from both systems is similar, i.e., drug diffusion through an outer membrane. Theoretically, SR aims to prolong drug release over a longer period of time, whereas CR aims to provide steady-state plasma concentrations. However, as outlined in the Section 2.1, in practice, it is often difficult to clearly differentiate between SR and CR delivery systems, and nowadays the terms are often used interchangeably. One distinguishing feature that can be applied between the two is that in CR systems, the dose of the drug is not used to characterize the system, but rather the *release rate* is the characteristic factor (e.g., µg/hour). Another feature is that the term CR is typically associated with an RCM comprising polymers such as microporous poly(ethylene-co-vinylacetate) and silicon rubber, i.e., different polymers than those described earlier for SR.

Examples of CR systems using an RCM are presented in Table 2.6 and are also given in the relevant chapters throughout this book. Reservoir CR systems for transdermal delivery are also known as membrane-controlled transdermal therapeutic systems (TTS membrane). These devices were developed as systems that release the drug to the skin at a controlled rate. In order to achieve this, the release of drug from the patch is slow, so that drug release from the patch is the actual rate-limiting step, not drug absorption through the stratum corneum of the skin. Suitable physicochemical properties of the drug include a low molecular weight and high lipophilicity. In most cases, drug molecules diffuse into the RCM prior to use and are more readily available than from the bulk of the transdermal therapeutic systems (TTS), providing a "burst effect," i.e., a fast initial release. Used patches will also still contain drug (as described earlier, an excess of drug is present initially, to provide the driving force for CR). This can present safety issues, as for example, used fentanyl TTS patches still contain notable fentanyl concentrations and require additional guidelines for their safe disposal. Other TTS membrane technologies are presented in Table 2.6 and discussed further in Chapter 9 (see Figure 9.4).

TABLE 2.6

Examples of Products Using a Rate-Controlling Membrane for Controlled Release

Product	Drug	Clinical Indication
Transdermal therapeutic systems		
Transderm-Scop	Scopolamine	Travel sickness
Durogesic	Fentanyl	Analgesic
Ocular inserts		
Ocusert	Pilocarpine	Glaucoma
Implants		
Norplant (s.c.)	Levonorgestrel	Contraceptive
Progestasert (intrauterine)	Progesterone	Contraceptive
Mirena (intrauterine)	Levonorgestrel	Contraceptive

CR systems based on an RCM are also widely used in ophthalmic delivery (see also Chapter 13). With ocular inserts, the drug is again delivered via diffusion out of a reservoir. Ophthalmic drugs (e.g., pilocarpine) are released at an almost constant rate (e.g., 20 or 40 μg/hour), which minimizes side effects caused by initial high peak concentrations of the drug and also significantly increases the dosing interval (e.g., to 4 or 7 days).

In parenteral delivery, CR systems using an RCM are used to provide constant drug release for prolonged periods. For example, the Norplant® contraceptive implant is a set of six flexible silicone rubber copolymer rods, containing levonorgestrel. The silicone polymer serves as the RCM and releases the drug continuously at the rate of 30 μg/day over a 5-year period. These and other long-acting parenteral systems are described in Chapter 6.

2.5.3 Controlled Release via Matrix Systems

Matrix diffusion systems release a drug via diffusion through a polymeric matrix, not through an outer RCM. As described earlier, the release profile of a drug from a matrix diffusion–controlled release system is typically not constant, but decreases with time (Equation 2.8). Thus, these systems provide release profiles that tail off with time, rather than a predetermined, zero-order rate. However, in some cases, it can be possible to control the release of drug molecules from a matrix system at a preprogrammed, more constant rate, for example, by carefully controlling the loading level and polymer solubility of the drug, drug diffusivity in the polymer matrix, and other factors. Examples include some matrix diffusion–controlled release TTS systems for transdermal drug delivery, e.g., Nitroderm® TTS (nitroglycerin) and Transtec® (buprenorphine) (see also Chapter 9, Figure 9.4). These TTS systems can be cut into smaller segments as no dose dumping occurs, allowing for specific drug treatment.

2.6 CONCLUSION

In this chapter we have discussed the different types of drug release from dosage forms or DDS, as a prerequisite for the drug to reach the blood stream. We have seen that the release of drugs from the delivery systems can be immediate, delayed, sustained, or controlled. The underlying mechanisms of drug release will therefore also be different, ranging from dissolution, to diffusion, erosion, and osmotic pressure; release is also described by different kinetics.

As a consequence of these different release mechanisms and kinetics, the plasma concentration–time curves for the same drug can be very different, allowing the formulation scientist to control the appearance of drug in the body, thus going a long way in controlling the delivery of the right amount of drug at the right time to the patient.

We should however be aware of the fact that these different systems, mechanisms, and kinetics often do not appear in a "pure form" for a given DDS and grey areas, such as a mixture of mechanisms and kinetics, are common. It is therefore necessary to experimentally, or increasingly also by computational methods, investigate the release mechanisms and kinetics for a developed delivery system. It should also be noted that we are using the plasma concentration–time curve as a proxy for the appearance of the drug in the body, but that this is not identical to appearance of the drug at the actual target site in the body, or even of the medications effectiveness. Just how we can improve the appearance of the drug at the right target site in the body, while minimizing drug distribution in other parts of the body (i.e., to maximize effectiveness while minimizing side effects), is the topic of Chapter 5.

FURTHER READING

Chien, Y.W. 1991. *Novel Drug Delivery Systems*. London, U.K.: Informa Healthcare.

Hadgraft, J. and M.S. Roberts. 2003. *Modified-Release Drug Delivery Technology*. New York: Marcel Dekker.

Kydonieus, A.F. 1991. *Treatise on Controlled Drug Delivery: Fundamentals, Optimization, Applications*. Boca Raton, FL: CRC Press.

Lee, V.H.L. and S.K. Mukherjee. 2006. Drug delivery: Oral colon-specific. In *Encyclopaedia of Pharmaceutical Technology*, ed. J. Swarbrick, pp. 1228–1241. London, U.K.: Informa Healthcare.

Li, X. 2005. *Design of Controlled Release Drug Delivery Systems*. New York: McGraw Hill.

Liu, F., S. Pygall, and E. McConnell. 2011. *Update on Polymers for Oral Drug Delivery*. Smithers Rapra Technology. Akron, OH: Smithers Rapra Technology.

McGinity, J. and L.A. Felton. 2008. *Aqueous Polymeric Coatings for Pharmaceutical Dosage Forms, Drugs and the Pharmaceutical Sciences*, 3rd ed. London, U.K.: Informa Healthcare.

Perrie, Y. and T. Rades. 2012. *Fasttrack: Pharmaceutics—Drug Delivery and Targeting*, 2nd ed. London, U.K.: Pharmaceutical Press.

Wen, H. and K. Park. 2010. *Oral CR Formulation Design and Drug Delivery Theory to Practice*. Hoboken, NJ: John Wiley & Sons, Inc.

Wilson, C. and P. Crowley. 2011. *Controlled Release in Oral Drug Delivery*. New York: Springer.

3 Improving the Water Solubility of Poorly Soluble Drugs

Kohsaku Kawakami and Anya M. Hillery

CONTENTS

3.1 INTRODUCTION

Ideally, drug development would involve the selection of active pharmaceutical ingredients (APIs) that possess ideal drug delivery characteristics, followed by their development using simple dosage forms. However, the reality is that increasingly formulators must work with APIs that have challenging physicochemical properties, including poor water solubility.

The increase in proportion of poorly soluble candidates is attributed to both improvements in synthesis technology, which has enabled the design of very complicated compounds, and also a change in focus in the discovery strategy of new APIs, from a so-called phenotypic approach to a target-based approach. The phenotypic approach involves trial-and-error methodology in which compounds are tested against cells, tissues, or the whole body. This approach takes into account various physicochemical and biological factors that may affect the efficacy of candidates, including solubility, protein binding, and metabolism. In the target-based approach, candidate compounds are screened against specific targets, based on hypotheses concerning action mechanisms. Lead compounds are typically dissolved in dimethyl sulfoxide for high-throughput screening (HTS), which means that even very poorly soluble drugs can be tested. Although the HTS approach provides a clear lead with respect to molecular design, compounds with poor aqueous solubility can progress to development after screening.

Poor water solubility has important ramifications for the drug discovery process, as poorly soluble lead compounds cannot be adequately formulated for subsequent preclinical studies in animals. Thus, it may not be possible to follow up potentially promising leads, which instead have to be dropped from the discovery process, never realizing their true potential. Although it may be possible to overcome the solubility problem by chemical modification of the drug, in many cases this is not feasible.

Poor water solubility also has important ramifications for drug bioavailability. In order to cross an epithelial interface, the drug must usually be dissolved in the biological fluids at that interface. For example, for the oral route, the first step in the oral absorption process is dissolution of the drug in the gastrointestinal (GI) lumen contents. A drug that is poorly soluble in the aqueous GI fluids will demonstrate poor and erratic dissolution, with concomitant low absorption and thus poor bioavailability—even if the drug possesses good intestinal permeability characteristics. Furthermore, the rate of intestinal absorption is driven by the concentration gradient between the intestinal lumen and the blood. A low concentration gradient is a poor driver for absorption, with a concomitant retarded flux across the intestinal epithelium.

As described in detail in Chapter 7, a significant hurdle associated with the oral route is the extreme variability in GI conditions, which can cause large intra- and interindividual variability in pharmacokinetic profiles. Poor water solubility exacerbates this variability, as there is a lack of dose proportionality for these compounds, as well as significant variability depending on the presence of food and fluids in the GI tract. The activity of bile salts on drug solubilization is a further important variable. These formulation and bioavailability concerns are equally relevant for poorly soluble drugs delivered via alternative epithelial routes, such as the pulmonary, topical, nasal, vaginal, and ocular routes.

3.1.1 Biopharmaceutics Classification System

The Biopharmaceutics Classification System (BCS) classifies drugs into four categories, based on their aqueous solubility and ability to permeate the GI membrane (Figure 3.1). (However, it should be noted that the BCS is relevant to permeation across all biological membranes, not just the GI tract.) Based on pioneering work by Gordon Amidon at the University of Michigan (Amidon et al. 1995), the system has been adopted by the U.S. Food and Drug Administration (FDA) to allow pharmaceutical companies a waiver of clinical bioequivalence studies (a biowaiver), when seeking regulation of postapproval changes and generics. Increasingly, the BCS is being used as a tool in product development, to flag up potential solubility and permeability difficulties that may be associated with lead compounds.

A drug is considered highly soluble if its highest dose strength is soluble in less than 250 mL of water, as tested over a pH range of 1–7.5. A drug is considered high permeable if the oral absorption compares favorably (i.e., higher than 90%) to an intravenous injection of the drug. Absorption in vivo can be carried out by monitoring the appearance of the drug in the plasma after oral administration. Intestinal permeability may also be assessed by other methods, including in vivo intestinal perfusions studies in humans, in vivo or in situ intestinal perfusion studies in animals, in vitro permeation experiments with excised human or animal intestinal tissue, and in vitro permeation experiments across epithelial cell monolayers, such as the Caco-2 cell line.

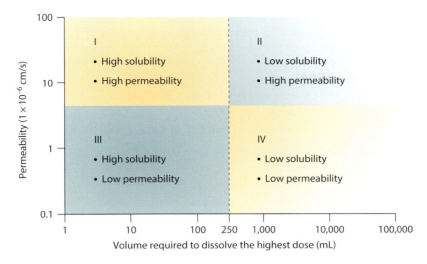

FIGURE 3.1   The Biopharmaceutics Classification System. (Courtesy of Particle Sciences, Inc.)

Using the BCS, four distinct classes of drug can be defined as the following:

- Class I drugs possess characteristics that ensure good bioavailability: they dissolve rapidly in the GI fluids and then rapidly permeate the epithelial barrier.
- Drugs that fall into Class II possess good permeability characteristics, but they have poor solubility, which limits their bioavailability. Approximately 35%–40% of the top 200 drugs listed in the United States and other countries as immediate-release oral formulations are practically insoluble (see also Chapter 20, Figure 20.6). The bioavailability of a Class II drug can be markedly improved by improving its solubility: various methods to improve drug solubility are the focus of this chapter. The fact that about 40% of the top marketed drugs are practically insoluble, yet are nevertheless used commercially, is a testimony to the success of current solubilization methods.
- Class III drugs, although highly soluble, possess poor permeability. Permeability across epithelial barriers and strategies to improve epithelial permeability are described in Chapter 4.
- Class IV drugs have both poor solubility and permeability. In the case of Class IV drugs, improving the solubility may help somewhat toward improving bioavailability, although poor permeability will still be an issue.

3.1.2   STRATEGIES TO IMPROVE WATER SOLUBILITY

Poor solubility and permeability problems may be addressed at the chemical level, via lead optimization: this approach is described in Chapter 20 (Section 20.8). This chapter describes approaches used to increase the solubility of a poorly soluble API. A wide range of approaches can be used, as summarized in Table 3.1.

Which approach to use is partly determined by the nature of the drug. Poorly soluble drugs, i.e., Class II and Class IV of the BCS classification situation, can be further subclassified into two types of molecules (Bergström et al. 2007):

1. "Grease ball": highly lipophilic compounds, with a high log P (>4) and a low melting point (<200°C). These compounds cannot form bonds with water molecules; thus, their solubility is limited by the solvation process.

2. "Brick dust": compounds usually with low energy, highly stable crystal forms, with a high melting point (>200°C), and with poor water and lipid solubility (log $P < 2$). The water solubility of such compounds is restricted due to strong intermolecular bonds within the crystal structure.

TABLE 3.1

Mechanisms to Improve the Solubility of Poorly Soluble Active Pharmaceutical Ingredients

Approach	Example
Physical modification	Micronization
	Nanosizing
Chemical modification	Prodrug formation
Crystal engineering	Salt formation
	Polymorphs
	Cocrystals
Formulation approaches	Solvent composition
	• Cosolvents
	• pH adjustment
	• Formulation excipients: surfactants, oils, etc.
	ASDs
	Cyclodextrin inclusion complexes
	Colloidal systems
	• Micelles
	• SMEDDS
	• Liposomes
	• Nano- and microparticulates

ASDs, amorphous solid dispersions; SMEDDS, self-microemulsifying drug delivery systems.

The solubility of "grease ball" molecules can be increased if appropriate formulation strategies are used to overcome, or at least improve, the solvation process. A traditional approach for parenteral formulations is to administer the drug with a cosolvent: as a mixture of water with a water-miscible solvent such as propylene glycol, ethanol, and poly(ethylene glycol) (PEG) 300. However, even with the use of a cosolvent, it may only be possible to achieve low drug loading. Additionally, harsh vehicles such as organic solvents carry the risk of toxicity, particularly cardiotoxicity, in vivo. Precipitation of the drug on dilution with the body fluids may also occur, causing pain and inflammation at the injection site, as well as the possibility of emboli. For intramuscular delivery, precipitation of the formulation may result in the formulation acting more like a depot injection, resulting in the delayed absorption of the drug. Other formulation strategies for "grease ball" molecules include the use of cyclodextrin (CD) inclusion complexes, and the use of micelle and emulsion-based delivery systems. In contrast, "brick dust" molecules are not only poorly soluble in water, but also in oils, rendering them unsuitable for many lipid-based formulation approaches. For "brick dust" molecules, the main strategies used tend to focus on crystal modification, including salt formation and cocrystals.

A further approach is to enhance drug dissolution kinetics. Drug solubility is an equilibrium measure; the rate at which solid drug, or drug in a formulation, passes into solution (i.e., the dissolution rate) is also a critically important parameter. Because intestinal transit time is relatively rapid, a drug with a very slow dissolution rate may not have sufficient time to dissolve in the GI fluids for absorption to take place. Increasing the dissolution kinetics can therefore result in an improved bioavailability for oral formulations. Mechanisms of improving drug dissolution include reducing particle size (e.g., NanoCrystals®), selecting a metastable polymorphic form, and using amorphous solid dispersions (ASDs).

All these approaches are described in this chapter, beginning with a discussion on particle-size reduction technologies.

3.2 CRYSTAL SIZE: NANOSIZING

According to the Noyes–Whitney equation (Chapter 2, Equation 2.1), the dissolution rate of drug particles is proportional to the surface area of the particles in contact with the dissolution medium. A decrease in drug crystal size results in an increased surface area to volume ratio (Figure 3.2);

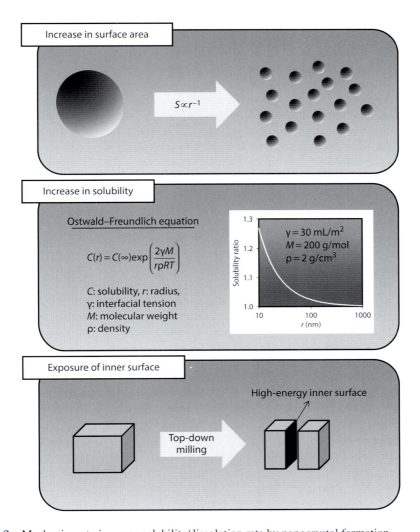

FIGURE 3.2 Mechanisms to increase solubility/dissolution rate by nanocrystal formation.

therefore, reducing the crystal size of a drug powder will increase its dissolution rate. For example, if the crystal size is reduced from 1 μm to 100 nm, the surface area increases 10-fold, which should lead to a 10-fold enhancement of the dissolution rate. For drugs where dissolution is the rate-limiting step, improved dissolution in vivo will translate to higher, and more uniform, bioavailability. Furthermore, as described by the Prandtl equation, the diffusion layer thickness around the drug crystal may also be decreased, thus resulting in an even faster dissolution rate.

The solubility per se also increases as a result of reducing crystal size. Assuming that a particle is spherical, dependence of solubility on particle size can be described by the Ostwald–Freundlich equation:

$$C(r) = C(\infty)\exp\left(\frac{2\gamma M}{r\rho RT}\right) \qquad (3.1)$$

where

    $C(r)$ and $C(\infty)$ are the solubilities of a particle of radius r and of infinite size

    γ, M, and ρ are the interfacial tension at the particle surface, the molecular weight of the solute, and the density of the particle, respectively

According to this equation, solubility increases with a decrease in particle size, i.e., an increase in surface curvature. However, an example of the calculation in Figure 3.2, in which typical values are substituted for Equation 3.1, shows that an increase in solubility is almost negligible for crystals of 100 nm (1.02-fold) and only 1.27-fold at 10 nm. A crystal size of 10 nm cannot be produced with current formulation technologies. Additionally, the API crystal may not be spherical, whereas this effect is only valid when an increase in surface curvature is achieved.

Nanocrystals produced by top-down technologies (see Section 3.2.2.1), should also expose high-energy surfaces to the outer environment. This further enhances the dissolution rate (Figure 3.2). This mechanism is not applicable for nanocrystals produced by bottom-up procedures.

An increase in the surface area is thus likely to be the dominant mechanism for the enhanced dissolution in most cases. Micronization, the process of reducing drug crystal to the micron size range via milling techniques, has long been used in the pharmaceutical industry as a means of improving drug formulation and oral bioavailability. Recent improvements in the technology now allows for size reductions to extend even further: "nanosizing" refers to API crystal size reduction down to the nanometer range, typically ca. 100–300 nm, thereby providing a considerable increase in surface area and thus dissolution rate. "Nanomaterials" are defined by the FDA as materials with a length scale of approximately one to one hundred nanometers in any dimension. Although current commercial nanosized preparations typically have a size range of 100–300 nm and are thus outside this range, exceptions are also accepted if the material exhibits dimension-dependent properties or phenomena.

In this chapter, the authors use the term "nanocrystal" in the general sense, to describe any drug crystal in the nanometer size range, whereas NanoCrystals® denote a patented technology of Elan Corporation, described in Section 3.2.2.1. Once a drug is nanosized, the nanocrystals can be formulated into various dosage forms, including injectables, tablets, capsules, and powders for inhalation; they are thus suitable for delivery via a wide variety of routes. Dispersions of nanocrystals liquid media are known as "nanosuspensions." Nanosuspensions prepared in water can be used as granulation fluids for the preparation of tablets; produced in oils, they can be used directly to fill capsules. Nanosuspensions may be stored in the liquid form, but postpreparation workup is also possible, such as spray-drying and freeze-drying, to obtain nanocrystals in a dry powder form. Sugars may also be added to formulations, to function as protectants during the drying process.

Nanosized formulations on the market include Avinza® (morphine sulfate), Focalin® XR (dexmethylphenidate hydrochloride), Megace ES® (megestrol acetate), Ritalin® LA (methylphenidate hydrochloride), Tricor® (fenofibrate), Triglide® (fenofibrate), and Zanaflex Capsules® (tizanidine hydrochloride).

3.2.1 Stabilization

Since nanocrystals have high-energy surfaces, stabilizers are needed to prevent irreversible aggregation. Steric stabilization can be achieved by the adsorption of hydrophilic polymers and/or surfactants onto the particle surface. Common polymers and/or surfactants used to provide steric stabilization include poly(vinylpyrrolidone) (PVP), hydroxypropyl methylcellulose (HPMC), hydroxypropylcellulose (HPC), α-tocopheryl PEG-1000-succinate (TPGS), polysorbate (Tween 80), and the pluronic surfactants F68 and F127. The repulsive steric layer prevents the particles from approaching each other. However, in most cases, the use of steric stabilization alone is not sufficient for nanosuspension stability. Further stability can be conferred by the adsorption of charged molecules, such as ionic surfactants (sodium lauryl sulfate and docusate sodium), onto the particle surface. In this case, electrostatic charge repulsion provides an electrostatic potential barrier to particle aggregation. Surfactants also help in the wetting and dispersion of the drug particles, which are usually very hydrophobic. Nanocrystals may have advantages over other surfactant-containing formulations such as emulsions and micelles, because the level of surfactant required is much lower, being only the amount necessary to stabilize the solid–fluid interface.

3.2.2 Manufacture

3.2.2.1 Top-Down Technologies

"Top-down" technologies involve disintegration methods, i.e., starting with coarse crystals and applying forces to reduce them to the nanocrystal range. As stated earlier, these top-down technologies expose high-energy surfaces to the outer environment, which further enhances the dissolution rate (Figure 3.2). The two most widely used technologies are media milling and homogenization.

Media milling. A milling chamber is initially charged with milling media: tiny balls, typically 1 mm or less, comprising materials such as ultradense ceramic media, or glass beads. A suspension of the API, with appropriate stabilizing agents, is then added to the chamber, and the chamber is rotated at a very high shear rate under a controlled temperature. The forces generated from the impaction of milling media with the drug cause crystal disintegration to the nanosize range. Processing time, as well as other operational parameters (milling speed, media load, media size, temperature, additives, etc.) can be tailored in order to maximize the crystal-size reduction process for each particular API. The process can produce stable, nanosized dispersions, with very tight, monodisperse, crystal-size distribution profiles. However, this method carries the risk of media ball erosion, which could result in the presence of unwanted media residues in the final product. For this reason, high-abrasion-resistance balls are used, and the final nanosuspension must be analyzed to ensure the absence of trace impurities.

NanoCrystals® are a proprietary wet milling technology from the Elan Corporation, which uses a highly crossed-linked polystyrene resin (PollyMill®) as the milling media (Merisko-Liversidge and Liversidge 2011). A crude slurry of the poorly water-soluble API, in a water-based stabilizer solution, is then added and subjected to shear forces. The NanoCrystal® particles of the drug are stabilized against agglomeration by the surface adsorption of patented, generally regarded as safe (GRAS), stabilizers.

The NanoCrystal® technology is used in a variety of commercially available preparations, where reformulation of the poorly soluble drug has resulted in many advantages. Sirolimus is a water-insoluble immunosuppressant drug, which was originally marketed as a self-emulsifying oral solution that required refrigeration and necessitated a complicated reconstitution procedure. In contrast, the NanoCrystal® tablet formulation (Rapamune®) offers improved bioavailability, less fluctuations in blood levels, easier storage (as no refrigeration is required), and improved palatability. Another example is the new chemical entity

MK-0869, which was developed to treat chemotherapy-induced nausea and vomiting. The NanoCrystal® formulation (Emend®) resulted in a 600% improvement in bioavailability. Furthermore, in contrast to the original formulation, there is no need to take the drug with food (an important issue for this patient group, who are suffering from nausea and vomiting). Invega Sustenna® is the first commercial depot formulation product using NanoCrystal® technology, for the delivery of paliperidone palmitate in the management of schizophrenia.

High-pressure homogenization. This process involves the application of high shear and impaction forces to drug suspensions in order to reduce their particle size to the nanoscale. A number of commercial technologies now exist, including SkyePharma's Dissocubes®, which can produce stable nanoparticle suspensions in water at room temperature. Triglide® (fenofibrate) is manufactured using this technology. The Nanopure® technology (PharmaSol GmbH) enables the production of nanosuspensions in nonaqueous media, for example with oils and PEG.

3.2.2.2 Bottom-Up Technologies

Nanocrystals may also be produced via "bottom-up" technologies, which involve assembly methods, i.e., starting from molecules in solution, then building up to form solid nanocrystals. Note that bottom-up procedures may produce metastable crystalline forms, including the amorphous state; the advantages and disadvantages of which are discussed in Section 3.6. Supercritical fluid technologies are being studied, although the technology is at an early stage. More developed methods include (1) precipitation and (2) emulsion as template.

Precipitation. Typically, the water-insoluble API is dissolved in an organic solvent, which is then mixed with an antisolvent, usually water. The addition of water causes a rapid supersaturation (nuclei formation) and growth of nanosized crystalline or amorphous drug. The limitation of this technique is that the drug needs to be soluble in at least one solvent, and the solvent needs to be miscible with the nonsolvent. This technology is available from DowPharma and BASF Pharma Solutions.

Emulsions as template. An emulsion is initially prepared comprising an organic solvent, or a mixture of solvents, loaded with the drug, which is dispersed in an aqueous phase containing suitable surfactants. The organic phase is then evaporated so the drug particles precipitate instantaneously to form a nanosuspension stabilized by surfactants. It is possible to control the particle size of the nanosuspension by controlling the size of the emulsion droplets. However, the possible use of hazardous solvents in this process raises safety and cost concerns.

A hybrid approach is also feasible: for example, Baxter's Nanoedge® technology employs both "bottom-up" and "top-down" approaches, through microprecipitation and also homogenization. Significant progress in nanocrystal preparation technology is being made, which should lead to substantially more products being brought to market in the near future.

3.3 SALT FORMATION

Salt formation, as described in Chapter 2 (Section 2.2.1), is the most common and effective method of increasing solubility and dissolution rates of acidic and basic drugs. In fact, more than 50% of the drugs currently listed in the USP are salt forms. The actual solubility of a salt, which is governed by the solubility products of the API and the counter salt, may not be much better than that of the free form. However, the *dissolution rate* is usually much faster, because of alterations in the microenvironmental pH. This phenomenon can be explained by considering that a weakly acidic drug is unionized in the stomach and therefore has a low dissolution rate. If the free acid is converted to the corresponding sodium or potassium salt, the strongly alkali sodium or potassium cations exert a neutralizing effect. Thus, in the immediate vicinity of the drug, the pH is raised to, for example, pH 5–6, instead of pH of 1–2 in the bulk medium of the stomach, thereby resulting in an alkaline microenvironment around the drug particle. This causes dissolution of the acidic drug in this localized region of higher pH, which gives rise to overall faster dissolution rates. When the dissolved drug diffuses away from the drug surface into the bulk of the gastric fluid where the pH is again lower, the free acid form may precipitate out. However, the precipitated free acid will be in the form of very fine, wetted, drug particles. These drug particles exhibit a very large total effective surface area in contact with the gastric fluids, much larger than would have been obtained if the free acid form of the drug had been administered. This increase in surface area results in an increased dissolution rate. Similarly, a strong acid salt of a weak base causes a localized drop in pH around the drug, which enhances the dissolution of weak bases.

An obvious limitation of this approach is that salt formation is limited to those APIs with at least one acidic or basic group. If the approach is feasible, sodium salts are most commonly used for acidic drugs and hydrochloride salts are most commonly used for basic drugs. However, this does not imply that these salts have necessarily the highest solubilizing potential; merely that the long history of their use means that there is correspondingly more information available about them. They also have low toxicity and a low molecular weight, thereby minimizing the overall mass of the dose as a salt form, compared to the free base/acid. It should be stressed though, that the most effective salt depends on the particular API under study—for optimal results, the choice of salt should be determined via a rigorous screening process.

Figure 3.3 shows an example of the effect of salts on solubility and dissolution rates of a basic compound, haloperidol, in 0.01 M HCl (Li et al. 2005). The fastest dissolution rate was observed for the mesylate salt, in accordance with its higher solubility than the hydrochloride or phosphate salts. The dissolution of the hydrochloride salt was suppressed because of the *common-ion effect*, i.e., release of hydrochloride ions was hindered because of the presence of the same ion in the testing solution (the experiment was carried out in 0.1 M HCl). The common-ion effect may be of particular relevance for the oral bioavailability for hydrochloride salts, as HCl is also present in the gastric fluids.

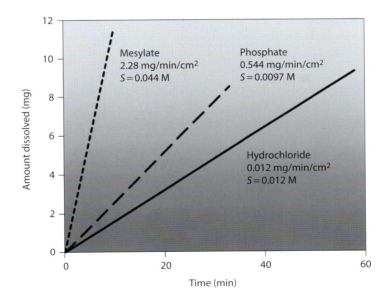

FIGURE 3.3 Dissolution profiles of haloperidol mesylate, hydrochloride, and phosphate at 37°C. The surface area of the test solids was 0.5 cm². The dissolution medium was 0.01 M HCl. (From Li, S., Doyle, P., Metz, S. et al.: Effect of chloride ion on dissolution of different salt forms of haloperidol, a model basic drug. *J. Pharm. Sci.* 2005. 94. 2224–2231. Copyright Wiley-VCH Verlag GmbH & Co. KGaA. Reproduced with permission.)

3.4 COCRYSTALS

Cocrystals have received much attention recently as a novel approach for overcoming low-solubility problems. The strategy centers on cocrystallizing the API with a crystalline solid, designated the coformer. This may produce a cocrystal with more favorable physicochemical properties, in which the API and coformer are connected by noncovalent interactions. Crystal engineering facilitates optimal coformer association for each particular API. Although there are various combination patterns of drug and coformer molecules, the "synthon approach" is frequently used in the design of a cocrystal (Thakuria et al. 2013), whereby particular functional groups of the API interact with complementary functional groups of the conformers, to form "supramolecular synthon" cocrystals (Figure 3.4).

Cocrystals are differentiated from salts by assessing the degree of proton transfer. In general, cocrystal formation is expected for $\Delta pK_a < 3$ and salt formation is expected for $\Delta pK_a > 3$. If one component is liquid at room temperature, then the crystals are designated solvates, whereas if both components are solids at room temperature, then the crystals are designated as cocrystals. Although solvates have the potential to enhance drug dissolution rate (for example, the solvated forms of spironolactone), hydrates generally exhibit slower dissolution rate relative to anhydrates. Also, they are often physically unstable, which can lead to desolvation during storage and possible crystallization into less soluble forms. High solvent levels may also cause toxicity problems. Cocrystals, in contrast, are typically produced by more rational design and so tend to be more stable.

As is the case for salt formation, the solubility of cocrystals is governed by the solubility product of the API and the coformer. Practically speaking, a coformer with high solubility tends to effectively increase cocrystal solubility. For carbamazepine, the solubility was increased by more than two orders of magnitude when it was combined with nicotinamide or glutamic acid, but the increase was only 2-fold with the aid of salicylic acid (Good and Rodríguez-Hornedo 2009).

On a laboratory scale, cocrystals can be prepared by either grinding or precipitation. The grinding method, notably solvent-assisted grinding, is believed to be better for estimating cocrystallization ability (Friscic et al. 2006), whereas the precipitation method is more suitable for screening. For industrial production, hot-melt extrusion may be used. Importantly, a recent Guidance for Industry issued by the FDA has specified that cocrystals need *not* be regarded as a novel API. This significantly reduces the regulatory hurdles needed for their licensing, which represents a distinct advantage for this approach.

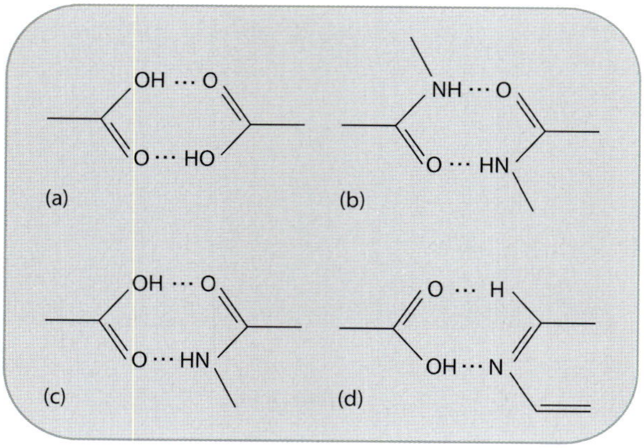

FIGURE 3.4 Representative hydrogen-bonding synthons for forming cocrystals, showing the (a) carboxylic acid dimer synthon, (b) the amide dimer synthon, (c) the acid-pyridine heterosynthon, and (d) the acid-amide heterosynthon.

3.5 POLYMORPHS

Depending on the conditions (temperature, solvent, time) and method used for crystallization, the molecules in a crystal can arrange in different ways: either they may be packed differently in the crystal lattice or there may be differences in the orientation, or conformation, of the molecules at the lattice sites (Florence and Attwood 2011). Different crystalline forms of the same compound are called "polymorphs." Although chemically identical, the different crystal lattices are at different free energy states. At a given temperature and pressure, only one of the crystalline forms is stable and the others are known as metastable forms. A metastable polymorph usually exhibits a greater aqueous solubility and dissolution rate, and thus greater absorption, than the stable polymorph.

Various physicochemical characteristics, including solubility and reactivity, can be correlated with their free energy difference, ΔG_{A-B}, according to the following equation:

$$\Delta G_{A-B} = RT \ln\left(\frac{r_A}{r_B}\right) = RT \ln\left(\frac{x_A}{x_B}\right) \qquad (3.2)$$

where

r_A and r_B are the rates of chemical reaction of forms A and B, respectively, x_A and x_B are their solubilities, respectively
R is the gas constant
T is the temperature

The actual performance may or may not be predicted from this equation. For example, the dissolution rate may be governed by particle size rather than solubility. Nevertheless, thermodynamics is the basis for understanding various physicochemical characteristics.

A comparison of the melting enthalpy is a particularly useful method for determining the thermodynamic relationship between different polymorphic forms. Figure 3.5 shows a plot of the free energy of two polymorphs against temperature. The most stable form at ambient temperature is

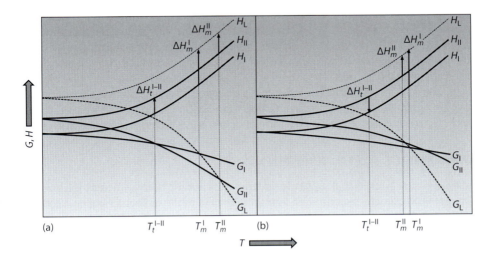

FIGURE 3.5 Free energy (G) and enthalpy (H)—temperature (T) diagrams of two crystal forms, which are in relationship of (a) enantiotropy or (b) monotropy. The superscripts I, II, I–II, and L represent Form I, Form II, Form I–II transition, and the liquid state, respectively. Form I is the stable form in the case of the monotropic relationship and the stable form at lower temperature in the case of the enantiotropic relationship. T_t and T_m are the transition and melting temperatures, respectively. ΔH_t and ΔH_m are the enthalpies of polymorphic transition and melting, respectively.

defined as Form I and the metastable form as Form II. For an enantiotropic system (Figure 3.5a), their free energies become equal at the transition temperature T_t^{I-II}. However, the experimental transition temperature may be different from the theoretical one. In many cases, the transition is observed at a higher temperature, because of the high-energy barrier of the transformation. It is even possible that Form I melts without transformation. In this case, the melting enthalpy of Form I is usually larger than that of Form II, as indicated by the difference in the length of the arrows in Figure 3.5a. Because the enthalpy of Form I is lower than that of Form II at that temperature, the transition from Form I to Form II is an endothermic process. Thus, it may be possible to convert reversibly between the two polymorphs on heating and cooling (Kawakami 2007). Conversely, a monotropic system is shown in Figure 3.5b: the stability order remains unchanged below their melting temperatures. Metastable Form II may or may not cause polymorphic transition to Form I upon heating. If this occurs, the transition is an exothermic process and irreversible. Unless the transformation occurs, the metastable Form II melts with a smaller melting enthalpy than that of the stable Form I.

Most modern drugs show polymorphism, frequently being able to crystallize in three or more forms. The most stable polymorph has the lowest solubility and slowest dissolution rate. Therefore, selecting the metastable form represents a potential strategy in order to improve the solubility of poorly soluble drugs. A review of the literature concluded that solubility ratios were less than 2 in most cases, with an average value of 1.7 (Pudipeddi and Serajuddin 2005) suggesting that in many cases, solubility gains are small. However, in some cases, the free energy differences between polymorphs are large enough to significantly affect solubility and hence bioavailability. An example is chloramphenicol palmitate, where one polymorphic form was shown to attain approximately seven times greater blood levels in comparison to another (Aiguiar and Zelmer 1969).

In addition to solubility considerations, the issue of polymorphism has profound implications for the stability of a formulation and thus must be studied very carefully in this context. Under a given set of conditions, the polymorphic form with the lowest free energy will be the most stable, and other polymorphs will tend to transform into it. The rate of conversion is variable and is determined by the magnitude of the energy barrier between the two polymorphs. If there is a high-energy barrier and the crystal is stored at a low temperature, it can be expected that the conversion rate will be slow. Occasionally, the most stable polymorph appears only several years after the compound was first marketed. A case in point is the development of the antiretroviral drug, ritonavir. The most stable crystalline form appeared 2 years after the initial product launch and demonstrated solubility approximately 4- to 5-fold less than the original crystal form of the drug. The original soluble crystal form could no longer be generated, so that the newly emerged, more stable form required reformulation into gelcaps and tablets, rather than the original capsules, to ensure adequate oral bioavailability (Bauer et al. 2001). With such possible risks, it is perhaps understandable that pharmaceutical companies may be reluctant to employ the metastable form for the purpose of solubility improvement. Nevertheless, in some cases, it can be an excellent option for overcoming the solubility problem.

3.6 AMORPHOUS SOLID DISPERSIONS

Crystalline solids are packed in a regularly ordered, repeating pattern, whereas the corresponding amorphous form is characterized by a random arrangement of the molecules and an absence of long range, 3D, order. Because the amorphous form of a drug has no crystalline lattice, dissolution is more rapid, as no energy is required to break up the crystal lattice during the dissolution process; wettability is also typically better for the amorphous form. Therefore, the bioavailability of the amorphous form of an API is generally greater than that of the crystalline form: for example, the amorphous form of novobiocin is at least 10 times more soluble than the crystalline form (Florence and Attwood 2011).

Figure 3.6 shows both the ideal dissolution profile expected from the energy state of the amorphous form according to Equation 3.2 and also a typical profile of a real amorphous solid. The discrepancy is usually because soon after suspending an amorphous form in aqueous media,

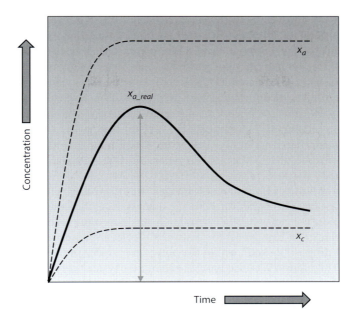

FIGURE 3.6 Dissolution profile of the crystalline state, ideal profile of the amorphous state, and a typical profile of the real amorphous solid. The theoretical crystalline and amorphous solubilities are represented by x_c and x_a, respectively. The real amorphous solids exhibit a peak at x_{a_real}, followed by a gradual decrease in the concentration due to the appearance of crystalline solids.

crystalline solids appear. Thus, the solution concentration does not reach the ideal amorphous solubility x_a but exhibits a peak at x_{a_real}, followed by a gradual decrease, until the concentration reaches the solubility of the crystal, x_c, after a sufficient period of time. Note that x_a is not supersaturation, but a solubility equilibrium for the amorphous form. However, once the crystalline drug appears, the solution with a solubility greater than the crystalline solubility can be regarded as being in a supersaturated state, relative to the crystalline solubility. This is the "spring-and-parachute" concept of amorphous dissolution and subsequent absorption: the formation of a supersaturated solution is "the spring," to drive up solubility in the GI tract; after which supersaturation must be maintained long enough to drive drug absorption—"the parachute" phase (Brouwers et al. 2009). The "parachute" phase must be maintained for a long time to achieve an improvement in the oral absorption behavior.

The dissolution advantage of amorphous solids can be negated by crystallization of the amorphous solid in contact with the dissolution medium, as well as rapid crystallization of the supersaturated solution. Furthermore, there is a risk that the high-energy amorphous state will convert to the crystalline state with time, leading to a decrease in the dissolution rate with ageing. There are also difficulties associated with processing the amorphous material. In order to overcome these disadvantages, much research has been focused on developing ASDs, in which the amorphous form is associated with a hydrophilic polymeric carrier, as described next.

3.6.1 Miscibility with Excipients

The term "solid dispersion" is a general one, denoting any formulation in the solid state, in which the API is dispersed in an inert matrix. Typically, a hydrophobic API is dispersed in a hydrophilic matrix carrier. Although several classifications of solid dispersions exist (including simple eutectic mixtures, solid solutions, glass solutions, and glass suspensions), it is ASDs that have received the most interest in drug delivery and are discussed here. In an ASD, an amorphous drug is *molecularly dispersed* within a solid matrix. The matrix comprises a polymeric carrier, that is typically hydrophilic, amorphous, and has a high glass transition temperature (T_g). Polymers such as PVP, HPMC, and its derivatives are commonly used. Novel polymers have also been especially designed for this purpose, such as Soluplus® (BASF), a polyvinyl caprolactam–polyvinyl acetate–PEG graft copolymer.

ASDs demonstrate enhanced solubility for a number of reasons: (1) the drug is present in the amorphous form and thus demonstrates more rapid dissolution in comparison to its crystalline counterpart (spring-and-parachute effect), (2) the drug is arranged within the carrier as a molecular dispersion, with a maximally reduced particle size, (3) the drug is intimately associated with a hydrophilic amorphous carrier, and (4) a recent important finding is that supersaturation can be achieved via the formation of nanoparticles, composed of the API and the polymeric excipients (Alonzo et al. 2010).

ASD formulations also have important advantages with respect to stability. ASDs significantly reduce the dangers of crystallization (of both the amorphous solid and the supersaturated solution) that can occur in contact with the dissolution medium, thus leading to the generation of supersaturated solutions that can persist for biologically relevant timeframes (Alonzo et al. 2010). Furthermore, as the drug is "locked" within the polymeric carrier in the solid state, its molecular motion is very low. Therefore ASDs also provide enhanced physical stability compared to the amorphous form alone, so that long-term storage stability is significantly improved.

In order to optimize the extension of parachute behavior, careful consideration of the excipient species and mixing ratio is required. Although a larger polymer/API ratio is preferred for physical stabilization and effective creation of supersaturated state, this also increases formulation volume. Thus, an optimum mixing ratio must be determined for each drug–polymer pairing. The free energy of mixing is described by the Flory–Huggins equation:

$$\frac{\Delta G_i}{kT} = \phi_d^i \ln \phi_d^i + \frac{\phi_p^i}{N} \ln \phi_p^i + \chi \phi_d^i \phi_p^i \tag{3.3}$$

where
   ΔG_i, ϕ_d^i, and ϕ_p^i are the mixing free energy, drug fraction, and polymer fraction of phase i, respectively
   χ is the interaction parameter between the drug and polymer, for which a value <0.5 indicates a miscible combination and a value greater indicates an immiscible combination
   k and N are the Boltzmann constant and segment number of the polymer molecule, respectively

The overall mixing free energy ΔG can be obtained from

$$\Delta G = X_d \Delta G_d + X_p \Delta G_p \tag{3.4}$$

where
   subscripts d and p represent the drug-rich and polymer-rich phases, respectively
   X is the fraction of each phase

Figure 3.7 shows examples of the phase diagram drawn by minimizing ΔG. A significant expansion of the two-phase region is observed with an increase in N. Thus, an increase in the molecular weight of a polymeric excipient can cause phase separation. Because the molecular weight of a drug molecule is usually in the range of 200–1000 Da, N is >100 in most cases, where the drug solubility rapidly decreases above $\chi = 1$. Because high miscibility cannot be expected for a combination of a hydrophilic polymer and a poorly soluble drug, χ is expected to be >1. Even for $\chi = 1$, phase separation is expected under drug-rich conditions, when N is >8 and the drug solubility in the matrix is almost constant at ca. 33% for $N > 100$. The expected solubility becomes only 15% when $\chi = 1.5$ and

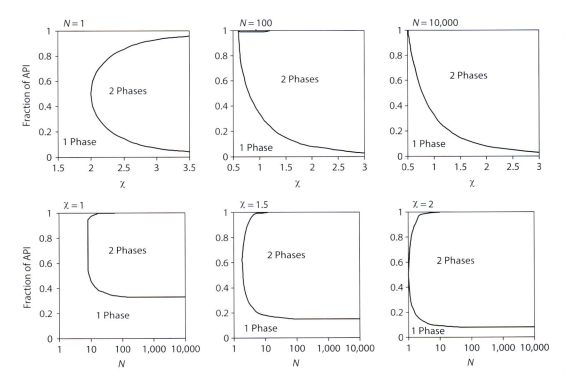

FIGURE 3.7   Theoretical phase diagrams of solid dispersions composed of polymeric excipients and the active pharmaceutical ingredients, which were established on the basis of the Flory–Huggins theory. The miscible system and the system that induces phase separation are described as "1 phase" and "2 phases," respectively.

is <10% when χ = 2.0. This theoretical calculation demonstrates why only a small amount of drug can be loaded into a stable solid dispersion.

3.6.2   Storage Stability

One of the significant issues that has hampered the development of ASDs until recently is their lack of physical stability. As outlined earlier, there is a risk that the amorphous state within the carrier will convert to the crystalline state, leading to a decrease in the supersaturation effect with ageing. X-ray powder diffraction (XRPD) is the most convenient method for evaluating whether a formulation is in the amorphous state. However, the XRPD technique cannot detect nuclei and small crystals that should enhance the crystallization. If the API and polymeric carrier are completely mixed, the formulation has only one T_g, which can be determined by DSC. A formulation with multiple T_gs indicates the existence of multiple phases. Although phase separation itself is not a fatal problem, the probability of crystallization of the API-rich phase is high.

Physical stability is governed by T_g of the system. A high T_g of the carrier increases the overall T_g of the mixture, resulting in a more stabilized amorphous form. Recent investigations have revealed that the amorphous form can be expected to be stable for 3 years at 25°C, if the system T_g is higher than 48°C (Kawakami et al. 2014). The absorption of moisture decreases the stability significantly, because molecular mobility is enhanced.

For crystalline formulations, accelerated studies for assuring chemical stability are based on the Arrhenius rule; however, the possibility of the crystallization of the amorphous form cannot be quantitatively predicted from accelerated studies. Furthermore, if phase separation or crystallization occurs at elevated temperatures, it may affect chemical stability. This unpredictability of the amorphous form stability leads to a prolongation of the drug developmental period. Currently, much research effort is directed at mechanisms to improve long-term stability predictions.

3.6.3 IMPROVEMENT IN ORAL ABSORPTION

In order to achieve an improvement in oral absorption, it is important to obtain a supersaturated state for a prolonged period ("spring and parachute"). The conventional dissolution test protocol utilizes sink conditions and focuses on the dissolution rate of the API, rather than providing an assessment of supersaturation. As such, it is not very effective in predicting the oral absorption enhancement, or if crystallization will occur in the GI tract.

Figure 3.8 compares three different griseofulvin (GF) formulations: a physical mixture, a solid dispersion containing both amorphous and crystalline GF, and an ASD. The ASD demonstrated the highest plasma concentration. The solid dispersion that contained both amorphous and crystalline GF displayed similar in vitro and in vivo profiles to that obtained for the physical mixture. Using the crystalline GF as a template, this can be explained by the immediate crystallization of the amorphous solid and/or dissolved solutes. This study demonstrates the importance of both complete amorphization of the API, and the maintenance of supersaturation, in order to improve oral absorption.

3.6.4 MANUFACTURE

In principle, most classes of material can be prepared in the amorphous state, if the rate at which they are solidified is faster than that at which their molecules can align themselves into a crystal lattice with 3D order. Small-scale production of solid dispersions can be achieved by the melt-quenching method, whereby the API is heated above its melting temperature, followed by rapid cooling. The required cooling rate depends on the glass-forming ability of the compound. Another simple method for obtaining amorphous solids is by precipitation, by adding a poor solvent to the

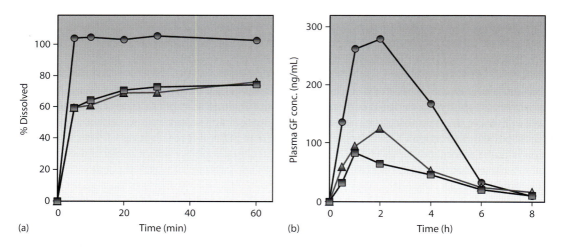

FIGURE 3.8 (a) In vitro dissolution profiles and (b) plasma concentrations after oral administration of griseofulvin (GF) formulations to rats (7.5 mg/kg). A physical mixture (square), a solid dispersion that contains both amorphous and crystalline GF (triangle), and an ASD (circles) are shown. All of the formulations were prepared with Eudragit L-100 as an excipient. The Japanese Pharmacopeia-specified simulated intestinal solution of pH 6.8 (JP-2 solution) with 20 mM taurocholic acid was used as a medium in the dissolution test. (Reprinted with permission from Zhang, S. et al., *Mol. Pharm.*, 8, 807. Copyright 2011 American Chemical Society.)

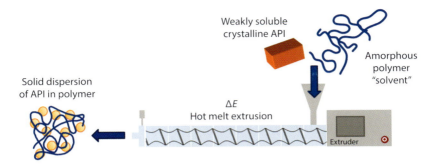

FIGURE 3.9 Hot-melt extrusion for the preparation of amorphous solid dispersions. (Courtesy of BASF SE, Ludwigshafen, Germany.)

API solution. Grinding, by supplying mechanical and thermal energy, can also convert a crystalline drug to the amorphous state. Although it is a simple procedure, it frequently leaves nuclei or small crystals, which can accelerate crystallization during storage.

A major problem with ASDs in the past, which significantly hampered progress in the field, was the difficulty in scaling up to large-scale industrial manufacturing. However, this problem has now been largely overcome due to the successful application of hot-melt extrusion technology, which can be a continuous process, in which equipment is available that allows solvent-free manufacture at temperatures above the relevant T_g (Figure 3.9). In this process, a crystalline API and an amorphous polymer are fed into the extruder, before being conveyed and exposed to shear stress. This transforms the API into its amorphous form, which is blended with the polymer and codispersed. Surfactants and other excipients may be added to aid in the extrusion process or to improve dissolution performance in vivo. The resulting solid (referred to as extrudate) is then pressed out and collected for further processing, e.g., into granules, spheres, powders, films, patches, and injections (Figure 3.9). Spray-drying is also frequently utilized for industrial production of ASDs.

3.6.5 COMMERCIAL EXAMPLES

ASDs can be prepared as tablets and as such offer significant advantages. For example, Kaletra®, an HIV protease inhibitor, was initially marketed as a liquid self-emulsifying formulation. Patients were required to take six large soft capsules once daily (or three capsules twice daily); moreover, the capsules required refrigeration for storage. This formulation was replaced by ASD tablets, prepared using hot-melt technology. The number of the tablets required was reduced to four a day (or two tablets twice a day) and the tablets could be stored at room temperature.

Although there has been over 40 years of research in the area, the number of commercial products using ASDs is not extensive (Table 3.2). Research has been hindered by the manufacturing and stability issues described earlier. However, these issues have been largely addressed and continue to be improved upon, so it is expected that more commercial products based on ASDs will soon follow.

3.7 CYCLODEXTRINS

CDs are cyclic oligosaccharides, based on α-D-glucopyranose units, which can improve water solubility by complexation of the API within a cavity (Figure 3.10). The hydrophilic exterior surface of the CD molecules makes them water soluble, but the hydrophobic parts of central cavity provides a microenvironment for the housing of the hydrophobic parts of APIs. A range of CDs are available (α-CD, β-CD, γ-CD), with varying numbers of sugar residues and thus different-sized cavities, to accommodate guest APIs with various structures. More water-soluble CD derivatives are also available, including the hydroxypropyl derivatives of β-CD and γ-CD and the randomly methylated β-CD (RMβCD).

TABLE 3.2

Examples of Commercial Products Using Amorphous Solid Dispersions

Commercial Preparation	Drug
Accolate®	Zafirlukast
Accupril®	Quinapril hydrochloride
Ceftin®	Cefuroxime axetil
Cesamet®	Nabilone
Certican®	Everolimus
Crestor®	Rosuvastatin calcium
Gris-PEG®	Griseofulvin
Intelence®	Etravirine
Isoptin®	Verapamil
Kaletra®	Lopinavir/ritonavir
Nivadil®	Nilvadipine
Prograf®	Tacrolimus
Rezulin®	Troglitazone
Sporanox	Itraconazole
Zelboraf®	Vemurafenib

CDs as solubility enhancers have been used in the pharmaceutical industry for many years: the first CD product was introduced in Japan in 1976 (Prostarmon E®) and comprised the prostaglandin E2 solubilized by β-CD as the molecular inclusion complex. The first U.S.-approved product was the antifungal itraconazole in a hydroxypropyl-β-CD complex (Sporanox® oral solution) and was introduced in 1984. Currently, 35 different drugs are marketed worldwide as either solid or solution-based CD complex formulations.

Biocompatible CDs are obviously safer and less toxic than harsh organic solvents for improving water solubility. They are also less likely to have problems of precipitation in vivo in contact with aqueous body fluids. This is because CDs solubilize compounds as a linear function of their concentration if they form 1:1 complexes, so that in contact with aqueous fluids in vivo, both the drug and CD concentration are reduced in a linear manner. In contrast, organic solvents solubilize solutes as a *log* function of their concentration, so that precipitation is more likely to occur with the rapid dilution that occurs in contact with the aqueous environment.

3.8 MICELLAR SOLUBILIZATION OF DRUGS

3.8.1 SURFACTANT MICELLES

Surfactants are amphiphilic molecules typically comprising a long-chain hydrocarbon tail and a head group that can be either (1) anionic, e.g., sodium dodecyl sulfate; (2) cationic, e.g., dodecyltrimethylammonium bromide; or (3) nonionic, e.g., *n*-dodecyl tetra(ethylene oxide). At low concentrations, surfactants are widely used as formulation excipients. For solid dosage forms, the addition of even a small amount of surfactant helps improve solubility, because it can improve wetting properties and aid in the rapid disintegration of the dosage form. Surfactants are also used as stabilizers in many other delivery systems, such as emulsions, microemulsions, nanocrystals, and ASDs. Other complementary effects of surfactants may play a role in enhancing oral bioavailability. For example, nonionic surfactants such as TPGS present in a formulation can inhibit P-glycoprotein efflux pumps. Furthermore, many surfactants function as absorption enhancers, promoting both transcellular and paracellular transport pathways across the GI epithelium (see also Chapter 7).

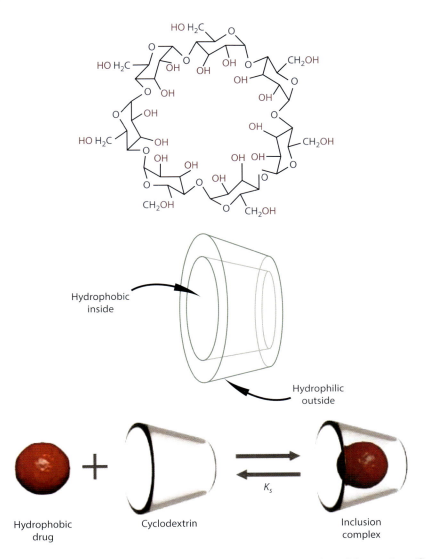

FIGURE 3.10 Schematic representation of formation of a cyclodextrin–drug 1:1 complex. (Courtesy of Pierre Fabre Medicament—Supercritical Fluids Division.)

Their role as solubilizers is due to their self-aggregation properties: when surfactant molecules are dissolved in water at concentrations above the critical micelle concentration (CMC), they form colloidal-sized aggregates known as micelles, in which the hydrophobic portions are driven inward, to form a hydrophobic core, while the hydrophilic portions face outward, toward the water. Surfactant micelles increase the solubility of poorly soluble substances in water, because the nonpolar drug molecules are solubilized within the hydrophobic micelle core. Polar molecules (or polar portions of a drug molecule) will be adsorbed on the micelle surface, and substances with intermediate polarity are distributed along surfactant molecules in certain intermediate positions (Florence and Attwood 2011). The capacity of surfactants to solubilize drugs depends on various factors, including the physicochemical nature of both the drug and surfactant, the temperature, pH, etc.

However, surfactant micelles are not static aggregates: they dissociate, regroup, and reassociate rapidly (hence, they are often referred to as "association colloids"). The solubilizing capacity of the surfactant can be lost on dilution with aqueous fluids in vivo. Micellar collapse can lead to drug precipitation, for example, in the lumen of the GI tract following oral administration, or in contact with the blood after injection. Furthermore, the solubilization capacity for poorly soluble drugs in surfactant micelles is usually less than 20 mg/g of surfactant, meaning that gram amounts of surfactant are usually required for complete solubilization, which is not often realistic practically. At high concentrations, surfactants may possibly cause damage to the GI epithelium, disrupting proteins in the plasma membrane (see also Chapter 7). If given intravenously, surfactants at high concentrations can cause anaphylactic reactions and other toxicity issues.

3.8.2 POLYMERIC MICELLES

The amphiphilic block copolymers are a newer class of surfactants, which are able to form stable micelles, known as polymeric micelles, at low CMC values (Torchilin 2004). The amphiphilic block copolymers typically consist of (1) a hydrophobic polyester block (for example, polylactic acid [PLA] or poly(lactic-*co*-glycolic acid) [PLGA]) and (2) a hydrophilic block comprising PEG or poly(ethylene oxide). When the length of the hydrophilic block exceeds the length of the hydrophobic block, these copolymers can form spherical micelles in aqueous solution (see also Chapter 1, Figure 1.14). Again, the hydrophobic blocks form the micellar core, which can accommodate a poorly soluble drug; the hydrophilic blocks form the outer shell. Polymeric micelles are being extensively investigated for drug delivery applications, in particular for parenteral administration. For example, Genexol-PM® is a polymeric micelle formulation of paclitaxel for the treatment of breast cancer. It is composed of a low-molecular-weight amphiphilic diblock copolymer, mono-methoxy PEG-block-poly(D,L-lactide). In addition to being able to solubilize poorly soluble drugs at low CMC values, the stability and outer hydrophilic layer of polymeric micelles promote long circulation times in the blood, which can allow time for their accumulation at sites of inflammation and infection (attributed to the enhanced permeability and retention [EPR] effect). Furthermore, they can be actively targeted to the site of action, by the attachment of a specific targeting vector to the outer surface. These drug delivery and targeting aspects of polymeric micelles are described further in Chapter 5 (Section 5.5.3).

3.9 OILS, EMULSIONS, AND COLLOIDAL CARRIERS

3.9.1 OILS AND COARSE EMULSIONS

A simple approach to enhancing the solubility of "grease ball" molecules is to dissolve the hydrophobic drug in an oily liquid; the solubilized drug can then be filled into capsules for oral delivery. Medium-chain mono-, di-, or triglycerides and their esters are convenient solvent choices, because of their large solubilization capacity and good compatibility with capsules. Oils can also be used to prepare oil-in-water (O/W) emulsions, to solubilize hydrophobic drugs within the oil phase. These systems are thermodynamically unstable, due to the large interfacial energy between the oil and water phase.

After oral administration, oils and coarse emulsions are subjected to the physiologically complex processes of digestion and absorption that exist for dietary lipids within the GI tract. Oil-based delivery systems initially form coarse droplets within the aqueous GI fluids, which then require the detergent action of bile salts to emulsify them into smaller, stabilized droplets. These smaller droplets are then subjected to the action of pancreatic lipase, which digests the oil into free fatty acids and monoglycerides, finally liberating the drug. The hydrophobic drug, with the free fatty acids and monoglycerides of the digested oil carrier, in addition to cholesterol and lecithin, are all subsequently incorporated into bile salt micelles. These micelles serve as carriers, to shuttle the hydrophobic API through the aqueous GI contents, to reach the absorbing surface of the enterocyte. The secretion and activity of bile salts demonstrate high inter- and intravariability and are profoundly affected by the fasting state of the patient, thus introducing significant variability into the drug absorption process for these drug delivery systems (DDS).

3.9.2 Self-Microemulsifying Drug Delivery Systems

A promising advance in emulsion technology is the development of self-emulsifying DDS (self-emulsifying drug delivery systems [SEDDS]), which comprise physically stable, isotropic mixtures of oils, surfactants, solvents, and cosolvents/surfactants, in very specific combinations, that require careful selection (Hauss 2007). Typical components may include

- Oils: Such as peanut oil and medium-chain triglycerides (e.g., Neobee® M5)
- Emulsifiers: Surfactants such as Labrafac® CM-10 and polyglycolyzed glycerides with varying fatty acid and PEG chain lengths
- Cosolvents: Such as propylene glycol and Transcutol®

SEDDS can be orally administered in soft or hard gelatin capsules. On dilution in the aqueous GI fluids and facilitated by the gentle agitation of the GI contents due to peristaltic activity, they spontaneously self-form into very fine, relatively stable O/W emulsions, with lipid droplets size of about 100–200 nm (SEDDS), or less than 100 nm for self-microemulsifying drug delivery systems (SMEDDS). SMEDDS thus differ from coarse emulsions in that they are a thermodynamically equilibrium solution and as such, form spontaneously, without energy input. A phase diagram may be used to obtain the optimum composition of the formulation, to promote optimal microemulsification.

Oral absorption is facilitated by the rapid release of drug from the high surface area of the small lipid droplets. SMEDDS do not require bile salts and other digestive processes for their digestion, thereby minimizing inter- and intrasubject variations. For SMEDDS, there is also the possibility that due to the lipidic nature of the delivery system, lymphatic absorption via the intestinal lacteals, or the intestinal Peyer's patches, is enhanced. Drugs absorbed via the lymphatics avoid the "first-pass" effects of the liver (see Chapter 7), which represents a significant advantage for the oral absorption of enzymatically labile APIs.

Supersaturatable SEDDS (S-SEDDS) are an extension of the original SEDDS, which incorporate less surfactant than original system, in order to avoid the adverse effects of surfactants in the GI tract. There is a risk that with less surfactant in the formulation, the system may precipitate out on dilution with the GI fluids. However, the surfactant concentration can safely be reduced by incorporating a polymeric precipitation inhibitor (PPI), most commonly HPMC. The PPI prevents precipitation of the system in the GI tract and also maintains a supersaturated state of the drug for extended periods, thereby providing a strong driving force for the absorption process. Figure 3.11 shows an example of the effect of supersaturation in a study of various self-emulsifying formulations containing ca. 60 mg/g paclitaxel (Gao et al. 2003). As shown in Figure 3.11a, when the formulation without HPMC was diluted 50-fold using simulated gastric fluid, precipitation occurred in 10 minutes and the paclitaxel concentration decreased to ca. 0.1 mg/mL, where the equilibrium solubility was expected to be 0.02 mg/mL. In contrast, the paclitaxel concentration from the formulation with HPMC was greater than 0.9 mg/mL at 10 minutes, and a high level of supersaturation was maintained for more than 2 hours. Oral absorption was greatly enhanced by the formulation with HPMC, as shown in Figure 3.11b, which can be explained by the maintenance of a high concentration of dissolved paclitaxel in the small intestine. Thus, addition of a PPI to self-emulsifying formulations appears to be a very useful approach. However, surfactant and polymer molecules usually have strong interactions and may form various types of complexes, which also require attention during the design of the formulation.

SMEDDS can be incorporated into soft capsules for oral administration and examples now on the market include Agenerase® (amprenavir), Aptivus® (tipranavir), Fortovase® (saquinavir), Kaletra®

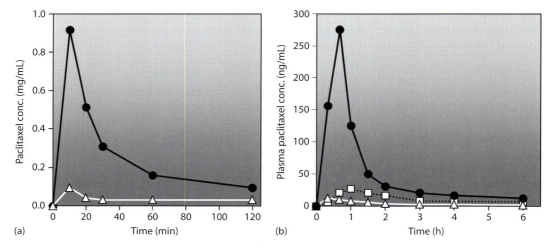

FIGURE 3.11   (a) Dissolution behavior of two self-emulsifying formulations of paclitaxel. A 50-fold dilution in simulated gastric fluid at 37°C was performed. Paclitaxel concentration is shown after dilution of a formulation containing hydroxypropyl methylcellulose (HPMC) (circles), or without HPMC (triangles). (b) An oral absorption study of paclitaxel formulations in fasted rats, showing plasma paclitaxel concentrations after dosing with a formulation containing HPMC (circles), without HPMC (triangles), or a control formulation of Taxol® (squares). (From Gao, P., Rush, B.D., Pfund, W.P. et al.: Development of a supersaturatable SEDDS (S-SEDDS) formulation of paclitaxel with improved oral bioavailability. *J. Pharm. Sci.* 2003. 92. 2386–2398. Copyright Wiley-VCH Verlag GmbH & Co. KGaA. Reproduced with permission.)

(lopinavir/ritonavir), Neoral® (cyclosporine A), Rapamune (sirolimus), and Xtandi® (enzalutamide). Neoral® is a SMEDDS reformulation of Sandimmune®, which was a SEDDS introduced in 1994. Neoral® emulsifies spontaneously into a microemulsion with a particle size smaller than 100 nm, which increases the bioavailability nearly by a factor 2 over the original emulsion formulation. In addition, Neoral® shows a much faster onset of action, a reduced inter-/intrasubject variability and a much lower impact of food intake on cyclosporin pharmacokinetics.

3.9.3   OTHER COLLOIDAL CARRIERS

As described in Chapter 5, a wide variety of micro- and nanoparticulate DDS have been developed, particularly for the parenteral route of delivery (see also Chapter 5, Figure 5.1). Such DDS include liposomes, niosomes, lipoprotein carriers, polymeric micro- and nanoparticles, and dendrimers. Many of these systems offer the advantage of increasing the solubility of an API. For example, hydrophobic APIs can be associated with the lipid bilayers of a liposomal DDS (see Chapter 5, Figure 5.9). Hydrophobic drugs can also be accommodated within the hydrophobic core region of micro- and nanoparticles and other colloidal carriers. As these DDS are predominantly investigated for their drug delivery and targeting purposes, rather than for their solubilization potential, the reader is referred to Chapter 5 for further information on these other colloidal carriers.

3.10   CONCLUSIONS

Enhancing the water solubility of poorly soluble drugs continues to be a pressing concern in the pharmaceutical industry, with important implications for the drug discovery process, as well as for the formulation, bioavailability, and therapeutic efficacy of poorly soluble APIs. There are now many different ways to improve drug solubility, as outlined in this chapter. However, there is no single all-encompassing solution to this problem, and each approach has associated advantages and limitations. Furthermore, in considering which approach to use, it is important that each potential drug candidate is considered on an individual basis. A number of interrelated factors must be taken into account, including the following:

- *The nature of the API*: What type of API needs to be solubilized? Is it a grease ball or a brick dust molecule? What solubilizes the API? Can it dissolve in an oily/lipidic medium, or does it need a more complex blend of oils, surfactants, and cosolvents? How stable is the API and can it withstand harsh industrial processes?
- *The delivery route*: Parenteral, oral, etc.
- *Therapeutic issues*: The disease, the desired therapeutic outcome, the dose, and the duration of drug administration.
- *Industrial issues*: The availability of backup candidates, the developmental timeline, the available resources for development and manufacture, and the relevant regulatory issues.

Whatever method for solubility improvement is chosen, meticulous characterization and stability analysis are required, including the establishment of rigorous protocols to predict physical stability. Stability will need to be ascertained both in vitro and also in vivo. For the oral route, this will require a study of the stability and performance of the formulation within the highly variable and complex GI milieu.

The field has seen recent novel and exciting advances to improving drug solubility, such as innovative lipid delivery systems like SMEDDS, the success of NanoCrystal® technology, and the reemergence of ASDs as a viable solubilization technology platform. Techniques such as hot-melt extrusion have allowed successful scale-up manufacturing of processes that were hitherto confined to the laboratory setting. There is currently intense research focus in this area, which is expected to yield further improvements in current methods, as well as the development of new approaches.

REFERENCES

Aiguiar, A.J. and J.E. Zelmer. 1969. Dissolution behaviour of polymorphs of chloramphenicol palmitate and mefenamic acid. *Journal of Pharmaceutical Science* 58:983–987.

Alonzo, D.E., G.G.Z. Zhang, D. Zhou et al. 2010. Understanding the behavior of amorphous pharmaceutical systems during dissolution. *Pharmaceutical Research* 27:608–618.

Amidon, G.L., H. Lennernä, V.P. Shah et al. 1995. A theoretical basis for a biopharmaceutic drug classification: The correlation of *in vitro* drug product dissolution and *in vivo* bioavailability. *Pharmaceutical Research* 12(3):413–420.

Bauer, J., S. Spanton, R. Henry et al. 2001. Ritonavir: An extraordinary example of conformational polymorphism. *Pharmaceutical Research* 18:859–866.

Bergström, C.A.S., C.M. Wassvik, K. Johansson et al. 2007. Poorly soluble marketed drugs display solvation limited solubility. *Journal of Medicinal Chemistry* 50:5858–5862.

Brouwers, J., M.E. Brewster, and P. Augustijns. 2009. Supersaturating drug delivery systems: The answer to solubility-limited oral bioavailability? *Journal of Pharmaceutical Science* 98:2549–2572.

Florence, A.T. and D. Attwood. 2011. *Physicochemical Principles of Pharmacy*, 5th ed. London, U.K.: Pharmaceutical Press.

Friscic, T., A.V. Trask, W. Jones et al. 2006. Screening for inclusion compounds and systematic construction of three-component solids by liquid-assisted grinding. *Angewandte Chemie International Edition* 45:7546–7550.

Gao, P., B.D. Rush, W.P. Pfund et al. 2003. Development of a supersaturatable SEDDS (S-SEDDS) formulation of paclitaxel with improved oral bioavailability. *Journal of Pharmaceutical Science* 92:2386–2398.

Good, D.J. and N. Rodríguez-Hornedo. 2009. Solubility advantage of pharmaceutical cocrystals. *Crystal Growth & Design* 9:2252–2264.

Hauss, D.J. 2007. Oral lipid-based formulations. *Advanced Drug Delivery Reviews* 59(7):667–676.

Kawakami, K. 2007. Reversibility of enantiotropically-related polymorphic transformations from a practical viewpoint: Thermal analysis of kinetically reversible/irreversible polymorphic transformations. *Journal of Pharmaceutical Science* 96:982–989.

Kawakami, K., T. Harada, K. Miura et al. 2014. Relationship between crystallization tendencies during cooling from melt and isothermal storage: Toward a general understanding of physical stability of pharmaceutical glasses. *Molecular Pharmaceutics* 11:1835–1843.

Li, S., P. Doyle, S. Metz et al. 2005. Effect of chloride ion on dissolution of different salt forms of haloperidol, a model basic drug. *Journal of Pharmaceutical Science* 94:2224–2231.

Merisko-Liversidge, E. and G.G. Liversidge. 2011. Nanosizing for oral and parental drug delivery: A perspective on formulating poorly-water soluble compounds using wet media milling technology. *Advanced Drug Delivery Review* 63(6):427–440.

Neslihan, R. and S. Benita. 2004. Self-emulsifying drug delivery systems (SEDDS) for improved oral delivery of lipophilic drugs. *Biomedicine & Pharmacotherapy* 58(3):173–182.

Pudipeddi, M. and A.T.M. Serajuddin. 2005. Trends in solubility of polymorphs. *Journal of Pharmaceutical Science* 94:929–939.

Thakuria, R., A. Delori, W. Jones et al. 2013. Pharmaceutical cocrystals and poorly soluble drugs. *International Journal of Pharmacy.* 453:101–125.

Torchilin, V.P. 2004. Targeted polymeric micelles for delivery of poorly soluble drugs. *Cellular and Molecular Life Sciences* 61(19–20):2549–2559.

Zhang, S., K. Kawakami, M. Yamamoto et al. 2011. Coaxial electrospray formulations for improving oral absorption of a poorly water-soluble drug. *Molecular Pharmaceutics* 8:807–813.

FURTHER READING

Brewster, M.E. and T. Loftsson. 2007. Cyclodextrins as pharmaceutical solubilizers. *Advanced Drug Delivery Review* 59:645–666.

Burger, A. and R. Ramberger. 1979. On the polymorphism of pharmaceuticals and other molecular crystals. I. Theory of thermodynamic rules. *Mikrochimica Acta [Wien]* II:259–271.

Kawakami, K. 2012. Modification of physicochemical characteristics of active pharmaceutical ingredients and application of supersaturatable dosage forms for improving bioavailability of poorly absorbed drugs. *Advanced Drug Delivery Review* 64:480–495.

Kesisoglou, F., S. Panmai, and Y. Wu. 2007. Nanosizing—Oral formulation development and biopharmaceutical evaluation. *Advanced Drug Delivery Review* 59:631–644.

Lawrence, M.J. and G.D. Rees. 2000. Microemulsion-based media as novel drug delivery systems. *Advanced Drug Delivery Review* 45:89–121.

Porter, C.J.H., C.W. Pouton, J.F. Cuine et al. 2008. Enhancing Intestinal drug solubilisation using lipid-based delivery systems. *Advanced Drug Delivery Review* 60:673–691.

Serajuddin, A.T.M. 1999. Solid dispersion of poorly water-soluble drugs: Early promises, subsequent problems, and recent breakthroughs. *Journal of Pharmaceutical Science* 88:1058–1066.

Strickley, R.G. 2004. Solubilizing excipients in oral and injectable formulations. *Pharmaceutical Research* 21:201–230.

Thakuria, R., A. Delori, W. Jones et al. 2013. Pharmaceutical cocrystals and poorly soluble drugs. *International Journal of Pharmacy* 453:101–125.

Williams, H.D., N.L. Trevaskis, S.A. Charman et al. 2013. Strategies to address low drug solubility in discovery and development. *Pharmacological Reviews* 65:315–499.

Overview of Epithelial Barriers

Floriane Laurent, Anya M. Hillery, and Randall Mrsny

CONTENTS

4.1   INTRODUCTION

Epithelial barriers are essential for life. These are cellular structures established and maintained at interfaces that separate the internal environment of an organism from the external world. For example, epithelia cover the external surface of the body and line the gastrointestinal (GI), respiratory, and reproductive tracts. All epithelia present an asymmetric complement of surface components, intracellular organelles, and cytoskeletal arrangements that are organized in accordance with the external environment they experience. Most importantly, all epithelia must maintain the homeostatic conditions of the body. Thus, epithelial barriers must perform a number of critical functions that are distinct for specific internal/external interfaces. Some of the best-known functions performed by epithelia relate to their ability to restrict the casual movement of water and solutes, unidirectional movement of specific molecules, sense changes at their surface, and transmit that information into the body (e.g., neuroepithelial cells), as well as exclude and respond to pathogens, toxins, and innocuous agents (e.g., myoepithelial cells). As one might expect, epithelia of the body are under constant stress by external challenges; epithelial cells have a relatively rapid turnover rate typically in the range of a few days to weeks.

The goal of this chapter is to provide an overview of the characteristics and properties for the various types of epithelia present in the body of interest for drug delivery strategies and describe factors that influence drug transport across epithelia. While all epithelial surfaces have the potential for drug delivery, poor accessibility (salivary glands, pancreatic ducts, etc.) or difficulty in sustaining drug exposure (e.g., esophagus) can limit such opportunities. Additionally, particularly harsh conditions (gastric mucosa) can limit the types of drugs to be considered for delivery. A thick mucus covering or high enzymatic burden at an epithelial surface can also limit drug delivery opportunities. In this overview, we will not suggest that any one epithelium is the best site for drug delivery. Instead, we will focus on how the properties of each make them more or less suitable for particular classes of drugs and delivery strategies based upon the physicochemical properties of these drugs and delivery systems. We will also make an effort to compare and contrast the biological and physiological properties of various epithelia to provide a rationale for optimal alignment of drug delivery strategies at particular epithelial surfaces.

4.2   EPITHELIA

4.2.1   GENERAL ORGANIZATION

Epithelia are organized into one layer (simple epithelia), or more than one layer (stratified epithelia) of closely associated cells, that are supported by a basement membrane (Paulsson 1992). Interaction with the basement membrane provides the basis for an asymmetric epithelial architecture, allowing each cell to establish specialized surfaces that face the external world (apical surface) or the subepithelial environment of the body (basal surface) and take on characteristic squamous, cuboidal, or columnar shapes. The basement membrane is composed of tissue-specific isoforms of laminin and collagen (commonly type IV), as well as particular proteoglycans, proteins, and glucosamines that are organized in a layered arrangement. These acellular structures not only provide a mechanism of support but also separate epithelial cells from underlying connective tissue that contains blood capillaries and lymph vessels; oxygen and nutrients must diffuse from the underlying capillaries. Interactions between surface-expressed elements on epithelial cells and specific basement membrane components influence their behavior, providing the basis for intracellular signaling that affects adhesion, shape, migration, proliferation, and differentiation. Desmosomes as well as hemidesmosomes and focal contacts provide a common basis for epithelial cells to maintain a close association with other epithelial cells or the basement membrane, respectively. Loss of these close contacts can induce a specialized form of epithelial cell apoptosis known as anoikis.

Epithelial barriers come in several forms, with their structure being dependent upon their required functions (Pakurar and Bigbee 2005). For example, the skin is a multilayered epithelium where the cells proliferate while in contact with the basement membrane and transition into a nonviable form as they migrate away from the basal cell layer. The resulting stratified epithelium, with a superficial layer that becomes morphologically and biochemically modified through the process of cornification, provides a barrier that limits water loss and provides protection from environmental challenges. Similar stratification of epithelia is observed in the majority of the buccal cavity and female reproductive tract, except in these cases the surface cells do not undergo cornification. Like the skin, stratified epithelia of the mouth and vagina present a multilayer format of cells that function to limit casual uptake of materials following topical exposure. Epithelia of internal tissues, such as the respiratory tract and most of the GI tract, have less issues with abrasion compared to the skin, mouth, and vagina, and are organized to focus more on absorption and secretion events rather than protection. These epithelia form more delicate single layers or pseudostratified layers of cells. The plasma membrane of these single layers or pseudostratified layers of cells in epithelia is organized into two distinct regions, namely, apical and basolateral domains, as a result of apical–basal polarization processes.

4.2.2 POLARITY PROPERTIES

The apical plasma membrane domain of simple epithelia faces the lumen of a tissue that directly communicates with the outside world, essentially equal to the epithelia of the skin. Although the cornified stratified squamous epithelium of the skin appears dramatically different from the simple columnar epithelium of the small intestine, all epithelia are structured to organize their functional properties in a polarized manner that is derived from specific interactions with a basement membrane present at their basal (inside the body) surface. One essential feature of this polarity is the communication between adjacent epithelial cells, which affects not only the apical to basal bias of epithelial cells but also their side-to-side (planar) arrangement: an organization that is essential for the control of epithelial cell proliferation. Lateral interactions between adjacent epithelial cells are facilitated by desmosomes, which bring cells in close approximation, and gap junctions (GJs) that directly connect the cytoplasm of two neighboring cells to enable the passage of ions and small molecules of <1 kDa.

While GJs and desmosomes promote intercellular communication, these structures are insufficient to establish apical to basal bias in epithelial cells since they do not form a barrier sufficient to restrict the movement of lipids and proteins between the apical and basal plasma membrane domains. The mixing of apical and basolateral domain plasma membrane components is restricted by an annulus of selected proteins and lipid structures known as the apical junctional complex (AJC). The AJC is formed by two structures, the adherens junction (AJ) and the tight junction (TJ), which bring two adjacent epithelial cells in extremely close approximation (Tsukita et al. 2001). These structures are organized in specific arrangement, are stabilized at the apical neck of polarized epithelial cells by a specialized annular ring of actin and myosin filaments, and are associated with an adjacent population of desmosomes (Figure 4.1).

There is a polarized distribution of lipids: the apical membrane is enriched in sphingolipids and cholesterol, while the basolateral membrane comprises mainly phospholipids (Figure 4.2).

AJ structures are important to maintain the integrity of the AJC and play a critical role in the regulation of epithelial cell proliferation. E-cadherin, the dominant AJ protein involved in establishing calcium-dependent cell–cell contacts, interacts with catenin proteins to signal AJ status. While AJ and TJ structures both form annular ring structures in polarized epithelial cells, it is the TJ that establishes the barrier that restricts lipid and protein movement between apical and basal plasma membrane domains. These membrane domains are organized by selected phosphorylation states of inositol lipids (Shewan et al. 2011). This restriction of plasma membrane component migration between apical and basal domains is referred to as the fence function of the TJ.

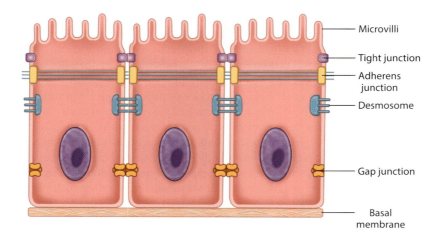

FIGURE 4.1 The apical junctional complex in intestinal epithelial cells is formed by two structures: the adherens junction and the tight junction. Note also the nearby association of desmosomes.

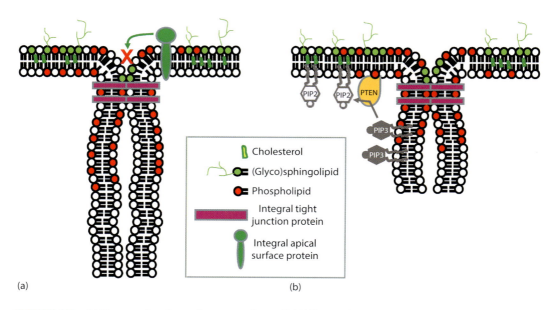

(a)   (b)

FIGURE 4.2 (a) Cartoon showing a plasma membrane lipid bilayer with tight junctions. An apical protein with significant extracellular domain cannot pass the tight junctions. (b) Phosphatase and tensin homolog (PTEN) generates polarised distribution of phosphoinositides in the inner plasma membrane leaflet. Sphingolipids (green) are enriched in the apical domain, whereas phospholipids (red) are not. PIP2, phosphatidylinositol 4,5-bisphosphate; PIP3, phosphatidylinositol (3,4,5)-trisphosphate; and PTEN, phosphatase and tensin homolog. (Reprinted from *Biochim. Biophys. Acta—Biomembr.*, 1788(4), Giepmans, B.N.G and Ijzendoorn, S.C.D., Epithelial cell–cell junctions and plasma membrane domains, 820–831. Copyright 2009, with permission from Elsevier.)

Due to the extremely close apposition of plasma membranes at the AJC, these structures also impede the movement of transport of water, electrolytes, and other small molecules between adjacent epithelial cells. This is different from the form of impediment established of the cornified layer of the skin. In simple epithelia, transmembrane proteins bring epithelial cells into such close approximation that it was initially suggested membranes of adjacent cells were fused

(Farquhar and Palade 1963). Our current understanding of TJ structures is that they are composed of family members from three classes of transmembrane proteins:

- Claudins
- TJ–associated marvel proteins, which include the protein occludin
- An immunoglobulin superfamily that includes the junctional adhesion molecules (JAM) and the lipolysis-stimulated lipoprotein receptor

Like the AJ, TJs organize with scaffold proteins that interact selectively with regulatory factors as well as cytoskeletal elements (Figure 4.3).

Claudins are thought to play a major role in the regulation of the paracellular transport of ions and molecules. Their two extracellular loops form pores in the paracellular space and, depending on the claudins, bear a variable population of positively and negatively charged amino acids; different epithelia express 3–5 discrete forms of the 27 currently known claudin family members. In this way, each epithelium has distinct permselective properties to affect/control/regulate the paracellular transport of ions. Various components of the TJ interact with specific polarity complexes that function to establish and stabilize the different phosphorylation states of inositol lipids present in the apical and basal plasma membrane domains of polarized epithelia. Defects or mutations in specific

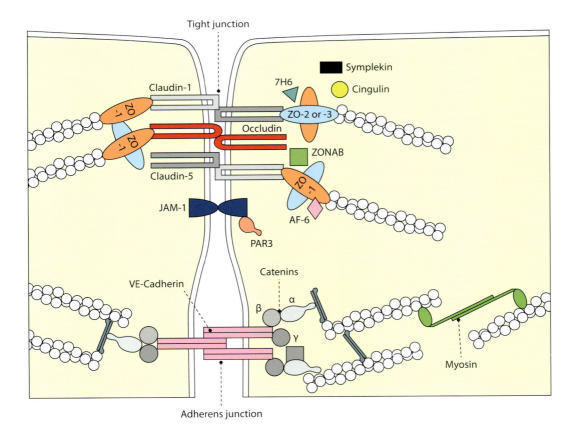

FIGURE 4.3 The molecular composition of the tight junction (TJ). TJs are constituted by the transmembrane proteins occludin, claudins, and junctional adhesion molecule 1 (JAM-1), which seal the paracellular space and connect TJ to the actin cytoskeleton via interaction with proteins from the zona occludens (ZO) family. (With kind permission from Springer Science+Business Media: *Histochem. Cell Biol.*, Tight junctions and the modulation of barrier function in disease, 130(1), 2008, 55, Förster, C., Copyright 2008.)

TJ components can disrupt their ability to maintain normal fence and barrier function, which have been linked to a variety of diseases.

4.2.3 REGULATION OF BARRIER PROPERTIES

Epithelial cells that make up the various epithelial barriers of the body are continuously being replaced due to the senescence, growth, or repair of frank lesions. Thus, the barrier function of an epithelium is constantly changing at a microscopic scale but will appear relatively constant macroscopically. Simultaneously, biological events may be dynamically altering the barrier properties of an epithelium. For example, proinflammatory cytokines (e.g., tumor necrosis factor alpha and interferon gamma) can incite TJ disorganization, resulting in reduced epithelial barrier function. At sites of infection, white cells such as neutrophils will be drawn between adjacent epithelial cells. Their ability to access sites of infection requires the local dismantling of TJs to allow for efficient transmigration, an event that results in only a very minor and transient local barrier loss. More importantly, these observations show that simple epithelial barriers, such as those in the gut and lung, can dynamically open and close to allow transit of AJC structures or sufficiently for an entire cell to pass through.

Dynamic opening and closing of AJCs is achieved through mechanisms that control intracellular signaling processes associated with phosphorylation states of key cytoplasmic regulators of this complex organization of proteins (Dorfel and Huber 2012). An elaborate cohort of kinases and phosphatases functions to stabilize TJ and AJ proteins in a specific manner; the outcome of these activities is the steady-state condition of a "closed" format of these structures. Transient "opening" of these structures involves an adjustment of phosphorylation status of a discrete set of AJ and TJ elements. There are several regulatory pathways involved in controlling phosphorylation events associated with dynamic AJC opening; these pathways appear to have the ability to sense the requirement to open the AJC, in particular with the TJ, to allow molecules of less than 10 kDa in size, versus entire cells, to pass (Madara 1998).

4.3 TRANSPORT ACROSS EPITHELIA

As discussed earlier, the various epithelia of the body perform different functions based upon their location and thus have distinct properties related to the potential for molecules to move across them. For example, properties of the skin epithelia are dominated by the role of providing a barrier that limits dehydration, while the respiratory epithelium must protect the body from inhaled particulates while performing efficient gas exchange. These barrier functions must not preclude essential functions required of these epithelia such as protecting the rest of the body from radiation energy or efficiently exchanging gases for respiration. Also, as discussed earlier, each of the epithelia of the body must try to retain these properties during periods of growth and following tissue repair in order to maintain homeostasis. In the context of this chapter, we will describe some of the general properties and the dynamic nature of various epithelia of the body that restrict the transport of materials of pharmaceutical interest. These properties will relate to three general pathways across epithelial cells (Figure 4.4):

1. Paracellular: transport between adjacent cells
2. Transcellular: transport through individual cells, using pathways that access the cell cytoplasm
3. Endocytosis/transcytosis: uptake and transport through individual cells using vesicles that do not involve access to the cytosol

4.3.1 PARACELLULAR PATHWAY

Based upon the surface area of epithelia, this potential space between adjacent epithelial cells represents only a fraction of the overall surface exposure to the external environment. Transport

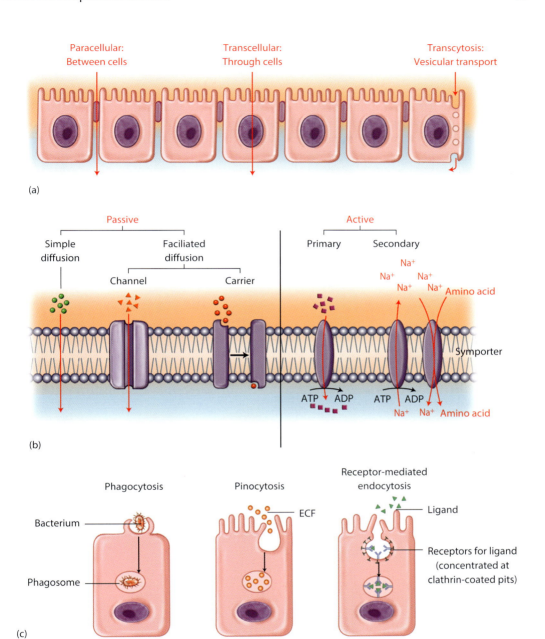

FIGURE 4.4 General transport pathways across epithelial cells/membranes. (a) General transport pathways: paracellular, transcellular, transcytosis. (b) Transcellular membrane transport: transport across the apical cell membrane can be via (1) passive transport, which can be via (I) simple diffusion and (II) facilitated diffusion. Facilitated diffusion can, in turn, be either *channel-mediated* facilitated diffusion or *carrier-mediated* facilitated diffusion. (2) Active transport, which can be (I) primary active transport and (II) secondary active transport. (c) Endocytosis/transcytosis: transport in vesicles that can be via (1) phagocytosis, via specialized cells of the reticuloendothelial system, e.g., neutrophils and macrophages, (2) pinocytosis: nonspecific internalization of extracellular fluid (ECF); any dissolved solutes that happen to be in the ECF also internalized and (3) receptor-mediated endocytosis/transcytosis, a highly selective type of endocytosis/transcytosis, by which cells take up specific ligands.

through the paracellular pathway is limited by the presence of TJs. These structures can be considered as a system of charge-selective small pores of approximately 4 Å and a second pathway created by larger discontinuities in the barrier that lack charge or size discrimination. This pore size translates to roughly compounds >200 Da being restricted from transporting across the epithelial barrier. There are, however, variations in TJ properties between different tissues, in particular related to the charge of the permeating molecule. As noted earlier, each epithelium or even the region within an epithelium (e.g., intestinal crypt versus villus tip) can express a different set of claudin proteins, each with its own sealing and charge-selective properties (Figure 4.5).

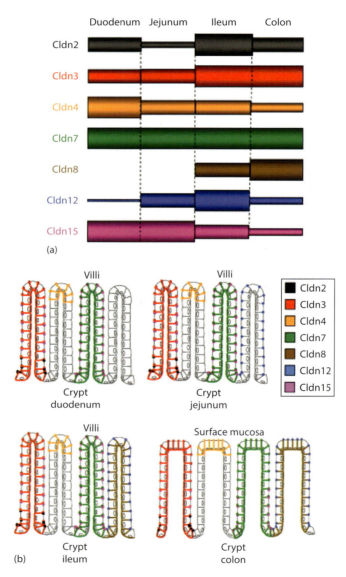

FIGURE 4.5   (a) Relative expression levels of different claudin proteins along the mouse intestine from duodenum to colon. (b) Claudin proteins expression differences as cells transition from the crypt to the villus tip in the small intestine and from the crypt to the luminal surface in the colon. These changes in claudin protein expression have been correlated with differences in solute permeability and perm-selective properties of that epithelium. (Reprinted from *Biochim. Biophys. Acta*, 1778(3), Chiba, H. et al., Transmembrane proteins of tight junctions, 588–600. Copyright 2008, with permission from Elsevier.)

Thus, the permeability and permselective properties of these epithelia can be distinct based upon the combination of expressed claudin proteins. In the following texts, we describe several exemplary epithelia to demonstrate these points.

Claudin-1, claudin-3, claudin-4, claudin-5, claudin-8, claudin-11, claudin-14, and claudin-19 each function to produce tight seals in an epithelium without establishing much permeability (Krause et al. 2008; Krug et al. 2012). For example, claudin-1–deficient mice die within 1 day of birth due to excessive transepidermal water through their skin. Claudin-2 forms channels permeable to small inorganic or organic cations and water. Claudin-10b is also cation selective but appears to produce smaller pores than that of claudin-2, since claudin-10b is not permeable to water and shows strong preference for Na^+ over Li^+ and K^+ ions, whereas claudin-2 is not selective between these cations. Claudin-15 is also a channel for cations and is indirectly involved in glucose absorption through its permselection for Na^+. Claudin-10a and claudin-17 are noted for their anion selectivity. The nature of this selectivity is dependent upon the other claudins present. For example, claudin-10a in Madin-Darby canine kidney (MDCK) II cells increases permeability for Cl^- and NO_3^-, but only NO_3^-, in MDCK C7 cells. Based upon these properties, it is not surprising to find that claudin-2, claudin-10b, and claudin-15 are found in leaky, cation-selective TJs like those of the kidney and small intestine and that these epithelia have slightly larger than average paracellular pores of roughly 6.5 Å.

Efforts have been made to enhance the paracellular transport of a pharmaceutical agent through modulation of TJ function. Some of these approaches selectively disrupt extracellular components of TJ and AJ proteins (peptides that emulate or antibodies that bind to these domains); others act to disorganize the specific membrane environment associated with the AJC (e.g., chitosan and thiolated polymers, EDTA). Still others appear to nonselectively disrupt the organization of functional TJ structures, such as sodium caprate. It has been suggested that sodium caprate increases paracellular permeability through decreasing the expression of the TJ protein tricellulin and thereby disorganizing tricellular contacts. Additionally, some approaches have copied methods used by certain pathogens that use secreted enzymes working at the external surface of the cell, as well as manipulation of intracellular signaling mechanisms that affect either TJs or AJs. It is important to note that opening the paracellular route can be achieved indirectly by affecting AJs, an event that leads to destabilized cell–cell contacts that result in TJ disorganization. More information on these various pathways and mechanisms are described in the relevant chapters dealing with each specific route of epithelial delivery (i.e., Chapters 7 through 13, inclusive).

In general, however, one can assume certain aspects of the paracellular pathway following its opening by such an agent. For example, if a transport enhancer causes TJs to open with the retention of the normal complement of claudin proteins at the plasma membrane, then one could envision a channel that is larger than the normal TJ pores and with charge properties similar to the native TJ structure. In this way, larger molecules could transport across the epithelia, but there may be preference to those with particular surface charge properties. If, however, the enhancing agents result in claudin internalization by the epithelial cell, the resulting paracellular pathway should not have charge preferences related to claudin properties. Making the assumption that the material accesses the entire epithelial surface area, transport across epithelial barriers via the paracellular route can be approximated similarly to transport via the transcellular route/pathway.

4.3.2 TRANSCELLULAR TRANSPORT

The majority of pharmaceuticals that have been developed have the ability to readily transport across epithelia to enter the systemic circulation of the body. Such agents are relatively small (<400 Da in mass) and, as described in Section 4.3.4, have certain physicochemical properties that allow for them to partition into and out of the various lipid and aqueous compartments of epithelial cells in a nonselective manner. This allows them to migrate across an epithelium based upon a chemical potential gradient. The flux of such molecules crossing the epithelium has been modeled according to Fick's first law of diffusion, which states that the rate of diffusion across a membrane is proportional to the difference in concentration on each side of the membrane:

$$\frac{dm}{dt} = \left(\frac{Dk}{h}\right) \cdot A \cdot \Delta C \tag{4.1}$$

where
   dm/dt is the rate of diffusion across the membrane
   D is the diffusion coefficient of the drug in the membrane
   k is the partition coefficient of the drug into the membrane
   h is the membrane thickness
   A is the available surface area
   ΔC is the concentration gradient, i.e., $C_o - C_i$, where C_o and C_i denote the drag concentrations on the outside and the inside of the membrane, respectively

When sink conditions occur, it ensures that a large concentration gradient is maintained throughout the absorption phase, thereby enhancing the driving force for absorption. The maintenance of sink conditions means that

$$C_o > C_i, \quad \text{thus } \Delta C \approx C_o$$

and thus Equation 4.1 is reduced to

$$\frac{dm}{dt} = \left(\frac{Dk}{h}\right) \cdot A \cdot C_o \tag{4.2}$$

Substituting further into Equation 4.2 gives

$$\frac{dm}{dt} = P \cdot A \cdot C_o \tag{4.3}$$

where P, the permeability constant, is defined as Dk/h and has the units cm/s.
   This can be simplified further to give

$$\frac{dm}{dt} = K_1 C_o \tag{4.4}$$

Equation 4.4 indicates that the rate of diffusion is proportional to drug concentration. K_1 is a pseudo-rate constant and is dependent on the factors D, A, k, and h.
   A further point here is that in order to reach the underlying blood capillaries to be absorbed, the drug must pass through at least two epithelial membrane barriers (the apical and basolateral epithelial cell membranes) and also the endothelial membrane of the capillaries. In some cases, for example, in stratified epithelia such as that found in the skin and buccal mucosa, the epithelial barrier comprises a number of cell layers rather than a single epithelial cell. Thus, the *effective* barrier to drug absorption is not diffusion across a single membrane as described earlier but diffusion across the entire epithelial and endothelial barrier, which may comprise several membranes and cells in series. The parameters ΔC, D, K, h, and A of Equation 4.1 refer to transport across the entire effective barrier, from the epithelial surface to the circulating blood, rather than simply the apical plasma membrane.

The use of endogenous pathways for the potential delivery of large hydrophilic drugs, such as proteins, into the body offers some exciting promise for the future. Such strategies are focused on simple epithelia, such as in the gut and lung, where these transcytosis pathways operate. Importantly, certain pathogens appear to hijack endogenous transcytosis pathways at these same simple epithelia to gain entry into a host. Some of these pathogen-targeted pathways have also been examined for the delivery of macromolecules.

4.3.4 PHYSICOCHEMICAL PROPERTIES OF A DRUG THAT INFLUENCE EPITHELIAL TRANSPORT

As described earlier, conventional drug molecules tend to be absorbed predominantly via transcellular passive diffusion across the epithelial interface, and the rate of diffusion is directly proportional to the concentration gradient across the epithelium (Equation 4.4). Diffusion is also critically dependent on the physicochemical properties of the drug, such as its lipid solubility, degree of ionization, size, and solubility (Florence and Attwood 2011); these factors are discussed here.

4.3.4.1 Lipid Solubility and Partition Coefficient

With respect to passive diffusion, the outer membrane of the epithelial cell may be regarded as a layer of lipid, surrounded on both sides by water (Figure 4.4b). Thus, for transport through the apical membrane, the drug must initially partition into the lipid membrane, according to its lipid/water partition coefficient, must then diffuse across the lipid phase at a rate determined by the concentration gradient, and finally, must be distributed out at the other side of the membrane, according to its lipid/water partition coefficient.

The degree to which a drug partitions into the lipid membrane is a measure of its lipid solubility and can be described by its oil/water equilibrium partition coefficient. This can be determined by adding the drug to a mixture of equal volumes of a lipophilic liquid (often octanol, but other solvents are also used) and water and shaking the mixture vigorously to promote partitioning of the drug into each phase. When equilibrium is attained, the phases are separated and assayed for drug. The partition coefficient (P) is given by

$$P = \frac{C_{oil}}{C_{water}} \tag{4.5}$$

where
C_{oil} is the concentration of drug in the oil phase
C_{water} is the concentration of drug in the water phase

Often, the logarithm of the partition coefficient, ($\log P$), of a compound is quoted. For a given drug, if $\log P = 0$, there is equal distribution of the drug in both phases; if $\log P > 0$, the drug is lipid soluble; and if $\log P < 0$, the drug is water soluble.

A plot of percentage absorption versus $\log P$ yields a parabolic curve, with peak absorption occurring within a narrow optimal range of $\log P$ values. Values of $\log P$ that are too high (>6) are associated with poor transport characteristics. This is because highly lipidic drugs also have poor aqueous solubility, which limits bioavailability. Furthermore, if a drug is too lipophilic, it will tend to remain in the lipidic membrane and never partition out again into the underlying aqueous environment. Values of $\log P$ that are too low (<3) are also associated with poor permeation ability. Very polar compounds are not sufficiently lipophilic to be able to pass through lipid membrane barriers. If a drug molecule forms hydrogen bonds with water, desolvation and breaking of the hydrogen bonds is required, prior to partitioning into the apical membrane of the epithelial cell. If the number of hydrogen bonds between the drug and water is >10, too much energy is required and there will be minimal drug transport across the membrane. Drugs that are ionized will also demonstrate poor permeation ability, as described next.

4.3.4.2 Degree of Ionization and Drug pK_a

Most drugs are weak electrolytes, existing in unionized and ionized forms in an aqueous environment. According to the pH-partition hypothesis, the unionized form of a drug is lipophilic and diffuses readily across cell membranes. Conversely, the ionized form is hydrophilic and thus cannot penetrate cell membranes easily. The degree of ionization (and therefore ability to cross a membrane) depends on both the pK_a of the drug and the pH of the biological fluid in which it is dissolved.

The extent of ionization can be quantified by the Henderson–Hasselbalch equation (Box 4.1), which shows that for weak acids (with a pK_a in the range 2.5–7.5), the unionized form of a weak acid predominates when the pH is lower than the pK_a. Thus, weak acids will be predominantly unionized in the stomach, which favors their absorption in this region. In contrast, a very low percentage is unionized in the small intestine, which suggests unfavorable absorption for weak acids in this region. However, for weak bases (with a pK_a in the range 5–11) the reverse is true: when the pH is lower than the pK_a, the ionized form predominates. Thus, weak bases are poorly absorbed, if at all, in the stomach since they are largely ionized at low pH, but they are well absorbed in the small intestine, where they are unionized. Strong acids, such as cromoglycate, as well as strong bases, such as mecamylamine, are ionized throughout the GI tract and are therefore poorly absorbed.

Although the pH-partition hypothesis is useful, it must be viewed as an approximation because it does not adequately account for certain experimental observations. For example, most weak acids are well absorbed from the small intestine. Similarly, quaternary ammonium compounds are ionized at all pHs but are readily absorbed from the GI tract. These discrepancies arise because the pH-partition hypothesis does not take into account factors such as

- The large mucosal surface area of the small intestine can compensate for ionization effects.
- The relatively long residence time in the small intestine again can compensate for ionization effects.
- A local pH exists at the membrane surface that differs from the bulk pH of the GI lumen, due to the attraction of hydrogen ions to negatively-charged membrane components.
- Even the unionized form of a drug displays limited absorption.
- Charged drugs, such as quaternary ammonium compounds, may interact with organic ions of opposite charge, resulting in a neutral species, which is absorbable.
- Bulk transport of water from the gut lumen to the blood, or vice versa, can drag water-soluble molecules with it, resulting in a respective increase or decrease in the absorption of water-soluble drugs.
- Some drugs interact with the overlying mucus layer of the epithelium, which can affect the overall transport properties.
- Some drugs are absorbed via active pathways.

For ionizable drugs, log P is pH-dependent. Therefore, log D, the log distribution coefficient of the drug at different pHs, can be employed as an estimation and/or prediction of the absorptive potential at a particular pH value. The pH at which the log D is measured should therefore be reported, but in practice, values normally correspond to determinations carried out at a physiological pH of 7.4. The relationship between the observed overall partition coefficient and the distribution coefficient is given by the following equation:

$$D = P(1-\alpha) \tag{4.6}$$

where α is the degree of ionization of drug.

Although described here with respect to the oral route of drug delivery, it should be noted that all epithelial interfaces (buccal, sublingual, dermal, vaginal, pulmonary, nasal, etc.) show variations in local pH, and so these factors must be considered with respect to each particular route, absorption site, and drug moiety.

BOX 4.1   HENDERSON–HASSELBALCH EQUATION QUANTIFIES THE DEGREE OF IONIZATION OF DRUG SPECIES AT A PARTICULAR pH

The Henderson–Hasselbalch Equation is

Acid   Base

$$pH = pK_a + \log \frac{[A^-]}{[HA]} \qquad pH = pK_a + \log \frac{[B]}{[BH^+]}$$

To illustrate the concept of the pH-partition hypothesis, consider the absorption of the weak electrolytes, aspirin, and codeine, from the GI tract.

WEAKLY ACIDIC DRUG ASPIRIN (pK_a = 3.5)

In the stomach, assume the pH of the gastric fluid = 1.
   The Henderson–Hasselbalch equation gives

$$1 = 3.5 + \log \frac{A^-}{[HA]} \Rightarrow \log \frac{[A^-]}{[HA]} = -2.5. \quad \text{Therefore } \frac{[A^-]}{[HA]} = 10^{-2.5} = 0.003$$

Thus, at the low pH in the stomach, [HA] is for greater than [A⁻], i.e., the drug is predominantly unionized, that is, in the absorbable form.
   In the intestine, assume the pH of the intestinal fluid = 6.5. The Henderson–Hasselbalch equation gives

$$6.5 = 3.5 + \log \frac{A^-}{[HA]} \Rightarrow \log \frac{[A^-]}{[HA]} = 3. \quad \text{Therefore } \frac{[A^-]}{[HA]} = 10^3 = 1000$$

Thus, at the high pH in the small intestine, [A⁻] is 1000 times greater than [HA], i.e., the drug is predominantly in the ionized, unabsorbable form.

WEAKLY BASIC DRUG CODEINE (pK_a = 8)

In the stomach, assume the pH of the gastric fluid = 1.
   The Henderson–Hasselbalch equation gives

$$1 = 8 + \log \frac{[B]}{[BH^+]} \Rightarrow \log \frac{[B]}{[BH^+]} = -7. \quad \text{Therefore } \frac{[B]}{[BH^+]} = 10^{-7} = 0.0000001$$

Thus, at the low pH in the stomach, [BH⁺] is 10^7 greater than [B], i.e., the drug is predominantly ionized, that is, in the unabsorbable form.
   In the intestine, assume the pH of the intestinal fluid is 6.5. The Henderson–Hasselbalch equation gives

$$6.5 = 3.5 + \log \frac{[A^-]}{[HA]} \Rightarrow \log \frac{[A^-]}{[HA]} = 3. \quad \text{Therefore } \frac{[A^-]}{[HA]} = 10^3 = 1000$$

Thus, at the high pH in the small intestine, the drug is much less ionized than in the stomach and is therefore more readily absorbed.

4.3.4.3 Molecular Weight and Molecular Volume

The size of a drug can limit its ability to cross an epithelium. As described in the following texts, a molecular weight threshold of 500 Da is normally the cutoff limit for transmembrane transport. Molecular volume is a further important consideration, as it affects the rate of drug diffusion. In the hypothesis of a spherical particle, the Stokes–Einstein equation is as follows:

$$D = \frac{RT}{N_A} \frac{1}{6\pi\eta a} \tag{4.7}$$

where
  D is the diffusion coefficient of a spherical particle (m²/s)
  a is the radius of the particle (m)
  R is the gas constant (=8.314 J/mol·K)
  T is the temperature (K)
  N_A is Avogadro's number (6.022×10^{23} mol^{-1})
  η is the viscosity of the fluid (Pa·s)

$$\text{Given that: } V = \left(\frac{4}{3}\right)\pi a^3 \tag{4.8}$$

where V is the volume and a is the radius; it can be seen from Equation 4.8 that drug diffusivity is inversely proportional to the cube root of the molecular volume. A more complex relationship pertains for more complex and organized structures such as lipid bilayers, but again, drug diffusivity is inversely proportional (probably by an exponential relationship) to the molecular volume. As molecular volume is related to molecular weight, large molecules will in general diffuse at a slower rate than small molecules.

Molecular volume is also determined by the overall conformation of the molecule and the heteroatom content that may be involved in inter- and intramolecular hydrogen bonding. Thus, molecules that assume a compact conformation will have a lower molecular volume and a corresponding higher diffusivity. An important consequence of this property is that even if such molecules have a high molecular weight (>500 Da), their high diffusivity may nevertheless be able to facilitate absorption. These parameters have particularly important implications for the paracellular route of drug absorption, as the TJs bind cells together very efficiently and so can block the paracellular passage of even relatively small molecules.

4.3.4.4 Solubility

A drug molecule must be in solution before it can cross an epithelial barrier; therefore drugs with limited aqueous solubility typically exhibit poor absorption behavior, as well as large inter- and intra-subject variation in bioavailability. Drug solubility can be critically dependent on the pK_a of the drug and the prevailing pH of the fluids at the epithelium interface in question. The ionized form of a drug molecule is the more water-soluble form, therefore the dissolution rate of weak acids increases with increasing pH, whereas the dissolution rate of weak bases decreases with increasing pH. Solubility requirements are thus diametrically opposed to permeability requirements: solubility favors charged species, whereas permeability and transmembrane transport favors neutral species. In practice, a balance between the lipid and aqueous solubility of a drug is required for successful absorption. Chapter 3 discusses in detail various strategies to improve the solubility of poorly soluble active pharmaceutical ingredients (APIs), which include reducing crystal size, cocrystal formation, using cosolvents, cyclodextrin inclusion complexes, amorphous solid dispersions, and colloidal carriers.

4.3.4.5   Lipinski's Rule of Five

The effects of the physicochemical properties of a drug discussed thus far have been summarized very effectively by Christopher Lipinski and colleagues, who studied the physicochemical properties of several thousands of drugs and drug candidates in clinical trials, in order to identify calculable parameters that were related to solubility and permeability (Lipinski et al. 2001). They devised a simple mnemonic, described as the "rule of five" (because all properties are a factor of five) that states that *poor* absorption or permeation are more likely when

- There are more than 5 H-bond donors (expressed as the sum of the O–H and N–H bonds on the molecule)
- The molecular weight is greater than 500
- The log P is over 5 (or multicore log P (M log P) is over 4.15)
- There are more than 10 H-bond acceptors (expressed as the sum of nitrogen and oxygen atoms on the molecule)

If two parameters are out of range, poor absorption or permeability is likely. Drugs that are substrates for biological transporters are exceptions to this rule: such drugs include certain antibiotics, antifungals, vitamins, and cardiac glycosides.

4.3.4.6   Stability

A further physicochemical property of the API with important implications for epithelial transport is its chemical stability. Degradative enzymes associated with epithelial interfaces can metabolize a labile drug, deactivating it prior to its absorption. Enzymes in the GI tract include proteases, glycosidases, lipases, and nucleosidases, which are highly efficient at breaking down proteins, carbohydrates, fats, and nucleic acids, respectively, from foodstuffs—so that they can be absorbed to make energy available to the body. However, these enzymes can also degrade drug molecules, resulting in poor bioavailability. Furthermore, drugs that are orally absorbed may subsequently be subjected to the "first-pass" effect of the liver. In some cases, this presystemic metabolism accounts for a significant, or even total, loss of drug activity.

The extremely high metabolic activity of the GI tract has been a major impetus for the exploration of alternative routes for systemic drug delivery. In comparison to the oral route, much less is known about the nature of the enzymatic barrier presented by the buccal, nasal, pulmonary, dermal, and vaginal routes. However, it is generally accepted that such routes have a lower enzymatic activity, particularly toward drugs such as peptides and proteins. Furthermore, such routes also offer the advantage of avoiding first-pass metabolism by the liver.

4.3.4.7   Implications of the Physicochemical Properties of a Drug for Epithelial Transport

The physicochemical properties of the drug will influence the route and mechanism of drug absorption through the mucosa. Thus, it would seem most likely that (1) low molecular weight hydrophilic compounds would tend to be absorbed via the paracellular route, rapidly moving through the aqueous compartment between the epithelial cells; (2) lipid-soluble drugs of low molecular weight would usually be absorbed via transcellular passive diffusion, diffusing through the lipidic membrane barrier; (3) transport of macromolecular drugs would be largely restricted to endocytic processes; and (4) drugs bearing structural similarities to endogenous nutrients may be absorbed via carrier-mediated mechanisms.

However, these considerations are obviously only broad generalizations. Thus, although a drug molecule may be *predominantly* absorbed via one particular route/mechanism, it is also likely that *suboptimal* transport will occur via other routes and mechanisms. In particular, drugs that are absorbed via active mechanisms are often also absorbed, to a (much) lesser extent, via passive diffusion mechanisms.

Poorly permeable drugs are defined as Class III and Class IV drugs of the biopharmaceutics classification system (see Chapter 3, Section 3.1.1 and Figure 3.1). Poorly permeable drugs include the new "biologics," i.e., peptides, proteins, and DNA-based medicines, produced through recent advances in biotechnology and molecular biology. These macromolecular drugs are not typically successful candidates for transport across epithelial barriers because they are (1) too large, (2) too charged, and (3) too labile, for successful epithelial transport.

Strategies to improve drug permeation include modifying the drug molecule per se, to confer enhanced permeability properties. For example, the lipid solubility of a drug molecule can be increased by blocking the hydrogen bonding capacity of the drug. This may be achieved by, for example, substitution, esterification, or alkylation of existing groups on the molecule, to decrease the drug's aqueous solubility, favoring partitioning of the drug into the lipid membrane. The development of clindamycin, which differs from lincomycin by the single substitution of a chloride for a hydroxyl group, is such an example. Alternatively, the drug may be covalently bound to a lipid carrier, such as long-chain fatty acids. These approaches involve modifying the existing structure of the drug, to form a new chemical entity. This raises regulatory hurdles, as well as running the risk of compromising the activity of the drug, increasing drug toxicity, and also increasing the molecular weight to such an extent that the molecule will be too large to cross the membrane barrier.

Alternatively, the prodrug approach can be used, which involves the chemical transformation of the active drug substance to an inactive derivative (prodrug), which is subsequently converted to the parent compound in vivo by an enzymatic or nonenzymatic process. A lipophilic prodrug of a drug can be prepared, to enhance membrane permeability in comparison to the parent drug. Enzymatic or chemical transformation converts the inactive prodrug to the pharmacologically active drug, after absorption has taken place.

Pharmaceutical scientists may also employ formulation strategies in order to optimize drug absorption. Such strategies include mechanisms to increase the solubility of the drug: a wide variety of mechanisms are available and are described in full in Chapter 3. Excipients may also be included in the formulation to enhance epithelial permeation—such excipients include absorption enhancers, enzyme inhibitors, mucoadhesive polymers, and CR technologies. This is a current area of active research and the reader is referred to the relevant chapters for each particular epithelial route, for further information. Finally, another approach to improving epithelial permeability is via the rational design of therapeutic entities that possess desirable properties for epithelial transport. Rational drug discovery, design, and development are discussed in Chapter 20.

4.4   FURTHER COMPONENTS OF EPITHELIAL BARRIERS

Depending upon their location in the body and required functional properties, each epithelium will have some sort of additional barrier strategy, which can provide further physical or biochemical barriers to drug transport, as described here.

4.4.1   KERATIN

Cornification of the stratified squamous epithelium of the skin is most noted for its keratin-containing surface. Formation of this keratin-laden barrier is achieved through the progressive alteration of epithelial cells through a series of biochemical events as they transition from a viable to nonviable form. While most notable in the skin, keratinization of epithelia at body orifices plays a critical role in the transition from epidermal to mucosal epithelial phenotypes. The process of keratinization is controlled by a variety of factors with frictional stress being one of the most prominent. For example, the buccal epithelium of the inner cheek vestibule is not keratinized, except in areas involved in mastication such as the dorsal tongue surface, hard palate, gingiva, and dentition plate where keratinization results in the structure known as the *linea alba*. Additionally, areas that experience less abrasive conditions have less keratinization, such as the postauricular region.

4.4.2 Mucus

Mucus is a highly viscous hydrogel. The main components of mucus are mucins; these are high molecular weight, heavily glycosylated proteins made of oligosaccharide side chains attached to a protein core. Roughly 75% of the protein core is glycosylated, providing mucus with its gel-like structure. The carbohydrate structures within mucins give mucus its water-retaining properties, make it resistant to proteolysis, and provide a mechanism where invading pathogens can become trapped to limit their infectivity. Thus, epithelial surfaces use mucus for at least three important functions: physical protection, retention of water to maintain a hydrated surface, and entrapment/inactivation of invading pathogens. Certain drugs and DDS can also become entrapped in the mucus layer of an epithelium.

The thickness, structure, and exact composition of mucus depend on its location. For example, proteins generated from the genes MUC5AC and MUC5B are the main constituents of airway mucus, while intestinal mucus is made mostly of protein products of the MUC2 and MUC3 genes. Mucus is often organized as a loosely associated layer ($\approx$450 µm thick) that lays atop a firmly adhering layer ($\approx$30 µm thick) that interacts with the glycocalyx ($\approx$1 µm thick) structure at the apical surface of mucosal epithelia. The process of how mucins and other components of mucus are secreted to reach the apical surface of an epithelium can play a significant role in dictating its properties. Mucins can be secreted from goblet cells present in the epithelium itself. Alternately, cells located in submucosal glands can secrete mucins that are transferred to the apical surface via ducts. During this transit, the mucin can be further processed and other mucus components can be added or removed. Importantly, mucus can function as a barrier to a variety of drugs, including those that are lipophilic (Sigurdsson et al. 2013).

4.4.3 Efflux Pumps

A challenging aspect of successful transcellular transport is the action of specific pumps (e.g., P-glycoprotein, MRP2) that effectively efflux molecules back to the apical surface once they have entered the cytoplasm of an epithelial cell. The P-glycoprotein is a 170–180 kDa membrane glycoprotein acting as an ATP-dependent efflux pump that reduces the intracellular accumulation and/or the transcellular flux of a wide variety of drugs, including peptides such as gramicidin D, valinomycin, and cyclosporin. As these efflux systems are located on the apical surface of the plasma membrane, it can be assumed that their physiological role is to restrict transcellular flux of some molecules.

Extensive work has been performed to identify these pumps and the characteristics of molecules that make them potential substrates. While an extensive discussion of these efflux pumps is beyond the scope of this chapter, the evaluation of all new prospective therapeutic entities as a potential substrate for these pumps is now a routine element of drug development. Further information is also given in the respective chapters describing each particular epithelial route.

4.5 PROPERTIES OF DISTINCT EPITHELIAL BARRIERS

Despite having the same fundamental role of protection of the organism against the penetration of foreign and potentially harmful particles, epithelia can be significantly different depending on their location in the body. The barrier they represent to the entry of macromolecular drugs is thus specific to the body area they cover. In the following sections, we will compare and contrast anatomical features and physiological functions of epithelia that could affect the delivery of drugs: skin epidermis, buccal mucosa, corneal epithelium, nasal epithelium, the various segments of the GI and respiratory tract and the vaginal cavity. Our goal is to highlight the basis for how each of these epithelia presents specific challenges as a barrier to drug delivery.

4.5.1 SKIN

The structure of the skin is shown in Figure 4.6. Keratinocytes are the primary cell type observed in human epidermis and these cells organize into at least five distinct layers as they mature from the basement membrane to the surface, where they slough off (see also Figure 4.7 for stratified squamous epithelium and Chapter 9, Figure 9.1 for the epidermal layers.). These layers are referred to as the *stratum basale*, *stratum spinosum*, *stratum granulosum*, *stratum lucidum*, and *stratum corneum*. While thickness and replicative function of the *stratum basale*, also known as the *stratum germinativum*, is fairly constant in the skin from various locations in the body, the other layers are present in varying thickness at discrete sites. For example, keratinocytes of the *stratum lucidum* are filled with an intermediate form of keratin known as eleidin, and this layer is prominent only in the palms of the hands and soles of the feet. By comparison, the skin on the backs of the hands and fingers are strikingly thinner and lack a defined *stratum lucidum*.

Besides keratinocytes, there are melanocytes in the *stratum basale* layer of the epidermis. Other nonkeratinocyte cells observed associated with the skin epithelium are those cells involved in the penetrations made by sweat gland ducts and hair follicles to reach the surface of the skin. These structures can provide sparse drug absorption pathways that are distinct from the keratinized skin surface that represents the majority of the skin's surface area. It should also be pointed out that hair follicle and sweat gland duct distributions are also regionally variable. Also, the amount of skin surface epithelium available for drug delivery is quite variable among the population; the realistic

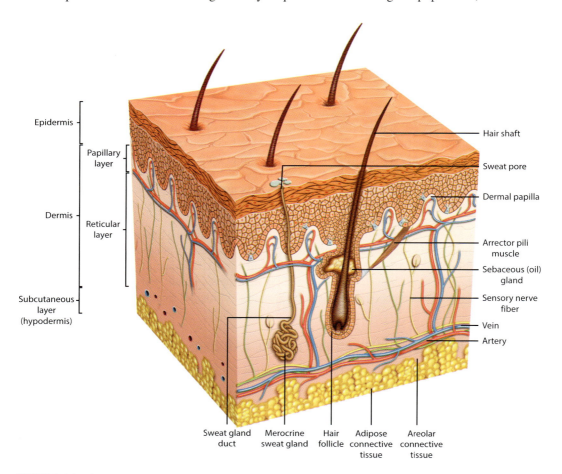

FIGURE 4.6   Structure of the skin. (Alexilusmedical/Shutterstock.com.)

issue for these approaches relates to the method of application (gels, lotions, adhesive patches, etc.) and the duration of exposure by that method.

The complex process of cell replication, maturation, and keratinization takes about 28 days in human skin. Despite these ongoing biochemical processes, skin epithelium is considered to be a fairly inert structure with regard to drug delivery. The primary structure involved in limiting drug transport is the *stratum corneum* layer of the epidermis. This dehydrated layer is best suited for the transport of small nonpolar molecules; larger, hydrophilic molecules fail to move effectively across this outermost layer of the skin. Transdermal absorption requires a drug to pass through the *stratum corneum* and other epidermal layers and reaches the blood capillaries of the underlying dermis (Figure 4.6).

4.5.2 ORAL CAVITY

The oral cavity consists of the lips, inside of the cheeks, soft palate, and lower surface of the tongue and floor of the oral cavity, hard palate, and gingiva. The mucosa of the inner cheek is referred to as the buccal mucosa. There are three major types of epithelium found in the oral cavity with differing degrees of keratinization: masticatory, specialized, and lining epithelium. The masticatory epithelium is dehydrated, mechanically tough and chemically resistant; it is keratinized, is 100–200 μm thick, and covers the gingiva and hard palate. Its outermost region is made of flattened hexagonal cells filled with cytokeratin aggregates and bound by a cell envelope, surrounded by a complex mixture of lipids. The presence of neutral lipids such as ceramides and acyl ceramides makes it relatively impermeable to water. The specialized epithelium is stratified, keratinized, and covers the dorsal surface of the tongue. Its main function is taste perception (presence of papillae). Finally, the lining epithelium protects the rest of the oral cavity. It is 500–600 μm thick and covers 60% of the oral mucosa. The lining epithelium is a nonkeratinized, stratified squamous epithelium composed of 10–50 cell layers (Figure 4.7). It is quite flexible and has a turnover time of 3–8 days. The oral cavity has only a small surface area available for absorption, 100 and 50.2 cm² for the oral and buccal mucosae, respectively. The majority of drug intake occurs through the lining epithelium. Its unique feature is the absence of TJs, resulting in a "leaky" epithelium. However, desmosomes,

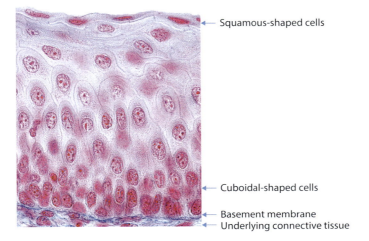

Squamous-shaped cells

Cuboidal-shaped cells

Basement membrane
Underlying connective tissue

FIGURE 4.7 Stratified squamous epithelium comprises several layers of cells. The cells have a cuboidal to columnar shape in the deep layers and a squamous shape in the apical layers. The nonkeratinized variety lines wet surfaces, such as the lining of the oral cavity and the vagina. (*Note:* The keratinized variety forms the epidermis, i.e., the superficial layer of the skin, shown in Figure 4.6 and Chapter 9, Figure 9.1.) (Jose Luis Calvo/Shutterstock.com.)

hemidesmosomes, and GJs still form loose intracellular links, and the presence of proteins and lipids in the intercellular space renders it relatively impermeable to drug penetration.

Moreover, the oral epithelium is covered by a 70–100 μm thick film of saliva, which contains 99% water and 1% proteins, glycoproteins, and electrolytes. Saliva has a pH of 5.6–7.4 and it lubricates the mouth, facilitates the swallowing process, and protects the teeth from demineralization. The daily volume of saliva secreted into the oral cavity varies from 0.5 to 2 L. Its constant production and elimination greatly reduces the drug residence time into the oral cavity. It can also result in uneven drug distribution after release from a solid DDS. Saliva also controls the bacterial flora in the oral cavity and initiates food digestion due to the presence of metabolic enzymes such as α-amylase, lysozyme, and lingual lipase that break 1–4 glycosidic bonds, cell wall, and fats, respectively. These enzymes initiate the digestion of food and impede bacterial infection, but they also metabolize some peptide and protein drugs, reducing their bioavailability after oral administration. Other proteases such as aminopeptidases, carboxypeptidases, and esterases are present either at the surface of the epithelium or in the mucosa. Although drugs enter in contact with only some of these enzymes, depending on the pathway they use to cross the epithelium, they can lead to peptide and protein degradation. Mucus in the oral cavity is mainly secreted by the sublingual and salivary glands. It is 30–40 μm thick and adheres strongly to the surface of buccal epithelial cells. At the pH of saliva, it is negatively charged due to the presence of charged sialic acid groups ($pK_a = 2.6$) and ester sulfates at the terminals of some oligosaccharide chains.

4.5.3 INTESTINES

The epithelium of the small intestine is organized in villi, approximately 1 mm long fingerlike projections, which add greatly to its functional surface area (Figure 4.8). The height and density of these villi decrease from the duodenum to the ileum; villi are lacking in the colon. The intestinal epithelium is characterized by a monolayer of columnar absorptive epithelial cells called enterocytes, interspersed with goblet cells, Paneth cells, M cells, Tufts cells, and a host of enteroendocrine cells. Tubular invaginations of the epithelium at the base of villi are called crypts of Lieberkühn and are the sites where stem cells reside that generate nascent forms of the cell types listed earlier. Lieberkühn crypts are lined with epithelial cells that are primarily involved in secretion, while cells that progress up the villus are more involved in absorptive processes. As cells progress up the villus, they elaborate hairlike projections of cell membrane called microvilli that cover the surface of these cells. Microvilli can be as numerous as 3000/cell and are organized to form a structure known as the brush border. They are the primary surface for nutrient absorption and also contain glycoprotein enzymes. Stem cells located in the crypts differentiate into cell types found in both the crypt and the villus that must be replaced due to senescence and desquamation.

Stem cells divide to form Paneth cells that migrate to the base of the crypt where they secrete antimicrobial peptides. Other forms of stem cell division lead to cells that migrate away from the crypt base, advancing up the villus in the small intestine or to the surface epithelium in the colon that lacks villi. Absorptive enterocytes make up the majority of these cells in both the small intestine and the colon. Goblet cells are responsible for the secretion of mucus into the GI tract lumen and are more abundant in the colon than in the small intestine. As goblet cells migrate from the bottom of the crypt, they develop granules filled with mucins and typically release their contents upon reaching the crypt opening. These cells also secrete enzymes involved in digestion such as peptidases, intestinal lipases, and intestinal amylase. Tufts cells are responsible for secreting a series of opioid molecules that can signal in a paracrine fashion. The small intestinal epithelium is also interspersed with specialized enteroendocrine cells referred to as K cells, I cells, S cells, M cells, N cells, and L cells, which respond to luminal nutrients by the basal secretion of glucose-dependent insulinotropic peptide, cholecystokinin, secretin, motilin, neurotensin, or a combination of peptide YY, glucagon-like peptides 1 and 2, and glicentin, respectively. Finally, microfold cells (also referred to as M cells) named for

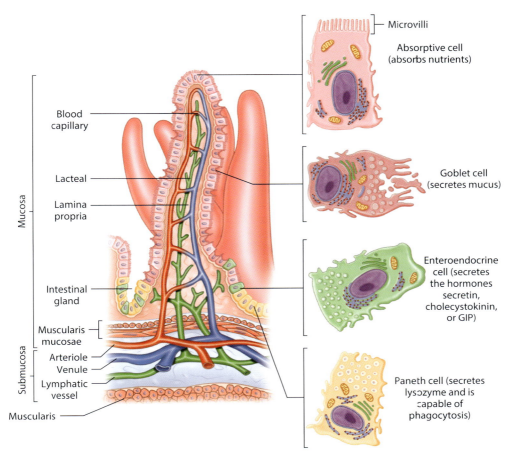

Enlarged villus showing lacteal, capillaries, intestinal glands, and cell types

FIGURE 4.8   Intestinal villus, showing intestinal glands and the different cell types: absorptive cells (entero-cytes) containing microvilli, goblets cells, enteroendocrine cells, and Paneth cells. (From Tortora, G.J. and Derrickson, B., eds.: *Principles of Anatomy and Physiology*, 14th ed. 2013. Copyright Wiley-VCH Verlag GmbH & Co. KGaA. Reproduced with permission.)

their apical membrane character are found in specialized regions known as lymphoid follicle-associated epithelium (FAE); these areas are specialized for endocytosis and vesicular transport to the underlying lymphoid tissue as part of the process to sense and process antigens found in the intestinal lumen. FAE regions are covered by only a thin layer of mucus and associated with underlying lymphoid cells (see also Chapter 7, Figure 7.9).

The GI tract contains a wide array of enzymes, which are present in a variety of locations, including the gut lumen, adsorbed to the mucus layer, as integral membrane proteins of the brush border (microvilli) of the enterocytes, within the enterocytes (both free within the cell cytoplasm and within cellular lysosomes), and the colonic microflora. The GI tract is also associated with extremes of pH, ranging from an acidic environment in the stomach (pH = 1.5–3.5) to much more alkaline conditions in the large intestine (pH = 7.5–8.5). Acid- or base-labile drugs are susceptible to degradation depending on the prevailing pH conditions (Woodley 1994).

Mucus in the intestines is secreted by goblet cells at the top of the crypts and progressively moves upward between the villi. As a consequence, villi tips are not always covered in mucus. In the small intestine, mucus is a 200 μm loosely adherent discontinued layer covering the surface of the epithelium (Johansenn et al. 2011). It can contain enzymes such as pepsin, trypsin, and chymotrypsin and metabolic enzymes (cytochrome P450 and P-glycoprotein) that are trapped in the viscous layer and remain concentrated close to the apical cell surface. Their presence can potentially result in the degradation of protein drugs.

In the large intestine, both absorptive and goblet cells are located in long, straight, tubular glands. Mucus in the colon does not contain the degradative enzymes found in the small intestine and is composed of two layers: a thin (≈100 μm) and dense inner layer that firmly adheres to the surface of the epithelium and is further converted into a second layer (≈700 μm), which is more loosely attached and has a more disorganized appearance. This outer layer turns out to be a suitable environment for bacteria, which use the glycans on the mucins as a source of energy and as attachment sites for the lectins on their surface. The inner layer is devoid of bacteria and keeps pathogens away from the surface of the epithelium (Kim and Ho 2010).

The glycocalyx forms a 500 nm thick coating layer between the apical surface of the cells and the mucus. It is composed of a negatively-charged network of proteoglycans, glycolipids, and glycoproteins, including densely packed mucins MUC-3, MUC-12, and MUC-17 as well as membrane-bound glycoprotein enzymes secreted by the enterocytes. In combination, mucus and the glycocalyx prevent autodigestion and diffusion of pathogens from the intestinal lumen. Unfortunately, they also keep particulate DDS away from the surface of the epithelium, as their size and surface properties can prove unsuitable for diffusion across such a viscous and tightly packed barrier. Despite the presence of these barriers hindering the penetration of drugs and a short transit time (2–4 hours), the small intestine remains a promising site for drug absorption due to its large surface area and high blood flow. Conversely, the colon is considered as a poor site for drug absorption because of its lower permeability and less extensive surface area; its long transit time (24–48 hours) can potentially make up for these deficits to result in increased drug absorption.

4.5.4 Respiratory Tract

The primary function of the respiratory tract is to perform the extraction of oxygen from the air and the elimination of carbon dioxide from the body. This is achieved in the deep lung in alveolar sacs. The other elements of the respiratory tract serve critical functions to allow this process to occur in a wide variety of conditions. The respiratory tract can be divided into a sequence of segments that provide a conduit to the deep lung: nasal cavity, pharynx, trachea and bronchioles. All of these segments have specialized epithelia (as summarized in Chapter 11, Table 11.1), which function to perform specific tasks that result in unique drug delivery opportunities and challenges. In general, the nasal cavity epithelium functions to humidify and warm entering air; the pharynx carries air from the nose and mouth and directs it toward the larynx that acts as an opening to the trachea; the trachea continues to humidify and warm the incoming air and traps large particulates that could compromise the function of the deep lung. As the surface area and residence time of the pharynx are not significant, we will focus on examining the epithelia of the nasal cavity, trachea, and deep lung.

4.5.4.1 Nasal Cavity

The total surface area available in the human nasal mucosa is estimated to be about 180 cm^2, of which 10 cm^2 is olfactory mucosa and 170 cm^2 is the richly vascularized respiratory mucosa. The effective surface area for drug absorption of the nasal epithelium is increased due to the presence of ≈400 microvilli on the apical surface of each cell. The main epithelium is pseudostratified columnar and contains many mucus-secreting goblet cells (Figure 4.9).

Mucus at the surface of the nasal epithelium is about 5 μm thick and composed of two distinct layers: periciliary and superficial layers. The periciliary layer directly covers the surface of the cells and is very viscous. The superficial layer lies on top of the periciliary layer and faces the lumen of

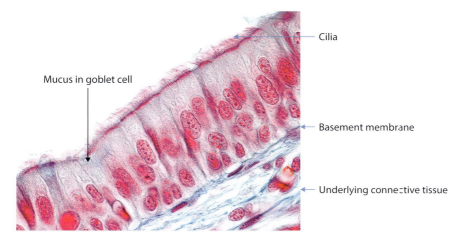

FIGURE 4.9 Pseudostratified ciliated columnar epithelium is not a true stratified tissue. It appears to have several layers, but actually only has a single layer (hence "pseudo"-stratified). The impression that there is more than one layer is because (1) the nuclei of the cells are at different heights and (2) although all the cells are attached to the basement membrane, not all of them reach the apical surface. This epithelium is found lining the airways of most of the upper respiratory tract. (Jose Luis Calvo/Shutterstock.com)

the nasal cavity. On the apical surface of epithelial cells, 4–6 μm long hairlike protrusions, called cilia, beat together at a frequency of 1000 strokes per minute to propel the mucus toward the mouth to clear dust and pathogens from the nasal cavity before they penetrate further into the body. DDS in the nose can be immobilized by the mucus and cleared toward the mouth by the constant ciliary beating. Mucus is secreted at a flow rate of ≈5 mm/min and this fast renewal rate means that particles are eliminated from the nasal cavity in less than 20 minutes. It also contains metabolic enzymes and peptidases, which together with the constant secretion and clearance mechanisms considerably reduce the amount of drug penetrating through the epithelium. Due to the permeable properties of the respiratory epithelia and highly vascularized nature of the adjacent submucosa, studies have shown that rapid systemic delivery of topically applied drugs can be achieved via this route.

The olfactory epithelium (≈60 μm thick) is also pseudostratified columnar and covers a bony structure known as the cribriform plate of the skull that separates the nasal cavity from the brain (see Chapter 10, Figure 10.4). This epithelium is composed of supporting, basal, and olfactory cells; it has been described as a gateway to the brain as a variety of materials, including biopharmaceuticals, have been shown to access the brain (cerebral spinal fluid) following topical administration. Multiple studies have now demonstrated a nose–brain pathway from this epithelium that bypasses the blood–brain barrier (see Chapter 10, Section 10.6).

4.5.4.2 Trachea and Bronchioles

The conducting airway is lined by a pseudostratified columnar epithelium (30–40 μm thick) that is composed of several distinct cell types that include absorptive cells, brush cells, goblet cells, and ciliated and serous cells (Figure 4.9). Mostly, this epithelium is known for its populations of ciliated cells interspersed with goblet and absorptive cells covered with microvilli. Basal cells in the tracheal epithelium, with their close approximation to the basement membrane, may function as a stem cell-like population. There is also an extensive organization of mucus-secreting glands with ducts that penetrate the tracheal epithelium. After division at the carina, the airway undergoes multiple additional branching to reach the alveoli of the lung. As these divisions occur, the epithelium transitions to a simple cuboidal format and ultimately a squamous format. During this transition, there is a progressive loss of goblet cells, a reduction of ciliated cells, and an increase in the population of

Clara cells, which are dome-shaped cells with short microvilli that secrete glycosaminoglycans to protect the epithelium. These cells may also function as a bronchiole progenitor cell type.

An airway surface liquid (ASL) covers the tracheal epithelium. This media contains protease inhibitors and is present as a double layer: the mucus layer is a 1–10 μm thick and viscous layer that traps inhaled particles. It lies on top of a 7 μm layer called periciliary liquid layer, which is in direct contact with the cell surface and keeps the mucus at an optimum distance from the epithelium. The ASL is a highly hydrated system, and its level of hydration is crucial in maintaining a good clearance mechanism as it influences the viscosity of the mucus layer and thus its capacity to be cleared from the airway (Hollenhorst et al. 2011). Each ciliated cell in the conducting airway has 200–300 cilia on its surface that beat together to propel the ASL and overlying mucus toward the pharynx and eventually to reach the orifice of the esophagus where they are swallowed. This mechanism, called mucociliary clearance, ensures that dust and pathogens are eliminated from the airway.

4.5.4.3   Respiratory Epithelium

While the surface area of the trachea and bronchioles is large, the surface area of the respiratory epithelium is much larger, comprising the majority of the 70 m² of surface area in human lungs. Type I pneumocyte cells cover ≈97% of the alveolar surface, with cuboidal type II pneumocytes covering the remainder (Figure 4.10). The type I pneumocyte cells are extremely thin (≈25 nm thick), a characteristic critical for efficient gas exchange between the alveoli and the blood. Thus, the main role of the alveolar epithelium is to provide a barrier of minimal thickness that is readily permeable to gases such as oxygen and carbon dioxide. This very thin epithelial barrier and its intimate association with vascular capillaries make it a promising site for rapid drug delivery into the systemic circulation. Within each alveolus are macrophages that perform the critical function of clearing debris and pathogens as there are no ciliated cells present to move these materials from these blind

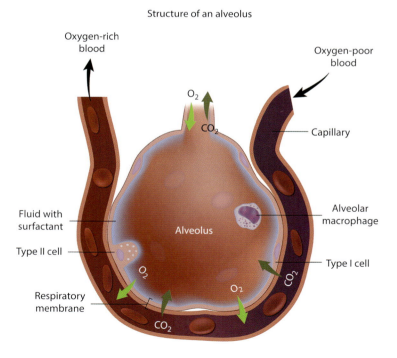

FIGURE 4.10   Alveolar epithelium, showing both type I and type II pneumocytes and fixed alveolar macrophages. (Alila Medical Media/Shutterstock.com.)

end sacs. Therefore, drugs delivered to the alveoli must dissolve quickly in the airway surface fluid or they will be engulfed by alveolar macrophages and directed to the macrophage lysosomes, where they will be degraded (Tam et al. 2011).

4.5.5 FEMALE REPRODUCTIVE TRACT

The female reproductive extends from the vagina to the uterus and oviducts (see Chapter 12, Figure 12.1). Restricted access to the uterus and oviduct makes drug delivery to these organs limited. As a consequence, the vagina is the main site that has been examined for drug absorption within the female genital system. The epithelium in the vagina is stratified squamous: it is composed of several layers of cells that are cuboidal close to the basement membrane and become very flattened as they migrate toward the luminal surface, where they disintegrate (Figure 4.7). New cells arise from division of cells close to the basement membrane and degenerated cells are continuously replaced by newly migrated. The cell turnover rate in the vaginal epithelium is 10–15 layers in 7 days.

The pH in the vagina is normally acidic (pH = 3.8–4.2) and this acidity can lead to rapid drug degradation. The acidic pH is usually maintained by glycogen conversion into lactic acid by lactobacilli, but its value is influenced by the presence of cervical mucus as well as the amount of vaginal fluid, infections, and sexual arousal. The cervix, positioned at the base of the uterus to interface with the vagina, secretes mucus from glandular crypts. Cervical mucus establishes a semipermeable viscoelastic barrier whose consistency varies depending on the stage of the menstrual cycle, changing from a fluid and clear consistency during midcycle to a thick and viscous material during the remainder of the cycle. Besides mucus, vaginal fluid is composed of transudates from vaginal and cervical cells. In total, vaginal fluid is complex, being composed of a mixture of enzymes and enzyme inhibitors, proteins and carbohydrates, amino acids, cervical mucus, and exfoliated epithelial cells, as well as endometrial and tubal fluids. The constant secretion of large volumes of fluid can modify the properties of DDS and further decrease their residence time on the site of action. For instance, the volume of vaginal fluid secreted daily by women of reproductive age has been estimated to be 0.75–1 g/h. This figure is reduced by half for postmenopausal women. Fluctuations in physiological factors can also alter the thickness of the epithelium, which can vary by 200–300 μm during the menstrual cycle due to variations in estrogen concentration.

Characteristics of the vaginal epithelium change depending on several factors such as age and hormonal activity. These parameters can affect both vaginal fluid production and enzyme concentration (particularly endopeptidases and aminopeptidases) present in this fluid. Indeed, the surface of the vaginal epithelium is coated with a great variety of enzymes, mostly present on the external and basal epithelial layers, which are able to degrade or metabolize drugs delivered. These proteases are the main barrier for absorption of intact peptide and protein drugs into the blood as they hydrolyze these molecules before they can cross the epithelium. Exopeptidases, endopeptidases, lysozyme, and enzymes having trypsin-like activity are also present. As a result of this variability, designing drugs and delivery devices that are efficient for large populations of women is problematic.

4.5.6 THE EYE

The anatomy of the eye is described in detail in Chapter 13 (Figures 13.1 and 13.2). To summarize here, the eye can be divided between the anterior segment, which includes the cornea, the aqueous humor, and the iris-ciliary body, and the posterior segment constituted by the retina, choroid, vitreous humor, and sclera. The two main ocular epithelia cover the cornea and the conjunctiva. The cornea itself can be divided between the corneal epithelium, stroma, and corneal endothelium. The corneal epithelium is a 50–90 μm thick stratified columnar epithelium. It is made of a basal layer of columnar cells, followed by a couple of layers of polyhedral cells and 2–4 layers of squamous, polygonal-shaped superficial cells with flattened nuclei. The stroma is formed of collagen fibrils running parallel to the surface of the cornea and embedded in an extracellular matrix. It is a hydrophilic layer that represents 90% of the corneal thickness. Finally, the corneal endothelium is a monolayer of hexagonal epithelial-like cells connected by TJs and desmosomes.

The cornea has a very small surface area for drug penetration ($\approx$1 cm²) and both the epithelium and endothelium limit the passage of foreign macromolecules due to the presence of TJs (see also Chapter 13, Figure 13.5). Moreover, due to the presence of successive hydrophobic (epithelium) and hydrophilic (stroma) layers, the cornea effectively restricts the passage of both types of drugs, resulting in poor drug permeation following topical delivery (Urtti 2006). Despite its poor penetrability, the cornea remains an important route for intraocular absorption.

The conjunctiva is a thin mucous membrane that lines the inner eyelid. The conjunctival epithelium is stratified columnar and contains mucus-secreting goblet cells and blood vessels. Together with the corneal epithelium, it forms the ocular surface epithelium. The conjunctival epithelium is thinner than the corneal epithelium, with loosely packed epithelial cells covered with microvilli, on top of which lies the glycocalyx, which is approximately 300 nm thick and is mainly composed of glycoproteins and glycolipids. The surface for drug absorption in the conjunctiva is 17 times larger than the surface on the cornea, and the conjunctiva is more permeable than the cornea. As a result, conjunctival absorption is an important limiting factor for corneal absorption.

The tear film is made of three coats: an outer layer made of both polar and nonpolar lipids (including sterol esters and wax) that acts as a thin polar surfactant, an intermediate aqueous layer, and a mucus layer at the base of the tear film (see also Chapter 13, Figure 13.3). This inner mucus layer prevents the attachment of bacteria and other foreign particles to the ocular surface. The continual replacement and removal of tears remove these elements from the conjunctival and corneal epithelia. The volume of tears in the eye is approximately 7 µL, and the eye can hold up to 30 µL without overflow or spillage. However, the typical volume of eyedrop dispensers for topical application is 50 µL. The excess volume triggers a reflex blinking of the eye, resulting in a major fraction of the drug being wasted. Once the excess solution has been removed, more than 90% of the drug is then drained through the nasolacrimal duct to the nasal cavities due to the tear turnover mechanism. It is then absorbed through the nasal epithelium into the systemic circulation. These two successive clearance mechanisms decrease the drug residence time on the site of delivery to only a few minutes, typically resulting in drug bioavailability of <5%.

4.6   CONCLUSIONS

Despite their common function as barriers, various epithelia present a wide range of properties that differ depending on the specific requirements of their location in the body. Each, however, provides a unique set of challenges to overcome for the efficient and effective delivery of drugs. The small surface area available for drug absorption in the oral-cavity and ocular epithelia contrasts with the large region covered by the intestinal and respiratory epithelia, which seem to provide great opportunities for drug delivery. It should be noted that these epithelia are, in general, absorptive rather than secretory in nature, providing a net flux in the direction that favors drug delivery. Some epithelial surfaces are not as easily accessible (salivary gland, pancreatic ducts, etc.) and delivering drugs to these locations may be too challenging to be practical. The secretion of large volumes of mucus, often coupled with efficient clearance mechanisms, is associated with a decreased bioavailability, especially in the conducting airway and reproductive tract. Finally, metabolic events can also contribute to reducing the amount of drug that crosses the epithelium.

REFERENCES

Anderson, J.M. and C.M. Van Itallie. 2009. Physiology and function of the tight junction. *Cold Spring Harb Perspect Biol* 1(2):a002584.
Chiba, H. et al. 2008. Transmembrane proteins of tight junctions. *Biochim Biophys Acta* 1778(3):588–600.

Dorfel, M.J. and O. Huber. 2012. Modulation of tight junction structure and function by kinases and phosphatases targeting occludin. *J Biomed Biotechnol* 2012:807356.

Farquhar, M.G. and G.E. Palade. 1963. Junctional complexes in various epithelia. *J Cell Biol* 17:375–412.

Florence, A.T. And Attwood, D. 2011. *Physicochemical Principles of Pharmacy*, 5th ed. London, U.K.: Pharmaceutical Press.

Förster, C. 2008. Tight junctions and the modulation of barrier function in disease. *Histochem Cell Biol* 130(1):55–70.

Giepmans, B.N.G. and S.C.D. Ijzendoorn. 2009. Epithelial cell–cell junctions and plasma membrane domains. *Biochim Biophys Acta—Biomembr* 1788(4):820–831.

Hollenhorst, M.I., K. Richter, and M. Fronius. 2011. Ion transport by pulmonary epithelia. *J Biomed Biotechnol* 1–16, Article ID 174306.

Johansson, M.E., D. Ambort, T. Pelaseyed et al. 2011. Composition and functional role of the mucus layers in the intestine. *Cell Mol Life Sci* 68(22):3635–3641.

Kim, Y.S. and S.B. Ho. 2010. Intestinal goblet cells and mucins in health and disease: Recent insights and progress. *Curr Gastroenterol Rep* 12(5):319–330.

Krause, G., L. Winkler, S.L. Mueller, et al. 2008. Structure and function of claudins. *Biochim Biophys Acta* 1778(3):631–645.

Krug, S.M., D. Günzel, M.P. Conrad, et al. 2012. Charge-selective claudin channels. *Ann N Y Acad Sci* 1257:20–28.

Lipinski, C.A. et al. 2001. Experimental and computational approaches to estimate solubility and permeability in drug discovery and development settings. *Adv Drug Deliv Rev* 46(1–3):3.26.

Madara, J.L. 1998. Regulation of the movement of solutes across tight junctions. *Annu Rev Physiol* 60:143–159.

Pakurar, A.S. and J.W. Bigbee. 2005. Epithelial tissues. In *Digital Histology: An Interactive CD Atlas with Review Text*. Hoboken, NJ: John Wiley & Sons, Inc.

Paulsson, M. 1992. Basement membrane proteins: Structure, assembly, and cellular interactions. *Crit Rev Biochem Mol Biol* 27(1–2):93–127.

Shewan, A., D.J. Eastburn, and K. Mostov. 2011. Phosphoinositides in cell architecture. *Cold Spring Harb Perspect Biol* 3(8):a004796.

Sigurdsson, H.H., J. Kirch, and C.M. Lehr. 2013. Mucus as a barrier to lipophilic drugs. *Int J Pharm* 453(1):56–64.

Tam, A., S. Wadsworth, D. Dorscheid et al. 2011. The airway epithelium: More than just a structural barrier. *Therap Adv Respir Dis* 5(4):255–273.

Tortora, G.J. and B.H. Derrickson, eds. 2013. *Principles of Anatomy and Physiology*, 14th ed. Hoboken, NJ: John Wiley & Sons, Inc.

Tsukita, S., M. Furuse, and M. Itoh. 2001. Multifunctional strands in tight junctions. *Nat Rev Mol Cell Biol* 2(4):285–293.

Urtti, A. 2006. Challenges and obstacles of ocular pharmacokinetics and drug delivery. *Adv Drug Deliv Rev* 58(11):1131–1135.

Woodley, J.F. 1994. Enzymatic barriers for GI peptide and protein delivery. *Crit Rev Ther Drug Carrier Syst* 11(2–3):61–95.

Section II

Parenteral Routes for Drug Delivery and Targeting

5 Nanotechnologies for Drug Delivery and Targeting
Opportunities and Obstacles

Alexander T. Florence and Daan J.A. Crommelin

CONTENTS

5.1 INTRODUCTION

Routine parenteral administration by injection serves to deliver drugs to body tissues. The most important routes of injection of these sterile products are intramuscular (i.m.), intravenous (i.v.), and subcutaneous (s.c.). The process of developing conventional parenteral formulations involves the selection of appropriate vehicles (e.g., aqueous, oily, and emulsions) to achieve the desired bioavailability following injection, as described in Chapter 3. The present chapter focuses on advanced drug delivery and targeting systems (DDTS), in the nanometer range, administered via the parenteral route and serves to provide the reader with a basic understanding of the principle approaches to drug targeting.

5.1.1 RATIONALE FOR THE DEVELOPMENT OF PARENTERAL DRUG DELIVERY AND TARGETING SYSTEMS

As introduced in Chapter 2, there are many limitations associated with conventional drug therapy. An i.v.-administered drug is subject to a number of pharmacokinetic processes in vivo, which can decrease the drug's therapeutic index, including the following:

- *Distribution*: Drugs administered i.v. distribute throughout the body and reach nontarget organs and tissues, resulting in drug wastage and consequently side effects and possibly toxicity.
- *Metabolism*: The drug may be rapidly metabolized in the liver or other organs.
- *Excretion*: The drug may be cleared rapidly from the body through the kidneys.

As a result of these processes, only a small fraction of the drug will reach the target tissue. Moreover, it may be cleared rapidly from this site and, therefore, not be available long enough to induce the desired effect. Reaching the target cell is often not the ultimate goal; in many cases, the drug has to enter the target cell to reach an intracellular target site. As discussed in Chapter 4, many drugs do not possess the required physicochemical properties to enter target cells; they may be too hydrophilic, too large, or labile or not transportable by the available active transport systems.

DDTS aim to overcome the limitations of conventional drugs and thus improve drug performance. An ideal DDTS should

- Specifically target the drug to target cells or target tissue
- Prevent the drug reaching nontarget organs, cells, or tissues
- Ensure minimal drug leakage during transit to target
- Protect the associated drug from metabolism
- Protect the associated drug from premature clearance
- Retain the drug at the target site for the desired period of time
- Facilitate transport of the drug into the cell
- Deliver the drug to the appropriate intracellular target site
- Be biocompatible, biodegradable, and nonantigenic

In certain situations, some of these requirements may be inappropriate. For example, the drug may work outside the cell; thus, cell penetration may not be necessary. In this chapter, there are also examples mentioned of passive targeting approaches (see Section 5.3.1), where the drug does not require to be specifically targeted to the cell or tissue.

The parenteral route of administration is associated with several major disadvantages. Parenteral administration is invasive and may require the intervention of trained medical professionals. Parenteral formulations often require refrigeration for storage. Strict regulations govern their use, which generally dictate that they are as simple as possible. The inclusion of excipients in the formulation is kept to an absolute minimum. Developing a DDTS requires an enormous amount of R&D investment in terms of cost, effort, and time, which can cause a significant delay in the development and marketing of a system and thus result in the final product being relatively expensive. Parenteral DDTS must, therefore, offer real therapeutic advantages to justify their use. Box 5.1 lists a number of pharmacokinetic considerations to decide if the use of DDTS is indicated for a particular drug.

**BOX 5.1   PHARMACOKINETIC CONSIDERATIONS
RELATED TO DRUG TARGETING**

- Drugs with high total clearance are good candidates for targeted delivery.
- Carrier-mediated transport is suitable for response sites with a relatively small blood flow.
- The higher the rate of elimination of free drug from either central or response compartments, the greater the need for targeted drug delivery; this also implies a higher input rate of the drug carrier combination to maintain the therapeutic effect.
- For maximizing the targeting effect, the release of drug from the carrier should be restricted to the response compartment.
- The rate of release and quantity of drug released should be at the appropriate therapeutic level.

Drugs used in the treatment of diseases that are life threatening or that dramatically affect the quality of life of the patient are prime candidates for inclusion in a DDTS. Such drugs include those used in the treatment of cancer, as well as life-threatening microbial, viral, and fungal diseases. Chronic diseases such as arthritis can also be found on the priority list.

5.1.2   GENERALIZED DESCRIPTION OF PARENTERAL DRUG DELIVERY AND TARGETING SYSTEMS

The technology used for targeted drug delivery with carrier systems differs from the technology to achieve prolonged release profiles for a drug. If prolonged release of a drug via the parenteral route is required, s.c. or i.m. injection of a controlled release system is the first option to consider. The relevant technology is already available and validated for many years. Long-acting injections and implants are the subject of Chapter 6; some examples include

- The long-, medium-, and short-acting insulin formulations, prepared by crystal manipulation or physical complex formation
- Depot injections (aqueous suspensions, oily injections) of contraceptives and psychotropic drugs (such as fluphenazine esters)
- Polymeric implants, for example, Zoladex®, a biodegradable matrix implant of the luteinizing hormone-releasing hormone (LHRH) agonist goserelin, for s.c. administration
- Infusion pumps

In contrast, a DDTS generally comprises three functionally specific units, as listed in Table 5.1: the drug or therapeutic agent, the carrier system, and a "homing" ligand.

A "homing" ligand is a target-specific recognition moiety. For example, galactose receptors are present on liver parenchymal cells; thus, the inclusion of galactose residues on a drug carrier can allow the interaction of the carrier to the target cells, once the two are in close proximity. A number of different target-specific recognition moieties are available and discussed further in the succeeding texts. However, an important point to note here is that target-specific recognition moieties are not idealized "magic bullets," capable of selectively directing the drug to the appropriate target and

TABLE 5.1

Components of Drug Delivery and Targeting Systems

Drug Delivery and Targeting Systems Component	Purpose
The active moiety	To achieve the therapeutic effect
The carrier system, which can be either soluble or particulate	To effect a favorable distribution of the drug
	To protect the drug from metabolism
	To protect the drug from early clearance
A "homing" ligand[a]	To specifically target the drug to the target cells or target tissue

[a] Not necessary when "passive" targeting approaches are used.

ignoring all other nontarget sites. Although the so-called homing ligand can increase the *specificity* of the drug for its target site, the process must rely on the (random) encounter of the homing ligand with its appropriate receptor during its circulation lifetime. This is a stochastic process, that is, one involving elements of chance rather than certainty; for example, particle escape from the circulation (extravasation) or a ligand–receptor interaction leading to cellular uptake. Box 5.2 lists the main relevant processes that have an element of uncertainty in terms of outcome.

The carrier systems that are presently on the market or under development can be classified into two groups based on their size:

1. Soluble macromolecular carriers
2. Particulate carrier systems

This classification is sometimes rather arbitrary, as some soluble carriers are large enough to enter the colloidal size range. A useful distinction is that with macromolecular carrier systems, the drug is "covalently attached" to the carrier and has to be released through a chemical reaction. In contrast, with colloidal carriers, the drug is generally "physically associated" and does not need a chemical reaction to be released. Here, diffusion barriers in the carrier comprise the major barriers for premature release from the system. Stimulus-sensitive systems are also being devised to ensure that

BOX 5.2   STOCHASTIC PROCESSES IN VECTOR-MEDIATED DRUG TARGETING

Definition: A stochastic process involves some random variable or variables, that is, it involves chance events or a probability, rather than a certainty, of an event occurring. Four examples are as follows:

1. Aggregation of nanoparticles under shear flow in the blood circulation or in static conditions, e.g., in cell culture. Aggregation can cause jamming of particles and would change the apparent size—biological action paradigm.
2. Extravasation. The escape from the circulation after i.v. administration.
3. Nanoparticle ligand–receptor interactions. Requiring both close approach (to within several nm) of particles and receptors and the correct orientation of vectors toward the receptor, both being random events.
4. Diffusion of vectors in tissues, which are complex environments involving iterative approaches to reach destinations.

drug is only released once the systems have reached the target. Release in this case is instigated by an external source, such as heat or magnetic force, the subject of Chapter 14.

Soluble carriers include antibodies and soluble synthetic polymers such as poly(hydroxypropyl methacrylate), poly(lysine), poly(aspartic acid), poly(vinylpyrrolidone), poly(*N*-vinyl-2-pyrrolidone-*co*-vinylamide), and poly(styrene-*co*-maleic acid anhydride).

Many particulate carriers have been designed for drug delivery and targeting purposes for i.v. administration. They usually share three characteristics:

1. Their size range: the minimum size is approximately 10 nm and the maximum size relevant for drug targeting is approximately 10–300 nm.
2. They are all biodegradable, but at different rates.
3. The drug is physically associated with the carrier, and in general, drug release kinetics are controlled by either diffusional transport or matrix degradation, or both.

Figure 5.1 shows a selection of some of the wide variety of both soluble and particulate DDTS that are currently under development or commercially available.

A full appreciation of the respective advantages and disadvantages of soluble and particulate carriers cannot be gained without first taking into account the relevant anatomical, physiological, and pathological considerations in vivo, described next.

5.2 ANATOMICAL, PHYSIOLOGICAL, AND PATHOLOGICAL CONSIDERATIONS

The body is highly compartmentalized and should not be considered as a large pool without internal barriers for transport. A schematic figure (Figure 5.2) illustrates the organs involved. The degree of body compartmentalization or, in other words, the ability of a macromolecule or particulate to move around, depends on its physicochemical properties, in particular

- Molecular weight/size
- Charge
- Surface hydrophobicity
- Surface attached ligands for interaction with surface receptors.

The smaller its size, the easier a drug carrier can passively move from one compartment to another. An important question is whether, and where, the carriers can pass through the endothelial lining of the blood circulation. Under physiological conditions, the situation exists as depicted in Figure 5.3. A continuous endothelium, as found in most parts of the body, comprises a continuous tube of endothelial cells, separated by intercellular clefts, i.e., narrow gaps between neighboring endothelial cells (Figure 5.3a). The cells are positioned on a continuous basal membrane. The exact characteristics of this barrier are still under investigation, but it is clear that particulate systems greater than 10 nm cannot pass through. Fenestrated capillaries, as found, for example, in the kidney and most endocrine glands, have many small pores (fenestrations) in the plasma membrane of the endothelial cells, ranging from 70 to 100 nm diameter (Figure 5.3b). Again, the subendothelial basement membrane is continuous. Sinusoidal capillaries (Figure 5.3c) are found in the liver, spleen, and bone marrow. The endothelial cells have large fenestrations, as well as large intercellular clefts, that allow the movement of molecules such as proteins through the capillary walls. The subendothelial basement membrane is either absent (as in the liver) or present as a fragmented interrupted structure (as in the spleen and bone marrow). Sinusoids in the liver contain phagocytic Kupffer cells, which remove bacteria and other particulates from the blood.

This anatomical information has important implications for the rational design of targeted carrier systems. If a therapeutic target is located outside the blood circulation and if normal anatomical

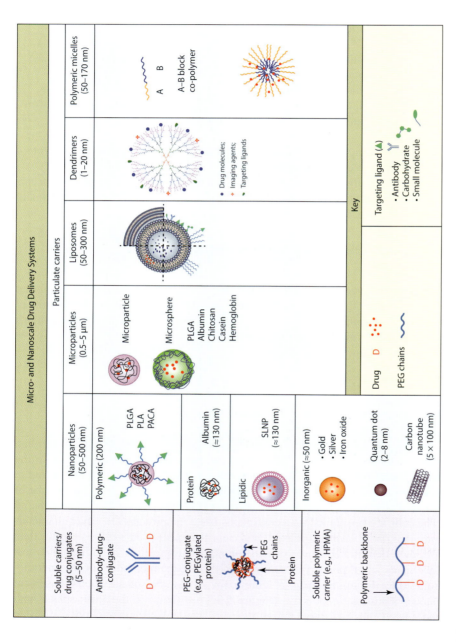

FIGURE 5.1 Schematic illustration of various types of nano-/micro-DDTS. Soluble carriers include antibodies, hydrophilic polymers such as PEG, and soluble synthetic polymers such as poly(hydroxypropyl methacrylate). Particulate carriers include nanoparticles, microparticles, liposomes, dendrimers, and micelles. Particulate carriers can also have covalently attached PEG chains (e.g., PEGylated liposomes, PEGylated microparticles) and contain targeting ligands or a targeting ligand–PEG conjugate. PLGA, poly(lactic-*co*-glycolic acid); PLA, poly(lactic acid); PACA, poly(alkyl cyanoacrylate); SLNP, solid–lipid nanoparticles. (*Note:* This figure illustrates some generalized drug carrier systems. Not all of these examples are necessarily injected intravenously and/or are intended for drug-targeting purposes. Many of the larger constructs are injected subcutaneously or intramuscularly for sustained release. Also, the size values quoted are indicative only; a wide variation in these values is possible).

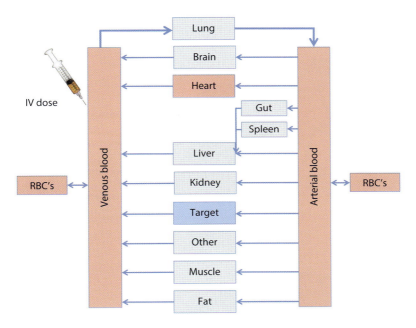

FIGURE 5.2   Simplified diagram showing the flow of the blood through the different organs of the body.

conditions exist around the target site, a small-sized macromolecular carrier must be selected, in order to achieve sufficient "escaping tendency" from the blood circulation. Particulate carriers will generally fail to extravasate, simply because there is no possibility for endothelium penetration in normal circumstances.

In addition to the issue of endothelial permeability, the effect of macrophages in direct contact with the blood circulation (e.g., Kupffer cells in the liver) on the disposition of carrier systems must be considered. Unless precautions are taken, particulate carrier systems are readily phagocytized by these macrophages and tend to accumulate in these cells. Phagocytic uptake by the cells of the mononuclear phagocyte systems (MPS), also known as the reticuloendothelial system (RES), has been introduced in Chapter 4 (Section 4.3.3). The MPS constitutes an important part of the body's immune system and comprises both fixed cells, such as macrophages in the liver (also known as Kupffer cells), spleen, lungs, bone marrow, and lymph nodes, and mobile cells such as blood mono-cytes and tissue macrophages. Its functions include

- The removal and destruction of bacteria
- The removal and metabolism of denatured proteins
- Antigen processing and presentation
- Storage of inert colloids
- Assisting in cellular toxicity

The cells of the MPS are always on the alert to phagocytize "foreign body–like material." Thus, aside from being responsible for the removal of particulate antigens such as microbes, other foreign particulates, such as microspheres, liposomes, and other particulate carriers, are also susceptible to MPS clearance.

Clearance kinetics by the MPS are highly dependent on the physicochemical properties of the "foreign" particulate; on particle size, charge, and surface hydrophobicity; and perhaps, on shape. These properties are considered in turn.

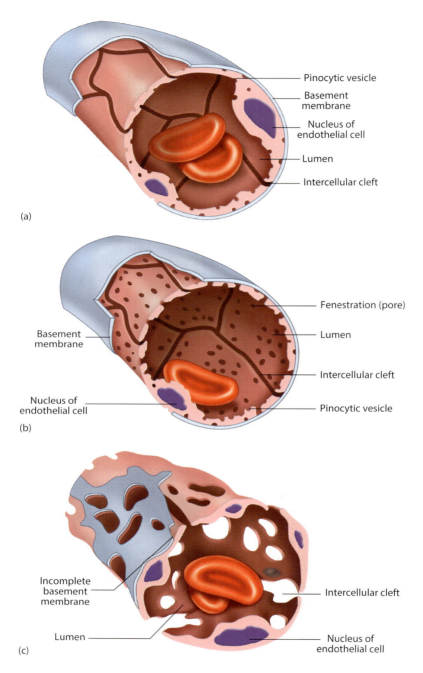

FIGURE 5.3 Schematic illustration of the structure of the three different classes of blood capillaries: (a) continuous, (b) fenestrated, and (c) discontinuous (sinusoidal) capillaries. (From Tortora, G.J. and Derrickson, B.H., eds.: *Principles of Anatomy and Physiology*, 14th ed. 2013. Copyright Wiley-VCH Verlag GmbH & Co. KGaA. Reproduced with permission.)

Particle size: Particulates in the size range of 0.1–7 μm tend to be cleared by the MPS, localizing predominantly in the Kupffer cells of the liver.

Particle charge: For liposomes, it has been shown that negatively charged vesicles tend to be removed relatively rapidly from the circulation, whereas neutral vesicles tend to remain in the circulation for longer periods.

Surface hydrophobicity: Hydrophobic particles are immediately recognized as "foreign" and are generally rapidly covered by plasma proteins known to function as opsonins, which facilitate phagocytosis. The extent and pattern of opsonin adsorption depends highly on surface characteristics such as charge and hydrophilicity. Strategies to decrease MPS clearance, by increasing the hydrophilicity of the particle surface, are described in Section 5.3.1.

A further consideration is that in some "pathological" conditions, the endothelium exhibits modified characteristics. In general, the permeability is enhanced; the phenomenon is called the "enhanced permeability and retention" (EPR) effect, coined by Hiroshi Maeda, as will be discussed in Section 5.2.1. For example, the endothelial fenestrations in inflammation sites can be as large as 20 nm. In tumor tissue, even larger fenestrations can be found. However, in this case, the pattern is not uniform and depends on the tumor type and stage of development. Even within one tumor, highly permeable sites can be identified in close proximity to sites of low permeability. Also, necrotic tissue affects tumor permeability. Because of the EPR effect, particles in the colloidal size range can "theoretically" enter tumor tissue or sites of inflammation. This phenomenon can be exploited for drug delivery. However, while the EPR effect has been demonstrated in animals, there has been no clear evidence of the effect in human tumors. The EPR effect has been invoked in many publications as a means of access of carrier systems to such sites, but the evidence for the mechanism, or mechanisms, of accumulation is often not to be found. Extrapolation of animal data to humans is less than clear, as it is with other phenomena too. The status of this putative mechanism and some of these questions are discussed further for example in Sections 5.2.1, 5.5.1.1 and 5.5.1.2.

Shape: The development of asymmetric carbon nanotubes has given the opportunity to investigate the influence of shape on many processes. Most current vector systems are spherical or quasi-spherical, with different degrees of flexibility. But shape clearly affects the movement and behavior of the systems, for example, their transport through pores (see Figure 5.4) and their interaction with biomembranes. Shape can affect endocytosis: elongated particles have been found to be more readily endocytosed than spherical equivalents. Cytotoxicity is also influenced by shape, as asbestos has taught us. Carbon nanotubes are said to have a propensity for membrane interactions, but more work is required to elucidate all the shape-related factors in attachment and uptake.

5.2.1 ENHANCED PERMEATION AND RETENTION EFFECT: CONCEPT OR REALITY?

Much of the research carried out on vector-mediated targeting is performed in small animal models, usually rodents. In these animal models, a certain degree of accumulation of vectors has been observed, for example, at tumor sites, with both active and passive carrier systems, a feature habitually ascribed to the EPR effect. A certain accumulation indeed occurs in rodents, but this has rarely exceeded about 5% of the vector. Imaging techniques show this modest accumulation but unless the drug itself can be visualized, the data are optimistic, as the drug may be trapped within the carrier particles. In addition, the literature is difficult to interpret as it has been found that different imaging techniques can lead to different data. The EPR effect has been considered recently by a number of experts in a consensus paper (Prabhakar et al. 2013) that points out the heterogeneity of the EPR effect and the limited evidence for the effectiveness of the mechanism, which was originally proposed, as stated in this chapter, for (flexible) macromolecules rather than (rigid) nanoparticles.

Nanoparticle uptake is often determined by radiolabeling of the vector and not the drug. The only sensible approach is to measure the drug in the tumor and to distinguish active free drug, from the drug that is still bound to the carrier. One problem in assessing true targeting is that the capacity of a vector system for the drug is crucial, and hence, the same vector carrying different drugs might

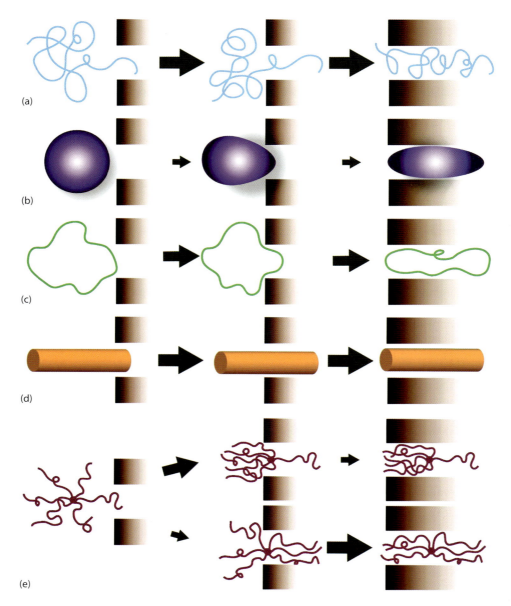

FIGURE 5.4 Polymer architecture and the passage of a polymer through a pore, for polymers with approximately equivalent hydrodynamic volume; (a) linear random coil polymer readily penetrates and reptates through a pore; (b) polymer with a rigid, globular conformation must deform to pass through. Beyond a threshold MW, both entry and passage through the pore could be difficult; (c) a cyclic polymer lacks a chain end for entering the pore and must deform to enter and pass through it; (d) polymer with a rigid elongated or tubular conformation easily enters and passes through; and (e) arm orientation and distance between chain ends of branched polymers impact the rate of entry and passage through a pore. While initial entry of only one chain end (top) may occur rapidly, passage of the entire polymer through the pore is sterically hindered, because the remaining arms must deform for the polymer to pass through; a symmetric conformation (bottom) is less likely since multiple chain ends must penetrate the pore at the same time. Once entry has been achieved, passage of the polymer would be less hindered than for the asymmetric distribution. (Reprinted with permission from Fox, M.E., Szoka, F.C., and Frechet, J.M.J., Soluble polymer carriers for the treatment of cancer: The importance of molecular architecture, *Acc. Chem. Res.*, 42(8), 1141–1151. Copyright 2009 American Chemical Society.)

produce quite different results. Drug load can be high, as in many liposomes, but in some nanoparticles, it can be as low as 5% w/w. Should 10% of such a carrier reach the target, and the available drug is only 5% of the vector mass, the so-called drug targeting represents only 0.5% of the administered dose. Many studies suggest that maximal vector uptake is in the vicinity of 5%: arithmetic matters!

The rate of release and diffusion of drug from the carrier is clearly another issue in effective therapeutic action. Distribution of drug throughout tumor tissue is not, unsurprisingly, homogeneous since tumors are not themselves homogeneous. In tumor tissue extracellular matrix barriers, necrotic tissue and pressure differences may form barriers for access to parts of the diseased tissue in spite of pore formation in the endothelium. Generally speaking, the larger the carrier, the more its penetration is hindered. And drugs such as doxorubicin (although a small molecule) do not necessarily diffuse freely through tumor tissue, not only because of the complex diffusion pathways but also as they can bind to cellular and tissue components.

And then the following question has to be answered: how predictive are animal models for the therapeutic effects in the human patient? As seen in a few observations, the tumor cell lines used in the most "popular" animals (rodents) are fast growing (days, weeks). These tumor cell lines are rather homogeneous compared to human tumors; furthermore, the immune systems in rodents differs from man. Not only that, but adsorption of proteins onto the surface of nanoparticles in the blood varies with residence time in the circulation. Both the plasma proteins in rodents and the dynamics of circulation are different from those in man. Therefore, direct extrapolations from findings in animals to humans can only be made after careful validation studies. In some patients, indeed a degree of accumulation in the primary tumor and metastases has been demonstrated by imaging techniques, but the same imaging techniques showed no EPR effect in the majority of other patients.

5.3   PASSIVE AND ACTIVE TARGETING

5.3.1   PASSIVE TARGETING

Passive targeting exploits the "natural" (passive) distribution pattern of a drug carrier in vivo and no homing ligand is attached to the carrier. For example, as described earlier, particulate carriers tend to be phagocytized by cells of the MPS. Consequently, the major organs of accumulation are the liver and the spleen, both in terms of total uptake and uptake per gram of tissue. An abundance of MPS macrophages and a rich blood supply are the primary reasons for the preponderance of particles in these sites. After phagocytosis, the carrier and the associated drug are transported to lysosomes, and the drug is released upon disintegration of the carrier in this cellular compartment. This passive targeting to the MPS (and particularly to the liver) is advantageous in a number of situations, including

- The treatment of macrophage-associated microbial, viral, or bacterial diseases (e.g., leishmaniasis)
- The treatment of certain lysosomal enzyme deficiencies
- The immunopotentiation of vaccines (further information on the use of nanoparticles for vaccines is given in Chapter 17)
- The activation of macrophages, by loading the carrier system with macrophage-activating agents such as interferon γ, to fight infections or tumors

If the drug is not broken down by the lytic enzymes of the lysosomes, it may be released in its active form from the lysosomal compartment into the cytoplasm and may even escape from the phagocyte, causing a prolonged release systemic effect.

Technology is available to reduce the tendency of macrophages to rapidly phagocytize colloidal drug carrier complexes. The process of "steric stabilization" involves the coating of the delivery system with synthetic or biological materials, which make it energetically unfavorable for other macromolecules to approach the surfaces carrying these chains. A standard approach is to graft hydrophilic, flexible polyethylene glycol (PEG) chains to the surface of the particulate carrier (termed "PEGylation"). This highly hydrated PEG layer reduces the adsorption of opsonins from the plasma and consequently slows down phagocytosis. The net effect of PEG attachment is that macrophage/liver uptake of the particles is delayed or reduced, thus increasing the circulation time.

In order to avail of the EPR effect, two conditions should be satisfied:

1. The size of the drug carrier system should exceed the size of normal endothelial fenestrations to ensure that the carrier system only crosses inflamed/tumor endothelium; a certain size range is preferred as there is an upper limit to the endothelial fenestration dimensions under pathological conditions.
2. The circulation time in the blood compartment should be long enough to allow the carrier systems to "escape" from the circulation at the pathological site, though not all will.

As described earlier, the circulation time of a particulate carrier in the blood can be prolonged using "stealth" technology to enhance particle hydrophilicity. If the circulation time is sufficiently prolonged and the particle size does not exceed, say, 200 nm, then accumulation in tumor and inflammation sites can be observed. The goal today is to enhance accumulation in targets, which in tumors in man has rarely exceeded around 5% of the vector.

5.3.2 ACTIVE TARGETING

In active-targeting strategies, a specific ligand is attached to the carrier system, to improve delivery to a specific cell, tissue, or organ. Thus, delivery systems designed for active-targeting are usually composed of three parts: the carrier, the ligand, and the drug. Preferably, the ligand is covalently attached to the carrier, although successful targeting attempts of noncovalently attached ligand–carrier combinations have also been described.

Target sites for active-targeting strategies can differ widely. A list of cell-specific receptors and their corresponding ligands, expressed under physiological conditions, is presented in Table 5.2. Thus, for example, galactose can be used to target a drug carrier to parenchymal liver cells. Receptors are not always exposed directly to the vector, which might thus have to access the target and then interact. In the future, it is expected that the rapidly growing field of genomics will be used to identify specific receptors for targeting purposes, which might enhance the effectiveness of targeting.

Other receptors may become available under pathological conditions. Such receptors include

- Antigenic sites on pathogens (bacteria, viruses, parasites)
- Infected host cells expressing specific antigenic structures
- Tumor-associated antigens (TAAs) (i.e., antigenic structures specifically occurring at the surface of tumor cells)
- Substances such as fibrin in blood clots (i.e., potential ligands for targeting of fibrinolytics)

TABLE 5.2

Examples of Cell-Specific Ligands/Carriers In Vivo

Cell	Cell-Specific Ligands/Carriers
Parenchymal liver cells	Galactose, polymeric IgA, cholesterol ester—VLDL, LDL
Kupffer cells	Mannose/fucose, galactose (particles), (oxidized) LDL
Liver endothelial cells	Mannose, acetylated LDL
Leucocytes	Chemotactic peptide, complement C3b

Abbreviations: VLDL, very-low-density lipoproteins; LDL, low-density lipoproteins.

Sometimes it is necessary for the carrier-bound drug to reach all target cells to be clinically successful, as is the case with antitumor therapy. The so-called "bystander" effects can help to achieve fully effective therapy. Bystander effects occur when the targeted drug carrier reaches its target site, and released drug molecules also act on surrounding nontarget cells. In other cases, not all target cells have to be reached, as is the case, for example, for targeted gene delivery for the local production of a therapeutic protein. However, the cell nucleus is a target that is difficult to reach, as several barriers inside the cell have to be overcome (e.g., cytoplasmic membrane, endosomal wall, and the nuclear membrane).

Antibodies raised against a selected receptor are extensively used as homing ligands, as described in the succeeding texts. Modern molecular biotechnology permits the production of large amounts of tailor-made material. Other potential candidates are also emerging, in the cytokine and the growth hormone family and, finally, among the adhesion molecules that play a role in the homing of inflammatory cells to inflammation sites.

5.4 SOLUBLE CARRIERS FOR TARGETED DRUG DELIVERY

As described earlier, the major advantage of soluble carriers over particulate carriers is their greater ability to extravasate; also, being soluble, they can be taken into cells via the process of pinocytosis. Active-targeting strategies for soluble carriers include attaching rather simple ligands such as galactose, for targeting to liver parenchymal cells (see Table 5.2); alternatively, more complicated structures, such as antibodies, or antibody fragments (Fab or single-chain antibodies [SCA]) can be used as targeting ligands.

However, a number of disadvantages are also associated with the use of soluble carriers:

- Limited drug-loading capacity: a low ratio of drug to carrier limits the mass transport mediated by the drug carrier.
- The drug is covalently bound to the carrier: this can mask the active site of the drug and the conjugation reaction may damage a labile drug moiety.
- The carrier confers limited protection on the drug.
- The rate of cleavage of the drug from the carrier may not be optimal for activity.

5.4.1 MONOCLONAL ANTIBODIES

The therapeutic antibodies that are on the market all come from the IgG family, with the IgG1 isotype being the most commonly used. Monoclonal antibodies (MAbs) belong to the largest molecules in our therapeutic arsenal with a molecular weight (MW) of 150,000 (1000 times the MW of paracetamol/acetaminophen). As shown in Figure 5.5, the structure consists of two heavy and two light chains. A sugar chain is connected to each heavy chain. The antigen-binding site of IgG molecules represents the homing part, which specifically interacts with the target (cells, pathogens, tissue). These antigen-binding sites are located at both tips of the Y-shaped molecules. The sites that are responsible for the pharmacological effects of IgG, such as complement activation and macrophage interaction, are located at the stem part of the Y. The rest of the molecule forms the connection between the homing ligand and the pharmacologically active sites, and also contributes to the long blood circulation characteristics of the IgG molecule, which has an elimination half-life much greater than 24 hours.

Humanized or human antibodies have replaced earlier generations of murine antibodies. There are four options in basic structure. The earliest generation was fully based on mouse hybridoma technology; then, the Fc part was mouse derived, with the Fab fragment comprising a human sequence (chimeric MAb), followed by humanized antibodies, where only the complementarity-determining region was mouse derived. Nowadays, fully human MAbs have entered the field. Mouse-derived antibodies can induce human antimouse antibodies (HAMA) in the patient. In multiple injection schemes,

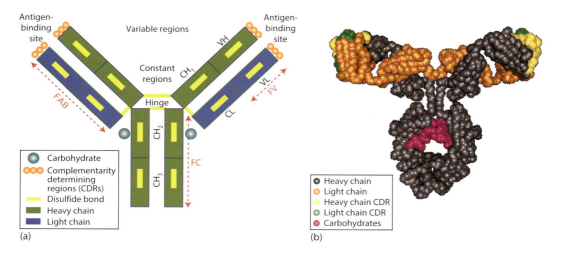

FIGURE 5.5 Molecular structure of IgG, Fab, and single-chain antibody: (a) cartoon and (b) a 3D representation. (With kind permission from Springer Science+Business Media: *Pharmaceutical Biotechnology: Fundamentals and Applications*, 4th ed., Springer, New York, 2013, Crommelin, D.J.A., Sindelar, R.D., and Meibohm, B., eds.)

this HAMA reaction can cause neutralization of the homing capacity of the homing ligand, loss of activity, and anaphylactic reactions. Chimeric, humanized, and human antibodies induce HAMA (much) less frequently but, for instance, can still raise anti-idiotypic antibodies against the binding site structure. These anti-idiotypic antibodies can also interfere with the targeting performance.

MAbs have rather complicated names as shown in Figure 5.6. The WHO International Nonproprietary Name (INN) system provides the logic behind these names (see WHO 2008). Murine MAb names end with –*momab*, chimeric with –*ximab*, humanized with –*zumab* and, finally, fully

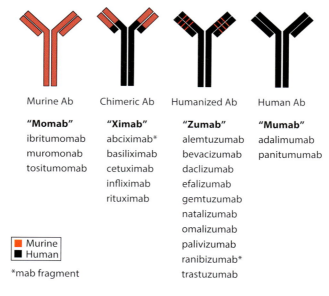

FIGURE 5.6 Monoclonal antibody (Ab) nomenclature. (With kind permission from Springer Science+Business Media: *Pharmaceutical Biotechnology: Fundamentals and Applications*, 4th ed., Springer, New York, 2013, Crommelin, D.J.A., Sindelar, R.D., and Meibohm, B., eds.)

human MAb with *–mumab*. The therapeutic indication is also referred to, e.g., *–li-* for an immuno-modulating and *–tu-* for an antitumor effect. This helps to understand INN such as trastuzumab and infliximab.

The large size of MAbs (around $15 \times 8 \times 5$ nm) and their hydrophilicity have some disadvantages. Penetration through endothelial and epithelial barriers and transport through extracellular spaces are slow. This leads to low uptake percentages at the target site if the target is located beyond the vascular endothelium (extravascular space). For instance, in cancer patients, the uptake of MAb in solid tumors is around 3% of the dose, which means that 97% ends up somewhere else in the body. Therefore, to facilitate (diffusional) transport, often the full antibody molecule (MW 150 kDa) is not utilized for target ligand (i.e., specific receptor) binding, but only the antigen-binding domain carrying the Fab (MW 50 kDa) fragment or even smaller fragments (SCA, MW 25 kDa).

In 1986, the first marketed MAb for therapeutic use was the mouse anti-CD3 antibody muro-monab (OKT3), for the prevention of rejection of kidney transplants. Since then, a long list of MAb has been approved mainly for oncology, autoimmune diseases, and transplant rejection. In 2012, 5 out of 10 on the list of the globally most successful medicines (in terms of revenue) were MAbs, a clear indication of the fast-growing success of this family of medicines.

MAbs against TAAs have been developed to assist in tumor imaging. The MAb is conjugated with a diagnostic imaging agent (e.g., ^{111}In). Commercial products include

- Satumomab pendetide (Oncoscint® CR/OV), for colorectal and ovarian adenocarcinomas
- Capromab pendetide (Prostascint®), for prostate cancer
- Arcitumomab (CeaScan®), for a number of carcinoembryonic antigen (CEA)-rich tumors

The therapeutic use potential of MAb as an "all-in-one" bioactive and targeting ligand molecule per se may be limited because of the limited immunological activity on the Fc part. Therefore, antibody–drug conjugates (ADC) have been developed. An example of an ADC is the MAb–drug conjugate, Adcetris® (brentuximab vedotin). This ADC is used for the treatment of Hodgkin lymphoma patients. Adcetris® consists of three components:

1. The chimeric IgG1 antibody cAC10, specific for human CD30 (exposed on lymphoma cells), which binds (random collision events, as well as access issues permitting, cf. stochastic paradigm, Box 5.2) to the tumor.
2. Monomethyl auristatin E (MMAE), a microtubule-disrupting agent (anticancer agent).
3. A protease-cleavable linker that covalently attaches MMAE to cAC10. It is stable outside the cell, and upon cell internalization, it is cleaved, releasing MMAE.

Bispecific antibodies are manufactured from two separate antibodies to create a molecule with two different binding sites. One binding site links the MAb to the target cell. The other site is chosen to bring T-lymphocytes or natural killer cells in close contact with the target site, in order to exert a pharmacological effect, for example, to kill the target cell. This approach is now in early stage clinical trials.

5.4.2 SOLUBLE POLYMERIC CARRIERS

Over the years, different soluble polymeric systems have been developed in attempts to enhance drug performance. Here again, the emphasis is on the improvement in drug disposition conferred by the carrier and ligand, as well as the protection offered by the system against premature inactivation. The strategy, as shown in Figure 5.7, involves the use of a soluble macromolecule, the molecular weight of which ensures access to the target tissue. The drug moiety can be bound via either a direct linkage or a short chain "spacer." The spacer overcomes problems associated with the shielding of the drug moiety by the polymer backbone and allows cleavage of the drug from the polymer.

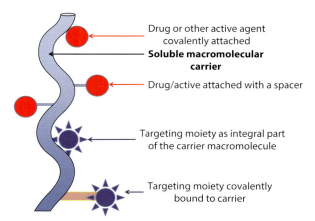

FIGURE 5.7   Components of a soluble macromolecular site-specific delivery system.

The spacer allows greater exposure of the drug to the biological milieu thereby facilitating drug release. A targeting moiety, which can be either an integral part of the polymer backbone or covalently bound, may also be incorporated into the system.

A crucial feature of such carrier systems is their solubility, which enables them to be taken up into target cells by the process of pinocytosis (see Chapter 4, Section 4.3.3.1). The intact carrier enters the target cell through pinocytotic capture. Through an endosomal sorting step, the carrier reaches the lysosomes where it is exposed to the actions of a battery of degradative enzymes. The drug–carrier linkage is designed to be cleaved by these enzymes, liberating free, active drug that can leave the lysosome by passage through its membrane, reaching the cytoplasm and other parts of the cell. Intralysosomal release of the drug from the carrier can also be achieved by making the drug–carrier linkage acid labile, as the lysosomal interior has a pH of approximately 4.5–5.5.

5.4.2.1   HPMA Derivatives

Poly(*N*-(2-hydroxypropyl)methacrylamide) has been investigated as a soluble macromolecular carrier system, using doxorubicin as the active drug. The bulk of the conjugate consists of unmodified HPMA units, which comprise about 90% of the carrier; the remaining units are derivatized with doxorubicin (see also Chapter 1, Figure 1.10). A tetrapeptide spacer (sequence Gly-Phe-Leu-Gly) connecting doxorubicin to the HPMA units proved to be cleavable by lysosomal thiol proteinases. Enzymatic cleavage breaks the peptide bond between the terminal glycogen and the daunosamine ring, liberating free doxorubicin, which can diffuse to the cytoplasm and nucleus where it (presumably) exerts its action.

Targeting moieties can also be incorporated into this delivery system. Targeting systems that have been investigated include

- *Galactose*: for targeting to parenchymal liver cells
- *Melanocyte-stimulating growth factor*: for targeting to melanocytes
- *MAbs*: for targeting to tumors

Interestingly, the doxorubicin–polymer conjugate alone, without a target-specific ligand, showed an enhanced therapeutic index in animal models and considerable accumulation of the drug in tumor tissue. The EPR effect, as discussed earlier, is held responsible for this phenomenon. After optimizing conjugate performance in terms of doxorubicin "payload" and desired molecular weight range of the polymer backbone, clinical grade material is now available and clinical trials are in progress to evaluate the potential of this concept.

5.4.2.2 SMANCS

The cytotoxic neocarzinostatin (NCS) is a small protein (MW 12 kDa) associated with a low-molecular-weight chromophore. NCS is rapidly cleared by the kidney and its cytotoxicity is non–cell specific. To modify its disposition, two poly(styrene-*co*-maleic acid anhydride) copolymers (MW 1500) have been coupled to one molecule of NCS, to give styrene-maleic-anhydride-neocarzinostatin (SMANCS) systems (see also Chapter 1, Figure 1.11). It was the work with SMANCS that led Hiroshi Maeda to develop the concept of the EPR effect.

SMANCS has been shown to retain nearly all the in vitro activity of NCS, with much improved pharmacokinetic properties. Tumor uptake has been shown to increase in animal models. Clinical successes have been reported with SMANCS formulated in Lipiodol® (a lymphographic vehicle) after intra-arterial administration in patients with unresectable hepatocellular carcinomas. The fact that the drug is administered in this lipid vehicle perhaps complicates interpretation, but clinical trials, mainly in Japan, are still being conducted some decades after the compound's first discovery.

5.4.2.3 Drug-Lipid Conjugates

One of the issues already touched upon is the need for optimization of drug loading and (too) early release from carrier systems. This can be an issue of polymer–drug compatibility in polymeric systems, although in liposomal systems interactions between drug and lipid are more conducive to high loadings (e.g., for doxorubicin liposomes discussed in Section 5.5.1). An alternative is to prepare drug–lipid conjugates, where the hydrophobic component is a compound such as squalene, which aggregates to form micelle-like structures with around 50% of drug content.

Coupling drugs to relatively low-molecular-weight insoluble or poorly soluble lipids, such as squalene—a natural lipid—has been proposed as a means of changing the biodistribution of the drug and to enhance activity. Using low-molecular-weight "carriers" means that the drug content is a significant element of the system. Gemcitabine, an anticancer nucleoside that is rapidly deactivated in vivo, displays an enhanced half-life and residence time when coupled to squalene. The gemcitabine–squalene compound (Figure 5.8) of which 40% is the drug payload, associates in aqueous solution because of its dual hydrophile–lipophile (surfactant) nature, forming aggregates of around 100–130 nm in diameter. These aggregates have a greater cytotoxic activity and accumulate to a higher extent in the liver and spleen than the drug itself. Other drugs have been coupled to squalene and clinical trials are in progress. Because of their structure, these squalene derivatives can also be readily incorporated into liposomes.

5.5 PARTICULATE CARRIERS FOR DRUG TARGETING

Advantages of particulate carriers include the following:

- The high drug loading that is possible with some systems.
- The drug does not have to be chemically attached to the carrier.
- A considerable degree of protection from degradation in vivo and control of drug release may be conferred on drug molecules encapsulated within the carrier.

The carrying capacity of a particulate is determined by the affinity of the drug for the carrier material: large proteins and peptides may have problems in mixing isotropically with polymeric core materials. This can lead to premature release of the drug.

However, a major limitation of these systems is their inability to cross continuous endothelial barriers (Figure 5.3) and leave the general circulation. In general, microparticulate carriers are phagocytized by the macrophages of the MPS, thereby rapidly localizing predominantly in the liver and spleen. However, sterically stabilized particulate carriers have extended circulation times and can remain in the blood; they either act as circulating drug reservoirs, or may slowly escape from the blood pool at pathological sites which exhibit increased vascular permeability.

Intravenously administered particles with dimensions exceeding a few μm (the diameter of the smallest capillaries) will be filtered by the first capillary bed they encounter, usually the lungs, leading

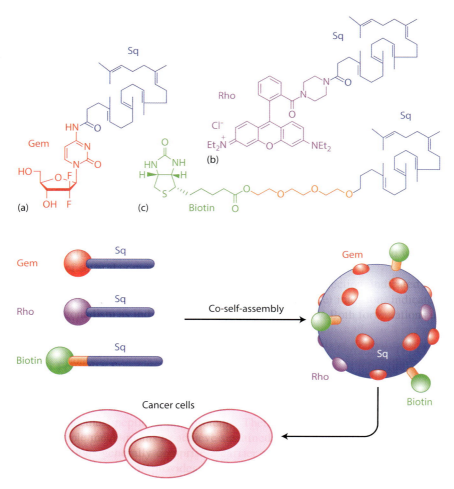

FIGURE 5.8 The structures (above) of (a) of gemcitabine-squalene (Gem-Sq), (b) the marker rhodamine-squalene (Rhod-Sq), and (c) targeting molecule biotin-squalene (Biotin-Sq) and (below) their assembly into multifunctional particles for cancer cell delivery. (From Bui, D.T., J.T. Nicolas, A. Maksimento, D. Desmaele, and P. Couvreur. 2014, *Chem.Comm.*, 50, 5336–5338. Reproduced by permission of The Royal Society of Chemistry.)

to embolism. Intra-arterially administered particles with dimensions exceeding the diameter of the smallest capillaries (around 7 μm) will be trapped in the closest organ located upstream; for example, administration into the mesenteric artery leads to entrapment in the gut and into the renal artery leads to entrapment in the kidney. This approach is under investigation to improve the treatment of diseases in the liver. Arterial chemoembolization involves the use of drug-eluting microspheres, which combine the action of blocking the arterial blood supply to hepatic tumors, causing ischemia, and delivering drugs such as doxorubicin and irinotecan, which then have prolonged contact with the tumor.

Active-targeting strategies for particulate systems are similar to those discussed for soluble macromolecular systems (see Table 5.2 and Section 5.4.1 on antibodies).

5.5.1 LIPOSOMES

Liposomes are vesicular structures based on one or more lipid bilayer(s) encapsulating an aqueous core (Figure 5.9). The lipid molecules are usually phospholipids, amphipathic moieties with a hydrophilic head group, and two hydrophobic chains ("tails"). Such moieties spontaneously orientate in water to give the most thermodynamically stable conformation, in which the hydrophilic head group faces out

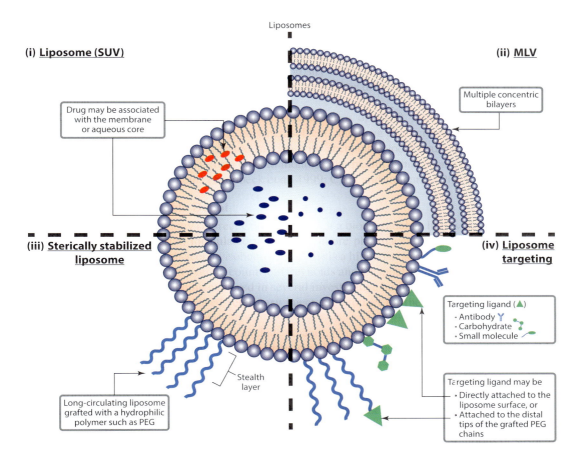

FIGURE 5.9 Schematic illustration of four possible different types of liposome: (i) small unilamellar vesicle (SUV), showing a water-soluble drug entrapped in aqueous interior and a water-insoluble drug incorporated into the liposomal membrane; (ii) multilamellar vesicle (MLV), with three concentric lipid bilayers; (iii) sterically stabilized liposomes, containing a "stealth" layer of polyethylene glycol (PEG), which shields the liposome from opsonization and uptake by the reticuloendothelial system; (iv) target-specific recognition ligands (including antibodies, sugars, and various other molecules) that can be coupled to liposomes, either to the liposome surface directly, or to the ends of grafted PEG chains.

into the aqueous environment and the lipidic chains orientate inward avoiding the water phase; this gives rise to bilayer structures. In order to reduce exposure at the edges, the bilayers self-close into one or more concentric compartments around a central discrete aqueous phase. Dependent on the preparation protocol used, liposome diameters can vary between 20 nm and 20 μm, thus encompassing the nano- and microrange of diameters. In general, they can be multilamellar or unilamellar, i.e., a multitude of concentrically orientated bilayers surrounds the aqueous core or only one bilayer surrounds an aqueous core, respectively. However, other structures have also been described.

If multilamellar structures are formed, water is present in the core of the liposome and also entrapped between the bilayers. Depending on the physicochemical nature of the drug, it can either

- Be captured in the encapsulated aqueous phase (i.e., the aqueous core and the aqueous compartments between the bilayers) (hydrophilic drugs) or
- Interact with the bilayer surface (e.g., through electrostatic interactions) or be taken up in the bilayer structure (lipophilic drugs)

Thus, liposomes can serve as carriers for both water-soluble and lipid-soluble drugs. The liposomal encapsulation of a wide variety of drugs, including antitumor and antimicrobial agents, chelating agents, peptides, proteins, and genetic material, have all been described.

Bilayer composition can be almost infinitely varied by choice of the constituent lipids. Phosphatidylcholine, a neutral phospholipid, has emerged as the major lipid component used in the preparation of pharmaceutical liposomes. Phosphatidylglycerol and phosphatidylethanolamine are also widely used. Liposomal bilayers may also accommodate sterols, glycolipids, organic acids and bases, hydrophilic polymers, antibodies, and other agents, depending on the type of vesicle required.

The rigidity and permeability of the bilayer strongly depend on the type and quality of lipids used. The alkyl-chain length and degree of unsaturation play a major role. For example, a C18 saturated alkyl chain produces rigid bilayers with low permeability at room temperature. The presence of cholesterol also tends to rigidify the bilayers. Such systems are more stable and can retain the entrapped drug for relatively longer periods, whereas more "fluid" bilayer systems can be prepared if a more rapid release is required.

Liposomes can be classified on the basis of their composition and in vivo applications:

- *Conventional liposomes*, which are neutral or negatively charged, are generally used for passive targeting to the cells of the MPS.
- *Sterically stabilized ("stealth") liposomes*, which carry hydrophilic coatings, are used to obtain prolonged circulation times.
- *Immunoliposomes ("antibody targeted")*, which can be either conventional or sterically stabilized, are used for active-targeting purposes.
- *Cationic liposomes*, which are positively charged, are used for the delivery of genetic material.

As phospholipid bilayers form spontaneously when water is added, the important challenge in liposome preparation is not the assembly of simple bilayers (which happens automatically), but in causing the bilayers to form stable vesicles of the desired size, structure, and physicochemical properties, with a high drug encapsulation efficiency. There are many different approaches to the preparation of liposomes; however, what they all have in common is that they are based on the hydration of lipids.

Liposomes represent highly versatile drug carriers, offering almost infinite possibilities to alter structural and physicochemical characteristics. This feature of versatility enables the formulation scientist to modify liposomal behavior in vivo and to tailor liposomal formulations to specific therapeutic needs. It took decades to develop the liposome carrier concept to a pharmaceutical product level, but a number of commercial preparations are now available in important disease areas and many more formulations are currently undergoing clinical trials. Examples of the different applications and commercial products of various types of liposomal systems are given in the following texts.

5.5.1.1 Conventional Liposomes

These can be defined as liposomes that are typically composed of only phospholipids (neutral and/or negatively charged) and/or cholesterol. Most of the early work on liposomes as a drug carrier system employed this liposomal type. These systems are rapidly taken up by the phagocytic cells of the MPS, localizing predominantly in the liver and spleen, and are therefore used when targeting to the MPS is the therapeutic goal. Conventional liposomes have also been used for antigen delivery and a liposomal hepatitis A vaccine (Epaxal®) has received marketing approval in Switzerland.

A commercial product based on conventional liposomes has been introduced for the parenteral delivery of the antifungal drug, amphotericin B, which is associated with a dose-limiting nephrotoxicity in conventional formulations. AmBisome®, a liposomal formulation of amphotericin B, comprises small unilamellar vesicles with diameters between 50 and 100 nm. Two other lipid-based formulations of amphotericin B are used in the clinic:

- Abelcet® consists of ribbonlike structures having a diameter in the 2–5 µm range.
- Amphocil® comprises a colloidal dispersion of disk-shaped particles with a diameter of 122 nm and a thickness of 4 nm.

In spite of the large differences in structural features (a further example of "liposomal" versatility), all formulations have been shown to greatly reduce the kidney toxicity of amphcatericin B, allowing higher doses to be given and thereby improving clinical efficacy.

5.5.1.2 Long-Circulating Liposomes

At present, the most popular way to produce long-circulating liposomes is to covalently attach the hydrophilic polymer, PEG, to the liposome bilayers. As discussed in Section 5.3.1, the highly hydrated PEG groups create a steric barrier against interactions with molecular and cellular components in the biological environment. Figure 5.10 shows how "PEGylation" of liposomes can extend their blood circulation profile.

Long-circulating liposomes can enhance their chances of extravasation and thus accumulate at sites where pathological reactions occur. For example, the commercial product Doxil® (marketed as Caelyx® in Europe) consists of small-sized PEGylated liposomes, encapsulating the cytostatic doxorubicin. The resulting long-circulation times and small size of the vesicles facilitate their accumulation in tumor tissue

DaunoXome® liposomes are also long-circulating liposomes, in this case encapsulating the cytostatic daunorubicin. Although a nonstealth system, long-circulation times are attained by using a particularly rigid bilayer composition, in combination with a relatively small liposome size. The encapsulation of these anthracycline cytostatics in liposomes affects a modified biodistribution of the drug; the drug is distributed away from the heart, where in free form it can exert considerable toxic effects, and is preferentially taken up by solid tumor tissue.

5.5.1.3 Immunoliposomes

Immunoliposomes have specific antibodies or antibody fragments on their surface to enhance target site binding. The primary focus of their use has been in the targeted delivery of anticancer agents.

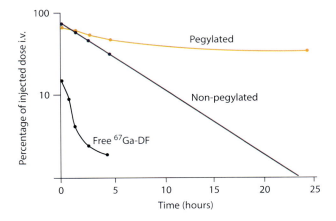

FIGURE 5.10 Comparison of the blood levels of free gallium desferal (^{67}Ga-DF), with ^{67}Ga-DF encapsulated in polyethylene glycol stabilized liposomes, and nonstabilized liposomes, upon intravenous administration in rats. (From Woodle, M. et al., Improved long circulating (Stealth) liposomes using synthetic lipids, *Proc. Int. Symp. Control. Release Bioact. Mater.*, 17, 77–78, 1990. With permission from the Controlled Release Society.)

Long-circulating immunoliposomes can also be prepared. The antibody can be coupled directly to the liposomal surface; however, in this case, the PEG chains may provide steric hindrance to antigen binding. Alternatively, a bifunctional PEG linker can be used, to couple liposomes to one end of PEG chains and antibodies to the other end of these chains (Figure 5.9). Steric hindrance is not a problem in the latter approach.

5.5.1.4 Cationic Liposomes

Cationic liposomes demonstrate considerable potential for improving the delivery of genetic material. The cationic lipid components of the liposomes interact with, and neutralize, negatively charged DNA, thereby condensing the DNA into a more compact structure. Depending on the preparation method used, the complex may not be a simple aggregate, but an intricate structure in which the condensed DNA is surrounded by a lipid bilayer. These systems are discussed further in Chapter 16.

5.5.2 Niosomes

An alternative to phospholipid-based liposomes can be found in niosomes (nonionic surfactant vesicles) developed by L'Oreal for cosmetic use. These systems, which can be uni- or multilamellar, are based on several different families of synthetic, nonionic low hydrophilic-lipophilic balance (HLB) amphipathic molecules, such as sorbitan monolaurate and others of this series, or alkyl glucosides, all formed with cholesterol. While there is considerable experimental experience with niosomes as parenteral delivery systems, there are no clear advantages over liposomal systems: although composed of nonionic surfactants/cholesterol mixtures rather than phospholipid/cholesterol mixtures and thus having a potential for greater variation in their composition, they turn out to have very similar physical and biological properties.

5.5.3 Polymeric Micelles

As described in Chapter 3 (Section 3.8), when low MW amphipathic (surfactant) molecules with distinct hydrophilic and hydrophobic sections are dispersed in water, spherical micelles in the nanometer size range are formed above a certain concentration, the critical micelle concentration (CMC). These micelles can solubilize drug molecules in their hydrophobic core, a useful formulation technique that has been used for the preparation of injections of paclitaxel, for example. There is a constant exchange between the surfactant monomers and their micelles. The more hydrophobic the surfactant, the lower the CMC. Nonionic surfactant systems generally have very low CMCs so that the micellar phase can exist even after dilution. However, micelles used as carriers for drug-targeting purposes must be stable in the blood circulation and should not disintegrate upon contact with blood components so that the drug load can be kept on board. In spite of the low CMCs of surfactant molecules, they are fragile. Using polymeric amphipathic materials to form the so-called polymeric micelles allows the formation of more stable systems suitable for targeting. The rate of exchange between the micellar and monomeric forms is slow because of the strong interaction forces between the molecules in the polymeric micelle.

There are several types of polymeric surfactant. Micellar systems based on amphipathic block copolymers have been explored as i.v.-administered drug carrier systems. These block copolymers can be, for example, composed of a hydrophilic PEG block (A) and a hydrophobic block (B) based on poly(aspartic acid) or poly(β-benzyl-L-aspartate). These form micelles in aqueous solution with spherical core/shell structures having diameters around 20–40 nm (see also Chapter 1, Figure 1.14). The hydrophobic core of these micelles can be loaded with lipophilic drugs such as doxorubicin. After i.v. administration, the micelles may accumulate at tumor sites. Some of the characteristics of these micellar systems are listed in Box 5.3. A subset of polymeric micelles is formed by lipid-core micelles, which comprise systems formed from conjugates of soluble copolymers with lipids. One example is the PEG–phosphatidylethanolamine conjugate.

BOX 5.3   POLYMERIC MICELLES AS DRUG CARRIERS

- Critical micelle concentration of the amphipathic copolymers is low; interaction between polymer units is strong, so that blood components cannot disrupt the aggregates.
- Molecular weight of the polymeric unit is small enough to allow clearance through glomerular filtration.
- Diameter is large enough to prevent penetration through intact endothelium.
- Diameter can be chosen in the range where the EPR effect is observed (<200 nm).
- Release kinetics of the drug depend on the selected polymer structure (hydrophilicity/hydrophobicity balance).
- Drug is in the hydrophobic core of the micelle and is protected from exposure to aqueous degradation processes.
- They have been shown to have a high drug–carrying capacity ("payload").

These systems are chemically versatile (using different AB or ABA copolymers), as it is possible by varying the copolymer characteristics to enhance payload and by using covalent binding strategies to further optimize their performance.

5.5.4   LIPOPROTEIN CARRIERS

Lipoproteins are nature's template for targeting growing cells, as they transport cholesterol to such cells by receptor-mediated endocytosis. These endogenous lipid carrier systems comprise a lipid core and a coat where the ligand apolipoproteins can be found. The lipid core material consists of cholesterol and other lipids (cholesterol esters, triacylglycerols and phospholipids), transported in plasma and other body fluids as lipoproteins, i.e., complexes of the lipid material bearing apolipoproteins as targeting moieties. This has given rise to their exploration as systems aimed at cancer cells. There are different types of lipoproteins: (1) high-density lipoproteins, size about 10 nm; (2) low-density lipoproteins (LDL), size about 25 nm; (3) very-low-density lipoproteins (VLDL), size about 30–90 nm; and (4) chylomicrons, size about 10–90 nm. Most studies have focused on LDL, which has a plasma half-life of 3–4 days. On some tumor cells, LDL receptor density is increased, making cytostatic-loaded LDL an interesting potential drug delivery system. With the Kupffer cells playing a major role in the uptake process, 90% of the LDL receptor activity is concentrated in the liver.

The obstacle to any clinical use is that LDL has to be extracted from patients, the cholesteryl esters in the interior of the particle removed, and the system processed to replace the interior with drug molecules. An alternative is to covalently attach drugs to the exterior, but this and other processing techniques can disrupt the conformation of the apolipoprotein and, hence, diminish or eradicate targeting properties. Some years ago, synthetic cholesterol ester–rich microemulsion systems were used to mimic LDL, but this work does not seem to have progressed.

5.5.5   POLYMERIC MICRO- AND NANOPARTICLES

Many polymers can be used in the preparation of micro- and nanoparticles (Figure 5.1). The choice of polymer is often based on the biodegradability of the polymer, the ease with which the polymer can be prepared as nanoparticles, and the loading capacity of the system for the chosen drug. The surface properties of the nanoparticles and the presence of reactive groups on the surface are clearly important in the ability to alter the behavior of these systems. The loading capacity of polymeric nanoparticles is determined by the affinity of the drug for the polymer: it cannot be assumed that all drugs will be freely miscible with the polymer in question, as mentioned earlier. The difficulty in the formulation of poly(lactic-*co*-glycolic acid) (PLGA) implants of LHRH is a case in point. Smaller molecules may mix (or dissolve) more readily, but any incompatibility will affect stability, capacity, and release rates. The nature of the drug dispersion within nanoparticles is thus crucial in terms of the subsequent rate of release of the active drug.

The polymers used for the formulation of nanoparticles include synthetic polymers such as poly(alkyl cyanoacrylates) (PACAs), PLGAs, and poly(caprolactones), as well as natural polymers such as albumin, gelatin, alginate, collagen, and chitosan. We discuss some of these in the following texts. Although we can generically refer to "polymeric nanoparticles," their surface properties and the nature of their interior structure will depend on the polymer employed. Particle interiors may be porous or solid and surfaces will have different functionalities and contours that will influence both drug release and the ability to functionalize the particles.

5.5.5.1 Poly(Alkyl Cyanoacrylates)

The PACAs, used widely as tissue adhesives, are biodegradable and well tolerated. PACA nanoparticles can be produced by an emulsion polymerization technique or by precipitation from the monomer in the presence of surfactants. The alkyl cyanoacrylate monomer polymerizes rapidly in the presence of water to form nanoparticles in vigorously stirred oil-in-water dispersion, the process of polymerization being initiated by anions (OH$^-$) at the oil–water interface. The drug to be encapsulated is incorporated in the particles as the oil phase is evaporated.

Degradation kinetics in vivo are controlled by the alkyl-chain length. Poly(butyl cyanoacrylate) nanoparticles are degraded fairly rapidly (1 day), while poly(hexyl cyanoacrylate) nanoparticles take a number of days to degrade in vivo. PACA nanoparticles accumulate in the liver (60%–90% of the injected dose) and the spleen upon i.v. injection. The macrophages in the liver are their major target. PACA nanoparticles loaded with doxorubicin have shown an increased therapeutic index in a number of animal tumor models, due to a reduction of the peak drug concentration in cardiac tissue, the organ most severely affected by doxorubicin upon injection. This is a common finding with doxorubicin delivery through liposomes, niosomes, and other carriers, so is not unique to PACA systems. The release of drug from the Kupffer cells upon breakdown of the nanoparticles in the lysosomal system may induce a slow release pattern that is still tumoricidal but lacks the cardiotoxic effects.

Sterically stabilized PACA nanoparticles can be prepared by adsorbing polyoxyethylene polymers of the poloxamer or poloxamine class. In addition, PEGylated nanoparticles can be prepared using block copolymers such as poly[methoxy-polyethylene glycol-cyanoacrylate-*co*-hexadecyl cyanoacrylate].

5.5.5.2 Poly(Lactic-*co*-Glycolic Acid)

The versatility of most polymers in the formation of nanoparticles lies in the ability to control the molecular weight of the polymer and also the ratio of the comonomers. This has been demonstrated strikingly with PLGA polymers. Figure 5.11 shows the relevant structures.

PLGAs have been a prominent material in the fabrication of experimental nanosystems. Changing the ratio of the monomers (lactic acid/glycolic acid) leads to different affinities for encapsulated drugs. The rate of release of drugs is also modulated by the molecular weight. One advantage of PLGA systems is that they degrade in vivo by hydrolysis into lactic acid and glycolic acid. The preparation, properties, and degradation of PLGA microspheres for use as long-acting injections are further discussed in Chapter 6 (Section 6.2.4).

5.5.5.3 Albumin and Other Protein Microspheres and Nanoparticles

In 2005, the FDA approved an albumin–paclitaxel formulation, Abraxane®. This formulation showed an improved therapeutic index in a number of solid tumors in the clinic. The exact nature of these albumin-based nanoparticles (size around 300 nm) is still under investigation,

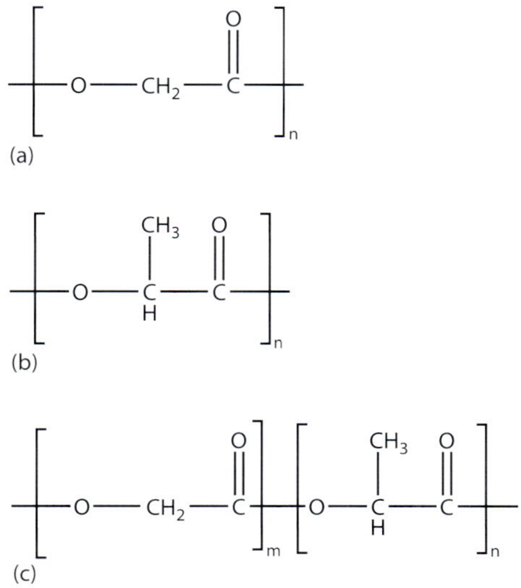

(a)

(b)

(c)

FIGURE 5.11   Structure of (a) poly(glycolic acid), (b) poly(lactic acid), and (c) poly(lactic-*co*-glycolic acid).

and several mechanisms have been suggested for the improved performance compared to conventional paclitaxel therapy. Among those is the possibility of increased uptake of albumin by fast growing solid tumors/tumor endothelia, as tumor cells need albumin-transported nutrients for their growth.

Albumin particles may also be prepared by stabilizing the structure by cross-linking the monomers. In the preparation of such albumin microspheres, an aqueous solution of albumin and the drug moiety is initially emulsified in oil, forming a water-in-oil emulsion. The protein can be chemically cross-linked by the addition of a cross-linking agent such as glutaraldehyde or butadione or thermally cross-linked. Either method produces stabilized particles for possible use in drug delivery. The size of the particles is based on the droplet size of the initial emulsion and can range from 15 nm to 150 μm. The rate of drug release has been shown in some cases to be dependent on the degree of cross-linking achieved. Hemoglobin, casein, and ferritin have been used as the basis for protein microspheres; there is no reason why such materials cannot be used as nanoparticles also.

5.5.6   DENDRIMERS

Dendrimers (treelike-branched polymers) are still a promising class of polymeric drug carrier system, some 30 years after their conception. They have several potential advantages over many other polymers in that they are synthesized in such a manner that each dendrimer molecule has exactly the same chemical composition and thus diameter, which can range from several nanometers upward. They are formed from a core multivalent molecule to which several dendrons (partial dendrimers) may be covalently attached, or other reactive groups, layer by layer or "generation" by "generation" (Figure 5.12). Research into all aspects of dendrimer chemistry and technology has escalated since the original work by Tomalia and others in the 1980s. The wide variety of chemical architectures of these spherical or quasi-spherical structures allows the construction of monodisperse particles in the 1–20 nm diameter range.

Dendrons are usually asymmetric partial dendrimers, which can be either used as such or formed covalently or noncovalently into complete dendrimers. This allows the formation of "Janus" forms with both a hydrophilic and hydrophobic surface. Because many such amphipathic dendrons and lipophilic dendrimers associate in solution, a range of supramolecular structures can be formed. These themselves may have potential as novel delivery systems. Drugs may be situated in the

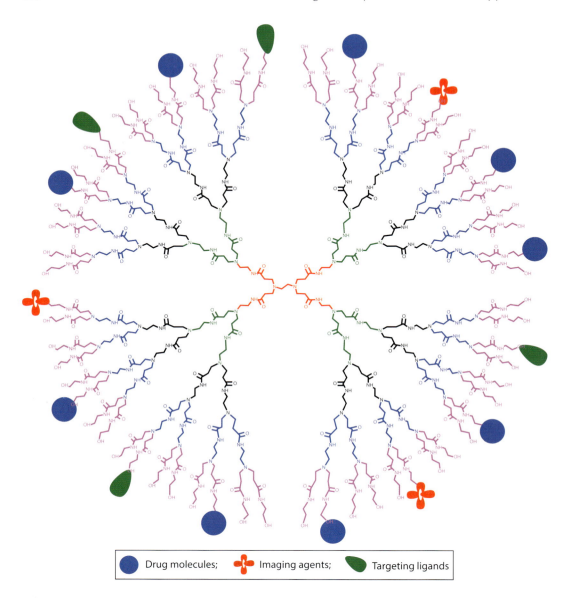

Drug molecules;   Imaging agents;   Targeting ligands

FIGURE 5.12 Polyamidoamine (PAMAM) dendrimer structure. Drug molecules may also be encapsulated in the interior of the structure. (Courtesy of the Kannan Group, Center for Nanomedicine, Johns Hopkins University, Baltimore, MD.)

interior of the dendrimer, on the dendrimer surface, perhaps covalently attached, or in intermediate positions. Hydrophobic dendrimers can associate into dendrimer aggregates 100–200 nm in diameter; hydrophilic dendrons can form vesicles—dendrisomes—so the family of systems based on these structures is immensely varied.

5.5.7 OTHER PARTICULATE CARRIER SYSTEMS

A variety of other nanoparticulate DDTS have been studied, including metallic (gold, silver, and iron oxide) nanoparticles, mesoporous silica nanoparticles, quantum dots, and carbon-based nano-materials, including nanotubes, nanorods, and nanocages (see also Chapter 18). Many new DDTS systems leverage techniques from the microelectronics industry, to precisely fabricate DDTS in the nanometer range; these nanofabrication techniques are described in Chapter 19.

5.6 PHARMACEUTICAL ASPECTS OF CARRIER SYSTEMS

In order for parenteral DDTS to become commercial products, certain pharmaceutical issues need to be addressed, including

- Purity of the carrier material
- Reproducibility of the characteristics of the drug carrier system
- Drug carrier–related safety aspects, including immunological responses
- Scaling-up possibilities
- Shelf life

Historically, DDTS were developed in environments where the primary goal was "proof of concept," rather than developing a commercial product. The typical pharmaceutical considerations described earlier were not dealt with seriously in the early days of drug carrier research; thus, early drug carrier systems were associated with long gestation periods from product development to product marketing.

The time frame associated with the development of a drug-targeting concept to a targeted drug product can be illustrated by the "liposome story." Liposomes were originally used as biochemical tools for the study of cell membrane behavior in the 1960s; the idea to use them as drug carriers was subsequently developed in the early 1970s. It took more than 20 years to develop the system from a concept to the first commercial parenteral liposome preparation carrying a drug (amphotericin B). Although this may seem like quite a long gestational period, it must be remembered that liposomes were one of the first colloidal carrier systems designed for targeted drug delivery. Comparatively, little was known about such systems and many technological and biopharmaceutical hurdles had to be overcome before marketing authorization for the first product could be obtained. Some of the hurdles encountered and solved over the years while developing liposomes as drug carriers are outlined here.

Poor quality of the raw material: In the early 1980s, the quality of lipids of several suppliers could vary considerably. Nowadays, a few suppliers provide the global market with high-quality products. Interestingly, over the years, the price per unit has dropped considerably while the quality has improved.

Poor characterization of the physicochemical properties of liposomes: Liposome behavior in vitro and in vivo is critically dependent on their physicochemical properties. Therefore, a full physicochemical characterization of pharmaceutical liposomes is required in early stages of a development program (Table 5.3). In later development stages, these quality control assays can be used to obtain regulatory approval and to ensure batch-to-batch consistency.

Shelf life: Shelf-life issues that need to be addressed include avoidance of preadministration leakage of the liposome-associated drug (retention loss), size stability (occurrence of fusion or aggregation), and phospholipid degradation (occurrence of peroxidation and hydrolysis).

Scaling-up problems: Several of the laboratory-scale liposome preparation methods were difficult to scale up to industrial scale.

Safety data: As these carriers are novel delivery systems, there initially existed a paucity of data on their safety during chronic use. However, their present safety record and the experience with marketed parenteral liposome preparations (e.g., amphotericin B, doxorubicin and daunorubicin) indicate that the safety of these systems is not a major limiting factor.

TABLE 5.3

Quality Control Assays of Liposomal Formulations

Assay	Methodology/Analytical Target
Characterization	
pH	pH meter
Osmolarity	Osmometer
Phospholipid concentration	Lipid phosphorus content/HPLC
Phospholipid composition	TLC, HPLC
Cholesterol concentration	Cholesterol oxidase assay, HPLC
Drug concentration	Appropriate compendial method
Chemical stability	
pH	pH meter
Phospholipid peroxidation	Conjugated dienes, lipid peroxides, FA composition (GLC)
Phospholipid hydrolysis	HPLC, TLC, FA concentration
Cholesterol autoxidation	HPLC, TLC
Antioxidant degradation	HPLC, TLC
Physical stability	
Vesicle size distribution	
Submicron range	DLS
Micron range	Coulter Counter, light microscopy, laser diffraction, GEC
Electrical surface potential	Zeta-potential measurements
Surface pH	pH sensitive probes
Number of bilayers	SAXS, NMR
Percentage of free drugs	GEC, IEC, protamine precipitation
Dilution-dependent drug release (simulating release upon injection)	Measure the extent of drug release from the drug–liposome product following dilution
Relevant body fluid–induced leakage	GEC, IEC, protamine precipitation
Biological characterization	
Sterility	Aerobic and anaerobic cultures
Pyrogenicity	Rabbit or LAL test
Animal toxicity	Monitor survival, histology, pathology

Source: Barenholz, Y. and Crommelin, D.J.A., Liposomes as pharmaceutical dosage forms, In *Encyclopedia of Pharmaceutical Technology*, eds. J. Swarbrick, J.C. Boylan, Vol. 9, pp. 1–39, Marcel Dekker: New York, 1991.

Abbreviations: FA, fatty acids; TLC, thin-layer chromatography; HPLC, high-pressure liquid chromatography; DLS, dynamic light scattering; GEC, gel exclusion chromatography; SAXS, small-angle x-ray scattering; IEC, ion exchange chromatography; LAL, limulus amebocyte lysate.

Biochemists, who worked with drug-loaded liposomes in the early days, had a completely different perception of "stability," reproducibility, upscaling, and toxicity than pharmaceutical scientists, who are familiar with the development of pharmaceutical formulations. For example, for a biochemist, a shelf life of a week at −70°C may be acceptable, whereas a pharmaceutical product would be expected to have a minimum shelf life of 2 years, preferably without refrigerator cooling. It took several years and considerable "mental adaptation" to bridge this cultural gap. Currently, quality is ensured by improved purification schemes, the introduction of validated analytical techniques, and a better insight into lipid degradation mechanisms, leading to better shelf-life conditions (Table 5.3). These quality control considerations are discussed further in Chapter 21.

The development of liposomal systems has thus contributed greatly to the development of drug carrier systems in general and has highlighted the various pharmaceutical hurdles that must be overcome before a DDTS can reach the marketplace. In addition, liposomal development has provided fundamental knowledge on the fate of particulate systems in vivo and how this fate can be manipulated for therapeutic gain.

5.7 OBSTACLES TO SUCCESSFUL TARGETING

We have discussed earlier in this chapter some of the exciting developments in pharmaceutical nanotechnology for drug delivery and targeting. For balance, it is essential to recount some of the features of nanosystems and their interaction with the biological environment that often impede *quantitative* delivery of drugs and other therapeutic agents to target organs, tissues, cells, or cell compartments. It is widely recognized that only maximally around 5% of the administered dose of drug in carrier systems reaches the desired target. Quantitative targeting signifies that losses to nontarget tissues and organs will be minimized. When this is realized, it will transform the field. Drug targeting through the use of nanotechnology is at the same time simple in concept but complex in reality. The route from the point of injection to the site of action in a target is not a straightforward one, as Figure 5.13 attempts to illustrate, recounting the possibility of particle aggregation, jamming in narrow pores and vessels, the lack of uptake into the target, the premature release of the drug, and other factors that are listed in Box 5.4. What cannot be shown in the figure are the dynamics of the processes involved.

One reason that we need to consider the delivery of drugs within nanoparticles in a different light from the delivery of free molecules is that once particulates are involved, we introduce the element of chance into the equation. If a small molecule drug is stable and soluble in biological fluids such as the blood, it will diffuse readily by way of the circulation throughout the body, not, of course, equally

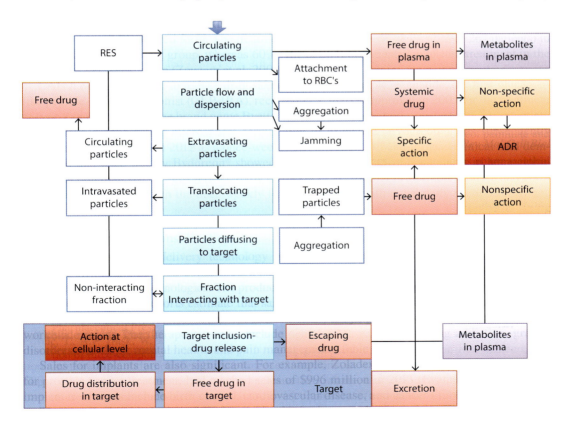

FIGURE 5.13 Diagrammatic (and simplified) illustration of the complex route from site of intravenous injection to target, with each potential stage labeled. Note the events that occur in the target as well as en route to the target tissue. RBC's, red blood cells; ADR, adverse drug reaction.

BOX 5.4   FACTORS REDUCING THE EFFECTIVENESS OF TARGETED SYSTEMS

- Premature release of drug load before the target is reached
- Problems in accessing the target site (size, charge, hydrophobicity)
- Adsorption of nanosystems to erythrocytes
- Complex flow patterns in the blood, which discourage epithelial contact
- Ligand loss during circulation
- Incomplete ligand–receptor interaction
- Poor diffusion of released drug into the target tissues
- Suboptimal levels of drug attained in the target
- Leakage of drug molecules from the target site
- Design of systems based on inappropriate cell choice (activity) or species (distribution)

to each site in the body, but sufficient at appropriate doses to affect specific receptors. The rates of metabolism and clearance are issues discussed elsewhere. The interaction of free drug molecules with their receptors is an equilibrium process, but because of the high number of free molecules, the probability of interaction is very high at the correct dose. One question to be posed is how we estimate the dose of nanoparticles to be delivered to, say, a tumor, as the delivered dose depends on the release of drug from the carrier as well as the degree of penetration of the particles into the target. With nanosystems, their small size is both vital for access to targets and for their accumulation in specific sites, say, through the EPR effect, but their trajectories are often complex. Fundamental to nanoparticle targeting is the behavior of the systems in vivo. Box 5.4 summarizes some of the issues that require control and further study to ensure that quantitative targeting is achieved. The use of external stimuli to activate the carriers to release their content is increasingly being studied, as described in Chapter 14.

While much effort is concentrated on the uptake of drug into target cells, consideration also has to be given to the loss of drug by leakage back into the circulation. There is, as discussed, no true "homing" device as biological interactions with nanoparticulates (ligand–receptor interactions) occur over very small distances (of the order of <10 nm). Even then, particles must be oriented in the optimum configuration and external forces can, of course, reduce the extent of interaction.

5.7.1 CHALLENGES

It is not only the properties of the carrier system that determine success. The nature of the drug itself and, of course, drug carrier interactions are also of great importance. If we consider the case of paclitaxel (Figure 5.14), once released from the vector, it has several characteristics that can influence the outcome: it has a low water solubility, which can lead to precipitation, it is a medium-sized molecule (MW 853.9 Da) with thus a lower diffusion coefficient than smaller drugs, it binds to albumin and is a P-gp activator. While the drug is in the carrier, the effects of efflux pumps are avoided, but once released, the drug, of course, can be acted on by cellular efflux systems.

There are pharmaceutical challenges to secure the production and design of the "perfect" or optimized nanosystem for targeting, which occur at several stages, including small-scale preparation in the laboratory (crucial for successful experiments), later production on a larger scale, instability during in vitro testing in cell culture systems, or during storage, and there are all the issues we have discussed that occur after administration. Different routes present different challenges. Intravenously, the flow of nanoparticles in the blood, their entrapment by the Kupffer cells, their interaction with red blood cells, their escape from the circulation, their diffusion in the extracellular matrix, their stability both physical and chemical, the potential loss of surface ligands, and the camouflage by PEG chains of the ligand, are indeed the gamut and mélange of chemical and biological problems which must be overcome.

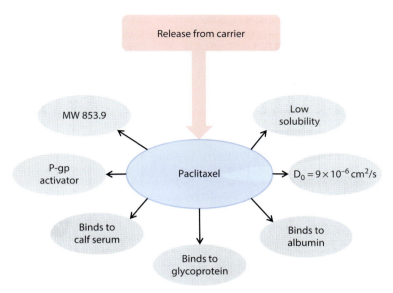

FIGURE 5.14 Paclitaxel properties that impact on its efficacy after its release from a carrier system. This illustrates the point that drug targeting needs not only appropriate delivery systems but also (more) appropriate drugs.

5.7.1.1 Particle Size and Particle Size Distribution

It is important to have systems with as narrow a size range as possible, as size is a parameter that determines biological activity and the fate of the particles. It is possible to prepare nanoparticles that are monosized, but often particles have a range of initial diameters. The size is then reported as a mean or median diameter by weight or number. The correct reporting of diameters and size distribution is crucial for comparison of data obtained in different laboratories. Both are valuable to explain outcomes. A system with 10% of the particles above the desired range may indeed have 90% of the drug load in these particles. The initial size of systems may not be the size once the system has "aged" or has been added to cell culture media or injected. Nanoparticles are colloids and the factors affecting colloid stability, of course, apply to them. Aggregation of particles through flocculation or coalescence results in an increase in the effective diameter of the product. Systems must be designed to prevent aggregation without affecting action. The adsorption of PEG chains by providing a hydrophilic barrier on particle surfaces achieves a degree of stability through steric or enthalpic stabilization but may also reduce the interaction of the particles with surfaces of cells by the same token or may mask the smaller surface ligands.

5.8 CONCLUSIONS AND PROSPECTS

Progress has been made toward the delivery of drugs more precisely in the body, but much remains to be done. Several targeted drugs have entered the marketplace successfully. They have in common that they are indicated for the treatment of life-threatening diseases like cancer, severe immunological diseases, and fungal infections and, therefore, contribute considerably to our therapeutic armamentarium. Targeted drug delivery concepts are still in some regard in statu nascendi some 40 years after Speiser advocated the use of nanoparticles and nanocapsules in therapy. Much has been learned along the way. Multidisciplinary approaches where different disciplines join forces (molecular biology, biotechnology, pathology, pharmacology, immunology, pharmaceutical sciences, engineering, clinical sciences, etc.) turned out to be the key to future success. In particular, our insights into the anatomical, physiological, and pathological constraints to the targeting concept have been growing fast over the last two decades. That know-how will help us to speed up new developments on a rational basis. Moreover, progress in molecular biology and biotechnology, enabling scientists to engineer protein structures and to produce them on a large scale, will have a great impact on drug-targeting concepts and the actual production of targeted drug delivery systems. Box 5.5 lists several attributes of ideal systems, indicating some of the complex sequence of events required to initiate a therapeutic response. In reading about the new developments in drug targeting, it would be wise to measure the claims against this list, which, although not exhaustive, is a guide. Doubt about any one of these attributes should be taken into account in evaluating the approach.

BOX 5.5   PRIMARY ATTRIBUTES OF TARGETED SYSTEMS

1. Appropriate chemical architecture: size, shape, and flexibility
2. Acceptability in terms of safety, compatibility, and biodegradability
3. Access to tissues and receptors, which might sometimes be hidden
4. Affinity to receptors sufficient to engage for sequential effects, such as cellular uptake
5. Accumulation in target tissues[a]
6. Absorption into target cells[a]
7. Activation and/or release of the drug at an appropriate rate[a]
8. Action of the drug at the cellular level and sometimes at the level of the nucleus
9. Adaptability in terms of facile modification to accommodate different surface ligands or more than one active agent to optimize efficacy

[a]   This list refers mainly to targeting to extravascular sites. Of course, sometimes the target may be a circulating molecule (such as with anti-TNF antibodies) or a biological structure (such as a thrombus in the targeted delivery of antithrombotics); hence, some of the aforesaid attributes such as 5, 6, and 7 would need modification.

The existing generation of targeted drug delivery systems contains active compounds that often have been used for many years in their "free" form before they were associated with a carrier. Examples are amphotericin B, doxorubicin, and daunorubicin. These drugs were chosen on the basis of their pharmacokinetic and pharmacodynamic profile while administered in the "free" form. It is expected that in the near future, "carrier-dependent drugs"—drugs that have not been used before in "free form" in therapy and indeed molecules that have been designed purposely for encapsulation—will enter the market.

In spite of thousands of papers in the field, there are still relatively few quantitative studies examining each part of the procession from the point of administration to the point of delivery of the payload at the target site. Systems have often been tested in cell culture systems and not in test animals. When the latter has occurred, the choice of animal model (mouse, rat, guinea pig, or other species), has not always been chosen well. It is well known that an individual delivery vector can induce markedly different effects in different cell lines. Of great importance too is the extrapolation of data obtained in small animals to the human situation, the ultimate test of most delivery systems. Scaling factors are bound to influence outcomes: if nanoparticle size is important, then the manner in which these nanoparticles circulate in vivo is crucial, as we have discussed earlier. Factors such as the time taken for systems to circulate, the different concentrations of nanoparticles used—measured in terms of mg/kg to achieve equal dosage and whether or not this is appropriate—will cause differences in flow and accumulation of carriers, shear forces acting on particles and receptors, and the nature of tumor models. As dynamic processes are involved, all these factors come into play in the ability to effectively extrapolate data.

Nonetheless, it is the complexity of the field that is the driving force for further research and it is imperative that such work continues, however, with greater attention to the factors discussed in this chapter.

FURTHER READING

Bae, Y.H., R.J. Mrsny, and K. Park, eds. 2013. *Cancer Targeted Drug Delivery*. New York: Springer.

Barenholz, Y. and D.J.A. Crommelin. 1991. Liposomes as pharmaceutical dosage forms. In *Encyclopedia of Pharmaceutical Technology*, eds. J. Swarbrick, J.C. Boylan, Vol. 9, pp. 1–39. Marcel Dekker: New York.

Bertrand, N. and J.C. Leroux. 2012. The journey of a drug-carrier in the body. An anatomo-physiological perspective. *J. Control. Release* 161:152–163. Marcel Dekker: New York.

Bui, D.T., J.T. Nicolas, A. Maksimento, D. Desmaele, and P. Couvreur. 2014, *Chem.Comm.*, 50, 5336–5338.

Crommelin, D.J.A. and A.T. Florence. 2013. Towards more effective advanced drug delivery systems. *Int. J. Pharm.* 454:496–511.

Crommelin, D.J.A., R.D. Sindelar, and B. Meibohm, eds. 2013. *Pharmaceutical Biotechnology: Fundamentals and Applications*, 4th ed. New York: Springer.

Davis, J.D., R. Deng, C.A. Boswell et al. 2013. Monoclonal antibodies: From structure to the therapeutic application. In *Pharmaceutical Biotechnology, Fundamentals and Applications*, eds. D.J.A. Crommelin, R. Sindelar, and B. Meibohm, pp. 143–178. New York: Springer.

Fox, M.E., F.C. Szoka, and J.M.J. Frechet. 2009. Soluble polymer carriers for the treatment of cancer: The importance of molecular architecture. *Acc. Chem. Res.* 42(8):1141–1151.

Okano, T., N. Yui, N. Yokoyama et al. 1994. *Advances in Polymeric Systems for Drug Delivery*, Vol. 4. Montreux, Switzerland: Gordon and Breach Science Publishers.

Prabhakar, U., H. Maeda, R.K. Jain et al. 2013. Challenges and key considerations of the enhanced permeability and retention (EPR) effect for nanomedicine drug delivery in oncology. *Cancer Res.* 73:2412–2417.

Seymour, L.W. 1992. Passive tumor targeting of soluble macromolecules and drug conjugates. *Crit. Rev. Ther. Drug Carrier Syst.* 9:135–187.

Storm, G. and D.J.A. Crommelin. 1998. Liposomes: Quo vadis? *Pharm. Sci. Technol. Today* 1:19–31.

Torchilin, V.P. and V. Weissig, eds. 2003. *Liposomes*, 2nd ed. Oxford, U.K.: Oxford University Press.

Tortora, G.J. and B.H. Derrickson, eds. 2013. *Principles of Anatomy and Physiology*, 14th ed. Hoboken, NJ: Wiley.

World Health Organisation. 2008. International Nonproprietary Names (INN). Working Group Meeting report on Nomenclature for Monoclonal Antibodies (mAb).

Woodle, M., M. Newman, I. Collins et al. 1990. Improved long circulating (Stealth) liposomes using synthetic lipids. *Proc. Int. Symp. Control. Release Bioact. Mater.* 17:77–78.

6 Long-Acting Injections and Implants

Jeremy C. Wright and James Matriano

CONTENTS

6.1   INTRODUCTION

Long-acting injectables (LAIs) and implants are designed to release a drug at a controlled rate for a specific duration and thereby achieve therapeutic drug concentrations for a prolonged period. This chapter describes a variety of LAIs and implant platforms that are currently commercially available, or at an advanced stage of development. This chapter is closely related to Chapter 5, which describes parenteral drug delivery from a targeting perspective, as opposed to the controlled-release perspective that is described here.

   In a general chapter such as this, it is not possible to include all the LAIs and implants that have been researched and/or developed. The authors chose systems that have been commercialized or seemed highly relevant to the discussion in the chapter, but in the end, the choice was somewhat arbitrary. The authors acknowledge the many investigators in this area whose work was not possible to include due to space limitations.

6.1.1   RATIONALE FOR LONG-ACTING INJECTABLES AND IMPLANTS

There are both therapeutic and commercial benefits associated with the use of LAIs and implants, listed in Box 6.1 and described further here.

6.1.1.1   Therapeutic Benefits

LAIs and implant drug delivery technology platforms offer significant benefits for the effective treatment of a broad range of medical conditions. These conditions include treatment for a variety of diseases (e.g., mental health, endocrinology, oncology, and pain management), infections, surgical procedures, local injuries, and prophylactic treatment regimens. The achievement of a safe, effective, and acceptable treatment requires expertise that spans multiple disciplines including formulation scientists, toxicologists, clinicians, and regulatory experts.

   Some specific examples are discussed further here. LAIs and implants are beneficial for patients that have trouble maintaining daily oral or injection regimens or have preference for injection over

---

BOX 6.1   RATIONALE FOR SUSTAINED-RELEASE INJECTABLES AND IMPLANTS

Reduction in fluctuation of drug concentrations
Reduction in the number of treatment administrations
Improved patient compliance
Potentially improved efficacy
Improved safety
Product differentiation
Life cycle management
    Reduced development time
    Ability to leverage safety data of approved active pharmaceutical ingredient
    Potential for additional patent claims

oral dosage forms. For example, LAIs have been useful for patients with mental disease that experience frequent relapses while on oral medications. Schizophrenia can be a severe and chronic illness and can be costly to treat. Relapse and rehospitalization significantly add to the economic burden of this disease. Further emphasizing the rationale for LAIs are studies that provide evidence that LAIs are underutilized, despite strong evidence-based findings that they have a favored impact on treatment adherence, costs, quality, and reduced hospitalizations (Gopalakrisna et al. 2013).

Studies by Louza et al. evaluated patients that were on oral antipsychotic medication who had a history of poor treatment adherence. Switching to a long-acting risperidone injection demonstrated significant clinical improvements, reinforcing that the long-acting antipsychotic is a positive alternative in patients in whom compliance is an issue (Louza et al. 2011).

Poor patient compliance is often associated with treatment of opioid addiction. The issue of treatment adherence is complex and limits treatment effectiveness. Naltrexone is a μ-opioid antagonist and has been used for treatment for opioid dependence. An attractive option to overcome poor treatment compliance is the use of sustained-release technology (Kjome and Moeller 2011). A retrospective review of two types of naltrexone implants, oral naltrexone and historical controls, demonstrated differences between the immediate-release naltrexone versus the implants. The opioid-free rates combined for the implants were 82%, compared to 58% in patients in the oral naltrexone and 52% for the historical control groups (Reece 2007).

LAIs have the potential for improved efficacy. A Phase 3 study compared the efficacy of treatment for patients with chronic hepatitis C infection, who were administered interferon alpha-2a (IFN-α2a) subcutaneously (s.c.) 3 times a week, to a once-weekly PEGylated IFN-α2a for 12 weeks. At week 72, patients receiving PEGylated IFN-α2a demonstrated a 69% response rate with undetectable levels of hepatitis C viral RNA (<100 copies/mL). The response rate was 28% in patients receiving the non-PEGylated form. The treatment completion rate was 84% (223 of 267 patients) in the once-weekly treatment group, whereas 60% (161 of 264) of patients completed treatment in the three times per week group (Zeuzem et al. 2000).

LAIs have the potential to reduce drug side effects. For example, exenatide, the first glucagon-like peptide-1 receptor agonist (GLP-1RA), received U.S. market approval in 2005 for the treatment of type 2 diabetes, under the brand name Byetta®. The formulation is administered by s.c. injection, twice daily. A long-acting microsphere formulation of exenatide (Bydureon®) was subsequently developed and approved in 2012 as a once-weekly s.c. injection. A head-to-head clinical trial demonstrated that once-weekly Bydureon® showed significantly greater improvements in hemoglobin A1c reduction than Byetta® given twice daily (Drucker et al. 2008). Furthermore, nausea and vomiting was less frequent with the once-weekly formulation than with twice daily version (DeYoung et al. 2011).

6.1.1.2 Commercial Considerations

LAIs and implant drug delivery technology platforms also offer significant commercial benefits (Box 6.1). A multibillion-dollar industry has developed that is provided by the benefits from extended drug release technologies. The products include both traditional pharmaceutical drugs and biotech drugs. Market reports estimate that sales for LAIs exceeded $8 billion (EvaluatePharma® 2013 worldwide sales; see also Table 6.5, which lists PEGylated drugs, with estimates of their worldwide sales). The therapeutic areas include dental disease, diabetes, drug addiction, growth disorders, oncology, mental health, and pain management (Table 6.1).

Sales for implants are also significant. For example, Zoladex® (goserelin acetate), an implant for prostate cancer treatment, had reached sales of $996 million in 2013 (EvaluatePharma). Other implant treatments include birth control, cardiovascular disease, and cancer (Table 6.2).

The complexities and costs of discovering and bringing new drugs to approval and to the market are ever increasing. The regulatory and drug safety requirements are also becoming more stringent. Life cycle management has become an integral part of the pharmaceutical industry to ensure maximal value for approved drugs. Life cycle management includes providing strategies to improve

TABLE 6.1

Examples of Marketed Sustained-Release Injectable Products

Area	Drug	Trade Name	Platform
Dental disease	Doxycycline hyclate	Atridox®	In situ forming depot
Diabetes	Exenatide	Bydureon®	Microspheres
Drug addiction	Naltrexone	Vivitrol®	Microspheres
Growth disorders	Octreotide	Sandostatin® LAR Depot	Microspheres
Oncology	Leuprolide acetate	Eligard®	In situ forming depot
Mental health	Paliperidone palmitate	Invega® Sustenna®	NanoCrystal®
Pain management	Bupivacaine	Exparel®	Liposome

TABLE 6.2

Examples of Marketed Implants

Therapeutic Area	Drug	Trade Name	Platform
Birth control	Levonorgestrel	Norplant® system	Silicone implant
Cardiovascular	Everolimus	Xience Xpedition™	Drug-eluting stent
Ocular	Dexamethasone	Ozurdex®	PLGA implant
Oncology	Carmustine	Gliadel® wafer	Polyanhydride copolymer
Oncology	Goserelin	Zoladex®	PLGA implant

efficiencies such as getting products to the market faster while reducing development costs. Included in the strategy for life cycle management is "upgrading" the existing product, such as daily oral or injectable, by developing a long-acting dosage form. Patent extensions can also be created through novel formulations as part of product life cycle management.

6.1.2   SELECTION OF THE TECHNOLOGY PLATFORM AND TYPE OF DELIVERY

A variety of different LAIs and implant technology platforms are available to effect controlled release of an active pharmaceutical ingredient (API) (Table 6.3).

TABLE 6.3

Examples of Parenteral Sustained-Release Technology Platforms

Platform	Technology
Long-acting injectables	Oil-based lipophilic depots
	In situ forming depots
	PEGylation
	PLGA microspheres
	Liposomes
	Polymeric nanoparticles
Implants	Pellets
	Implants based on an RCM
	Polymeric erodible compositions
	Osmotic pumps
	Drug-eluting stents

Each platform has unique properties. There are a number of factors that influence the choice of the drug delivery platform. Much of platform selection depends on the product requirement needs. In some cases, more than one type of platform is capable of meeting product requirement needs. However, care and attention to the capability and limits of each technology is critical for selection. For example, drug delivery duration requirements can limit the type(s) of technology that would be suitable:

- Liposomes and micelles have been generally limited to providing circulating drug for up to a few days.
- In situ forming depots can be tailored to provide release rates from a few days to months.
- Microsphere compositions can provide sustained drug release from a week to several months.
- Implants can provide sustained release ranging from months to years.

Other factors affecting the selection of the appropriate platform include the API type (small molecule, peptide, protein), chemical compatibility of the API with the platform, drug solubility, therapeutic indication, patient population, cost-benefits of the platform, understanding the patient and physician needs, manufacturing complexity, and commercial manufacturing costs. An additional decision criterion can be the development stage of the platform, i.e., platforms utilized in marketed products with regulatory approval vs. earlier-stage technologies. The final selection should be based on the knowledge and experience of the development team and other decision makers.

LAIs and implants are delivered via the parenteral route. They can be used for systemic, local, and targeted delivery. The majority of marketed LAIs and implants release drug from the administration site, to be ultimately absorbed into systemic circulation. The administration of the LAIs or implants into the s.c. or intramuscular (i.m.) space avoids the first-pass metabolism effect. This is particularly desirable for drugs that have significant susceptibility to first-pass metabolism that results in poor bioavailability.

In cases where local drug concentrations are required and the anatomical site is accessible, LAIs and implants have the ability to maintain high local concentrations for prolonged periods. High local drug concentrations with reduced systemic drug levels can potentially reduce undesirable systemic toxicity. Examples include treatments for ocular diseases, release of chemotoxic agents to tumor sites, inhibition of cardiovascular restenosis, and management of local pain. Several products for localized delivery have been commercialized, such as a dexamethasone implant for macular edema (Ozurdex®), implants delivering carmustine for malignant gliomas (Gliadel® wafer), and drug-eluting stents (DES, multiple active agents).

In situations where access to the target tissue/organ is not available or not optimal, technologies that provide drug targeting, such as the incorporation of antibodies or receptors for target ligands, offer additional multifunctional advantages. In cases where the therapeutic agents can be highly toxic or potency is limited, targeted delivery offers improved specificity and efficacy. As an example, antibody drug conjugates (ADCs) can widen the therapeutic window by targeting to tumor-specific or overexpressed cell surface antigens. ADCs have been marketed for Hodgkin lymphoma and anaplastic large cell lymphoma (Adcetris™) and HER2-positive metastatic breast cancer (Kadcyla™). Targeting ligands are described further in Chapter 5 (Section 5.3.2).

6.1.3 History

The history of LAIs and implants constitutes an important part of the general history of the field of controlled release, and the information in this section complements that of Chapter 1.

6.1.3.1 Compressed Steroid Pellets

While sustained parenteral drug delivery began to emerge as a clearly defined subarea of pharmaceutics in the middle of the twentieth century, there was appreciation of the utility of extended delivery via the parenteral route as early as the 1800s. By the 1930s, it was recognized that implanted pellets containing hydrophobic compounds could provide sustained release (Deanesly and Parkes 1937). Examples included pellets containing estradiol for the treatment of prostate cancer and pellets containing testosterone for the treatment of hormone deficiency (Chien 1982). A significant improvement occurred with advances in polymer technology, specifically the development of polymeric materials that possessed significant biocompatibility. Further, the development of the field has been significantly influenced by advances in pharmacokinetics (PK) and pharmacodynamics, which served to highlight the need for controlled, extended drug delivery and sustained drug plasma/tissue levels in achieving desired therapeutic responses.

6.1.3.2 Higuchi Model

The Higuchi model (Higuchi 1963; refined by Paul and McSpadden 1976) provided insight into the release kinetics of drug dispersed in an ointment base or dispersed in a polymer matrix. The model predicts that the drug delivery rate will decline with the inverse square root of time. This model was later applied to other matrix-based delivery systems.

6.1.3.3 Folkman Experiments

In the 1960s, while circulating rabbit blood inside an arteriovenous shunt of silicone rubber, Folkman discovered that if the tubing was exposed to anesthetic gases on the outside, the rabbits would fall asleep (Folkman et al. 1966). He proposed that short, sealed segments of such tubing containing a drug could form the basis of a constant-rate drug delivery system (DDS) and be implanted (Folkman and Long 1964). Further work in the 1960s and 1970s led to the establishment of the concept of the rate-controlling membrane (RCM)/drug reservoir delivery system as yielding a constant delivery rate and producing a zero-order, flat PK profile. A number of systems based on this underlying concept are described in Section 6.3.2.

6.1.3.4 Invention of PLGAs

Biodegradable polymers of poly(hydroxy acids) were investigated for use as sutures in the 1960s (Schmitt and Polistina 1967). This group of polymers encompasses poly(lactides) (PLA), poly(glycolides), and their copolymers, poly(lactide-*co*-glycolides) (PLGA). The relevant structures are given in Chapter 5 (Figure 5.11). PLGA systems degrade in vivo by hydrolysis of ester linkages into lactic acid and glycolic acid. The in vivo hydrolysis of PLGA polymers depends on the ratio of lactide to glycolide monomer units (Lactel 2014). The rate is fastest at approximately a 50:50 ratio and then decreases as either the proportion of glycolide or lactide unit increases (see, for example, Tables 6.4 and 6.6).

Investigation of PLGAs for drug delivery applications followed and has had a major impact on the field. Biodegradable polymers are attractive for drug delivery applications because of two potential major attributes: (1) if the polymer erodes only at the surface, then it would seem possible to design systems exhibiting a steady drug release rate, and (2) for parenteral applications, the system can be expected to completely erode, thereby eliminating the need for a procedure to remove the

TABLE 6.4
Eligard (Leuprolide Acetate) Drug and Polymer Composition

Dosing Interval	API Dose (mg)	Polymer DL-Lactide/Glycolide Molar Ratio
Monthly	7.5	50:50
3 months	22.5	75:25
4 months	30	75:25
6 months	45	85:15

system at the end of the delivery lifetime. In the late 1960s, workers at DuPont developed micropar-ticle and pellet depot delivery systems by mixing drugs with PLA (Boswell and Scribner 1973). In parallel, in the 1970s, researchers at Southern Research Institute and the University of Alabama at Birmingham were investigating steroid-containing PLGA microparticles for contraceptive drug delivery (Hoffman 2008). Currently, PLGAs are used for both LAIs (as in situ forming depots and microparticles) and bioerodible implants, as described in Sections 6.2 and 6.3.3, respectively.

6.1.3.5 Controlled-Release Era

The history of drug delivery was dramatically changed when Alza Corporation was founded in 1968 by Alejandro Zaffaroni (Hoffman 2008). At the time, Alza Corporation was unique because its business model was not to create new drugs but to better enable drug therapy by delivering the drug at the right place and at the right time. The ability to control drug release provided multiple benefits including reduced side effects, improved compliance, and better efficacy. The combination of a drug and delivery system had a profound impact not only in medicine but in all the ancillary sci-entific disciplines that were involved as an integral part of each delivery platform. Alza developed controlled-release systems based on RCM designs, which found applications in a variety of dif-ferent platform configurations. Ocusert® (pilocarpine) was the first system designed with an RCM and was approved in 1974. A second RCM system, Progestasert® (progesterone) was approved in 1976 as an intrauterine contraceptive (Barnhart 1985). A family of osmotic systems was developed by Dr. F. Theeuwes and colleagues at Alza in the 1970s (Theeuwes 1975, 1978). The ALZET®-implantable osmotic pump was developed for preclinical research (Theeuwes and Yum 1976), and DUROS®, an osmotically driven titanium implant, the size of a matchstick, was later developed for human use. Both technologies are described in detail in Section 6.3.4.

The OROS® (osmotic [controlled] release system) platform was also developed by Theeuwes and coworkers in the 1970s and 1980s. The OROS® is a solid oral tablet with a water-permeable membrane and laser-drilled hole (see also Chapter 1, Figure 1.7). As the tablet passes through the gastrointestinal tract, water is absorbed, creating osmotic pressure that pushes the active drug out. Approved products include Alpress® (prazosin), Acutrim® (phenylpropanolamine), Adalat® CR (nifedipine), Cardura® XL (doxazosin), Concerta® (methylphenidate HCl), Covera HS® (verapamil HCl), Ditropan XL® (oxybutynin chloride), DynaCirc CR® (isradipine), Efidac 24® (pseudoephed-rine), Glucotrol XL® (glipizide), Invega® (paliperidone), Jurnista® (hydromorphone), Metoros® (metoprolol), Procardia XL® (nifedipine), Sudafed® 24 Hour (pseudoephedrine), and Volmax® (sal-butamol). A number of other innovative controlled drug delivery technologies were also developed by Alza, including passive transdermal delivery, transdermal delivery by iontophoresis (E-Trans) (Phipps et al. 2002), and microneedles (Macroflux®) (Matriano et al. 2002; Cosman et al. 2010).

Several companies spun out from Alza, including Durect, Alexza, and Zosano Pharma. Durect Corporation was founded by Drs. F. Theeuwes and J. Brown and initially focused on applications of the DUROS® system. The ALZET® product line was subsequently acquired from Alza; other Durect Corporation drug delivery platforms include DURIN® (implant) (see Section 6.3.3), the sucrose ace-tate isobutyrate extended-release (SABER®) and CLOUD® injectable depot DDS (see Sections 6.2.2.3 and 6.2.2.4), and the nonparenteral DDS, ORADUR® (oral-controlled release with abuse resistance) and TRANSDUR® (transdermal drug delivery). Other major contributions to the field of controlled drug delivery with emphasis on LAIs and implants can be found in subsequent sections of this chapter.

6.2 LONG-ACTING INJECTABLES

6.2.1 Oil-Based Lipophilic Depots

If a drug can be dissolved (or dispersed) in an oil, then an injectable extended-release dosage form can potentially be designed and developed. Oil-based (lipophilic) depots tend to exhibit maximum utility with hydrophobic small molecule drugs, as there is correlation between drug release and the partitioning (and solubility) of drug between the oil phase and the external aqueous phase (i.e., the in vitro release medium or tissue fluid) (Larsen and Larsen 2009; Larsen et al. 2012).

Oil-based depots have been developed for the extended release of steroids (e.g., testosterone, estradiol) and antipsychotic drugs (e.g., fluphenazine, haloperidol) (Chien 1982; Larsen and Larsen 2009). Oftentimes, a hydrophobic ester of a drug has achieved success, especially for steroids. Oil-based depots are generally administered i.m., although s.c. administration and local administration are also possible.

Oleaginous vehicles utilized for these depots have included castor oil, corn oil, cottonseed oil, medium-chain triglycerides, olive oil, peanut oil, sesame oil, and fractionated coconut oil. Vehicles have also been gelled by the addition of aluminum monostearate and benzoic acid derivatives to the vehicles. Duration of drug release from oil-based depots ranges from several days to 6 weeks or longer.

The partitioning of drug between the oil-based depot and the external aqueous phase is described in terms of oil/water partition coefficients or distribution coefficients (Larsen et al. 2012). The release of drugs from oil-based depots has been modeled by Hirano and coworkers (Hirano et al. 1981). The release was analyzed in terms of the oil/water partition coefficient and mass transfer (boundary layer) coefficients. An exponential relationship with time resulted and was compared to the in vivo release of several compounds.

Several recent oil-based depots include Faslodex® (fulvestrant) injection, indicated for treatment of metastatic breast cancer, and Nebido® (testosterone undecanoate) injection, indicated for treatment of male hypogonadism as indicated by testosterone deficiency. Both formulations use a castor oil–based depot.

6.2.2 In Situ Forming Systems

In situ forming sustained-release formulations are liquids at ambient conditions that transform into a semisolid when injected into a target tissue site. These in situ forming systems have been used to deliver small molecules, peptides, and proteins. There are multiple formulation platforms that correspond to multiple mechanisms to achieve sustained drug release (Heller 2009; Wright et al. 2012). The compositions generally comprise a carrier and solvent(s) and are formulated as solutions or suspensions. This section provides examples of products that utilize several formulation platforms.

6.2.2.1 Atrigel®

Dunn and coworkers, at the Southern Research Institute in Birmingham, Alabama, developed some of the first marketed in situ forming systems (Dunn 2003; Dadey 2008). The technology was licensed as the Atrigel system (Atrix Laboratories) and comprises PLGA that is dissolved in *N*-methyl-2-pyrrolidone (NMP); drug is added to form a solution or suspension.

There are a number of marketed products based on the Atrigel technology. Atridox® (doxycycline hyclate) provides site-specific delivery of the antibiotic doxycycline hyclate, for the treatment of chronic adult periodontitis. Upon injection into the periodontal pocket, the liquid product solidifies, allowing for controlled release of doxycycline for a period of 7 days. Atrisorb® GTR barrier does not contain a drug but allows for the provision of a custom-fitted gel barrier, which promotes regeneration of periodontal tissue.

The Atrigel technology platform is also used in Eligard® s.c. injection, which is a suspension of leuprolide acetate for the treatment of prostate cancer. By varying the ratio of DL-lactide to glycolide of the PLGA copolymer, four different controlled-release profiles are possible, allowing a dosing interval that can range from 1 to 6 months (Table 6.4).

6.2.2.2 Alzamer® Depot™ Technology

The Alzamer formulation comprises a drug mixed with a PLGA polymer, dissolved in a low-water-miscible solvent. On administration, the gradual loss of the solvent from the depot and the concurrent ingress of water or interstitial fluid change the physical form in situ, from a viscous gel to a finely porous spongelike structure. Initial drug release occurs through diffusion from the depot. The solvent systems selected have low water miscibility (e.g., benzyl benzoate), which slows down phase inversion and thus significantly reduces porosity of the depot. The reduced porosity slows down drug release, thereby minimizing burst effects. Further drug release is facilitated by PLGA degradation.

PLGA polymers are favored with lactide/glycolide ratios of 50:50. Drug loading ranges from 5% to 30%. By careful selection of the polymer, solvent, and drug characteristics, near zero-order delivery profiles have been demonstrated in vivo with acceptable local and systemic tolerability. Drug delivery durations can vary from days to months. The platform has been evaluated for small molecules, peptides, and proteins (Brodbeck et al. 1999; Chen and Junnarkar 2008).

6.2.2.3 SABER® Injectable Depot

The SABER® depot technology (Durect Corp.) consists of a biodegradable, high-viscosity, nonpolymeric liquid carrier material, formulated with one or more pharmaceutically acceptable solvents and other excipients. Sucrose acetate isobutyrate (SAIB) is a high viscosity agent that has been used as a food additive approved in over 40 countries (Reynolds and Chappel 1998). The use of SAIB as a pharmaceutical excipient is being explored in several investigative human drug products, and SAIB is a component in an FDA-approved veterinary injectable product for the controlled release of the peptide deslorelin (Sucromate™).

Solvents in the SABER® system may include ethanol, N-methyl-2-pyrrolidone (NMP), dimethyl sulfoxide (DMSO), benzyl alcohol, benzyl benzoate, and others. Upon injection, the water-miscible solvents (e.g., ethanol or DMSO) diffuse out of the depot, resulting in an increased in situ viscosity. Water-immiscible solvents (e.g., benzyl benzoate) diffuse at a slower rate: the selection of appropriate solvents thus allows flexibility in the control of the drug release rate. Additives such as biodegradable polymers (e.g., PLA and PLGA) further moderate the levels of drug release. SABER® formulations are compatible with small molecules, peptide, and proteins, for durations ranging from days to months (Okumu et al. 2002; Shah et al. 2014). The formulations can be solutions or suspensions and are used for both systemic and local delivery (Wright et al. 2008).

POSIMIR™ (Durect Corp.) is a SABER® formulation for the local administration of the anesthetic agent bupivacaine, for postoperative pain treatment. Administered to the surgical site, it is designed to provide 72 hours of local anesthesia/analgesia. POSIMIR™ comprises bupivacaine, SAIB, and benzyl alcohol. Clinical studies demonstrated 72 hours of sustained bupivacaine delivery, without initial drug burst, in a variety of surgical models (Figure 6.1) (Shah et al. 2014). SABER®-bupivacaine was administered to a variety of injection/instillation sites, in a dose-linear fashion. The absorption of bupivacaine was rapid, measurable within 0.5–1 hour, with a gradual increase in concentration, without initial burst. The differences in the PK profiles were attributable to the different surgeries evaluated and are thought to be associated with differences in local blood perfusion, type of tissue, and patient-to-patient variability. PK was dose linear with administration volumes ranging from 2.5 to 7.5 mL. The bioavailability of bupivacaine was 100%. The product is currently in Phase 3 clinical trials.

Relday™ (Zogenix 2013) is being developed as a once-monthly SABER® depot injection of risperidone, for the management of schizophrenia. Positive single-dose safety and PK profiles from a Phase 1 clinical trial of Relday have been reported. Therapeutic blood concentrations were achieved on the first day of dosing and were maintained throughout the 1-month period. Dose proportionality was demonstrated (Zogenix 2013).

Intra-articular injection of a recombinant human growth hormone SABER® formulation in dogs demonstrated sustained high local concentrations of the protein in the synovial joint for at least 14 days as a potential treatment for osteoarthritis or knee trauma (Sekar et al. 2009). The SABER® system has been evaluated for a variety of other compounds, including small molecules such as steroids and neoplastic agents, peptides, and proteins (Wright et al. 2008).

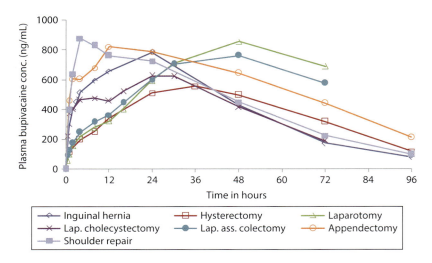

FIGURE 6.1 Pharmacokinetics of SABER®-bupivacaine across different surgery types. (Courtesy of Durect Corporation.)

6.2.2.4 CLOUD® Injectable Depot

Durect Corporation is also developing the CLOUD® injectable depot platform. This technology involves complexation of the API as an additional rate-controlling element. It is well suited for APIs with good aqueous solubility, confers additional API stability, provides a broader control over biodegradation characteristics, and involves low-formulation viscosities, which potentially allows for injection with finer needles.

6.2.3 PEGYLATED PEPTIDES/PROTEINS

PEGylation is the conjugation of hydrophilic polyethylene glycol (PEG) with the API, e.g., a protein, peptide, or small molecule. The goal of PEGylation is to modify the drug PK, i.e., extending the circulating half-life while maintaining the biological activity. The technology was pioneered in the 1950s and 1960s for protein modifications (Davis 2002). PEG is water soluble, nontoxic, and FDA approved. PEGylation increases the hydrodynamic volume, slows proteolysis, and reduces renal drug clearance. Drug solubility in aqueous and organic environments can also be increased with PEG, and toxicity and immunogenicity to the API is generally reduced (Veronese and Pasut 2005; Zhang et al. 2012).

PEG-adenosine deaminase (Adagen®) was the first approved product, developed by Enzon. PEG is covalently modified to the adenosine deaminase enzyme for the treatment of severe combined immunodeficiency disease. A variety of PEGylated therapeutic agents have now been commercialized, for a variety of treatments including immunodeficiency, cancer, inflammation, and ocular disease; they account for greater than $7 billion in sales (Table 6.5). The current PEGylated products are all large molecules. There are ongoing efforts for PEGylation of small molecules, and although no product has yet been commercialized, several late-stage clinical trials are underway (Riley and Riggs-Sauther 2008; Li et al. 2013; Nektar 2014). PEGylation technology can also be applied to the preparation of long-circulating liposomal and other nanoparticle DDS, as described in Sections 6.2.5 and 6.2.6.

6.2.3.1 PEG Conjugation

This section provides some of the main considerations on the technology. PEG can be linear or branched and sizes have ranged from 5,000 to greater than 30,000 Da. PEGylation can be

TABLE 6.5
FDA-Approved PEGylated Drugs

Trade Name	Drug Name	Company	PEG Shape/ Size	Indication[a]	Year of Approval	Sales ($,B) 2013 WW[b]
Adagen®	Pegademase	Enzon	Multiple linear 5,000	SCID	1990	0.048
Oncaspar®	Pegaspargase	Enzon	Multiple linear 5,000	Leukemia (ALL, CML)	1994	0.055
PEG-Intron®	Peginterferon alpha-2b	Schering-Plough/ Merck & Co.	Linear 12,000	Hepatitis C	2000	0.496
Pegasys®	Peginterferon alpha-2a	Hoffmann-La Roche	Branched 40,000	Hepatitis C	2001	1.146
Neulasta®	Pegfilgrastim	Amgen	Linear 20,000	Neutropenia	2002	4.392
Somavert®	Pegvisomant	Pharmacia & Upjohn/Pfizer	4–6 linear 5,000	Acromegaly	2003	0.217
Macugen®	Pegaptanib	Pfizer/Valeant Pharmaceuticals International	Branched 40,000	Age-related macular degeneration	2004	0.009
Mircera®	mPEG-epoetin	Hoffmann-La Roche/Chugai	Linear 30,000	Anemia associated with chronic renal failure	2007	0.459
Cimzia®	Certolizumab pegol	UCB	Branched 40,000	Reducing signs and symptoms of Crohn's disease	2008	0.829
Puricase1®/ Krystexxa®	PEG-uricase	Savient	10,000	Gout	2010	0.026

Source: Modified from Li, W. et al., *Progr. Polym. Sci.*, 38, 421, 2013.

[a] ALL, acute lymphoblastic leukemia; CML, chronic myeloid leukemia; SCID, severe combined immunodeficiency disease.

[b] Worldwide sales.

heterogeneous, i.e., multiple sites on the molecule can be PEGylated. PEGylation can be reversible or nonreversible. Peptides or proteins that are covalently bound with PEG are considered to be new chemical entities (NCEs).

PEGylation can be achieved through various platforms such as chemical modification, enzymatic methods, and genetic engineering. A critical factor of PEGylation technology is to maintain biological activity of the API. As a consequence, the type of PEGylation platform to use should be carefully considered in development. Although PEGylation can impact the potency of the molecule, this can be offset by the benefit of the increased circulating half-life. There are a number of good reviews on this topic (Veronese and Pasut 2005, 2012).

Chemical modification of the API with PEG, under mild conditions, is relatively well established and can produce high yields. However, the reactions are not highly specific and can lead to losses in bioactivity. The common reactive groups in polypeptides are nucleophiles (thiols, alpha-amino groups, epsilon-amino groups, carboxylate and hydroxylate groups) (Veronese and Pasut 2005). Nonspecific PEGylation runs the risk of a potential loss of biological activity. Site-specific PEGylation methods/technologies are thus emerging, in order to improve bioactivity. Site-specific conjugations avoid random PEGylation, resulting in a single species. The process can be optimized for both functional activity and half-life. Examples include a transglutaminase method and functionalization of maleimide PEG (Gong et al. 2011).

6.2.4 PLGA MICROSPHERES

As outlined in Section 6.1.3.4, research into PLGA microspheres began in the late 1960s and 1970s. The development of biodegradable microspheres based on PLGA has been one of the major successful applications of LAIs and implants. The literature on microspheres is quite extensive and the reader is referred elsewhere for more detailed discussions (Wang and Burgess 2012; Gombotz and Hoffman 2013).

Microspheres can range in size from 1 to 999 μm and may have either a continuous polymeric matrix structure wherein the drug is homogeneously dispersed or a shell-like wall surrounding the drug reservoir/core. PLGA microspheres may have some inherent porosity, and porosity can develop during drug delivery. They are almost always suspended in an aqueous vehicle, prior to administration through a small gauge needle.

As noted earlier, PLGA systems degrade in vivo by hydrolysis into lactic acid and glycolic acid. While initial expectations were that PLGA microspheres would surface erode and therefore release drug at a nearly constant rate, for a number of systems, it was found that surface-available drug was released quickly in an initial burst. The drug would then not diffuse appreciably through the dense PLGA matrix ("lag time") until the PLGA molecular weight declined to a critical value. At this point, the diffusivity in the microspheres increased substantially, coupled with the loss of physical integrity, resulting in a second, large period of drug release. PLGA degradation can be autocatalytic, as acid moieties accumulate during polymer hydrolysis. Drug release is considered to be a complex phenomena involving diffusion of the drug in a manner dependent on the degradation of the PLGA, polymer erosion, water sorption into the microspheres, diffusion of polymer degradation products, and (aqueous) pore formation in the microspheres. Mathematical modeling of drug delivery from PLGA microspheres has recently been reviewed (Versypt et al. 2013).

6.2.4.1 Manufacture of PLGA Microspheres

There are a number of processes for the manufacture of PLGA microspheres (Wang and Burgess 2012) including the following:

Solvent evaporation/extraction: In this process, the PLGA is dissolved in an organic solvent, and the drug is then dissolved or dispersed in the polymer/solvent solution. This mixture is emulsified in an aqueous solution (or other nonsolvent), and the solvent is removed by evaporation/extraction. The resultant microspheres may be harvested by filtration or centrifugation and further processed. Particle size of the microspheres can be controlled by processing parameters (Freitas et al. 2005).
Coacervation: In this process, drug particles are dispersed in a PLGA/solvent solution. Coacervation can be induced by several methods, usually by addition of a nonsolvent, resulting in deposition of the polymer around the drug particle.
Prolease process: In this process, the drug particles are dispersed into a PLGA solution, which is then atomized into a liquid nitrogen/ethanol bath. The droplets are transformed into microspheres, which can be further processed (Johnson and Herbert 2003).

The processes for microsphere manufacturing do not result in the encapsulation of all the drug, and for commercialization, attention must be paid to the overall process yield of encapsulated drug.

6.2.4.2 PLGA Microsphere Commercial Products

A number of PLGA-based LAI microsphere products have been developed and commercialized for applications ranging from prophylactic treatment of prostate cancer, to growth hormone deficiency,

TABLE 6.6
Examples of Marketed Poly(Lactide-*co*-Glycolide) Microsphere Products

Product	Company	Drug	Indication	Duration
Lupron Depot®	Abbvie	Leuprolide	Prostate cancer	1, 3, 4, and 6 months
Trelstar® Depot	Debiopharm/Actavis	Triptorelin	Prostate cancer	4, 12, and 24 weeks
Sandostatin® LAR	Novartis	Octreotide	Acromegaly	4 weeks
Vivitrol®	Alkermes	Naltrexone	Alcohol dependence and opioid dependence relapse	1 month

to alcohol and opioid dependency (Table 6.6). By modifying polymer composition, molecular weight, and ratio of lactide to glycolide monomers, a wide range of release profiles can be obtained, from weekly, to as long as 6 months (Table 6.6).

Risperdal® Consta® is a further commercial example. This PLGA microsphere formulation is a prolonged-release suspension for the i.m. delivery of risperidone, for the treatment of schizophrenia and bipolar disorder. This formulation exploits the PLGA degradation pattern of initial release of surface drug (first day), followed by a lag time of approximately 3 weeks, followed by a period of substantial release of drug (approximately 3 weeks). By timing injections at 2-week intervals and administering oral drug for the first several weeks of therapy, sustained plasma levels are achieved in patient populations (Ereshefsky and Mascarenas 2003).

As outlined in Section 6.1.1.1, exanatide was originally marketed as a twice daily s.c. injection. A long-acting microsphere formulation of exenatide was subsequently developed (Bydureon®), based on 50:50 PLGA. This PLGA microsphere formulation only requires a once-weekly s.c. injection. A further PLGA microsphere product, Nutropin® depot, delivered the protein human growth hormone (somatropin) for 2–4 weeks. It was approved by the FDA and marketed by Genentech, but distribution was ceased for commercial reasons.

PLGA microspheres are also used for targeted or local delivery of drugs to specific body sites, if the site is accessible for injection or other administration. One such example is in the dental area for the treatment of periodontitis. Arestin® microspheres (Orapharma) are PLGA microspheres that contain the antibiotic, minocycline HCl. The microspheres are for subgingival administration into periodontal pockets around affected teeth, as an adjunct with scaling and root planning.

6.2.5 Liposomes

As described in detail in Chapter 5 (Section 5.5.1), liposomes are colloidal suspensions consisting of vesicular bilayer structures, which are generally composed of phospholipids and cholesterol. Liposomes offer drug delivery and drug-targeting capabilities. Although discussed here as LAIs, it should be remembered that liposomes are generally limited to providing circulating drug for up to a few days. They are currently unable to provide therapeutic drug levels for a prolonged period (for example, a week or longer).

Their classification can be based on structure, method of preparation, composition, and application. They can be processed to different sizes, composition, charge, and lamellarity (Immordino et al. 2006; Xu and Burgess 2012). Because of their chemical properties, they can incorporate lipid- and water-soluble drugs, including small molecules, peptides, and proteins. In some cases, liposomes can improve drug solubility. Liposomes can also confer (1) protection from degrading enzymes, improving drug stability, (2) improved drug potency for therapy, (3) reduced toxicity, (4) drug-targeting possibilities, and (5) modified PK (e.g., long-acting delivery properties). However, additionally, liposomes can be disrupted by high shear (injection through very fine needles) and temperature excursions during storage. Application of liposomes includes oncology, infectious disease vaccine delivery/adjuvants, pain management, dermatology, and eye disorders. Marketed liposome–based products include treatments for fungal infections, vascular disease, cancer, and pain management. The clearance of liposomes is determined by a number of factors, including their physicochemical properties, membrane fusion events, rate of drug leakage, uptake of liposomes by the reticular endothelial system (RES; also called the mononuclear phagocyte system), mechanical filtration, and interactions with specific serum proteins (e.g., opsonins) (Ishida et al. 2002).

A challenge of liposomes injected intravenously (i.v.) is that they are phagocytosed by the RES and accumulate in the liver and spleen. Circulating liposomes interact with high- and low-density lipoproteins and also result in rapid release of the encapsulated drug into the blood. The addition of saturated phospholipids and cholesterol helps to overcome some of the limitations (Immordino et al. 2006). Further research on the technology led to a significant step in longer-circulating liposomes by the incorporation of PEG into the composition, resulting in sterically stabilized, PEGylated liposomes. PEG has been widely used as a steric stabilizer (Chiu et al. 2001) and, as described in Section 6.2.3 for protein PEGylation, confers other useful properties such as good biocompatibility, improved solubility, low toxicity, and low immunogenicity (Veronese and Pasut 2012; Zhang et al. 2012).

Woodle and Martin developed PEGylated liposomal doxorubicin (Doxil® in the United States, Caelyx® in Europe) (Hoffman 2008). Doxil® is the only approved Stealth (avoids detection by immune system) liposomal product for treatment of ovarian cancer, AIDS-related Kaposi's sarcoma, and multiple myeloma. The product is administered by slow i.v. infusion. Stealth liposomes can passively target tumor sites by preferential accumulation into the tumor sites that have higher vascular permeability than normal tissue.

A local sustained-release liposome formulation delivering bupivacaine (Exparel®) was approved by the U.S. FDA in 2011 for local postsurgical anesthesia based on clinical studies in bunionectomy or hemorrhoidectomy. This liposomal formulation of bupivacaine is injected into soft tissue to produce local analgesia. The reported half-life of liposomal bupivacaine after a single dose was 23.8 hours in hemorrhoidectomy and 34.1 hours in bunionectomy clinical studies (Exparel 2012), as compared to 2.7 hours for bupivacaine HCl that is formulated in an isotonic aqueous buffer.

6.2.6 LONG-ACTING NANOPARTICLES

Nanoparticles, i.e., particulate drug carriers that have mean particle diameters below 1000 nm, are described in detail in Chapter 5 (Section 5.5.5). Such systems include those made from synthetic polymers such as PLGA (e.g., the Accurins™ technology; see Table 6.8) and poly(alkyl cyanoacrylates), as well as natural polymers such as albumin, gelatin, and chitosan. Similar to liposomal DDS, many are used in conjunction with PEG technology, in order to achieve extended circulation times, avoid RES capture, and facilitate uptake in sites exhibiting enhanced permeability, such as tumor sites. Although polymeric nanoparticles do provide extended circulation times in comparison with a free drug, the times are still limited to hours and days, rather than the longer time frames possible with microsphere-based systems. Thus, polymeric nanoparticles tend to be studied more for their ability to facilitate drug targeting to an active site, rather than to achieve extended release.

As described in Chapter 3 (Section 3.2), nanocrystals are drug crystals that have been size reduced to the nanometer range (*Note*: NanoCrystals® specifically denote a patented size-reduction technology of Elan Corporation). Dispersions of nanocrystals in liquid media are known as nanosuspensions; the drug crystals are stabilized by polymers and/or surfactants. Production of nanocrystals utilizes the basic principles of milling, precipitation, homogenization, or a combination(s) of these processes. This perceived simple process addresses a major issue for many drugs that are poorly soluble. Particle size reduction increases surface area for greater dissolution, increased drug solubility, and enhanced bioavailability. The particle characteristics are optimized based on target parameters including the desired PK profile and product shelf life. The first marketed products utilizing nanocrystal structures were oral dosage forms (Junghanns and Müller 2008). Parenteral application followed, including Invega® Sustenna® for schizophrenia in 2009. The product composition is a suspension of paliperidone palmitate NanoCrystals® that slowly dissolve after i.m. injection and then hydrolyze into the active moiety, paliperidone, followed by systemic absorption. Initial dosing is required on day 1 and a week later; maintenance doses are monthly (Kim et al. 2012).

6.3 IMPLANTS

Implantable systems can deliver either systemic or local therapy. The systems can be either nonerodible or bioerodible. While nonerodible systems generally require removal or explantation at the end of the delivery lifetime, this attribute is, conversely, a safety feature in that the system can be removed if the need arises to terminate therapy prior to the end of the delivery lifetime.

6.3.1 PELLET SYSTEMS/ANIMAL HEALTH

As outlined in Section 6.1.3, implants of steroid compounds were early sustained-release systems. There are a number of compressed pellet systems that have been utilized in animal health, mainly steroid based, including

- Ralgro® (zeranol)
- Synovex® (estradiol benzoate)
- Revalor® (estradiol-17beta + trenbolone acetate)
- Finaplix® (trenbolone acetate)

These implanted pellet systems have durations of 60–140 days (Ferguson 2001), due to the very slow dissolution and uptake of the compressed steroids.

An additional animal health product is the Compudose® implant that delivers estradiol over periods in excess of 200 days. The Compudose® implant comprises a cylindrical implant that consists of an outer layer of silicone rubber matrix, containing dispersed estradiol, overlaying an inner inert core of silicone rubber (Ferguson et al. 1988; Ferguson 2001). This design, featuring a thin layer of drug-containing matrix and a relatively thick drug-free core, minimizes the tailing in the drug release profile that is typically seen with matrix-release systems.

6.3.2 IMPLANTS BASED ON A RATE-CONTROLLING MEMBRANE

RCM implants consist of an inner core in which drug is dissolved or dispersed and an outer RCM layer, through which drug can diffuse (see also Chapter 1, Figure 1.3). As long as the drug concentrations at the inner and outer membrane surfaces are constant, a constant rate of drug release can be achieved. The first RCM products were insertable DDS including the Ocusert® (delivery of pilocarpine for the treatment of glaucoma from a system placed in the cul-de-sac of the eye) and the Progestasert® (delivery of progesterone to the uterus for contraception), both of which were developed by Alza Corporation (Hoffman 2008).

The Vitrasert® implant releases ganciclovir for the treatment of cytomegalovirus retinitis. The system is implanted intravitreally. The Vitrasert® consists of a compressed tablet of ganciclovir, overcoated with poly(vinyl alcohol) (PVA), further partially overcoated with poly(ethylene vinyl acetate) (PEVA) and then affixed to a PVA suture stub. The system is designed to release the drug over 5–8 months (Lee et al. 2008). An extensive array of other ocular implants is described in Chapter 13.

The Norplant system is a 5-year contraceptive system consisting of the hormone levonorgestrel encapsulated into six thin, flexible silicone capsules (Silastic tubing) for s.c. implantation on the inside of a woman's upper arm. The silicone rubber copolymer serves as an RCM. The system has been approved in many countries and was approved by the U.S. FDA in 1990. Later in the United States, the 6-rod Norplant system became associated with removal problems and marketing in the United States ceased in 2002 (Kleiner and Wright 2013).

The Implanon® system consists of a single-rod system, 4 cm long and 2 mm wide, which makes for easier insertion and removal than the Norplant system. The system has a PEVA core containing 68 mg of etonogestrel (reservoir). The core is covered with a PEVA RCM and is designed to release drug to prevent ovulation for a 3-year period. The Implanon® system was first approved (ex-US) in 1998 and received U.S. approval in 2006 (Alam et al. 2008).

6.3.3 ERODIBLE IMPLANTS

The advantage of using erodible implants is that a surgical procedure is not required to remove the implant at the end of the product lifetime. The PLGAs are widely used as bioerodible implants. The Zoladex® implant is indicated for the palliative treatment of advanced carcinoma of the prostate and breast cancer. The implant contains the peptide goserelin acetate, a potent synthetic peptide analog of gonadotropin-releasing hormone (GnRH), dispersed in a matrix of PLGA (Hutchinson and Furr 1985, 1990; Dijkman et al. 1990). The Zoladex® implant is in the form of a cylinder. The Zoladex® 3.6 mg implant is implanted s.c. with a 16 G needle and releases drug over a period of 28 days; the Zoladex® 10.8 mg implant is implanted s.c. with a 14 G needle and releases drug over a 3-month period. The encapsulated drug is released by a combination of diffusion- and erosion-controlled mechanisms.

DURIN® (Durect Corp.) is an implant that incorporates either biodegradable (PLGA polymers) or nonbiodegradable polymers. The implant either can be a drug-loaded monolithic design or can incorporate an RCM around a drug reservoir core. The RCM is created through a coextrusion manufacturing process, consisting of different polymer types between the inner core and outer RCM, of a uniform rod-shaped implant. In addition to the RCM, drug diffusion and release is controlled by drug content, polymer type and composition, and other excipients such as permeation enhancers and stabilizers (Gibson et al. 2012).

A biodegradable polyanhydride copolymer was developed for the Gliadel® wafer implant, for the treatment of brain tumors (Brem and Langer 1996). Gliadel wafers are small biodegradable polyanhydride disks, 1.45 cm in diameter and 1.0 mm thick, designed to deliver the chemotherapeutic drug, BCNU (carmustine), directly into the surgical cavity created after the tumor (high-grade malignant glioma) has been surgically excised. Up to eight wafers may be implanted along the walls and floor of the cavity that contained the tumor. This system was approved by the FDA in 1996. The biodegradable polyanhydride copolymer in the Gliadel® wafer is trade named Polifeprosan 20 and consists of poly[bis(p-carboxyphenoxy)propane/sebacic acid] in a 20:80 molar ratio. The chemical structure of Polifeprosan 20 is shown in Figure 6.2.

Upon exposure to the aqueous environment in the surgical cavity, the anhydride bonds hydrolyze to release carmustine and the polymer constituents carboxyphenoxypropane and sebacic acid. More than 70% of the copolymer biodegrades within 3 weeks back to constituent monomers. The Gliadel® wafer was investigated clinically by Brem and coworkers (Brem et al. 1991) and is associated with an approximately 2-month improvement in median survival in the treatment of newly diagnosed malignant glioma (Perry et al. 2007).

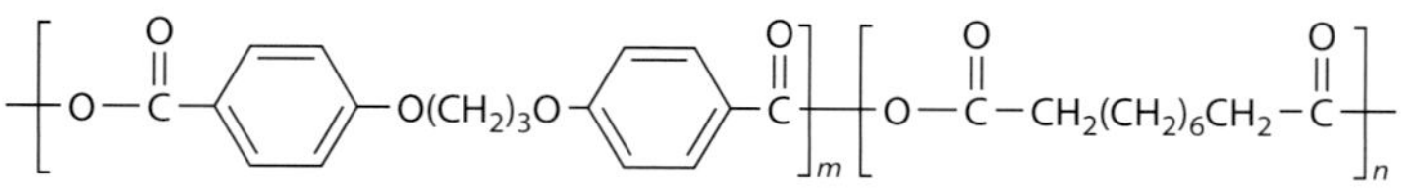

FIGURE 6.2 Polyanhydride random copolymer (Polifeprosan 20) chemical structure (ratio $n:m$ = 20:80).

6.3.4 OSMOTIC PUMPS

Osmosis is the movement of water through a semipermeable membrane in response to high solute (usually ionic) concentration on the opposing side of the membrane. If the influx of water displaces (pumps) an equal volume of drug solution or suspension from the delivery system, steady, zero-order delivery can be achieved. Osmotic drug delivery systems for oral, animal research, veterinary, and human delivery have been intensively investigated and developed (Eckenhoff et al. 1987; Zingerman et al. 1997; Wong et al. 2003; Wright et al. 2003).

The osmotic transport of water can be described by the following equation:

$$\frac{dV}{dt} = \frac{A}{l} L_p [\sigma \Delta \pi - \Delta P] \tag{6.1}$$

where
   dV/dt is the volume flux of water across the membrane
   $\Delta \pi$ and ΔP are, respectively, the osmotic and hydrostatic pressure differences across the semipermeable membrane
   L_p is the membrane mechanical permeability coefficient
   σ is the reflection coefficient (usually ≈ 1)
   A and l are, respectively, the membrane area and thickness (Theeuwes and Yum 1976)

Equation 6.1 can be utilized to mathematically describe the delivery rate from an osmotic drug delivery system (with assumptions including $\sigma = 1$ and negligible ΔP):

$$\frac{dM}{dt} = \left(\frac{A}{l}\right) L_p [\Delta \pi] c \tag{6.2}$$

where
   dM/dt is the delivery rate of drug
   c is the drug concentration

From Equation 6.2, it can be appreciated that if the osmotic pressure and the drug concentration are held constant, a constant rate of drug delivery will result. Osmotic pressure can be held constant by utilizing saturated solute solutions with excess solid solute. Some examples of osmotically driven implants are described below.

6.3.4.1 ALZET® Implantable Osmotic Minipumps

The ALZET® osmotic pump was developed for implantable drug delivery to laboratory animals (Theeuwes and Yum 1976; Culwell et al. 2012). It has become a popular research tool for continuous delivery of a wide range of test agents, including small molecules, peptides, proteins, nucleic acids, and lipids, and is cited in over 15,000 research publications. The ALZET® pump has a cylindrical shape and is composed of three concentric layers: the inner drug reservoir, the osmotic layer, and the outer RCM (Figure 6.3). The flexible, inner reservoir is molded from a synthetic elastomer. The osmotic layer contains sodium chloride (NaCl). The outermost, rate-controlling semipermeable membrane is formulated with a blend of cellulose esters. An additional component of the ALZET® system is the flow moderator, a 21-gauge stainless steel tube with a plastic end cap, which prevents random drug diffusion.

Operationally, the pump is filled with drug solution and the flow moderator is inserted through the delivery port. When placed in an aqueous medium (or implanted in an animal), water is transported across the semipermeable membrane into the osmotic layer. The osmotic layer expands,

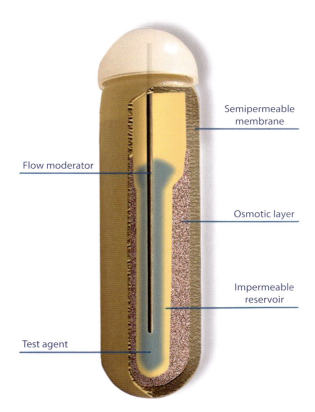

Semipermeable membrane

Flow moderator

Osmotic layer

Impermeable reservoir

Test agent

FIGURE 6.3 Cross section of the ALZET® osmotic pump. (Courtesy of Durect Corporation.)

exerting pressure on the flexible reservoir and pumping the drug solution through the flow moderator and out via the delivery portal. Various ALZET® pump sizes have been developed, to provide
researchers with a wide range of delivery rates to choose from (ALZET 2014).

The ALZET® pump can be implanted s.c. or intraperitoneally in unrestrained animals as small
as mice and young rats; it has also been used in a variety of other animal species. The pump is particularly useful for continuous delivery of compounds with short half-lives, which are notoriously
difficult to reliably deliver by conventional dosing regimens. The pump eliminates the need for
repeated animal handling for dosing, thus minimizing unnecessary animal stress and experimental
variables, which can interfere with study results. Local or targeted delivery can be achieved either
by implantation directly at the desired site of action or via attachment to a catheter delivering to
the desired site of action, such as blood vessels, the spinal cord, or into the cerebral ventricles and
brain parenchyma. Another important capability of the ALZET® pump is the possibility of delivering solutions in an intermittent fashion, which is achieved by loading alternating series of drug and
placebo solutions within the catheter (Culwell et al. 2012).

Some interesting examples of the use of the ALZET® pump in preclinical research include their
use in the evaluation of the following:

- The impact of dosing schedules on the anticancer effects and pulmonary toxicity of bleomycin, in an animal model of lung cancer (Sikic et al. 1978).
- The in vivo efficacy of the angiogenesis inhibitor, anginex, to inhibit tumor growth in
  mouse xenograft models of ovarian carcinoma (Dings et al. 2003).
- The anti-obesity effects of continuous infusion of fibroblast growth factor 21 in diet-
  induced obese mice (Coskun et al. 2008).

- Intracerebroventricular (ICV) infusion of the tricyclic pyrone CP2 in a transgenic mouse model of Alzheimer's disease. The 5X FAD mouse strain is known to manifest the Alzheimer's phenotype with generous deposition of β-amyloid. A 2-week ICV infusion of CP2 with Alzet pumps reduced β-amyloid 1–42 in brain tissue by 40%–50% in comparison with the control (Hong et al. 2009).
- The role of the cerebellum in a rat model of noise-induced chronic tinnitus. Animals with established tinnitus had the paraflocculus of the cerebellum either surgically removed or inactivated by chronic lidocaine infusion into the subarcuate fossa using ALZET® pumps. Lidocaine treatment via ALZET® pumps gradually eliminated tinnitus over a 14-day infusion period, but these effects were reversed after a 6-week washout period (Bauer et al. 2013).

6.3.4.2 DUROS® Implantable Osmotic System

A schematic of the DUROS® implantable osmotic system is shown in Figure 6.4. The system consists of an outer cylindrical reservoir, an RCM at one end, the "osmotic engine" (tablets containing primarily NaCl, combined with other pharmaceutical excipients), an elastomeric piston, the drug formulation in the drug reservoir, and an exit port at the other end (see diffusion moderator/orifice in Figure 6.4). The outer reservoir can be constructed from titanium alloy or a biocompatible rigid polymer. The RCM is composed of a specially designed semipermeable polyurethane polymer. The drug formulation consists of the drug dissolved or suspended in pharmaceutically acceptable vehicles or solvents. The drug must remain stable in the formulation for the duration of delivery; accordingly, the formulation components are specific to the properties of the drug being delivered.

DUROS® dimensions can range from as small as several millimeters in outside diameter (OD) to 10 mm OD × 60 mm in length. Resulting volumes available for the drug formulation (drug reservoir volume) range from less than 100 µL to slightly more than 1 mL.

When implanted, water is osmotically imbibed from extracellular fluid into the osmotic engine at a rate controlled by the hydraulic permeability of the semipermeable membrane. The osmotic engine expands at a controlled rate, resulting in the displacement of the piston and delivery of the drug formulation at a controlled rate through the diffusion moderator and orifice, as described by Equation 6.2. Sufficient NaCl is usually present in the osmotic engine so that a saturated solution of NaCl is present in the osmotic engine compartment throughout the full delivery stroke of the piston.

The volumetric delivery rate from the DUROS® system depends on the membrane permeability, which, in turn, depends on the choice of the material used for the membrane or, in the case of a copolymer such as polyurethane, the copolymer composition. Additionally, the volumetric delivery rate of the pump can be adjusted by the membrane surface area and the membrane thickness. The resulting delivery rate is then fixed for a given pump design; different doses can be provided by different formulation concentrations.

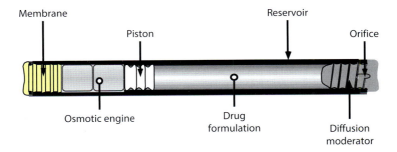

FIGURE 6.4 Schematic of DUROS® implantable osmotic system. (Reprinted from *J. Control. Release*, 75, Wright, J.C., Leonard, S.T., Stevenson, C.L. et al., An in vivo/in vitro comparison with a leuprolide osmotic implant for the treatment of prostate cancer, 1–10. Copyright 2001, with permission from Elsevier.)

DUROS® systems can be implanted s.c. in body sites where there is sufficient extracellular water and where there are no biomechanical limitations on implant placement or patient comfort restrictions. For systemic delivery, a preferred site is the s.c. space of the inside of the upper arm. Implantation is accomplished under local anesthesia; a small skin incision is made and a trocar, or a custom designed implanter can be used for implant insertion (Wright et al. 2003). For site-specific delivery, the DUROS® system can be implanted directly at the site, or the system can be implanted at an accessible location and a catheter tunneled to the desired site of action. Implant removal (explantation) is accomplished under local anesthesia. A small incision is made at one end of the implant site and in the thin capsule surrounding the implant, and the implant is pushed out.

The Viadur® implant was the first approved product, in March 2000, to incorporate the DUROS® implant technology. The implant contains the peptide drug leuprolide acetate and was developed for the palliative treatment of advanced prostate cancer (Wright et al. 2001). Suppression of circulating testosterone levels is the primary therapeutic approach to the management of advanced prostate cancer (Crawford et al. 1995). Sustained administration of leuprolide, a GnRH agonist, acts on the pituitary–testicular axis to produce an initial, transient increase in circulating testosterone levels, followed by long-term suppression to low concentrations.

The drug reservoir of the Viadur® implant contains a concentrated solution of leuprolide acetate dissolved in DMSO. The system was designed to deliver leuprolide for 1 year; the steady serum leuprolide levels observed during clinical trials are shown in Figure 6.5. In the clinical trials, the expected initial surge in testosterone levels was observed, followed by sustained suppression of testosterone levels (Fowler et al. 2000).

In vivo/in vitro delivery rate correlation was investigated for the Viadur® implant in rats, dogs, and humans. Based on analysis of explanted systems, there was a good correlation between the amount of drug delivered in vitro and the amount delivered in vivo (Wright et al. 2001).

The DUROS® technology is also being investigated for exenatide, a GLP-1RA for the treatment of type 2 diabetes. Twelve-month stability data of an exenatide DUROS® implant (Intarcia) at human body temperature have been reported (Yang et al. 2009). A Phase 2 study with a 3-month s.c. implant that provided steady delivery of exenatide has been completed (Dahms et al. 2011), and a Phase 3 study is in progress. The DUROS® implant delivery system has also been investigated for the s.c. delivery of interferon (Omega DUROS®), sufentanil (Chronogesic® system), and the site-targeted delivery of other opioids (Wright et al. 2003; Yang et al. 2008).

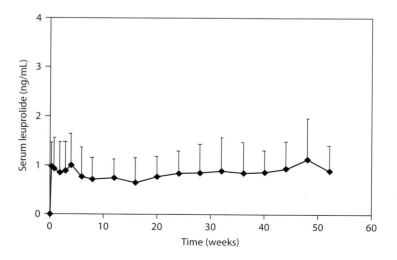

FIGURE 6.5 Serum leuprolide levels for patients treated with the Viadur® implant. (Reprinted from *J. Control. Release*, 75, Wright, J.C., Leonard, S.T., Stevenson, C.L. et al., An in vivo/in vitro comparison with a leuprolide osmotic implant for the treatment of prostate cancer, 1–10. Copyright 2001, with permission from Elsevier.)

TABLE 6.7

Examples of Drug-Eluting Stents

Trade Name	Drug	Stent	Polymer
Cypher® stent	Sirolimus	Stainless steel	PEVA and poly(butyl methacrylate)
Taxus® stent	Paclitaxel	Stainless steel	Copolymer of poly(styrene-isobutylene-styrene)
Endeavor® stent	Zotarolimus	Cobalt/chromium alloy	Copolymer of 2-methacryloyloxyethyl phosphorylcholine, lauryl methacrylate, hydroxypropyl methacrylate, and trimethoxysilylpropyl methacrylate
Xience V® stent	Everolimus	Cobalt/chromium alloy	Copolymer of vinylidene fluoride and hexafluoropropylene

6.3.5 SURFACE-RELEASING SYSTEMS: DRUG-ELUTING STENTS

Coronary artery disease is a major health problem worldwide. Narrowing of arteries can be treated by balloon angioplasty followed (optionally) by placement of a stent, a bare metal wire device intended to prevent restenosis of the treated artery. However, restenosis is observed in a significant number of patients treated with bare metal stents (Zhao and Alquier 2012).

A major implantable drug delivery system application is the DES, which has revolutionized the treatment of vascular disease by reducing restenosis rates by 60%–75% across all patients when compared to bare-metal stents (Kukreja et al. 2009). DES are classified as combination drug-device products. A DES is normally a three-component system consisting of a scaffold (i.e., the stent that serves to keep the artery patent), a polymer to control the release of drug, and an antiproliferative drug, to inhibit neointimal growth.

A DES is generally designed as a matrix delivery system; drug particles are blended with the polymer and the mode of release from these coatings is diffusion controlled. A successful DES requires careful attention to all three components of the system. Drug release kinetics are governed by the choice of polymer, the specific drug incorporated in the system, and, as matrix systems, a number of other design variables. Drugs are released from DES designs over periods ranging from approximately 10 days to over 30 days (Zhao and Alquier 2012).

The first DES to obtain U.S. approval was the Cypher® stent in 2003 (note that Johnson & Johnson ceased manufacturing the Cypher stent in 2011). The Taxus® stent from Boston Scientific obtained U.S. approval in 2004. Other DES with U.S. approval include the Endeavor® stent (Medtronic) and the Xience V® stent (Abbott Vascular). Their properties are summarized in Table 6.7.

The commercial success of the DES has led to the involvement of other companies in the field and to the investigation of refined design options. Of special note is the concept of a fully bioresorbable DES, which would perform all the needed scaffolding and drug release functions and then be resorbed. Such a design would eliminate any potential for late-stage thrombosis events. A bioresorbable DES design has been tested in clinical trials (Kleiner and Wright 2013). The literature on DES is extensive; more detailed overviews can be found in Zhao and Alquier (2012) and in Kleiner and Wright (2013).

6.4 EMERGING TECHNOLOGIES

There are a number of emerging platforms that are being developed, summarized in Table 6.8. These technologies include drug encapsulation with targeting capabilities; the addition of a polymer of amino acids to therapeutic peptides and proteins; bioerodible polymer and solvent formulations; incorporation of nonnatural amino acids that allow site-specific chemical modification; and nanoparticle manufacturing based on high-resolution micromolding fabrication. Table 6.8 presents examples of emerging technologies; we have selected several examples of platform technologies

TABLE 6.8

Examples of Emerging Technologies

Platform Name	Company	Technology	Reference
Accurins™	BIND Therapeutics	Drug-encapsulated targeted nanoparticles with a controlled-release polymer (PLA, PLGA), a stealth layer (PEG), and targeting ligands to bind to cell surface or tissue markers.	Valencia et al. (2011) Hrkach et al. (2012)
XTEN®	Amunix	Hydrophilic and unstructured polypeptide polymers using natural amino acids designed to mimic properties of PEG (e.g., to increase hydrodynamic volume). XTEN is added to proteins or peptides by genetic fusion or by chemical conjugation.	Schellenberger et al. (2009) Podust et al. (2013)
PASylation®	XL-Protein	Conformationally disordered polypeptide sequence consisting of proline, alanine, and/or serine (PAS) that are used for attachment to a therapeutic protein.	Schlapschy et al. (2013) Di Cesare et al. (2013)
Medusa®	Flamel Technologies	Amphiphilic polymer comprising a poly-L-glutamate backbone grafted with alpha-tocopherol (vitamin E) molecules. The active agent binds via hydrophobic or electrostatic interactions within the polymer.	Chan et al. (2007)
Re-CODE™	Ambryx	Enabling placement of a chemically reactive amino acid into a specific position in coding DNA sequence. The amino acids are nonnative and have selective linkable bioorthagonal chemical groups that allow site-specific modification of therapeutic peptides and proteins.	Xie and Schultz (2006) Cho et al. (2011)
Fluid Crystal™	Camurus	Injectable depot is based on lipid liquid crystal (LLC) phase formation. Composition includes long-chain lipids such as soy phosphatidyl choline (SPC) and glycerol dioleate (GDO) and solvent (e.g., ethanol, propylene glycol) for viscosity control.	Tiberg and Joabsson (2010)
PRINT®	Liquidia Technologies	A continuous, roll-to-roll, high-resolution molding technique is being developed to fabricate and define micro- and nanoparticles.	Mack et al. (2012) Perry et al. (2012)

and corresponding companies that have advanced their platforms into preclinical proof of concept, clinical testing, or near-term commercialization.

6.5 DEVELOPMENT CONSIDERATIONS FOR LAIs AND IMPLANTS

A number of factors that influence the development of LAI and implantable DDS are discussed here.

6.5.1 TARGET PRODUCT PROFILE

The development and establishment of a target product profile (TPP) is an important element that defines the drug product. The TPP should be based on the unmet medical need, which can be derived from several sources including clinical and scientific analysis, market research, and physician and patient surveys. The TPP should include product description, indication and usage, dosage and administration, dosage forms and strengths, contraindications, warnings and precautions, adverse reactions, drug interactions, use in specific populations, drug abuse and dependence, overdosage, clinical pharmacology, nonclinical toxicology, clinical studies, patient counseling, how supplied, and storage and handling (FDA Guidance TPP 2007).

For controlled-release dosage forms, the TPP should indicate the delivery platform (e.g., LAIs vs. implant, erodible vs. nonerodible implant). The selection of appropriate excipients or materials of construction is a key element to support the TPP requirements. Nonerodible components should exhibit biocompatibility consistent with the intended use. For biodegradable systems, biocompatibility is also very important; careful consideration should be given to the degradation and/or diffusion characteristics of each excipient in the LAI or implant. Disappearance of the depot or biodegradable implant in situ should be consistent with the duration of the drug release kinetics. The depot/implant constituents must be present to provide the functional attributes for sustained release and should disappear in a reasonable time thereafter. With regard to implants, setting specification limits on dimensions can drive specifications on drug loading dose, drug concentration, and formulation excipient composition.

TPPs may evolve as a program develops. Having a TPP helps to set goals and target specifications for formulation attributes for LAIs or implants. Additional benefits from a TPP include assuring that perspectives from all relevant expertise areas (e.g., medical, formulation, regulatory, and marketing) contribute to establish desired product performance characteristics. The TPP provides regulatory agencies the therapeutic indication and the performance attributes for the product (FDA Guidance TPP 2007).

It is important to include additional details on PK and efficacy target duration within the TPP. Drugs that have known therapeutic windows provide an advantage to define the drug release rate targets for controlled-release formulations. However, as with the case of NCEs, the desired therapeutic concentration is often not determined until clinical efficacy studies are performed. The development challenge is whether the delivery system will be able to meet target drug PK parameters.

The injection volume will vary from product to product and is influenced by the drug concentration and total dose in the formulation. The volume administered should be within the maximal allowed volume for s.c., i.m., or other administration routes. Drugs with higher potency generally require lower doses per treatment and allow flexibility in drug-loading dose in the formulation. Drugs with lower potency generally require higher doses to achieve efficacy. It is important to understand the range in loading capacities for the drug delivery platform technology. The TPP should also consider the target needle gauge size such that the formulation can be easily injected. The TPP should also include information on the target product shelf life and indicate storage condition requirements. Additional information on product shelf life is described in Section 6.5.7.

6.5.2 BIOLOGY OF THE TREATMENT

The physicochemical aspects of the formulation may be fairly developed. It is very important that the biology of the disease also be well understood. In addition to the biocompatibility considerations discussed later, the pharmacodynamics and desired temporal pattern of drug delivery should be matched to the underlying medical need. For example, treating a disease with a sustained-release GnRH agonist works well for treating prostate cancer when chronic suppression of testosterone levels is desired. In contrast, sustained release of parathyroid hormone would not be beneficial for osteoporosis treatment because daily pulsatile administration would be warranted (Potter et al. 2005). A perspective of the biology of chronic versus acute treatment and severity of the disease state should also be considered. These attributes can affect product risk/benefit analysis, product design, and the overall development program.

6.5.3 HOST RESPONSE/BIOCOMPATIBILITY

A marketed LAI or implant provides the desired PK and efficacy for a given therapeutic. A successful commercialization outcome may fail if the local tissue response (biological interface) is too severe and other key safety parameters are not met. Biocompatibility is an essential element in the success of LAIs and implants. Thus, a fundamental understanding of what reactions may occur, as well as the severity and the duration of the reaction, is important. Table 6.9 highlights some of the key assessments in addressing biocompatibility. Local site assessments are critically important for biocompatibility testing. The evaluations should include histological as well as clinical findings. Toxicological evaluation of systemic toxicity often focuses on the active agent. It is also important to be able to distinguish drug-related toxicities versus toxicities that may be contributed by the delivery system.

It is vital for scientists and product developers to recognize that host responses (foreign body reactions) to LAI and implants may be elicited and that these reactions can be acceptable transient responses. The perspective of the acceptability of biocompatibility of the LAI and implant varies and should include assessments such as treatment indication, disease severity, acute vs. chronic treatment, and patient population. It should be also noted that go/no-go decisions based on biocompatibility assessment can occur at all stages of development. As each delivery technology is unique, it is not possible to include all the details on this topic in this section, but the discussion is intended to highlight some of the key issues that investigators seek to understand and address for developing LAIs and implants.

The biological response resulting from a LAI or implant results in a series cascade of complex events. The host's foreign body response includes recognition of the LAI or implant and induction of innate and specific immune inflammatory processes and tissue healing. Work by Anderson and Shive (1997) provides valuable information surrounding the events following microsphere implantation. The temporal change in the local tissue environment after implantation is characterized as a continuum of cellular responses. The formulation composition and residence time can strongly influence the magnitude of the response. The active agents (e.g., cytotoxic anti-inflammatory agents) as well as the concentrations that are incorporated into the delivery system can also influence the

---

TABLE 6.9

Biocompatibility Assessment

Host Response	Assessment
Local tissue response	Clinical visibility at administration site
	Redness, induration, swelling, pain
	Level of severity
	Duration of response (transient vs. prolonged)
	Tissue recovery
	Temporal relationship of the foreign body response vs. the erosion kinetics of depot or implant
	Potential effect on drug release kinetics
	Response to therapeutic agent and/or excipient(s)
Systemic response (hypersensitivity, immunogenicity)	Clinical safety
	Level of severity
	Duration of response
	Response to therapeutic agent and/or excipient(s)
	Immunologic mechanism(s)
	Effect on drug release kinetics
	Effect on efficacy

foreign body responses (Spilizewski et al. 1985; Ike et al. 1992). Over time, with biocompatible polymers, the fibrosis becomes residual, leaving minimal scarring and fibrotic tissue.

It is important to evaluate biocompatibility preclinically. However, the findings and interpretations from animal studies should be tempered by the understanding that the preclinical models chosen for evaluation have limits on predicting the clinical outcome. A review by Engelhardt (2008) assessed the predictive capability of nonclinical studies that evaluated injection-site reactions from the administration of biopharmaceuticals to the clinical outcome. Based on the available literature and personal observations for some therapeutic agents, the interspecies correlation was poor. In some cases, profound inflammation was observed in animals, whereas relatively mild cutaneous responses were observed in humans. In other cases, minimal inflammation was found in animals, but significant inflammation was observed in humans. In the case of biologic agents, it is important to be aware whether the agent has biologic activity in the host species.

Evaluation of biocompatibility should include experimental designs that allow evaluation of injection site reactions that can discriminate the contribution(s) of the active agent versus delivery system excipients or a combination of multiple components. It is also important to determine the appropriate rodent and nonrodent species used in the conduct of biocompatibility studies. The assessments should include the presence or absence of a response, incidence, onset, duration, severity, and recovery. In some cases, the relationship between observations of a local response versus drug delivery performance (e.g., PK) may be warranted. Biocompatibility studies may include in vitro cytotoxicity studies (ISO 1992).

6.5.4 IMMUNOGENICITY

As part of the development of controlled-release delivery systems, the evaluation of immunogenicity may be relevant. This will depend on a number of factors including the type of therapeutic agent (e.g., small molecule vs. peptide vs. protein), formulation composition, route of administration, duration of treatment, stage of development, therapeutic indication, and patient population.

Immunogenicity reactions are antigen-specific immune responses. The activation of the immune response involves both innate and specific immune systems. The factors that influence the induction and elicitation of unwanted immunogenicity (humoral or cellular) to therapeutic compounds are complex. They include host/patient factors (e.g., genetics, age, sex, and disease state), the drug (e.g., protein structure, sequence homology to host, complexity, purity, and stability), route of administration, dose, dosing frequency, and formulation composition (Hermeling et al. 2004; Schellekens 2005). The FDA and European Medicines Agency have provided guidance on immunogenicity testing for large molecules (EMEA 2007; FDA 2013). The guidance focuses on the testing strategy and the analytical methods to assess anti-drug responses with an emphasis on therapeutic proteins. Because of the regulatory requirements of testing and monitoring for anti-drug responses, appropriate risk assessments should be conducted as part of development activities for sustained-release platforms.

With regard to LAIs and implants, developers should also determine the influence of sustained-release systems on immunogenicity. For example, patients treated with exenatide twice daily in an aqueous formulation (Byetta®) reported 38% incidence of low antibody titers and 6% had high titers. Patients receiving a once-weekly microsphere formulation of exenatide (Bydureon®) reported a 45% incidence of low antibody titers to exenatide and a 12% incidence in patients having high antibody titers. Although a higher incidence was observed with the microsphere-based long-acting formulation, the immunogenicity testing results were acceptable from a drug-approval perspective.

The value of preclinical immunogenicity testing to predict the human response can be questionable but may have merit in some cases. For example, immunogenicity testing in support of good laboratory practice toxicokinetic studies provides value in the interpretation of the PK data that are generated. Studies conducted by Alza Corporation provide a good example of the influence of excipients for a sustained-release dosage form (Alzamer®) on immunogenicity to a therapeutic protein in rats. Rat growth hormone (rGH) was chosen as the model protein so that immunogenicity was not due to sequence homology differences between the animal host and the protein. Antibody responses were not observed in animals receiving injections of rGH formulated in an aqueous bolus or in an Alzamer depot containing benzyl benzoate as the solvent. In contrast, antibodies to growth hormone were observed in rats injected with an rGH Alzamer depot with benzyl alcohol as a cosolvent (Chen and Junnarker 2008). Studies by Schellenberger conducted immunogenicity testing as a screening and selection mechanism to determine amino acid polymer constructs (called XTEN) that demonstrate minimal or no antibody responses in a mouse model. These constructs were then further developed (Schellenberger et al. 2009).

6.5.5 Manufacturing

Initially, LAIs and implants are produced on the laboratory bench using research processes where the objective is to produce sufficient amounts of the LAI for preliminary testing and cost is not a primary consideration. It is often possible to use the laboratory bench process for initial development work. However, the laboratory bench process may not scale up easily, may be expensive, and may be difficult to validate.

Ideal processes for LAI's involved unit operations that are standard for the pharmaceutical industry or the medical device industry, such as lyophilization, fluid mixing, dispensing of solutions into unit-dose containers, and molding and extrusion. Novel delivery systems can involve entirely new process steps that require extensive process development. The investment in process development may be justified based on the therapeutic need or market opportunity.

In the authors' experience, some manufacturing problems with novel delivery systems do not become evident to the development team until processing at larger scale or until multiple batches have been made at commercial scale. This occurrence can be especially true for raw materials, where variations in the characteristics of the raw material may occur over several years of production by the outside supplier.

6.5.6 Sterilization/Aseptic Processing

Implantable or injectable drug delivery systems must be sterile and nonpyrogenic and pass compendial tests for sterility and pyrogens/endotoxins (FDA 2012; USP 2013). While nonsterile products can be evaluated in rodents during development, the final commercial product must be able to withstand a sterilization process, or the product must be manufactured via an aseptic process. Common methods of sterilization include steam, dry heat, ethylene oxide, gamma irradiation, and electron beam sterilization. Regulatory agencies prefer terminal sterilization over aseptic processing because of the higher level of sterility assurance (10^{-6} vs. 10^{-3}), and use of aseptic processing will need to be justified in the regulatory approval process (Yaman 2012).

6.5.7 Product Shelf Life

For commercial products, a 2-year shelf life at the indicated storage temperature is generally desired from distribution and economic considerations. Storage temperatures include refrigerated, frozen, and controlled room temperature. It should be noted that actual storage conditions in other climatic zones (III and IV) may be different. Product stability is usually tested in the final container/closure system at a range of temperatures and humidities (including a temperature greater than the intended storage condition) so that the shelf life can be extrapolated to longer time periods. The impact of the method of sterilization on product degradation must be considered, as often initial stability experiments are performed on nonsterile product, and sterilization can result in accelerated degradation kinetics or new routes of degradation. Regulatory agencies generally request 6–12 months of stability data in the marketing application (FDA 1996, 2003).

6.6 CONCLUSIONS

Significant strides have been made in developing effective LAIs and implants that utilize a range of underlying delivery technologies. Therapeutic agents married with liposomes, microspheres, in situ forming depots, PEGylation, nanoparticle, and implant platforms have been successfully developed and commercialized. Clinical uses of products leveraging these technologies have improved effectiveness of therapies for the patient, doctor, and healthcare system.

The challenges to develop safe, biocompatible, therapeutic, and cost-effective delivery systems can be similar or unique with each platform. Continued development, refinement, and optimization will be required for each technology. While cost considerations may favor LAIs and implants that are nontargeted, improvements in efficacy may justify the economics of targeted delivery systems. New product opportunities leveraging LAIs and implant technologies are expected to further improve drug efficacy and overall patient care, as well as to continue to benefit the healthcare system.

ACKNOWLEDGMENTS

We thank Dr. Felix Theeuwes and Dr. WeiQi Lin for their helpful review and insights and Jose Gadea for the assistance with the ALZET® section.

REFERENCES

Alam, S., J.B. Baldwin, K.A. Tombros et al. 2008. The single-rod contraceptive implant. In *Clinical Proceedings® from Association of Reproductive Health Professionals*, pp. 7–9.

ALZET 2014, ALZET® Osmotic Pumps–Implantable pumps for research, Accessed January 19, 2014. http://www.alzet.com.

Anderson, J.M. and M.S. Shive. 1997. Biodegradation and biocompatibility of PLA and PLGA microspheres. *Adv. Drug Del. Rev.* 28:5–24.

Barnhart, E. ed. 1985. *Alza Corporation: Progestasert*, pp. 590–602. Oradell, NJ: Medical Economics Company, Inc.

Bauer, C.A., K. Wisner, L.T. Sybert et al. 2013. The cerebellum as a novel tinnitus generator. *Hear. Res.* 295:130–139.

Boswell, G. and R. Scribner. 1973. Polylactide-drug mixtures. US Patent 3,773,919, Issued November 1973.

Brem, H. and R. Langer. 1996. Polymer-based drug delivery to the brain. *Sci. Med.* 3:52–56.

Brem, H., M.S. Mahaley Jr., N.A. Vick et al. 1991. Interstitial chemotherapy with drug polymer implants for the treatment of recurrent gliomas. *J. Neurosurg.* 74:441–446.

Brodbeck, K.J., S. Pushpala, and A.J. McHugh. 1999. Sustained release of human growth hormone from PLGA solution depots. *Pharm. Res.* 16(12):1825–1829.

Chan, Y.-P., R. Meyrueix, R. Kravtzoff et al. 2007. Review on Medusa®: A polymer-based sustained release technology for protein and peptide drugs. *Expert Opin. Drug Deliv.* 4:441–451.

Chen, G. and G. Junnarkar. 2008. ALZAMER® Depot™ Bioerodible technology. In *Modified Release Drug Delivery Technology*, 2nd ed., Vol. 2, M.J. Rathbone, J. Hadgraft, M.S. Roberts, and M.E. Lane (eds.), pp. 215–225. New York: Informa Healthcare Inc.

Chien, Y.W. 1982. *Novel Drug Delivery Systems*, pp. 247–280. New York: Marcel Dekker, Inc.

Chiu, G.N.C., M.B. Bally, and L.D. Mayer. 2001. Selective interactions with phosphatidylserine containing liposomes alter the steric stabilization properties of poly(ethylene glycol). *Biochim. Biophys. Acta* 1510:56–59.

Cho, H.S., T. Daniel, Y.J. Buechler et al. 2011. Optimized clinical performance of growth hormone with an expanded genetic code. *Proc. Nat. Acad. Sci.* 108:9060–9065.

Coskun, T., H.A. Bina, M.A. Schneider et al. 2008. Fibroblast growth factor 21 corrects obesity in mice. *Endocrinology* 149:6018–6027.

Cosman F., N.E. Lane, M.A. Bolognese et al. 2010. Effect of transdermal teriparatide administration on bone mineral density in postmenopausal women. *J. Clin. Endocrinol. Metab.* 95:151–158.

Crawford, E.D., E.P. DeAntonio, F. Labrie et al. 1995. Endocrine therapy of prostate cancer: Optimal form and appropriate timing. *J. Clin. Endocrinol. Metab.* 80:1062–1078.

Culwell, J.A., J.R. Gadea, C.E. Peer et al. 2012. Implantable drug delivery systems based on the principles of osmosis. In *Long Acting Injections and Implants*, J. Wright and D. Burgess (eds.), pp. 335–357. New York: Springer.

Dadey, E.J. 2008. The Atrigel® drug delivery system. In *Modified-Release Drug Delivery Technology*, 2nd ed., M.J. Rathbone, J. Hadgraft, M.S. Roberts, and M.J. Lane (eds.), pp. 183–189. New York: Informa Healthcare Inc.

Dahms, J., Y. Chandrasekher, R. Fielding et al. 2011. Pharmacokinetic and pharmacodynamic assessments with ITCA 650, continuous subcutaneous delivery of exenatide, in Type 2 diabetes. Presented at the *European Association for the Study of Diabetes*, September 14, 2011, Lisbon, Portugal.

Davis, F.F. 2002. The origin of pegnology. *Adv. Drug Deliv. Rev.* 54:457–458.

Deanesly, R. and A.S. Parkes. 1937. Biological properties of some new derivatives of testosterone. *Biochem. J.* 31:1161–1164.

DeYoung, M.B., L. MacConell, V. Sarin et al. 2011. Encapsulation of exenatide in poly-(D,L-lactide-co-glycolide) microspheres produced and investigational long-acting once-weekly formulation for type-2 diabetes. *Diab. Technol. Therap.* 13:1–10.

Di Cesare, S., U. Binder, T. Maier et al. 2013. High-yield production of PASylated human growth hormone using secretory *E. coli* technology. *BioProcess Int.* 11:30–38.

Dijkman, G.A, P.F. del Moral, J.W. Plasman et al. 1990. A new extra long acting depot preparation of the LHRH analogue Zoladex. First endocrinological and pharmacokinetic data in patients with advanced prostate cancer. *J. Steroid Biochem. Mol. Biol.* 37:933–936.

Dings, R.P.M., D.W.J. van der Schaft, B. Hargittai et al. 2003. Anti-tumor activity of the novel angiogenesis inhibitor anginex. *Cancer Lett.* 194:55–66.

Drucker, D.J., J.B. Buse, K. Taylor et al. 2008. Exenatide once weekly versus twice daily for the treatment of type 2 diabetes: A randomised, open-label, non-inferiority study. *Lancet* 372:1240–1250.

Dunn, R.L. 2003. The Atrigel drug delivery system. In *Modified-Release Drug Delivery Technology*, M.J. Rathbone, J. Hadgraft, and M.S. Roberts (eds.), pp. 647–655. New York: Marcel Dekker, Inc.

Eckenhoff, B., F. Theeuwes, and J. Urquhart. 1987. Osmotically actuated dosage forms for rate-controlled delivery. *Pharm. Tech.* 11:96–105.

EMEA Guidance. 2007. Guideline on immunogenicity assessment of biotechnology-derived therapeutic proteins. Accessed May 19, 2014. http://www.ema.europa.eu/docs/en_GB/document_library/Scientific_guideline/2009/09/WC500003946.pdf.

Engelhardt, J.A. 2008. Predictivity of animal studies for human injection site reactions with parenteral drug products. *Exp. Toxicol. Pathol.* 60:323–327.

Ereshefsky, L. and C.A. Mascarenas. 2003. Comparison of the effects of different routes of antipsychotic administration on pharmacokinetics and pharmacodynamics. *J. Clin. Psychiatry* 64:18–23.

EvaluatePharma®. 2013. Evaluate Ltd. www.evaluate.com.

Exparel 2014. Package Insert. Pacira Pharmaceuticals Inc. San Diego, CA. 2012. Accessed June 2, 2014. http://www.exparel.com/pdf/EXPAREL_Prescribing_Information.pdf.

FDA. 1996. Guidance for industry—Q1C stability testing for new dosage forms. Accessed June 2, 2014. http://www.fda.gov/downloads/Drugs/GuidanceComplianceRegulatoryInformation/Guidances/ucm073374.pdf.

FDA. 2003. Guidance for industry Q1A(R2) stability testing of new drug substances. Accessed June 2, 2014. http://www.fda.gov/downloads/drugs/guidancecomplianceregulatoryinformation/guidances/ucm073369.pdf.

FDA. 2007. Guidance for industry and review staff, target product profile—A strategic development process tool. Accessed June 2, 2014. http://www.fda.gov/downloads/Drugs/GuidanceCompliance RegulatoryInformation/Guidances/ucm080593.pdf.

FDA. 2012. Guidance for industry—Pyrogen and endotoxins testing: Questions and answers. Accessed January 26, 2014. http://www.fda.gov/downloads/Drugs/GuidanceComplianceRegulatoryInformation/Guidances/UCM310098.pdf.

FDA. 2013. Draft guidance for industry—Immunogenicity assessment for therapeutic protein products. Accessed May 19, 2014. http://www.fda.gov/downloads/Drugs/GuidanceComplianceRegulatoryInformation/Guidances/UCM338856.pdf.

Ferguson, T.H. 2001. Implants: Overview of technology. In *AAPS Workshop, Assuring Quality and Performance of Sustained and Controlled Release Parenterals*, April 19, 2001, Washington, DC.

Ferguson, T.H., G.F. Needham, and J.F. Wagner. 1988. Compudose®: An implant system for growth promotion and feed efficiency in cattle. *J. Control. Release* 8:45–54.

Folkman, J. and D.M. Long. 1964. The use of silicone rubber as a carrier for prolonged drug therapy. *J. Surg. Res.* 4:139–142.

Folkman, J., D.M. Long, and R. Rosenbaum. 1966. Silicone rubber: A new diffusion property useful for general anesthesia. *Science* 154:148–149.

Fowler, J.E., M. Flanagan, D.M. Gleason et al. 2000. Evaluation of an implant that delivers leuprolide for 1 year for the palliative treatment of prostate cancer. *Urology* 55:639–642.

Freitas, S., H.P. Merkle, and B. Gander. 2005. Microencapsulation by solvent extraction/evaporation: Reviewing the state of the art of microsphere preparation process technology. *J. Control. Release* 102:313–332.

Gibson, J.W., A.J. Tipton, R.J. Holl et al. 2012. Zero-order prolonged release coaxial implants. US 8,263,108 B2, Issued September 2012.

Gombotz, W.R. and A.S. Hoffman. 2013. Injected depot DDS. In *Biomaterials Science*, 3rd ed., B.B. Ratner, A.S. Hoffman, F.J. Schoen, and J.E. Leomons (eds.). Waltham, MA: Elsevier/Academic Press.

Gong, N., A.I. Ma, L.J. Zhang et al. 2011. Site-specific PEGylation of exenatide analogues markedly improved their glucoregulatory activity. *Brit. J. Pharmacol.* 163:399–412.

Gopalakrishna, G., A. Aggarwal, and J. Lauriello. 2013. Long-acting injectable Aripiprazole: How might it fit in our tool box? *Clin. Schizophr. Relat. Psychoses* 7:87–92.

Heller, J. 2009. Patient-friendly bioerodible drug delivery systems. *J. Control. Release* 133:88–89.

Hermeling, S., D.J.A. Crommelin, H. Schellekens et al. 2004. Structure–immunogenicity relationships of therapeutic proteins. *Pharm. Res.* 21:897–903.

Higuchi, T. 1963. Mechanism of sustained-action medication. *J. Pharm. Sci.* 52:1145–1149.

Hirano, K., T. Ichihashi, and H. Yamada. 1981. Studies on the absorption of practically water-insoluble drugs following injection. I. Intramuscular absorption from water-immiscible oil solutions in rats. *Chem. Pharm. Bull.* 29:519–531.

Hoffman, A.S. 2008. The origins and evolutions of controlled drug delivery systems. *J. Control. Release* 132:153–163.

Hong, H.-S., S. Rana, L. Barrigan et al. 2009. Inhibition of Alzheimer's amyloid toxicity with a tricyclic pyrone molecule *in vitro* and *in vivo*. *J. Neurochem.* 108:1097–1108.

Hrkach, J., D. Von Hoff, M. M. Ali et al. 2012. Preclinical development and clinical translation of a PSMA-targeted Docetaxel nanoparticle with a differentiated pharmacological profile. *Sci. Transl. Med.* 4:1–11.

Hutchinson, F.G. and B.J. Furr. 1985. Biodegradable polymers for the sustained release of peptides. *Biochem. Soc. Trans.* 13:520–523.

Hutchinson, F.G. and B.J. Furr. 1990. Biodegradable polymer systems for the sustained release of polypeptides. *J. Control. Release* 13:279–294.

Ike, O., Y. Shimizu, R. Wasa et al. 1992. Controlled cisplatin delivery system using poly(DL-lactic acid). *Biomaterials* 13:230–234.

Immordino, M.L., F. Dosio, and L. Cattel. 2006. Stealth liposomes: Review of the basic science, rationale, and clinical application, existing and potential. *Int. Natl. J. Nanomed.* I:297–315.

Intarcia Therapeutics, Inc. 2014. ICTA 650. Accessed June 3, 2014. http://www.intarcia.com/products-technology/type2-diabetes.html.

Ishida, T., H. Harashima, and H. Kiwada. 2002. Liposome clearance. *Biosci. Rep.* 22:197–224.

ISO Document 10993. 1992. Biological compatibility of medical devices. Part 5. Tests for cytotoxicity: *In vitro* methods.

Johnson, O.L. and P. Herbert. 2003. Long-acting protein formulations—ProLease technology. In *Modified-Release Drug Delivery Technology*, M.J. Rathbone, J. Hadgraft, and M.S. Roberts (eds.), New York: Marcel Dekker, Inc., pp. 671–677.

Junghanns, J.-U.A.H. and R.H. Müller. 2008. Nanocrystal technology, drug delivery and clinical applications. *Int. J. Nanomed.* 3:295–309.

Kim, S., H. Solari, P.J. Weiden, and J.R. Bishop. 2012. Paliperidone palmitate injection for the acute and maintenance treatment of schizophrenia in adults. *Patient Prefer. Adherence* 6:533–545.

Kjome, K.L. and F.G. Moeller. 2011. Long-acting injectable naltrexone for the management of patients with opioid dependence. *Subst. Abuse: Res. Treat.* 5:1–9.

Kleiner, L.W. and J.C. Wright. 2013. Implants and Inserts. In *Biomaterials Science: An Introduction to Materials in Medicine*, 3rd ed., B.D. Ratner, A.S. Hoffman, F.J. Schoen, and J.E. Lemons (eds.), pp. 1062–1071. New York: Academic Press/Elsevier.

Kukreja, N., Y. Onuma, and P.W. Serruys. 2009. Future directions of drug-eluting stents. *J. Interven. Cardiol.* 22:96–105.

Lactel 2014. Lactel® Absorbable Polymers. Accessed January 20, 2014. http://www.absorbables.com/technical/biodegradation.html.

Larsen, S.W. and C. Larsen. 2009. Critical factors influencing the *in vivo* performance of long-acting lipophilic solutions—Impact on *in vitro* release method design. *AAPS J.* 11(4):762–770.

Larsen, S.W., A. Mette, M.A. Thing et al. 2012. Oily (lipophilic) solutions and suspensions. In *Long Acting Injections and Implants*, J. Wright and D. Burgess (eds.). New York: Springer.

Lee, S.S., P. Yuan, and M.R. Robinson. 2008. Ocular implants for drug delivery. In *Encyclopedia of Biomaterials and Biomedical Engineering*, 2nd ed., G.E. Wnek and G.L. Bowlin (eds.), pp. 1981–1995. New York: Informa Healthcare.

Li, W., P. Zhan, E. De Clerq et al. 2013. Current drug research on PEGylation with small molecular agents. *Progr. Polym. Sci.* 38:421–444.

Louza, M.R., H. Elkis, S. Ruschel et al. 2011. Long-acting injectable risperidone in partially adherent and nonadherent patients with schizophrenia. *Neuropsychiatric Dis. Treat.* 7:391–398.

Mack, P., K. Horvath, A. Garcia et al. 2012. Particle engineering for inhalation formulation and delivery of biotherapeutics. Accessed June 3, 2014. http://www.liquidia.com/Publications.html.

Matriano, J., M. Cormier, J. Johnson et al. 2002. Macroflux® microprojection array patch technology: A new and efficient approach for intracutaneous immunization. *Pharm. Res.* 19:63–70.

Nektar 2014. Accessed June 3, 2014. http://www.nektar.com/product_pipeline/all_phases.html.

Okumu, F.W., L.N. Dao, P.J. Fielder et al. 2002. Sustained delivery of human growth hormone from a novel gel system: SABER™. *Biomaterials* 23:4353–4358.

Paul, D.R. and S.K. McSpadden. 1976. Diffusional release of a solute from a Polymer Matrix. *J. Membr. Sci.* 1:33–48.

Perry, J., A. Chambers, K. Spithoff et al. 2007. Gliadel wafers in the treatment of malignant glioma: A systematic review. *Curr. Oncol.* 14:189–194.

Perry, J.L., K. Reuter, M.P. Kai et al. 2012. PEGylated PRINT Nanoparticles: The impact of PEG density on protein binding, macrophage association, biodistribution, and pharmacokinetics. *Nano Letters* 12:5304–5310.

Phipps, J.B., R. Padmanabhan, W. Young et al. 2002. E-TRANS technology. In *Modified-Release Drug Delivery Technology*, M.J. Rathbone, J. Hadgraft, and M.S. Roberts (eds.). New York: Marcel Dekker.

Podust, V., B.C. Sim, D. Kothari et al. 2013. Extension of *in vivo* half-life of biologically active peptides via chemical conjugation to XTEN protein polymer. *Protein Eng. Des. Sel.* 26:743–753.

Potter, L.K., L.D. Greller, C.R. Cho et al. 2005. Response to continuous and pulsatile PTH dosing: A mathematical model for parathyroid hormone receptor kinetics. *Bone* 37:159–169.

Reece, A.S. 2007. Psychosocial and treatment correlates of opiate free success in a clinical review of a naltrexone implant program. *Subst. Abuse Treat. Prev. Policy* 2:35–49.

Reynolds, R.C. and C.I. Chappel. 1998. Sucrose acetate isobutyrate (SAIB): Historic aspects of its use in beverages and a review of toxicity studies prior to 1988. *Food Chem. Toxicol.* 36:81–93.

Riley, T. and J. Riggs-Sauther. 2008. The benefits and challenges of PEGylating small molecules. *Pharm. Technol.* 32. Accessed June 3, 2014. https://www.nektar.com/pdf/PharmTechnologySmallMoleculePEG.July2008.pdf.

Schlapschy, M., U. Binder, C. Börger et al. 2013. PASylation: A biological alternative to PEGylation for extending the plasma half-life of pharmaceutically active proteins. *Protein Eng. Des. Sel.* 26:489–501.

Schmitt, E. and R. Polistina. 1967. Surgical sutures. US Patent 3,297,033, Issued 1967.

Schellekens, H. 2005. Factors influencing the immunogenicity of therapeutic proteins. *Nephrol. Dial. Transplant.* 20:vi3–vi9.

Schellenberger, V., C.W. Wang, N.C. Geething et al. 2009. A recombinant polypeptide extends the *in vivo* half-life of peptides and proteins in a tunable manner. *Nat. Biotechnol.* 27:1186–1190.

Sekar, M., F. Okumu, W. van Osdol et al. 2009. SABER™ formulation for intra-articular delivery of recombinant human growth hormone. In *AAPS National Biotechnology Conference*. Poster #NBC-09-00476.

Shah, J., D. Ellis, and N. Verity. 2014. Pharmacokinetic characteristics of SABER®-Bupivacaine in humans demonstrate sustained drug delivery for up to 72 hours in a variety of surgical models. In *39th Annual American Society of Regional Anesthetic and Pain Medicine Meeting*, April 3–6. Accessed June 5, 2014. http://www.durect.com/pdf/101100947-03_PK_Poster_L1d.pdf.

Sikic, B.I., J.M. Collins, and E.G. Mimnaugh. 1978. Improved therapeutic index of bleomycin when administered by continuous infusion in mice. *Cancer Treat. Rep.* 62:2011–2017.

Spilizewski, L.K., R.E. Marchant, C.R. Hamlin et al. 1985. The effect of hydrocortisone acetate loaded poly(DL-lactide) films on the inflammatory response. *J. Control. Release* 2:197–203.

Theeuwes, F. 1975. The elementary osmotic pump. *J. Pharm. Sci.* 64:1987–1991.

Theeuwes, F. 1978. Osmotic system for the controlled and delivery of agent over time. US Patent 4,111,202, Issued September 5.

Theeuwes, F. and S.I. Yum. 1976. Principles of the design and operation of generic osmotic pumps for the delivery of semisolid or liquid drug formulations. *Ann. Biomed. Eng.* 4:343–353.

Tiberg, F. and Joabsson, F. 2010. Lipid liquid crystals for parenteral sustained-release applications: Combining ease of use and manufacturing with consistent drug release control. On Drug Delivery. Accessed June 5, 2014. http://www.ondrugdelivery.com/publications/Injectable%20Formulations%202010/Camurus.pdf.

United States Pharmacopeia 37/National Formulary 32. 2013. General requirements for tests and assays, <1> Injections. 1:33–38. Rockville, MD: United States Pharmacopeial Convention.

Valencia, P.M., M.H. Hanewich-Hollatz, W. Gao et al. 2011. Effects of ligands with different water solubilities on self-assembly and properties of targeted nanoparticles. *Biomaterials* 32: 6226–6233.

Veronese, F.M. and G. Pasut. 2005. PEGylation, successful approach to drug delivery. *Drug Discov. Today* 10:1451–1458.

Veronese, F.M. and G. Pasut. 2012. Protein PEGylation. In *Long Acting Injections and Implants*, J. Wright and D. Burgess (eds.). New York: Springer.

Versypt, A.N.F., D.W. Pack, and R.D. Braatz. 2013. Mathematical modeling of drug delivery from autocata-lytically degradable PLGA microspheres—A review. *J. Control. Release* 165:29–37.

Wang, Y. and Burgess D.J. 2012. Microsphere technologies. In *Long Acting Injections and Implants*, J. Wright and D. Burgess (eds.). New York: Springer.

Wong, P.S.L., S.K. Gupta, and B.E. Stewart. 2003. Osmotically controlled tablets. In *Modified-Release Drug Delivery Technology*, 1st ed., M.J. Rathbone, J. Hadgraft, and M.S. Roberts (eds.). New York: Marcel Dekker Inc.

Wright, J.C., R.M. Johnson, and S.I. Yum. 2003. DUROS® osmotic pharmaceutical systems for parenteral and site-directed therapy. *Drug Del. Tech.* 3:3–11.

Wright, J.C., S.T. Leonard, C.L. Stevenson et al. 2001. An *in vivo/in vitro* comparison with a leuprolide osmotic implant for the treatment of prostate cancer. *J. Control. Release* 75:1–10.

Wright, J.C., M. Sekar, W. van Osdol et al. 2012. *In Situ* forming systems (Depots). In *Long Acting Injections and Implants*, J. Wright and D. Burgess (eds.). New York: Springer.

Wright, J.C., A.N. Verity, and F.W. Okumu. 2008. The SABER™ delivery system for parenteral administration. In *Modified-Release Drug Delivery Technology*, 2nd ed., M.J. Rathbone, J. Hadgraft, M.S. Roberts, and M.E. Lane (eds.), pp. 151–158. New York: Informa Healthcare.

Xie, J. and P.G. Schultz. 2006. A chemical toolkit for proteins—An expanded genetic code. *Nat. Rev. Mol. Cell Biol.* 7:775–782.

Xu, X. and D.J. Burgess. 2012. Liposomes as carriers for controlled drug delivery. In *Long Acting Injections and Implants*, J. Wright and D.J. Burgess (eds.), pp. 195–220. New York: Springer.

Yaman, A. 2012. Methods of sterilization for controlled release injectable and implantable preparations. In *Long Acting Injections and Implants*, J. Wright and D. Burgess (eds.), pp. 459–473. New York: Springer.

Yang, B., C. Negulescu, R. D'vaz et al. 2009. Stability of ITCA 650 for continuous subcutaneous delivery of exenatide at body temperature for 12 months. In Presentation at the *Ninth Annual Diabetes Technology Meeting*, San Francisco, CA. Accessed January 2013. http://www.intarcia.com/documents/11609DURO SPosterDiabetesTechMtg.pdf.

Yang, B., C. Rohloff, R. Mercer et al. 2008. Presentation of continuous delivery of stabilized proteins and peptides at consistent rates for at least 3 months from the DUROS® device. In *American Association of Pharmaceutical Scientists Annual Meeting*.

Zeuzem, S., S.V. Feinman, J. Rasenack et al. 2000. Peginterferon alfa-2a in patients with chronic hepatitis C. *N. Engl. J. Med.* 343:1666–1672.

Zhang, C., X.L. Yang, Y.H. Yuan et al. 2012. Site-specific PEGylation of therapeutic proteins via optimization of both accessible reactive amino acid residues and PEG derivatives. *Biodrugs* 26:209–215.

Zhao, J. and L. Alquier. 2012. Drug-eluting stents. In *Long Acting Injections and Implants*, J. Wright and D. Burgess (eds.). New York: Springer.

Zingerman, J.R., J.R. Cardinal, R.T. Chern et al. 1997. The *in vitro* and *in vivo* performance of an osmotically controlled delivery system-IVOMEC SR® bolus. *J. Control. Release* 47:1–11.

Zogenix 2013. Zogenix Reports Positive Top-Line Results From Extended Relday™ Phase 1 Clinical Trial. Accessed May 19, 2014. http://ir.zogenix.com/phoenix.zhtml?c=220862&p=irol-newsArticle&ID= 1814238&highlight.

Section III

Nonparenteral Routes for Drug Delivery and Targeting

7 Oral Drug Delivery

Anya M. Hillery and David J. Brayden

CONTENTS

7.1 INTRODUCTION

For patients, the oral route represents the easiest and most practical method of taking a drug; it is also usually the safest and least expensive route available. Commercially, as discussed in Chapter 21, the oral drug delivery market comprises the largest slice of the overall drug delivery market. However, the oral route is not without significant challenges. We begin our discussion with a consideration of the anatomy and physiology of the gastrointestinal (GI) tract, the implications therein for successful oral drug delivery, and the factors that must be taken into account in designing appropriate oral drug delivery systems (DDS). This is followed by a discussion on oral controlled-release (CR) technology, beginning with the most widely used and successful technology in oral CR: the hydrophilic matrix tablet. Dissolution- and/or diffusion-controlled matrix and reservoir constructs, osmotic-controlled devices, and microelectronic devices are also described. Drug targeting is discussed with respect to regional targeting within the GI tract: via gastroretention in the stomach and drug targeting to the small intestine (SI) and colon, respectively.

As described in Chapter 3 (Section 3.1.1), Class I molecules of the biopharmaceutical classification system (BCS) dissolve rapidly in the GI fluids and are highly permeable, so that they rapidly cross the epithelial barrier and are taken into the systemic circulation. It has proven much more difficult to achieve satisfactory oral absorption of larger, more polar active pharmaceutical ingredients (APIs), particularly those that are enzymatically labile. For example, the protein insulin, a 5808 Da protein that was introduced for the management of diabetes nearly 100 years ago (in 1923 by Lilly), is still not delivered via the oral route, despite intensive worldwide research efforts. Other poorly bioavailable APIs include polypeptides, proteins, and nucleotide-based drugs: the new "biologics" arising from advances in biotechnology and genomics. Currently, these entities must be delivered parenterally. Their successful oral administration offers many tantalizing rewards, including increased patient acceptance and compliance, reduced costs because sterile manufacturing and specialized health-care professionals are not required, expanded markets for drugs that could become major products if converted from daily or weekly injections, the potential for life cycle management of older injectable products, and the possibility of new intellectual property for oral systems with such molecules.

Strategies directed at enhancing oral bioavailability are considered from a formulation perspective via the use of excipients, including absorption enhancers (AEs), enzyme inhibitors (EIs), and solubilizers. Nanoparticulate DDS are described, as well as chemical conjugation approaches and the possibility of systemic delivery via the colon. Finally, the potential of the oral route for vaccination and the induction of mucosal immunity are discussed.

7.2 ANATOMY AND PHYSIOLOGY OF THE GI TRACT: IMPLICATIONS FOR ORAL DRUG DELIVERY

The anatomy of the GI tract is shown in Figure 7.1. It comprises a continuous tube of about 4.5 m in length in its normal contractile state, which extends from the mouth to the anus (Tortora and Derrickson 2013). It includes the organs of the mouth, most of the pharynx, the esophagus, the stomach, the SI (which in turn is divided into the duodenum, jejunum, and ileum) and the large intestine (which includes the ascending, transverse, descending, and sigmoid colon) (Figure 7.1). Secretions from accessory structures (including the salivary glands, liver, gallbladder, and pancreas) flow into the GI lumen via ducts and aid in the enzymatic breakdown of both food and drug molecules. A full description of the epithelium lining the GI tract is given in Chapter 4 (Section 4.5.3).

The mouth and drug delivery in the oral cavity (buccal and sublingual routes) is the subject of Chapter 8, and is not discussed further here. The stomach, a saclike chamber for the storage and digestion of ingested food, affords a hostile environment for many drug molecules, being

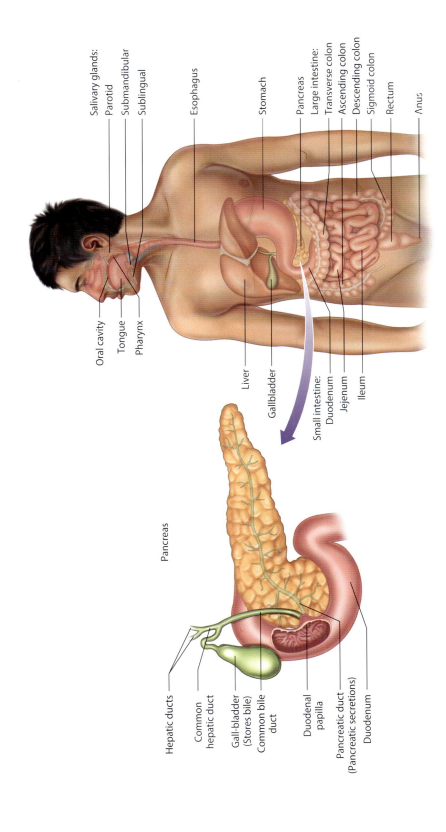

FIGURE 7.1 The gastrointestinal tract and (arrow) the accessory structures of the small intestine.

both strongly acidic (with a pH in fasting conditions between 0.8 and 2), as well as containing many degrading enzymes (renin, pepsin, amylases). While the stomach does not comprise absorbing epithelia particularly appropriate for systemic administration, it is a target for retained local therapies as described in Section 7.5.1. The SI is concerned with completing the process of digestion started in the stomach, principally via the action of pancreatic enzymes secreted into the duodenal lumen. Bile salts, produced in the liver and stored in the gallbladder, aid in fat digestion and absorption. As described in Chapter 4 (Section 4.5.3), a number of structural features in the SI greatly increase the surface area available for absorption, (the presence of villi [see also Chapter 4, Figure 4.8], microvilli, *plicae circulares*, and intestinal glands) making this area of crucial importance to the absorption of digested nutrients and ingested drug molecules. The final stage of digestion occurs in the colon, through bacterial action. The colonic lumen is the home of 10^{12} FU/mL bacteria: this unique feature can be exploited for targeting to this region, as described in Section 7.5.3.

7.2.1 ADVANTAGES OF THE ORAL ROUTE FOR DRUG DELIVERY

The oral route offers many advantages for drug delivery, including the following:

1. *An expanded surface area*: The surface area of the SI in humans is approximately 200 m², larger than the surface area of a tennis court and considerably larger than the surface area of other transmucosal routes such as the nose or oral cavity.
2. *Absorptive epithelium*: The specialized columnar epithelium of the GI tract is highly conducive to the rapid absorption of drugs of the appropriate physicochemical characteristics (see also Chapter 4, Section 4.5.3).
3. *Rich blood supply*: The highly vascular surface of the GI mucosa ensures rapid absorption and onset of action, as well as the maintenance of sink conditions.
4. *Acceptability and compliance*: The oral route allows for self-administration and is noninvasive. For patients, swallowing a pill or other oral DDS is simple, fast, and convenient. In comparison to all other possible routes of drug delivery, peroral drug administration demonstrates the highest patient acceptability and compliance. Compliance can be further improved by using CR dosage forms, which can convert three or four times daily-dosage regimens into much simpler, minimal-dosing schedules.
5. *Amenable to CR and targeted delivery*: The oral route is amenable to a wide range of drug release profiles, including immediate release (IR), delayed release (DR), sustained release (SR), CR, and pulsatile release (PR). Drug targeting to different regions of the GI tract is feasible, as is cell-specific targeting, albeit to a lesser degree.

7.2.2 LIMITING FACTORS FOR ORAL DRUG DELIVERY

A considerable number of factors in the GI tract may also limit oral drug delivery; these factors are considered here.

7.2.2.1 Epithelial Barrier

Transport across epithelial barriers is discussed in detail Chapter 4 (Section 4.3). To summarize here with respect to the oral route, paracellular transport is, under normal conditions, severely restricted by the presence of epithelial tight junctions between adjacent enterocytes and so is confined to small hydrophilic molecules, such as simple sugars like mannitol. Transcellular passive diffusion is usually the predominant pathway for low molecular weight, lipophilic drug molecules. The rate of absorption is governed by Fick's law and is determined by the physicochemical properties of the drug, as well as the concentration gradient across the cells. The physicochemical properties of the drug that favor epithelial transport via transcellular passive diffusion have been described in detail in Chapter 4 and are summarized here:

- *Molecular weight*: Oral absorption is optimal for drugs with a molecular weight less than 500 Da.
- *Lipid solubility and partition coefficient*: The drug must be sufficiently lipid soluble to partition into the lipid bilayers of the plasma membrane of the enterocytes. Ionized, polar molecules have poor lipid solubility and thus exhibit poor transcellular transit. The optimal log P value for passive transcellular permeation is in the range of -0.4 to 3.5.
- *Solubility*: The drug must be soluble in the GI fluids for transcellular permeation. A major challenge of oral drug delivery is to increase the water solubility of poorly soluble BCS Class II and IV drugs. Strategies to increase drug solubility are discussed in Chapter 3.
- *Drug pK_a*: This will affect the charge on weak acid and weak bases, as predicted by the Henderson–Hasselbalch equation (see Chapter 4, Box 4.1). The extent of ionization of a drug molecule varies according to the prevailing pH; it affects both solubility (ionized drugs are more soluble) and transepithelial permeability (unionized drugs are more permeable) properties.
- *Chemical stability*: The GI tract carries out biochemical breakdown of structurally complex foodstuffs of the diet into smaller, absorbable units. An API must be able to withstand the onslaught of digestive enzyme activity directed at foodstuffs. Acid- and base-labile drugs may be vulnerable to degradation, depending on the prevailing pH of the GI region.

A further mechanism of transcellular absorption is via carrier-mediated transport. A wide variety of nutrient carriers (including amino acid transporters, oligopeptide transporters, glucose transporters, lactic acid transporters, monocarboxylic acid transporters, phosphate transporters and bile acid transporters) exist on the luminal membrane of the enterocytes and serve as transporters to facilitate the absorption of digested foodstuffs by the intestine, using either active or passive mechanisms (see also Chapter 4, Figure 4.4). A drug molecule that possesses a similar structure to a nutrient may thus be transported by the nutrient carrier. A drug molecule may also be conjugated to a carrier substrate, such as a vitamin or sugar, so that the drug–substrate conjugate can be transported. However, many of these transport carriers are saturable, thereby limiting the amount of drug absorption possible. Competition between the drug and nutrient substrate may occur, with the concomitant risk of impaired nutrient absorption. Furthermore, nutrient transporters have evolved for the absorption of the digested products of protein nutrients, i.e., amino acids and di- and tripeptides, so that most peptide and protein drugs are too large for transporter uptake.

7.2.2.2 Enzymatic Barrier

Enzymatic degradation occurs along the GI tract, beginning with the action of salivary amylase in the mouth. Pepsin and gastric lipase are found in the stomach. Enzymatic degradation in the SI is principally due to digestive enzymes secreted by the pancreas into the duodenal lumen (Figure 7.1). These pancreatic enzymes are capable of digesting all three categories of foodstuffs: (1) proteins, via serine proteases including trypsin, chymotrypsin, elastase, and procarboxypeptidases; (2) carbohydrates, via pancreatic amylase; (3) fats, via pancreatic lipase; and (4) nucleic acids, via pancreatic ribonuclease and deoxyribonuclease.

Drugs that structurally resemble nutrients, such as polypeptides, proteins, and DNA-based therapies, are highly susceptible to pancreatic enzyme degradation. Further enzymatic breakdown is carried out via the membrane-bound enzymes of the brush border of the enterocytes, which include disaccharidases and aminopeptidases. Enzymatic degradation in the enterocytes can occur either within the cell cytoplasm or within cellular lysosomes. In the large intestine, colonic microflora may also be involved in presystemic drug metabolism. Finally, venous blood draining from the digestive tract is first carried, via the hepatic portal vein, to the liver. An API absorbed from the GI tract may be metabolized by this *first-pass effect* (i.e., pre-systemic hepatic metabolism), prior to reaching the systemic circulation.

7.2.2.3 Mucus Barrier

GI mucus comprises a complex mucus layer that coats the epithelium, protecting the GI tract. Several hundred microns thick, it presents a physical barrier that hinders the diffusion of drug molecules to the absorbing surface. A drug may also bind with the mucus layer, via hydrogen bonding, covalent interactions, etc. As mucus is alkaline, it may also affect the ionization (and by extension, absorption) properties of the API.

7.2.2.4 P-Glycoprotein Efflux Pump

This 1280 amino acid, 170 kDa protein functions as an energy dependent, drug efflux pump at the apical surface of cells. There is a high level of expression of P-glycoprotein (P-gp) in the epithelial cells of the SI. Substrate molecules that diffuse into enterocytes can be extracted directly back to the lumen via the pump. Thus, P-gp can play an important role in limiting the oral bioavailability of certain drugs, e.g., many anticancer drugs including vincristine and taxol are substrates of P-gp and not optimally available by the oral route.

7.2.2.5 pH Range

The pH along the GI tract ranges from acidic to basic, with values ranging from 1 to 3 in the stomach, 4 to 5 in the duodenum, 5 to 6 in the jejunum, 6 to 7 in the ileum, and 7 to 8 in the colon. The pH of the fluids along the GI tract plays a critical role in the dissolution, solubilization, and intestinal permeability of ionizable drugs (see also Chapter 4, Box 4.1). Furthermore, acid- or base-labile drugs are susceptible to pH-mediated breakdown. For example, the extreme acidity possible in the stomach can rapidly degrade many peptide- and protein-based drugs, as well as drugs such as erythromycin, penicillin, and omeprazole.

7.2.2.6 Variability in Conditions in the GI Tract

The GI tract is characterized by a wide variability in the conditions that can exist along its length, which can affect consistency and reproducibility via this route. The large variation in pH along the GI tract has already been described. Another source of variability is the intestinal transit time. GI motility propels the luminal contents forward through the digestive tract, so that typically, small intestinal transit time is between 2 and 4 hours, although large variations are possible. Optimal drug absorption typically occurs in the upper jejunal region of the SI. A relatively short intestinal transit time can limit the effective contact time of an API with the absorbing surface and may thus compromise drug absorption. Posture, age, sleep, body mass index, physical activity, the presence or absence of food and fluid, overall diet, and disease states are all factors that can affect intestinal transit times. Similarly, there is a large inter- and intrasubject variation in the rate of gastric emptying, which can range from approximately 30 minutes to several hours.

Food in the GI lumen generally results in less efficient drug absorption due to a number of mechanisms. By slowing down gastric emptying, molecules unstable in gastric fluids are rendered more susceptible to metabolism. Food also provides a rather viscous environment, which may retard drug dissolution, as well as drug diffusion to the absorptive surface. Certain drugs may bind to food constituents forming a nonabsorbable complex (e.g., the chelation of tetracycline antibiotics with calcium ions in milk). GI fluids are secreted in response to food: enzymes present in the fluids may deactivate a drug moiety. Bile salts, secreted in order to aid in the emulsification and absorption of lipids, may also destabilize lipophilic drugs and lipid-based drug carriers, such as liposomes and emulsions. The secretion and activity of bile salts exhibits high intra- and intervariability, making it difficult to anticipate these effects. Food constituents may also compete with drugs for carrier-mediated absorption mechanisms. A classic example is the breakdown products of dietary proteins, which can compete with L-dopa for oral uptake via an active transport mechanism. Further variability in the GI tract is afforded by individual subject variations such as gender, race, age, and disease state.

7.3 MECHANISMS OF CONTROLLED RELEASE FOR ORAL DRUG DELIVERY

Conventional oral dosage forms (solutions, suspensions, capsules, and tablets) feature immediate release (IR) of the API in the GI lumen, allowing the drug to dissolve in the GI contents and subsequently to be absorbed across the epithelium. Such dosage forms are useful when a rapid onset of action is required. There are many other types of drug release profile possible, as defined in Chapter 2 (Section 2.1), including sustained release (SR), controlled release (CR), modified release (MR), delayed release (DR), pulsatile release (PR), enteric release (ER) and stimuli-sensitive release (SSR). However, it should be remembered that the situation in vivo is highly complex, variable, and difficult to precisely ascertain. In particular, it can be difficult to differentiate accurately between SR and CR delivery systems and in this chapter, the terms CR and SR are used interchangeably.

The first commercial oral SR formulations date back to the early 1950s, with the introduction of the Spansule® system by Smith, Kline & French Ltd. Each Spansule® consisted of a capsule, loaded with hundreds of beads. Each bead comprised a drug core, surrounded by a waxy coating. Upon ingestion, the outer capsule rapidly disintegrated, liberating the coated beads. SR was facilitated by the gradual dissolution of the waxy coating as the beads traveled down the GI tract. The SR profile could be manipulated by mixing pellet populations of different coating thicknesses (see Chapter 1, Figure 1.2). Since the 1950s, oral SR technologies have become increasingly sophisticated (Omidian et al. 2011), to include a wide range of specialist technologies, as described next.

7.3.1 Matrix Systems

In a matrix system, the API is dispersed within a polymer network that retards drug release. Drug release from "swell-and-gel" hydrophilic matrices, hydrophobic matrices, and geometric matrices are described here.

7.3.1.1 Hydrophilic Swellable Matrix Systems

More than 75% of all oral CR delivery products currently marketed are based on hydrophilic matrix tablets. Their widespread use can be attributed to their simplicity, robustness, and versatility. Their method of manufacture is similar to that for conventional tablet formulation (i.e., granulation, blending, compression, and coating), thus established and cost-effective process technology is used, without the need for specialist technology or equipment (Rajabi-Siahboomi et al. 2008).

In hydrophilic matrices, the API is dispersed within a water-swellable polymer, known as a glassy hydrogel. On ingestion, aqueous fluids penetrate the matrix, lowering the glass transition temperature of the polymer, so that it converts from the glassy state to the rubbery state. The polymer chains hydrate, relax, and swell, thereby forming an outer, rubbery, pseudo-gel surface layer (Figure 7.2). In this "swell-and-gel" system, the mechanism of CR is via drug diffusion: the drug dissolves in the incoming fluids and then diffuses out through the viscous, swollen, polymer gel network (Rajabi-Siahboomi et al. 2008; Siepmann and Siepmann 2008; Liu et al. 2011).

Three distinct regions exist, which are defined within the matrix system during drug release (Siepmann and Siepmann 2008; Tu et al. 2010):

- *Zone* 1: The innermost area, where the polymer is in the dry, glassy state. The polymer is nonswollen. The drug is present as undissolved drug particles.
- *Zone* 2: A relatively thin layer, where the polymer chains are undergoing hydration and relaxation from the glassy state. The polymer is beginning to swell and disentangle; there is a mixture of both dissolved and dispersed drug.

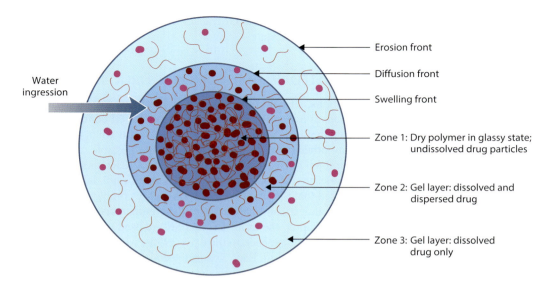

FIGURE 7.2   Hydrophilic "swell and gel" matrix drug release.

- *Zone* 3: The outermost area, where the polymer is completely swollen, due to maximal elongation of the polymer chains. In this region, the drug is dissolved; the polymer per se may also be undergoing erosion and dissolution.

Three fronts can be defined, at the corresponding interface of each zone (Figure 7.2): (1) the swelling front, where nonswollen polymer is starting to swell with water; (2) the diffusion front, where the disentangling polymer molecules start to maximally extend; and (3) the erosion front, which exists at the interface between the GI fluids and the polymer.

   The rate of drug release is complex and is determined by the rate of drug diffusion through the gel barrier, which in turn is affected by many factors, including drug solubility and the degree of polymer hydration, swelling, and porosity. In many cases, erosion constitutes a possible further release mechanism: the swollen hydrated chains display weakened mechanical properties, which may cause the hydrated polymer to break away from the system. This polymer disintegration leads to further dispersion and dissolution of the drug. Polymer dissolution may also occur, thereby facilitating the formation of pores and channels in the system and further modulating release rates.

   From Figure 7.2, it can be seen that as water diffuses into the matrix, the drug diffuses out. Drug in the outer boundary will leave first, resulting in the formation of an outer layer of drug-free matrix. As more drug diffuses out, the layer of drug-free matrix (the depletion zone) becomes larger (see also Chapter 14, Figure 14.11). Therefore, the drug has an increasingly longer distance to travel before it is released from the system, resulting in a decreased amount of drug released over time (see also Chapter 2, Section 2.4.2.2). Although there has been considerable research effort directed toward developing zero-order, constant release, oral delivery systems, in the belief that "flatter is better," in practice, this is not always necessary. A gradual tailing off in the release profile is perfectly acceptable, as long as the drug is maintained within the therapeutic range (i.e., below the toxic concentration and above the minimum effective concentration) for an extended period. While plasma levels are within the therapeutic range, there is enough drug present to produce a therapeutic response—it is not therapeutically necessary to have constant drug release.

TABLE 7.1

Oral Sustained-Release Formulations Based on Hydrophilic Matrix Tablets Using Hydroxypropyl Methylcellulose

Product (Active)	Company	Indication
Tegretol® XR (carbamazepine)	Novartis	Anticonvulsant
Ambien® CR (zolpidem)	Sanofi	Sedative
Xanax® XR (alprazolam)	Pfizer	Anxiolytic
Glucophage® XR (metformin)	BMS	Antihyperglycemic
Wellbutrin® XR (bupropion)	Valeant	Antidepressant
Keppra® XR (levetiracetam)	UCB	Anticonvulsant
Advicor® (niacin XR and lovastatin)	Abbott	Anticholesterol
Simcor® (simvastatin and niacin)	Abbott	Anticholesterol
Niaspan® (niacin)	AbbVie	Antihyperlipidemic
Intuniv® (guanfacine)	Shire Pharma	ADHD

Most water-swellable polymers used in hydrophilic matrices are based on cellulose and include hydroxypropyl methylcellulose (HPMC) and methylcellulose (MC). HPMC is by far the most widely used, forming a component of many commercially available preparations (Table 7.1). HPMC offers many advantages for hydrophilic matrix formulation, including fast and uniform gel formation, which is crucial to protect the matrix from disintegration and premature drug release (Rajabi-Siahboomi et al. 2008). The gel formed by HPMC is strong and viscous, which provides the necessary diffusional barrier to control drug release. Different chemistries (degree of methoxyl and hydroxypropyl substitution) and viscosities of HPMC allow customization of the desired release profile. It may also be combined with other polymers to offer further flexibility and control of release. Combinations with water-insoluble ethyl cellulose (EC) polymers can be used to delay release rates, whereas combinations with soluble polyvinylpyrrolidone (PVP), and poly(ethylene oxide) (PEO) polymers can be used to enhance solubility and dissolution rate. Incorporation of the anionic polymer sodium carboxymethylcellulose promotes drug release via erosion-mediated mechanisms.

Alternatives to cellulose as the matrix material release include xanthan and locust bean gums, as used in the Timer$_x$® system. This technology is used in Opana® ER for the extended release of the opiate analgesic, oxymorphone hydrochloride. Contramid® uses a chemically and physically modified high-amylose cornstarch: the release rate can be customized by altering the degree of starch cross-linking. Commercial preparations include once-daily Ryzolt® (tramadol) for pain management and Oleptro® (trazodone) for depression.

7.3.1.2 Hydrophobic Matrices

Much less widely used perorally than their hydrophilic counterparts, hydrophobic matrices can be formed from water-insoluble polymers such as (1) EC; (2) PVC; (3) poly(methacrylates), e.g., the Eudragit® NE and NM series; and (4) fatty acids, alcohols, and waxes. The matrix can be either homogeneous, where the drug is partly dissolved and evenly distributed throughout the matrix, or porous, in which the matrix contains an additional soluble polymer, which dissolves quickly in the GI fluids, leaving pores in the release matrix to facilitate drug release. Although inert matrix technologies have been largely confined to parenteral implantable systems (see also Chapter 6), there has been recent interest in their use for peroral extended release, due to their ease of manufacturing and as an alternative to hydrophilic matrices. As per hydrophilic matrices, the rate of drug release typically tails-off with time. Hydrophobic matrices may be excreted intact ("exhausted ghosts") in the stools following drug release in the GI tract.

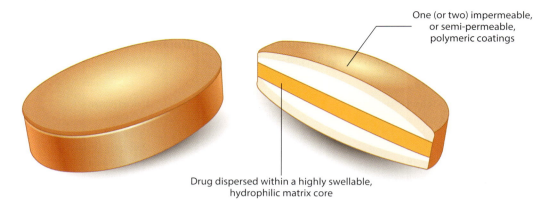

FIGURE 7.3   Geomatrix™ technology.

7.3.1.3   Geomatrix™ System

In recent years, research has been directed toward modifying the drug release profile from matrix tablets, to try and convert the declining drug release rates typically obtained with matrix systems to a constant release rate over time. In the Geomatrix™ multilayered tablet, the geometry of the tablet is modified to enhance zero-order release (Conte et al. 1993; Moodley et al. 2012). The technology comprises an inner core layer, in which the drug is dispersed within a highly swellable hydrophilic polymer matrix (e.g., HPMC). The core is flanked by one or two impermeable, or semipermeable, polymeric coatings, applied to one, or both, bases of the core (Figure 7.3). On ingestion, the external flanking layers initially physically restrict the core layer from exposure to the aqueous fluids of the GI tract, thereby reducing the available surface area of the core for drug release. However as time progresses, the hydrophilic core swells, thereby increasing the exposed surface area available. This progressive increase in surface area with time compensates for the increasing path length, so that zero-order release can be achieved (Conte et al. 1993) (see also Chapter 1, Section 1.6).

Drug-specific customized release can be achieved by varying the number and type of flanking polymers and the core polymer characteristics. For example, zero-order release for the highly soluble drug, diltiazem, was achieved by using blocking polymers of hydrophobic EC on either side of the core. In contrast, low-viscosity hydrophilic HPMC polymers were required for zero-order release of the sparingly soluble calcium blocker nicardipine. Currently, the Geomatrix™ platform is used commercially to provide zero-order extended release in Paxil CR™ (paroxetine), Requip® (ropinirole), and Coruno® (molsidomine). The NSAID diclofenac (Diclofenac-ratiopharm®) is delivered in a biphasic profile with Geomatrix™: an initial burst release of the active drug followed by sustained release of the anti-inflammatory drug over several hours. More than one drug can be delivered with the system: Madopar® DR is a three-layered matrix tablet for the management of Parkinson's disease, which releases a fraction of levodopa and benserazide within an hour and then further amounts of levodopa continuously over 6 hours. This is designed to have a steadying effect, minimizing the troubling "on–off" phenomenon that is seen with conventional Parkinson's drug therapy.

7.3.2   Reservoir Drug Delivery Systems

In reservoir DDS, a drug reservoir is surrounded by a polymeric coating, which functions as a rate-controlling membrane (RCM; see also Chapter 1, Figure 1.3). Release can be controlled by modifying the coating composition, including the type and thickness of the coating polymer, the number of polymer layers, and the presence of excipients such as pore formers and fillers. Reservoir systems for oral drug delivery typically use water-insoluble polymers, in order to provide a constant diffusional barrier to drug release. Widely used polymers include EC, as well as polymers from the Eudragit® NE and NM series. Hydrophilic polymers such as HPMC, poly(vinyl alcohol) (PVA), and polyethylene glycol (PEG) can also be included as pore formers in the membrane.

Drug release from reservoir systems is based on diffusion of the drug through the coating membrane and will follow Fick's laws of diffusion. If both the drug permeability and the drug concentration in the reservoir are constant, zero-order drug release will, in principle, be attained. In practice, this is not typically so—for example, water ingression on contact with the GI fluids may cause swelling of the polymer coating, which will affect coating thickness and the partition coefficient of the drug into the coating. Furthermore, as time proceeds after ingestion, the reservoir may become exhausted of drug. Zero-order release is further confounded by *lag effects* (an initial delay in drug release, as drug initially diffuses through the coating, before attaining equilibrium release conditions) and *burst effects* (an initial burst of release, which may occur if the drug has already diffused through the coating during storage). Nevertheless, although perfect zero-order release is not achieved, a "near-zero" release profile is possible.

Although "near-zero" release is possible, a number of disadvantages are also associated with reservoir DDS. First, there are manufacturing capability complications, as the knowledge, expertise, and equipment to apply specialist coatings are required. Second, there is the susceptibility of these systems to lag times and burst effects. Third, reservoir systems run the risk of dose dumping if the surrounding coat is damaged—tablets must carry a warning for patients not to crush or chew the tablet. Finally, as described earlier, if the polymer coating is not dissolved or otherwise degraded, the drug-voided shell will remain intact and be excreted by the patient. Many of these disadvantages can be overcome by using reservoir DDS as multiparticulate dosage forms, described in Section 7.4.

7.3.3 ORALLY ACTIVE OSMOTIC PUMPS

OROS™ (i.e., *OR*al *OS*motic) technology uses osmotic pressure as the mechanism to control drug release from the delivery system. (*Note:* OROS™ was the original trademark of the Alza Corporation, however, the patent has expired and the term is now used generally, to describe oral osmotic release systems.) Importantly, the osmotic pressure of the GI fluid remains *constant* throughout the GI tract, regardless of the formulation used, or the surrounding environmental conditions. Thus, the extreme variability in conditions that characterizes the GI tract is circumvented, allowing for much more reliable and reproducible oral drug delivery.

Examples of OROS technology include the bilayer push–pull osmotic pump (PPOP) (see also Chapter 2, Section 2.5.1), as used in Glucotrol XL®, for the once-daily administration of glipizide, a blood-glucose-lowering drug for diabetic patients, as well as Procardia XL®, a CR formulation of nifedipine for the treatment of hypertension. Since their introduction by Pfizer in the 1980s, these two drugs have become top sellers in their respective market segments. Other highly successful examples of the PPOP include Ditropan XL® (oxybutynin) and Cardura XL® (doxazosin).

The Oros-Concerta® is a trilayer push–pull system for the delivery of methylphenidate in the treatment of ADHD in children (Rosen and Abribat 2005). Conventional IR preparations of methylphenidate necessitate twice daily dosing, with one of the doses required during school hours, which can be stigmatizing for affected children. To complicate matters further, tolerance to methylphenidate is associated with constant plasma levels of the drug. Oros-Concerta® is a trilayer tablet consisting of one push layer and two separate drug layers: (1) a lower layer, containing a high concentration of methylphenidate, and (2) an upper layer, with a lower drug concentration (Figure 7.4a). An outer IR layer of methylphenidate provides an initial quick onset of active. Then, as the osmotic mechanism begins to operate, drug is released via the orifice. The ascending concentration release profile overcomes the development of drug tolerance associated with constant plasma levels of methylphenidate (Rosen and Abribat 2005; Katzman and Sternat 2014).

The liquid OROS (L-Oros®) system provides for the osmotic-powered delivery of drugs in a liquid form, thus combining the advantages of CR with the high bioavailability of a liquid

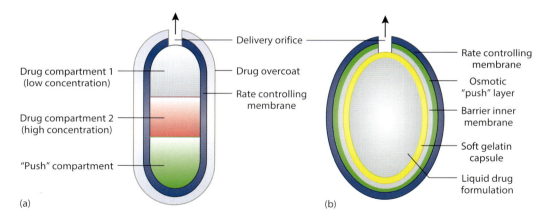

FIGURE 7.4 Oral osmotic drug delivery systems: (a) trilayer system and (b) liquid OROS system.

formulation (Rosen and Abribat 2005). A soft capsule, containing the liquid drug, is in turn surrounded by three further layers: a barrier layer, an osmotic engine layer, and a rate-controlling semipermeable membrane (Figure 7.4b). A delivery orifice is drilled through the outer three layers. When the osmotic engine expands, it compresses the soft gelatin capsule, pushing the drug formulation out through the delivery orifice. The barrier layer ensures the capsule is separated from the osmotic engine, thereby minimizing hydration of the capsule and mixing with the drug layer. The system is under study for the delivery of emulsions and microparticulate suspensions of highly insoluble drugs and polypeptides.

Andrx™ is an oral osmotic system in which the semipermeable membrane contains *two* laser-drilled exit ports, one on either side of the tablet, which facilitates an enhanced drug release from the system. Furthermore, greater dissemination of the drug through the GI lumen is expected on release, in comparison to a single exit point. The technology is used for the once-daily delivery of metformin (Fortamet®) and lovastatin (Altocor®).

Although a highly successful technology, OROS is not without drawbacks, chief among these being the high associated costs of manufacture—as very precise laser drilling is required, necessitating specialist knowledge and equipment. Dose dumping is again an issue, as the entire dose is contained in the drug reservoir. Also, the biologically inert components of the delivery system remain intact during GI transit and are eliminated in the stools as an insoluble shell. Finally, although theoretically perfect zero-order CR can be achieved with this system, it should be remembered that different segments of the GI tract typically demonstrate different absorption capacities. Optimal absorption usually occurs in the upper SI, and the drug absorption efficacy of the tissue typically decreases as the formulation transits along the intestine. Thus, even a zero-order drug release system such as OROS may not actually result in a constant drug concentration in the blood.

7.3.4 MICROELECTRONIC CONTROLLED RELEASE

Newer CR technologies are emerging based on microelectronics, which provide exceptionally precise control over drug release in the GI tract. For example, the InteliSite® Companion Capsule is a radio-frequency activated, nondisintegrating drug delivery device for oral drug delivery. A radio-label included in the drug reservoir permits determination of the capsule location within a specific region of the GI tract via gamma scintigraphy. When the capsule reaches the desired location in the GI tract, the capsule contents are expelled via an active spring-loaded release mechanism. The subsequent release, degree of dispersion, and GI transit of the drug can be followed via radioimaging. The InteliSite® capsule is currently used in drug and formulation development, to precisely determine the rate and extent of drug absorption within specific regions of the GI tract.

The IntelliCap® system is a microelectronic drug delivery pill, comprising a drug reservoir and delivery pump, used in conjunction with an electronic microcontroller, wireless communicator, and sensors. The delivery device is mounted within a small pill-shaped capsule that is swallowed and moves through the GI tract. Within the GI tract, the capsule can take measurements of local pH and temperature. Analysis of the physiological measurements allows accurate localization of the capsule within a specific region of the GI tract. Once the desired site has been identified (e.g., the SI, colon), the capsule dispenses a metered dose of the therapeutic agent, according to a pre-programmed release profile and/or a command issued from an operator. The onboard electronics enable both extremely accurate, and highly adaptable, delivery release patterns.

Although the present generation of microelectronic oral delivery devices are for research purposes only, they do show great promise for the future provision of personalized care. These devices could facilitate the administration of the correct dose, at the correct time, in precisely the desired location in the GI tract. Drug release can be adjusted to the specific needs, and also physiological conditions, of each individual patient.

7.4 MULTIPARTICULATES

In multiparticulate dosage forms, the drug dose is divided out into multiple mini-unit dosage forms, most commonly using beads ("nonpareil" cores, 1–2 mm in diameter), which are then loaded into a gelatin capsule (Figure 7.5).

Multiparticulates can be used in combination with matrix and/or reservoir architecture, to achieve many different drug release options (Skalsky and Stegemann 2011). For example, the spheroidal oral drug absorption system (SODAS®, originally developed by Elan) is based on nonpareil cores, onto which a solution of the active drug is applied. The drug layer is covered by further layers of product-specific functional coating polymers, which provide the RCM (Figure 7.5). Drug release is controlled by diffusion across the membrane; membrane erosion, dissolution, and pore formation may also occur. In the SODAS® formulation for methylphenidate, each Ritalin® LA capsule contains two populations of beads: half of the dose as uncoated beads, for immediate effect; the remainder comprise drug cores layered with enteric-coated, delayed-release polymers for extended release. Numerous other products have been relaunched using the SODAS® technology, all of which permit once-daily dosing with minimal side effects compared to the conventional formulations, including Cardizem® LA (diltiazem), Avinza® (morphine), and Focalin® XR (dexmethylphenidate).

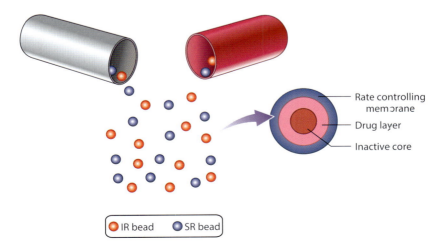

FIGURE 7.5 Multiparticulate drug delivery system for oral drug delivery.

Other examples illustrate the versatility of multiparticulate technology. Diffucaps® multiparticulates provide a delayed-onset release for the antihypertensive drug, propranolol. Innopran® XL is administered at bedtime, and the Diffucaps® polymer-coated beads ensure a delay in propranolol release for 4 hours, ensuring that maximum propanolol plasma levels occur in the early morning: the time when patients are most at risk of adverse cardiovascular events. A further example is Moxatag®, a once-daily extended release amoxicillin tablet formulation, based on the Pulsys™ multiparticulate platform. This technology utilizes different pellet populations, coated with enteric and CR polymers, to provide a triple-pulse delivery of amoxicillin along the GI tract. It is known that pulsatile antibiotic therapy can be a more effective and efficient modality than standard antibiotic regimens. Finally, as an alternative to using beads as the cores, Minitabs® are tiny (approximately 2 mm in diameter) cylindrical tablets, which can be coated with RCMs. This modality combines the simplicity of tablet formulation with the versatile CR capabilities of multiparticulates.

Multiparticulate dosage forms are technically more complicated, time consuming, and expensive to produce, requiring specialist know-how and equipment. However, they offer several advantages over their single-unit dosage counterparts:

1. *CR*: Multiparticulates can be used to provide a wide range of drug release patterns (e.g., IR, SR, CR, ER, delayed-onset release, pulsatile biphasic and triphasic release). The final capsule can be loaded with a mixture of different bead populations, thereby facilitating multiple release profiles and/or multiple active ingredients—all within a single dosage form.
2. *Reduction in GI variability*: Multiparticulates are not subject to the vagaries of gastric emptying. Gastric emptying is an "all-or-nothing" process, so that a single-dose solid dosage form such as a tablet is either all in the stomach or all in the duodenum. Due to the large inter- and intrasubject variation in the rate of gastric emptying, there can be extreme variability in the bioavailability with single-dose forms (SDFs). In contrast, when the gelatin capsule of a multiparticulate formulation dissolves in the stomach, hundreds of coated pellets are released, all small enough to pass through the pyloric sphincter, even if the sphincter is closed. Variability issues, such as whether the patient is in a fasted or fed state or the patient's posture, are also minimized. Furthermore, the pellets are widely dispersed throughout the GI tract, which tend to reduce the effect of variations in GI motility, and GI pH, along the tract.
3. *Avoid dose dumping*: In a multiparticulate system, the dose is distributed among hundreds of mini-units. Many hundreds of pellets would have to fail before dose dumping could occur, making it a far less likely event than for a single unit Solid Dosage Form (SDF).
4. *Convenience*: Multiparticulates are more convenient for the patient, as the capsules can be opened and sprinkled onto food or beverages ("sprinkle dosing"), a particular advantage for pediatric and geriatric patients, who often experience difficulty swallowing tablets.

7.5   REGIONAL TARGETING FOR ORAL DRUG DELIVERY

Microelectronic-controlled drug release has been described in Section 7.3.4 and features oral capsules that, via remote signaling, can release their payload at the desired site in the GI tract; as such, they are capable of very precise regional targeting. Here, we describe other mechanisms to achieve regional drug targeting, to the stomach, SI and colon, respectively. Targeting oral vaccines to the Peyer's patches (PP) is a further area of research and is described in Section 7.7.

7.5.1   Regional Targeting to the Stomach: Gastroretention

For drugs with a narrow absorption window, there is a risk that they may transit their optimal absorbing region too rapidly, limiting the extent of absorption possible. Gastroretention (GR) strategies

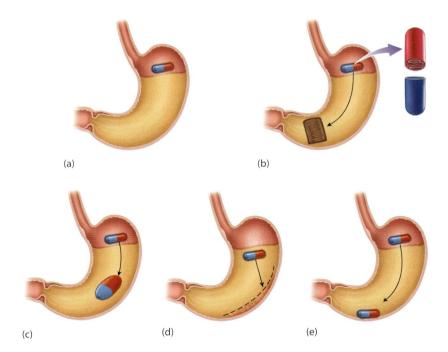

(a)   (b)

(c)   (d)   (e)

FIGURE 7.6 Gastroretentive drug delivery systems: (a) floating (effervescent, hydrodynamically balanced system, porous foam powders), (b) expandable, (c) swelling (superporous hydrogels, swelling polymers), (d) bioadhesive, and (e) high density.

retain a drug in the stomach for an extended period, allowing it to be slowly and gradually released, for subsequent optimal absorption in the SI. GR strategies are also useful for drugs absorbed via saturable transport mechanisms in the upper SI (e.g., L-dopa, gabapentin): a slow, extended presentation of the drug to its transporter minimizes the possibility of transporter saturation. GR devices can also be used to retain drugs in the stomach for local action (e.g., antacids, antibiotics to treat *Helicobacter pylori* for ulcer therapy, cytotoxics for gastric cancer) and for drugs that have poor solubility at small intestinal pH, e.g., cinnarizine.

To achieve successful GR, formidable physiological challenges must be overcome; in particular, the dosage form must be retained in the stomach during the very powerful housekeeper wave of the migrating motor complex. A wide variety of GR systems have been developed, using many different approaches (Streubal et al. 2006; Pawar et al. 2011). The principal approaches are described here and summarized in Figure 7.6.

7.5.1.1 Floating Drug Delivery Systems

In this approach, the DDS floats on top of the stomach contents, retarding transit to the pylorus and prolonging gastric residence time (Pawar et al. 2011; Prajapati et al. 2013). Obviously, sufficient fluid must be present in the stomach to support these systems. Buoyancy can be achieved in a number of ways (Figure 7.6a). Effervescent systems are based on the inclusion of bicarbonate, citrate, or tartrate and cause the release of CO_2, which becomes partially entrapped in the dosage form and provides sufficient buoyancy for floatation. Cifran OD® uses a gas-generating system for the gastroretention of ciprofloxacin. Liquid Gaviscon® contains sodium alginate, which reacts with HCl in the stomach to form a thick, alginate, raftlike layer. HCl reacts with calcium carbonate and sodium hydrogencarbonate in the formulation, to form bubbles of CO_2, which become trapped in the alginate raft, enabling it to float on the surface. The raft forms a strong physical barrier against reflux, while antacids in the formulation neutralize excess HCl.

A hydrodynamically balanced system (HBS) comprises a floating capsule, based on the use of gel-forming hydrocolloids, which then entrap air for buoyancy. The drug is mixed with swellable, gel-forming, hydrophilic polymers such as HPMC and then encapsulated. On ingestion, the capsule disintegrates and water starts to penetrate the system, causing swelling and hydration of the polymers. The air trapped by the swollen polymers allows the system to float. In addition, the gel layer provides diffusion-controlled sustained release. The HBS system is available commercially for the delivery of levodopa (Madopar® HBS), a drug that is absorbed by a saturable carrier system from the SI and thus benefits from a slower arrival at the SI. Madopar® shows less side effects and less "on–off" phenomenon than conventional formulations. Valrelease® also uses an HBS system for the sustained delivery of diazepam. Diazepam is preferentially absorbed in the stomach, due to its increased solubility in acidic pH, making it an ideal candidate for gastric retention.

Porous foam powder systems incorporate a highly porous, polypropylene foam powder in the formulation, to provide an inherent low density. Both single-unit systems and also floating microparticles have been shown to have excellent in vitro floating properties and a broad spectrum of drug release patterns.

7.5.1.2 Size-Increasing Delivery Systems

These DDS are initially small, to facilitate swallowing, but then they massively, and rapidly, increase in size in the stomach, so that the system becomes physically too large to pass through the pyloric sphincter. The products must not increase so much in size that they adversely affect GI function. An example of an expandable system is the Accordion Pill™, in which the drug, or combination of drugs, is loaded onto a polymeric film, which is then folded into an accordion-like shape and packed into a gelatin capsule (Figure 7.6b). After the capsule dissolves in the stomach, the accordion unfolds and is retained in the stomach for up to 12 hours. Drug is released from the expanded system via diffusion. A variety of other expandable systems are currently being investigated, which expand into various complex shapes in vivo. A major challenge with this approach is the difficulty in the manufacture of such complex constructs.

Superporous hydrogels are a class of hydrogels with, as the name implies, very large pore sizes (more than 100 µm, which contrasts with the nanometer size range of conventional hydrogels). The large pore size allows them to swell much faster, and to a greater extent, than conventional hydrogels, so that the systems become too large to pass through the pyloric sphincter (Figure 7.6c) (Chen et al. 2000). The addition of excipients such as croscarmellose sodium confers high mechanical strength on the system, enabling it to withstand the mechanical force of the peristaltic waves.

Acuform™ is a tablet that absorbs water from the gastric fluids and expands to 150% its original size. Patented swelling polymers facilitate the uptake of water, and the enlarged tablet is retained in the stomach for approximately 8–10 hours, while drug slowly diffuses out. Acuform™ technology is used for the once-daily oral administration of gabapentin (Gralise®), which is absorbed via the L-amino acid carrier in the jejunum. The IR formulation results in carrier saturation, hence multiple daily doses are required. Gralise® appears to cause less dizziness and sleepiness than the IR gabapentin formulation, so its advantages go beyond a simple convenient argument for a once-a-day tablet.

7.5.1.3 Bioadhesive Systems

Mucoadhesive microparticles based on the hydrophilic, gel-forming polymer Carbopol® 934P have shown promise in preclinical studies for the GR of furosemide, amoxicillin, and riboflavin, compared to suspensions. Other microparticles and minitablets designed for gastric retention use chitosan, thiolated chitosan, and polycarbophil as the mucoadhesive polymer, with a range of payloads including tetracyclines and low-molecular-weight heparin (Figure 7.6d). Still, none have reached the market and there are major issues concerning high mucus turnover and weak gastric mucus adherence, leading to unpredictable retention times. Mucoadhesive polymers are discussed further in Section 7.6.1.4, as formulation excipients to improve oral bioavailability.

7.5.1.4 High-Density Systems

High-density systems incorporate excipients such as barium sulfate, titanium oxide, and iron salts, with a higher density than gastric acid (usually >3 g/cm^2) to provide a sinking system (Figure 7.6e). Intuitively, this would seem to be an obvious approach, but there are technical difficulties in compacting high concentrations of active and preventing premature release: no system has yet reached the market.

7.5.1.5 Other Approaches

Other research approaches include magnetic systems with external remote control of transit time. Ion-exchange resins can also be formulated to have gastroretentive properties. Resin beads can be entrapped with bicarbonate, an anionic drug is then attached to the resin and the composition is loaded into a semipermeable construct. When it reaches gastric juice, bicarbonate and chloride are exchanged, but the resulting gas is withheld in the membrane and causes the system to float. However, these approaches are still at the research stage.

Of interest is the development of gastroretentive intraruminal devices in large animal veterinary medicine. These large devices are designed to release antiparasitic molecules including ivermectin and morantel over the 9-month growing season in cattle and are designed to stay in the rumen after administration with a balling gun. Elegant intraruminal device strategies comprise principles of expansion, osmotic release, and rolled-up polymers with drug embedded in a porous matrix. Due to the different physiology between the monogastric human and ruminal stomachs, technology development has been parallel, but distinct, in respect of engineering designs, payloads, and clinical indications.

7.5.2 REGIONAL TARGETING FOR SMALL INTESTINAL DELIVERY

An enteric coating is designed to prevent drug release in the stomach; instead, release is delayed until the SI. Such coatings have a long history in pharmaceutical practice and have traditionally been used either to prevent the drug from causing stomach irritation (e.g., aspirin) or to prevent the acidic environment of the stomach from causing premature drug degradation (e.g., enteric coatings for penicillins). Many oral peptide formulations currently in clinical trials have enteric coatings designed to undergo initial dissolution at the pH of duodenal and jejunal fluid, thereby protecting the labile peptide from acid degradation in the stomach (see Section 7.6).

Almost all enteric coatings currently in use are synthetic, or semisynthetic, pH-sensitive polymers containing ionizable carboxylic groups. These remain unionized (and thus insoluble) in the acidic environment of the stomach, but become ionized (and thus soluble) in the less acidic conditions of the SI, thereby allowing the coating to dissolve and the drug to be released.

The most widely used group of polymers to achieve enteric coating are the poly(methacrylates), marketed as the Eudragit® L and Eudragit® S series. A wide range of copolymers within this series is available, allowing drug release across a wide range of pH conditions (Table 7.2). Cellulose-based derivatives are a further option, from either cellulose itself, such as cellulose acetate phthalate (CAP), or derivatives of the hydrophilic polymer HPMC, such as HPMC phthalate (HPMCP). Poly(vinyl) derivatives include poly(vinylacetate phthalate) (PVAP).

An extension of this idea is the use of pH-sensitive hydrogels, whereby the drug is trapped within a 3D hydrogel matrix that exhibits negligible swelling in the acidic conditions of the stomach, but drastically increased swelling in the alkaline environment of the SI, releasing drug (Figure 7.7). One of the most extensively studied of these systems comprises an anionic hydrogel comprising PEG grafted on poly(methacrylic acid) (Peppas et al. 2006).

TABLE 7.2

Threshold pH of Commonly Used Polymers for Enteric Coatings

Polymer	Threshold pH
Eudragit® L100	6.0
Eudragit® S100	7.0
Eudragit® L30D	5.6
Eudragit® FS30D	6.8
Eudragit® L100-55	5.5
Polyvinyl acetate phthalate	5.0
HPMCP	4.5–4.8
HPMCP 50	5.2
HPMCP 55	5.4
Cellulose acetate trimellitate	4.8
Cellulose acetate phthalate	5.0

Abbreviation: HPMCP, hydroxypropyl methylcellulose phthalate.

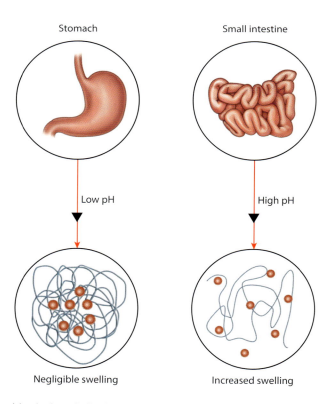

FIGURE 7.7 pH-sensitive hydrogels for drug targeting to the small intestine.

7.5.3 REGIONAL TARGETING FOR COLONIC DELIVERY

Targeting drugs for local delivery to the colon is particularly appropriate for the topical treatment of inflammatory bowel disease (IBD), including ulcerative colitis (UC) and Crohn's disease. First-line drug treatments for UC include (1) the anti-inflammatory drug, mesalazine (also known as 5-aminosalicylic acid [5-ASA]), which acts locally on the colonic mucosa to reduce inflammation, and (2) the anti-inflammatory corticosteroid, budesonide. A number of mechanisms are used to achieve drug targeting to the colon (Lichtenstein 2009; Omidian and Park 2010; Perrie and Rades 2012) and are summarized here:

1. *Rectal delivery*: A cohort of UC patients have restricted inflammation limited to the sigmoid colon, which is amenable to rectally delivered foams, liquid enemas, and suppositories. Foams incorporating mesalazine, or budesonide, are low in volume, adhesive, and viscous; they disperse as far as the descending colon. Scintigraphy studies with mesalazine reveal that foams can disperse more widely than enemas, for left-sided UC. Foams are well tolerated by patients and their local delivery means that systemic side effects are minimized.

2. *pH-triggered colonic delivery*: The most commonly used pH-dependent coating polymers for colonic targeting are Eudragit® L100 and Eudragit® S100, which dissolve at pH 6.0 and 7.0, respectively (Table 7.2). Drug release within this range can be tailored by using polymer combinations in various ratios. A number of commercial preparations based on pH-sensitive Eudragit-S resin coatings are available for the colonic delivery of mesalazine (e.g., Asacol®, Claversal®, and Mesasal®) and also budesonide (e.g., Entocort® and Budenofalk®).

3. *Time-delay colonic delivery*: The drug can be incorporated within a matrix- or reservoir-CR system comprising poorly soluble, slowly eroding polymers so that drug release is delayed until the colon is reached. A lag time of 5 hours is usually considered sufficient, given that gastric emptying takes about 2 hours and SI transit is about 3–4 hours. An example is the Pentasa® technology, based on the delayed release of mesalazine, using pellets coated with EC. More experimental time-delay systems include the following: (1) The Pulsincap™ uses a swellable hydrogel plug to initially seal an API within an insoluble capsule body. As the system passes through the GI tract, the hydrogel plug swells until eventually it is ejected from the capsule, thereby liberating the drug. (2) The OROS-CT™ is an oral osmotic device for colon targeting, based on as many as five to six push–pull osmotic units filled into a hard gelatin capsule (see Section 7.3.3). This allows for a lag time of 3–4 hours, thereby facilitating colonic targeting. However, neither of these approaches has been commercialized.

4. *Combination approaches*: Both the pH and time-delay methods outlined earlier are limited by the wide variability that exists in the GI tract with respect to GI pH and transit time. This results in a lack of consistency and reproducibility with these approaches. In order to improve colon-targeting efficacy, these strategies can be combined. Lialda® delivers mesalazine to the colon using a MultiMatrix (MMX®) technology. This sophisticated system uses an outer enteric layer of Eudragit-S film, which delays initial release of the active drug until the tablet is exposed to a pH of 7.0 or higher. As the coating disintegrates, intestinal fluids interact with an inner hydrophilic matrix, resulting in the formation of a viscous rate-controlling gel, which provides CR via diffusion. An additional lipophilic matrix, interspersed within the hydrophilic matrix, slows down the penetration of aqueous fluids into the tablet core, reducing the rate of drug dissolution, thereby prolonging drug release and therapeutic activity. The hydrophilic matrix may also adhere to the colonic mucosa, this bioadhesion can further facilitate targeted drug delivery.

Recent studies in man have revealed that nanoparticles have an inherent attraction to the inflamed colon in UC (Schmidt et al. 2013). The prospect of a combination approach, using pH-dependent coated capsules which release targeted nanoparticles for local delivery of payload in the inflamed colon, is therefore under intense research scrutiny. Research has focused on the use of polymeric nanoparticles based on poly(lactide-co-glycolide) (PLGA), although trimethylated chitosan, solid–lipid nanoparticles (SLNs), and lipid-based microemulsions are also being investigated.

7.5.3.1 Microbially-Triggered Colonic Delivery

Although bacteria are present throughout the GI tract, the colonic region is much more heavily colonized than the upper regions, due to the decreased peristalsis, presence of feces, and lower oxidation-reduction potentials of this region. These resident microflora are a highly specific environmental feature of the colon, thus making it a more selective approach for colon-targeting than methods such as pH-sensitivity or time-control.

One of the earliest prodrug approaches for local colonic delivery for the treatment of IBD was using sulphasalazine, which comprises a dimer of the anti-inflammatory 5-ASA (mesalazine) linked via an azo bond to a carrier molecule, sulphapyridine. The majority of the sulphasalazine reaches the colon intact, where the conjugate is cleaved by colonic microflora to the active metabolite, 5-ASA, and the by-product, sulphapyridine. Colon-specific targeting is achieved due to the absence of azoreductase activity in the upper parts of the GI tract and the limited absorption of the prodrug there. As it is the sulphapyridine by-product that seems to be responsible for most of the side effects of sulphasalazine (including hypersensitivity reactions), various new azo-bond prodrugs have been developed, substituting sulphapyridine for non-toxic molecules. Dipentum® (olsalazine) comprises an azo-linked dimer of 5-ASA molecules, which is cleaved in the colon to release two molecules of 5-ASA for topical action. Colazal® (balsalazide), a prodrug of 5-ASA attached via an azo bond to 4-aminobenzoyl-β-alanine, releases 99% of the 5-ASA in the colon.

Naturally-occurring polysaccharides which are susceptible to colon-specific microfloral degradation are also being investigated, including amylose, chitosan, guar gum and pectin. The Colal-Pred® delivery system comprises small, prednisolone-containing pellets, coated with a combination of amylase (for colon-specific targeting) and the water insoluble polymer EC (to minimize over-swelling of the highly water-soluble amylase, thereby avoiding premature prednisolone release).

7.6 IMPROVING ORAL BIOAVAILABILITY

7.6.1 Formulation Strategies to Improve Oral Bioavailability

The use of formulations strategies has been investigated in particular for the oral delivery of therapeutic peptides and proteins, such as insulin. The subject has been extensively reviewed (Maher and Brayden 2011; Maher et al. 2014; Park et al. 2011); here, only the most clinically advanced approaches are described.

7.6.1.1 Absorption Enhancers

Low concentrations of surfactants (including sodium lauryl sulfate, polyoxyethylene-9-lauryl ether, and polyoxyethylene-20-cetyl ether) have been studied as absorption enhancers (AEs). Being amphipathic molecules, they can associate with the amphipathic bilayers of the plasma membranes of epithelial cells, thereby enhancing the fluidity and permeability of the bilayer and promoting transcellular transport. Bile acid salts (e.g., sodium taurocholate), and their derivatives, are also thought to function in part via membrane-solubilizing effects.

Other AEs work by transiently loosening the tight junctions (TJs) between cells, thereby facilitating paracellular transport. Many candidates are under study, with medium chain fatty acids such as sodium caprate (a C10 fatty acid) and sodium caprylate (a C8 fatty acid) showing particular promise. At low concentrations, sodium caprate causes a contraction of scaffolding proteins that maintain TJ integrity; this action causes the removal of claudin-5 and tricellulin, thus enhancing paracellular permeability. At higher concentrations, however, its surfactant properties results in mild membrane perturbation, and this may be the initiating event that causes intracellular enzyme cascades that impact the tight junction especially. Sodium caprate is a formulation excipient in GIPET™ (Gastrointestinal Permeation Enhancement Technology), which also uses other AEs, including other medium chain fatty acid derivatives, as well as microemulsion systems based on medium chain fatty acid glycerides. The drug and different AEs are formulated as enteric-coated tablets or capsules. Sodium caprate has also been studied in an enteric-coated oral formulation of antisense oligonucleotides in the 1990s.

Sodium caprylate also enhances absorption via tight junction disruption/paracellular pathway. In the transient permeability enhancer™ (TPE™) technology, the drug is initially solubilized with sodium caprylate and PVP in water. The resulting solution is then lyophilized and the powder is dispersed in a lipophilic medium. This drug/oil suspension is filled into enteric-coated capsules. Trials are ongoing using the TPE™ technology for the oral delivery of octreotide acetate, which is currently given via long-acting injection.

In the Peptelligence™ approach, a number of complementary factors combine to enhance absorption for peptides. An enteric coating on the tablet provides initial protection in the acidic stomach. In the duodenum, the enteric coating dissolves and a disintegrant in the formulation ensures the tablet contents are rapidly released. Citric acid, contained within the core, quickly reduces the intestinal pH in the immediate vicinity. Pancreatic proteases, which have a neutral/alkaline pH range for activity, are thus prevented from degrading the API in this acidic microenvironment. The simultaneous release of lauroyl-L-carnitine, a C12 fatty acid acylcarnitine, further enhances peptide absorption, again by transiently loosening the tight junctions and increasing paracellular transport. The technology is used in Oracal®, an oral formulation of calcitonin for the treatment of osteoporosis, which has recently completed a Phase 3 clinical trial.

A protein oral delivery (POD™) technology for the oral delivery of peptides and proteins uses EDTA as an AE, in combination with a protease inhibitor (e.g., soybean trypsin inhibitor) and a lipoidal carrier (vegetable or fish oil or a synthetic omega-3 fatty acid), administered as enteric-coated capsules. The chelating agent EDTA forms complexes with calcium ions and thus ruptures tight junctions, again facilitating absorption via the paracellular pathway. Listed in the USP, EDTA is considered nontoxic and nonirritant at "allowable" levels.

To be successful, the membrane-perturbing effects of AEs must be local, rapid, and mild—otherwise, there could be a risk of more permanent membrane damage, especially for drugs requiring repeat administration. Given the possible safety risks, as well as the cost of working with new chemical entities as AEs, the pharmaceutical industry has focused on simple enhancers that have a history of safe use and are designated generally regarded as safe (GRAS) status based on prior use as food additives, or FDA "allowed" excipients that can be manufactured cheaply (Maher et al. 2009). Multiple clinical trials of a range of enhancers suggest that the most advanced candidates have good safety profiles. It should also be remembered that the GI tract is sufficiently robust to withstand the daily onslaught and damage afforded by food and fluid intake, as well as bile and gastric secretions. The robustness of the GI tract can be attributed to a number of factors, including the rapid and efficient repair of the GI epithelium within hours, rapid turnover of the epithelial lining in 4–5 days, and dilution effects in the aqueous GI contents. A further issue regarding the safety of AEs is the risk of possible bystander absorption, whereby breaching the membrane barrier could inadvertently facilitate the entry of potentially harmful agents such as bacteria and viruses. This risk is unproven to date, but seems unlikely, given that the extent of tight junction opening and the induced membrane porosity is orders of magnitude smaller than the size of bacteria and viruses.

7.6.1.2 Enzyme Inhibitors

Research into the use of EIs has particularly focused on the use of proteolytic EIs, in order to enhance the oral absorption of peptide and protein drugs (Bernkop-Schnürch 1998). A wide variety of peptidase inhibitors (PIs) have been studied in man, including soybean trypsin inhibitor and aprotinin, although both have been associated with pharmacology and safety issues. Some recently identified EIs include chicken and duck ovomucoids, derived from the egg white of avian species, which show promise in the protection of insulin from pancreatic enzymes, but do not have associated safety problems.

There is, however, a lack of specificity when using EIs: an incorporated EI may inhibit the breakdown of proteins from dietary sources, thereby decreasing their absorption, with possible toxic consequences. Efforts have been directed toward improving EI specificity. Examples include the Peptelligence™ technology, outlined in Section 7.6.1.1, whereby a localized release of citric acid confines the inhibition of pancreatic enzyme activity to the immediate vicinity surrounding the tablet. Similarly, the protease inhibitor used in the POD™ technology (also outlined in Section 7.6.1.1) is contained within a lipoidal carrier, again to enhance specificity.

A further approach to protect labile molecules involves the physical complexation of an API with an absorption-promoting carrier molecule. N^{α}-deoxycholyl-L-lysyl-methylester (DCK), a dihydroxyl bile acid derivative, is synthesized by conjugating the amino acid, L-lysine, with deoxycholate. Being positively charged, DCK can physically form a complex in solution with the anionic residues of a drug molecule. The reversible association of DCK with insulin offers protection to insulin from intestinal protease degradation. Absorption may also be enhanced due to the increased lipophilicity of the complex and (given that the carrier is a bile acid derivative) the possibility of enhanced interaction with the bile acid uptake transporters present in the GI tract.

7.6.1.3 Solubility Enhancers

As described in Chapter 3, the solubility of an API is a crucially important factor for oral bioavailability. A drug must initially dissolve in the GI lumen contents, prior to permeating the epithelial layer. Poorly soluble drugs will thus demonstrate limited oral absorption. Furthermore, poor solubility results in a low concentration gradient across the enterocytes, which is a poor driver for transepithelial transport. Poorly water-soluble drugs also demonstrate a lack of dose proportionality, as well as significant intra- and intersubject variability, which is exacerbated by the presence, or absence, of food and fluid in the GI lumen. A variety of formulation approaches are used to improve the solubility of poorly water-soluble drugs and these are discussed fully in Chapter 3.

7.6.1.4 Mucoadhesives and Mucolytics

Widely studied mucoadhesives include Carbopol® (carboxypolymethylene), a synthetic copolymer of acrylic acid and allyl sucrose, which has shown enhancing effects for oral insulin in rodents. Chitosan, a cationic polysaccharide comprising copolymers of glucosamine and N-acetylglucosamine, is strongly mucoadhesive, due to the formation of hydrogen and ionic bonds between the positively charged amino groups of chitosan and the negatively charged sialic acid residues of mucin glycoproteins. This polymer also has some absorption-enhancing effects, as it causes opening of tight junctions, which facilitates the paracellular transport of hydrophilic compounds. Thiolated polymers (thiomers) are mucoadhesive polymers with thiol-bearing side chains, capable of not only physical interaction with mucus but also actual covalent bond formation between the thiol groups of the polymers and cysteine-rich subdomains of mucus glycoproteins. These polymers also have permeation enhancement action and can inhibit P-gp efflux pumps. Preclinical studies have shown the potential of thiomers for oral insulin enhancement. ThioMatrix is exploiting this potential for the noninvasive delivery of a variety of APIs, using their proprietary Thiomer Technology™.

There is again, however, a lack of specificity with this formulation approach: mucoadhesion cannot be targeted to a specific area of the SI, but may occur much higher in the GI tract, for example, in the stomach, where the mucus lining is much thicker than in the SI. A further limitation is the rapid turnover of intestinal mucus, which will limit the amount of time the drug can be realistically retained at the absorbing surface. A mucoadhesive system may be sloughed off and released into the lumen, without ever gaining access to the enterocyte surface (Ensign et al. 2012).

A different line of research is focused on the use of mucopenetrating polymers and mucolytics, in order to minimize the mucus diffusional barrier and thus improve access to the apical membrane of the enterocytes. Mucolytics, such as acetylcysteine, can minimize the diffusional barrier of the overlying mucus gel layer of the GI epithelium. Hydrophilic PEG, hydroxypropyl methacrylate (HPMA), and poly(sialic acid) confer a "slippery surface" on nanoparticles, so that entrapment in the mucus glycoprotein meshes can be avoided. However, this hydrophilic surface coating may subsequently limit uptake via passive diffusion through the hydrophobic cell membranes of the enterocytes.

7.6.2 Micro- and Nanoparticulate Drug Delivery Systems

From the early 1980s, a wide array of micro- and nanoparticulate DDS have been intensively investigated as carriers to facilitate oral drug delivery and targeting; some excellent reviews are available (Pouton and Porter 2008; Bakhru et al. 2013; Yun et al. 2013). Studies have focused in particular on the oral delivery of peptides and proteins, especially insulin; the current state of research spans from proof-of-concept studies to late-stage clinical studies.

Nanocarriers investigated for the oral route include microparticles, microspheres, nanoparticles, nanocapsules, liposomes, polymeric micelles, niosomes, dendrimers, SLNs, and microemulsions. A wide range of polymers, including physiological lipids, as well as synthetic, semisynthetic, and natural polymers (including poly(alkyl cyanoacrylates), PLGA, chitosan, alginate, and poly(ethylenimine)), have been studied for the fabrication of nanocarriers. Micro- and nanoparticulate DDS can enhance oral bioavailability via a number of different mechanisms, as outlined in Box 7.1.

Targeting ligands can be associated with nanocarriers, to facilitate cell-specific uptake (Box 7.1). Coupling the targeting vector to a nanoparticulate carrier, rather than an API per se, has the advantage that the conjugation in no way interferes with the biological activity of the API; it also allows for high drug loading. Many types of ligands for epithelial targeting have been described, including lectins, sugars, vitamins (in particular, vitamin B_{12}), peptides, cholesterol, folic acid, and albumin (Li et al. 2014). However, it should be remembered that targeting vectors are only beneficial if the delivery system can actually gain proximity to the corresponding cellular target, but this is in no way guaranteed. Many barriers (including the GI luminal contents, the mucus layer, the presence of enzymes and bile salts) can prevent access of the DDS to the enterocyte surface. Other problems

BOX 7.1   MECHANISMS BY WHICH DRUG DELIVERY SYSTEMS CAN ENHANCE ORAL BIOAVAILABILITY

- Improved drug solubilization
- Protection from enzymatic degradation
- Enhanced interfacial area
- Increased lipidity
- Enhanced transepithelial transport
- Possibility of enhanced lymphatic uptake
- The capacity to incorporate:
  - Targeting ligands
  - Mucoadhesives/mucolytics
  - Enzyme inhibitors
  - Absorption enhancers
  - Solubility enhancers

with the approach include low, or variable, receptor expression; possible carrier saturation; and interference with dietary absorption mechanisms.

A full description of these delivery systems is given in Chapter 5 (Sections 5.4 and 5.5), and the reader is referred there for further information on the delivery systems per se; the focus here is on specific micro- and nanoparticulate DDS showing potential for improved oral bioavailability. The Lipid Polymer Micelle™ (LPM) technology uses reverse micelles to enhance oral bioavailability. A hydrophilic micelle core protects a water-soluble peptide, leuprolide, from degradation. The surfactants (short- and medium-chain fatty acid esters) used to form the micelle may also assist in transepithelial permeation. The reverse micelle is stabilized by either the inclusion of a synthetic/semisynthetic polymer (gelatin, poly(lactide)) or interfacial polymerization, in order to prevent phase inversion of the system on dilution in the aqueous GI contents. LPM™ technology has shown improved oral bioavailability for leuprolide in animal studies.

CobOral™ comprises a dextran nanoparticle for the oral delivery of peptides. The drug-loaded nanoparticle is coated with cobalamin, a vitamin B_{12} analog, to exploit receptor recognition by intrinsic factor–B12 receptors on enterocytes. Entrapment within nanoparticles affords protection of the peptide from enzymatic degradation, while targeting to the receptor facilitates oral absorption. Similarly, Oradel™ is a vitamin B_{12}-coated carbohydrate-based nanoparticle containing insulin. It is unclear if these approaches are still in development.

Serum-specific nanoencapsulated particles™ (SSNe) are composed of a natural polymeric backbone (carrageen or alginate), with covalently linked C20–C40 alkyl side chains, which form multiple double-helix aggregates, approximately 35 nm in size, containing the drug within the aqueous helical core. The alkyl chains are exposed as "spikes" that protrude externally from the nanoparticle surface and are claimed to offer protection to the entrapped drug from enzymatic degradation, as well as promoting absorption due to their pronounced lipidity.

Liposomal formulations have not yet demonstrated success by the oral route, and progress in this field is hindered by their poor GI stability and poor transepithelial permeation. Solid lipid nanoparticles (SLNs) represent a more stable alternative, and a variety of constructs are under study by different research groups, including SLNs containing mineral oil, for the oral delivery of leuprolide; ethyl palmitate SLNs for oral insulin; and stearic acid and tripalmitin SLNs for the oral delivery of salmon calcitonin.

7.6.3 CHEMICAL CONJUGATION APPROACHES TO IMPROVING ORAL BIOAVAILABILITY

The chemical conjugation of a drug molecule to another molecule can improve oral bioavailability via a number of mechanisms (Al-Hilal et al. 2013). Conjugation of the API with cell surface–specific ligands offers the potential for drug targeting to enterocytes and/or uptake via specific carrier systems. Increasing the lipophilicity of a drug by conjugating it with lipidic moiety may enhance transcellular passive absorption. Drug conjugation with hydrophilic polymers, such as low-molecular-weight chitosan or PEG, can increase a drug's solubility and hence its bioavailability, as well as offering protection from enzymatic attack.

The chemical conjugation approach is limited by the chemical nature of the drug and the availability of reactive groups on it, as this determines the type of chemical linkages possible. It may be possible to add functional groups to the native drug, to facilitate subsequent conjugation. Crucially, chemical conjugation must not compromise the biological activity of the original drug molecule. The resulting chemical conjugate constitutes a new chemical entity, which raises significant regulatory challenges. The nature of the conjugating link is important: if the resulting linkage is stable in vivo, the entire drug conjugate can be absorbed. Significant amounts of the carrier may also be absorbed, which could result in safety issues. An alternative approach is to use labile drug-conjugate links, from which the drug is subsequently released in vivo. Although chemical conjugation approaches have been very successful for drugs intended for local action in the GI tract (see, for example, Section 7.5.3.1 for the use of azo prodrugs for the local delivery of anti-inflammatories to the colonic mucosa), problems arise when systemic delivery is the aim. This is because a drug released from its carrier in the lumen of the GI tract is then subject to enzymatic degradation and other harsh environmental conditions; the epithelial barrier also remains to be surmounted.

7.6.4 SYSTEMIC DELIVERY VIA THE COLON

The colon appears to be more amenable to permeation by poorly absorbed molecules than higher up in the GI tract, the implication being that it may be a more suitable site to enable systemic absorption of therapeutic peptides and proteins. The increased permeability is thought to be in part because the apical membrane of the colonic epithelium is more fluidic than in the jejunum, therefore basal peptide flux, and sensitivity to permeation enhancers, is higher. Furthermore, although peptidases are present in colonic enterocytes, the challenge of luminal pancreatic peptidases is greatly reduced. The longer residence time in the colon increases contact time of the API with the absorbing surface, again favoring absorption. Finally, the colonic blood supply reaches the liver indirectly, not via the hepatic portal vein, negating the first-pass effect.

However, in order for systemic delivery via the colon to be successful, several problems need to be addressed. These include the minimum 4-hour delay that exists, from swallowing a dosage form to the appearance of the drug in the blood. There is also the safety issue of increased permeation in the presence of a high microfloral population. Variable gastric emptying and small intestinal transit times result in highly unpredictable delivery. Finally, accurate dissolution of formulations can be problematic, due to the presence of gas, feces, and unpredictably located pockets of aqueous solution. This is largely why local delivery *to* the colon has been more successful than systemic delivery *from* the colon to date.

Approaches to deliver APIs to the colon for systemic absorption are similar to those used for regional targeting for local colonic delivery (Section 7.5.3) and are predominantly based on pH-triggered delivery, time-delay methods, microbially triggered delivery, and the use of absorption enhancers and EIs (Maroni et al. 2012). Advanced concepts are based on the premise that even if colonic regional targeting of a solid-dosage form is achieved, a peptide still requires protection and transepithelial flux upon release from the dosage form, and that this can be afforded by nanoencapsulation, with an option to include peptidase inhibitors (PIs) and AEs. PIs may be safer in the colon, as necessary peptide digestion for nutrition is not relevant here, in comparison to the jejunum. Evidence suggests that nanoparticles made from chitosan, polycarbophil, and dextran also achieve this peptidase inhibition aim. While there is little doubt that systemic oral peptide delivery from the colon can be achieved if the formulation can be regionally delivered in a predictable fashion, there is little information on the risks of repeat-dose administration of enhancer-based formulations, nor if pathogens can be activated by such systems to cause IBD-like dysfunction via access to immunocompetent cells underlying the epithelium.

7.7 ORAL VACCINES

One of the most important attempts at regional targeting in the GI tract is for oral vaccination, with subunit antigens designed to be taken up by ileal PP. PP are the main gut-associated lymphoid tissue and approximately 150 line the SI, with a predominance in the ileum. Disorganized lymphoid follicles occur in the colon, but their role is uncertain.

The role of PP is one of immune surveillance, sampling, and presenting particulate antigen via microfold (M) cells to the underlying professional antigen presenting B lymphocytes, macrophages, and dendritic cells. When pathogens including *Salmonella*, *Yersinia*, reovirus, and poliovirus are presented, the outcome is generation of mucosal immunity, whereby secretory IgA is transported from the blood to the gut lumen, to attach and remove specific pathogens upon future exposure (see also Chapter 17, Figure 17.3).

In man, approximately 5% of M cells comprise the follicle-associated epithelium (FAE) in each follicle. Unlike intestinal enterocytes, M cells of the PP are capable of extensive endocytic uptake of macromolecules and microparticles. They are designed to access particulates in the 200 nm–3 μm

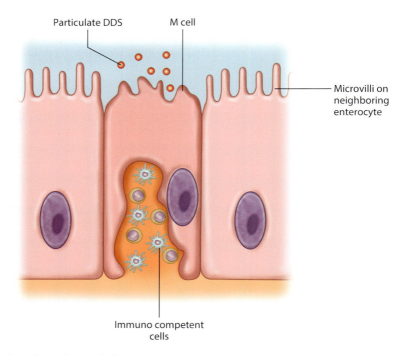

Particulate DDS   M cell

Microvilli on
neighboring
enterocyte

Immuno competent
cells

FIGURE 7.8   Peyer's patch M cell, flanked by enterocytes on either side.

diameter range, primarily by the process of macropinocytosis, assisted by a reduced surface mucus layer and stunted microvilli. M cells are devoid of lysosomes and have a short journey to the lymphocytes invaginating the basolateral pocket, leading to enabled antigen presentation and subsequent mucosal immunity. Figure 7.8 shows these unique features of PP M cells.

The ability of M cells to extensively endocytose macromolecules and nanoparticles was initially exploited for drug delivery. Many microparticulate DDS (including PLGA microparticles, polymerized liposomes, poly(alkylcyanoacrylate) nanoparticles, and microparticles composed of polyanhydride copolymers of fumaric and sebacic acid) have been investigated as drug carriers for the oral delivery of poorly absorbed APIs, including peptides and proteins. However, research into drug delivery via the PP has tailed off because effective targeting of particles to M cells in patches has proved elusive, and the main focus has now switched to oromucosal vaccination. Mucosal immunity is especially important for enteric pathogens, and for subunit vaccines, which generate weak local and systemic immunity unless in a live attenuated form.

Understanding M cell differentiation is crucial to the design of effective nanoparticulate oral vaccines. M cell differentiation pathways are better understood since the advent of knockout mice models to examine intestinal epithelial cell lineage. All enterocytes, as well as the M cells in the FAE and those (newly discovered) in the villous epithelium, are initially derived from leucine-rich repeat containing G protein–coupled receptor (Lgr5+) stem cells in the dome crypts. Additional signals from the dome crypts selectively induce early M cell differentiation via the NF receptor activator of NFkB ligand, which is selectively expressed in cells below the FAE, thereby localizing its signaling effect. Finally, the result is production of a transcription factor, *Spi-B*, which activates the final step of rapid M cell maturation. Functional maturation is also assisted by B lymphocytes expressing the chemokine, CCR-6, which is attracted by CCL20 elaborated from the FAE when under challenge.

M cells in the FAE are the focus of targeting concepts, as this is the gateway to particle entry. Several tenets have emerged from the study of PLGA, polystyrene latex, liposome, and chitosan-based nanoparticles in PP-containing intestinal loops from mice and rabbits, as well as the extensive use of various nanoparticles as DDS (Des Rieux et al. 2006): (1) increased particle hydrophobicity promotes uptake, (2) the ideal diameter is 200–500 nm, and (3) specific targeting ligands can enhance uptake. However, there is little predictability of particle uptake in M cells between different species. Major issues are that a quantifiable relationship between particle uptake and mucosal immunity is unknown and that several hundredfold higher doses of vaccine are needed for oral immunization compared to parenteral comparisons. The limited clinical trial data in the public domain showed equivocal antibody production from oral administration of enterotoxigenic *Escherichia coli* antigens in PLG microparticles designed as a vaccine for traveler's diarrhea. The impact was that if nanoparticulate oral vaccines are to succeed, then incorporation of adjuvants and/or M cell targeting agents is required.

On the adjuvant side, toll-like receptor agonists, CpG and monophosphoryl lipid A are routinely included in mice oral vaccine studies, either mixed or coentrapped with vaccine-loaded nanoparticles. None are yet approved for man as part of a vaccine formulation, and regulatory hurdles are extremely high.

M cell targeting can be achieved in mice with several approaches (Devriendt et al. 2012; Woodrow et al. 2012). First, M cells express unique glycoprotein surface receptor signatures and, for example, the α-L-fucose receptor can be targeted with nanoparticles decorated with the lectin *Ulex europaeus* agglutinin-1 (UEA-1), stable small molecule analogs of UEA-1, or antibodies to the receptor. This approach has led to increased oral vaccine responses in mice with targeted particles, in some, but not all, studies. Unfortunately, the α-L-fucose receptor is not well expressed on human M cells, whereas its expression on mucus provides unwanted competition prior to epithelial receptor binding.

Second, mimicry of adhesion mechanisms used by microbes to attach to M cells has been exploited. A notable example includes attachment of the tripeptide, Arg-Gly-Asp (RGD) and other peptidomimetics to nanoparticles, to yield increased uptake by human PP compared to untargeted particles. RGD is a motif that binds overexpressed $\alpha_5\beta_1$ integrin receptor on M cells; this receptor is the target of *invasin* from *Yersinia* as well as endogenous fibronectin. Third, enterotoxins can bind M cells, and a peptide motif from *Clostridium perfringens* enterotoxin attaches to the claudin-4 receptor, which is overexpressed on M cells. Conjugation of influenza vaccine antigens to the motif is being tested in mice for intranasal vaccination via M cells. It is worth noting that our current understanding of how targeting ligands on nanoparticles, or as drug-ligand conjugates, interact with M cell receptors in the intestine luminal milieu is limited and that it is accepted that such targeting is only relevant when the formulation is very close to the receptor target. Thus, achieving mucus penetration of nanoparticles consisting of hydrophilic coatings and near-neutral zeta potentials may be a more important outcome for oral vaccine particle research than ligand targeting per se, since these particles can move unhindered through the meshes of mucus and reach the vicinity of the FAE. Table 7.3 highlights current targeting approaches for oral vaccine antigens.

TABLE 7.3
Targeting[a] Approaches with Potential for Oral Vaccine Conjugation or Particle Entrapment via Primate Peyer's Patch Microfold Cells

Receptor	Delivery System
Complement C4a receptor	Co1 ligand from phage display library, conjugated to antigen
Glycoprotein 2	Binds bacterial pili on Type I fimbriae
Tetragalloyl-D-lysine stable lectin mimetic	Enteric-coated capsules with multiantigens conjugated to ligand and tested in rhesus macaques
$\alpha_5\beta_1$ integrin	Bacterial *invasin* as a particle coating entrapping an antigen or as an antigen conjugate
Dectin-1	Complexed HIV antigen with secretory IgA
Claudin-4	Receptor for *Clostridium perfringens* enterotoxin, potential for a peptide motif conjugated to an antigen

[a] Targeting is only relevant if the ligand system can gain proximity to M cell target.

Finally, oral delivery of vaccine antigen–nanoparticle formulations that can reach colonic lymphoid follicles may be a feasible vaccine approach for protection against genital virus invasion. Recent designs include PLG nanoparticles with HIV antigens and toll-like receptor agonists. These nanoparticles were entrapped in large micron-sized PLG particles coated with Eudragit FS30D, which dissolves at pH > 7.0, thus ensuring both colonic delivery in mice and lack of uptake by nanoparticle-preferring M cells in the SI.

7.8 CONCLUSIONS

The oral route continues to be the most important route for drug delivery and targeting, in terms of both research effort and commercial success. Various mechanisms (principally matrix and reservoir diffusion-based systems, and oral osmotic devices) can be used to successfully control drug release in the GI tract, and a wide variety of drug release profiles is possible. New technologies are emerging, based on modern microelectronics, which allow extremely precise control of drug release within a specific area of the GI tract.

Regional targeting of drug formulations in the GI tract has become more sophisticated, as a result of better understanding of GI physiology in different species. In the 1990s, the concepts of gastroretentive devices and colon-specific formulations were thought to be unfeasible due to large intrasubject variation, as well as the lack of suitable polymer constructs that could reproducibly release in target regions. Advances in formulation design to produce tablet swelling in the stomach, to model time-based surface erosion between the duodenum and colon, and to trigger bacterial enzymatic activation in the colon have all been relatively successful for selected drug types. These have led to highly profitable ranges of products that maximize the therapeutic index of the payload, due to local release and optimized pharmacokinetics.

Some progress has been made in improving the oral bioavailability of poorly absorbable drugs such as therapeutic peptide and proteins. What is striking is the relative simplicity of many of the formulations that have advanced clinically, which typically include formulation excipients such as an absorption enhancer and an EI that have a proven safety record in food or pharmaceutical formulations, incorporated in an enteric-coated tablet or capsule.

Despite the considerable research effort directed toward using micro- and nanoparticulate DDS to enhance oral bioavailability, real progress remains disappointingly slow. Nanocarriers have not shown significant enhancement of oral bioavailability thus far—and it is estimated that even the best oral peptide formats will likely have relatively low bioavailabilities of $\leq 5\%$. Additionally, complex delivery systems are expensive; they require specialized manufacturing know-how and facilities and possibly difficult reconstitution procedures. Nanoparticles are showing considerably more promise as carriers for oral vaccines, and the focus is to promote antigen-loaded nanoparticle uptake by M cells of the PP. However, progress toward exploiting receptor signatures on M cells also remains slow.

Improving our understanding of how drug molecules and nanoparticles interact with, and permeate, mucus *en route* to the intestinal wall, and the intensive study of enterocyte drug and particle uptake pathways, will bring about the next phase in successful oral drug delivery: the advance from current intestinal regional delivery, to efficient and accurate epithelial cell targeting.

REFERENCES

Al-Hilal, T.A., F. Alam, and Y. Byun. 2013. Oral drug delivery systems using chemical conjugates or physical complexes. *Adv. Drug Deliv. Rev.* 65(6):845–864.

Bakhru, S.H., S. Furtado, A.P. Morell et al. 2013. Oral delivery of proteins by biodegradable nanoparticles. *Adv. Drug Deliv. Rev.* 65(6):811–821.

Bernkop-Schnürch, A. 1998. The use of inhibitory agents to overcome the enzymatic barrier to perorally administered therapeutic peptides and proteins. *J. Control. Release* 52:1–16.

Chen, J., W.E. Blevins, H. Park et al. 2000. Gastric retention properties of superporous hydrogel composites. *J. Control. Release* 64(1–3):39–51.

Conte, U., L. Maggi, P. Colombo et al. 1993. Multi-layered hydrophilic matrices as constant release devices (Geomatrix™ Systems). *J. Control. Release* 26(1):39–47.

des Rieux, A., V. Fievez, and M. Garinot. 2006. Nanoparticles as potential oral delivery systems of proteins and vaccines: A mechanist approach. *J. Control. Release* 116:1–27.

Devrclient, B., B.G. De Geest, B.M. Goddeeris et al. 2012. Crossing the barrier: Targeting epithelial receptors for enhanced oral vaccine delivery. *J. Control. Release* 160:431–439.

Ensign, L.M., R. Cone, and J. Hanes. 2012. Oral drug delivery with polymeric nanoparticles: The gastrointestinal mucus barriers. *Adv. Drug Deliv. Rev.* 64:557–750.

Katzman, M.A. and T. Sternat. 2014. A review of OROS methylphenidate (Concerta®) in the treatment of attention-deficit/hyperactivity disorder. *CNS Drugs* 28(11):1005–1033.

Li, X., M. Yu, W. Fan et al. 2014. Orally active-targeted drug delivery systems for proteins and peptides. *Expert Opin. Drug Deliv.* 11(9):1435–1447.

Lichtenstein, G.R. 2009. Mesalamine in the treatment of ulcerative colitis: Novel therapeutic options. *Gastroenterol. Hepatol.* 5:66–73.

Liu, F., E. McConnell, and S. Pygall. 2011. *Update on Polymers for Oral Drug Delivery*. Palo Alto, CA: Smithers Rapra Technology.

Maher, S. and D.J. Brayden. 2011. Overcoming poor permeability: Translating permeation enhancers for oral peptide delivery. *Drug Discov. Today Technol.* 9(2):13–19.

Maher, S., T.W. Leonard, J. Jacobson, and D.J. Brayden. 2009. Safety and efficacy of sodium caprate in promoting oral drug absorption: from in vitro to the clinic. *Adv. Drug Deliv. Rev.* 61(15):1427–1449.

Maher, S., B. Ryan, A. Duffy et al. 2014. Formulation strategies to improve oral peptide delivery. *Pharm. Pat. Anal.* 3(3):313–336.

Maroni, A., L. Zema, M.D. Del Curto et al. 2012. Oral colon delivery of insulin with the aid of functional adjuvants. *Adv. Drug Deliv. Rev.* 64(6):540–556.

Moodley, K., V. Pillay, Y.E. Choonara et al. 2012. Oral drug delivery systems comprising altered geometric configurations for controlled drug delivery. *Int. J. Mol. Sci.* 13(1):18–43.

Omidian, H. and K. Park. 2010. Oral Targeted drug delivery systems: Gastric retention devices. In *Oral Controlled Release Formulation Design and Drug Delivery*, eds. H. Wen and K. Park. Hoboken, NJ: Wiley.

Omidian, H., S. Fesharaki, and K. Park. 2011. Oral CR mechanisms and technologies. In *Controlled Release in Oral Drug Delivery: Advances in Delivery Science and Technology*, eds. C.G. Wilson and J. Patrick, p. XIV. New York: Springer.

Park, K., I.C. Kwon, and K. Park. 2011. Oral protein delivery: Current status and future prospect. *Reactive Funct. Polym.* 71:280–287.

Pawar, P.K., S. Kansal, G. Garg et al. 2011. Gastroretentive systems: A review with special emphasis on floating drug delivery systems. *Drug Deliv.* 18 (2):97–110.

Peppas, N.A., J.Z. Hilt, A. Khademhosseini et al. 2006. Hydrogels in biology and medicine: From molecular principles to bionanotechnology. *Adv. Mater.* 18(11):1345–1360.

Perrie, Y. and T. Rades. 2012. Delayed release drug delivery systems. In *Pharmaceutics: Drug Delivery and Targeting.* eds. Y. Perrie and T. Rades. London, U.K.: Pharmaceutical Press.

Pouton, C.W. and C.J. Porter. 2008. Formulation of lipid-based delivery systems for oral administration: Materials, methods and strategies. *Adv. Drug Deliv. Rev.* 60(6):625–637.

Prajapati, V.D., G.K. Jani, T.A. Khutliwala et al. 2013. Raft forming system—An upcoming approach of gastroretentive drug delivery system. *J. Control. Release* 168(2):151–165.

Rajabi-Siahboomi, A.R. and B.S. Tiwari. 2008. Modulation of drug release from hydrophilic matrices. *Pharm. Technol. Europe.* 1:1–8.

Rosen, H. and T. Abribat. 2005. The rise and rise of drug delivery. *Nat. Rev. Drug Discov.* 4:381–385.

Schmidt, C., C. Lautenschlaeger, E.M. Collnot et al. 2013. Nano- and microscaled particles for drug targeting to inflamed intestinal mucosa: A first *in vivo* study in human patients. *J. Control. Release* 165(2):139–145.

Siepmann, J. and F. Siepmann. 2008. Mathematical modeling of drug delivery. *Int. J. Pharm.* 364(2):328–343.

Skalsky, B. and S. Stegemann. 2011. Coated multiparticulates for controlling drug release. In *Controlled Release in Oral Drug Delivery Series: Advances in Delivery Science and Technology*, eds. Wilson, G.C. and J. Patrick, p. 414. New York: Springer.

Streubal, A., J. Siepmann, and R. Bodmeir. 2006. Drug delivery to the upper small intestine window using gastroretentive technologies. *Curr. Opin. Pharmacol.* 6(5):501–508.

Tortora, G.J. and B.H. Derrickson, eds. 2013. *Principles of Anatomy and Physiology*, 14th ed. Hoboken, NJ: Wiley.

Tu, J., Y. Shen, R. Mahalingam et al. 2010. Polymers in oral modified release systems. In *Oral Controlled Release Formulation Design and Drug Delivery*, eds. H. Wen and K. Park, pp. 71–88. Hoboken, NJ: Wiley.

Woodrow, K.A., K.M. Bennett, and D.D. Lo. 2012. Mucosal vaccine design and delivery. *Annu. Rev. Biomed. Eng.* 14:17–46.

Yun, Y., Y.W. Cho, and K. Park. 2013. Nanoparticles for oral delivery: Targeted nanoparticles with peptidic ligands for oral protein delivery. *Adv. Drug Deliv. Rev.* 65(6):822–832.

8 Buccal and Sublingual Drug Delivery

Marc B. Brown and Viralkumar F. Patel

CONTENTS

8.1 INTRODUCTION

Topical delivery to the oral cavity is used to treat localized conditions of the mouth, such as aphthous ulcers, fungal infections, and periodontal disease. However, the oral cavity can also be used to achieve the systemic delivery of a drug, i.e., oral transmucosal delivery. As described in Chapter 7, the peroral (i.e., via the gastrointestinal [GI] tract) route remains the preferred route for the administration of therapeutic agents because of its low cost, ease of administration and high level of patient compliance. However, this route of administration also has disadvantages, such as hepatic first-pass metabolism and acidic and enzymatic degradation within the GI tract, which often prohibits its use for certain drug classes, including peptides and proteins. Consequently, other absorptive mucosa (i.e., the mucosal linings of the nasal, rectal, vaginal, ocular, and oral cavity) are often considered as an alternative site for drug administration. One such alternative route is oral transmucosal drug delivery, which offers many distinct advantages over peroral administration for systemic drug delivery, including the avoidance of the hepatic first-pass effect and presystemic elimination within the GI tract.

Oral transmucosal delivery is further subdivided into the following:

1. Sublingual delivery: the systemic delivery of drugs through the mucosal membranes lining the floor of the mouth. This route is typically used when a rapid onset of action is required.
2. Buccal delivery: drug administration through the mucosal membranes lining the inner cheeks (buccal mucosa). Buccal delivery can additionally be used to prolong drug retention in the oral cavity, which is advantageous for both systemic and local drug delivery.

A further type of drug delivery to the oral cavity involves orally disintegrating tablets (ODTs), also known as "fast melts," e.g., Zydis® ODT fast-dissolve formulation. These dosage forms are designed to dissolve rapidly (i.e., less than 30 seconds) in the mouth, in contrast to conventional tablets that must be swallowed whole. ODTs can be used as an alternative for patients who experience dysphagia (difficulty in swallowing), such as in pediatric and geriatric populations, or where compliance is an issue. However, in this case, drug absorption actually takes place in the GI tract after swallowing the dissolved active pharmaceutical ingredient (API), rather than from the oral cavity. As such, ODTs involve GI absorption rather than absorption from the oral cavity. Therefore they are not considered further here—they are discussed instead in Chapter 2 (Section 2.2.2).

8.1.1 Overview of the Structure of the Oral Mucosa

The oral cavity comprises the lips, cheek, tongue, hard palate, soft palate, and floor of the mouth (Figure 8.1).

The lining of the oral cavity, referred to as the oral mucosa, includes the buccal, sublingual, gingival, palatal, and labial mucosa. The oral mucosa is a stratified squamous epithelium, comprising many cell layers (see Chapter 4, Figure 4.7). The epithelium sits on an underlying connective tissue

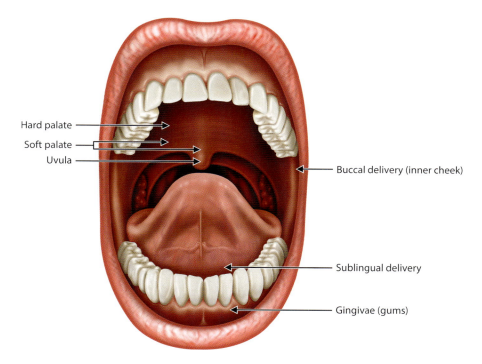

FIGURE 8.1   The oral cavity. (Modified from Alexilusmedical/Shutterstock.com.)

layer (the lamina propria), which contains a rich blood supply. An API is absorbed through the blood capillaries in the lamina propria and gains access to the systemic circulation (Squier and Wertz 1996).

The oral mucosa varies depending on its location in the oral cavity, so that three distinct types are described (see also Chapter 4, Section 4.5.2):

1. The lining mucosa: found in the outer oral vestibule (the buccal mucosa) and the sublingual region (floor of the mouth). It comprises approximately 60% of the total surface area of the oral mucosal lining in an adult human.
2. The specialized gustatory (taste) mucosa: found on the dorsal surface of tongue, specifically in the regions of the taste buds on the lingual papillae, located on the dorsal surface of the tongue. These regions contain nerve endings for general sensory reception and taste perception. This specialized mucosa comprises approximately 15% of the total surface area.
3. The masticatory mucosa: found on the hard palate (the upper roof of the mouth) and the gingiva (gums) and comprises the remaining approximately 25% of the total surface area.

The specialized mucosa is dedicated to taste perception. The masticatory mucosa is located in the regions particularly susceptible to stresses and strains resulting from masticatory activity. The superficial cells of the masticatory mucosa are keratinized, to help withstand the physical stresses of this region. The multilayered barrier of the masticatory mucosa, reinforced with keratin in the surface layers, presents a formidable barrier to drug permeation.

In contrast, the lining mucosa is subject to much lower masticatory stress and consequently has a nonkeratinized epithelium, which sits on a thin and elastic lamina propria, and submucosa. It is this lining epithelium that is the primary focus for drug delivery.

8.2 PHYSIOLOGICAL BARRIERS TO DRUG DELIVERY ACROSS THE ORAL MUCOSA

The environment of the oral cavity presents some significant challenges for systemic drug delivery. Certain physiological aspects of the oral cavity play significant roles in this process, including its pH, fluid volume, enzyme activity, and permeability. Table 8.1 provides a comparison of the physiological characteristics of the oral mucosa in comparison with the mucosa of the GI tract (Patel et al. 2011).

8.2.1 SALIVA

The principal physiological environment of the oral cavity, in terms of pH, fluid volume, and composition, is shaped by the secretion of saliva. Saliva is secreted by three major salivary glands

TABLE 8.1

Comparison of Oral and Gastrointestinal Mucosa

Absorption Site	Estimated Surface Area	Percent Total Surface Area	Local pH	Mean Fluid Volume (mL)	Relative Enzyme Activity	Relative Drug Absorption Capacity
Oral cavity	100 cm² (0.01 m²)	0.01	5.8–7.6	0.9	Moderate	Moderate
Stomach	0.1–0.2 m²	0.20	1.0–3.0	118	High	Moderate
Small intestine	100 m²	98.76	3.0–4.0	212	High	High
Large intestine	0.5–1.0 m²	0.99	4.0–6.0	187	Moderate	Low
Rectum	200–400 cm² (0.04 m²)	0.04	5.0–6.0	—	Low	Low

(parotid, submaxillary and sublingual). The parotid and submaxillary glands produce a watery secretion, whereas the sublingual glands produce mainly viscous saliva with limited enzymatic activity. The main functions of saliva are to lubricate the oral cavity, to facilitate swallowing and to prevent demineralization of the teeth. It also contributes to carbohydrate digestion and regulates oral microbial flora by maintaining the oral pH and enzyme activity. The daily total salivary volume is between 0.5 and 2.0 mL. However, the volume of saliva constantly available is around 1.1 mL with a pH of $\approx$5.5–7.6, thus providing a relatively low fluid volume available for drug release from dosage forms, when compared to the GI tract. Overall, the pH and salivary compositions are dependent on the flow rate of saliva, which in turn depends upon three factors: the time of day, the type of stimulus and the degree of stimulation. For example, at high flow rates, the sodium and bicarbonate concentrations increase, leading to an increase in the pH. Such changes in pH can affect the absorption of ionizable drugs. For example, drugs such as midazolam, buprenorphine, nicotine, fentanyl, and lamotrigine are reported to have pH-dependant drug absorption across the oral mucosa (Mashru et al. 2005; Myers et al. 2013; Nielsen and Rassing 2002; Streisand et al. 1995).

Nevertheless saliva provides a water-rich environment of the oral cavity, which can be favorable for drug release from delivery systems, especially those based on hydrophilic polymers. However, saliva flow decides the time span of the released drug at the delivery site. This flow can lead to premature swallowing of the drug before effective absorption occurs through the oral mucosa and is a well-accepted concept known as "saliva washout."

8.2.2 EPITHELIAL BARRIER

Drug permeability through the oral cavity mucosa represents another major physiological barrier for oral transmucosal drug delivery. The oral mucosal thickness varies depending on the site, as does the composition of the epithelium. The characteristics of the different regions of interest in the oral cavity are shown in Table 8.2 (Patel et al. 2011).

As outlined earlier, the areas of the mucosa subject to mechanical stress (i.e., the masticatory mucosa of the gingiva and hard palate) are keratinized (similar to the epidermis), which makes drug permeation difficult. They also contain neutral lipids like ceramides and acylceramides, making them relatively impermeable to water. Any formulation adhering to these areas can also present a problem in swallowing. For these reasons, the masticatory mucosa has not been used widely for drug delivery. In contrast, the nonkeratinized epithelia do not contain acylceramides and have only small amounts of ceramides, and also contain polar lipids, mainly cholesterol sulfate and glucosyl ceramides. These epithelia are therefore considerably more permeable to water than the keratinized epithelia and are consequently more widely explored for drug delivery.

TABLE 8.2
Characteristics of Oral Mucosa

Tissue	Structure	Thickness (µm)	Turnover Time (Days)	Surface Area (cm²) ± SD	Permeability	Residence Time	Blood Flow[a]
Buccal	NK	500–600	5–7	50.2 ± 2.9	Intermediate	Intermediate	20.3
Sublingual	NK	100–200	20	26.5 ± 4.2	Very good	Poor	12.2
Gingival	K	200	—	—	Poor	Intermediate	19.5
Palatal	K	250	24	20.1 ± 1.9	Poor	Very good	7.0

Abbreviations: K, keratinized tissue; NK, nonkeratinized tissue.

[a]   In rhesus monkeys (mL/min/100 g tissue).

8.2.3 MUCUS BARRIER

The apical cells of the oral epithelia are covered by mucus layer; the principal components of which are complexes made up of proteins and carbohydrates; its thickness ranges from 40 to 300 µm. In the oral mucosa, mucus is secreted by the major and minor salivary glands as part of saliva. Although most of the mucus is water ($\approx$95%–99% by weight), the key macromolecular components are a class of glycoprotein known as mucins (1%–5%). Mucins are large molecules with molecular masses ranging from 0.5 to over 20 MDa, containing large amounts of carbohydrate. They are made up of basic units ($\approx$400–500 kDa) linked together into linear arrays, which are able to join together to form an extended 3D network, which acts as a lubricant and may also contribute to cell–cell adhesion. At physiological pH, the mucus network carries a negative charge due to the sialic acid and sulfate residues and forms a strongly cohesive gel structure that binds to the epithelial cell surface as a gelatinous layer. This gel layer is believed to play a role in mucoadhesion for drug delivery systems (DDS), which work on the principle of adhesion to the mucosal membrane and thus extend the dosage form retention time at the delivery site (described further in Section 8.3.2).

8.3 ORAL TRANSMUCOSAL DRUG DELIVERY CONSIDERATIONS

Despite the physiological challenges, the oral mucosa, due to its unique structural and physiological properties, offers several opportunities for drug delivery. As the mucosa is highly vascularized, any drug diffusing across the oral mucosa membranes has direct access to the systemic circulation via capillaries and venous drainage and will bypass hepatic first-pass metabolism. The rate of blood flow through the oral mucosa is substantial and is generally not considered to be the rate-limiting factor in the absorption of drugs by this route (Table 8.2).

In contrast to the harsh environment of the GI tract, the oral cavity offers relatively consistent and mild physiological conditions for drug delivery that are maintained by the continual secretion of saliva. Compared to secretions of the GI tract, saliva is a relatively mobile fluid with less mucin and has limited enzymatic activity and virtually no proteases, which is especially favorable for protein and peptide delivery. The enzymes that are present in the buccal mucosa are believed to include aminopeptidases, carboxypeptidases, dehydrogenases, and esterases.

Within the oral cavity, the buccal and sublingual routes are the focus for drug delivery because of their higher overall permeability, compared to the other mucosa of the mouth (Table 8.2). The buccal and sublingual mucosa are also approximately 13 and 22 times more permeable to water, respectively, in comparison with the skin. Based on relative thickness and their epithelial composition, the sublingual mucosa has the highest potential drug permeability of the oral mucosa and thus is suitable for systemic drug delivery with rapid onset of action. For rapid oral transmucosal delivery, a drug can be presented as lozenges, films or patches, sprays or compressed tablets having fairly rapid disintegration in the mouth (3 minutes or less). The buccal mucosa has moderate permeability and is suitable for both local and systemic drug delivery. The drug can be presented as a mucoadhesive formulation (patch or tablet) and can be released slowly, either to achieve a sustained release systemic absorption profile or to achieve sustained release effects locally in the oral cavity.

A further possibility for drug delivery to the oral cavity is that of vaccine delivery. The oral mucosa has a number of nonspecific mechanisms to protect against invading pathogens, including (1) salivary secretions, which keep the epithelial surface moist, inhibiting bacterial colonization; (2) a process of continuous shedding of the stratified squamous epithelium from the apical surface layer, therefore minimizing bacterial colonization; and (3) a highly resilient underlying lamina propria, which ensures that tissue integrity is maintained. In addition, the oral mucosa contains various types of specific immune-competent cells, as well as specific immune-competent tissue, particularly in the oropharyngeal region. Thus, the oral cavity could offer a potential route for noninvasive vaccine delivery. Promising data are emerging, in particular with respect to vaccine delivery via the sublingual route; this research is described in Chapter 17 (Section 17.4.4).

8.3.1 Transport Mechanisms across the Oral Mucosa

As described in Chapter 4, drugs can be transported across epithelial membranes by passive diffusion, carrier-mediated transport, or other specialized mechanisms (see Chapter 4, Section 4.3 and Figure 4.4). Most studies of buccal absorption indicate that the predominant mechanism is *passive diffusion* across lipid membranes. Passive diffusion can occur via paracellular transport between the epithelial cells, as shown, for example, for flecainide, sotalol, and metformin (Deneer et al. 2002). Passive diffusion may also occur via transcellular transport through the epithelial cells, as shown, for example, for buspirone and lamotrigine (Birudaraj et al. 2005; Mashru et al. 2005). Drugs such as nicotine are believed transverse both pathways simultaneously (Nielsen and Rassing 2002). Specialized transport mechanisms have also been reported for a few drugs and nutrients, including monocarboxylic acids and glucose. Usually, one route is predominant, depending on the physicochemical properties of the drug. The relevant physicochemical properties of the API that favor epithelial absorption have been described in detail in Chapter 4 (Section 4.3.4).

8.3.2 Mucoadhesion Theory

Mucoadhesive systems are used in the oral cavity to maintain an intimate and prolonged contact of the formulation with the oral mucosa, allowing a longer duration for drug absorption. Mucoadhesion is a complex process and numerous theories have been presented to explain the mechanisms involved. The wettability theory (Ugwoke et al. 2005) is mainly applicable to liquid or low viscosity mucoadhesive systems and is essentially a measure of the spreadability of the DDS across the biological substrate. The electronic theory (Dodou et al. 2005) describes adhesion by means of electron transfer between the mucus and the mucoadhesive system arising through the differences in their electronic structures. The electron transfer between the mucus and the mucoadhesive results in the formation of a double layer of electrical charges at the mucus and mucoadhesive interface. The net result of such a process is the formation of attractive forces within this double layer. According to the fracture theory (Ahagon and Gent 1975), the adhesive bond between systems is related to the force required to separate both surfaces from one another and relates the force for polymer detachment from the mucus to the strength of their adhesive bond. The work of fracture has been found to be greater when the polymer network strands are longer, or if the degree of cross-linking within such a system is reduced. According to the adhesion theory (Jiménez-Castellanos et al. 1993), adhesion is defined as being the result of various surface interactions (primary and secondary bonding) between the adhesive polymer and mucus substrate. Primary bonds due to chemisorption result in adhesion due to ionic, covalent, and metallic bonding, which is generally undesirable due to their permanency. The diffusion-interlocking theory (Lee et al. 2000) proposes the time-dependent diffusion of mucoadhesive polymer chains into the glycoprotein chain network of the mucus layer. This is a two-way diffusion process with the penetration rate being dependent upon the diffusion coefficients of both interacting polymers. In practice however, the mechanism by which mucoadhesion occurs will depend on many factors including the nature of the mucus membrane, the mucoadhesive material(s) used and the type of formulation, so that is difficult to predict or assign a single mechanism (Boddupalli et al. 2010).

Significant research has focused on the development of mucoadhesive delivery systems (including tablets, films, wafers, and patches) that contain different components to extend the residence time of dosage forms at the site of application. The most widely investigated macromolecules are those containing numerous negative groups, including hydroxyl (OH^-) and carboxyl (COO^-) groups, which permit hydrogen bonding with the cell surface. Although hydrogen bonds are weak, numerous bonds are possible with such macromolecules, ensuring a firm attachment to the buccal epithelium. Examples include hydroxypropyl cellulose (HPC), sodium carboxymethyl cellulose (SCMC), hydroxyethyl cellulose (HEC), hydroxypropyl methylcellulose (HPMC), polyvinylpyrrolidone (PVP), polyvinyl alcohol (PVA), and polyacrylic acid (PAA), either alone or in combination. Polycarbophil is a synthetic polymer of polyacrylic acid, lightly cross-linked with divinyl glycol. It is a weak polyacrylic acid (pKa = 4.2) containing multiple negative carboxyl groups (COO⁻), for bonding to the buccal surface. Other mucoadhesive polymers used in commercially available formulations include xanthan and locust bean gums. Thiolated polymers (thiomers) comprise a new class of mucoadhesives, derived from hydrophilic polymers such as polyacrylates. The thiol groups are capable of forming covalent bonds with cysteine-rich subdomains of the mucus gel layer, leading to increased residence time and improved bioavailability (Albrecht et al. 2006).

8.4 LOCAL/TOPICAL DRUG DELIVERY TO THE ORAL CAVITY

Local/topical delivery to the oral cavity is of use in the treatment of local conditions of the mouth, such as aphthous mouth ulcers, oral inflammatory lesions, fungal infections (e.g., oral candidiasis), viral infections (e.g., herpes simplex virus) and bacterial infections.

Liquid dosage forms are not readily retained in the oral cavity, cannot easily achieve drug targeting within the mouth and deliver relatively uncontrolled amounts of drug. For these reasons, liquid formulations for topical delivery tend to be confined to antibacterial mouthwashes, such as chlorhexidine gluconate mouth rinse (Periogard®), benzydamine hydrochloride mouthwash (Oroeze®), and antiseptic essential oil mouth rinse (Listerine®).

Semisolid dosage forms (SSDFs) for topical delivery include gels, creams, paste and ointments. These typically contain a hydrophilic polymer and drug, plus any required excipient dissolved or suspended as a fine powder, in an aqueous or nonaqueous base. Semisolid formulations can be applied using the finger (or syringe) to a target region and tend to be more acceptable in terms of "mouth feel" to patients, relative to a solid dosage form (SDF). However, they may deliver variable amounts of active ingredients. They are further limited by their short contact time with the oral mucosa and the need for multiple doses each day.

In order to improve contact time, more advanced technologies for local therapy typically incorporate a mucoadhesive polymer into the formulation. Kenalog® in Orabase® dental paste contains the steroidal drug triamcinolone acetonide, for the temporary relief of symptoms associated with oral inflammatory lesions and ulcerative lesions resulting from trauma. This paste formulation contains hydrophilic polymers such as gelatin, pectin, and SCMC, in Plastibase®: a plasticized hydrocarbon gel. Upon application, the paste forms a thin film covering the lesions; it is recommended for use at bedtime to ensure intimate contact between the formulation and mucosal site.

Similarly, Gelclair® contains PVP and sodium hyaluronate in a liquid gel, which, in contact with the oral mucosa, forms a protective film that adheres to the mucosa, offering rapid and effective pain management of lesions of the oral mucosa. PVP hydrate acts as a bioadhesive polymer and retains the formulation at the site of action for a prolonged period of time.

Lauriad™ technology utilizes hypromellose and a proprietary milk protein concentrate, in order to achieve mucoadhesion. The technology is used for the buccal delivery of the antifungal drug miconazole (Loramyc® in Europe, Oravig® in the United Kingdom). The buccal tablet provides sustained local release of miconazole over several hours, with just one daily application. Lauriad™ technology is also used in Sitavig® buccal tablets, for the prolonged local delivery of the antiviral drug acyclovir, in the treatment of oral herpes.

8.5 ORAL TRANSMUCOSAL DRUG DELIVERY SYSTEMS

Continuous research into the improvement of oral transmucosal drug delivery has resulted in the development of several conventional and novel dosage forms like solutions, tablets/lozenges, chewing gums, sprays, patches and films, hydrogels and microspheres. These dosage forms can be broadly classified into liquid, semisolid, solid or spray formulations. An overview of the different types of dosage forms currently available and in development is provided in the following text.

8.5.1 Sublingual Drug Delivery

As described earlier, the sublingual mucosa is more permeable and thinner than the buccal mucosa and, because of the considerable surface area and high blood flow, it is a favorable site when a rapid onset is desired. The sublingual route is thus generally used for drug delivery in the treatment of acute conditions, such as in pain relief, antimigraine treatment or relief from an angina attack. Limitations of the route include that its surface is constantly washed by saliva; the short residence time at the site of absorption may result in incomplete absorption. Also tongue activity makes it difficult to keep a dosage form in contact with the mucosa for an extended period of time.

Glyceryl trinitrate (GTN) provides an excellent example of an API that benefits from the avoidance of first-pass metabolism when administered via the sublingual route. It was one of the first drugs successfully developed for oral transmucosal delivery, with a sublingual form introduced as early as 1847. GTN sublingual tablet (Nitrostat®) is delivered rapidly via the sublingual route, providing rapid relief/prophylaxis in angina. Other commercial examples of sublingual tablets include pain relief formulations such as buprenorphine hydrochloride (Temgesic®), buprenorphine hydrochloride in combination with naloxone (Suboxone®), and fentanyl citrate (Abstral®). Further sublingual tablets include zolpidem (Edluar®), lorazepam (Ativan®), nicotine (Nicorette Microtab®), and asenapine maleate (Sycrest®).

A major limitation of sublingual tablets is that dissolution can vary considerably, depending on the type, size, and shape of the tablet. Thus, sublingual dosage forms generally have a high inter- and intraindividual variation in absorption and bioavailability. Also, such types of systems are not able to provide unidirectional drug release. A sublingual aerosol spray is an alternative to sublingual tablets, which can deliver the API more rapidly and uniformly into the salivary fluid, or onto the mucosal surface, where it is readily available for absorption. Nitroglycerin (Nitrolingual® pump spray, Nitromist®, and Glytin®), zolpidem (Zolpimist®), nicotine (Nicorette Quickmist®), and flurbiprofen (Benactiv®), are the currently marketed, spray-type formulations.

8.5.2 Buccal Drug Delivery

The buccal mucosa can be used for both local and systemic therapies. As outlined earlier, the buccal mucosa is relatively permeable, robust and, in comparison with other mucosal tissues, more tolerant to potential allergens and has a reduced tendency to irreversible irritation or damage. When rapid buccal delivery is required, the drug may be delivered as buccal tablets that have a relatively rapid in-mouth disintegration time (15 minutes or less). Sprays, films, wafers, lozenges, and lollipop-like systems may also be used for this purpose. The buccal mucosa, in contrast to the sublingual mucosa, also offers a smooth and relatively immobile surface: this facilitates the placement of a retentive DDS for prolonged exposure, such as a buccal tablet, film or wafer. In this case, the drug is released slowly, for continuous absorption throughout the oral mucosa. In some cases, some of the dose may be swallowed, with subsequent absorption occurring from the GI tract.

8.5.2.1 Buccal Tablets

Buccal tablets can be formulated so that the drug is dispersed within a bioadhesive matrix. This type of system facilitates multidirectional release. Unidirectional release may be achieved via the incorporation of a second impermeable backing layer, typically an insoluble polymer layer, which covers the underlying drug/adhesive layer. In addition to ensuring unidirectional flow across the buccal mucosa, the outer impermeable layer may prevent overwetting of the tablet and subsequent formation of slippery mucilage, which can limit bioadhesive retention. In some cases, the outer layer can be a slowly dissolving layer, so that after drug absorption has taken place, the DDS dissolves and does not have to be physically removed.

Disadvantages of buccal tablets can include poor patient acceptability (due to uncomfortable mouth feel, taste, irritation, and discomfort) and the nonubiquitous distribution of drug within saliva for local therapy. There is also the risk that the dosage form could separate from the oral mucosa, be swallowed, and then adhere to the wall of the esophagus.

Some commercially available buccal tablets are described here. Buccastem® tablets are a long-established buccal delivery system for the antiemetic prochlorperazine, which contains PVP, xanthan gum, and locust bean gum, to achieve bonding with the buccal mucosa. On application, the tablet softens and adheres to the gum where it forms a gel, from which the prochlorperazine is released and absorbed. One 3 mg buccal tablet twice a day results in steady-state plasma levels bioequivalent to those achieved by the standard oral dosage of one 5 mg tablet taken three times a day.

Suscard® buccal tablets are prolonged release mucoadhesive buccal tablets for the delivery of GTN. The Suscard tablet is placed high up between the upper lip and gum to either side of the front teeth; hypromellose in the formulation ensures mucoadhesion. Once in place, the duration of action of the tablet correlates with its dissolution time and is normally 3–5 hours. However, the first few doses may dissolve more rapidly until the patient is used to the presence of the tablet. During the dissolution time, the tablet softens and adheres to the gums; in practice the presence of the tablet is not noticeable to the patient after a short time.

Striant® SR tablets contain testosterone in a mucoadhesive system comprising a combination of carbopol, polycarbophil, and hypromellose designed to adhere to the gum or inner cheek. It provides a controlled and sustained release of testosterone through the buccal mucosa as the buccal system gradually hydrates. Application of Striant® twice a day, in the morning and in the evening, provides continuous systemic delivery of testosterone.

An interesting formulation approach is used in the OraVescent® tablet, which exploits a localized, transient pH change that occurs over the course of tablet disintegration and dissolution, in order to facilitate drug dissolution and buccal absorption. As the tablet disintegrates in the oral cavity, a CO_2-generating reaction produces a modest initial decrease in the pH of the tablet microenvironment. For weakly basic drugs, this lower pH (below its pKa) favors the ionized form of the drug, thereby accelerating drug dissolution. A pH-modifying substance present in the tablet (e.g., sodium carbonate) then begins to dissolve, so that the pH microenvironment subsequently increases with time, causing ionized drugs to convert predominantly to the unionized form, thereby increasing transmembrane permeability. This technology is used for the buccal delivery of the opioid analgesic, fentanyl (Fentora®), to treat "breakthrough" cancer pain. Following buccal administration of Fentora®, fentanyl is readily absorbed with an absolute bioavailability of 65%. The absorption profile of Fentora® is largely the result of an initial absorption from the buccal mucosa, with peak plasma concentrations generally attained within an hour after buccal administration. Approximately, 50% of the total dose administered is absorbed transmucosally and becomes systemically available. The remaining half of the total dose is swallowed and undergoes more prolonged absorption from the GI tract (Darwish et al. 2007).

8.5.2.2 Buccal Patches/Films/Wafers

Buccal patches/films/wafers are emerging as alternative formulation choices to buccal tablets. Their thinness and flexibility means that they are less obtrusive and therefore more acceptable to the patient. Similar to buccal tablets, they can provide multidirectional or unidirectional drug release. They are usually prepared by casting a solution of the polymer, drug, and any excipients (such as a plasticizer) on to a surface and allowing it to dry. They can be up to 10–15 cm^2 in size, but are more usually 1–3 cm^2, often with an ellipsoid shape to fit comfortably into the center of the buccal mucosa. The relative thinness of the films, however, means that they are

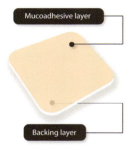
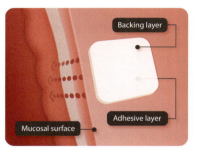

BEMA technology breakdown   BEMA drug delivery   BEMA film placement

FIGURE 8.2 Buccal delivery using a BioErodible MucoAdhesive Film showing unidirectional drug delivery. (Courtesy of BioDelivery Sciences International, Inc.)

more susceptible to overhydration and loss of the adhesive properties. Such formulations may also suffer from low drug loading.

BioErodible MucoAdhesive Film (BEMA™) features a bilayered buccal film technology, in which the active drug is dissolved within a mucoadhesive layer; a backing layer then facilitates unidirectional flow of drug. On application, saliva moistens the mucoadhesive layer, ensuring adhesion of the film to the buccal mucosa within seconds; rapid systemic drug absorption then follows (Figure 8.2). The film is bioerodible and dissolves completely within 15–30 minutes of application.

Onsolis® (fentanyl buccal soluble film) uses the BEMA™ bilayer delivery technology for the buccal delivery of the potent opioid analgesic, fentanyl citrate. The API is incorporated into a bio-adhesive layer comprising carboxymethyl cellulose and HPC. An insoluble backing layer of HEC promotes unidirectional drug release and drug absorption across the oral mucosa. The amount of fentanyl delivered transmucosally is proportional to the film surface area. The absorption pharmacokinetics of fentanyl from Onsolis® is a combination of an initial rapid absorption from the buccal mucosa (about 50% of the dose) and a more prolonged absorption of swallowed fentanyl from the GI tract. Of the swallowed fentanyl, about 20% of the total dose escapes hepatic and intestinal first-pass elimination and becomes systemically available. The BEMA™ platform is also used for the buccal delivery of buprenorphine with naloxone (Bunavail®) and is under study for a variety of other drugs.

8.5.2.3 Buccal Sprays

Buccal sprays offer the advantage of a rapid onset of action: a spray delivers the dose in fine particulates or droplets; thus, the lag time for the API to be available at the site of absorption is reduced. For example, a pharmacokinetic study of buccal insulin spray in patients with Type I diabetes revealed no statistical difference in glucose, insulin, and C-peptide plasma levels, compared to insulin administered subcutaneously (s.c.) (Pozzilli et al. 2005).

Oral-lyn™ is an oral spray for the systemic delivery of insulin, for the treatment of Type I and II diabetes, based on the RapidMist™ technology platform. Regular recombinant human insulin is delivered to the membranes of the oral cavity by a simple asthma-like inhaler device, and Oral-lyn™ is offered as a pain-free alternative to s.c. injections of prandial insulin. It is on the market in some regions and is preparing for FDA approval.

The RapidMist™ technology is based on the formation of microfine micelles, loaded with insulin. A combination of different proprietary absorption enhancers encapsulates and protects insulin within the micelles. The device produces a high-velocity (100 mph) aerosol, which is claimed impels the insulin-containing micelles across the superficial layers of the buccal mucosa; rapid transmucosal absorption is further facilitated by aerosol particle size and the absorption enhancers in the formulation. Following oral administration, insulin appears in the blood within 5 minutes, peaks at 30 minutes, and is back to baseline at 2 hours. This rapid absorption profile means that even though the formulation contains regular recombinant human insulin, its delivery profile is more akin to that of the synthetic fast-acting insulin analogs (lispro, aspart, and glulisine) administered subcutaneously.

The spray orifice in the actuator of the device is designed for maximum impact with the buccal cavity. It is claimed that the size of the insulin-containing micelles that are formed (85% having mean size >10 μm) ensures that absorption is limited to the mouth, with no entry of product into the lungs, thereby avoiding pulmonary side effects.

The technology is designed so that one spray delivers approximately 10 U of insulin; thus, approximately 1 U of insulin is absorbed systemically. Application of >10 U of insulin (e.g., after a meal) therefore requires more than 10 puffs, which is time-consuming and not very user-friendly. The insulin is claimed to be released from the device as a metered dose, so that the delivered dose should be identical from the first puff to the last. However, effective dosing requires some patient education and training. Notwithstanding, the RapidMist™ technology is in clinical development for the buccal delivery of a variety of other drugs, including fentanyl citrate, morphine, and low-molecular-weight heparin.

8.5.3 Other Oral Transmucosal Drug Delivery Systems

A lozenge-on-a-stick unit offers nonspecific drug delivery to oral mucosa, i.e., drug release from the formulation is not specific to either the sublingual or buccal mucosa and a drug can be absorbed both from these mucosa as well as from the GI tract once swallowed by the patient. An example of this type of delivery device is Actiq®, for the oral transmucosal delivery of the potent opioid analgesic fentanyl citrate. The advantage of this system is that the stick handle allows the Actiq® unit to be easily removed from the mouth if signs of excessive opioid effects appear during administration. Normally, approximately 25% of the total dose of Actiq® is rapidly absorbed from the buccal mucosa and becomes systemically available. The remaining 75% of the total dose is swallowed with the saliva and then is slowly absorbed from the GI tract. About 1/3 of this amount (i.e., 25% of the total dose) escapes hepatic and intestinal first-pass elimination and becomes systemically available. Thus, the generally observed 50% bioavailability of Actiq® is divided equally between rapid transmucosal, and slower GI, absorption.

Medicated chewing gum is another means of delivering drugs nonspecifically to the oral cavity. The approach is widely used for the delivery of nicotine as an aid in smoking cessation. Nicotine, a small, lipophilic molecule, is readily absorbed from the oral mucosa when administered in chewing gum. Blood levels are obtained within 5–7 minutes and reach a maximum about 30 minutes after the start of chewing. Blood levels are roughly proportional to the amount of nicotine chewed.

Nicotrol® Inhaler (nicotine inhalation system) consists of a mouthpiece and a plastic cartridge delivering 4 mg of nicotine from a porous plug containing 10 mg nicotine. The cartridge is inserted into the mouthpiece prior to use. Most of the nicotine released from the Nicotrol® Inhaler is deposited in the mouth, with only a fraction (less than 5%) of the dose reaching the lower respiratory tract. An intensive inhalation regimen (80 deep inhalations over 20 minutes) releases on average 4 mg of the nicotine content of each cartridge, of which about 2 mg is systemically absorbed. Peak plasma concentrations are typically reached within 15 minutes of the end of inhalation. Absorption of nicotine through the buccal mucosa is relatively slow, so that the rapid peaks and troughs obtained with cigarette smoking are not achieved.

DDS such as micro- and nanoparticles and liposomes are also being investigated for oral transmucosal delivery (Heanden et al. 2012). For example, polymeric microparticles (23–38 μm) of Carbopol®, polycarbophil and chitosan or Gantrez® (copolymers of monoalkyl esters of poly (methyl vinyl ether/maleic acid)) were found to adhere to porcine esophageal mucosa, with particles prepared from the polyacrylic acids exhibiting greater mucoadhesive strength during tensile testing studies. In contrast, in elution studies, particles of chitosan or Gantrez® were seen to persist on mucosal tissue for longer periods of time (Kockisch et al. 2004).

8.5.3.1 Pediatric Transmucosal Formulations

Sublingual and buccal DDS are a particularly attractive choice for children, as this target group typically experiences problems in swallowing SDFs. However, relatively few products have been approved for pediatric indications at the current time. Pediatric gel formulations are typically used for topical oral treatment, e.g., teething and antifungal gels. For systemic delivery, SDFs such as lozenges, tablets, capsules, and films/wafers are preferred to liquid formulations, because of their improved drug stability, ease of manufacture, and less bulky nature, all of which increases pediatric patient compliance and decreases the cost of goods. As such, lozenges and tablets are some of the more common types of dosage forms for the buccal/sublingual route in pediatric patients. However, where prolonged contact with the mucosa is required, semisolid formulations may offer greater pediatric acceptability than SDFs, as the former can be spread evenly and thinly over the mucosa, rather than having to be deliberately retained and thus potentially obstructing swallowing, eating, and drinking.

8.6 IN VITRO AND IN VIVO ASSESSMENT OF ORAL TRANSMUCOSAL SYSTEMS

Despite the intensive research efforts dedicated to developing oral transmucosal DDS, relatively few formulations have made the successful translation to commercial product. This can partly be attributed to the current lack of standardized methodology or guidance available for the evaluation and optimization of such delivery systems in vitro and in vivo, prior to performing much more expensive and time-consuming clinical evaluations. Table 8.3 lists the typical in vitro and in vivo methods available for the assessment of current oral transmucosal dosage forms (Patel et al. 2012).

TABLE 8.3
Types of Oral Transmucosal Dosage Forms and Their Testing Requirements

Test Methods	Tablets/Lozenges	Films/Wafers/Patches	Liquids/Gel/Cream/Ointments	Spray
In vitro evaluation				
Weight variation	√	√		
Uniformity of content	√	√	√	√
Friability	√			
Resistance to crushing	√			
Tensile strength	√	√		
Thickness	√	√		
Film endurance		√		
Viscosity			√	
Droplet size				√
Disintegration test	√	√		
Dissolution test	√	√	√	
Residence time	√	√	√	
Mucoadhesion strength	√	√	√	
Permeability study	√	√	√	√
In vivo evaluation				
Buccal absorption test	√	√	√	√
Perfusion study	√	√	√	√
Residence time	√	√	√	√
Pharmacokinetic study	√	√	√	√

Drug permeation across the oral mucosa is one of the key determinants of the effectiveness of a DDS. This can be assessed using isolated mucosa in a diffusion cell. Buccal epithelium from pigs, dogs, monkeys, rats, hamsters, rabbits, and primates are the most frequently used in such permeation studies. Of these, pig buccal mucosa is the most common, due to its close resemblance to human mucosal tissue with respect to lipid composition, keratinization and overall thickness. The drawbacks of using tissue models include the difficulty in maintaining tissue viability and integrity, as well as the complexity involved in tissue excision. Increasingly, buccal epithelial cell culture models are being used as an alternative.

In vitro–in vivo correlation (IVIVC) data are often used during pharmaceutical development in order to optimize the formulation, while reducing product development time and costs. A good correlation is a tool for predicting in vivo results based on in vitro data, and it allows dosage form optimization with the fewest possible trials in human, fixes drug release acceptance criteria, and can be used as a surrogate for further bioequivalence studies. Very few attempts have been made so far to obtain IVIVC for oral transmucosal DDS, and significant research effort is needed in this area.

8.7 CONCLUSIONS

Oral transmucosal drug delivery has attracted significant attention from academic and industrial researchers due to a number of advantages the route offers, such as avoidance of first-pass metabolism, improving medication compliance, rapid drug response, and possibility of controlled release. Many formulation approaches have been explored for buccal and sublingual routes, although the number of commercially available formulations is limited. A commercially available buccal spray for the oral transmucosal delivery of insulin has been developed. Oral transmucosal dosage forms will continue to be an exciting research focus for achieving the systemic absorption of drugs that are unsuitable for delivery via the oral route, especially for the new "biologics," such as peptides, proteins, and DNA-based medicines.

As regards future directions, problems that need to be addressed include those associated with palatability, as well as irritancy caused by DDS retention at the site of application. Robust and validated in vitro and in vivo methods are essential tools that need to be developed and standardized, in order to assess the performance of oral transmucosal DDS and to predict their in vivo behavior.

BIBLIOGRAPHY

Ahagon, A. and A.N. Gent. 1975. Effect of interfacial bonding on the strength of adhesion. *Journal of Polymer Science Part B: Polymer Physics* 13: 1285–1300.

Albrecht, K., M. Greindl, C. Kremser et al. 2006. Comparative *in vivo* mucoadhesion studies of thiomer formulations using magnetic resonance imaging and fluorescence detection *Journal of Controlled Release* 115: 78–84.

Birudaraj, R., B. Berner, S. Shen, and X. Li. 2005. Buccal permeation of buspirone: Mechanistic studies on transport pathways. *Journal of Pharmaceutical Sciences* 94: 70–78.

Boddupalli, B.M., Z.N.K. Mohammed, R.A. Nath, and D. Banji. 2010. Mucoadhesive drug delivery system: An overview. *Journal of Advance Pharmaceutical Technological Research* 1: 381–387.

Darwish, M., M. Kirby, P. Robertson Jr., W. Tracewell, and J.G. Jiang. 2007. Absolute and relative bioavailability of fentanyl buccal tablet and oral transmucosal fentanyl citrate. *Journal of Clinical Pharmacology* 47: 343–350.

Deneer, V.H.M., G.B. Drese, P.E.H. Roemele, J.C. Verhoef, L. Lie-A-Huen, J.H. Kingma, J.R.B.J. Brouwers, and H.E. Junginger. 2002. Buccal transport of flecainide and sotalol: Effect of a bile salt and ionization state. *International Journal of Pharmaceutics* 241: 127–134.

Dodou, D., P. Breedveld, and P. Wieringa. 2005. Mucoadhesives in the gastrointestinal tract: Revisiting the literature for novel applications. *European Journal of Pharmaceutics and Biopharmaceutics* 60: 1–16.

Gilhotra, R.M., M. Ikram, S. Srivastava, and N. Gilhotra. 2014. A clinical perspective on mucoadhesive buccal drug delivery systems. *Journal of Biomedical Research* 28: 81–97.

Hearnden, V., V. Sankar, K. Hull, D.V. Juras, M. Greenberg, A.R. Kerr, P.B. Lockhart, L.L. Patton, S. Porter, and M.H. Thornhill. 2012. New developments and opportunities in oral mucosal drug delivery for local and systemic disease. *Advance Drug Delivery Review* 64: 16–28.

Jiménez-Castellanos, M.R., H. Zia, and C.T. Rhodes. 1993. Mucoadhesive drug delivery systems. *Drug Development Industrial Pharmacy* 19: 143–194.

Khutoryanskiy, V.V. 2011. Advances in mucoadhesion and mucoadhesive polymers. *Macromolecular Bioscience* 11: 748–764.

Kockisch, S., G.D. Rees, S.A. Young, J. Tsibouklis, and J.D. Smart. 2004. In-situ evaluation of drug-loaded microspheres on a mucosal surface under dynamic test conditions. *International Journal of Pharmaceutics* 276: 51–58.

Lee, J.W., J.H. Park, and J.R. Robinson. 2000. Bioadhesive-based dosage forms: The next generation. *Journal of Pharmaceutical Sciences* 89: 850–866.

Li, B. and J.R. Robinson. 2005. Preclinical assessment of oral mucosal drug delivery systems. In *Drug Delivery to the Oral Cavity: Molecules to Market*, eds. T.K. Ghosh and W.R. Pfister, pp. 41–66. Boca Raton, FL: Taylor & Francis.

Li, H., Y. Yu, S. Faraji Dana, B. Li, C.Y. Lee, and L. Kang. 2013. Novel engineered systems for oral, mucosal and transdermal drug delivery. *Journal of Drug Targeting* 21: 611–629.

Mashru, R., V. Sutariya, M. Sankalia, and J. Sankalia. 2005. Transbuccal delivery of lamotrigine across porcine buccal mucosa: *In vitro* determination of routes of buccal transport. *Journal of Pharmacy and Pharmaceutical Science* 8: 54–62.

Morales, J.O. and J.T. McConville. 2011. Manufacture and characterization of mucoadhesive buccal films. *European Journal of Pharmaceutics and Biopharmaceutics* 77: 187–199.

Myers, G.L., D.H. Samuel, B.J. Boone, B.A. Bogue, P. Sanghvi, and M. Hariharan. 2013. Sublingual and buccal film compositions. U.S. 8475832 B2.

Nair, A.B., R. Kumria, S. Harsha, M. Attimarad, B.E. Al-Dhubiab, and I.A. Alhaider. 2013. *In vitro* techniques to evaluate buccal films. *Journal of Controlled Release* 166: 10–21.

Nielsen, H.M. and M.R. Rassing. 2002. Nicotine permeability across the buccal TR146 cell culture model and porcine buccal mucosa *in vitro*: Effect of pH and concentration. *European Journal of Pharmaceutical Sciences* 16: 151–157.

Patel, V.F., F. Liu, and M.B. Brown. 2011. Advances in oral transmucosal drug delivery. *Journal of Controlled Release* 153: 106–116.

Patel, V.F., F. Liu, and M.B. Brown. 2012. Modeling the oral cavity: *In vitro* and *in vivo* evaluations of buccal drug delivery systems. *Journal of Controlled Release* 161: 746–756.

Pather, S.I., M.J. Rathbone, and S. Senel. 2008. Current status and the future of buccal drug delivery systems. *Expert Opinion on Drug Delivery* 5: 531–542.

Pozzilli, P. et al. 2005. Biokinetics of buccal spray insulin in patients with type 1 diabetes. *Metab Clin Exp* 54(7): 930–934.

Preis, M., C. Woertz, P. Kleinebudde, and J. Breitkreutz. 2013. Oromucosal film preparations: Classification and characterization methods. *Expert Opinion on Drug Delivery* 10: 1303–1317.

Senel, S., M.J. Rathbone, M. Cansız, and I. Pather. 2012. Recent developments in buccal and sublingual delivery systems. *Expert Opinion on Drug Delivery* 9: 615–628.

Shojaei, A.H. 1998. Buccal mucosa as a route for systemic drug delivery: A review. *Journal of Pharmaceutics and Pharmaceutical Sciences* 1: 15–30.

Smart, J.D. 2005a. Buccal drug delivery. *Expert Opinion on Drug Delivery* 2: 507–517.

Smart, J.D. 2005b. The basics and underlying mechanisms of mucoadhesion. *Advanced Drug Delivery Review* 57: 1556–1568.

Squier, C.A. and P.W. Wertz. 1996. Structure and function of the oral mucosa and implications for drug delivery. In *Oral Mucosal Drug Delivery*, ed. M.J. Rathbone, pp. 1–26. New York: Taylor & Francis.

Streisand, J.B., J. Zhang, S. Niu, S. McJames, R. Natte, and N.L. Pace. 1995. Buccal absorption of fentanyl is pH-dependent in dogs. *Anesthesiology* 3: 759–764.

Sudhakar, Y., K. Kuotsu, and A.K. Bandyopadhyay. 2006. Buccal bioadhesive drug delivery—A promising option for orally less efficient drugs. *Journal of Controlled Release* 114: 15–40.

Ugwoke, M.I., R.U. Agu, N. Verbeke, and R. Kinget. 2005. Nasal mucoadhesive drug delivery: Background, applications, trends and future perspectives. *Advance Drug Delivery Review* 57: 1640–1665.

9 Transdermal Drug Delivery

Simon R. Corrie and Mark A.F. Kendall

CONTENTS

9.1 INTRODUCTION

Transdermal drug delivery is an emerging field with key applications in local delivery of molecules to local tissue sites, through to systemic delivery applications under sustained-release conditions. Historically, transdermal delivery has been limited to molecules that fit a narrow physicochemical profile (low molecular weight, adequate solubility in both oil and water, high partition coefficient); however, the range of deliverable drugs is rapidly expanding on all fronts, thanks to advances primarily in (1) enhancing the permeability of the outermost skin layer (stratum corneum [SC]), (2) increasing the driving force for drug transport across the SC, (3) physical approaches that bypass the SC altogether, and (4) novel combinations of these methods. With ≈20 small molecule drug delivery systems (DDS) currently FDA-approved involving transdermal approaches, based on current trials in small animals and humans, it is likely that delivery of genetic and cellular therapies, vaccines, therapeutic proteins, and nanoparticle-encapsulated systems will be available in the near future.

9.1.1 ANATOMY AND PHYSIOLOGY OF THE SKIN: IMPLICATIONS FOR TRANSDERMAL DRUG DELIVERY

The skin is the largest organ of the human body (1.5–2.5 m^2 surface area); its structure is described in Chapter 4 (Section 4.5.1 and Figure 4.6). It is composed of three principal skin layers: the SC, 10–20 µm; the viable epidermis (VE), 50–100 µm; and the dermis, 1–2 mm. The dermis is highly vascularized, while the SC/VE is not, yet capillary loops in the reticular dermis provide a route for nutrient supply and waste removal from the live cells in the VE. Traveling deeper into the dermis, hypodermis, and subcutaneous tissue, the blood vessels become larger and less dense, draining into the superficial venous plexus and the deep venous plexus. Stem and progenitor cells, located at the dermoepidermal junction, are responsible for skin renewal, continually pushing dead and dying cells into the SC and regenerating the underlying viable tissue.

The SC is the outermost layer, composed of corneocytes arranged in a "bricks and mortar" structure; it comprises the major physical barrier function of the skin (Figure 9.1). The corneocytes are terminally differentiated and flattened keratinocytes, rich in keratin filaments and embedded in a dense matrix of proteins to restrict transport of large molecules into the epithelial tissues beneath. The physicochemical properties of this layer are primarily responsible for limiting the free diffusion of molecules into the skin to those species <500 Da (approximately), which are moderately soluble in both water and oil phases.

The VE, i.e., the layers of the epidermis deep to the SC, is primarily composed of a densely packed layer of keratinocytes (≈95%) interspersed with small numbers of highly specialized cell types, including Langerhans cells (3%–5%; dendritic cells) and melanocytes (≈8%; melanin production to prevent UVB-associated damage) (Figure 9.1). Langerhans cells are potent antigen-presenting cells (APCs) and are thought to be key mediators of the improved immune responses observed in some transdermal vaccination techniques (see Section 9.3.3).

The dermis is a highly hydrated tissue composed largely of a collagenous extracellular matrix, containing a capillary network and a relatively lower density of cells, primarily fibroblasts and mast cells. Dermal dendritic cells (dDCs) have relatively recently been identified as a distinct APC population native to this layer, and their principal functions are still being determined, along with

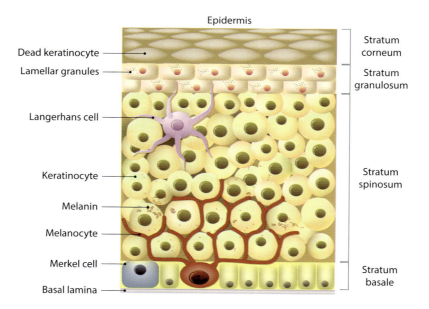

FIGURE 9.1 Epidermal cell layers. (Designua/Shutterstock.com.)

their interactions with other cells, tissue layers, and other components of the immune system. The vasculature of the dermis provides routes for systemic drug delivery, via circulating blood or lymphatic fluid.

9.1.2 ADVANTAGES OF THE TRANSDERMAL ROUTE FOR DRUG DELIVERY

In general terms, transdermal delivery methods are considered to be more acceptable to patients in comparison to more invasive delivery routes, including needle/syringe methods. This is because transdermal methods are relatively noninvasive, do not necessarily require trained health staff for administration, and "access" to viable delivery sites is rarely a concern due to the high number of available skin sites. Importantly, a number of transdermal techniques also allow the user or medical practitioner to control release rates into the tissue, from rapid delivery applications (e.g., local anesthetics) through to sustained-release periods (up to 1 week, e.g., pain medications).

For the delivery of drugs to *local* skin epithelial tissues, transdermal methods can clearly target the tissue of interest without off-target systemic effects. Transdermal delivery also has advantages for *systemic delivery* applications due to direct and/or indirect uptake of drugs from skin tissue fluid into the blood or lymphatic circulation, respectively. This potentially avoids first-pass liver metabolism, thereby requiring lower initial concentrations for therapeutic effects. Controlled-release strategies can also be used to control the drug delivery rate and pharmacodynamic profile. Furthermore, in the last decade or so, a clear narrative is emerging that transdermal methods provide unique access to a unique population of immunologically sensitive cells, which can be exploited for immunotherapeutic and vaccine applications.

9.1.3 LIMITING FACTORS FOR TRANSDERMAL DRUG DELIVERY

The key limitation in transdermal drug delivery, particularly for diffusion-based mechanisms, is that efficient delivery is limited to those compounds that possess the physicochemical properties that favor transepithelial transport. These properties are described in detail in Chapter 4 (Section 4.3.4); to summarize here for the transdermal route, this refers to low-molecular-weight (<500 Da approximately) compounds, with high partition coefficients and moderate solubility in both oil and water phases. This is due primarily to the physicochemical properties of the SC, which are crucial to maintain the skin's barrier function. To overcome this limitation, a range of methods have been developed to enhance diffusion through the SC, or breach this outer layer completely, to gain direct access to the epithelial tissues underneath.

The other limitation, again particularly related to diffusion-based methods, is that attempts to increase delivery rates through the SC (direct breaching, enhanced permeability, etc.) are generally accompanied by local tissue irritation and/or inflammation. This has been problematic for the health-care providers, FDA, and patients, although it is becoming apparent that a minor degree of local irritation may enhance some therapeutic strategies, especially those that target the immune system.

9.2 MECHANICAL AND BIOLOGICAL PROPERTIES OF THE PRINCIPAL SKIN LAYERS

9.2.1 STRATUM CORNEUM: THE KEY MECHANICAL BARRIER

The SC is a semipermeable barrier that, owing to its variable mechanical properties, is challenging to breach in a minimally invasive manner, to target the underlying viable skin strata (whether it is the VE, dermis, or a combination of both). Mechanically, the SC is classified as a bioviscoelastic solid and shows highly variable properties. Obvious differences include the huge variation in

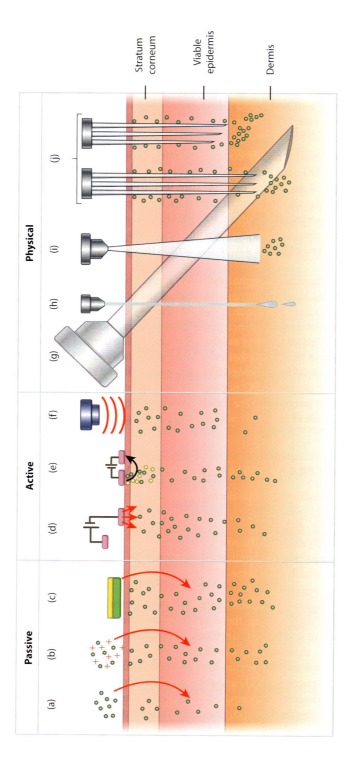

FIGURE 9.3 Summary of approaches to achieving transdermal drug/vaccine delivery. *Passive approaches*: (a) drug, (b) drug + penetration enhancer, (c) transdermal patch. *Active approaches*: (d) electroporation, (e) iontophoresis (arrow = induced convective flow, could be electroosmosis or electrophoresis), (f) ultrasonic wave. *Physical approaches*: (g) intradermal injection, (h) liquid jet injector, (i) gene gun. (j) microneedle arrays, delivering to either the dermis or the viable epidermis.

of a metered syringe); however, the same cannot yet be stated for transdermal techniques. This is particularly important in terms of potential toxic effects related to overdose of some medications (e.g., fentanyl is a commonly studied drug for transdermal delivery systems in controlling pain medication, but as an opiate, it can also cause respiratory depression), based on patient-to-patient variability, and the differences between adults and children in pharmacokinetics and metabolism.

9.3.1 PASSIVE APPROACHES RELYING ON DIFFUSION

9.3.1.1 Transdermal Therapeutic Systems

Transdermal patches, also called transdermal therapeutic systems (TTS), comprise medicated skin patches that facilitate controlled release of the API (Figure 9.3c). They are designed so that the drug diffuses through the patch at a rate that is much slower than its diffusion-rate through the SC. Therefore drug is delivered at a controlled rate that is dictated by the patch, and not the skin.

The first TTS, developed by Alza (see Chapter 1, Section 1.4.1), was FDA-approved for use in humans in 1979 and comprised a skin patch that delivered scopolamine for up to 3 days (Transderm Scop®), for the treatment of nausea related to motion sickness. Nicotine patches followed approximately a decade later and remain the most well-known transdermal system for the general public. Structurally, a TTS contains a number of layers, including (1) an impermeable backing layer, (2) a layer (or layers) that provide the controlled-release mechanism, and (3) a peel-away strip, which is removed prior to application. Various mechanisms are used to provide controlled release, including (1) reservoir systems, whereby drug release is controlled by a rate-controlling membrane (RCM), which may be porous or microporous (Figure 9.4a), and (2) matrix systems, whereby drug release is controlled by diffusion through an adhesive matrix (Figure 9.4b).

Many other designs are used to facilitate controlled release, including, for example, a multilayered matrix TTS, which contains more than one drug-in-adhesive layer, the individual layers may be separated by a membrane. A reservoir-matrix TTS contains a drug reservoir layer with an associated RCM and an additional drug-in-adhesive matrix layer. More recently, TTS designs have moved away from reservoir systems to other CR mechanisms, thus avoiding adverse events related to rapid "dumping" of drug molecules into the skin, should the RCM become damaged.

The patches are recommended only for use for up to 7 days at the same location, to avoid local adverse skin responses mainly related to skin stripping upon removal (Wohlrab et al. 2011). All of the drugs approved for delivery using TTS technology (for example, nicotine, clonidine, scopolamine, nitroglycerin, and fentanyl) fit the required physicochemical profile described earlier, i.e., small and lipophilic molecules. The great challenge is that only a limited number of drugs are amenable to administration using this approach, so large and/or hydrophilic drugs are generally not deliverable using a TTS.

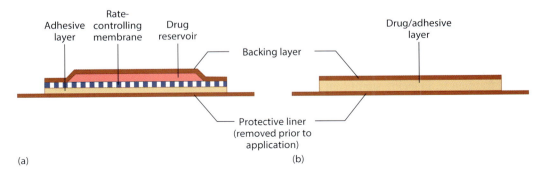

FIGURE 9.4 Transdermal therapeutic systems. Mechanisms of achieving controlled release: (a) reservoir system and (b) matrix system.

9.3.1.2 Penetration Enhancers

To improve the passive diffusion rates across the skin for a range of molecules, chemical enhancers have been introduced in order to increase skin permeability by disrupting the lipid structure of the SC (Figure 9.3b). However, increased permeation enhancement is also related to increased irritation and injury to the underlying layers. This approach provides enhanced permeability, and sometimes driving force, for transdermal transport. Hundreds of such enhancers have been trialed, with several specifically designed and approved for human use (1-dodecyl-azacycloheptan-2-one [Azone] and 2-n-nonyl-1,3dioxolane [SEPA™]). While a key challenge is identifying useful chemical enhancers for a wide range of different drugs, high-throughput screening approaches have been developed, using spectroscopic markers that correlate with (1) lipid permeability and (2) skin irritation. Amphiphiles containing either long-carbon tails or multiple aromatic rings were predicted to be efficient and indeed have been proven so experimentally, particularly by Mitragotri and colleagues.

A key limitation with conventional chemical enhancers is that their effects are often not limited to the SC layer itself; hence, irritation or injury to the underlying tissues often results. A range of alternative or combination approaches has been used with some success. Packaging of drugs inside micro- and nanosized carriers, such as liposomes, dendrimers, and nanoparticles, has the advantage of limiting delivery of the carrier to the SC alone so that enhancers do not reach the underlying tissue. Such encapsulation methods can have other advantages as well, including protective effects, additional control over release profiles, etc.

Biochemical enhancers have also been introduced, which comprise polypeptides that have pore-forming or pore-enhancing biological activity. These can often be formulated with conventional chemical enhancers, using the latter to increase lipid disruption at the SC.

Other combinations of enhancers have also been trialed in order to balance the permeability/irritation aspects of individual enhancers. In a high-throughput screening approach, one study examined 500 pairs of chemical enhancers in over 5000 individual compositions, finding a combination that provided significant permeability enhancement with low skin irritation (a combination of sodium laureth sulfate and phenylpiperazine at 0.35% and 0.15% by weight, respectively, in a 1:1 mixture of ethanol and phosphate buffer), validated in a rat model of leuprolide acetate peptide delivery (Karande et al. 2004).

Alternatively, some investigators have made use of the appendage delivery pathways to bypass the need to interact with the SC altogether—particularly for skin-specific and cosmetic treatments; however, this area is in its infancy (Chourasia and Jain 2009).

9.3.2 ACTIVE APPROACHES REQUIRING APPLICATION OF ENERGY

While conventional and novel chemical enhancers have resulted in significantly improved drug delivery in preclinical and clinical studies, resulting in FDA-approved therapies, this approach is still fundamentally limited to small molecule delivery. The application of electrical or ultrasonic energy to the skin has been shown to significantly increase the permeability of the SC above and beyond the activity of enhancers, often with minimal injury to underlying tissues. The high electrical and thermal resistance of the SC layer means that relatively high currents or temperature can be applied transiently, inducing significant permeability in the SC alone, with the effect lasting up to several hours following the initial energy input.

9.3.2.1 Electroporation

High voltage pulses (100–1500 V), applied transiently to the SC (0.01–10 ms), have been shown to disrupt lipid structure in the SC allowing drug transport via diffusion through "electropores" and electrophoresis-driven flow (Figure 9.3d). The level of enhancement in terms of delivered drug can be several orders of magnitude higher than control skin, depending on physicochemical properties of the drug of interest. However, this technique is still sensitive to the molecular weight of the drug to be delivered, its lipophilicity, competitive ions present in the formulation, and also the viscosity of the formulation. To date, most studies have been limited to preclinical animal studies due to the complexities involved in designing the devices (including proteins and DNA); however, both ascorbic acid and lidocaine have been successfully delivered to human volunteers via this method. Concern over safety and pain related to this technique are largely offset by the acceptance of related techniques (e.g., electrochemotherapy—8×1000 V cm^{-1}, 100 μs with plate electrodes); however, further design improvements are required to limit electrical activity to the SC, thus avoiding excitation of sensory nerves in the deeper dermis (Denet et al. 2004).

9.3.2.2 Iontophoresis

Iontophoresis involves application of a low-voltage current in order to increase permeability but also provides an electrical driving force for transport (Figure 9.3e). An Ag/AgCl electrode set is favored due to the pH stability, as it does not produce the sharp decrease in pH that can be seen with Pt electrodes. The driving force is distinct to that induced by electroporation because it facilitates either direct transfer of charged moieties (electrophoresis) or mobile ion flows (electroosmosis), causing convective transport of weakly charged, or neutral, molecules. Interestingly, the delivery rate is a function of the current, therefore complex delivery profiles can be achieved for controlled-release purposes. Indeed, low-cost approaches to iontophoresis are available: simply linking the patch to a disposable, constant-voltage, disposable battery, avoiding the need for a microprocessor. For this reason, iontophoresis is considered suitable for clinical use because the process can easily be controlled by untrained individuals. While a range of devices have been commercialized and approved for local treatments (e.g., lidocaine for anesthesia, tap water for hyperhidrosis), there is still concern over systemic drug delivery (e.g., fentanyl for pain relief) due to potential side effects based on the variation in the delivered dose between patients, due to age or skin type (Kalia et al. 2004).

9.3.2.3 Ultrasound

Ultrasound (also referred to as sonophoresis or phonophoresis) was first used as a skin permeabilization method for the local delivery of corticosteroids to the skin in the 1950s. The ultrasound wave is a compression wave (frequency >20 kHz), in which an electrical signal is transmitted to a "horn" and converted into a sound wave via piezoelectric crystals (change static dimensions in response to electric field) (Figure 9.3f). This causes the horn to be displaced in a cyclic fashion, with characteristic amplitude and frequency. Early studies utilized high-frequency sonophoresis (HFS; $\approx$0.7–16 MHz), resulting in up to 10-fold increase in local drug delivery through the SC. However, it was later recognized that low-frequency sonophoresis (LFS; $\approx$20–100 kHz) led to significant improvements in local delivery rates (up to three orders of magnitude better than HFS) due to the effects of generating, and collapsing, air bubbles near the SC (cavitation).

Key variables in ultrasound-induced drug delivery include the type of pulse program (e.g., continuous vs. timed pulses), the horn-to-skin distance, the overall treatment time, and the ultrasound medium composition. The medium is crucial: it contains the drug to be delivered, generally a chemical enhancer, and is usually formulated in aqueous solutions for LFS and gels for HFS, both of which are designed to have acoustic impedance similar to that of the skin. Pulsing is often used to reduce local heating of the tissue, although a small degree of heating can also act to help permeabilize the SC. The horn-to-skin distance is generally very close to (if not touching) the skin for HFS treatments, but often up to 1 cm from the skin for the LFS treatments. This is because the cavitation-associated effects of LFS treatment are dependent on the generation of gas bubbles in the medium above the skin, an effect diminished if the probe and skin are in direct contact. Treatment times can be wide ranging, from transient delivery regimes on the order of seconds/minutes up to steady-state delivery regimes, which can be on the order of hours/days. Synergistic effects have been observed when combining chemical enhancers into the medium (e.g., surfactants).

HFS treatments have found utility in the local delivery of low-molecular-weight (<1000 Da), skin-specific drugs, but generally are not favored for systemic delivery. In comparison, LFS treatments are still relatively new (only 20-year history of research so far); however, a number of groups have demonstrated that macromolecules, including proteins (e.g., interferon-γ $\approx$17 kDa, erythropoietin $\approx$48 kDa), can be effectively delivered via this method, paving the way for systemic delivery. Vaccines have also been delivered via this route (e.g., tetanus toxoid), with LFS treatment clearly showing activation of skin-resident APCs, paving the way for future innovations in transdermal vaccination. However, the prospect of systemic effects is also a safety issue; hence, the development pathway is expected to be longer for LFS treatments.

9.3.3 Physical Approaches That Breach the Stratum Corneum

While the techniques described earlier have shown utility in permeabilizing the SC, allowing transport of mainly low-molecular-weight compounds via different mechanisms, the size and other physicochemical properties of the drug and/or formulation media are still key parameters that limit delivery and uptake by the key epidermal and dermal layers. To address this problem, physical methods of breaching the SC in order to deliver drug payloads directly in the epidermis and dermis have been developed. The key advantages of this approach include (1) the variety of drugs based on size/charge/polarity/hydrophilicity that can be delivered, due to the lack of direct interaction with the SC, and (2) in terms of vaccine delivery, compounds that can be delivered directly into the vicinity of the potent APCs present in the VE and dermis. For these reasons, physical approaches typically address the therapeutic challenge of vaccine delivery (with or without immune-boosting adjuvants) to the epidermal layers. Importantly, as the methods earlier have proven in many ways well suited to the local delivery of small, lipophilic compounds, physical approaches have historically been developed for a different challenge.

9.3.3.1 Intradermal Injections

Conceptually, the simplest method for targeting the epidermal layers of the skin with a drug is to simply inject the compound into that layer via a standard needle/syringe (Figure 9.3g). However, this approach proves to be extremely challenging from a technical point of view, due to the submillimeter precision required to target the appropriate layers. Furthermore, variations with individuals (injection site) and between individuals based on age, gender, levels of sun exposure, etc., can have dramatic effects on the hydration levels and hence mechanical properties of the skin layers, which cannot be predicted or controlled for, using such an approach. However, the intradermal injection route has certainly proven the concept that targeting immune cells in the skin results in significantly higher antibody titers for a range of vaccines (e.g., rabies, influenza), in comparison to the more traditional intramuscular or subcutaneous injection routes.

9.3.3.2 Liquid Jet Injectors

Interest in using high-speed liquid jet injectors arose in the mid-twentieth century because of its needle-free approach (Furth et al. 1995). Mitragotri provides an excellent review on the status of the liquid injector field (Mitragotri 2006). This technique has seen resurgence, with liquid delivered around the Langerhans cells in gene transfer and DNA vaccination experiments, and the delivery of drugs. Since then, there have been many developments in the field. This includes the licensing of new liquid jet injector devices for vaccine delivery and, separately, research investigations into microscale liquid jet injectors. As shown in Figure 9.3h, current liquid jet injectors typically disrupt the skin in the epidermal and dermal layer. To target exclusively the viable epidermal cells, such as Langerhans cells (as one example), the challenge of more controlled delivery needs to be addressed. With the dermal disruption induced by administration, liquid jet injectors are also reported to cause pain to patients.

9.3.3.3 Gene Gun

With gene guns, pharmaceutical or immunomodulatory agents formulated as particles are accelerated in a supersonic gas jet to sufficient momentum to penetrate the skin (or mucosal) layer and to achieve a pharmacological effect (Figure 9.3i). Klein et al. pioneered this innovation with systems designed to deliver DNA-coated metal particles (of diameter of the order of 1 μm) into plant cells for genetic modification, using pistons accelerated along the barrels of adapted guns (Klein et al. 1987). The concept was extended to the treatment of humans, with particles accelerated by entrainment in a supersonic gas flow. Prototype devices embodying this concept have been shown to be effective, painless, and applicable to pharmaceutical therapies ranging from protein delivery to conventional vaccines and DNA vaccines.

Prior to operation, the gas canister is filled with helium or nitrogen to 2–6 MPa, and the vaccine cassette, comprising two 20 mm diaphragms, is loaded with a powdered pharmaceutical payload of 0.5–2 mg. The pharmaceutical material is placed on the lower diaphragm surface. Operation commences when the valve in the gas canister is opened to release gas into the rupture chamber, where the pressure builds up until the two diaphragms retaining the vaccine particles sequentially burst. The rupture of the downstream diaphragm initiates a shock that propagates down the converging–diverging nozzle. The ensuing expansion of stored gas results in a short-duration flow (often <1 ms) in which the drug particles are entrained and accelerated through the device. After leaving the device, particles impact on the skin and penetrate to the epidermis, to induce a pharmacological effect.

9.3.3.4 Microneedles, Microprojections, and Arrays Thereof

Microprojection arrays (MPAs), or "microneedles," comprise arrays of sharp-tipped projects that are designed to pierce the tough SC layer and rapidly deliver vaccine payloads to the underlying epidermal and dermal tissues (Figure 9.3j). These devices emerged in the late 1990s and have now been fabricated from a range of materials (silicon, metal, polymer, ceramic, and hybrids/composites), in a range of different geometries (density of projections 10^1–10^4 cm^{-2}; length 0.03–3 mm). Key design parameters include the shape, density, length, and tip sharpness of the projections, along with the application velocity, as these directly affect the skin penetration depth achieved by the array. The different materials of fabrication allow for different delivery modes, including (1) dissolution of a vaccine coating from a solid projection, (2) injection of a liquid vaccine through a hollow projection, and (3) projections designed to dissolve upon skin insertion, releasing the vaccine payload. Furthermore, coating strategies aim to optimize formulations for long-term thermostability (e.g., incorporating "glassy" sugars to stabilize biomolecules when dried), controlled release of vaccine payload following skin insertion, and high release efficiency into the skin (Prausnitz and Langer 2008; Kim et al. 2012).

Compelling data have been published showing that MPAs can elicit protective immune responses in comparison to standard intramuscular or subcutaneous injection in animal models, and in many cases with lower vaccine dosage required (e.g., Fernando et al. 2010). Influenza vaccine administration has been the most investigated test case using coated projections, showing protective immunity in comparison with the needle, on the basis of total IgG antibody levels and functional assays involving hemagglutination inhibition and neutralizing antibody activity (Kim et al. 2012). However, a range of different vaccines (including split virion, protein subunit, DNA plasmids, siRNA, etc.) targeting different diseases (influenza, human papillomavirus, West Nile virus, hepatitis B, herpes simplex virus, chikungunya virus, etc.) have been investigated.

To date, clinical trials have shown early promise both in terms of pain reduction and immunogenicity, in comparison to standard intramuscular injections for a range of approved and emerging vaccines. Most patients and clinicians report significantly lower pain scores for microneedle devices in comparison with standard hypodermic needles; there is minimal skin reactivity beyond a mild erythema, which resolves in minutes or hours in the majority of cases. Transdermal vaccination is discussed further in Chapter 17 (Section 17.4.2).

9.3.3.5 Nanopatch™: A Case Study

While early results in preclinical and clinical trials investigating microneedle technology are encouraging, key challenges remain to be solved, including minimizing delivered dose to increase distribution and availability and the significant improvement of vaccine formulation thermostability, for storage at ambient conditions for long periods. The Nanopatch™ (Vaxxas) was designed as an ultrahigh density MPA with vaccine formulated into a dry-coating layer over solid projections (Figure 9.5). The array design was fabricated in silicon, using standard semiconductor processing techniques. The coating rapidly dissolved once hydrated in the skin, for rapid delivery in seconds. In order to target the immune APCs in the mouse epidermis and dermis, a 21,000 cm^{-2} density array was designed, with very short tips (0.03–0.3 mm) in order to maximize the delivered dose at the target site. In a key proof-of-concept study, the authors showed that Nanopatches™ indeed targeted ≈50% of available APCs in mice (both in the VE and dermis), resulting in less than 100th of the standard intramuscular dose required for equivalent protective immune responses using an

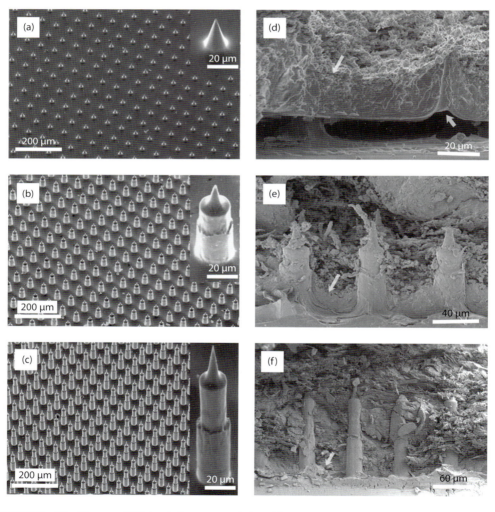

FIGURE 9.5 The Nanopatch™. (a–c) Arrays containing silicon projections of increasing length; (d–f) corresponding penetration of each array into mouse abdominal skin (cryo-SEM). (Adapted from *Biomaterials*, 34(37), Coffey, J., Corrie, S.R., and Kendall, M.A.F., Early circulating biomarker detection using a wearable microprojection array skin patch, 9572–9583. Copyright 2013, with permission from Elsevier.)

influenza-based mouse model (Fernando et al. 2010). Investigations have found that localized death of skin cells induced by the dynamic application of the Nanopatch plays an important role in generating potent immunity: a "physical immune enhancer" (Depelsenaire et al. 2014).

The Nanopatch™ has been used to formulate delivery strategies for a range of vaccines, and along the way, a number of novel coating technologies and formulations were developed to support long-term thermostability. Using a "jet-coating" approach, the Nanopatch™ technology has delivered many different classes of vaccine, including inactivated whole virus vaccines (e.g., FluVax®: commercially available seasonal influenza vaccine), viruslike particles (VLPs, e.g., Gardasil®: commercially available tetravalent HPV vaccine), and DNA plasmids (e.g., preclinical HSV2 vaccine), among others. Immune adjuvants, designed to boost the host immune response to a vaccine formulation, have also been incorporated into the coating mixtures to further enhance immune responses, e.g., from 100-fold to 900-fold dose reduction in a FluVax® model using Quil-A adjuvant (Fernando et al. 2010). Long-term thermostability has also been demonstrated using the jet-coating approach, with comparative immunogenicity observed with freshly coated devices or those coated and stored for over 6 months at 23°C prior to skin application.

9.4 CONCLUSIONS

In conclusion, transdermal drug delivery methods show great promise for both localized and systemic delivery of therapeutic molecules. Advances in technology over the past several decades have overcome the key problems encountered when attempting to deliver drugs through the SC, leading to the delivery of molecules with a wide variety of size, charge, and physicochemical properties. Most recently, technologies developed to breach the SC (while causing minimal discomfort) have allowed the delivery of vaccines and related macromolecules with/without adjuvants to the immunologically sensitive skin layers, leading to significantly improved vaccine responses in comparison to traditional methods (e.g., needle/syringe). These technologies (including gene guns, liquid jet injectors, and microneedles) will potentially allow more widespread distribution of life-saving vaccines and therapies to those who most need them, without the current problems arising from the need for needles, cold chains, and trained health-care professionals.

REFERENCES

Banchereau, J. and R.M. Steinman. 1998. Dendritic cells and the control of immunity. *Nature* 392(6673): 245–252.

Boyer, G., L. Laquieze, A. Le Bot et al. 2009. Dynamic indentation on human skin *in vivo*: Ageing effects. *Skin Research and Technology* 15(1):55–67.

Coffey, J., S.R. Corrie, and M.A.F. Kendall. 2013. Early circulating biomarker detection using a wearable microprojection array skin patch. *Biomaterials* 34(37):9572–9583.

Chourasia, R. and S.K. Jain. 2009. Drug targeting through pilosebaceous route. *Current Drug Targets* 10(10):950–967.

Crichton, M.L., X.F. Chen, H. Huang et al. 2013. Elastic modulus and viscoelastic properties of full thickness skin characterised at micro scales. *Biomaterials* 34(8):2087–2097.

Crichton, M.L., B.C. Donose, X. Chen et al. 2011. The viscoelastic, hyperelastic and scale dependent behaviour of freshly excised individual skin layers. *Biomaterials* 32(20):4670–4681.

Denet, A.R., R. Vanbever, and V. Preat. 2004. Skin electroporation for transdermal and topical delivery. *Advanced Drug Delivery Reviews* 56(5):659–674.

Depelsenaire, A.C., S.C. Meliga, C.L. McNeilly et al. 2014. Colocalization of cell death with antigen deposition in skin enhances vaccine immunogenicity. *Journal of Investigative Dermatology* 134(9):2361–2370.

Fernando, G.J.P., X.F. Chen, T.W. Prow et al. 2010. Potent immunity to low doses of influenza vaccine by probabilistic guided micro-targeted skin delivery in a mouse model. *PLOS ONE* 5(4):11.

Furth, P.A., D. Kerr, and R. Wall. 1995. Gene-transfer by jet injection into differentiated tissues of living animals and in organ-culture. *Molecular Biotechnology* 4(2):121–127.

Huzaira, M., F. Rius, M. Rajadhyaksha et al. 2001. Topographic variations in normal skin, as viewed by *in vivo* reflectance confocal microscopy. *Journal of Investigative Dermatology* 116(6):846–852.

Kalia, Y.N., A. Naik, J. Garrison et al. 2004. Iontophoretic drug delivery. *Advanced Drug Delivery Reviews* 56:619–658.

Karande, P., A. Jain, and S. Mitragotri. 2004. Discovery of transdermal penetration enhancers by high-throughput screening. *Nature Biotechnology* 22(2):192–197.

Kendall, M.A.F., Y.-F. Chong, and A. Cock. 2007. The mechanical properties of the skin epidermis in relation to targeted gene and drug delivery. *Biomaterials* 28(33):4968–4977.

Kim, Y.-C., J.-H. Park, and M.R. Prausnitz. 2012. Microneedles for drug and vaccine delivery. *Advanced Drug Delivery Reviews* 64:1547–1568.

Klein, T.M., E.D. Wolf, R. Wu et al. 1987. High-velocity microprojectiles for delivering nucleic-acids into living cells. *Nature* 327(6117):70–73.

Mitragotri, S. 2006. Current status and future prospects of needle-free liquid jet injectors. *Nature Reviews Drug Discovery* 5:543–548.

Prausnitz, M.R. and R. Langer. 2008. Transdermal drug delivery. *Nature Biotechnology* 26(11):1261–1268.

Sullivan, S.P., D.G. Koutsonanos, M.D. Martin et al. 2010. Dissolving polymer microneedle patches for influenza vaccination. *Nature Medicine* 16(8):915-U116.

Wohlrab, J., B. Kreft, and B. Tamke. 2011. Skin tolerability of transdermal patches. *Expert Opinion on Drug Delivery* 8(7):939–948.

10 Nasal Drug Delivery

Per Gisle Djupesland and Anya M. Hillery

CONTENTS

10.1 INTRODUCTION

Nasal drug delivery is ideally suited to the topical treatment of local nasal conditions, such as the common cold, allergic and nonallergic rhinitis, nasal polyps, and chronic nasal and sinus inflammations. Drugs for topical delivery include antihistamines (e.g., azelastine), anti-inflammatory corticosteroids (e.g., budesonide and fluticasone), and topical nasal decongestants (e.g., oxymetazoline and xylometazoline). The idea is that delivered locally, these drugs are effectively targeted to their site of action, which should maximize their therapeutic effect while minimizing unwanted side effects.

However, as described in this chapter, current delivery devices are in fact not very effective at targeting to the posterior nasal cavity; drug delivery to this region requires optimization.

The nasal route is also increasingly used as a noninvasive route for systemic delivery, particularly when rapid systemic absorption and clinical effect are desired, for example, in the rapid relief of a migraine attack. Marketed nasal antimigraine drugs include sumatriptan (Imitrex® nasal spray), zolmitriptan (Zomig® nasal spray), and dihydroergotamine mesylate (Migranal® nasal spray). A further example is the intranasal (IN) delivery of opiates, such as fentanyl (Lazanda® and Instanyl®), when rapid pain relief is required. In addition, nasal delivery has become a useful alternative for systemic drug absorption in situations where the gastrointestinal (GI) route is unfeasible, such as for patients with nausea, vomiting, and gastric stasis (frequent in migraine patients); or patients with swallowing difficulties, such as children and elderly; or those who suffer from dry mouth.

The nasal route is suitable for drugs with poor oral bioavailability, due to, for example, GI instability, poor and delayed oral absorption, and drugs that undergo extensive first-pass effects in the gut wall or liver. A variety of peptide and protein drugs that demonstrate poor oral bioavailability are capable of systemic absorption via the nasal route, and a number of commercially available preparations are on the market, including for nafarelin (Synarel®), salmon calcitonin (Miacalcin®, Roritcal®), oxytocin (Syntocinon®), desmopressin (Desmospray®), and buserelin (Suprecur®). Similarly, "biologic" drugs such as monoclonal antibodies and antisense DNA demonstrate poor oral bioavailability and currently must be given by injection. These molecules are unlikely to realize their full clinical potential unless the patient can easily and conveniently self-administer the drug. The nasal route has emerged as a highly promising alternative epithelial route for the systemic delivery of these drugs.

The nasal route may also be used as an alternative to injections for the administration of vaccines, potentially including immunotherapeutics. For example, FluMist®, a nasal vaccine to protect against influenza, has been commercially available since 2003. A further possibility of the nasal route is as a portal of entry for drugs into the central nervous system (CNS). The barrier between the blood and the brain (the blood–brain barrier [BBB]) plays a vital role in protecting the delicate milieu of the brain, but also prevents CNS therapeutics from gaining access. Exploitation of a direct "nose-to-brain" (N2B) pathway, which bypasses the BBB, could therefore facilitate the treatment of numerous disabling psychiatric and neurodegenerative disorders, as well as brain cancers.

10.2 ANATOMY AND PHYSIOLOGY OF THE NASAL CAVITY: IMPLICATIONS FOR DRUG DELIVERY

The anatomy of the nose is shown in Figure 10.1. It extends 6–9 cm from the nostrils to the nasopharynx (throat) and is subdivided into left and right sides by a vertical partition, the nasal septum. The anterior portion of the nasal cavity, the nasal vestibule, then narrows into a triangular-shaped slit, the nasal valve, located approximately 1.5–2.5 cm from the nostril. Beyond the valve region, the posterior cavity is characterized by groove-like air passages (meatuses), formed by the scroll-like projections of the superior, middle, and inferior turbinates (conchae), which extend out from the lateral walls, almost reaching the septum.

Specific anatomical and physiological features of the nasal cavity have important implications for nasal drug delivery and targeting, and are discussed in more detail here.

10.2.1 Nasal Geometry

The complex anatomical features of the nose are designed to facilitate its role in protecting the lower airways by filtering, warming, and humidifying the inhaled air. When air enters the nostrils, it passes first through the nasal vestibule, which is lined by the skin containing vibrissae (short, coarse hairs) that filter out large dust particles. The nasal valve is the narrowest portion of the nasal passage, with a mean cross-sectional area of only about 0.6 cm^2 on each side. It is the primary

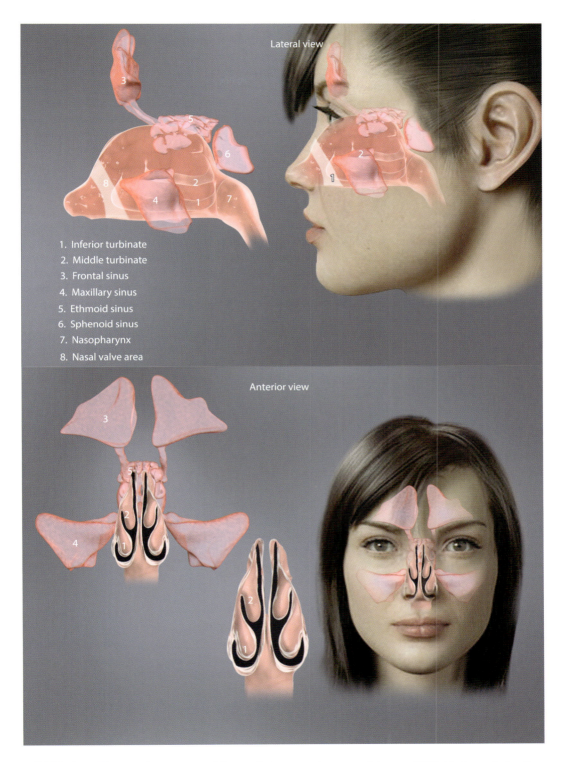

Lateral view

1. Inferior turbinate
2. Middle turbinate
3. Frontal sinus
4. Maxillary sinus
5. Ethmoid sinus
6. Sphenoid sinus
7. Nasopharynx
8. Nasal valve area

Anterior view

FIGURE 10.1 The complex anatomy of the nasal airways and paranasal sinuses. (With kind permission from Springer Science+Business Media: *Drug Deliv. Transl. Res.*, Nasal drug delivery devices: Characteristics and performance in a clinical perspective—A review, 3, 2013, 42, Djupesland, P.G., Copyright 2012.)

regulator of airflow and resistance, accounting for up to 80% of nasal resistance, and almost half of the total resistance of the entire respiratory system. With increasing inspiratory flow rate, the action of Bernoulli forces progressively narrow the valve. The valve can even close completely with vigorous sniffing. Beyond the valve, the nasal cavity comprises a set of narrow, warrenlike passageways, the meatuses. The turbinates introduce turbulence to the airflow, forcing it through the narrow meatuses and ensuring maximal contact between the incoming air and the mucosal surface. The dense vascular capillary bed directly beneath the mucosal surface, in addition to the mucus layer and nasal secretions, ensures that the incoming air is warmed and humidified. The arrangement of conchae and meatuses also increases the available surface area of the internal nose and prevents dehydration by trapping water droplets during exhalation.

10.2.2 MUCUS AND THE MUCOCILIARY ESCALATOR

Goblet cells in the mucosa secrete mucus, a complex, thick, viscoelastic gel, containing specific glycoproteins known as mucins, which give the secretions their characteristic viscosity and elasticity. Mucos also contains proteins, lipids, and antibacterial enzymes (such as lysozyme and immunoglobulins). The secreted mucosal blanket lies over the nasal epithelium and functions as a protective barrier against the entry of pathogens. The underlying epithelial cells are ciliated, and the mucus works in tandem with the ciliated cells, in an extremely efficient self-cleansing mechanism known as the "mucociliary escalator." In this process, inhaled particulates (dust, pollutants, microbes, etc.) become trapped in the sticky mucus layer. The underlying cilia beat in unison, via an ATP energy–dependent process, propelling the mucus layer toward the throat. The ciliary movement can be considered a form of rhythmic waving, which enables hooklike structures at the ciliary tips to propel mucus along. The ciliary beat frequency is in the range of about 100 strokes per minute. At the throat, entrapped particles are removed, either by being swallowed or expectorated.

10.2.3 NASAL EPITHELIUM

The nasal vestibule is lined by nonciliated, stratified squamous epithelium, which gradually transitions in the valve region into a ciliated, pseudostratified, columnar epithelium with goblet cells (see also Chapter 4, Figure 4.9). Mucus secreted by the goblet cells lies over the epithelium as a protective layer, and the mucociliary escalator filters the incoming air. In addition to cilia, the epithelial cells contain microvilli, which increase the available surface area.

Another important function of the nose is olfaction and, as such, the nasal cavity also contains a specialized olfactory epithelium, located in the roof of the nasal cavity. The olfactory epithelium and a direct olfactory pathway to the brain are described further in Section 10.6, in the context of N2B delivery (Figure 10.4).

10.3 OVERVIEW OF NASAL DRUG DELIVERY

As outlined in Section 10.1, nasal drug delivery is highly versatile and can be used for a variety of delivery options, including (1) for local delivery to the nose and (2) as a noninvasive route for systemic delivery, (3) for nasal immunization and vaccine delivery, and (4) for delivery of drugs to the CNS, via N2B delivery. The nasal route offers a number of advantages for these various delivery options, outlined here.

1. *An expanded surface area*: The area available for absorption is enhanced by the labyrinthine passages of the nasal cavity, as well as the microvilli present on the apical surface of the epithelial cells. The total surface area of the nasal cavity is about 160 cm^2 (although the *effective* surface area for absorption is influenced by the type of dosage form used to deliver the drug, and other factors).

2. *Permeable epithelium*: The epithelium of the nose is described in Chapter 4 (Figure 4.9). It is more permeable than that of the GI tract. For example, molecules larger than 500 Da are too large to permeate the GI tract via transcellular passive diffusion, whereas the cut-off molecular weight for nasal transcellular diffusion seems to be about 1000 Da. This enhanced permeability, coupled with reduced metabolic activity, accounts for the number of commercially available preparations for the systemic delivery of peptide drugs via the IN route, as listed in Section 10.1.

3. *Rich blood supply*: The nasal mucosa is highly vascularized, with specialized capacitance vessels that facilitate heat exchange and other physiological roles. The rich blood supply promotes the rapid absorption of drug molecules and is especially useful when rapid onset is desirable, for example, in the treatment of migraine attacks or breakthrough pain (BTP) in cancer.

4. *Relatively low enzymatic activity*: Nasal secretions possess a wide range of enzymes, including proteases such as neutral endopeptidase, aminopeptidase peroxidase, and carboxypeptidase N. Furthermore, intracellular enzymes are found in the epithelial cells lining the cavity. Nevertheless, the metabolic activity of the nose is relatively low, in particular in comparison with other epithelial sites such as the GI tract, so that enzymatic activity is not considered a substantial obstacle to nasal drug delivery. Hepatic first-pass metabolism is also avoided via this route.

5. *Accessibility and compliance*: The nasal route is easily accessible and nasal delivery devices (sprays, drops, etc.) tend to be unobtrusive and relatively easy to use. The route is noninvasive, with all the associated advantages over parenteral delivery, e.g., no needles/sharps and associated biohazard/disposal issues, less need for specialized medical personnel to administer the dose, nasal formulations do not normally require cold storage or have to be sterile.

6. *Suitability for controlled release*: Various types of nasal drug delivery systems (DDS) and formulation additives can allow for a sustained release of drug over time. Although in general, it should be noted that the nose is not particularly suited for sustained-release purposes.

7. *Delivery to the CNS*: The specialized olfactory epithelium of the nose offers a potential route for direct N2B drug delivery.

8. *Delivery to the lymphoid tissue*: As described in Section 10.7, the nasal-associated lymphoid tissue (NALT) makes the nose a highly effective immunological site, making the route an attractive one for the delivery of vaccines.

10.3.1 LIMITING FACTORS FOR NASAL DRUG DELIVERY

The same anatomical and physiological features that enable the many vital functions of the nose also impose substantial hurdles for efficient nasal drug delivery (Djupesland 2013; Djupesland et al. 2014). These barriers are considered here.

10.3.1.1 Small Size of the Nasal Cavity

The small size of the nasal cavity limits the volume of a liquid formulation that can be administered. Typically, volumes about 100–200 μL in each nostril are possible. A larger volume can drip back out of the nostril, which can cause discomfort and embarrassment. Drip-out also reduces the actual fraction of dose retained in the cavity, thereby introducing variability in the administered dose. Throat run-off is a further problem: high liquid volumes can "flood" the nasal cavity so that the dose runs off to the throat and is swallowed, resulting in drug loss and a bitter aftertaste, which can adversely affect patient compliance. Drugs with low aqueous solubility and/or requiring high doses can therefore present a problem for IN delivery.

10.3.1.2   Constraints of Nasal Geometry: Access and Penetration Difficulties

The complex geometry of the nose and the associated turbulent airflow present multiple barriers to the access and penetration of drug molecules. Initial entry is limited by the small size of the nasal vestibule. Beyond the vestibule, the narrow, slit-like, nasal valve constitutes a major, and often ignored, challenge to successful nasal drug delivery and targeting. As described earlier, the narrow valve becomes even narrower during inhalation, and it can close entirely during sniffing. Beyond the valve, the posterior nasal cavity comprises a complex labyrinthine series of tunnels and passageways, with turbulent airflow, which makes deep penetration further into the cavity very difficult.

Drug delivery from conventional nasal delivery devices (sprays, drops, pressurized metered-dose inhalers [pMDIs], nebulizers, and powder sprayers) is described in detail in the following texts, but it should be pointed out here at the outset that drug delivery from all the currently available conventional devices is suboptimal. A common feature of all these devices is their limited ability to deliver drug past the nasal valve, resulting in a large fraction of the dose being deposited in the anterior region of the nose (Aggarwal et al. 2004; Djupesland et al. 2014). Drug deposited in the anterior cavity is subject to loss via drip out and run off to the throat; further elimination occurs via sneezing, mechanical wiping, and ingestion. Anterior deposition also causes patient annoyance and discomfort or, more seriously, irritation and crusting of the tissue (Waddell et al. 2003). From a drug delivery perspective, deposition in the anterior cavity means that the drug is delivered to a limited, restricted portion of the nose, but fails to reach the much larger, highly vascular, expanded surface area of the posterior cavity.

Drug molecules that do succeed in penetrating the valve tend to enter the posterior cavity via the wider, lower part of the triangular-shaped valve, thereby gaining access to the floor of the nasal cavity. As such, the formulation tends to run off along the cavity floor to the throat but has limited exposure to the rest of the posterior cavity (Djupesland 2013).

Access and penetration difficulties are compounded in those pathological states that are associated with a hypersecretion of mucus, such as in inflammatory conditions and allergies. As described in Section 10.6, the specialized olfactory region of the nose associated with N2B delivery is located far beyond the valve, up in the roof of the nasal cavity (Figure 10.4). Accessing this area for the delivery of CNS therapeutics is also a formidable challenge.

10.3.1.3   Mucus and Mucociliary Clearance

The mucus layer and the process of mucociliary clearance present a number of obstacles for nasal drug delivery. For systemically acting drugs, the mucus layer presents a diffusional barrier for a drug in transit to the epithelial surface. The rate of diffusion of a drug through mucus depends on a number of factors, including the thickness and viscosity of the mucus layer, as well as the physico-chemical properties of the drug. However, nasal mucus is only a few microns thick (in contrast, for example, to the GI tract, which has a mucus thickness of about 500 μm) and so does not present as substantial diffusional barrier here as compared to the GI tract and other mucosal sites.

Mucociliary clearance may limit the contact time of a drug molecule with the epithelial surface. The drug, instead of settling locally, can instead be removed to the throat, to be swallowed or expectorated. The mucociliary escalator moves the mucus blanket toward the nasopharynx at an average speed of 6 mm/minute, so that a particle deposited in the valve region of the nose is cleared to the nasopharynx within 15–20 minutes. Limited contact time with the absorbing surface may compromise drug absorption for a systemically active drug, or shorten drug–receptor interaction time for a locally acting drug.

10.3.1.4   Epithelial Barrier

If systemic absorption is the aim of therapy, the epithelial barrier must also be considered. Transport across epithelial barriers is described in detail in Chapter 4. To summarize here with respect to the nasal route, drug permeation via the paracellular route (i.e., between epithelial cells) is limited because of the tight junctional complexes that are present between cells. The predominant transport pathway across nasal epithelium is usually via the transcellular route, by means of transcellular passive diffusion. The rate of absorption is governed by Fick's law (Chapter 4, Equation 4.1). The most important physicochemical drug factors affecting nasal absorption are as follows: (1) molecular weight (nasal absorption drops off sharply for drugs with a molecular weight > 1000 Da) and (2) lipophilicity (polar, hydrophilic, or ionized molecules exhibit poor permeability).

Thus, drugs currently administered nasally for systemic action generally comprise low molecular weight, lipophilic compounds. Although some peptides do show some systemic absorption (calcitonin, oxytocin, etc., see Section 10.1), the absolute bioavailability of peptide drugs via the nasal route is still relatively low. For example, the IN bioavailability of salmon calcitonin (MW 3432 Da) is only about 3%. Low bioavailability is compensated for by the extremely high potency of these drugs, which can produce therapeutic effects even at low plasma concentrations.

10.3.1.5 Mucosal Sensitivity

Mucosal sensitivity is a natural component of the nasal defense mechanism, but it also complicates nasal drug delivery, by making this area highly susceptible to irritation and injury. Exposure to chemicals, gases, particles, temperature, and pressure changes, as well as direct tactile stimuli, may cause nasal irritation, secretion, tearing, itching, sneezing, and severe pain. Nosebleeds, crusting, and potentially erosions or perforations may arise because of factors such as direct contact of the tip of a nasal spray nozzle during actuation and/or localized concentrated drug deposition on the septum. Formulation additives such as absorption enhancers (AEs) may also cause mucosal irritation or damage (see Section 10.5.1).

10.3.1.6 Nasal Cycle

The nasal cycle is an alternating cycle of congestion and decongestion that occurs every 1–4 hours and is observed in at least 80% of healthy humans. This reciprocal autonomic cycling of mucosal swelling means that at any given time, even though the total combined resistance remains fairly constant, one of the nostrils is generally considerably more congested than the other, with most of the airflow passing through the passage of lesser congestion. This may be a challenge to efficient drug delivery. Therefore, for most indications, it would seem prudent to deliver the drug to both nasal passages, when administering a given dose (Djupesland 2013).

10.3.1.7 Variability

As described further in Section 10.3.2, nasal deposition and clearance are dependent on a number of complex, interrelated factors, including the type of nasal delivery device, drug physicochemical factors, formulation factors, and physiological, anatomical and pathological factors. These factors combine to introduce considerable variability with respect to the emitted dose, the site of deposition, the resulting clearance, and ultimately, the clinical response. The variability is particularly serious for drugs with a narrow therapeutic index (e.g., opiates and hormonal drugs), and represents a limitation of the route.

10.3.1.8 Patient Acceptability and Compliance

Nasal delivery devices can be difficult and uncomfortable to use, which can compromise patient acceptability and compliance. These problems are described further in Section 10.4, with respect to each type of device. To summarize briefly here, compliance issues include drip-out and run-off problems for nasal drops and high volume nasal sprays; discomfort when using aqueous sprays; cold shock sensations associated with pMDIs; nebulizers that can be difficult and cumbersome to use; and nasal drops that require extreme head positions for their correct administration.

10.3.2 Nasal Deposition and Clearance

For IN administered drugs, the site of drug deposition and the resulting clearance from that site are important determinants of the clinical response. Drug deposition and clearance, in turn, are dependent on a number of other interrelated factors, outlined here and discussed further in the following sections.

The device: A wide variety of devices (including sprays, pMDIs, nebulizers, and drops) can be used to effect nasal drug delivery. The type of device used profoundly affects characteristics such as the velocity and size of the emitted particles, the plume characteristics, the volume emitted, and the angle of entry.

The drug: The physicochemical properties of the drug, such as molecular weight and volume, lipophilicity, solubility, and susceptibility to enzyme degradation, affect its deposition and clearance in vivo.

The formulation: A liquid formulation will result in different deposition and clearance patterns to a solid formulation; other factors affecting deposition and clearance include the formulation viscosity and the effects of formulation excipients such as AEs, enzyme inhibitors, mucoadhesives, and gel formers.

The patient: Patient factors that influence nasal deposition and clearance include the patient's skill at using the device, and whether the instructions for administration are followed correctly. Further factors include inter-individual anatomical and physiological variability, as well as the impact of airflow and breathing patterns and the presence of any pathologies (e.g., mucus hypersecretion).

A careful consideration of these interrelated factors is necessary to optimize the route. Nasal deposition and clearance patterns can be studied in vitro using nasal casts made of silicone or other materials. It is essential for interpretation that the nasal cast geometry and dimensions are realistic and validated, which unfortunately is not always the case. Furthermore, caution is necessary because such casts do not characterize physiological factors such as nasal valve dynamics, or mucociliary clearance. The field of computational fluid dynamics is a further important in vitro tool, which will play an increasingly important role as the quality and capabilities of the simulations increases.

Colored dyes offer a quick and inexpensive semiquantitative assessment of nasal deposition and clearance in vivo. More detailed information is achieved using gamma-deposition studies, where the fate of a radiolabel can be tracked in vivo. In particular, recent studies have carefully assessed regional differences in tissue attenuation in different nasal segments and also between the nose and lungs (Djupesland and Skretting 2012). These sophisticated studies have provided greater insight into regional nasal deposition and clearance, showing more precisely where drug deposits from different types of delivery devices. These and similar studies are helping to elucidate how drug delivery via the nasal route can be optimized.

10.4 NASAL DRUG DELIVERY DEVICES

A summary of the most widely used nasal devices is given here. Further information on a wide variety of nasal drug delivery devices is given in a recent extensive review (Djupesland 2013).

10.4.1 Spray Pumps

Metered-dose spray pumps are the most widely used nasal delivery devices and have been for several decades. Their popularity is attributed mainly to their ease of use. They comprise a container, a pump with a valve, and an actuator. Actuating the pump creates a force that drives the liquid through a swirl chamber at the tip of the applicator and out through the circular nozzle

FIGURE 10.2   Nasal spray pump. (Markus Gann/Shutterstock.com.)

orifice (Figure 10.2). The pumps typically deliver volumes of 100 μL (range 25–200 μL) per spray, and they offer high reproducibility with respect to the emitted dose, particle size, and plume geometry.

Traditional spray pumps replace the emitted liquid with air, and preservatives are therefore required to prevent contamination. Driven by early concerns that preservatives may impair muco-ciliary function, various preservative-free spray pumps have been developed. For example, Aptar Pharma has developed a pump that incorporates an aseptic microfilter membrane and a ball valve at the tip. Single-dose spray devices offer the advantages of being preservative free, as well as being easily portable and offering high accuracy of dosing. A Pfeiffer/Aptar single-dose device is used for the nasal delivery of the antimigraine drugs sumatriptan and zolmitriptan. However, initial fears about possible adverse effects of preservatives have proven unwarranted and many long-term studies have concluded that preservatives do not represent a safety concern (Marple et al. 2004). Their safety is evidenced, for example, by the fact that preservatives are found in all of the current top-selling formulations for topical steroids, i.e., in preparations which are intended for chronic use.

Future directions in pump technology include the development of pumps that incorporate pres-sure-point features, to improve dose reproducibility. This is because mechanical pumps rely on hand actuation, which inevitably provides a variable force that can influence particle size distribu-tion and plume characteristics. Other optimization approaches involve the development of pumps that need less priming (which causes drug wastage), and which feature dose counters.

As introduced in Section 10.3.1.2, a major limitation associated with nasal sprays is their sub-optimal delivery efficiency. Nasal sprays generate an expanding conical-shaped plume, with the majority of the particles at the periphery (Figure 10.2). Even if the spray pump is inserted as deep as 10–15 mm into the nostril, there is an obvious mismatch between the dimensions and shape of the expanding circular plume (diameter ≈ 2 cm), compared to the constricting dimension of the nasal vestibule, and the confined nasal geometry beyond. The mechanical action of inhalation further narrows the nasal valve, making plume penetration even more difficult. Sniffing, either intention-ally or reflexively to avoid drip-out, will further narrow the nasal valve and limit access. Thus, the majority of the drug particles, particularly those traveling at high speed, will impinge on the walls of the anterior nasal cavity and deposit there.

The sensation of high-speed particles impacting on the nasal walls can be unpleasant and uncomfortable for the patient. Drug formulation deposited in the anterior cavity may drip out of the nostril or run off to the throat. The small fraction of drug that actually penetrates the valve gains access predominantly to the floor of the nasal cavity, but fails to reach the majority of the rest of the nasal cavity, the ideal target site.

Nasal sprays can also be generated using squeeze bottles. The patient squeezes a plastic bottle partly filled with air, and the drug is atomized when delivered from a jet outlet. However, delivery efficacy is poor: in addition to the access and penetration issues common to all nasal sprays, the dose and particle size will vary considerably according to the squeezing force applied, causing dose reproducibility issues. Furthermore, nasal secretions and microorganisms may be sucked back into the bottle when the pressure is released.

10.4.2 Pressurized Metered-Dose Inhalers

pMDIs are used widely in pulmonary drug delivery and the reader is referred to Chapter 11 (Section 11.7) for a detailed description of this type of device. For the nasal route, conventional pMDIs based on the use of chlorofluorocarbon (CFC) propellants were associated with a number of disadvantages, including (1) very high particle speeds, which caused discomfort on impaction with the walls of the nasal cavity, (2) the unpleasant "cold freon effect" (the cold blast of propellant hitting the walls of the nose), and (3) nasal irritation and dryness. Following the ban on ozone-depleting CFC propellants, the number of pMDI products for both pulmonary and nasal delivery diminished rapidly, and they were removed from the U.S. market in 2003.

3M has pioneered the development of environmentally friendly hydrofluoroalkanes (HFA) as an alternative to the CFC propellants. HFA is now used as the propellant in pMDIs for the nasal delivery of a variety of drugs, including triamcinolone acetonide (Nasacort HFA®), ciclesonide (Omnaris HFA®), and beclomethasone (Qnasl®). These new-generation devices are associated with many advantages. HFA-based pMDIs do not cause as much drip-out or run-off as traditional liquid spray pumps, which aids patient acceptance and compliance, and reduces drug loss. Also known as "slow-mist" devices, HFA-based pMDIs produce much slower particle speeds than the old CFC versions. The slower speeds reduce particle impaction against the nasal walls, thereby enhancing nasal drug delivery, as well as patient comfort.

However, the speed of the emitted particles with the new HFA pMDIs is still the same, or even higher, than the speed of particles emitted for spray pumps: patients still describe the cold shock feeling on impact, and how they have to steel themselves prior to dosing. Furthermore, although the aerosol-generating mechanism is different, a similar mismatch exists between the expanding conical-shaped plume produced by a pMDI and the dimensions of the narrow nasal valve. Therefore, access and penetration difficulties remain an important issue. Gamma-deposition studies with these devices confirm this pattern of limited posterior disposition, with a very distinct anterior "hotspot" (Djupesland 2013).

It has been claimed that anterior deposition provides enhanced efficacy for the IN delivery of topical anti-inflammatory steroids using HFA pMDIs (Righton 2011). We question the validity of this idea. Drug deposited in the anterior cavity is subject to all the disadvantages outlined earlier (e.g., drip-out and elimination). Furthermore, nasal inflammatory diseases (e.g., rhinitis and sinusitis) are not associated with the anterior cavity but are localized instead in the posterior cavity, so that topical steroid therapy would be optimized by targeting this region instead (Djupesland 2013).

10.4.3 Powder Devices

Dry powder formulations may also be administered intranasally. Nasal powder sprayers utilize a pressure gradient to force out a fine plume of powder particles, similar to that of a liquid spray. The Becton Dickinson and Aptar group both offer powder sprayers that utilize a plunger technology: pressing the plunger ruptures a membrane to expel the powder. In the Fit-lizer™ multidose system, a capsule is first inserted into a holding chamber, which slices off the capsule ends. The patient then compresses the plastic bottle: the compressed air passes up through the sliced-open capsule, forcing out a fine spray of powder particles. In common with liquid sprays, there remains a mismatch between an expanding powder plume and the constricted nasal geometry, so that most of the dose is deposited in the anterior cavity.

Rhinocort Turbuhaler® is a newer breath-actuated nasal inhaler, based on a modification of the Turbuhaler® device for pulmonary delivery. To use, the patient is required to sniff quickly and forcefully through an adaptor: the resulting negative pressure pulls the powder formulation into the nose. However, as described earlier, sniffing causes constriction (and even possible closing) of the nasal valve, resulting in unwanted deposition in the anterior cavity. Furthermore, many rhinitis patients have nasal congestion, which can impede the flow rates required for efficient delivery. Nasal inhalation for the Turbuhaler® was also shown to produce significant drug deposition in the lungs (Thorsson et al. 1993).

10.4.4 Nebulizers

Nebulizers use compressed gases (air, oxygen, nitrogen), or ultrasonic or mechanical power, to break up medical solutions and suspensions into small aerosol droplets that can be directly inhaled into the mouth or nose over a period of minutes. The aerosol can be administered passively or assisted by active nasal inhalation, or even assisted by suction from the contralateral nostril. Nebulizers are primarily used to delivery topically acting drugs such as antibiotics or steroids, in patients with chronic rhinosinusitis (CRS).

Compared to other nasal delivery devices, nebulizers produce aerosols with a smaller particle size (<10 μm), traveling at slower speeds. These features are associated with less impaction against the nasal walls and greater drug deposition in the posterior nasal cavity. However, particles less than 5–10 μm can evade the normal filtration and cleaning mechanisms of the nose and may be inhaled into the lungs. For example, studies have shown that using a nebulizer resulted in as much as 33%–58% of the administered dose resulting in unwanted lung deposition (Suman et al. 1999; Djupesland et al. 2004). Lung deposition results in drug waste, as well as possible pulmonary irritation and unwanted systemic absorption. Unwanted lung deposition is an area of increasing concern, reflected by the most recent Food and Drug Administration (FDA) guidelines for nasal devices, which recommend minimizing the fraction of respirable particles below 9 μm. A further disadvantage of traditional nebulizers is that they cannot provide the dose reproducibility required for most active drugs.

A new generation of nasal nebulizers is under development, in order to achieve improved delivery profiles. ViaNase® is a handheld, battery-driven device that atomizes liquids by a process of Controlled Particle Dispersion™ and produces a vortical flow on the droplets as they exit the device. Flow characteristics (circular velocity and direction) can be altered to achieve different droplet trajectories and customized delivery. The ViaNase® device has been used to deliver nasal insulin in patients with early Alzheimer's disease, and clinical benefit has been demonstrated (Craft et al. 2012). However, unwanted lung deposition of up to 9% of the delivered dose was also reported for this device (Reger et al. 2008).

The Vibrent® pulsation nebulizer generates a fine aerosol mist of an aqueous liquid via a perforated pulsating membrane. Breath holding during delivery is recommended, in order to reduce the risk of lung inhalation. A further device is the Aeroneb Solo® (Aerogen), which uses a vibrating mesh technology to produce a low-velocity aerosol. The aerosol is delivered into one nostril, while a pump simultaneously aspirates at the same flow rate from the other nostril; meanwhile, the subject is instructed to avoid nasal breathing.

Although the new generation of nebulizers (ViaNase®, Vibrent®, Aeroneb Solo®) improves nasal deposition into the posterior cavity and sinuses, and reduces unwanted lung deposition, gamma-deposition studies have demonstrated that deposition still occurs mainly at the nasal valve, with a substantial fraction of the drugs delivered outside the target regions (Vecellio et al. 2011). Furthermore, these new devices are relatively complex in operation; the recommended delivery procedures require patient cooperation and take several minutes. The special breathing patterns may be challenging for some patients, especially when they are sick.

10.4.5 Nasal Drops

Nasal drops are administered by drawing liquid into a glass/plastic dropper, inserting the dropper into the nostril with an extended neck, and then squeezing the rubber top to emit the drops. Nasal drops are commercially available, for example, for the topical delivery of decongestants. A variant of nasal drops is the rhinyle catheter, where drops are filled by the patient into a thin flexible tube. One end of the tube is inserted in the mouth and the other is placed at the entrance of the nostril. Drops are delivered to the nose by blowing through the tube. The rhinyle catheter is used in some countries for the IN delivery of desmopressin.

Nasal drops, when delivered correctly, can increase penetration beyond the nasal valve—this facilitates deposition in the upper posterior nasal segments. This improved deposition is associated with improved clinical outcome. For example, a study showed significant benefits of fluticasone drops, compared with nasal spray, in avoiding surgery in patients with CRS and polyps (Aukema et al. 2005). Improved deposition was attributed to the gravitational forces in operation when the drops were administered using the requisite head maneuvers.

However these requisite head maneuvers constitute a major limitation of nasal drops: for their correct administration, they require the patient to assume an extreme head extension or adopt the "Mekka position," or similar. In this way, gravity helps to get the drops to the middle part of the nose and reduce complications of run-off and drip-out. But most patients do not want, or are unable, to perform the necessary head extensions required, as they are cumbersome and can be very uncomfortable, especially for patients with sinusitis and headaches. When nasal drops are administered without assuming a correct position, drip-out and run-off problems arise.

10.4.6 Breath Powered™ Bi-Directional™ Nasal Device: A Case Study

The Breath Powered™ Bi-Directional™ device by OptiNose is designed to optimize nasal drug delivery by exploiting functional aspects of the nasal anatomy. The device has a mouthpiece, connected to a delivery unit and a sealing nosepiece (Figure 10.3a). The user slides the nosepiece into one nostril until it forms a seal at the nostril opening—at which point it also mechanically expands the narrow slit-shaped part of the nasal valve. The user exhales into the mouthpiece, and the exhaled breath carries medication through the nosepiece into one side of the nose (Figure 10.3b). The pressure of the patient's exhaled breath automatically elevates the soft palate, sealing off the nasal cavity completely, thereby preventing unwanted lung deposition (Figure 10.3b). The breath pressure also gently expands the narrow nasal passages. Due to the sealing nosepiece, the force of the air exhaled into the mouthpiece balances the pressure across the closed soft palate so that an open flow path between the two nostrils is maintained. Thus, liquid or powder drug particles are released into an airstream that enters one nostril, passes entirely around the nasal septum posteriorly, and exits through the opposite nostril, following a "Bi-Directional" flow path (Figure 10.3c). The OptiNose technology can be combined with a variety of dispersion technologies (sprays, pMDIs, nebulizers) for the delivery of both powders and liquids.

In operation, the device ensures a mechanical expansion of the narrow nasal valve and also gently expands the narrow nasal passageways, thereby improving nasal deposition in the posterior cavity. Human studies of deposition patterns using gamma scintigraphy have shown that the OptiNose device achieves less deposition in the vestibule and significantly greater deposition to the upper posterior target regions beyond the nasal valve, when compared to conventional nasal spray devices, for both liquid and powder formulations (Djupesland et al. 2006; Djupesland and Skretting 2012;

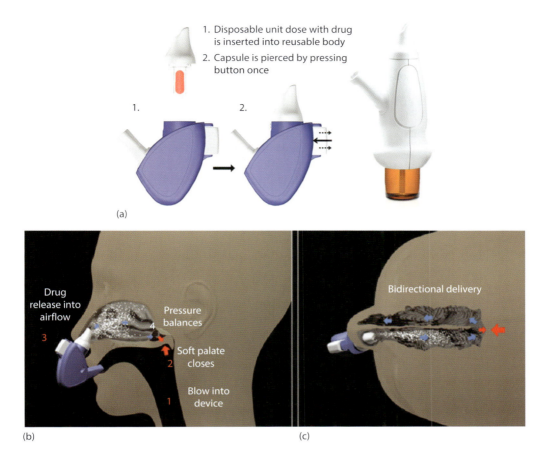

1. Disposable unit dose with drug is inserted into reusable body
2. Capsule is pierced by pressing button once

1.
2.

(a)

Drug release into airflow

3

Pressure balances

4

Soft palate closes

2

Blow into device

1

Bidirectional delivery

(b) (c)

FIGURE 10.3 (a) The OptiNose Breath Powered™ Bi-Directional™ devices. Multi-use powder device (blue) with disposable nosepiece on the left and the multi-dose liquid device (white) on the right. (b) Using the OptiNose devices automatically causes soft palate closure, thereby sealing off the nasal cavity and avoiding unwanted lung deposition. (c) Bidirectional flow: drug particles enter one nostril, pass posteriorly around the nasal septum, and exit via the opposite nostril. (Courtesy of OptiNose, Inc.)

Djupesland 2013). Comparisons using nebulized formulations with the OptiNose devices have shown that unwanted lung inhalation is prevented, even when small respirable particles are delivered.

Furthermore, the OptiNose device is simple to use, which enhances patient compliance. It does not require extreme head positions, and reduces problems of drip-out and run-off. The system uses warm, moist air to carry the particles, which enhances comfort and avoids cold shock sensations. Nebulizers can be adapted to the device with a breath-triggered mechanism, thereby avoiding the breath coordination issues and suction mechanisms required for the new-generation nasal nebulizers recently developed. The OptiNose powder device delivering sumatriptan has recently received US marketing approval for the treatment of acute migraine in adults (Onzetra™, Xsail™). A multi-dose liquid device version delivering fluticasone propionate has recently completed Phase 3 trials in patients with CRS with nasal polyps.

10.5 NASAL FORMULATION FACTORS

Liquid formulations, including solutions, suspensions, and emulsions, currently dominate the nasal drug market. Liquid formulations provide a humidifying effect, which is useful as many allergic and chronic diseases are associated with crusting and drying out of mucous membranes.

Disadvantages of liquid formulations include (1) their susceptibility to microbial contamination, (2) potentially reduced chemical stability of the active, and (3) potentially short residence time in the nasal cavity, unless countermeasures are taken. Powder formulations can offer the advantage of greater stability than liquid formulations and the possibility of avoiding the use of preservatives. Powder formulations may also offer an increased residence time in the nasal cavity and also possibly sustained-release effects, as the drug must initially dissolve in the nasal fluids, prior to being absorbed.

A variety of different formulation excipients can be included in a formulation, and many different agents are being investigated, in order to optimize nasal delivery and targeting (Illum 2003, 2012). For optimal results, the formulation should be tailored to the specific drug, therapeutic indication (i.e., for local, systemic, immunologic, or N2B delivery), and type of delivery (spray, nebulizer, dry powder, etc.) required.

10.5.1 ABSORPTION ENHANCERS

AEs can be included in formulations intended for systemic delivery in order to increase nasal bioavailability. They can potentially facilitate the systemic absorption of a wide range of drugs, including large biologics such as polypeptides, proteins, and antisense molecules (Merkus et al. 1993). As described in detail in Chapter 7 (Section 7.6.1.1), AEs generally act via a combination of complementary mechanisms. Many AEs are amphipathic in nature and thus associate with the amphipathic bilayers of the plasma membranes, increasing the fluidity and permeability of membranes and promoting transcellular transport. Some also interact with the junctional complexes between epithelial cells, facilitating paracellular transport. These agents may also promote nasal absorption via other effects, including increasing drug solubility, enzyme inhibition, and mucoadhesion.

However, AEs are also associated with safety and toxicity concerns. AEs can directly damage the delicate nasal epithelium. Early AEs, including surfactants such as laureth-9, bile salts, and their derivatives such as sodium taurodihydrofusidate and fatty acids, were associated with mucosal damage and are now used less frequently. The nasal epithelium provides a protective barrier defense mechanism against the entry of harmful agents. Interfering with this protective barrier could potentially facilitate the bystander absorption of harmful, irritating, and infectious agents. These safety concerns are reflected in the difficulties in getting regulatory approval for a new AE—which is almost as difficult as getting approval for a new therapeutic agent. Due to the regulatory hurdles, what may have originally been intended as an inexpensive formulation additive can actually become a very expensive, time-consuming approach, which may even require a completely new set of clinical trials. Therefore, the focus in this field is on investigating "generally regarded as safe" (GRAS) agents, rather than discovering new types of AEs.

Chitosan is being investigated as an AE for the nasal delivery of a number of drugs (its use in mucoadhesion is discussed in the following texts), including peptides and proteins (including leuprolide, salmon calcitonin, and parathyroid hormone), and most successfully for the IN delivery of morphine in BTP for cancer (Casettari and Illum 2014). The IN administration of morphine with a chitosan formulation facilitated up to sixfold increase in bioavailability compared to controls, with maximum plasma levels reached within minutes, and is currently showing promise in clinical trials.

Intravail® comprises a class of alkylsaccharides that have demonstrated impressive absorption enhancement for a range of peptides in a selection of animal models (Maggio and Pillion 2013). Surfactant in nature (the molecules contain a polar sugar head group, such as maltose or sucrose, esterified with a hydrophobic alkyl chain; the lead compound is tetradecyl maltoside), they are thought to interact with the amphipathic membranes of the nasal epithelial cells, inducing membrane-permeabilizing effects.

It should also be remembered, however, that although many AEs have demonstrated pronounced enhancement effects in animal studies, these effects can be a significant overestimation of what can realistically be achieved in humans, given the very distinct architectures and morphologies of the nasal cavity in different species.

10.5.2 Gel Formers

A drug for IN delivery may be administered as a nasal gel (typically a high-viscosity thickened solution or suspension), or an in situ gel-forming agent may be included in the formulation. The high viscosity of the formulation can facilitate a longer residence time in the nasal cavity and allows more time for drug–receptor interactions (for local effects) or transepithelial flux (for systemic absorption). High-viscosity formulations can also reduce postnasal drip-out and run-off, with all the associated advantages. Soothing/emollient excipients in the gel may also reduce irritation. Many viscosity enhancers used for gel formation also function as bioadhesives, as they complex with the mucus layer, increasing its viscoelasticity and reducing mucociliary clearance.

Pectins are widely used in the food industry as gelling agents; their gel-forming properties have also been investigated for nasal drug delivery. PecSys® technology uses a pectin derivative, low methoxyl (LM) pectin—which has a low degree of esterification of the galacturonic acids—as an in situ gel former. The product is administered to the nasal cavity as a solution and forms a gel on contact with the nasal mucosal surface, due to the interaction of the LM pectin with the calcium ions of the mucosal fluid. PecFent® uses the PecSys® technology for the IN delivery of the opioid analgesic fentanyl, indicated for management of BTP for cancer (Fallon et al. 2011).

10.5.3 Mucoadhesives and Ciliastatics

Mucoadhesives are incorporated into formulations to attach onto the nasal mucus layer, thereby prolonging the effective contact time of the drug in the nasal cavity. Mucoadhesives for oral drug delivery are described in detail in Chapter 7 (Section 7.6.1.4), and the same principles and mechanisms apply here. As noted earlier, many gel-forming excipients also demonstrate mucoadhesive properties. Thiomers are mucoadhesive polymers being investigated as formulation excipients for a variety of epithelial routes, including for nasal, oral, buccal, and vaginal delivery. Thiol side chains can form disulfide bridges between cysteine groups on mucus, which can improve mucoadhesion up to 100-fold (Lehr 2000). The mucoadhesive effects are complemented by their in situ gelling properties, due to the oxidation of thiol groups at physiological pH values, which results in the formation of inter- and intramolecular disulfide bonds. This increased viscosity in situ further increases nasal cavity residence time. Thiomers also confer AE effects, via reversible opening of the tight junctions between cells. They further offer the potential safety advantage of being too large to be absorbed through the nasal mucosa per se, in contrast to other low-molecular-weight AEs under investigation.

An acrylated poly(ethylene glycol)-alginate copolymer has been developed for IN delivery (Davidovich-Pinhas and Bianco-Peled 2011). The presence of PEG increases the viscosity of the formulation, and also PEG chains have the ability to penetrate the mucus surface and to form hydrogen bonds with sugars on glycosylated mucus proteins. Poly(acrylic acid) also forms hydrogen bonds between its carboxylic acid groups and the sialic acid–carboxylic acid groups present in the mucus. These attributes, combined with the gelation ability of the alginate component, make the copolymer an attractive candidate for nasal mucoadhesion. Further important mucoadhesives that have demonstrated promise for IN delivery include microcrystalline cellulose, chitosan, and poloxamer 407 (Ugwoke et al. 2005). Some of these polymers have been incorporated into micro- and nanoparticles, as described in Section 10.5.5. A mucoadhesive cyclodextrin-based drug powder formulation (μCo™ Carrier) is being developed in conjunction with the Fit-lizer™ powder device (Section 10.4.3), for the nasal delivery of a variety of drugs and influenza vaccine.

A further approach to enhancing residence time in the nasal cavity is to use a reversible ciliostatic in the formulation. If the cilia stop beating, mucociliary clearance is compromised, and the drug substance remains longer in the nasal cavity. Some AEs have been shown to cause irreversible ciliostasis, e.g., laureth-9 (0.3%) and sodium deoxycholate (0.3%), although at lower concentrations the effect is reversible. Others, such as sodium glycocholate, are well tolerated.

10.5.4 Other Formulation Excipients

Solubility enhancers: The inclusion of solubility enhancers can allow a more concentrated drug formulation to be administered, in a reduced volume. A reduced volume lessens the problems of run-off and drip-out from the nasal cavity. Increasing the solubility of the API also increases the concentration gradient across the nasal epithelium, providing a larger driving force for drug permeation via transcellular diffusion. Many AEs, being surfactants, also provide a solubilizing effect. Various formulation excipients that are used to increase drug solubility, such as cosolvents, cyclodextrins, and polymeric micelles, are described in Chapter 3.

Buffers: The normal pH in the nasal cavity is between 5.5 and 6.0. Transcellular passive diffusion across membranes is facilitated when the drug is in the unionized state; therefore, depending on the pKa of the drug, a buffer may be used to adjust the pH and facilitate the unionized form (see also Chapter 4, Box 4.1).

Enzyme inhibitors: Enzyme inhibitors may be added to a formulation to protect a labile drug such as a peptide or protein. For example, bestatine and comostate amylase have been studied in IN formulations as aminopeptidase inhibitors for calcitonin. Bacitracin and puromycin have been used to minimize enzymatic degradation of human growth hormone.

10.5.5 Micro- and Nanoparticulate Drug Delivery Systems

Micro- and nanoparticulate DDS, described in detail in Chapter 5, include microspheres, nanospheres, and liposomes. All of these systems have been extensively investigated for nasal drug delivery but with limited success to date. Micro- and nanoparticles can protect an API from enzymatic degradation and provide sustained release of the drug. Many particulate systems for IN delivery incorporate a mucoadhesive polymer, such as chitosan, gelatin, or alginate, to enhance retention in the nose. Solubility enhancers and AEs can also be included within the nanoparticle construct. Targeting ligands can be attached to the surface of the carrier to enhance nasal delivery and retention. Targeting vectors specifically studied to enhance nasal delivery include lectins extracted from *Ulex europaeus* I, soybean, peanut, and wheat germ agglutinin.

Starch microspheres, as well as being bioadhesive, seem to draw up moisture from the surrounding cells, resulting in the nasal mucosa becoming dehydrated. This results in reversible "shrinkage" of the cells, providing a temporary physical separation of the tight junctions and thus facilitating paracellular absorption. Starch microspheres have been shown in animal studies to enhance the absorption of insulin and other proteins. Mucoadhesive multivesicular liposomes have demonstrated promise for the nasal absorption of insulin, providing protective and sustained-release effects. Many different types of liposomal gels have also been investigated for nasal delivery. A further application, introduced in Section 10.7, is the use of micro- and nanoparticles for IN vaccination.

10.6 NOSE-TO-BRAIN DRUG DELIVERY

Chapter 15 describes in detail the anatomical and physiological barrier that comprises the BBB, as well as a number of strategies to cross or bypass the BBB. Such strategies include surgical methods, as well as strategies that involve reengineering of drug molecules to exploit endogenous transport mechanisms for BBB uptake. Carrier-mediated transport mechanisms, receptor-mediated transfer systems, and molecular-based ("Trojan horse" delivery) and nanotechnology-based approaches to cross the BBB have shown promising results in animals and some are being tested in human trials. However, these strategies are often complex and molecule specific. They may also require that the

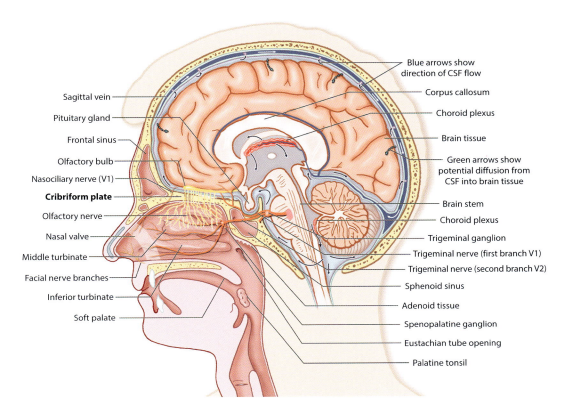

FIGURE 10.4   The olfactory epithelium is located on the inferior surface of the cribriform plate and superior nasal conchae. The axons of the olfactory receptor cells collectively form the olfactory nerve. Nose-to-brain delivery involves the transport of drug molecules along the olfactory nerve. (From Djupesland, P.G., Messina, J., and Mahmoud, R., The nasal approach to delivering treatment for brain diseases: An anatomic, physiologic, and delivery technology overview, *Ther. Deliv.*, 5(6), 709, 2014. With permission, Future Science, Copyright 2015; Illustrator Copyright: K.C. Toverud CMI.)

engineered molecules are first delivered into the systemic circulation, which has its own challenges and limitations in terms of absorption, bioavailability, and safety.

The nose offers a direct route to the CNS, in which the BBB can be circumvented in a noninvasive manner (Figure 10.4). N2B transport involves the direct transport of substances into the brain, via the olfactory nerve (and possibly in some cases, via the trigeminal nerve). The olfactory epithelium is located toward the roof of the nasal cavity and contains 10–100 million olfactory receptors, i.e., bipolar neurons, which react to molecules in the inspired air and initiate impulses in the olfactory nerve. The axons of the olfactory nerve pass through tiny holes in the flat bone plate (cribriform plate) that separates the olfactory mucosa from the overlying brain tissue and immediately synapse in the olfactory bulb. Axons from the secondary neurons make up the olfactory tracts, which extend posteriorly and end in the cerebral cortex.

It was initially believed that N2B transport occurred along the olfactory and trigeminal nerves as a slow, intra-axonal process. However, emerging research shows that the transport from the nose to CNS occurs within minutes, which would suggest that the most likely mechanism is actually bulk transport, propelled by arterial pulsations, along the ensheathed channels surrounding the axons (Iliff et al. 2012). A recent review describes a number of animal and human studies confirming that a variety of biologics, including peptides and proteins, are capable of reaching the brain following nasal administration (Lochhead and Thorne 2012).

Optimized N2B delivery would thus seem to require targeted delivery to the olfactory region. However, this region, located as it is in the roof of the posterior cavity, presents significant access difficulties—the big challenge for N2B delivery is to successfully target this area. Impel NeuroPharma is developing a HFA-driven pressurized delivery device for N2B delivery, which produces a narrower plume than that of the standard nasal sprays, in order to enhance drug deposition beyond the nasal valve. A study in rats found that 50% more of the dose was deposited in the region of the rat olfactory epithelium using this device, than when using nasal drops deposited at the nares (Hoekman and Ho 2011). Evidence of enhanced transport of morphine and fentanyl from the nose to the brain was also shown. While encouraging, it should be remembered that the olfactory region in rats covers a very large fraction of the nasal mucosal surface; further, drug delivery to the nares of anesthetized rats may not be an appropriate representation of drug delivery in nonanesthetized humans.

An adaptation of the Breath Powered™ OptiNose device has been developed to optimize N2B delivery, which uses a specialized nosepiece for better insertion into the narrower, upper part of the nasal valve. Gamma-deposition studies in humans confirm enhanced delivery in the superior olfactory region. Recent pharmacokinetic and pharmacodynamic studies with this device in humans strongly suggest direct N2B delivery of oxytocin (Quintana et al. 2015).

10.7 NASAL VACCINES

As described in Chapter 17, the NALT forms part of the body's common mucosal immune system. Infective agents that enter the body via the nose are presented to the abundant local nasal immune system, which comprises (1) various specialized immunocompetent cells distributed both in the mucus blanket and the mucosa per se and (2) organized lymphatic structures situated mainly in the pharynx, as a ring of lymphoid tissue known as Waldeyer's pharyngeal ring, which includes the adenoids and palatine tonsils.

In order to achieve an optimal immune response, a vaccine should be delivered to the respiratory mucosa rich in antigen-presenting dendritic cells and to M cells located in the organized lymphatic tissues. Nasal vaccination induces both mucosal (sIgA) and systemic (IgG) immune responses, making it a highly effective immunological site. Furthermore, the mucosal sIgA response is rapid and raised not only locally in the nose, but may also induce protection in distant mucosal sites. Nasal vaccines also offer the advantages over parenteral vaccines of being more acceptable to patients, cheaper, and easier to use.

Flumist® is a single-dose nasal spray from Becton Dickinson containing a live, attenuated influenza vaccine. Flumist® is sprayed into the nose as a needle-free way to protect against influenza. The volume administered (0.5 mL) is much larger than typically used for nasal delivery (100 μL), suggesting that a substantial amount of the drug dose is actually swallowed, i.e., oral delivery rather than IN; drip-out from the nasal cavity also occurs.

Advances in nasal vaccine technologies are directed towards optimizing nasal delivery devices, as well as improving nasal formulations. The Breath Powered™ OptiNose liquid device has shown preliminary promise in improving both local and systemic immune responses in humans for influenza vaccine, in comparison with a traditional nasal spray pump (Djupesland 2013). On the formulation front, the AE chitosan is also being investigated as a vaccine adjuvant to enhance the immune response. Chitosan nasal vaccines have been studied for influenza, pertussis, and diphtheria vaccines and shown enhanced immune responses in various animal models, in comparison to nasal vaccine formulations without chitosan (Illum 2012). Due to its positive charge, chitosan can complex with negatively charged DNA plasmids. Self-assembling nanoparticle complexes of chitosan–DNA for vaccination against respiratory syncytial virus have shown promise in preliminary studies. Other nanoparticulate adjuvants include proteasome-based adjuvants for the nasal immunization of human influenza virus and human streptococcus A, which are currently in Phase 2 clinical trials. Further nasal vaccine delivery systems are described in Chapter 17.

10.8 CONCLUSIONS

It can be concluded that drug delivery to the nose is a complex, challenging science, dependent on many interrelated factors, including the type of nasal delivery device, drug factors, formulation factors, and physiological and pathological factors. Understanding and addressing all of these issues is crucial to optimizing the potential of this route. In vitro studies using casts, and in vivo studies using gamma scintigraphy, continue to develop our understanding of nasal deposition and clearance processes and the factors that affect these processes, thereby allowing further optimization of the route.

Overcoming the various anatomical and physiological barriers associated with the nose is essential to optimizing therapy. Many commonly used nasal delivery devices are associated with suboptimal delivery, whereby a large fraction of the dose is deposited in the anterior cavity. Progress in the field is twofold, centering on improvements in (1) device design and (2) formulation parameters. On the device front, a new generation of nasal nebulizers offer improved drug deposition in the nose while reducing undesirable lung inhalation. HFA-based pMDIs are associated with improved delivery properties compared to the old CFC-based systems. Various new designs of nasal spray pumps are improving dose reproducibility and delivery, and the OptiNose bidirectional device provides improved nasal depostition.

Advances in formulation technology are directed toward the use of AEs to promote systemic delivery, gel-forming agents to promote drug retention within the nose, and micro- and nanoparticulate DDS to enhance the delivery of peptide and protein drugs. These improvements are driving the potential of this route forward, both for locally acting drugs and for the systemic delivery of a wide range of APIs.

The nose is continuing to show considerable promise as a site for mucosal immunization, using nasal vaccine DDS. Finally, although it is still too early to predict outcomes, the nose also offers the tantalizing possibility that it could be utilized as a direct portal of entry to the CNS, bypassing the BBB and so improving treatment for a wide range of CNS disorders.

REFERENCES

Aggarwal, R., A. Cardozo, and J.J. Homer. 2004. Clinical otolaryngology: The assessment of topical nasal drug distribution. *Allied Sci* 29:201–205.

Aukema, A.A., P.G. Mulder, and W.J. Fokkens. 2005. Treatment of nasal polyposis and chronic rhinosinusitis with fluticasone propionate nasal drops reduces need for sinus surgery. *J Allergy Clin Immunol* 115:1017–1023.

Casettari, L. and L. Illum. 2014. Chitosan in nasal delivery systems for therapeutic drugs. *J Control Release* 190:189–200.

Craft, S., L.D. Baker, T.J. Montine et al. 2012. Intranasal insulin therapy for Alzheimer disease and amnestic mild cognitive impairment: A pilot clinical trial. *Arch Neurol* 69(1):29–38.

Davidovich-Pinhas, M. and H. Bianco-Peled. 2011. Alginate-PEGAc: A new mucoadhesive polymer. *Acta Biomater* 7(2):625–633.

Djupesland, P.G. 2013. Nasal drug delivery devices: Characteristics and performance in a clinical perspective—A review. *Drug Deliv Transl Res* 3:42–62.

Djupesland, P.G., J.C. Messina, and R.A. Mahmoud. 2014. The nasal approach to delivering treatment for brain diseases: An anatomic, physiologic, and delivery technology overview. *Ther Deliv* 5(6):709–733.

Djupesland, P.G. and A. Skretting. 2012. Nasal deposition and clearance in man: Comparison of a bidirectional powder device and a traditional liquid spray pump. *J Aerosol Med Pulm* 25(5):280–289.

Djupesland, P.G., A. Skretting, M. Winderen et al. 2004. Bi-directional nasal delivery of aerosols can prevent lung deposition. *J Aerosol Med* 17:249–259.

Djupesland, P.G., A. Skretting, M. Winderen et al. 2006. Breath actuated device improves delivery to target sites beyond the nasal valve. *Laryngoscope* 116(3):466–472.

Fallon, M., C. Reale, A. Davies et al. 2011. Efficacy and safety of fentanyl pectin nasal spray compared with immediate-release morphine sulfate tablets in the treatment of breakthrough cancer pain: A multicenter, randomized, controlled, double-blind, double-dummy multiple-crossover study. *J Support Oncol* 9(6):224–231.

Hoekman, J.D. and R.J. Ho. 2011. Enhanced analgesic responses after preferential delivery of morphine and fentanyl to the olfactory epithelium in rats. *Anesth Analg* 13(3):641–645.

Illiff, J.J., M. Want, Y. Liao et al. 2012. A paravascular pathway facilitates CSF flow through the brain parenchyma and the clearance of interstitial solutes, including amyloid β. *Sci Transl Med* 4(147):147ra111.

Illum, L. 2003. Nasal drug delivery: Possibilities, problems and solutions. *J Control Release* 87:187–198.

Illum, L. 2012. Nasal drug delivery—Recent developments and future prospects. *J Control Release* 161(2):254–263.

Lehr, C.M. 2000. Lectin-mediated drug delivery: The second generation of bioadhesives. *J Control Release* 65(1–2):19.

Lochhead, J. and R. Thorne. 2012. Delivery of therapeutics to the central nervous system. Intranasal delivery of biologics to the central nervous system. *Adv Drug Deliv Rev* 64(7):614–628.

Maggio, E.T. and D.J. Pillion. 2013. High efficiency intranasal drug delivery using Intravail® alkylsaccharide absorption enhancers. *Drug Deliv Transl Res* 3:16–25.

Marple, B., P. Roland, and M. Benninger. 2004. Safety review of benzalkonium chloride used as a preservative in intranasal solutions. *Otolaryngol Head Neck Surg* 130:131–141.

Merkus, F.W.H.M., N.G.M. Schipper, W.A.J.J. Hermens et al. 1993. Absorption enhancers in nasal drug delivery: Efficacy and safety. *J Control Release* 24(1–3):201–208.

Quintana, D.S., L.T. Westlye, Ø.G. Rustan et al. 2015. Low-dose oxytocin delivered intranasally with Breath Powered device affects social-cognitive behavior: A randomized four-way crossover trial with nasal cavity dimension assessment. *Transl Psychiatry* 5:e602.

Reger, M.A., G.S. Watson, P.S. Green et al. 2008. Intranasal insulin improves cognition and modulates beta-amyloid in early AD. *Neurology* 71(11):866.

Righton, L. 2011. Bringing the patient's voice to nasal drug delivery. *Pulmonary and Nasal Drug Delivery*. *ONdrugDelivery*. Accessed January 8, 2015. www.ondrugdelivery.com.

Suman, J.D., B.L. Laube, and R. Dalby. 1999. Comparison of nasal deposition and clearance of aerosol generated by nebulizer and an aqueous spray pump. *Pharm Res* 16(10):1648–1652.

Thorsson, L., S.P. Newman, A. Weisz et al. 1993. Nasal distribution of budesonide inhaled via a powder inhaler. *Rhinology* 31:7–10.

Ugwoke, M., U. Remigius, N.V. Agu et al. 2005. Nasal mucoadhesive drug delivery: Background, applications, trends and future perspectives. *Adv Drug Deliv Rev* 57(11):1640–1665.

Vecellio, L., S. Le Guellec, D. Le Pennec et al. 2011. Deposition of aerosols delivered by nasal route with jet and mesh nebulizers. *Int J Pharm* 407:87–94.

Waddell, A.N., S.K. Patel, A.G. Toma et al. 2003. Intranasal steroid sprays in the treatment of rhinitis: Is one better than another? *J Laryngol Otol* 117:843–845.

11 Pulmonary Drug Delivery

Heidi M. Mansour, Paul B. Myrdal, Usir Younis,
Priya Muralidharan, Anya M. Hillery, and Don Hayes, Jr.

CONTENTS

11.1 INTRODUCTION

Historically, inhalation aerosol therapy dates back to ancient times when the ancient Greeks (including Hippocrates) and Egyptians used "vapors" to treat asthma and tuberculosis (TB) (Murphy 2007; O'Callaghan et al. 2002; Patton and Byron 2007). In 1778, an English physician by the name of John Mudge invented his own pewter tank inhaler for the treatment of a cough (Sanders 2007). Starting in the early twentieth century, bronchodilators and biologics (antibiotics and insulin) were commonly being delivered via inhalation devices (Newman et al. 2009). Vast advancements have recently been made in inhalation therapy. In 2013, the global pulmonary drug delivery market reached $32.4 billion and is expected to reach $43.9 billion in 2018 (Marketwatch 2014).

Pulmonary drug delivery is primarily used to treat local conditions of the airways, thereby delivering drugs directly to their site of action, for the treatment of respiratory disorders such as asthma and chronic obstructive pulmonary disease (COPD). Such drugs include, for example, short-acting β-agonists (e.g., albuterol), short-acting corticosteroids (e.g., beclomethasone), short-acting anticholinergics (e.g., ipratropium), mast cell inhibitors (e.g., cromolyn sodium), long-acting β-agonists (e.g., salmeterol and formoterol), long-acting corticosteroids (e.g., fluticasone propionate), and long-acting anticholinergics (e.g., tiotropium). Pulmonary drug delivery is also used for the local delivery of destructive mucolytics (e.g., dornase alpha) and nondestructive mucolytics (e.g., hypertonic saline solution and mannitol) in the treatment of cystic fibrosis (CF), as well as antibiotics (e.g., tobramycin and aztreonam) and antivirals (e.g., ribavirin and zanamivir) for respiratory infections.

An emerging application of pulmonary drug delivery is for systemic delivery, providing a painless, needle-free, noninvasive route of entry for the drug to the blood; very rapid absorption is possible. The pulmonary route is showing considerable promise for the systemic delivery of small molecules but also macromolecules like proteins, peptides, and vaccines, as well as gene-based therapies. Both local and systemic drug delivery applications are discussed in this chapter. The invasive method of endotracheal instillation, for localized lung delivery, is also described.

Given the multifactorial nature of pulmonary diseases, combination drug aerosol products (containing drugs from different therapeutic classes in the same aerosol) continue to grow and are the top-selling aerosol products worldwide. For superior disease state management and prevention, codeposition of inhaled pulmonary drugs in the same aerosol to the lung has been shown clinically to be superior than inhaling two separate single-drug aerosol products where codeposition of both drugs does not occur. In addition, the aerosol treatment burden on the diseased lungs is less with a combination drug aerosol product. Currently, several dual-drug combination aerosol products containing two drugs from two different therapeutic classes are commercially available as liquid aerosols and inhalable dry powders for the treatment and management of airway inflammatory disorders, including asthma, CF, and COPD. These top-selling dual-drug combination inhalation products are DuoNeb®, Advair®, Spiriva®, Dulera®, and Symbicort®. The upcoming sections will present these aspects in detail.

This chapter will describe pulmonary drug delivery in the context of treatment of human lung diseases and approved respiratory products used by patients. In addition, the different inhalation aerosol devices approved for human use and respiratory pharmaceutical products will be described. Finally, future trends and outlook will be presented.

11.2 ANATOMY AND PHYSIOLOGY OF THE RESPIRATORY TRACT: IMPLICATIONS FOR PULMONARY DRUG DELIVERY

The anatomy of the respiratory system is shown in Figure 11.1. The respiratory tract begins at the nose (drug delivery to the nasal cavity is the focus of Chapter 10). The nasal passageways open into the pharynx (throat), which serves as a common passageway between the respiratory and digestive

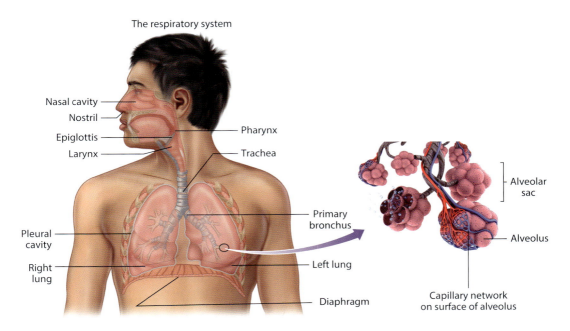

The respiratory system

Nasal cavity
Nostril
Epiglottis
Larynx
Pharynx
Trachea
Alveolar sac
Primary bronchus
Pleural cavity
Alveolus
Right lung
Left lung
Diaphragm
Capillary network on surface of alveolus

FIGURE 11.1 Anatomy of the lungs. (Detail of the alveoli, Alex Mit/Shutterstock.com.)

systems. The larynx (voice box, which houses the vocal cords) is located at the entrance of the trachea. Beyond the larynx, the trachea (windpipe) divides into the left and right primary bronchi, which enter the left and right lungs, respectively. Upon entering the lungs, the primary bronchi divide to form the secondary (lobar) bronchi, which in turn branch into the tertiary (segmental) bronchi and then the bronchioles. These further branch repeatedly, forming a fine network of bronchioles, which end in the terminal bronchioles. Each successive branching results in progressively narrower, shorter, and more numerous airways; the overall effect resembles an inverted tree, described as the bronchial tree. Clustered at the ends of the terminal bronchioles are the alveoli, tiny air sacs where gas exchange between air and blood takes place (see Chapter 4, Figure 4.10).

Anatomically, the respiratory tract can be divided into two sections (Tortora and Derrickson 2013), as shown in Figure 11.1: (1) the upper respiratory system, which includes the nose, pharynx, and associated structures; and (2) the lower respiratory system, which includes the larynx, trachea, bronchi, and lungs. Functionally, the respiratory tract can also be divided into two parts: (1) The conducting zone is a series of interconnected cavities and passageways, which includes the nose, pharynx (throat), larynx, trachea, bronchi, bronchioles, and terminal bronchioles. These serve to prepare ambient air for respiration by warming and humidifying the incoming air, as well as filtering the air of foreign particles and pathogens. (2) The respiratory zone is the tissues within the lungs where gas exchange between air and blood occurs. These tissues include the respiratory bronchioles, alveolar ducts, alveolar sacs, and alveoli.

Specific anatomical and physiological features of the respiratory system have important implications for pulmonary drug delivery and are discussed in more detail here.

11.2.1 RESPIRATORY GEOMETRY

The basic structure of the respiratory tract can be considered as a series of dichotomous branches (Figure 11.1). Each time the bronchial tree divides, it is referred to as a new generation; thus, the trachea (generation 0) divides into the primary bronchi (generation 1), then the secondary bronchi (generation 2), etc. There are 23 generations of airways; gas exchange occurs in the last 8.

This branching of the bronchial tree is associated with a number of key features. First, airway branching causes a reduction in the airflow velocity. Within the bronchial tree, airflow is highest in the bronchi and rapidly decreases as it descends into the lower airways. Airflow velocities in the main bronchi are estimated to be 100-fold higher than in the terminal bronchioles and 1000-fold higher than in the alveoli. In the alveoli, the air velocity is near zero, to facilitate respiration by passive diffusion of dissolved gases. Second, with the increasing number of bifurcations, the total cross-sectional area is increased significantly, so that the overall surface area of the alveolar region is very large. Third, the branching of the bronchial and bronchiolar tree is associated with significant changes in the epithelial lining, as described in Section 11.2.3.

11.2.2 MUCUS AND THE MUCOCILIARY ESCALATOR

With the exception of the alveolar region, mucus is secreted in the respiratory tract by goblet cells in the mucosa, as well as by submucosal glands. The mucus of the respiratory tract has already been described for nasal drug delivery (Chapter 10, Section 10.2.2). A ciliated epithelium lines most of the airways and is described in detail in Chapter 4 (Section 4.5.4). Each ciliated cell has about 200 cilia, as well as many interspersed microvilli. The cilia are surrounded by a low-viscosity fluid, known as the periciliary fluid layer; they project into the more viscous, overlying, mucus layer. As described in Chapter 10 (Section 10.2.2), mucus functions as a protective barrier against the entry of pathogens and also helps to filter the incoming air via the "mucociliary escalator." Dust particles, microbes, allergens, and other pollutants become trapped in the viscoelastic mucus layer. Rhythmic beating of the cilia propels the mucus, with any entrapped pollutants, against gravity toward the throat. The ciliary beat frequency is in the range of 1000–1200 beats/min. At the throat, entrapped particles are then either swallowed or expectorated, thereby removing them from the air-ways. Coughing and sneezing speed up the movement of cilia and mucus, accelerating the removal process. Soluble particles can also dissolve in mucus and may then be removed via absorption from the airways. A further lung clearance mechanism involves phagocytosis by alveolar macrophages (see Section 11.3.1.4).

11.2.3 RESPIRATORY EPITHELIUM

As the branching becomes more extensive in the bronchial tree, the respiratory epithelium undergoes considerable structural changes, as outlined in Table 11.1 (see also Chapter 4, Figures 4.9 and 4.10). In summary, the thick, pseudostratified, ciliated epithelium (with overlying mucus) found in the upper airways gradually gives way to a progressively thinner epithelium with less mucus and cilia

TABLE 11.1

Structural Characteristics of the Respiratory Epithelium Associated with Different Regions of the Respiratory Tract

Respiratory Region	Epithelium
Primary, secondary, and tertiary bronchi	Pseudostratified ciliated columnar epithelium, with many goblet cells (see Chapter 4, Figure 4.9)
Larger bronchioles	Ciliated simple columnar epithelium with some goblet cells
Smaller bronchioles	Ciliated simple cuboidal epithelium with no goblet cells
Terminal bronchioles	Nonciliated cuboidal epithelium
Alveoli	Simple squamous epithelium (see Chapter 4, Figure 4.10)

in the bronchioles and eventually becomes an extremely thin, nonciliated, single layer of squamous cells, with no overlying mucus layer, at the level of the alveoli.

The structure of the alveoli has been described in detail in Chapter 4 (Section 4.5.4.3 and Figure 4.10). The walls of the alveoli consist of two types of epithelial cells: (1) type 1 alveolar cells, comprising simple squamous epithelial cells, are the main sites for gas exchange. Type 1 cells cover 90% of the entire alveolar surface. (2) Type 2 alveolar (also called septal) cells secrete alveolar fluid, which comprises a lung surfactant monolayer on a thin aqueous layer. Pulmonary surfactant lowers the surface tension within the alveoli, thereby maintaining alveolar structural integrity. Also associated with the alveolar wall are the alveolar macrophages, which account for about 3% of the cells in this region.

11.3   OVERVIEW OF PULMONARY DRUG DELIVERY

As outlined in Section 11.1, the respiratory system can be used both for localized drug delivery to the lungs and as a noninvasive route for systemic delivery. Drugs used to exert a local effect, such as anti-asthmatics, anti-inflammatories, and mucolytics, are delivered directly to their site of action in the airways. Thus, therapeutic agents are effectively targeted to their site of action, maximizing therapeutic effect. Lower doses are therefore required for effective therapy, reducing the risk of systemic side effects and toxicity, as well as costs. By delivering the drug directly to the site of action, a more rapid onset of action is also possible, in comparison to the oral route.

For systemically acting drugs, pulmonary delivery also offers a number of potential advantages:

1. *An expanded surface area*: Physiologically, the lungs are designed for the efficient exchange of respiratory gases. The extensive branching results in a large total cross-sectional area. The surface area of the adult lung is in the range of 80–150 m^2, equivalent to the surface area of a tennis court. In addition, it is estimated that the lungs contain 300 million alveoli for gas exchange. These features also offer great potential for the delivery of systemically acting drugs.

2. *Thin and highly permeable epithelium*: The alveolar epithelial membrane is very thin and permeable, to facilitate its physiological role of rapid passive gas exchange (see also Chapter 4, Figure 4.10). A network of capillaries spreads over the outer surface of the alveoli (Figure 11.1). The exchange of oxygen and carbon dioxide takes place by gas diffusion across the alveolar and capillary walls, which together form the respiratory membrane, which comprises merely: a single epithelial cell (i.e., a type 1 pneumocyte), a single endothelial cell (which lines the blood capillary), and a common basement membrane between the two (see also Chapter 4, Figure 4.10). The respiratory membrane is at least an order of magnitude thinner than other epithelial membranes, such as in the intestines, and several orders of magnitude thinner than, for example, the skin epidermis. Thus, the alveolar epithelium has a higher (in many cases, much higher) permeability than alternative mucosal routes, which facilitates the absorption of large molecular weight APIs (including insulin), which are usually too large to permeate other epithelial interfaces. Drug absorption from this region is also very rapid, so that rapid onset of therapeutic activity is possible. The epithelium of the higher airways, such as the trachea and bronchi, is thicker than the alveolar region, but is still very thin in comparison to other epithelial interfaces and these regions also demonstrate enhanced permeability.

3. *Rich blood supply*: The lungs receive blood via the pulmonary arteries (deoxygenated blood) and the bronchial arteries (oxygenated blood from the aorta). The network of capillaries that covers the alveoli is so dense that each alveolus is effectively encircled by an almost continuous sheet of blood. This highly vascular surface further promotes the rapid absorption of drug molecules and ensures a rapid onset of action.

4. *Relatively low-enzymatic activity*: The enzymatic activity associated with respiratory secretions is considerably lower than that of the GI tract. Furthermore, drugs absorbed

via the respiratory route avoid the intestinal and hepatic first-pass metabolism associated with oral drug delivery. Pulmonary drug delivery is therefore an attractive option
for enzymatically labile drugs, such as therapeutic peptides, proteins, and DNA-based
medicines.

5. *Accessibility and compliance*: As described later, a large number of different types of
inhalable devices allow for the relatively easy self-administration of drugs, for either local
or systemic effects. The pulmonary route offers a painless, needle-free, noninvasive alternative to injections for systemic drug delivery.

11.3.1 LIMITING FACTORS FOR PULMONARY DRUG DELIVERY

The anatomy and physiology of the lungs also afford a number of obstacles to successful drug
delivery. As described in Section 11.6 *et seq*, pulmonary drug delivery is typically achieved via
the administration of an aerosol, i.e., a suspension of fine solid or liquid particles, dispersed in
a gas. The respiratory system is characterized by a variety of efficient defense mechanisms to
both minimize the entrance of particulates into the airways and ensure the effective clearance of
any particles that have been deposited. Although these mechanisms protect the airways from the
entrance of contaminants such as dust, fumes, pollutants, and pathogens, they also present significant barriers to effective drug delivery. These, and other, barriers to pulmonary drug delivery are
described here.

11.3.1.1 Respiratory Geometry

The branching architecture of the airways is such that it is very difficult for drug molecules to access
and penetrate deep into the lungs. To travel down the airways, the drug particles (which are often
traveling at high speeds) must pass through a successive series of branching tubes, of progressively
decreasing size. As described in Section 11.2.1, there are about 23 generations of branching, *en
route* to the alveolar space. Once the aerosol droplet/particle has deposited in the airways, there are
a number of further obstacles to be surmounted, as described next.

11.3.1.2 Mucus and Mucociliary Clearance

The mucus layer and the process of mucociliary clearance present a number of obstacles for pulmonary drug delivery. Drugs delivered as dry powders must first dissolve in the mucus secretions.
Typically, the high water content and the presence of surfactant (i.e., fatty acids and phospholipids)
in mucus aid the dissolution process. However, dissolution can be an issue for poorly soluble APIs
such as the anti-inflammatory corticosteroids, which are often poorly soluble and administered as
dry powders.

The mucus also presents a significant diffusional barrier for a drug in transit to the epithelial surface. The thickness of the mucus layer is about 5–8 μm in the larger airways (although
this can increase significantly in disease states, see below), then it progressively decreases in
thickness, until it disappears in the respiratory region. The rate of diffusion of an API through
mucus depends on a number of factors, including the thickness and viscosity of the mucus layer,
as well as the physicochemical properties of the drug. Interactions between the API and the
mucus layer may also occur, leading to mucoadhesion, via hydrogen bonding; charge interactions
can also occur, between a positively charged drug and negatively charged sialic acid residues of
the mucus layer.

Mucociliary clearance may limit the contact time of a drug molecule with the epithelial
surface. The drug, instead of settling locally, can instead be removed to the throat, to be swallowed or expectorated. Limited contact time with the absorbing surface can compromise drug
absorption for a systemically active drug, or shorten drug–receptor interaction time for a locally
acting drug.

11.3.1.3   Epithelial Barrier

As can be seen from Table 11.1, on penetrating deeper into the airways, the epithelium presents a progressively thinner diffusional barrier. Also, mucociliary clearance declines as the number of ciliated cells decreases and less mucus is produced. At the level of the alveoli, the epithelium comprises merely a single layer of highly permeable squamous cells, with no overlying mucus layer, or mucociliary clearance process. This region is also characterized by very large surface area and rich blood supply. For these reasons, delivery systems designed for systemically acting drugs should target the alveolar region. However, gaining access to this region is difficult, due to the various barriers and clearance mechanisms present.

As described in detail in Chapter 4 (Section 4.3), drug absorption across any epithelial interface can be via the paracellular route (i.e., between epithelial cells), or the transcellular route (i.e., through the cells). The presence of tight junctions between epithelial cells significantly limits paracellular transport, typically confining transit via this route to small hydrophilic molecules such as mannitol. There is evidence that the tight junctions of alveolar epithelium are leakier than at other epithelial interfaces, so that more paracellular transport is known to occur in this region (Chandra et al. 1992).

In the larger airways, the predominant pathway across the respiratory epithelium is via transcellular passive diffusion. The rate of absorption is determined by the physicochemical properties of the drug, as well as the drug concentration gradient across the cells. The various drug physicochemical factors that affect transepithelial transport have been described in Chapter 4 (Section 4.3); to summarize here, important drug properties include (1) molecular weight: large molecules (greater than 500 Da) demonstrate poor permeation, (2) lipophilicity: lipidic drugs exhibit good permeation properties, whereas polar, hydrophilic, or ionized molecules demonstrate poor permeability, and (3) drug solubility: a drug must dissolve in the respiratory fluids and be in solution for epithelial absorption to occur.

Pulmonary drug delivery may also result in absorption of an API from the GI tract, either due to direct swallowing of inhaled dose that deposits in the mouth and back of the throat, or due to secondary swallowing: drug deposited in the airways is moved, via mucociliary clearance and coughing, to the throat where it is then swallowed.

11.3.1.4   Alveolar Clearance

The alveolar macrophages represent a further clearance mechanism that exists in the lungs. These phagocytic cells are found in the lumen of the alveolar sacs and remove foreign particulates (e.g., fine dust, viruses, mycobacteria such as in TB, and bacteria causing pneumonia) via the process of phagocytosis (see also Chapter 4, Figures 4.4 and 4.10). Large molecular weight drugs, drug particles, and nanoparticulate drug delivery systems (DDS) can also be phagocytosed via these cells. Following phagocytosis, the drug is internalized within a phagocytic vesicle (phagosome), which subsequently fuses with a lysosome, which may lead to destruction of the API on exposure to lysosomal enzymes. An important consequence is that in order to avoid phagocytic capture, a macromolecular drug (or nanoparticulate DDS) must dissolve quickly in the alveolar fluid and be absorbed systemically.

11.3.1.5   Enzyme Activity

While the lung is regarded as a low metabolic organ compared to the hepatic system and GI tract, a low number of metabolizing enzymes (e.g., phospholipases, esterases, and cytochrome p450) and transporters (e.g., p-glycoprotein) have been discovered in certain types of pulmonary cells (Bend et al. 1985; Devereux et al. 1989). This enzymatic activity could be detrimental to the absorption of enzymatically labile drugs.

11.3.1.6   Variability

Drug delivery to the lungs is a complex process and dependent on many interrelated elements, including pharmaceutical, drug, physiological, and pathological factors. This introduces a variability that may be particularly problematic for drugs with a narrow therapeutic index.

11.3.1.7 Patient Acceptability and Compliance

Aerosol devices can be difficult and unpleasant to use, which can compromise patient compliance. These difficulties are described later, with respect to each type of device, and include issues such as the difficulty in coordinating actuation with inhalation for pressurized metered-dose inhalers (pMDIs), the unpleasantness of the "cold Freon" effect (the cold blast of pMDI propellant hitting the back of the throat) for pMDIs, the long inhalation times required for nebulizer therapy, the cumbersomeness and lack of portability of nebulizer devices, the high inspiratory effort required to generate an effective aerosol for dry powder inhalers (DPIs), and the unpleasantness of dry powder hitting the back of the throat for DPIs.

11.4 PARTICLE DEPOSITION IN THE AIRWAYS

The site of particle deposition is a crucial factor affecting pulmonary drug delivery and is considered in detail here.

11.4.1 Mechanisms of Inhaled Particle Deposition in the Airways

Particles deposit in the airways by five modes of deposition mechanisms (Altiere and Thompson 2007; Hickey and Thompson 2004; Stocks and Hislop 2002), which are comprised of three major modes and two minor modes. The three major modes are the following: (1) inertial impaction due to inertial force, (2) sedimentation due to gravitational force, and (3) diffusion due to Brownian motion. The two minor modes are: (1) electrostatic attraction due to attractive coulombic forces and (2) interception. The three major modes are described here:

Inertial impaction: Large particles, traveling at high velocity, tend to collide with the airway walls in the upper airway regions. Due to particle inertia, the particles continue on their original course and are unable to follow the sharp change in direction of the inspired air, as it weaves its way through the extensive branching of the oropharyngeal and bronchial tree. Thus, the particles impact on the airway walls, with impaction usually occurring near the sharp-angled bifurcations. For rapidly traveling particles of large size, inertial impaction is particularly relevant in the oropharyngeal region. This is as a result of the sharp change in airflow direction from the mouth to throat region, which is a $\approx 90°$ angle (Figure 11.1).

Inertial impaction is dependent on particle momentum, which in turn is dependent on particle size and velocity. Particles with larger diameters, higher densities, and higher velocities, will result in greater impaction. These properties are in turn critically influenced by the type of device used to generate the aerosol, as discussed later.

Sedimentation: Sedimentation is seen with smaller airborne particles that succumb to gravity, i.e., those particles that have successfully transited the upper airways, to reach the bronchiolar and alveolar regions. In these regions, the airstream velocity is relatively low and so deposition is primarily the result of sedimentation due to gravitational settling. The percentage of particles that deposit via this mechanism is dependent on the amount of time the particles are in these regions. If the time between the end of inspiration and the start of exhalation is extended, there is more time for sedimentation to occur. Therefore, introducing a breath-hold step after inspiration is an important mechanism to optimize the administration of many types of aerosol.

Brownian motion: Much smaller-sized particles (less than 500 nm) can deposit due to Brownian motion. The random bombardment of gas molecules can result in particle collision with the airway walls and thus particle deposition. Deposition due to Brownian motion is particularly associated with the alveolar spaces, where air velocity is near zero. It is of interest to note that inhaled cigarette smoke particles are typically in the aerodynamic size range of approximately 100–200 nm. These particles readily deposit in the alveolar region for fast systemic absorption, facilitating rapid nicotine action on the central nervous system (CNS).

11.4.2 Pharmaceutical Factors Affecting Particle Deposition

Particle deposition is influenced by the type of delivery device used and the specifics of the drug formulation, as these factors affect aerosol velocity, as well as the aerodynamic particle size, aerodynamic particle size distribution, particle shape, particle surface charge, particle density, and hygroscopicity of the inhaled particles. Here, we will discuss two important pharmaceutical variables, namely, aerosol velocity and aerodynamic particle size. Further information is given in the discussion later with respect to each type of aerosol device.

With regard to aerosol velocity, aerosols generated by pMDIs comprise particles of large size with extremely high velocities (up to 100 km/hour), which can lead to extensive impaction in the oropharyngeal region. Estimates have suggested as much as 80% of the dose from a pMDI can be lost to oropharyngeal impaction (Dolovich et al. 1981). The aerosol particles produced by DPIs travel within the inspired air, and as such, their velocity is determined by the inspiratory flow rate (IFR) (which can vary considerably, particularly in disease states). The generation of an aerosol via a nebulizer is not dependent on the patient's IFR, and these devices produce smaller liquid aerosol droplets with less oropharyngeal deposition. Soft mist inhalers (SMIs), by definition, produce a slow-moving aerosol mist without a propellant. The liquid aerosol is generated independently of the patient's IFR and is associated with less oropharyngeal deposition than propellant-based liquid aerosol systems.

Aerodynamic particle size is another important determinant of particle deposition in the airways. Aerosol particle size is expressed as aerodynamic diameter (D_a), defined as the equivalent diameter of a spherical particle of unit density having the same settling velocity from an air stream as the particle of interest (Suarez and Hickey 2000). It is an important measured parameter required by federal regulating agencies for the approval of any inhaled therapeutic aerosol product to be used in humans. D_a has been used for several decades to quantify an aerosol particle's inherent propensity to deposit in the lungs (Edwards et al. 1998).

Particles with $D_a \geq 10$ μm will not enter the lungs, but will instead exhibit extrathoracic deposition in the oropharyngeal region (Dalby et al. 2007; Murphy 2007; Suarez and Hickey 2000). Oropharyngeal impaction results in drug wastage: drug deposited here is either swallowed or expectorated and is not available to exert its therapeutic effect. It can cause local side effects, for example, the oropharyngeal candidiasis associated with inhaled corticosteroid use. Drug deposition here also feels unpleasant, can induce a cough reflex, and adversely affect patient compliance. Particles with $D_a < 10$ μm can readily enter the lungs and exhibit thoracic deposition. Particles with $D_a \leq 5$ μm can readily reach the mid-lung region and smaller airways. Particles with $D_a \leq 2$ μm can readily reach the respiratory bronchiolar and alveolar regions.

Aerosol dispersion performance by in vitro inertial impaction testing is required for the approval of inhalation aerosol products, as described by the U.S. Pharmacopeia (USP) Chapter <601>. The aerosol dispersion performance parameters that are required are as follows (United States Pharmacopeia 2006; Xu et al. 2011):

- The emitted dose is the total mass of drug emitted from the device.
- The fine particle fraction (FPF) is the mass fraction of aerosol particles that is less than a certain aerodynamic cutoff size to the total mass of particles emitted from the device. The aerodynamic cutoff size used in calculating FPF is at the discretion of the scientist carrying out the measurements, as it depends on the type of inertial impactor used, testing conditions such as airflow, and the specific lung disease of interest.
- The mass median aerodynamic diameter (MMAD) is a calculated statistical population value based on the entire aerodynamic size distribution based on mass, below which 50% of the total aerosol mass falls under.

Taking into account the effects of aerodynamic particle size on regional lung deposition patterns, it can be seen that MMAD $\leq$ 5 µm of the aerosol is desirable, to facilitate predominant deposition targeted to the smaller airways. Aerosols with larger MMADs will deposit higher in the respiratory tract. It is important to note that the MMAD is a calculated statistical value based on the entire aerosol size distribution plotted on a log–probability scale. The underlying assumption is that the mass-weighted aerodynamic particle size distribution data is log-normally distributed. The population variability spread of MMAD is described by the geometric standard deviation (GSD), which is also determined from the same plot as the MMAD. An aerosol with a GSD < 1.22 is described as monodisperse, whereas a GSD 1.22 is polydisperse. A polydisperse aerosol (GSD 1.22) will also show greater deposition than a monodisperse aerosol of equivalent MMAD.

The efficacy of an aerosol device to target the airways is of course related to the aerodynamic particle size generated by the device and the formulation physicochemical properties. Different types of aerosol-generating devices generate different MMAD values, for a given formulation. Droplets leaving pMDIs can be too large with a high velocity, which tends to cause unwanted oropharyngeal deposition in the mouth and throat via inertial impaction.

11.4.3 DRUG FACTORS AFFECTING PARTICLE DEPOSITION

Particle deposition is also affected by the physicochemical properties of the drug, such as its molecular weight, solubility, lipophilicity, and charge. These factors can affect stability in vitro, for example, a hygroscopic drug, formulated as a dry powder, tends to aggregate, which can make the generation of a dry powder aerosol more difficult. Drug interactions with mucus can affect residence time and the ability of a drug to permeate through the mucus layer. Drug properties will also influence the drug's ability to be absorbed across the respiratory epithelium, with low molecular weight, lipophilic drugs showing enhanced absorption properties (Section 11.3.1.3). For pMDIs, the importance of drug solubility in the formulation was highlighted by the difficulties encountered on switching from CFA to HFA propellants, as many drugs were insoluble in the new systems.

11.4.4 PHYSIOLOGICAL FACTORS AFFECTING PARTICLE DEPOSITION

Physiological factors affecting particle deposition include features such as the IFR, breathing patterns, thickness of the mucus blanket, mucociliary clearance, lung architecture, and patient skills. Increasing the IFR will increase airstream velocity, which will in turn increase deposition via inertial impaction in the upper airways. It will also increase airstream turbulence, which further enhances deposition, especially in the oropharyngeal region and the first few generations of the bronchial tree.

Some aerosol devices, such as DPIs, use the energy of inspiration to generate the drug aerosol. The drug in a DPI is in the form of a finely milled powder in large aggregates, either alone or in combination with a carrier such as lactose monohydrate. Turbulence generated inside the DPI device resulting from the patient's IFR and device anatomical features breaks up the aggregates into smaller respirable particles that can penetrate the smaller airways. Therefore, depending on the DPI powder formulation properties and DPI device anatomy, an increase in a patient's IFR can lead to enhanced penetration, due to the generation of smaller-sized particles. However, it is important to note that the process is complex, because higher air velocity can also increase inertial impaction, causing aerosol particles to deposit in the oropharyngeal and larger airways.

The type of breathing pattern is also important. Reference has already been made to the importance of the breath-hold step in the administration of many types of aerosol products, because it enhances gravitational settling time. A breath-holding time of 5–10 seconds postinspiration is recommended for maximum effect (Suarez and Hickey 2000). Rapid breathing, in contrast, is associated with increased deposition of larger particles in the larger airways (Valberg et al. 1982).

11.4.5 PATHOLOGICAL FACTORS AFFECTING PARTICLE DEPOSITION

Respiratory conditions such as asthma result in bronchoconstriction: the corresponding narrowing of the airways enhances deposition via inertial impaction. Furthermore, deposition via sedimentation may also be enhanced, as the bronchoconstriction of asthma has a greater effect on exhalation than inhalation.

Many disease states, including bronchitis, CF, and asthma, are associated with a hypersecretion of mucus, so that the thickness of the mucus layer can increase from the normal 8 µm, up to as much as 50 µm. This represents a much greater barrier for drug diffusion than in the normal physiological state. A thick mucus layer also narrows the airways, making aerosol impaction more likely.

Mucus hypersecretion can also cause an overloading of the mucociliary transport process, so that the cilia become paralyzed and the mucus is not cleared, but instead builds up as a thick, highly viscous layer. Mucociliary clearance is also known to be impaired in smokers, CF patients, patients with chronic bronchitis, and acute asthmatics. Packed-down, lodged mucus must be shifted by coughing, which can rapidly remove any deposited drug, thereby reducing the residence time of the API.

11.5 ENDOTRACHEAL INSTILLATION

Endotracheal instillation is an invasive method of pulmonary drug delivery, whereby formulations are administered directly to the lungs via endotracheal tubing. The procedure is used for the pulmonary delivery of lung surfactant. Physiologically, lung surfactant is synthesized and secreted by the alveolar type 2 cells (Wright and Clements 1987) upon stretching. It is essential for normal breathing, decreasing the work of breathing by significantly reducing the surface tension of the aqueous layer that lines the alveoli, from 72 mN/m to slightly above 0 mN/m. Lung surfactant is also critical in regulating normal pulmonary immunity in the deep lung region and modulating alveolar macrophage phagocytic activity in the alveolar respiratory region. Alveolar stability depends on variable surface tension where the surface area is low during lung compression and increases during surface expansion, as discovered by Clements (Clements et al. 1961; Schurch et al. 1976).

Structurally, lung surfactant is a thin monomolecular layer comprising primarily of essential phospholipids with a specific polar headgroup/hydrophobic diacyl chain configuration required for activity and a small amount of four essential lung surfactant–specific proteins (i.e., SP-A, SP-B, SP-C, and SP-D). SP-A and SP-D are large hydrophilic calcium-dependent proteins involved in stimulating alveolar macrophages and modulating pulmonary immunity. SP-B and SP-C are smaller hydrophobic proteins that stabilize the lung surfactant monolayer through molecular ordering. Children lacking in SP-B die of respiratory failure (Nogee et al. 1993). The entire process of surfactant production is not fully understood. There are other proteins involved as well, such as ABCA3 protein. ABCA3 transporter protein is found in the membrane that surrounds lamellar bodies, where it is involved with the transport of phospholipids into the lamellar bodies and interactions with lung surfactant–specific proteins to form lung surfactant.

Infants born prematurely (aka "preemies") suffer from respiratory distress syndrome (RDS), a fatal lung condition due directly to the lack of innate lung surfactant, or dysfunctional lung surfactant, as first reported by Avery and Mead (1959). Moreover, RDS can occur in full-term infants from surfactant dysfunction due to genetic ABCA3 transporter dysfunction (Anandarajan et al. 2009; Bullard et al. 2006; Shulenin et al. 2004). RDS can also occur in the elderly as adult RDS (Lewis and Jobe 1993) and secondary to other lung diseases such as lung injury, respiratory infections, COPD, and asthma. It is quickly fatal if not immediately treated.

Exosurf® was the first FDA-approved artificial lung surfactant replacement therapy that saved many premature infants from death due to RDS. This product was discontinued by the manufacturer relatively recently due to more advanced lung surfactant replacement therapeutic products. These include Surfaxin®, which is the first and only humanized synthetic lung surfactant replacement product available. The other current lung surfactant replacement therapeutic products are animal derived, as bovine- and porcine-derived nanocarrier suspensions. These are comprised of self-assembled multilamellar phospholipid nanocarriers and lung surfactant–specific proteins. Surfaxin® contains lucinactant, which is a recombinant human (rh) mimic of the essential human lung surfactant–specific protein, SP-B (Cochrane and Revak 1991). Lucinactant consists of a 21–amino acid cationic hydrophobic polypeptide (i.e., KL$_4$) consisting of repeating units of one lysine (a cationic amino acid "K") molecule and four leucine (a hydrophobic amino acid "L") molecules (Cochrane 1998; Cochrane et al. 1996; Ma et al. 1998; Mansour et al. 2008). Surfaxin® also contains multilamellar liposomes consisting of the specific phospholipids essential in lung surfactant function.

These liquid products are administered directly to the lungs by endotracheal instillation. As these consist of nanocarrier self-assemblies with significant surface activity, they spread at the air–water interface following installation and coat the airways down to the deep lung alveolar region. Pulmonary surfactant turnover and clearance in the alveolar region have been described in detail (Wright and Clements 1987; Wright and Dobbs 1991). The mucociliary escalator (in the conducting zone) and phagocytic alveolar macrophages (in the respiratory zone) are also sources of clearance. Studies of delivering dry powder artificial lung surfactant to preemies with RDS have been reported by the University of Cambridge (Morley et al. 1981) and the University of Oxford (Wilkinson et al. 1985). However, no dry powder products for lung surfactant replacement therapy currently exist.

The following sections describe various noninvasive pulmonary drug delivery approaches. These approaches all involve the generation of an aerosol, which is then inhaled into the lungs. The different devices currently in use, or under development, for aerosol generation are described in detail.

11.6 NEBULIZERS

In 2013, the nebulizers' market reached $685.7 million and is forecasted to grow to $893.5 million in 2018 (Marketwatch 2014). Nebulized liquid aerosols provide a continuous aerosol mist of droplets whose generation is not dependent on the patient's IFR, as the liquid aerosol droplets are generated by a compressor (powered by electricity or a battery) that entrains and disperses the liquid into droplets (Knoch and Finlay 2008; Niven and Hickey 2007). Inhalation via a face mask (e.g., infants and young children) or mouth port is achieved by the patient's normal tidal breathing (Murphy 2007). The mouthpiece or face mask and the auxiliary tubing should be washed after each use. Nebulizers, which were first developed in the late 1820s (Murphy 2007), can be noisy, costly, and not conveniently portable. Nebulizers are rather bulky; even the handheld ones are larger than inhaler devices used in pMDIs, DPIs, and SMIs. Treatment administration times can range in the range of 10–20 minutes per dose, and along with several treatments per day, the nebulization treatment burden on patients can be time consuming. It is important to note that nebulized solutions are required to be sterile, hence the unit-dose packaging. Some nebulized solutions used in marketed nebulizer products require refrigeration for proper stability.

Nebulizers are mostly used for acute rescue treatment and in niche patient populations, i.e., pediatric and geriatric patients, where ease of administration greatly outweighs convenience. Following FDA regulations in 2002 requiring sterility (Myrdal and Angersbach 2008), nebulizer inhalation solutions are currently packaged solely as unit-dose plastic (or glass) ampules with no preservative (Hickey and Mansour 2009). Historically, preservatives (e.g., benzalkonium chloride [BAC] and disodium ethylenediamine tetraacetic acid [EDTA]) were used in nebulized inhalation solutions but are no longer acceptable as they can cause airway irritation such as bronchoconstriction and coughing, especially in inflamed airways such as in asthma and COPD.

Aerosol droplets that are dispersed from nebulizer nozzles are based on the Bernoulli (Venturi) effect (Hickey and Mansour 2009). As described earlier (Section 11.4), the physicochemical properties of the liquid formulation impact the aerosol properties. Surface tension, sterility, density, viscosity, isotonicity, pH, and drug solubility are all important considerations in liquid formulations used in nebulizer therapies. In particular, viscosity, density, and surface tension all govern droplet formation. Point of breakup is under the condition where the surface tension forces are equal to the aerodynamic forces (Niven and Hickey 2007) and occur at the critical Weber number (W_e) (Hickey and Mansour 2009; Niven and Hickey 2007). Drug solubility and stability in aqueous and aqueous/alcohol environments can limit the types of drugs that can be formulated into nebulized solutions and suspensions.

The two delivery types are air-jet and ultrasonic (i.e., electronic) (Dalby et al. 2007). Air-jet nebulizers utilize compressed air generated from the compressor at a high air velocity to create the aerosol droplets. Figure 11.2 shows example air-jet nebulizer devices that are used clinically.

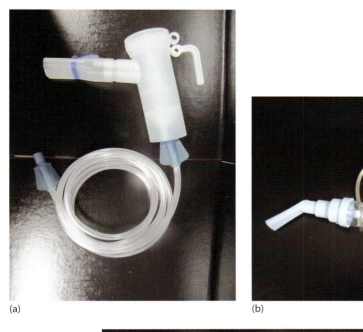
(a)

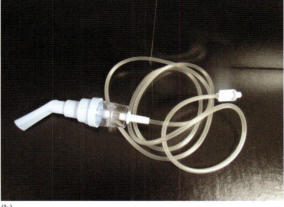
(b)

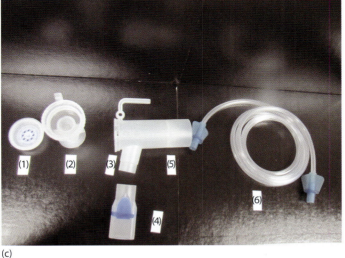
(c)

FIGURE 11.2 Examples of marketed air-jet nebulizers. (a) Pari LC® Plus nebulizer. (b) Respironics® nebulizer. (c) Pari LC Plus nebulizer with its components: (1) inspiratory valve cap, (2) nebulizer insert, (3) nebulizer outlet, (4) mouthpiece, (5) nebulizer cup, and (6) Wing-tip™ tubing (connection to compressor).

Air-jet nebulizers can be divided into the following three types (Knoch and Finlay 2008): (1) constant output (i.e., unvented), (2) breath enhanced (i.e., vented), and (3) breath activated. With constant output nebulizers, at least 50% of the aerosol is lost to the environment during exhalation, and higher than 6 L/minute compressed gas airflow rate is necessary for administration. A breath-enhanced nebulizer provides 70% of the aerosol (i.e., less loss), by using a compressed gas airflow rate (in the range of 3–6 L/minute) and a valve system that enables more aerosol to be generated during inhalation, and less aerosol during exhalation. A breath-activated nebulizer only delivers the aerosol during inhalation and hence nearly 100% delivery is achieved.

Ultrasonic nebulizer involves a rapidly vibrating piezoelectric transducer, which is located at the bottom of the drug liquid reservoir that generates high-frequency ultrasonic wave vibrations, which in turn, provide the energy to generate aerosol droplets. Ultrasonic vibration frequency influences the droplet size generated (Murphy 2007). Heat can be generated during ultrasonic nebulization, which can cause protein denaturation (Knoch and Finlay 2008).

Adrenaline (i.e., epinephrine) mixed with water and glycerol was first used to treat asthma by nebulization in 1911 (Murphy 2007). In 1951, due to adrenaline's systemic side effects following inhalation, it was replaced in the clinical treatment of asthma with isoprenaline (Murphy 2007). In addition to water, ethanol and glycerin along with propylene glycol can be used in nebulizer solutions. Table 11.2 lists the currently FDA-approved nebulizer products. These successful products are used clinically to treat patients with asthma (Barnes et al. 2009; Murphy 2007), CF,

TABLE 11.2
FDA-Approved Nebulizer Aerosol Pharmaceutical Products

Nebulizer Product Name	Drug(s), USP	Therapeutic Drug Class	Manufacturer
Accuneb™	Albuterol sulfate	SABA	Dey, L.P.
DuoNeb (Combivent®) (dual-drug combination)	Albuterol sulfate + ipratropium bromide	SABA + short-acting anticholinergic	Dey, L.P.
Combivent UDV® (dual-drug Combination)	Albuterol sulfate + ipratropium bromide	SABA + short-acting anticholinergic	Boehringer Ingelheim (approved in Canada)
Xopenex®	Levalbuterol hydrochloride	SABA	Sepracor, Inc.
Perforomist™	Formoterol fumarate	LABA	Dey, L.P.
Brovana®	Arformoterol tartrate	LABA	Sepracor, Inc.
Pulmicort Respules®	Budesonide	Corticosteroid (anti-inflammatory)	AstraZeneca
Intal®	Cromolyn sodium	Mast cell inhibitor	King Pharmaceuticals, Inc.
Pulmozyme®	Dornase alpha	rhDNase (destructive mucolytic in CF)	Genentech, Inc.
HyperSal®	Hypertonic saline solution	Osmotic hydrating agent (nondestructive mucolytic in CF)	Pari Respiratory Equipment
TOBI®	Tobramycin	Aminoglycoside antibacterial[a]	Novartis Pharmaceuticals Corporation
Cayston®	Aztreonam lysine	Monobactam antibacterial[a]	Gilead Sciences
Virazole®	Ribavirin	Nucleoside antiviral[b]	Gilead Sciences

Abbreviations: SABA, short-acting β_2 adrenergic agonist (short-acting bronchodilator); LABA, long-acting β_2 adrenergic agonist (long-acting bronchodilator); rhDNase, rh deoxyribonuclease I.

[a]   *Pseudomonas aeruginosa* secondary lung infections in cystic fibrosis (CF).
[b]   Viral lung infection due to respiratory syncytial virus (RSV).

COPD, pulmonary bacterial infections that occur secondary to CF, and pulmonary viral infections due to respiratory syncytial virus (RSV).

11.7 pMDIs

Riker Laboratories (later becoming 3M Pharmaceuticals) generated and marketed the first pMDI in 1956, which established the beginning of the rapidly progressing field of inhalation therapy (Newman et al. 2009). It was not until the invention of the pMDI that successful, cost-effective, and convenient pulmonary drug delivery was possible.

Since its development, the majority of pMDIs have been used for the treatment of patients suffering with asthma and COPD (Newman 2005). Changes in formulation requirements, such as the transition from chlorofluorocarbon (CFC) propellants to hydrofluoroalkane (HFA) propellants, have sequentially caused the evolution of pMDI formulations and hardware (Myrdal et al. 2014; Stein and Stefely 2003; Stein et al. 2014). These innovations have led to current pMDIs becoming an affordable, convenient, and robust method of pulmonary drug delivery.

Four major components comprise the pMDI, which include the canister, metering valve, actuator, and formulation. The formulation consists of drug, propellant, and excipients. These combined components affect product efficacy, performance, and patient compliance. In this section, the current and future prospects of MDI technology will be discussed.

11.7.1 CFC to HFA Propellant Transition

The propellant comprises the majority of the MDI formulation; therefore, its chemical inertness, toxicity, and flammability are requisite considerations (McCulloch 1999; McDonald and Martin 2000). MDIs were previously formulated with CFC propellants (Smyth 2003). However, CFC propellants were found to deplete the stratospheric ozone layer, leading to the search for a more environmentally friendly alternative (Molina and Rowland 1974). In 1989, the Montreal Protocol was signed to initiate the phasing out of destructive CFC propellants and reformulate with an appropriate substitute (USEPA 2014). Extensive research led to the development of two HFA propellants, HFA 134a (1,1,1,2-tetrafluoroethane) and HFA 227 (1,1,1,2,3,3,3-heptafluoropropane) (Wallington et al. 1994). Both are toxicologically safe, chemically inert, and volatile and have a decreased capacity to destroy the ozone layer (Smith 1995). Unfortunately, HFA propellants are not identical thermodynamically to the CFC precursors. This chemical change resulted in solubility and formulation challenges and in turn affected hardware compatibility (Stein and Stefely 2003). Indeed, the evolution of the MDI to HFA propellants has proven to be challenging but necessary.

11.7.2 Hardware

The packaging of a pMDI consists of three different components: the canister, a metering valve, and an actuator. The formulation is contained within a sealed canister, most commonly made out of aluminum alloy. The use of plasticized glass vials is limited to research settings due to their heavier weight and higher cost, but with the benefit that the formulation can be visualized. Coating the interior of the canister with chemically inert polymers can afford the reduction of drug adhesion on the canister wall, increasing dose reproducibility (Newman et al. 2009).

A key component of the pMDI is the metering valve, which is responsible for delivering a consistent amount of formulation each time it is actuated. pMDIs that are currently on the market possess varying metering valve designs, with metering chambers usually ranging from 25 to 100 µL per actuation (Stein et al. 2014). Metering valves require elastomeric seals for performance, which also protect the contents of the formulation and minimize moisture from entering the canister (Newman et al. 2009).

The metering valve is securely crimped to the canister, and the whole unit is fitted within a plastic actuator mouthpiece. When the pMDI is actuated, a predetermined volume of formulation is released from the canister by the metering valve and exits out of the spray nozzle (also referred to as the atomization orifice) within the actuator. This process then allows the formulation to flow through the expansion chamber forming an aerosol to be inhaled by the patient (Dunbar et al. 1997). Typically, spray nozzle diameters vary from 0.3 to 0.6 mm, depending on the type of formulation and the desired product performance characteristic (Myrdal et al. 2014; Newman et al. 2009; Stein et al. 2014). Proper alignment of the canister assembly within the actuator, and cleanliness of the spray nozzle, is vital to proper pMDI function (Stein et al. 2014).

Patient coordination is a very important aspect for optimal drug delivery via an MDI; the timing between actuation and inhalation determines the success of this delivery system (Blaiss 2007; Larsen et al. 1994; Newman et al. 1991). Advancements to packaging hardware have been developed to increase patient compliance and aid the continuing success of pMDIs, including technologies developed for those who have difficulty with the synchronization required to work the inhaler (Crompton 1982). A breath-actuated pMDI has been shown to benefit those patients with poor "press and breath" technique and is preferred over conventional pMDIs, since they avoid the need for patient coordination (Newman et al. 1991). Another innovation is the spacer or holding chamber, an extension of the plastic actuator mouthpiece, which removes atomized particles that would otherwise deposit in the oropharynx (Ahrens et al. 1995; Brown et al. 1990). The continuing improvement in hardware technology provides ease of use for the patient, making the pMDI such a useful form of drug delivery.

11.7.3 FORMULATION

pMDIs currently on the market are either formulated as solution or suspension products and contain drug, propellant, and other various excipients. Whether a pMDI is a solution or suspension is determined on the solubility characteristics of the active ingredient. Importantly, the physical and chemical stability of the formulation must also be considered (Myrdal et al. 2014).

Solution pMDIs have the advantage of providing a homogeneous formulation, allowing for consistent dose sampling and, subsequently, increased patient compliance (Smyth 2003). However, the transition from CFC to HFA propellants has rendered several drugs and excipients insoluble within the new formulation parameters (Noakes 2002). To bypass this difficulty, cosolvents are frequently added to HFA propellant formulations (Vervaet and Byron 1999). The most commonly utilized cosolvent in HFA propellant MDI formulations is ethanol (Myrdal et al. 2014). However, the amount of ethanol used must be taken into account. Although ethanol may be able to significantly affect solubility, it can also negatively affect the product performance of the aerosol (Gupta et al. 2003). In addition to ethanol, several other solubilization aids have been investigated to increase the solubility of a liquid formulation, including surfactants, micellar solubilization, hydrophobic and hydrophilic counter ions, and cyclodextrins (Myrdal et al. 2014).

Suspension HFA formulations are heterogeneous formulations that require the drug to have ideally very low solubility within the propellant system (Myrdal et al. 2014). The potential disadvantage of these formulation systems is the possibility of inconsistent dosing. The contents of a suspension formulation can phase separate, flocculate, and/or agglomerate if the canister has been left untouched for a prolonged period of time (O'Donnell and Williams 2013). Therefore, patients are required to shake their inhaler prior to each use to increase drug uniformity, decreasing patient compliance in comparison to solution formulations (Smyth 2003). Surfactants can be used in suspension formulations for the prevention of valve sticking and decreasing the interaction between drug particles that causes drug particle agglomeration (Myrdal et al. 2014). Besides surfactants, bulking agents used in combination with cosolvents provide another method to prevent dosing variability (Myrdal et al. 2014).

Although the CFC propellant to HFA propellant transition has produced several formulation challenges, current marketed pMDIs are a more efficient DDS. A future avenue of pMDI delivery lies in the development of combination therapy, where drugs in solution and suspension are both contained in one formulation. Figure 11.3 shows the components of two example marketed pMDIs.

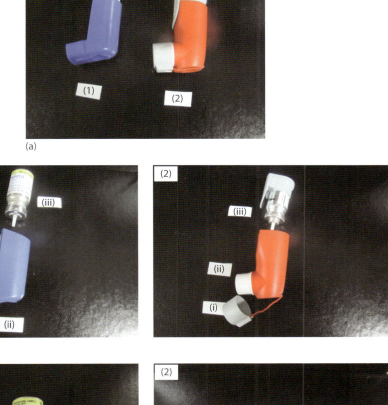

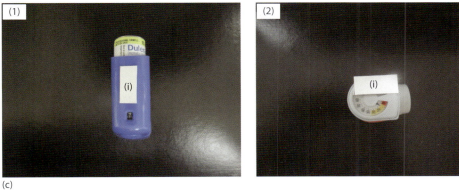

FIGURE 11.3 Examples of pressurized metered-dose inhalers. (a)(1) Dulera® pMDI and (a)(2) Symbicort® pMDI. (b)(1) Dulera® pMDI and (b)(2) B-Symbicort® pMDI with component parts: (i) cap, (ii) device mouthpiece, and (iii) canister containing formulation. Dose counter (i) for (c)(1) Dulera® pMDI and (c)(2) Symbicort® pMDI.

11.7.4 FUTURE DIRECTIONS IN pMDIs

Looking forward, advances in the field of pMDI drug delivery lie in the development of formulation and hardware technologies, combination formulations, and dose counters, to increase patient compliance. An example of formulation advances is PulmoSpheres®, which form a suspension of porous microspheres that are permeated by the propellant, creating a more evenly dense and stable formulation, in comparison to conventional micronized drug suspension formulations (Dellamary et al. 2000; Tarara et al. 2004). The permeation of the propellant into the core of the particles allows for a more homogeneously dense formulation, creating a formulation that is more stable and less prone to creaming or particle settling by reducing attractive forces between particles (Dellamary et al. 2000; Tarara et al. 2004; Weers et al. 2004). Increased formulation stability affords an increased dose uniformity for longer periods of time between vial agitation and shaking (Duddu et al. 2002; Tarara et al. 2004).

Current therapeutic guidelines suggest the simultaneous use of multiple drugs from different classes for patients suffering from asthma and COPD to maximally treat their clinical conditions (Nannini et al. 2007). It is recommended that patients use corticosteroids for long-term relief, along with quick acting beta-agonists to manage their chronic pulmonary disease (Nannini et al. 2007; O'Callaghan et al. 2002). The problem with the addition of multiple inhaled drug therapies to a patient's medication regimen is that it adds confusion and increased opportunity of poor adherence. Therefore, the ability to combine drug therapies in one pMDI provides great benefit to patients. Difficulties arise when attempting to formulate multiple drugs in one formulation (drugs in both solution and suspension, or two drugs in suspension). Solubility, chemical and physical stability, and drug–drug interactions are only a few of the challenges faced with combination therapy (Adi et al. 2012; Rogueda et al. 2011). Combination therapy is still being extensively researched but is a potentially monumental advancement in inhalation therapy.

Another difficulty arises when patients must determine when to replace their inhaler (Holt et al. 2005; Sheth et al. 2006). Not only is it impossible to visually determine how much of the formulation is remaining in an aluminum canister, but the last few doses of a suspension formulation can also be highly variable (Given et al. 2013). For those patients suffering with an acute asthma exacerbation, an empty or erratic inhaler could be a potentially life-threatening problem. This concern has led to the innovation of the dose counter, which is built into the plastic actuator and is a physical indicator of how many doses are left within each device (Sheth et al. 2006). The addition of the dose counter to the inhaler allows the patient ample time to obtain a replacement inhaler (Ogren et al. 1995). Other content indicators have been developed, including plastic glass vials (similar to those used in the research setting), but are not as precise or easily interpreted by the patient in comparison to dose counters (Stein et al. 2014). Table 11.3 lists the currently FDA-approved pMDI products.

11.8 DPIs

DPIs are widely used for the localized delivery of various types of drugs to the airways, for the treatment of specific disorders of the airways. These devices are also increasingly being investigated as a needle-free way of delivering therapeutics to the systemic circulation. Both applications are described here.

11.8.1 DPIs FOR LOCAL LUNG ACTION

In 2010, the DPI market was $6.6 billion (BBC 2014), reached $17.5 billion in 2013 (Marketwatch 2014), and is forecasted to reach $31.5 billion in 2018 (Marketwatch 2014). Pulmonary diseases that are currently treated by marketed DPI products are primarily asthma and COPD but also include pulmonary bacterial infection (e.g., TOBI® Podhaler®), pulmonary viral infection (e.g., Relenza® Diskhaler®), and mucolytic osmotic hydration in CF (e.g., Bronchitol®). Most DPI marketed products

TABLE 11.3

Currently Marketed HFA pMDI Products (FDA Approved)

pMDI Product Name	Drug(s), USP	Propellant/Excipients	Manufacturer
Ventolin HFA®	Albuterol sulfate	HFA 134a	GlaxoSmithKline
Symbicort® (dual-drug combination)	Budesonide + formoterol fumarate dihydrate	HFA 227/povidone K25, polyethylene glycol 1000NF	AstraZeneca
Advair HFA® (dual-drug combination)	Salmeterol xinafoate + fluticasone propionate	HFA 134a	GlaxoSmithKline
Dulera® (dual-drug combination)	Mometasone furoate + formoterol furoate dihydrate	HFA 227/anhydrous alcohol, oleic acid	Merck
Proventil HFA®	Albuterol sulfate	HFA 134a/ethanol, oleic acid	Schering-Plough
Flovent HFA®	Fluticasone propionate	HFA 134a	GlaxoSmithKline
QVAR®	Beclomethasone dipropionate	HFA 134a/ethanol	Ivax
Alvesco®	Ciclesonide	HFA 134a/ethanol	Nycomed
ProAir HFA®	Albuterol sulfate	HFA 134a/ethanol	Teva
Atrovent HFA®	Ipratropium bromide	HFA 134a/water, dehydrated alcohol, anhydrous citric acid	Boehringer Ingelheim
Xopenex HFA®	Levalbuterol tartrate	HFA 134a/ethanol, oleic acid	Sunovion Pharmaceuticals, Inc.
Aerospan®	Flunisolide hemihydrate	HFA 134a/ethanol	Forest Pharmaceuticals, Inc.
Asmanex HFA®	Mometasone furoate	HFA 227/ethanol, oleic acid	Merck & Co., Inc.

Sources: Chan, J.G. et al., *AAPS PharmSciTech*, 15(4), 882, 2014; Sheth, P., Theoretical and experimental behavior of suspension pressurized metered dose inhalers, Dissertation, The University of Arizona, Tuscon, AZ, 2014.

are single-drug formulations that consist of jet-milled drug that is physically blended with large nonrespirable lactose monohydrate crystalline carrier particles, which are typically in the aerodynamic size range of 70–100 μm. The process of jet-milling renders the drug particles into the respirable aerodynamic size range, typically in the range of 2.5–5 μm for currently marketed DPI jet-milled products. During aerosolization, the large nonrespirable lactose monohydrate carrier particles impact the back of the throat through inertial impaction and are then swallowed to the GI tract. Interfacial forces acting between DPI particles (i.e., interparticulate interactions) include mechanical interlocking (i.e., structural cohesion due to solid surface irregularities), van der Waals interactions, electrostatic interactions, and capillary force due to capillary condensation (Hickey et al. 2007; Xu et al. 2011).

Lactose monohydrate carrier-free DPI formulations are made when chemical incompatibility between the drug and lactose monohydrate exists, such as chemical degradation through the Maillard solid-state decomposition reaction. In addition, lactose monohydrate carrier-free formulations are made for patients who are lactose-intolerant or who have an allergic reaction to lactose, which can be attributed to the presence of residual protein on the surface of lactose particles.

The number of dual-drug DPI products continues to grow. There are several marketed dual-drug DPI products (Advair® Diskus®, Breo® Ellipta®, Anoro® Ellipta®, and Symbicort® Turbuhaler®) for asthma and COPD, which consist of individually jet-milled drug particles from two different therapeutic classes that are physically blended with large nonrespirable lactose monohydrate crystalline carrier particles, to form interactive physical mixtures. Advair® Diskus® dual-drug DPI has been the top-selling DPI product for a number of years, used for the management of asthma and COPD. Dual-drug DPIs have been shown to be clinically superior to inhaling two separate single-drug DPI aerosols. This has been shown to be attributed to therapeutic synergy resulting from dual-drug colocalization within the same site of the lungs that dual-drug DPI aerosol products provide.

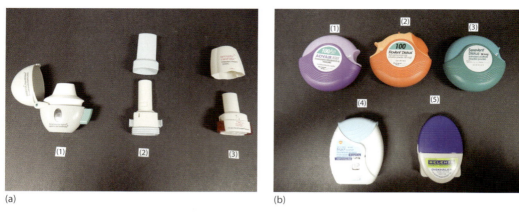

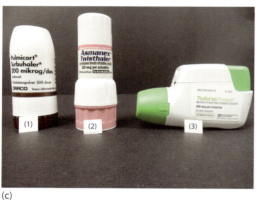

FIGURE 11.4 Images of DPI devices. (a) Unit-dose devices and capsules: (1) Spiriva® Handihaler®, (2) Foradil® Aerolizer®, (3) Arcapta™ Neohaler™. (b) Multi-unit dose devices: (1) Advair® Diskus®, (2) Flovent® Diskus®, (3) Serevent® Diskus®, (4) Breo® Ellipta®, (5) Relenza® Diskhaler®. (c) Multidose devices: (1) Pulmicort® Turbuhaler®, (2) Asmanex® Twisthaler®, (3) Tudorza™ Pressair™. (d) DPI dose counters (arrowed).

DPI devices are versatile due to the various types of metering systems available (Muralidharan et al. 2015a). DPI devices are classified based on their metering system (Figure 11.4), which can be divided into three main categories: unit-dose (Figure 11.4a), multi-unit-dose (Figure 11.4b), and multidose (Figure 11.4c) DPI devices. Dose counters are also engineered into DPI devices to aid patients (Figure 11.4d).

A unit-dose DPI device (Figure 11.4a) is a unit-dose capsule-based DPI device that employs a capsule that is premetered and preloaded with the powder formulation. The patient inserts the pre-loaded premetered capsule into the capsule chamber of the DPI device immediately prior to administration. The capsule is then pierced by needles existing within the DPI device itself. The resulting small holes in the pierced capsule provide the exit points for the aerosolized powder formulation. The presence of 1–2 grids composed of stainless steel or plastic within the device provides additional shearing forces for drug stripping. The mass range is typically 25–30 mg of drug-excipient formulation that is preloaded into one capsule. The preloaded and premetered capsules are individually sealed in individual aluminum foil blister packs by the manufacturer. The capsules used are inhalation-grade capsules. The polymer used in inhalation-grade capsules is either HPMC (hydroxypropyl methylcellulose) or hard gelatin. Example unit-dose capsule-based DPI devices include the Podhaler®, Rotahaler®, Inhalator®, Handihaler®, Aerolizer®, Cyclohaler®, and Neohaler®.

A multi-unit-dose DPI device (Figure 11.4b) consists of a foiled blister strip that is preloaded with individually sealed premetered doses. Example DPI devices using this mechanism include the Diskus®, Ellipta®, and Diskhaler®. Breo Ellipta® DPI contains individually foiled blister unit doses and excipients of nonrespirable lactose monohydrate carrier and magnesium stearate. Advair Diskus® contains individually foiled blister strip unit doses and nonrespirable lactose monohydrate carrier excipient but no magnesium stearate.

A multidose DPI (Figure 11.4c) device consists of a powder reservoir (i.e., powder bed) that is repeatedly sampled during dose metering with each use by the patient. Since the powder bed is repeatedly exposed to moisture ingress during patient use, multiuse powder bed devices contain desiccant within the dial base of the device to absorb moisture introduced through patient use. Examples of multidose powder reservoir DPI devices include the Twisthaler®, Flexhaler®, Pressair®, and Turbuhaler®.

11.8.2 DPIs for Systemic Delivery

Using the pulmonary route for systemic action is associated with the key advantages that it is non-invasive and can produce nearly instantaneous absorption. The first DPI device to be introduced for systemic action, Exubera® (developed by Nektar and marketed by Pfizer), was used for the delivery of rh insulin, stabilized in a trehalose amorphous glass inhalable powder. The launch of Exubera marked the first FDA-approved inhaled protein pharmaceutical product, as well as the first FDA-approved inhaled product for systemic action via the pulmonary route of delivery. The DPI device was a compressed-air active device that generated the aerosol independent of the patient's inspiratory force. The product was voluntarily withdrawn by Pfizer due to insufficient profit margins and the large cumbersome "bong" device perceived by patients.

Table 11.4 lists the currently FDA-approved DPI products for systemic action, administered via the pulmonary route as inhaled aerosols. It is important to note that these formulations do not contain nonrespirable lactose monohydrate carrier but are "carrier-free."

In 2014, a new inhaled rh insulin DPI product was approved by the FDA for the indication of both type I and II diabetes (Afrezza®). The onset was clinically shown to be faster than conventional subcutaneous (s.c.) insulin injection. In addition, glucose control was clinically shown to be longer than for conventional s.c. insulin injection. The unit-dose cartridge DPI device, the Dreamboat®, is a breath-actuated passive DPI device, of very small in size (Figure 11.5a). The insulin formulation uses Technosphere® particle technology (Figure 11.5b), which utilizes fumaryl diketopiperazine (FDKP), a small organic molecule that forms crystalline plates that self-assemble into microparticles bearing a net negative electrostatic surface charge (Leone-Bay and Grant 2008). Monomeric insulin is precipitated and has a net positive charge, which interacts with the surface of the FDKP particles through electrostatic attraction (Leone-Bay and Grant 2008). These particles efficiently

TABLE 11.4
Pulmonary Products for Systemic Action by Pulmonary Inhalation Aerosol Delivery

DPI Product Name	Device	Drug(s), USP	Therapeutic Drug Class	Manufacturer
Adasuve®	Staccato® device (thermally generated drug vapor condensation aerosol)	Loxapine	Antipsychotic	Teva/Alexza Pharmaceuticals
Afrezza®	Dreamboat DPI device (unit-dose cartridge)	rh Insulin with Technosphere® technology	Ultrarapid-acting mealtime endocrine hormone for diabetes mellitus I and II	Mannkind Corp

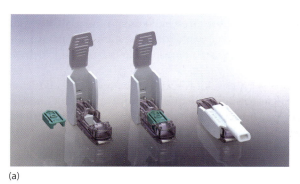
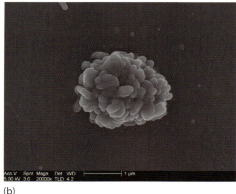

(a)   (b)

FIGURE 11.5   (a) The Dreamboat® passive DPI device and (b) an electron micrograph of Technosphere® particles technology. (Courtesy of Mannkind Corporation.)

reach the deep lung where they rapidly dissolve and are quickly absorbed (faster than injection) as monomeric insulin into the systemic blood circulation (Leone-Bay and Grant 2008).

Adasuve® is a new aerosol product for the noninvasive systemic delivery to the CNS, via the pulmonary route. This FDA-approved DPI utilizes the Staccato® device, which is a breath-actuated passive DPI device that generates the aerosol particles based on the patient's inspiratory force. The device contains a thin solid film of pure drug (the antipsychotic, loxapine) that is rapidly heated to vaporize (i.e., drug sublimation). It then rapidly condenses to inhalable particles. The inhalable particles readily reach the deep lung alveolar region efficiently, where they dissolve and are rapidly absorbed into the systemic circulation for subsequent delivery to the CNS for antipsychotic therapy.

11.9   SMIs

SMIs actively generate a slow-moving aerosol mist that is propellant-free. The power generated is mechanical energy, produced by a compressed spring mechanism inside the device. In addition to being an active device, the device is very small and conveniently portable for patients. Generation of the aerosol is independent of the patient's IRF. The aerosol is a slow-moving aerosol cloud of small fine droplets (i.e., "soft mist") with greater than 60% weight smaller than 5 µm (Wachtel and Moser 2008). Consequently, enhanced lung deposition can be achieved with less oropharyngeal (i.e., mouth/throat) deposition and higher FPFs than with pMDIs and DPIs (Dalby et al. 2004).

The long generation time of the aerosol cloud over 1.5 seconds facilitates inhalation–hand coordination (Dalby et al. 2004). The soft mist is first generated by a capillary tube that meters and draws the drug solution upward to the nozzle outlets that produce two liquid jets (Wachtel and Moser 2008). Two impinging fast-moving liquid jets collide at a fixed angle, which then disintegrate into inhalable fine droplets with reduced velocity (Wachtel and Moser 2008).

Water, ethanol, or a water/ethanol cosolvent can be used for dissolving the drug used in SMIs (Wachtel and Moser 2008). Higher FPF values and lower MMAD values are achieved with ethanol solutions in the SMI, whereas aqueous solutions result in relatively lower FPF values and higher MMADs (Wachtel and Moser 2008). Since the SMI is a multidose device containing an aqueous liquid, preservatives such as BAC and EDTA are necessary to prevent bacterial growth. Table 11.5 lists the SMI products that are currently FDA-approved for the treatment of asthma COPD.

TABLE 11.5

Soft Mist Inhaler Pharmaceutical Products (FDA Approved) for Asthma and Chronic Obstructive Pulmonary Disease Therapy

Product Name	Drug(s), USP	Therapeutic Drug Class	Manufacturer
Combivent® Respimat® (dual-drug combination)	Albuterol sulfate + ipratropium bromide	SABA + short-acting anticholinergic	Boehringer Ingelheim
Spiriva® Respimat®	Tiotropium bromide	Long-acting anticholinergic	Boehringer Ingelheim
Striverdi® Respimat®	Olodaterol	LABA	Boehringer Ingelheim
Stiolto® Respimat®	Tiotropium bromide/olodaterol	Long-acting anticholinergic + LABA	Boehringer Ingelheim

Abbreviations: SABA, short-acting β_2 adrenergic agonist (short-acting bronchodilator); LABA, long-acting β_2 adrenergic agonist (long-acting bronchodilator).

11.10 PULMONARY NANOMEDICINE

The importance of particle size and size distribution of inhaled aerosol has been emphasized in this chapter. It is only logical to consider submicron-sized aerosol particles in drug delivery to the lungs. The incorporation of nanotechnology into inhalation aerosol technologies has given rise to the developing field of pulmonary nanomedicine.

Commercial examples include the surfactant replacement products (for example, Surfaxin®, discussed in Section 11.5), which comprise phospholipid self-assembly nanocarriers (Mansour and Park 2013). In addition to RDS, the targeted treatment of many other pulmonary diseases (e.g., bronchiolitis obliterans syndrome in lung transplant recipients, pulmonary hypertension, pulmonary fibrosis, radiation-induced pneumonitis seen in lung cancer patients, certain lung cancers, and highly resistant lung infections) is becoming a clinical reality through pulmonary nanomedicine.

Pulmonary nanomedicine offers the possibility of improved systemic delivery of therapeutics, particularly for challenging drugs such as the new "biologics," including proteins and gene-based therapies. A commercial example is the Technosphere® technology (Section 11.8.2), which creates inhalable nanoaggregate aerosol particles in the solid-state. Utilizing a breath-actuated DPI device, these particles efficiently reach the alveolar region producing rapid systemic absorption of insulin, for the noninvasive treatment of both type I and II diabetes.

Various types of micro- and nanocarriers are described in detail in Chapter 5 for parenteral drug delivery and targeting. Many of these nanocarrier DDS are also being investigated for the pulmonary route (Mansour et al. 2009, 2010, 2011, 2013; Muralidharan et al. 2015b; Willis et al. 2012). In particular, polymeric- and lipid-based formulations have gained popularity for pulmonary drug delivery, due to their increased biocompatibility, sustained-release properties, and decreased tissue toxicity. However, the choice of polymer used in the carrier is critical, in order to decrease bioaccumulation and burden on excretion.

Lipids endogenous to the lungs, such as certain types of phospholipids, are a biomimetic choice for nanocarriers for pulmonary drug delivery. Phospholipid nanocarriers, which are inherently biocompatible and biodegradable, can be administered as instilled suspensions, as, for example, in the various lung surfactant replacement products for RDS (Section 11.5). Liposomes are a thermodynamically stable system that can also be formulated as a dry powder, for use in DPIs. Solid-state liposome, where it loses water, is called liposphere or proliposome; several drugs have now been successfully formulated into inhalable proliposomal dry powders (Willis et al. 2012). PEGylated liposomes have been developed as DPIs for the delivery of various drugs, including paclitaxel, a first-line lung cancer chemotherapeutic drug, for use in targeted lung cancer treatment

(Meenach et al. 2013; Muralidharan et al. 2014). Proteins such as insulin can also be conjugated to PEG and used for inhalation as a dry powder. Inhalation of PEGylated insulin has been shown to achieve both efficient absorption across lung epithelia and extended serum concentrations (Adis International 2004).

11.11   PULMONARY VACCINATION

There has been growing public health interest in recent years in using inhaled aerosols as needle-free vaccines for the prevention of infectious diseases, particularly for bacterial and viral infectious diseases that readily spread and infect via the respiratory route. Specifically, the rationale is respiratory mucosal immunization by directly targeting the respiratory-associated lymphoid tissues (RALT) and the avoidance of systemic side effects that have been associated with injection-based vaccines. In particular, the use of DPIs has been investigated for pulmonary vaccine delivery, due to the ease of mass administration without needing to reconstitute the formulation; enhanced stability, particularly at relatively high temperatures, of a solid powder formulation; mass distribution capability; lack of cold chain manufacturing; and the ability to protect the formulation from moisture ingress through DPI device innovation.

Pulmonary vaccination has been demonstrated for the prevention of TB using the DPI BCG (Bacillus Calmette–Guérin) vaccine. TB mycobacteria infect through the lungs and, in particular, remain latent within alveolar macrophages. Hence, targeting the lungs and the alveolar macrophages, in particular, is a viable strategy. The efficacy of inhaled aerosols for pulmonary vaccination has been demonstrated for other infectious diseases including RSV, malaria, pertussis, diphtheria, hepatitis, measles, influenza, tularemia, plague, pneumonia, and anthrax. Pulmonary vaccination is discussed further in Chapter 17 (Section 17.4.3.2).

11.12   CONCLUSIONS AND FUTURE DIRECTIONS

Inhalation aerosol medicine continues to grow and future forecast is predicted to be ≈$44 billion globally by 2018. Scientific advancements have enabled a wide variety of pulmonary diseases beyond asthma and COPD to be effectively treated in a targeted manner through inhalation aerosol delivery. Such pulmonary diseases include pulmonary infectious diseases, pulmonary fibrosis, pulmonary hypertension, CF, lung transplant immunosuppression, and lung cancer. The pulmonary route is also emerging as a viable route for systemic delivery; commercial products are now on the market for the systemic delivery of APIs (including insulin) via pulmonary inhalation. In addition, pulmonary vaccination using inhalation aerosols is a growing public health interest and an active area of promising research.

Given the multifactorial nature of pulmonary diseases, combination drug aerosol products containing drugs from different therapeutic classes in the same aerosol continue to grow and continue to be the top-selling aerosol products due to the demonstrated clinical superiority in lung disease management. Triple drug combination DPIs (GlaxoSmithKline/Theravance) are currently in clinical trials and more are being planned. In addition, "fusion drugs" are being developed as "first-in-class" therapeutics and tested as inhaled DPIs for the treatment and prevention of asthma and COPD.

New long-acting pulmonary therapeutics are currently available as once-daily inhaled aerosol products, leading to improved patient compliance and pulmonary disease state management, as well as a reduction in overall health-care costs. More once-daily inhalation aerosol products are expected to be approved in the near future. In addition, controlled release formulations for sustained drug release can be expected. Nanotechnology aerosol products are currently on the market for local and also for systemic action, noninvasively through the pulmonary route. Given the advantages, technological advancements, and medical needs, one can expect more such products to appear in the future.

Optimal performance of the pMDI requires both hardware and formulation compliance to produce dose uniformity, an ideal particle size distribution for lung deposition and chemical and physical stability. Since its development in 1956, several innovations to the hardware and formulation of the pMDI have occurred leading to its increasing popularity as a route for pulmonary therapy. The transition of CFC propellant to HFA propellant has provided a more environmentally friendly device, along with providing a better understanding of solubility and stability of various drugs and excipients within the HFA propellant system and how these changes affect hardware functionality. Other innovations include formulation advances (for example, PulmoSpheres®, which provide a more stable and uniform formulation), while combination drug therapy allows multiple therapy in one device, both aiding to increase patient compliance. Modifications to the MDI actuators such as spacers have provided an improvement in drug delivery efficiency, and the addition of dose counters now allows patients a better understanding of when to replace their inhaler. These innovations will continue to contribute to the affordability, convenience, and success of the pMDI.

REFERENCES

Adi, H., P.M. Young, and D. Traini. 2012. Co-deposition of a triple therapy drug formulation for the treatment of chronic obstructive pulmonary disease using solution-based pressurised metered dose inhalers. *J Pharm Pharmacol* 64:1245–1253.

Adis International Ltd. Insulin inhalation—Pfizer/Nektar therapeutics: HMR 4006, inhaled PEG-insulin—Nektar, PEGylated insulin—Nektar. 2004. *Drugs R&D* 5(3):166–170.

Ahrens, R., C. Lux, T. Bahl et al. 1995. Choosing the metered-dose inhaler spacer or holding chamber that matches the patients need—Evidence that the specific drug being delivered is an important consideration. *J Allergy Clin Immun* 96:288–294.

Altiere, R.J. and D.C. Thompson. 2007. Physiology and pharmacology of the airways. In *Inhalation Aerosols: Physical and Biological Basis for Therapy*, 2nd ed., A.J. Hickey (ed.), pp. 83–126. New York: Informa Healthcare USA, Inc.

Anandarajan, M., S. Paulraj, and R. Tubman. 2009. Abca3 deficiency: An unusual cause of respiratory distress in the newborn. *Ulster Med J* 78:51–52.

Avery, M.E. and J. Mead. 1959. Surface properties in relation to atelectasis and hyaline membrane disease. *Am J Dis Child* 97:517–523.

Barnes, P., J. Drazen, S.I. Rennard et al. 2009. *Asthma and COPD: Basic Mechanisms and Clinical Management*, 2nd ed. San Diego, CA: Academic Press/Elsevier.

BCC market research report. 2014. P.D.D.S.T.a.G.M. http://www.bccresearch.com/market-research/healthcare/pulmonary-drug-delivery-systems-hlc094a.html. Accessed July 1, 2014.

Bend, J.R., C.J. Serabjit-Singh, and R.M. Philpot. 1985. The pulmonary uptake, accumulation, and metabolism of xenobiotics. *Ann Rev Pharmacol Toxicol* 25:97–125.

Blaiss, M.S. 2007. Part II: Inhaler technique and adherence to therapy. *Curr Med Res Opin* 23 (Suppl. 3):S13–S20.

Brown, P.H., G. Blundell, A.P. Greening et al. 1990. Do large volume spacer devices reduce the systemic effects of high-dose inhaled corticosteroids? *Thorax* 45:736–739.

Bullard, J.E., S.E. Wert, and L.M. Nogee. 2006. Abca3 deficiency: Neonatal respiratory failure and interstitial lung disease. *Semin Perinatol* 30:327–334.

Chan, J.G., J. Wong, Q.T. Zhou et al. 2014. Advances in device and formulation technologies for pulmonary drug delivery. *AAPS PharmSciTech* 15(4):882–897.

Chandra, T., I.F. Miller, and D.B. Yeates. 1992. A pore transport model for pulmonary alveolar epithelium. *Ann Biomed Eng* 20:481–494.

Clements, J.A., R.F. Hustead, R.P. Johnson et al. 1961. Pulmonary surface tension and alveolar stability. *J Appl Physiol* 16:444–450.

Cochrane, C. 1998. Surfactant protein b and mimic peptides in the function of pulmonary surfactant. *FEBS Lett* 430:424.

Cochrane, C., S. Revak, T. Merritt et al. 1996. The efficacy and safety of kl4-surfactant in preterm infants with respiratory distress syndrome. *Am J Respir Crit Care Med* 153:404–410.

Cochrane, C.G. and S.D. Revak. 1991. Pulmonary surfactant protein b (sp-b): Structure–function relationships. *Science* 254:566–568.

Crompton, G.K. 1982. Problems patients have using pressurized aerosol inhalers. *Eur J Respir Dis: Suppl* 119:101–104.

Dalby, R.N., M. Spallek, and T.A. Voshaar. 2004. Review of the development of respimat® soft mist™ inhaler. *Int J Pharm* 283:1–9.

Dalby, R.N., S.L. Tiano, A.J. Hickey et al. 2007. Medical devices for the delivery of therapeutic aerosols to the lungs. In *Inhalation Aerosols: Physical and Biological Basis for Therapy*, 2nd ed., A.J. Hickey (ed.), pp. 417–444. New York: Informa Healthcare USA, Inc.

Dellamary, L.A., T.E. Tarara, D.J. Smith et al. 2000. Hollow porous particles in metered dose inhalers. *Pharm Res* 17:168–174.

Devereux, T.R., B.A. Domin, and R.M. Philpot. 1989. Xenobiotic metabolism by isolated pulmonary cells. *Pharmacol Therap* 41:243–256.

Dolovich, M., R.E. Ruffin, R. Roberts et al. 1981. Optimal delivery of aerosols from metered dose inhalers. *Chest* 80:911–915.

Duddu, S.P., S.A. Sisk, Y.H. Walter et al. 2002. Improved lung delivery from a passive dry powder inhaler using an engineered pulmosphere® powder. *Pharm Res* 19:689–695.

Dunbar, C.A., A.P. Watkins, J.F. Miller et al. 1997. An experimental investigation of the spray issued from a pMDI using laser diagnostic techniques. *J Aerosol Med—Depos Clearance Effects Lung* 10:351–368.

Edwards, D.A., A. Ben-Jebria, and R. Langer. 1998. Recent advances in pulmonary drug delivery using large, porous inhaled particles. *J Appl Physiol* 85(2):379–385.

Given, J., H. Taveras, H. Iverson et al. 2013. Prospective, open-label assessment of albuterol sulfate hydrofluoroalkane metered-dose inhaler with new integrated dose counter. *Allergy Asthma Proc* 34:42–51.

Gupta, A., S.W. Stein, and P.B. Myrdal. 2003. Balancing ethanol cosolvent concentration with product performance in 134a-based pressurized metered dose inhalers. *J Aeros Med: Offic J Int Soc Aerosols Med* 16:167–174.

Hickey, A.J. and H.M. Mansour. 2009. Delivery of drugs by the pulmonary route. In *Modern Pharmaceutics*, 5th ed., A.T. Florence and J. Siepmann (eds.), pp. 191–219. New York: Taylor & Francis.

Hickey, A.J., H.M. Mansour, M.J. Telko et al. 2007. Physical characterization of component particles included in dry powder inhalers. I. Strategy review and static characteristics. *J Pharm Sci* 96:1282–1301.

Hickey, A.J. and D.C. Thompson. 2004. Physiology of the airways. In *Pharmaceutical Inhalation Aerosol Technology*, A.J. Hickey (ed.), pp. 1–29. New York: Marcel Dekker, Inc.

Holt, S., A. Holt, M. Weatherall et al. 2005. Metered dose inhalers: A need for dose counters. *Respirology* 10:105–106.

Knoch, M. and W.H. Finlay. 2008. Nebulizer technologies. In *Modified—Release Drug Delivery Technology*, 2nd ed., M.J. Rathbone, J. Hadgraft, M.S. Roberts et al. (eds.), pp. 613–621. New York: Informa Healthcare USA. Inc.

Larsen, J.S., M. Hahn, B. Ekholm et al. 1994. Evaluation of conventional press-and-breathe metered-dose inhaler technique in 501 patients. *J Asthma* 31:193–199.

Leone-Bay, A. and M. Grant. 2008. Technosphere/insulin: Mimicking endogenous insulin release. In *Modified-Release Drug Delivery Technology*, 2nd ed., M.J. Rathbone, J. Hadgraft, M.S. Roberts et al. (eds.), pp. 673–679. New York: Informa Healthcare USA, Inc.

Lewis, J.F. and A.H. Jobe. 1993. Surfactant and the adult respiratory distress syndrome. *Am Rev Respir Dis* 147:218–233.

Ma, J., S. Koppenol, H. Yu et al. 1998. Effects of a cationic and hydrophobic peptide, kl4, on model lung surfactant lipid monolayers. *Biophys J* 74:1899–1907.

Mansour, H.M., P. Chun-Woong, and D. Hayes, Jr. 2013. Nanoparticle lung delivery and inhalation aerosols for targeted pulmonary nanomedicine. In *Nanomedicine in Drug Delivery*, p. 445. Boca Raton, FL: CRC Press, Taylor & Francis.

Mansour, H.M., S. Damodaran, and G. Zografi. 2008. Characterization of the *in situ* structural and interfacial properties of the cationic hydrophobic heteropolypeptide, kl4, in lung surfactant bilayer and monolayer models at the air–water interface: Implications for pulmonary surfactant delivery. *Mol Pharm* 5:681–695.

Mansour, H.M. and C.W. Park. 2013. Therapeutic applications and targeted delivery of nanomedicines and nanopharmaceutical products. In *The Clinical Nanomedicine Handbook*, S.A. Brenner (ed.), pp. 321–338. London, U.K.: CRC Press/Taylor & Francis.

Mansour, H.M., Y.S. Rhee, C.W. Park et al. 2011. Lipid nanoparticulate drug delivery and nanomedicine. In *Lipids in Nanotechnology*, A. Moghis (ed.), pp. 221–268. Urbana, IL: American Oil Chemists Society (AOCS) Press.

Mansour, H.M., Y.S. Rhee, and X. Wu. 2009. Nanomedicine in pulmonary delivery. *Int J Nanomed* 4:299–319.

Mansour, H.M., M. Sohn, A. Al-Ghananeem et al. 2010. Materials for pharmaceutical dosage forms: Molecular pharmaceutics and controlled release drug delivery aspects. Invited paper. *Int J Mol Sci: Spec Issue— Mater Sci Nanotechnol Sect—Biodegrad Mater* 11:3298–3322.

Marketwatch. 2014. BCC market research report. P.D.D.S.T.a.G.M. http://www.marketwatch.com/story/pulmonary-drug-delivery-systems-technologies-and-global-markets-2014-06-10. Accessed July 1, 2014.

McCulloch, A. 1999. Cfc and halon replacements in the environment. *J Fluorine Chem* 100:163–173.

McDonald, K.J. and G.P. Martin. 2000. Transition to cfc-free metered dose inhalers—Into the new millennium. *Int J Pharm* 201:89–107.

Meenach, S.A., K.W. Anderson, Z. Hilt et al. 2013. Characterization and aerosol dispersion performance of advanced spray-dried chemotherapeutic PEGylated phospholipid particles for dry powder inhalation delivery in lung cancer. *Eur J Pharm Sci* 49:699–711.

Molina, M.J. and F.S. Rowland. 1974. Stratospheric sink for chlorofluoromethanes: Chlorine atom-catalysed destruction of ozone. *Nature* 249:810–812.

Morley, C.J., A.D. Bangham, N. Miller et al. 1981. Dry artificial lung surfactant and its effect on very premature babies. *Lancet* 1:64–67.

Muralidharan, P., D. Hayes, Jr., and H.M Mansour. 2015a. Dry powder inhalers in COPD, lung inflammation and pulmonary infections. *Expert Opin Drug Deliv* 12:947–962.

Muralidharan, P., M. Malapit, E. Mallory et al. 2015b. Inhalable nanoparticulate powders for respiratory delivery. *Nanomed: Nanotechnol Biol Med* 11:1189–1199.

Muralidharan, P., E. Mallory, M. Malapit et al. 2014. Inhalable pegylated phospholipid nanocarriers and pegylated therapeutics for respiratory delivery as aerosolized colloidal dispersions and dry powder inhalers. *Pharmaceutics* 6:333–353.

Murphy, A. 2007. Drug delivery to the lungs. In *Asthma in Focus*, A. Murphy (ed.), pp. 113–132. London, U.K.: Pharmaceutical Press/Royal Pharmaceutical Society of Great Britain Publishing.

Myrdal, P.B. and B.S. Angersbach. 2008. Pulmonary delivery of drugs by inhalation. In *Modified-release Drug Delivery Technology*, 2nd ed., M.J. Rathbone, J. Hadgraft, M.S. Roberts et al. (eds.), pp. 553–562. New York: Informa Healthcare USA, Inc.

Myrdal, P.B., P. Sheth, and S.W. Stein. 2014. Advances in metered dose inhaler technology: Formulation development. *AAPS PharmSciTech* 15:434–455.

Nannini, L., C.J. Cates, T.J. Lasserson et al. 2007. Combined corticosteroid and long-acting beta-agonist in one inhaler versus placebo for chronic obstructive pulmonary disease. *Cochrane Database Syst Rev* Issue 4. Art no.CD003794. DOI:10.1002/14651858.CD003794.pub3.

Newman, S.P. 2005. Inhaler treatment options in COPD. *Eur Res Rev* 14:102–108.

Newman, S., P. Anderson, P. Byron et al. 2009. *Respiratory Drug Delivery: Essential Theory and Practice*, p. 388. Richmond, VA: Respiratory Drug Delivery Online/Virginia Commonwealth University.

Newman, S.P., A.W.B. Weisz, N. Talaee et al. 1991. Improvement of drug delivery with a breath actuated pressurized aerosol for patients with poor inhaler technique. *Thorax* 46:712–716.

Niven, R.W. and A.J. Hickey. 2007. Atomization and nebulizers. In *Inhalation Aerosols: Physical and Biological Basis for Therapy*, 2nd ed., A.J. Hickey (ed.), pp. 253–283. New York: Informa Healthcare USA, Inc.

Noakes, T. 2002. Medical aerosol propellants. *J Fluorine Chem* 118:35–45.

Nogee, L., D. de Mello, L. Dehner et al. 1993. Deficiency of pulmonary surfactant protein b in congenital alveolar proteinosis. *N Engl J Med* 328:406–410.

O'Callaghan, C., M.T. Vidgren, and O. Nerbrink. 2002. The history of inhaled drug therapy. In *Drug Delivery to the Lung*, H. Bisgaard, C. O'Callaghan, and G.C. Smaldone (eds.), pp. 1–20. New York: Marcel Dekker, Inc.

O'Donnell, K.P. and R.O. Williams, 3rd. 2013. Pulmonary dispersion formulations: The impact of dispersed powder properties on pressurized metered dose inhaler stability. *Drug Dev Ind Pharm* 39:413–424.

Ogren, R.A., J.L. Baldwin, and R.A. Simon. 1995. How patients determine when to replace their metered-dose inhalers. *Ann Allergy Asthma Immunol: Offic Public Am Coll Allergy Asthma Immunol* 75:485–489.

Patton, J.S. and P.R. Byron. 2007. Inhaling medicines: Delivering drugs to the body through the lungs. *Nat Rev Drug Discov* 6:67–74.

Rogueda, P.G.A., R. Price, T. Smith et al. 2011. Particle synergy and aerosol performance in non-aqueous liquid of two combinations metered dose inhalation formulations: An AFM and Raman investigation. *J Colloid Interf Sci* 361:649–655.

Sanders, M. 2007. Inhalation therapy: An historical review. *Prim Care Respir J: J Gen Pract Airways Group* 16:71–81.

Schurch, S., J. Goerke, and J.A. Clements. 1976. Determination of surface tension in the lung. *Proc Natl Acad Sci USA* 73:4698–4702.

Sheth, K., R.L. Wasserman, W.R. Lincourt et al. 2006. Fluticasone propionate/salmeterol hydrofluoroalkane via metered-dose inhaler with integrated dose counter: Performance and patient satisfaction. *Int J Clin Pract* 60:1218–1224.

Sheth, P. 2014. Theoretical and experimental behavior of suspension pressurized metered dose inhalers. Dissertation. Tuscon, AZ: The University of Arizona.

Shulenin, S., L.M. Nogee, T. Annilo et al. 2004. Abca3 gene mutations in newborns with fatal surfactant deficiency. *N Engl J Med* 350:1296–1303.

Smith, I.J. 1995. The challenge of reformulation. *J Aeros Med: Offic J Int Soc Aeros Med* 8 (Suppl. 1):S19–S27.

Smyth, H.D.C. 2003. The influence of formulation variables on the performance of alternative propellant-driven metered dose inhalers. *Adv Drug Deliv Rev* 55:807–828.

Stein, S.W., P. Sheth, P.D. Hodson et al. 2014. Advances in metered dose inhaler technology: Hardware development. *AAPS PharmSciTech* 15:326–338.

Stein, S.W. and J.S. Stefely. 2003. Reinventing metered dose inhalers: From poorly efficient CFC MDIs to highly efficient HFA MDIs. *Drug Deliv Technol* 3:46–51.

Stocks, J. and A.A. Hislop. 2002. Structure and function of the respiratory system: Developmental aspects and their relevance to aerosol therapy. In *Drug Delivery to the Lung*, eds. H. Bisgaard, C. O'Callaghan and G. C. Smaldone, pp. 47–104. New York: Marcel Dekker, Inc.

Suarez, S. and A.J Hickey. 2000. Drug properties affecting aerosol behavior. *Respir Care* 45:652–666.

Tarara, T.E., M.S. Hartman, H. Gill et al. 2004. Characterization of suspension-based metered dose inhaler formulations composed of spray-dried budesonide microcrystals dispersed in HFA-134a. *Pharm Res* 21:1607–1614.

Tortora, G.J. and B.H. Derrickson (eds.). 2013. *Principles of Anatomy and Physiology*, 14th ed. Hoboken, NJ: Wiley.

United States Pharmacopoeia and the National Formulary: The Official Compendia of Standards. 2006. Chapter <601>. Aerosols, nasal sprays, metered-dose inhalers, and dry powder inhalers monograph, Vol. 29/24, pp. 2617–2636. Rockville, MD: The United States Pharmacopeial Convention.

U.S. Environmental Protection Agency. 2014. The process of ozone depletion. http://www.epa.gov/ozone/science/process.html. Accessed January 18, 2014.

Valberg, P.A., J.D. Brain, S.L. Sneddon et al. 1982. Breathing patterns influence aerosol deposition sites in excised dog lungs. *J Appl Physiol: Respir Environ Exer Physiol* 53:824–837.

Vervaet, C. and P.R. Byron. 1999. Drug-surfactant-propellant interactions in HFA-formulations. *Int J Pharm* 186:13–30.

Wachtel, H. and A. Moser. 2008. The respimat, a new soft mist inhaler for delivering drugs to the lungs. In *Modified-Release Drug Delivery Technology*, 2nd ed., M.J. Rathbone, J. Hadgraft, M.S. Roberts et al. (eds.), pp. 637–645. New York: Informa Healthcare USA, Inc.

Wallington, T.J., W.F. Schneider, D.R. Worsnop et al. 1994. The environmental-impact of CFC replacements—HFCs and HCFCs. *Environ Sci Technol* 28:A320–A326.

Weers, J.G., A.R. Clark, P. Challoner et al. 2004. High dose inhaled powder delivery: Challenges and techniques. In *Respiratory Drug Delivery IX*, R.N. Dalby, P.R. Byron, J. Peart et al. (eds.), pp. 281–288. River Grove, IL: Davis Healthcare International Publishing, LLC.

Wilkinson, A., P.A. Jenkins, and J.A. Jeffrey. 1985. Two controlled trials of dry artificial surfactant: Early effects and later outcome in babies with surfactant deficiency. *Lancet* 10:287–291.

Willis, L., D. Hayes, Jr. and H.M. Mansour. 2012. Therapeutic liposomal dry powder inhalation aerosols for targeted lung delivery. *Lung* 190:251–262.

Wright, J.R. and J.A. Clements. 1987. Metabolism and turnover of lung surfactant. *Am Rev Respir Dis* 135:426–444.

Wright, J.R. and L.G. Dobbs. 1991. Regulation of pulmonary surfactant secretion and clearance. *Ann Rev Physiol* 53:395–414.

Xu, Z., H.M. Mansour, and A.J. Hickey. 2011. Particle interactions in dry powder inhaler unit processes: A review. *J Adhes Sci Technol* 25:451–482.

FURTHER READING

Bisgaard, H., C. O'Callaghan, and G.C. Smaldone. 2002. *Drug Delivery to the Lung*. New York: Marcel Dekker, Inc.

Finlay, W.H. 2001. *The Mechanics of Inhaled Pharmaceutical Aerosols*. London, U.K.: Academic Press.

Hickey, A.J. 2004. *Pharmaceutical Inhalation Aerosol Technology*, 2nd ed., Revised and Expanded. New York: Marcel Dekker, Inc.

Hickey, A.J. 2007. *Inhalation Aerosols: Physiological and Biological Basis for Therapy*, 2nd ed. New York: Informa Healthcare USA, Inc.

Hickey, A.J. and H.M. Mansour. 2008. Formulation challenges of powders for the delivery of small molecular weight molecules as aerosols. In *Modified-Release Drug Delivery Technology*, Vol. 2, Drugs and the Pharmaceutical Sciences Series, 2nd ed., M.J. Rathbone, J. Hadgraft, M.S. Roberts et al. (eds.), pp. 573–602. New York: Informa Healthcare.

Hickey, A.J. and H.M. Mansour. 2009. Delivery of drugs by the pulmonary route. In *Modern Pharmaceutics*, Vol. 2, Applications and Advances, Drugs and the Pharmaceutical Sciences Series, 5th ed., A.T. Florence and J. Siepmann (eds.), pp. 191–219. New York: Informa Healthcare.

Mansour, H.M., C.W. Park, and D. Hayes, Jr. 2013. Nanoparticle lung delivery and inhalation aerosols for targeted pulmonary nanomedicine. In *Nanomedicine in Drug Delivery*, A. Kumar, H.M. Mansour, A. Friedman et al. (eds.), pp. 43–74. London, U.K.: CRC Press/Taylor & Francis.

Patton, J.S. and P.R. Byron. 2007. Inhaling medicines: Delivering drugs to the body through the lungs. *Nat Rev Drug Discov* 6:67–74.

Zeng, X.M., G.P. Martin, and C. Marriott. 2000. *Particulate Interactions in Dry Powder Formulations for Inhalation*. London, U.K.: CRC Press/Taylor & Francis.

CONTENTS

12.1 INTRODUCTION

Delivery of drugs to the vagina was thought to be limited to topical administration until 1918, when Macht reported systemic absorption of vaginally dosed morphine, atropine, and potassium iodide. Nearly a century later, drug delivery to the female reproductive tract (FRT), most notably the uterus and vagina, has become an essential component of female reproductive health. Previous reviews largely focused on vaginal delivery; however, due to the connectivity of the reproductive organs, we believe the whole FRT should be considered.

A diverse array of active pharmaceutical ingredients (API) has been delivered via the FRT for both topical and systemic effect. Clinically, these most commonly include estrogens, progestins, and anti-infective agents. The HIV prevention field has extensively investigated the prophylactic delivery of antiviral drugs to the FRT since the turn of the twenty-first century, catalyzing innovation in vaginal dosage form design and development. Vaginal delivery has many advantages, including ease of use, painlessness, privacy, reversibility, avoidance of hepatic first-pass metabolism in the case of systemic delivery, and direct drug elution to the site of action in the case of topical delivery. Despite its importance to women's health, FRT drug delivery has not been a popular area of study for pharmaceutical scientists, likely because of an increasingly outdated social stigma surrounding women's health and sexuality. Nonetheless, several important historical "firsts" in pharmaceutical science have been achieved in response to problems in women's health, including the first commercially available zero-order release implant (Norplant®) and the first melt-extruded controlled-release device (NuvaRing®).

In this chapter, we provide the reader with broad overview of the principles and applications of drug delivery to the FRT, including a brief overview of anatomy and physiology, a description of drug transport principles and physiological effects, a summary of the benefits afforded by FRT delivery, and a survey of existing and upcoming drug delivery systems (DDS) and technologies. We focus largely on systems that have entered clinical development or have a large body of preclinical data to support clinical viability.

12.2 ANATOMY AND PHYSIOLOGY OF THE FRT

The FRT is a continuous passageway from the ovaries to the vaginal introitus and consists (traveling distally) of the fallopian tubes, uterus, cervix, and vagina (Figure 12.1).

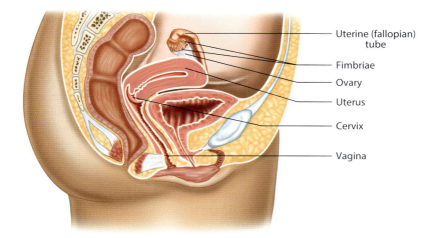

FIGURE 12.1 Sagittal view of the human female reproductive tract (FRT). (Modified from Tortora, G.J. and Derrickson, B.H., eds., *Principles of Anatomy and Physiology*, 14th ed. 2013. Copyright Wiley-VCH Verlag GmbH & Co. KGaA. Reproduced with permission.)

Gonadal differentiation occurs during the eighth to the ninth week of gestation. In the female embryo, the mesonephric (Wolffian) ducts completely regress and the paramesonephric (Müllerian) ducts develop into the proximal vagina, uterus, and fallopian tube. The two paramesonephric ducts meet the posterior wall of the urogenital sinus to form the sinovaginal bulbs, which, in turn, form the lower vagina.

As a result of the different embryonic origin of the lower and upper vagina, both blood supply and innervation of the upper and lower vagina differ. The lower one-fifth to one-quarter of the vagina is innervated from the deep perineal nerve, which is a branch of the pudendal nerve. The remainder of the vagina is innervated by the uterovaginal nerve plexus. Similarly, blood supply to the upper vagina is supplied from the uterine artery and vaginal artery, while the lower vagina is supplied by the pudendal artery and middle rectal artery.

The epithelium of the upper and the lower portions of the FRT differs, again as a result of differing embryonic origin. The Müllerian epithelium in the upper FRT differentiates into a single layer of columnar epithelial cells, while the vaginal plate differentiates into stratified squamous epithelium, meeting at the transformation zone, whose location starts on the ectocervix but migrates proximally with age.

12.2.1 OVARIES

Ovaries are solid, hormone-secreting structures located inside the peritoneal sac that serve as the proximal termini of the FRT. The adult ovary is 2.5–5.0 cm long, 1.5–3.0 cm wide, and 0.6–2.2 cm thick. Ovaries have four layers:

1. Germinal epithelium, the outer layer (neither germinal nor an epithelium)
2. Capsule (white coat), located underneath the germinal epithelium, composed of fibrous tissue
3. Cortex, comprising the bulk of the ovary, where the ova develops
4. Medulla, located inside the cortex, containing blood vessels and connective tissue

The ovaries are generally not considered an important site for drug delivery, but many ovarian pathologies, particularly ovarian cancer, might benefit from local or targeted drug delivery.

12.2.2 FALLOPIAN TUBES

The fallopian tubes are a symmetric pair of seromuscular channels, approximately 7–14 cm in length and 0.01–1 cm in diameter, which run medially from each ovary and attach to the uterus. The fallopian tubes transport eggs from the ovary into the uterine cavity through a combination of peristaltic contraction and flow induced by ciliary motion. Thus, occlusion of the tubes results in sterilization.

12.2.3 UTERUS

The uterus is a roughly pear-shaped organ approximately 8 cm long and 5 cm wide, with involuntary smooth muscle walls approximately 1.25 cm thick. These smooth muscle cells are capable of hypertrophy and recruitment of new myocytes from stem cells within the connective tissue, in order to accommodate expansion during pregnancy. The walls of the uterus are lined with the endometrium, a glandular epithelium that changes during menstruation and serves as the site of ovum implantation. Importantly, the pH of the intrauterine environment is typically near neutral, in contrast to the typically (but not always) acidic pH of the vagina (discussed in detail in the following text).

Uterine drug delivery is primarily used for long-term contraception, but there is a need for development of FRT delivery systems to target other pathologies, including uterine fibroids and endometriosis. Uterine fibroids are benign tumors that arise from smooth muscle cells in the myometrium of the uterus and are a common indication for hysterectomy. They are quite prevalent and can result in painful menstruation, bleeding, and reduced fertility in women. Endometriosis occurs when endometrial tissue grows in locations outside the uterine cavity, resulting in a chronic immune response, and symptoms including pain, infertility, and fatigue.

12.2.4 CERVIX

The opening of the uterus forms the cervix, a narrow channel composed of connective tissue and smooth muscle that protrudes slightly into the upper vagina, terminating at the cervical os. The exterior epithelial surface of the vaginal canal inside the vagina is referred to as the ectocervix, and the luminal surface is referred to as the endocervix. Importantly, goblet cells exist in the cervical canal and secrete mucin-containing mucus that can act as a barrier to sperm and infectious pathogens. Mucus is secreted cyclically into the vagina, where it can also act as a barrier to infection. As sperm must traverse the cervix to achieve fertilization, all contraceptive barrier devices function by preventing the passage of sperm from the proximal vagina into the cervix. Some contraceptive barriers also incorporate a drug delivery component and are discussed later in the chapter.

12.2.5 VAGINA

The vaginal tract of an adult female is a collapsed pseudocavity with a slight "S" shaped structure approximately 2 cm wide and 6–12 cm long. A common misconception is that the vaginal tract is a straight tube, which often leads to a fear that items placed in the proximal vagina can easily fall out. However, there are two distinct portions of the vagina, where the proximal axis, nearly horizontal when a woman is standing, forms a 130° angle with the distal axis. The walls of the vagina are generally formed into laterally oriented 2–5 mm thick folds (rugae) that are in close apposition. The vagina is capable of holding a few milliliter of fluid or gel without leakage.

Histologically, the vaginal wall consists of the following:

- A superficial nonsecretory stratified squamous epithelial layer (with underlying basement membrane)
- The lamina propria
- A muscular layer consisting of smooth muscle fibers
- The tunica adventitia, consisting of areolar connective tissue

12.2.5.1 Vaginal Epithelium

The vaginal epithelium is made up of noncornified, stratified squamous cells (see Chapter 4, Figure 4.7), is on average 50–200 μm thick, and consists of five cell layers:

1. Superficial (approximately 10 rows of cells)
2. Transitional (about 10 rows of cells)
3. Intermediate (about 10 rows of cells)
4. Parabasal (about 2 rows of cells)
5. Basal (a single row of cells)

Intercellular channels pass through the vaginal epithelium, allowing for the transport of leukocytes and large proteins, as well as fluid exudate from the vesicular plexus. Immune system cells can be found below the lamina propria and bridging the basal layer.

12.2.5.2 Lamina Propria

The lamina propria is a dense layer of vascularized, collagenous connective tissue beneath the vaginal epithelium. A diverse array of cells can be found in this layer, including fibroblasts, macrophages, mast cells, lymphocytes, Langerhans cells, plasma cells, neutrophils, and eosinophils. Drugs delivered to the vagina gain access to the systemic circulation through blood vessels in the lamina propria. Additionally, the lamina propria contains a lymphatic drainage system, which feeds into four lymph nodes: the iliac sacral, gluteal, rectal, and inguinal.

12.2.5.3 Cervovaginal Fluid

The interior of the vagina is coated with a thin layer of a complex biological media known as cervovaginal fluid (CVF). Approximately 2 g of CVF is produced per day in the absence of sexual arousal. CVF contains secretions from the cervix in the form of mucins and secretions from the Bartholin's and Skene's glands, endometrial fluid, and fluid transuded from the vascular bed of the vaginal tissue. It also contains a large number of squamous epithelial cells and commensal bacteria. During the menstrual cycle, this fluid increases in volume through mixing with fluid from the uterus, ovaries, follicles, and peritoneum, permitting movement of spermatozoa.

CVF acts as a barrier against infections, not only by directly binding to microorganisms, but by maintaining an acidic pH, which inhibits pathogenic bacterial proliferation. Additionally, CVF contains antimicrobial substances such as defensins, lysozyme, lactoferrin, fibronectin, spermine, and secretory IgA and IgG. Naturally, the vaginal fluid may contain highly variable amounts of semen and spermatozoa following intercourse.

A wide range of aerobic and anaerobic bacteria normally are found in the vagina with concentrations of 10^8–10^9 colonies/mL of vaginal fluid. Lactobacilli (found in 62%–88% of asymptomatic women) that normally reside in the vagina break down secreted glycogen to form lactic acid, resulting in an acidic vaginal environment. Many reviews give healthy CVF pH ranges between 3.5 and 5.5. However, studies have shown an even larger range in women who show no clinical pathologies, perhaps indicating there is not a "normal" vaginal pH as commonly thought. The pH of the vagina differs depending on age, ranging from a more acidic level during sexually mature childbearing years to neutral or alkaline before puberty and after menopause. Shifts in the

CVF pH can be, but are not always indicative of vaginal infection: a more alkaline environment can indicate bacterial vaginosis or trichomoniasis, while a more acidic environment can indicate fungal infection.

12.3 DRUG TRANSPORT IN THE FRT AND THE IMPACT OF VAGINAL PHYSIOLOGY

12.3.1 COMPARTMENTAL DESCRIPTION OF DRUG TRANSPORT

There are three basic compartments of importance when considering spatial drug pharmacokinetics (PK) in the FRT (Figure 12.2): (1) cervovaginal/uterine fluid, (2) epithelial and underlying tissues, and (3) systemic circulation. Cervovaginal/uterine fluid generally acts as a dissolution medium that relays drug molecules from dosage form to tissue. The fluid provides a sink-type boundary in the case of water-soluble drugs or can become a saturated (or partially saturated) rate-controlling impediment for release, in the case of less water-soluble drugs. In either case, the fluid serves as a source for drug transport throughout the tissues. Fluid convection in the FRT serves to effectively spread drugs distally from the device to the introitus, which is particularly advantageous for devices that provide a constant drug source from a fixed point in the FRT (such as an intravaginal ring [IVR] or intrauterine system). Also, it is likely that a fraction of any dose delivered vaginally remains dissolved in the fluid and is lost through the introitus, which could lead to a reduction in dose availability for bolus-type doses (such as gels and tablets). It has also been suggested that drugs can partition directly from dosage form to epithelium, thus some fraction of released drug molecules can bypass the fluid compartment completely.

The tissue compartment can be further stratified into various subcompartments. When considering continuum-scale drug transport, the tissue can be divided into the epithelium (nonvascularized) and the underlying lamina propria (vascularized). For vaginal delivery, a linear concentration reduction in drug concentration can be expected across the epithelium by Fick's

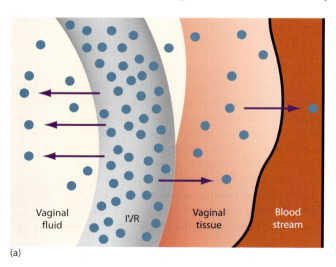
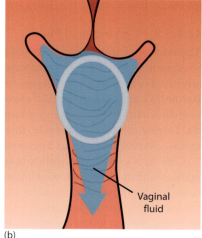

(a)   (b)

FIGURE 12.2 Generalized anatomy of the female reproductive tract and display of the various pharmacokinetic compartments. (a) An intravaginal ring is shown as a model dosage form and can elute drug into (1) vaginal fluid, (2) epithelial and underlying tissues, and (3) systemic circulation. (b) The vaginal ring is in contact with the walls of the vagina and is in contact with vaginal fluid. The virtual canal is anatomically connected through the cervical os to the uterus and fallopian tubes and ovaries. (Reprinted with permission from Kiser, P.F., Johnson, T.J., and Clark, J.T. State of the art in intravaginal ring technology for topical prophylaxis of HIV infection, *AIDS Rev.* 2012; 14:62–77. Copyright 2012 Permanyer Publications.)

laws, approximating the epithelium as a thin plane, while combined diffusion and elimination to the bloodstream should result in an exponential decay of drug concentration throughout the vascularized tissues. For topical prophylactic applications, a sufficient drug source must be maintained in the fluid compartment to ensure an effective drug level surrounding infectable cells at various depths in the tissue. For drugs that are activated intracellularly (such as many antiretrovirals), the tissue can be further stratified at the cellular level and chemical reactions must be considered along with diffusion and elimination.

Blood circulation serves as an intermediate sink for topical FRT delivery, in contrast to its function as a source for oral and parenteral delivery. Once in circulation, drugs delivered via the FRT are distributed similarly throughout the body but likely eliminated from the body via the same hepatic and renal clearance mechanisms.

From a pharmacokinetic/pharmacodynamic (PK/PD) perspective, intravaginal delivery can have several advantages over delivery from other routes. Intravaginal delivery requires reduced systemic drug concentrations to achieve effective levels in the FRT fluid and tissue, in comparison to oral and transdermal delivery. For systemic delivery, the hepatic first-pass effect is avoided, reducing presystemic drug loss. Furthermore, the FRT provides multiple sites for the noninvasive placement of controlled-release devices (e.g., IVRs and intrauterine systems), thus allowing for constant or sustained drug concentration in the FRT and the blood plasma, which is impossible to achieve through intermittent oral dosing. In a study directly comparing the systemic absorption of estrogen following vaginal, oral, or transdermal delivery over a 21-day period, vaginal delivery provided comparable efficacy to the other routes while avoiding unnecessarily high plasma concentrations (Figure 12.3).

Finally, we have observed that drug delivered vaginally in nonhuman primates (NHP) can be found at lower concentrations in proximal parts of the FRT, including the uterus and ovaries, despite distal fluid convection. Some have proposed a counter current exchange mechanism

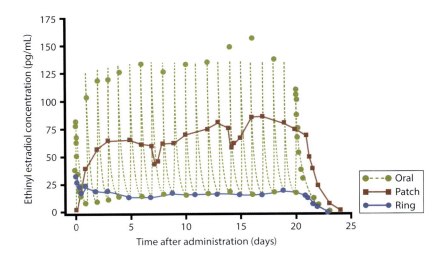

FIGURE 12.3 Comparison of systemic plasma ethinyl estradiol (EE) concentrations for 21 days via (1) vaginal delivery (NuvaRing®; shown in blue, 15 µg/day vaginal EE), (2) transdermal delivery (Evra™ transdermal patch; shown in red, 20 µg/day transdermal EE), and (3) oral delivery (Microgynon® combined oral contraceptive [COC] pill; shown in green, 30 µg/day oral EE). Systemic C_{max} values following vaginal delivery were 78% and 65% lower than via the oral and transdermal routes, respectively; furthermore, the characteristic peaks and troughs associated with oral delivery were avoided. (Modified from *Contraception*, 72, van den Heuvel, M.W., van Bragt, A.J.M., Mohammed, A.K. et al., Comparison of ethinyl estradiol pharmacokinetics in three hormonal contraceptive formulations: The vaginal ring, the transdermal patch and an oral contraceptive, 168–174, Copyright 2005, with permission from Elsevier.)

whereby drug could exchange from the venous bed in the vagina to the uterus. It is also possible that material is moved into the upper reproductive tract by diffusion down a concentration gradient in the cervical canal.

12.3.2 DRUG PROPERTIES

Several physicochemical properties are of interest for drugs delivered via the FRT, including aqueous solubility, tissue permeability, and hydrophobicity (see also Chapter 4, Section 4.3.4). Solubility in the vaginal fluid is important, since the concentration dissolved in the vaginal fluid functions as the boundary condition for transport through the tissues. Compounds also must have sufficient diffusivity in the tissue compartment for them to reach their potential sites of action or to reach circulation. Although hydrophobic compounds generally have better permeability across epithelial cell membranes, small hydrophilic compounds can rapidly diffuse in the extracellular space, quickly reaching continuum-level steady-state concentrations. The extracellular concentration at any point then serves as a drug source across individual cell membranes. Aqueous solubility is potentially more important for long-term delivery from a controlled-release device because local tissue and systemic concentrations should be directly related to the source concentration dissolved in the fluid and/or the device matrix.

Chemical stability of drugs dissolved in the CVF is an important factor in selection of appropriate drugs for vaginal delivery. As discussed previously, the CVF pH can exhibit intra- and interpatient variability, which must be considered for drugs with pH-dependent stability.

12.3.3 PHARMACOKINETIC MODELS

Several models of local PK for vaginal delivery have been reported. Saltzman used a steady-state solution to a one-dimensional diffusion–elimination model to predict the vaginal transport of ^{125}I-labeled IgG antibodies released from poly(ethylene-*co*-vinyl acetate) (EVA) matrix devices and also measured the concentration as a function of tissue depth experimentally. This treats the CVF as a thin conducting surface with constant API concentration. However, this model neglects the advective transport of drugs longitudinally through the vaginal tract. Saltzman later published an improved compartmental PK model for vaginal delivery that also considered fluid advection. Geonnotti and Katz constructed a finite-element model of a two-dimensional cross section of an IVR and the vaginal tract and surrounding tissues. The results of this model indicated that the thickness and fluid velocity of the vaginal fluid boundary layer have much greater impact on drug distribution through tissue than effective drug–tissue diffusivity.

12.3.4 CYCLIC AND INTERMITTENT CHANGES IN CERVICOVAGINAL MUCUS

The effects of menstrual cycle variation and sexual intercourse on vaginal drug delivery are often not considered. In cases where the flux of a drug from a device surface is limited by its solubility in the surrounding fluid, changes in the volume and/or composition of cervicovaginal mucus (CVM) could affect drug release. Some long-acting FRT dosage forms are left in place for one or more menstrual cycles. During this time, drug release rates could be modulated by the composition of CVM, which may contain various cyclically present drug-solubilizing factors, as well as by an increase in the fluid volume.

The epithelium of the vagina changes based on response to hormones (estrogens, progesterones, luteinizing hormone, and follicle-stimulating hormone), aging, sexual cycling, and pregnancy. These changes can all affect the vaginal absorption of drugs, rendering difficulty in achieving consistent drug levels. In the follicular phase, the basal and parabasal layers increase mitosis in response to higher levels of estrogen, resulting in increased epithelial thickness. Intercellular channels also increase in frequency, resulting in a more connected epithelium. Changes in drug transporter expression and distribution may also occur.

In contrast, during the luteal phase, desquamation occurs, causing epithelial thinning, loss of structure, and increased porosity of the epithelium as well as widening of the intercellular channels. After menopause, the vaginal epithelium becomes very thin, with indistinct cell boundaries and decreased microridging.

When selecting an API for formulation in an FRT-based dosage form, these changes should be considered as they can alter drug solubility, permeability, and absorption. This can especially confound the formulation of drugs with a low therapeutic index.

The drastic increase in vaginal pH following intercourse could also result in a temporary modulation of drug solubility, concentration, and transport.

12.3.5 Pregnancy

During pregnancy, the epithelial wall of the vagina becomes thicker and more vascularized, and blood flow slows. Recovery from this state takes several weeks postpartum. The effects of pregnancy on FRT drug transport are not often discussed, as many systems are intended to prevent pregnancy, to aid in achieving or maintaining pregnancy, are intended for use in postmenopausal women, and/or are contraindicated in pregnant women. Although intrauterine drug delivery systems (IUS) are highly effective in preventing pregnancy, pregnancies can still occur, and in these cases, IUS use is associated with a higher risk of ectopic pregnancy. Many API that could be delivered by the vaginal route are contraindicated with pregnancy. For instance, topical delivery of the chemotherapeutic sensitization agent, 5-fluorouracil, is regarded as unsafe for use during pregnancy by the FDA.

12.4 WHY DELIVER DRUGS TO THE FRT?

FRT DDS are naturally suited for women's health applications as they are only available to women. As discussed thus far, these systems can be designed for topical and/or systemic effect. Most FRT DDS to date are indicated for elimination of infection, postmenopausal hormone replacement, and contraception. As discussed in the previous section, a key limitation is that cyclical changes in the vaginal environment across multiple time scales can complicate drug dissolution/release and tissue absorption/transport.

12.4.1 Utility for Topical Dosing and Feasibility for Systemic Delivery

Direct drug delivery to the FRT is particularly advantageous when a local effect is desired and high systemic drug levels are not needed. This is the case with anti-infectives, HIV/sexually transmitted infection (STI) prophylactics, and some contraceptives (particularly low-dose, progestin-only formulations). Systemic delivery through the FRT also has many advantages, including low metabolic activity, relatively high permeability, ease of administration, prolonged retention, avoidance of hepatic first-pass metabolism, and potential for sustained and controlled release from long-acting devices.

12.4.2 Reduction/Elimination of User/Patient Interventions

Poor compliance/adherence to traditional oral drug therapies is a universal detractor of effectiveness, especially for prophylactic applications. Delivery to the lower FRT with longer acting devices, such as IVR, minimizes user/patient intervention, while delivery to the upper FRT with intrauterine devices/systems (IUD/IUS) effectively eliminates these interventions for the total device duration (up to 10 years for some contraceptive IUD). Poor adherence is also associated with frequency of dosing and duration of treatment across all routes of administration, strengthening the case for FRT delivery using long-acting devices when appropriate.

12.4.3 FRT INFECTIONS

A good portion of vaginal dosage forms comprise an anti-infective. The most common sources of infection are trichomonal, bacterial, candidal (yeast), and gonococcal. *Candida albicans* is responsible for approximately 25% of cases of vaginitis, which can be treated with local intravaginal azole agents (clotrimazole, miconazole, butoconazole, terconazole, and tioconazole). Additionally, modification of vaginal pH using a boric acid suppository has been found successful in treatment of candida, though it is not advised for pregnant women.

Trichomoniasis is a common STI caused by the anaerobic protozoan *Trichomonas vaginalis*, responsible for 25% of all cases of vaginitis. Because the infection resides in the vagina as well as the urethra, Skene's glands, and bladder, an oral dose of 2 g of metronidazole or tinidazole is typically prescribed, although these compounds are also administered vaginally.

Approximately, 40% of vaginosis is bacterial in nature and caused by an overgrowth of the *lactobacilli* responsible for maintaining vaginal pH. Bacterial vaginosis is not often treated, since it usually can resolve on its own, but several treatments are available, including metronidazole intravaginal gels, clindamycin intravaginal creams, and a combination of oral and intravaginal clindamycin.

12.4.4 MENOPAUSAL SYMPTOMS

Surrounding and following menopause, the concomitant decrease in estrogen levels may result in atrophic vaginitis, whereby the vulvar and vaginal tissue can become pale, thin, and dry, resulting in general discomfort and painful intercourse. An estrogen is often prescribed in the form of a vaginal cream, tablet, or ring, as local delivery is particularly desirable for estrogen to avoid endometrial stimulation.

12.4.5 CONTRACEPTION

Contraceptive hormones can be absorbed efficiently to systemic circulation from the vagina. Since low-dose progestin-only contraceptives are effective primarily through local effects, such as thickening of CVM, very low doses of progestins can be applied directly to the FRT to eliminate the need for estrogen delivery and greatly reduce systemic progestin exposure. This strategy is employed by the Mirena® and Skyla® IUS and in levonorgestrel (LNG)-releasing IVR. Reducing the dose also allows for higher duration in IVR and IUD, which have inherent size limitations.

12.4.6 TOPICAL PREEXPOSURE PROPHYLAXIS OF HIV

The continued absence of an HIV vaccine has motivated the investigation of woman-controlled strategies that can interrupt the early events of male-to-female sexual transmission. Early studies employed vaginal gels formulated with nonspecific polymeric agents (e.g., carrageenans, cellulose sulfate) designed to prevent infection and were categorically unsuccessful. More recent efforts have proven the concept of preexposure prophylaxis (PrEP), whereby ARV drugs are administered prior to sexual exposure. To date, clinical studies of vaginal HIV PrEP have reported mixed results, but overall this strategy shows promise to impact the pandemic if products are developed that women are highly motivated to use. As discussed in the following texts, the PrEP effort has spurred innovation in nearly every vaginal product category, most notably in IVR and vaginal gels. Vaginal delivery of ARV is thought to be advantageous because drugs can be delivered directly to the site of cells that are infectable by HIV with less systemic exposure. Nonetheless, only oral tablets have been approved by the FDA for prophylactic use to date. It remains uncertain whether oral delivery of ARV prevents HIV transmission in the mucosa or later events in the immune system required for dissemination of HIV.

The terms "PrEP" and "microbicides" are often used interchangeably. Here, we distinguish that PrEP refers to the administration of agents with specific pharmacological activity against HIV, whereas microbicides are a broad category of prophylactic antivirals and anti-infectives.

12.5 DRUG DELIVERY TO THE UPPER FRT

We will next discuss existing products and emerging technologies that utilize the FRT as a route for both topical and systemic drug absorption. A pictorial survey of commercially available vaginal delivery systems is shown in Figure 12.4.

12.5.1 DELIVERY TO THE OVARIES

Because the ovaries are located within the peritoneal sac, intraperitoneal injection of chemotherapeutics has been used to treat ovarian cancer. Additionally, ^{99m}Tc-labeled human albumin microspheres deposited in the vaginal fornix have been found to concentrate in the ovaries. This has been used as a method of evaluating fallopian tube function and illustrates another potential route of ovarian administration. Little has been explored in the field of ovarian drug delivery, and it remains a potentially interesting field for study.

12.5.2 DELIVERY TO THE FALLOPIAN TUBES

Pellets of quinacrine were first used by Jaime Zipper in Chile as a form of nonsurgical sterilization in the late 1970s. Using a modified IUD insertion device, seven 36 mg pellets of quinacrine hydrochloride were placed close to the entry to the fallopian tubes. These pellets caused chemical burns and scarring, resulting in occlusion and permanent sterilization. While over a hundred thousand women were sterilized using quinacrine, primarily in developing countries, ethical controversy in the late 1990s over inadequate testing and possible carcinogenicity resulted in its declining use.

The Essure® system is a permanent birth control option with a success rate of 99.8%. An Essure micro-insert is a stainless steel inner coil surrounded by a nickel–titanium outer coil and polyethylene terephthalate (PET) fibers. Upon release of each micro-insert into the proximal section of each fallopian tube, the outer coil expands from 0.8 to 2.0 mm diameter to anchor the device. The local inflammatory response to the PET fibers causes fibrosis and occlusion of the fallopian tubes, resulting in permanent sterilization.

12.5.3 INTRAUTERINE DRUG DELIVERY

IUD and IUS are T-shaped polymeric devices inserted in the uterus that can provide highly effective contraception for up to several years from a single device (Figure 12.4a). Traditionally, IUD did not contain API, but provided contraception solely by exhibiting a spermicidal foreign body response in the uterus. The modern IUD shape is likely derived from attempts in the early 1900s to create a stem-type pessary that extended from the vagina into the cervix. These devices were eventually abandoned due to their high rate of infection, as well as the controversial nature of contraception at the time. In fact, two scientists, Grafenburg of Germany and Oda of Japan, were exiled from their respective countries as a result of their research into contraceptives in the 1930s. In the late 1960s, the safety of combined oral contraception ("the pill") had come into question, prompting the investigation and clinical use of IUD as an alternative. By the early 1970s, 17 IUD or IUS were under development by 15 different companies, including the progesterone-releasing Progestasert® IUS, developed by Alejandro Zaffaroni and colleagues at the eponymous ALZA Corporation. Progestasert® was one of the first drug/device combinations to be used clinically. Another of these IUD, the Dalkon Shield®, was associated with extensive safety concerns and may have caused at least 18 deaths. It was eventually removed from the market in 1974. As one would

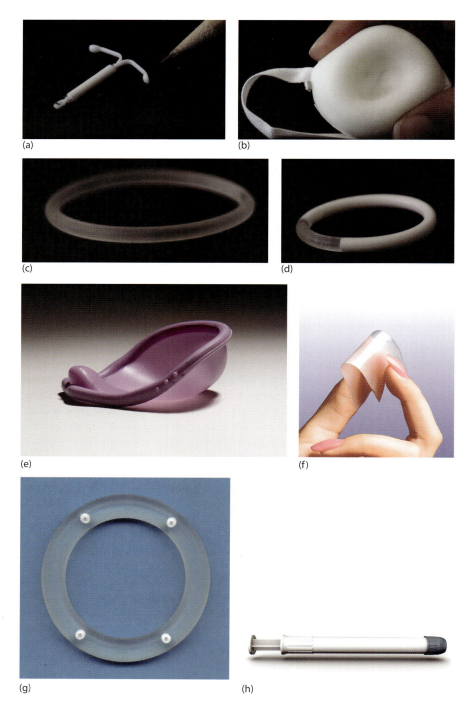

FIGURE 12.4 Pictorial survey of some female reproductive tract drug delivery systems. (a) Mirena®
intrauterine drug delivery system, (b) Today® Sponge; (c) NuvaRing®, (d) segmented tenofovir/levonorg-
estrel intravaginal ring; (e) Caya Contoured diaphragm (SILCS diaphragm), (f) vaginal contraceptive
film; (g) pod-insert intravaginal ring, (h) Crinone® gel applicator. (e: Photo courtesy of PATH. All rights
reserved, Seattle, Washington; f: Courtesy of Apothecus Pharmaceutical Corp., New York. Copyrighted by
Apothecus Pharmaceutical Corp., 2015; g: Photo courtesy of Marc Baum, Oak Crest Institute of Science,
Pasadena, CA; h: Photo courtesy of Actavis plc, Parsippany, NJ.)

expect, this resulted in a very sour public opinion of IUD. The reader is reminded that the Dalkon Shield® was sold at a time when FDA approval was required only for drugs, and not for medical devices.

After a lull in IUD/IUS use in the 1980s and 1990s, primarily resulting from issues with the Dalkon Shield®, IUD/IUS use has resurfaced in the past 15 years, resulting in FDA approval of two Bayer-marketed drug-releasing IUS. The Mirena® and Skyla® IUS, FDA approved in 2000 and 2013, respectively, both deliver "microdoses" (up to 14 and 20 µg/day, respectively) of the progestin LNG in the uterus for up to 3 and 5 years, respectively, and achieve contraception largely through local effects (e.g., cervical mucus thickening and inhibition of sperm function). Both Mirena® and Skyla® are over 99% effective in preventing pregnancy. Mirena® is also coindicated for treatment of heavy menstrual bleeding.

An alternative to progestin-releasing IUS is the Paragard® IUD, also known as the "Copper T." Paragard® was approved in 1981 but has seen a recent resurgence in popularity along with the IUD/IUS platform as a whole. Paragard® is a copper-releasing IUD that is also over 99% effective and can remain in place for up to 10 or 12 years. Paragard® use results in high concentrations of copper in the cervicovaginal fluids that are toxic to sperm cells and may aid in the prevention of implantation. Although Paragard® is primarily used as a long-acting contraceptive, it has clinically been shown to be effective as a method for emergency contraception (EC). However, hormonal EC methods, such as high oral doses of LNG, are more commonly used.

Both LNG-releasing IUS and the Paragard® IUD consist of a T-shaped inert (nondrug releasing) polyethylene frame varying slightly in overall dimensions, where Paragard® is the largest (32 mm wide and 36 mm long) and Skyla® is the smallest (28 mm wide and 30 mm long). In all three devices, barium sulfate is compounded into the polyethylene matrix as a radiopacifier. Furthermore, a polyethylene monofilament is tied to a loop at the end of the vertical stem, resulting in two threads that allow the user to confirm that her IUD/IUS is still in place. A silver ring was placed around the top of the vertical stem in Skyla® to render the device visible by ultrasound. In Mirena® and Skyla®, the vertical stem is covered with a cylindrical polydimethylsiloxane (PDMS or "silicone elastomer") reservoir-type cylinder, where the reservoir core contains a mixture of PDMS and LNG. Skyla® was introduced in 2013 as a smaller, low-dose variant of Mirena® specifically marketed for nulliparous women, whereas Mirena® had been originally indicated only for use in women who had at least one child. Skyla® has a lower overall drug load (52 versus 13 mg), shorter duration (3 versus 5 years), and lower final release rate (5 versus 11 µg/day). The lower release rate was likely achieved by using a thicker outer, rate-controlling membrane for Skyla® and/or is due to the lower drug concentration in the core (assuming the drug is mostly dissolved in the core). Nonetheless, Skyla® is nearly as effective as its predecessor for the indicated duration, with a cumulative 3-year pregnancy rate of 0.9% (compared to a 5-year pregnancy rate of 0.7% for Mirena®). The horizontal and vertical stems of Paragard® are partially wrapped with a total of 176 mg of copper wire, resulting in an exposed copper surface area of 380 mm². Frameless intrauterine implants also have been developed as alternatives to traditional T-shaped designs. These systems consist of a straight cylindrical segment containing an anchor system that is kept in place by suturing to the uterine fundus. A 5-year, copper-releasing frameless IUD (GyneFix®) is available in Europe but has not been approved in the United States, and a 20 µg/day LNG-releasing variant (FibroPlant™) is in development.

12.6 DRUG DELIVERY TO THE LOWER FRT

12.6.1 First Vaginal Drug Delivery Device: The Pessary

"Pessary" is a term used for devices placed into the vagina for structural support of the uterus of parous women but has been used also to refer to vaginal DDS. Pessaries have been used since the time of the ancient Greeks; Polybus (≈400 BC) recommended the use of a half pomegranate, and Soranus (110 AD) recommended a linen tampon soaked in vinegar with a piece of beef.

While pessaries have primarily been used for treatment of pelvic organ prolapse and urinary incontinence, the term pessary has also been used to refer to IVR that deliver estrogen for the management of urogenital atrophy. While contemporary vaginal DDS have been described as pessaries, the term is archaic and should only be used to refer to devices for pelvic floor disorders and not in the description of drug formulations or devices.

12.6.2 Vaginal Absorption of Toxins

Prior to 1918, it was widely believed that any compounds administered to the vagina were capable of only local effect. In 1918, Macht reported systemic absorption of molecules through the vagina by administration of a number of compounds to dogs and cats. He was able to demonstrate that compounds could quickly produce effects, including vomiting and death, upon administration to the vagina, thereby demonstrating systemic absorption.

12.6.3 Cervix and Combination Barrier Devices

Female barrier contraceptives are designed to prevent sperm transport from the proximal vagina to the cervix. However, these devices do not form a perfect barrier and have often been formulated with spermicides to increase efficacy. Common designs of such devices are the diaphragm, and cervical cap, and the contraceptive sponge.

The diaphragm is a dome-shaped cup with a steel flexible rim that holds the device in place against the vaginal walls, thereby sealing the cervix (Figure 12.4e). The cervical cap, first introduced in 1838 by Friedrich Wilde, is comprised entirely of latex and is intended to fit snugly around the base of the cervix. Diaphragms and cervical caps are generally intended for use with a spermicide. Currently, the only available cervical cap is branded as FemCap™.

The Today™ sponge is a sustained-release system consisting of a circular polyurethane disc impregnated with nonoxynol-9 (N9), intended to release approximately 125 mg of N9 over 24 hours (Figure 12.4b). N9 is a nonionic surfactant that immobilizes spermatozoa by removing lipid components of the midpiece and tail. The sponge is inserted against the cervix up to 24 hours before intercourse and is left in place for 6 hours following intercourse. Despite the inclusion of N9, the sponge has a failure rate of 8%–13%, considerably higher than many modern contraceptive options.

The combination barrier devices discussed earlier can be modestly effective in preventing pregnancy but are limited in that, with the exception of the female condom, they offer no protection against HIV/STI transmission, in contrast to the male condom. In fact, N9 is now thought to carry increased risk for HIV/STI acquisition (discussed in the following texts). To address this gap, strategies are under investigations that form a physical barrier to sperm, as well as a chemical barrier to infection, using modifications of SILCS diaphragm (Figure 12.4e). SILCS, also known as the Caya Countoured diaphragm, is a modified silicone diaphragm designed for easy insertion/removal and better anatomical fit. Key features are the thermoplastic nylon 6,6 spring forming a cervical rim, as well as a recessed "removal dome" to aid in easy removal. One incarnation for drug delivery is a SILCS variant designed to carry a dose of the 1% tenofovir (TFV) gel, which has shown effectiveness against HIV infection. Major and Malcolm reported a long-acting combination SILCS device, whereby the nylon spring was replaced by an elastic polyoxymethylene matrix containing the anti-HIV drug dapivirine (DPV). If effective, this variant could protect against pregnancy and HIV infection for 6 months.

12.6.4 Intravaginal Rings

12.6.4.1 Technical Description of Intravaginal Rings

IVRs are torus-shaped drug/medical device combinations used clinically for long-acting contraception and hormone replacement therapy and have garnered recent interest as HIV prophylactics (Figure 12.4c, d, and g). IVR dimensions are loosely constrained by the width of the vaginal canal, observed to vary by as much as 48–63 mm, although the fit and comfort of IVRs are also constrained by the overall balance of elastic properties between the compressible ring and the vaginal musculature. Clinical IVR outer diameters (OD) do not vary greatly, ranging from 54 mm (NuvaRing) to 56 mm (Femring®); however, OD of 50 to 61 mm are thought to be well tolerated based on clinical data. Interestingly, the first silicone IVR tested by Mishell had OD of up to 80 mm. The cross-sectional diameters of clinical IVR vary significantly, ranging from 4 mm (NuvaRing) (Figure 12.4c), to 9 mm (Estring®).

As with other drug/device combinations, most IVR can be classified as either matrix or reservoir devices. Matrix IVRs are simple, monolithic devices where drug is dissolved and/or dispersed evenly throughout the polymer matrix, whereas reservoir IVRs employ an outer polymeric membrane to control the rate of drug release (see also Chapter 2). Drug release from matrix IVR usually decreases significantly with time, due to an increase in the diffusion path length following progressive inward drug depletion. Drug release rates from reservoir IVR are linearly dependent on the dissolved core drug concentration and thus can be constant (if an excess of undissolved drug is present) or can decay in a first-order exponential fashion (if the entirety of drug is completely dissolved). Most clinical IVRs are reservoir devices, but matrix IVRs have seen significant investigation by the HIV PrEP field, likely due to their potential manufacturing simplicity and reduced cost. More advanced designs have been investigated that address the challenge of delivering multiple drugs simultaneously. A fixed-dose combination was achieved for the dual hormone contraceptive IVR NuvaRing by dissolving different concentrations of two drugs with similar chemical properties in the core. Advanced designs include (1) the segmented or multisegment IVR (Figure 12.4d), which theoretically can incorporate both matrix and reservoir components, and (2) the pod-insert IVR, wherein an injection-molded ring is fitted with several reservoir-controlled release inserts (Figure 12.4g). Teller et al. recently described an IVR containing a "flux-controlled pump" (FCP) for zero-order release of hydrophobic and macromolecular compounds. In an FCP, drug tablets formed from a hydrophilic gel-forming polymer matrix are incorporated in a hollow-tube segment. Upon hydration, a drug-containing gel is extruded through one or more orifices in the tubing wall of a rigid hollow segment, resulting in the pressure-driven release of hydrophobic drugs at rates independent of their molecular properties.

All commercially available IVRs are fabricated from either thermosetting silicone elastomers or thermoplastic EVA. The DPV IVR under clinical investigation for HIV PrEP is also synthesized from a silicone elastomer. Kiser and colleagues have extensively investigated thermoplastic polyurethane elastomers as IVR matrices, particularly poly(ether urethanes), which can be synthesized in water-swellable and nonswellable variants. Most IVR designs are limited to the delivery of small molecule drugs due to similar solubility/permeability constraints; however, several designs have been investigated, including the aforementioned pod-insert ring that could allow for delivery of larger molecules.

Drug release from IVR under sink conditions has been mechanistically described by solutions to diffusion equations, which assume a sink boundary between the device and CFV/surrounding tissues. It is unclear whether a biological sink is maintained for some hydrophobic antiviral compounds in the vaginal tract. It is thought that solubility-limited release occurs (i.e., CVF does not provide a sink condition for IVR drug release) in vivo for certain device/drug combinations where the vaginal fluid concentration remains saturated or supersaturated with drug throughout the dose duration.

In general, the efficacy of long-term controlled release drug–device combinations is dependent on their ability to remain in place for the duration of use. IVR differ from other long-acting drug/device combinations in that they are not surgically implanted and are designed to be inserted and removed by the user. If an IVR is too easily deformed, the ring may be expelled as a result of the user's day-to-day activities. However, if the force required to compress an IVR is too great, it may be difficult for the user to compress the ring prior to insertion, or the presence of the ring may cause damage to the vaginal epithelium.

12.6.4.2 Clinical History of Intravaginal Rings

IVRs were first investigated for the sustained delivery of contraceptive hormones to systemic circulation in the late 1960s. Mishell and colleagues were the first to clinically demonstrate the potential of contraception via IVR by testing a silicone IVR impregnated with the progestin medroxyprogesterone acetate (the drug now used in the Depo-Provera® contraceptive injection) in a small PK/PD pilot trial. In a series of seminal papers, Chien and colleagues used similar silicone IVR as a case study for the discussion of controlled drug release in the context of vaginal delivery. The WHO later developed a silicone IVR that released approximately 20 µg/day of LNG for topical contraception for up to 3 months. This LNG IVR was tested in a large clinical multicenter clinical trial and was shown to be over 95% effective in preventing pregnancy; however, development was discontinued, in part due to use of the IVR being associated with the formation of vaginal lesions, potentially due to excessive device stiffness. The first contraceptive IVR to market was the progestin/estrogen NuvaRing®, an EVA-based reservoir ring that releases the progestin etonogestrel and ethinyl estradiol (EE) at near constant rates over 21 days. The inclusion of EE and the 3-week-in/1-week-out pattern generally maintains a normal menstrual cycle, as in progestin/estrogen oral contraceptives that adhere to a similar 3 week/1 week cycle. Several important insights to IVR design and development gained from the NuvaRing® work were published by van Laarhoven and colleagues in the early 2000s.

Four silicone-based IVRs are currently available worldwide for various indications other than general contraception. Estring® and Femring®, both available in the United States, deliver estrogen analogs (estradiol and estradiol acetate, respectively) for postmenopausal hormone replacement. Progering® and Fertiring® both release natural progesterone and are approved only in a few Central and South American countries. Progering® is indicated for contraception in lactating women, while Fertiring® is indicated for hormone supplementation and pregnancy maintenance during in vitro fertilization.

IVR for HIV PrEP can be traced back to the N9-releasing silicone IVR reported by Malcolm and Woolfson at the Queen's University Belfast in the early 2000s. This group went on to formulate DPV into a silicone IVR that lead to phase 3 trials of this antiviral IVR in African women. Use of the DPV IVR demonstrated a 27% overall efficacy in preventing sexual transmission of HIV and 65% efficacy in regular users.

In the mid- to late 2000s, Kiser and colleagues published several papers reporting the in vitro and in vivo investigation of polyurethane IVR for the delivery of several ARV, including multisegment IVR designs for the simultaneous delivery of multiple API with disparate properties. One of these designs was used in the development of the first clinically tested IVR for multipurpose prevention (discussed in the following texts). ARV-releasing IVR have achieved high rates of efficacy in NHP challenge studies, and have demonstrated adherence dependent efficacy in recent dapivirine IVR trials. An IVR-releasing MIV-150 (a nonnucleoside reverse transcriptase inhibitor) demonstrated 80% efficacy in preventing simian–human immunodeficiency virus (SHIV) infection in rhesus macaques, while another IVR-releasing tenofovir disoproxil fumarate (TDF) was 100% effective in preventing SHIV infection in pig-tailed macaques challenged weekly for 4 months during continuous IVR use.

IVRs have also been investigated for delivery of lidocaine as a cervical anesthetic and oxybutynin as a treatment of overactive bladder syndrome, with the latter recently completing a Phase II clinical study. Bayer is currently developing a combination anastrozole/LNG 28-day IVR for treatment of endometriosis. Anastrozole is a third-generation aromatase inhibitor that has been used for management of severe endometriosis through suppression of estrogen synthesis. LNG, which causes the endometrium to become atrophic and inactive, has also been reported to improve endometrial pain. This ring is currently in Phase III trials.

12.6.5 Vaginal Semisolids (Gels and Creams)

Vaginal semisolids, known colloquially as vaginal gels or creams, may be the broadest category of vaginal drug delivery system with respect to their formulation, physical properties, and clinical indication. Semisolid formulations are an attractive option for drug delivery as they are typically self-administered by a patient or user and afford the potential for on-demand or as-needed use. Vaginal semisolids have many drawbacks as well, such as messiness and, as with traditional oral formulations, a tendency for poor user adherence/patient compliance. Here, we use the word vaginal "gel" in a colloquial and not scientific sense; a true gel must contain long-lasting cross-links and therefore must retain a predominantly elastic character regardless of strain rate. Vaginal semisolids made from hydrocolloids generally do not have these properties and are therefore not strictly gels but are rather viscoelastic fluids.

A majority of vaginal semisolid formulations are prepared from viscoelastic water-soluble polymers with appreciable viscoelasticity. As with other dosage form classes, semisolids can contain dispersed and/or dissolved drugs. Viscoelastic semisolid "gels" are formed by dissolution of a relatively small amount of solid (hydrophilic polymers) in large amounts of liquid, resulting in "gels" with physically or chemically cross-linked, 3D polymer matrices that exhibit solid-like behavior. A majority of clinically available vaginal gels are formed from either a cellulose derivative (e.g., methylcellulose, hydroxyethylcellulose [HEC], carboxymethyl cellulose), a Carbopol® brand cross-linked poly(acrylic acid), or poly(carbophil). The literature is replete with other examples of natural and synthetic polymers investigated as hydrophilic polymers used in these formulations. Typically, these matrix-forming polymers are classified as "excipients" as they do not contribute to the activity of the drug product. However, as discussed in the following texts, early HIV/STI prevention ("microbicide") research employed vaginal gels where the gel-forming polymer was also the active agent. There are many claims stating that gelling polymers such as Carbopol and poly(carbophil) can also serve a bioadhesive function to aid in gel retention, but there is not a single convincing paper in the literature showing improved performance of these systems, and thus the concept of "bioadhesive vaginal gels" should be treated with skepticism.

Several additional excipient classes are commonly used in vaginal gels, including gelling agents, humectants, and preservatives. In general, excipients are considered to be pharmacologically and toxicologically inert, but this is not always the case; for example, excipients such as sodium dodecyl sulfate, carrageenans, cellulose acetate phthalate (CAP), and benzalkonium chloride may have antimicrobial properties. Important tests for excipient biocompatibility include cell growth/cell viability assays, cell proliferation assays, and cytotoxicity assays.

Vaginal semisolids are typically applied by the user with little or no instruction from a clinician. Disposable applicators are used to deliver a single dose, typically around 3 or 4 mL, to the proximal vagina. Ideally, this gel dose will spread and coat the vaginal mucosa to result in even drug distribution throughout the tract, although this property is more important in some applications than in others. Gel distribution can occur passively via gravity or through ambulation and even intercourse. Retention in the vaginal tract is also of key importance, both from the standpoint of PK and compliance/acceptability, as excessive gel leakage will result in large drug fractions being lost, as well as being an inconvenience to the user. Several semisolid characteristics affect distribution and retention. First of all, the appropriate volume of gel must be applied, as an excess may result in leakage, but a deficiency may result in inadequate distribution. Most importantly, the bulk viscoelastic properties of the gel formulation (e.g., viscosity, elastic modulus, or storage/loss modulus, depending on the mechanical behavior of the formulation) will influence distribution and retention. Ultimately, these properties are products of the strength and temporal nature of the physical and/or chemical interactions within the polymer. Traditionally, new vaginal semisolids were evaluated comparatively to existing products, but as with vaginal rings, the HIV prevention effort of the past two decades has catalyzed more rigorous study of composition–structure–property–performance relationships of vaginal semisolids.

As with traditional matrix and reservoir IVRs, the mechanical properties of semisolids are intimately coupled to their drug dissolution/release behavior. Vaginal semisolids are perhaps one of the more complicated dosage forms for the mechanistic description of drug release, as the time-dependent size and shape of the formulation confounds the already complex interplay between vaginal diffusion and convection. Changes to the nature of the polymer network are likely to simultaneously affect the viscoelastic behavior of the formulation, as well as the drug diffusivity in the formulation matrix, which in turn affects drug release rates.

Perhaps the most widely utilized and investigated agent in vaginal semisolid delivery systems is the aforementioned N9. An effective spermicide, N9 is included in many over-the-counter, vaginally administered, sodium carboxymethylcellulose-based contraceptives, such as Gynol® and Conceptrol®. However, the use of N9-containing contraceptives is considered by some to be controversial, especially in populations at high risk for STI acquisition. This is because some studies have implicated N9, which can weaken the natural epithelial barrier of the vagina, in increased risk of susceptibility to such STI. N9 was one of the first agents investigated as an HIV microbicide and was at one time lauded as a potential cure-all multipurpose prevention agent due to its activity against HIV and other STI. Subsequent HIV microbicide trials investigated vaginal semisolids formulated with nonspecific polymeric agents (e.g., carrageenans, cellulose sulfate) designed to prevent infection. These formulations were determined to be clinically safe for vaginal administration but were categorically unsuccessful in preventing HIV infection, leading to the investigation of antiretroviral-eluting prophylactics as already discussed. Most notably, a HEC-based vaginal gel containing the anti-HIV compound TFV has been both successful and unsuccessful in preventing HIV infection in clinical studies. Mixed results are thought to be due, in part, to low clinical trial adherence. Several ARVs, including the integrase inhibitor raltegravir, are still under preclinical investigation for delivery via vaginal gels. PrEP via a vaginal gel has proven successful in multiple SHIV (recombinant simian HIV) challenge studies in macaques, results that still continue to motivate the investigation of vaginal gels for HIV PrEP and the investigation of HIV PrEP in general.

Although contraception and microbicides have gathered most of the recent attention in regarding vaginal semisolids, several other products are available for a diverse array of indications. For example, semisolids are used for the delivery of antibiotics, such as metronidazole and clindamycin, for the treatment of bacterial vaginosis. Polycarbophil-based vaginal gels containing natural progesterone (i.e., Crinone®, Prochieve®) are indicated both for fertility assistance and secondary amenorrhea (Figure 12.4h). Prochieve® has also been clinically studied for prevention of preterm birth but failed to receive FDA approval following lack of statistical significance in a clinical study. Prostaglandins are administered to induce labor via vaginal semisolids, such as the dinoprostone-releasing, silicon dioxide–based Prostin E2. Vaginal "creams" containing estrogens are available for management of vaginal atrophy and associated symptoms (e.g., vaginal dryness, painful intercourse) in postmenopausal women; however, there is increasing evidence that chronic dosing of unopposed estrogen (without a progestin) carries an increased risk of a variety of serious medical outcomes including endometrial and breast cancer, cardiovascular events (e.g., heart attack, deep venous thrombosis), and dementia.

Various vaginal gel formulations have been investigated for insulin delivery, but it has been difficult to achieve the sufficiently high systemic insulin levels necessary to warrant clinical development. An interesting platform under clinical investigation is a vaginal gel containing the dendrimer SPL7013, which has demonstrated in vitro inhibition of HIV, herpes simplex virus (HSV), and bacterial replication. This dendrimer has shown clinical efficacy in treating bacterial vaginosis and is also being considered for use as a condom coating.

Vaginal creams are employed in the treatment of vaginal intraepithelial neoplasia (VAIN), a premalignant lesion involving cells in the vaginal epithelium, thought to be associated with human papillomavirus (HPV) infection. Topical delivery is advantageous to avoid systemic exposure to compounds used in the treatment of VAIN, which are associated with a wide array of side effects. 5-fluorouracil is formulated in a topical cream for the treatment of diffuse VAIN, though it has been associated with ulceration and poor tolerance. It is also used in conjunction with surgery for vaginal cancer, both pre- and postoperatively. Aldara™ (Imiquimod) is another topical cream for treatment of diffuse VAIN. Unlike 5-fluorouracil, Imiquimod is an immune response modulator that recruits cytokines with antiviral and tumoricidal effects.

12.6.6 Erodible Vaginal Solids (Tablets, Suppositories, Inserts, and Films)

Thus far, we have discussed nonerodible solid devices (IUS/IUD, IVR, and cervical barriers) and semisolids (gels and creams). Between these two categories lie erodible/dissolvable solid formulations, which include vaginal tablets and suppositories, and the emerging vaginal film platform. In theory, vaginal solids are not fundamentally different from oral solid dosage forms, although the slow convection of the FRT (in comparison to the GI tract) lends the possibility for sustained-release tablets with high residence times, including the incorporation of bioadhesives to enhance retention, as with vaginal gels.

Solid vaginal tablets and suppositories are available for a variety of indications, somewhat mirroring the list earlier for vaginal semisolids. The majority of these formulations are designed for delivery of antifungals and/or antibiotics, including the Candizole-T® tablet, which is a fixed-dose combination of two antifungals (miconazole and tinidazole) and an antibiotic (neomycin). Vaginal solids are also employed for vaginal delivery of progesterone, estradiol, and the spermicide N9 for similar indications as discussed earlier for gels.

There are many reports of off-label vaginal application of oral solid dosage forms in the clinic for various reasons. Many clinical studies have demonstrated the vaginal administration of the abortifacient misoprostol for cervical ripening and labor induction. Also, sildenafil (Viagra®) has been used in fertility maintenance, and the NSAID indomethacin has been dosed vaginally for prevention of preterm birth. Other oral formulations, including oral contraceptives and hormone therapies, have been dosed vaginally in cases of intolerance to oral medication.

As with many of the platforms discussed thus far, the HIV PrEP effort has spurred innovation within the vaginal tablet platform. The quick-dissolve tablet platform has been adapted for vaginal use and is currently under clinical investigation for delivery of the HIV RT inhibitors TFV and emtricitabine.

Vaginal films have recently emerged as an alternative to vaginal gels. Vaginal films are typically thin, solid polymeric sheets formed from water-soluble polymers and can incorporate drug substances (Figure 12.4f). In contrast to semisolid formulations, films are less messy to apply and provide a formulation strategy for compounds that are unstable in aqueous solution. The concept of thin-film drug delivery is thought to have evolved from consumer products for the oral cavity, particularly the breath mint "patch" (see also Chapter 8, Section 8.5.2.2).

Vaginal films can be designed for rapid disintegration and complete drug release, or to be slow dissolving for more sustained release. The majority of vaginal films currently in use or development are based on poly(vinyl alcohol) (PVA) and hydroxypropylmethyl cellulose, whose mass fraction and molecular weight can control disintegration time. Films can include plasticizers such as glycerol to reduce mechanical rigidity. Films are typically formed by solvent casting but can also be formed by hot-melt extrusion and incorporate thermoplastic polymers. Again, drugs can be dissolved and/or dispersed in the polymer matrix but should, if possible, be completely dissolved in the solvent casting or extrusion phase, to ensure even distribution.

Vaginal drug delivery applications for thin films have already been discussed, such as contraception, treatment of bacterial vaginosis, and HIV/STI prevention. To date, the only vaginal film to market has been the Vaginal Contraceptive Film®, which is another N9-containing contraceptive product marketed as a hormone-free contraceptive. Films have also been investigated for delivery of itraconazole for candidiasis and clindamycin phosphate for bacterial vaginosis. Rohan et al. have recently published several papers describing the in vitro evaluation of several antiretroviral-releasing vaginal films for HIV PrEP.

12.6.7 VAGINAL "INSERTS"

The term vaginal "insert" is somewhat ambiguous and has been used to describe vaginal solids not readily placed in other categories. For instance, the Endometrin® "insert" is a progesterone tablet placed in the proximal vagina, near the cervix, with the aid of a specialized applicator. Cervidil®, also called an "insert," is a swellable, nondissolving, dinoprostone-releasing slab, composed of a cross-linked poly(ether urethane)-based copolymer. The slab is encased in a knitted polyester bag with a string for retrieval.

12.7 NEW TECHNOLOGIES FOR FRT DELIVERY

In the last few decades, a large body of work has been reported in the area of DDS in the FRT; in particular, vaginally applied concepts are at the cutting edge of this area of study.

12.7.1 MULTIPURPOSE PREVENTION TECHNOLOGIES

Biomedical interventions designed to simultaneously address multiple reproductive health needs are commonly termed "multipurpose prevention technologies" (MPT). MPT can include products that protect against multiple STIs or against unintended pregnancy and an STI. The only commercially available MPT are barrier contraceptives (e.g., male and female condoms, cervical caps), which exhibit reasonable efficacy in providing a barrier to both agents of infection (i.e., viruses and bacteria) and sperm cells, but are prone to misuse and inconsistent use. Specifically, a single product that combines HIV PrEP and contraception has been identified as an urgently needed MPT by the global health community. It has been hypothesized that the inclusion of a contraceptive into an HIV PrEP could bolster demand for unfamiliar PrEP products due to the widespread use of contraceptives.

Conceptually, MPT can take the form of any of the dosage forms already discussed. The first long-acting MPT to be tested clinically is the TFV/LNG IVR, which is a segmented dual-reservoir design that delivers both drugs for up to 90 days. Other MPT IVR are under investigation for delivery of similar combinations, including DPV/LNG (also for HIV prevention and contraception) and a pod-insert style TFV/acyclovir ring (for HIV and HSV-2 prevention). Antiviral drugs, such as DPV and TFV, have been added to barrier contraceptives to form MPT, such as the SILCS diaphragm variants discussed earlier. Antivirals can also be added to an EC to form an on-demand MPT, which could theoretically be administered before or after intercourse. These on-demand MPT could be formulated as any of the semisolid or dissolvable solid dosage forms already discussed.

The Population Council has investigated MPT gels based on the carrageenan family of sulfated polysaccharides. The Carraguard gel was one of the first microbicide products tested clinically and was not effective in preventing HIV infection. However, carrageenans have shown activity against HSV-2 and HPV, which, along with their usefulness as gel-forming polymers, has led to their investigation as MPT. Zinc acetate, which is thought to have activity against HIV and HSV-2, has been formulated in a carrageenan gel, along with the HIV nonnucleoside inhibitor MIV-150, in the hope of an on-demand MPT capable of HIV, HSV-2, and HPV prevention. A MIV-150/zinc acetate/carrageenan (MZC) gel formulation was completely protective against SHIV transmission in macaques, and although the relative contributions to efficacy of MIV-150 and zinc acetate is not clear, a variant without MIV-150 was partially protective, confirming in vivo activity of the zinc salt. In mouse models, zinc acetate/carrageenan gels have demonstrated efficacy against HSV-2 challenge, and carrageenan-only gels have protected against HPV challenge. A MZC IVR is also in development, with the eventual goal of including a contraceptive (e.g., LNG). If such a device was feasible to manufacture and regulatory approval was possible, it could potentially offer long-acting protection against three STIs and pregnancy.

12.7.2   Mucus-Penetrating Particles

The efficient trapping of particles by mucus through steric and adhesive interactions is a barrier to delivery to the vagina using nanoparticles. Densely coated PEG-covered PLGA and PS nanoparticles have been demonstrated to have improved vaginal distribution as compared to uncoated nanoparticles. When these mucus-penetrating particles were formulated with an acyclovir monophosphate (ACVp) core in mice, they showed improved protection against HSV-2 challenge (46.7% infected) as compared to soluble ACVp (84.0% infected).

12.7.3   Small-Interfering RNA and Gene Silencing

Small-interfering RNAs (siRNA) are short (21–23 nucleotide) strands capable of inducing RNA interference and gene silencing. Vaginally delivered siRNA-based therapeutics has been delivered using PLGA nanoparticles (100–300 nm) in mice by Woodrow et al. The small size of the nanoparticles was thought to allow for effective tissue penetration and cell entry, and gene silencing was found throughout the uterine horns, cervix, and vagina.

12.7.4   Antibody Delivery

Challenges in vaccine delivery to the vaginal environment include the need to overcome degradative enzymatic activity, continual mucus turnover, and a fluctuating microenvironment due to cyclic changes in the menstrual cycle. In order to address these changes through sustained release of antigen, Kuo-Haller et al. created PLGA microparticles and 4 mm diameter discs of EVA, for introduction of the model antigen ovalbumin into mouse reproductive tracts, and obtained equivalent antibody response as compared to vaccines delivered via a different mucosal site.

12.7.5   Electrospun Fiber Mats

Electrospun fibers have also been studied for vaginal drug delivery. Electrospun fiber meshes can be made out of a variety of synthetic and biological polymers (PLLA/PEO, PEG-PLLA, PVA/PVAc, PLGA, PCL/PGC-C18) and encapsulate a diverse array of drugs, including azidothymidine, ACV, and maraviroc. Erodible solid dosage forms constructed from this platform could circumvent the leakage and messiness associated with vaginal gels, but the advantage over films and tablets is yet unproven.

12.7.6   Semen-Triggered Systems

A number of strategies have been developed for semen-triggered vaginal delivery systems, primarily to prevent infection with HIV. Through triggered delivery of large amounts of anti-HIV drugs, these systems have the potential to deactivate virus before an infection event can occur. Three systems are in development: vaginal meshes, osmotic pump tablets, and microgel particles.

Vaginal meshes represent another novel approach to vaginal drug delivery. Huang et al. developed electrospun CAP fibers loaded with either TMC-125 or TDF. Because the CAP fibers have a pH-dependent solubility, they remain intact in the normally acidic pH environment of the vagina. Upon introduction of semen, the fibers dissolve, releasing the anti-HIV drugs.

Rastogi et al. recently reported a vaginal osmotic pump tablet formed from a hydroxypropylcellulose matrix and a pH-sensitive CAP coating. These tablets function as a reservoir at "normal" vaginal pH ($\approx$4–5), providing sustained drug release to the proximal vagina for up to 10 days, but the outer membrane rapidly dissolves at neutral pH, thus potentially providing a semen-triggered bolus of anti-HIV drugs.

An enzymatically triggered microgel system containing the HIV-1 entry inhibitor sodium poly(styrene-4-sulfonate) (PSS) has also been presented. Microgel particles containing PSA peptide substrates and PSS were created; upon exposure to seminal plasma, the microgel degraded and released PSS. A pH- and thermosensitive microgel composed of N-isopropylacrylamide, butyl methacrylate, and acrylic acid has also been developed, capable of burst release upon exposure to semen; initial tests for release of acid orange dye and fluorescein isothiocyanate–dextran proved promising (see also Chapter 14, Section 14.5.1).

12.7.7 PERMEABILITY ENHANCERS AND ENZYMATIC INHIBITORS

There are several instances in literature where permeability enhancers have been investigated in vitro and in vivo to enhance the vaginal absorption of large molecular therapeutics, such as peptides and proteins. These include simple organic acids (e.g., citric acid), cationic surfactants (e.g., benzalkonium chloride), cyclodextrins, and even hydrogen peroxide. However, any permeability enhancers, however safe or biocompatible, will likely weaken the epithelial barrier to infection, thus lessening the usefulness of such an approach.

Other studies are based on the use of aminopeptidase inhibitors, including bestatin, leupeptin, pepstatin, sodium glycocholate, aprotinin, and p-chloromercuribenzenesulfonic acid, to enhance the stability of proteins when delivered vaginally. An interesting approach demonstrated that a thiolated variant of carbopol 974P (a common vaginal gel polymer) inhibited aminopeptidase activity against LHRH in vitro. Again, the effect of any such inhibition on the effectiveness of the epithelial barrier must be considered in a drug delivery strategy.

12.7.8 ADDITIONAL FUTURE TECHNOLOGIES

In addition to those described in detail earlier, the following technologies are under investigation at the cutting edge of FRT drug delivery:

- Genetically modified lactobacillus that excrete antigen-binding fragments against STD
- IVR systems that deliver lactobacillus to modify commensal bacteria populations
- IVR that deliver moisturizers to treat vaginal dryness
- Mucosal vaccines that seek to elicit a focused immune response in the genital mucosa

12.8 OTHER SYSTEMS WITH THE FRT AS THE PRIMARY SITE OF ACTION

There are several non-FRT-contacting dosage forms that have the FRT as the primary site of action. One example is the oral PrEP (see earlier), which includes products such as Truvada®, a pill that combines 300 mg TDF and 200 mg emtricitabine. Other examples include HPV vaccination (Gardasil®) and antibiotics for treatment of infections including syphilis, cervicitis, gonorrhea infection, chancroid, chlamydia, salpingitis, and genital tuberculosis. Antivirals for HSV-2 (acyclovir, famciclovir, and valacyclovir) are given systemically as well.

Combined oral contraceptives (COC), containing both estrogen and progestin, are a common form of oral contraceptive, with effectiveness based on adherence of scheduled use. COCs work by suppression of both the follicle-stimulating hormone and luteinizing hormone as well as thickening of the cervical mucus and possibly altering tubal transport. Oral progestin-only contraceptive pills work through affecting the cervical mucus and endometrium and are also approximately as effective as COCs. Low-dose COCs and progestin-only contraceptives (oral, injectable, and intrauterine form) may also be used in the treatment of abnormal uterine bleeding.

Women experiencing polycystic ovary syndrome (PCOS) desiring to become pregnant may be treated with the selective estrogen receptor modulator, clomiphene citrate. PCOS may also be treated with injectable human gonadotropins. Abnormal uterine bleeding can also be managed with one of low-dose combination oral contraceptives, cyclic oral progestin, injectable progestin (Depo-Provera®), parenteral estrogen, and gonadotropin-releasing hormone agonists.

12.9   IN VIVO AND EX VIVO MODELS FOR FRT DRUG DELIVERY

Several animal models are used in the pharmacokinetic, safety, and efficacy testing of FRT dosage forms. However, anatomical and physiological differences between the human FRT and that of these animals should be considered when interpreting any results obtained. A key difference is the vaginal pH, which is neutral or near neutral in all animal models presented here (between 6 and 8), in contrast to the typical acidic pH (discussed earlier) in premenopausal women.

12.9.1   Rabbit

The New Zealand white rabbit (NZWR) model has been used extensively in the preclinical testing of vaginal DDS. Pharmacokinetic evaluations of vaginal dosage forms in NZWR date back to testing of early silicone IVR segments by Chien et al. in the 1970s. These studies also demonstrated a method for establishing empirical correlations between drug release rates from topical devices and blood–plasma concentrations. This practice continues today with the testing of IVR segments for HIV PrEP and contraception. Although such studies can provide valuable insight into the performance of vaginal formulations, care should be taken when interpreting results due to key differences between human and rabbit vaginal anatomy. First, two distinct epithelial types are found in the rabbit vagina. The lower third of the tract contains a stratified squamous epithelium similar to the human tract but the upper two-thirds of the tract, proximal to the urethral opening, are lined with a single layer of ciliated columnar cells, similar to those found in the human (and rabbit) endocervix. This difference is likely to affect the vaginal absorption and distribution of drugs, but such an effect has not been quantified.

NZWR are also used as a standard model for testing the safety of vaginally applied products, including drug formulations/drug-containing devices. A rabbit vaginal irritation test is specified in the ISO 10993 ("Biological Evaluation of Medical Devices"). The irritation test score is a composite of four histological examinations of necropsied vaginal tissue: epithelial integrity, leukocyte infiltration, vascular congestion, and edema.

NZWR have also been used to demonstrate the efficacy of vaginally administered contraceptives. Another interesting application of NZWR is the use of cutaneous transmission of cottontail rabbit papillomavirus as a model for vaginal HPV transmission.

12.9.2   Nonhuman Primates

NHP, most commonly pig-tailed and rhesus macaques, are used as a model for PK and efficacy testing for HIV-prevention formulations and devices. As HIV only infects humans, there is no way to demonstrate the prevention of vaginal transmission of HIV in an animal model. However, in many cases, antiretroviral drugs effective against HIV are also active against some strains of simian immunodeficiency virus (SIV) and/or SHIV. The potential efficacy of HIV PrEP products can be evaluated through vaginal SIV and SHIV challenge in macaques during or after dosing. This practice effectively proved the concept of PrEP through challenge studies in several vaginal gels and IVR, even in the face of several clinical failures now attributed to poor product adherence. For NHP, a reduced-diameter variant of IVR (approximately half of that of a human-sized IVR) must be manufactured, due to the smaller size of the macaque vagina.

12.9.3 SHEEP

The sheep model has become increasingly popular for PK and safety testing of vaginal products. The sheep vaginal tract is similar in size to the human tract allowing for the testing of full-sized nonerodible devices, especially for IVR. Interestingly, the rabbit vaginal irritation evaluation and score, described earlier, have been adapted for safety evaluations of ARV-releasing IVR in sheep.

12.9.4 MICE

Mice are susceptible to multiple STI, including HSV, allowing for the early assessment of the potential efficacy of PrEP products. However, the physiological nature of the murine vaginal mucosa confounds these assessments. The murine estrous cycle is rapid (4–5 days), necessitating hormonal synchronization of the mucosa. The epithelium is keratinized in the estrus phase, making mice less susceptible to infection, but thins in diestrus to more closely resemble the human epithelium. Similar to NHP, dosing mice with progestin can induce diestrus allowing for transmission events to be studied in the context of topical prevention. However, dosing mice with estradiol can induce estrus, wherein the mouse CVF more closely resembles human CVF and becomes useful in the study of drug and particle transport in mucus. Humanized mouse models can be susceptible to HIV infection and have been used in challenge studies to demonstrate potential efficacy.

Human endometrial grafts have also been used to create a model that allows for study of the effects of local progestin delivery, such as from the Mirena® and Skyla® IUS, in immunocompromised mice.

A major limitation for any mouse model is the size difference between the human and mouse vagina, especially for large devices such as IVR.

12.9.5 IN VITRO AND EX VIVO MODELS FOR FRT DELIVERY

Several ex vivo and in vitro models are used to evaluate permeability of drugs in the vaginal mucosa. The full porcine vaginal mucosa is often used to assess the permeability of vaginal drug delivery candidates. Drug diffusivity has been observed to be similar, but not always identical between the two tissues, meaning that data obtained can be a good first approximation. However, as with all explanted tissues, the exact thickness of tissue samples can be difficult to determine, hampering the quantitative evaluation of drug permeability. The EpiVaginal™ series of models are a multilayered in vitro culture of human-derived vaginal/ectocervical cells and dendritic cells that approximates the human vaginal mucosa. EpiVaginal™ tissues can be grown with a more reproducible thickness in transwells for comparative assessments of drug permeability and formulation toxicity.

To assess the potential efficacy of antiviral formulations ex vivo, macaque tissues can be biopsied following dosing and challenged with SIV or SHIV.

12.10 MAJOR QUESTIONS FOR FUTURE RESEARCH

A number of major questions in the field of FRT drug delivery have yet to be addressed thoroughly and will play an important role moving forward with systems designed to impact women's health. We include here several areas of investigation that, if addressed, would advance the field:

- The role of CVF in drug delivery
- The importance of mucus in the FRT as an innate immune barrier and a barrier to drug delivery
- The role of epithelial structure in the lower human FRT and how cyclical changes in the epithelium affect drug absorption

- Development of organ-wide transport models that describe how drugs are transported throughout the FRT
- Development of better tools to understand user perceptions, biases, and desires around FRT-targeted products, to inform the design of better systems with larger impact on women's health
- Development of more predictive in vitro and animal models, to evaluate the many systems being studied

12.11 CONCLUSIONS

Drug delivery to the FRT has proven useful for several applications, most notably for contraception, hormone replacement, fertility maintenance, and topical treatment and prophylaxis of infection. These interventions have been achieved through a wide variety of dosage forms, including tablets, gels, creams, films, and innovative long-acting devices, such as IVR and IUD/IUS, which have been a major driving force in the evolution of controlled-release science. The major advantages to delivery by this route are the reduction of unnecessarily high drug concentrations when a topical effect is desired, the avoidance of first-pass hepatic metabolism when a systemic effect is desired, and the potential for long-acting prophylaxis and therapy to increase adherence and compliance. Delivery by the FRT also has several drawbacks, most notably the intra- and interuser/patient variability in the vaginal environment, which can induce high variability in PK, and thus confounding accurate assessments of PK, PD, and efficacy. Naturally, it is most logical to investigate this route for indications pertaining to women's health, regardless of the potential feasibility for other applications.

FURTHER READING

Alexander, N.J., E. Baker, and M. Kaptein. 2004. Why consider vaginal drug administration? *Fertil Steril* 82(1):1–12.

Chien, Y.W., S.E. Mares, J. Berg et al. 1975. Controlled drug release from polymeric delivery devices. III: In vitro-in vivo correlation for intravaginal release of ethynodiol diacetate from silicone devices in rabbits. *J Pharm Sci* 64(11):1776–1781.

Clark, J.T., M.R. Clark, N.B. Shelke et al. Engineering a segmented dual-reservoir polyurethane intravaginal ring for simultaneous prevention of HIV transmission and unwanted pregnancy. *PLOS ONE* 9(3):e88509.

das Neves, J. 2014. Vaginal delivery of biopharmaceuticals. In *Mucosal Delivery of Biopharmaceuticals: Biology, Challenges and Strategies*, J. das Neves (ed.), pp. 261–280. New York: Springer.

das Neves, J. and M.F. Bahia. 2006. Gels as vaginal drug delivery systems. *Int J Pharm* 318(1–2):1–14.

Friend, D.R., J.T. Clark, P.F. Kiser et al. 2013. Multipurpose prevention technologies: Products in development. *Antiviral Res* 100(S):S39–S47.

Garg, S., D. Goldman, M. Krumme et al. 2010. Advances in development, scale-up and manufacturing of microbicide gels, films, and tablets. *Antiviral Res* 88(S1):S19–S29.

Geonnotti, A.R. and D.F. Katz. 2010. Compartmental transport model of microbicide delivery by an intravaginal ring. *J Pharm Sci* 99(8):3514–3521.

Gibbs, R.S., B. Karlan, A. Haney et al. 2008. *Danforth's Obstetrics and Gynecology*. Philadelphia, PA: Lippincott Williams & Wilkins.

Harwood, B. and D.R. Mishell Jr. 2001. Contraceptive vaginal rings. *Semin Reprod Med* 19(4):381–390.

Hoffman, A.S. 2008. The origins and evolution of "controlled" drug delivery systems. *J Control Release* 132(3):153–163.

Kiser, P.F., T.J. Johnson, and J.T. Clark. 2012. State of the art in intravaginal ring technology for topical prophylaxis of HIV infection. *AIDS Rev* 14(1):62–77.

Kuo, P.Y., J.K. Sherwood, and W.M. Saltzman. 1998. Topical antibody delivery systems produce sustained levels in mucosal tissue and blood. *Nat Biotechnol* 16(2):163–167.

Malcolm, R.K. 2008. Vaginal rings for controlled-release drug delivery. In *Modified-Release Drug Delivery Technology*, M.J. Rathbone (ed.), pp. 499–510. New York: Informa Healthcare.

Mishell, D.R. Jr. and M.E. Lumkin. 1970. Contraceptive effect of varying dosages of progestogen in silastic vaginal rings. *Fertil Steril* 21(2):99–103.

Nel, A., S. Smythe, K. Young et al. 2009. Safety and pharmacokinetics of dapivirine delivery from matrix and reservoir intravaginal rings to HIV-negative women. *J Acquir Immune Defic Syndr* 51(4):416–423.

Okada, H. and A.M. Hillery. Vaginal drug delivery. In *Drug Delivery and Targeting: For Pharmacists and Pharmaceutical Scientists*, 1st ed., A.M. Hillery (ed.), pp. 301–328. Boca Raton, FL: CRC Press.

Rivera, R., I. Yacobson, and D. Grimes. 1999. The mechanism of action of hormonal contraceptives and intra-uterine contraceptive devices. *Am J Obstet Gynecol* 181(5):1263–1269.

Saltzman, W.M., J.K. Sherwood, and D.R. Adams. 2000. Long-term vaginal antibody delivery: Delivery systems and biodistribution. *Biotechnol Bioeng* 67(3):253–264.

Sloane, E. 2002. *Biology of Women*, 4th ed. New York: Delmar.

Smith, J.M., R. Rastogi, R.S. Teller et al. 2013. Intravaginal ring eluting tenofovir disoproxil fumarate completely protects macaques from multiple vaginal simian-HIV challenges. *Proc Natl Acad Sci USA* 110(40):16145–16150.

Tortora, G.J. and B.H. Derrickson (eds.). 2013. *Principles of Anatomy and Physiology*, 14th ed. Hoboken, NJ: Wiley.

van den Heuvel, M.W., A.J.M. van Bragt, A.K. Mohammed et al. 2005. Comparison of ethinylestradiol pharmacokinetics in three hormonal contraceptive formulations: The vaginal ring, the transdermal patch and an oral contraceptive. *Contraception* 72:168–174.

13 Ophthalmic Drug Delivery

Clive G. Wilson, Mariam Badawi, Anya M. Hillery,
Shyamanga Borooah, Roly Megaw, and Baljean Dhillon

CONTENTS

13.1   INTRODUCTION

Vision problems constitute a major burden to society, although some sight-threatening diseases (such as cataracts) are now largely successfully treated. The leading causes of blindness and low vision in the Western world are primarily diseases of the elderly population, including age-related macular degeneration (AMD), diabetic retinopathy, and glaucoma. As the population demographic changes, treating these diseases becomes increasingly important as a health-care priority.

Effective drug delivery to the eye is extremely challenging, and unlike the other epithelial routes described in this book, ophthalmic drug delivery is used only for the treatment of *local* conditions of the eye and is not considered as a portal of drug entry to the *systemic* circulation. Significant advances have been made to optimize the localized delivery of medication to the eye, so that the route is now associated with highly sophisticated drug delivery technologies, some of which are unique to the eye. We begin this chapter with a consideration of the anatomical and physiologic barriers of the eye and the concomitant limitations they impose for successful ophthalmic delivery. The pathology of the most important ophthalmic diseases is described, as well as their location in the eye, as this has important implications for the delivery route chosen. This is followed by a discussion of the various routes of ophthalmic drug delivery and the associated drug delivery and targeting technologies for each route.

13.2   ANATOMY AND PHYSIOLOGY OF THE EYE: IMPLICATIONS FOR OPHTHALMIC DRUG DELIVERY

The function of the eye is to produce a clear image of the external world and to transmit this to the visual cortex of the brain. In order to do this, the eye must have constant dimensions, an unclouded optical pathway, and the ability to focus light on the retina. These requirements, and the need for protection of the globe, determine the special structure of the eye and its associated physiological responses.

13.2.1   ANATOMY OF THE EYE

The eye is a spherical fluid-filled structure composed of three layers: (1) the sclera/cornea, (2) the uvea, and (3) the retina (Figure 13.1). The sclera, the tough outer "white" of the eye, is a layer of dense connective tissue that covers the entire eyeball, except for the cornea. Its stiff nature helps maintain the globe's shape, making it more rigid, and protecting its inner parts. Extraocular muscles attach to the sclera through tendons, allowing the eye to move. The cornea is a transparent avascular layer at the front of the eye (FOTE), which covers the colored iris. Its curved structure helps to focus light onto the retina (Elkington and Frank 1999).

The anterior cavity (anterior to the lens; FOTE) is subdivided into two chambers: (1) the anterior chamber, between the cornea and the iris, and (2) the posterior chamber, which lies behind the iris, but in front of the lens. The lens is composed of proteins called crystallins, arranged like the layers of an onion; it lacks blood vessels and is completely transparent. Both chambers of the anterior cavity are filled with circulating aqueous humor, described further in the following texts.

The vitreous chamber (posterior to the lens; back of the eye [BOTE]) is the larger, posterior cavity of the eye. It contains a clear, jellylike, viscoelastic gel, the vitreous humor, which is formed during embryonic development and is not replaced thereafter. The character of this gel changes through life, becoming less structured, and therefore more liquid, in the elderly.

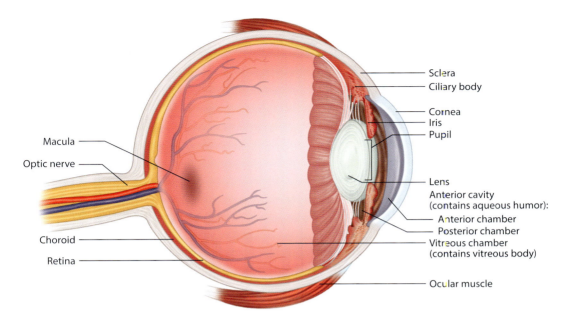

Sclera
Ciliary body
Cornea
Iris
Pupil
Macula
Optic nerve
Lens
Anterior cavity
(contains aqueous humor):
Anterior chamber
Posterior chamber
Vitreous chamber
(contains vitreous body)
Choroid
Retina
Ocular muscle

FIGURE 13.1 Sagittal section of the eye.

The second layer is the uvea or uveal tract, which is composed of the choroid, iris, and ciliary body. The choroid is highly pigmented (the melanin limits reflections within the eye) and contains many blood vessels that supply the retina. Anteriorly, the choroid meets the iris and the ciliary body. The iris is the thin, pigmented structure in the eye, the shape of a flattened doughnut, responsible for controlling the diameter of the pupil and thus the amount of light reaching the retina. The ciliary body comprises both the ciliary muscle and the ciliary processes. Contraction of the ciliary muscle relaxes the zonular fibers, resulting in a change in the curvature of the lens and facilitating fine focus (accommodation).

The ciliary processes are capillary-rich protrusions on the internal surface, which produce and secrete the aqueous humor. This transparent, slightly alkaline fluid contains less protein and glucose than plasma, but more ascorbic acid, an important antioxidant (Civan and Macknight 2004). Aqueous humor nourishes the lens and cornea, carrying oxygen and nutrients and removing waste, as it circulates through the anterior cavity. It subsequently drains at sclerocorneal trabecula into Schlemm's canal, to be resorbed into the blood. Normally, aqueous humor is completely replaced every 90 minutes.

The third layer of the eye is the retina, which contains a pigmented layer and a neural layer. The neuroretina comprises three distinct layers of neurons, interconnected by synapses: the photoreceptor layer, the bipolar cell layer, and the ganglion cell layer (Figure 13.2). Visual data are transduced by the photoreceptors and processed through the other neural layers before being transmitted as nerve impulses along the axons that form the optic nerve. There are two types of photoreceptor: rods (especially important for "gray" night vision) and cones (daytime vision). The larger cones only make up 5% of the total population of visual receptors and thus the retina is dominated by rods (about 100 million in man). The 5–6 million cones comprise S-type cones detecting blue light, M-type cones most sensitive to green light, and L-type cones detecting red light. The central retina, known as the macula, has a high density of cones, thus allowing for high-resolution central vision.

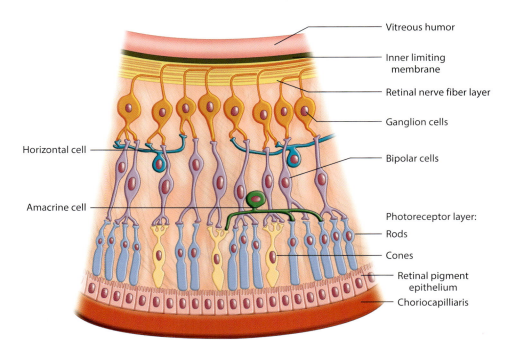

FIGURE 13.2 Microscopic structure of the retina.

13.2.2 Limiting Factors for Ophthalmic Delivery

A considerable number of anatomical and physiological barriers exist to protect the eye from the entry of foreign substances. These barriers act as major limiting factors for ophthalmic drug delivery, which will be discussed in the following texts. Further reviews of the barrier properties of the eye are discussed in Wilson and Tan (2012) and Mains and Wilson (2013).

13.2.2.1 Visibility Requirements

More than half the sensory receptors in the human body are located in the eyes and a large part of the cerebral cortex is devoted to processing visual information. Transduction of the light energy into a receptor potential occurs in the outer segment of rods and cones. Visual signals are subsequently transmitted through the neuronal cell layers to the retinal ganglion cells, which provide retinal output to the brain. Light must have an unobstructed path to reach the photoreceptors; subsequent neuronal processing and nerve transmission should also be unimpeded.

The eye then affords a particularly challenging route for drug delivery, as a drug delivery system (DDS) cannot interfere with this vital sensory role. For example, other epithelial routes such as the skin, or oral cavity, can use sustained-release technology such as an adhesive patch to provide drug release for a prolonged period of time. The situation is difficult for the eye as even if the device is transparent and well tolerated, any resulting refractive change would cause blinking, possible lacrimation, and attempted removal. This places design limitations for a DDS in the central axis of vision. Similar limitations are encountered within the eye. Delivery systems formulated as colloids and particulates such as nanoparticles and liposomes can be used for controlled release and drug targeting via many epithelial routes. However, within the eye, there is a risk that movement of fluid would transport particulate systems posteriorly, where they could adhere to the lens, causing opacification, or appear as free-floating objects in the visual path, scattering light and decreasing visual acuity.

The extreme sensitivity and delicacy of the eye is a further difficulty and contrasts with more robust drug delivery sites such as those of the skin or the oral cavity. For example, the cornea is one of the most sensitive surfaces in the body, containing 300–600 more pain receptors than the skin (Belmonte and Cervero 1996). Thus, some chelating agents and surfactants, widely explored at high concentrations as a means of enhancing epithelial permeability via the oral, nasal, pulmonary, and buccal routes cannot be used in ocular topical preparations.

13.2.2.2 Tear Film, Nasolacrimal Drainage, and Reflex Lacrimation

The tear film comprises a three-layer structure: an inner mucin layer produced by conjunctival goblet cells that adheres to the cornea, a middle aqueous layer secreted by the lacrimal gland, and an external lipid layer produced by meibomian glands of the eyelid (Figure 13.3).

In the physiological process of nasolacrimal drainage, the tear film spreads over the surface of the eyeball due to the blinking of the eyelids. The tears are then cleared away as rapidly as they are produced, draining first into the lacrimal canals, then the lacrimal sac, and finally into the nasal cavity (Figure 13.4). The tear film and the process of nasolacrimal drainage combine to protect the eye by removing foreign material and form the first optical surface. Tears also lubricate and moisten the eyeball and carry nutrients to the cornea and conjunctiva. Each blink partially redistributes the tear film over the surface of the cornea and conjunctiva, and nutrients are replenished.

This process has important implications for drug delivery to the eye surface. Any topical medication must be retained on the surface of the eye despite a constant flow of tears and continuous eye blinking. In practice, nonviscous topical medications are rapidly cleared by this mechanism: up to 90% of the dose administered from conventional eyedrops is lost in the first 15 seconds after administration. Clearance by nasolacrimal drainage results in drug loss and also raises the possibility of systemic side effects, due to drug absorption through the mucous membrane of the nasolacrimal duct. This represents a particular concern for drugs such as the beta-blockers used in the management of glaucoma: unwanted systemic absorption can result in adverse cardiovascular and respiratory effects.

Under normal conditions, the human tear volume is about 7–9 μL and is relatively constant. The maximum amount of fluid that can be held in the lower eyelid sack is 15–20 μL, but only 3 μL

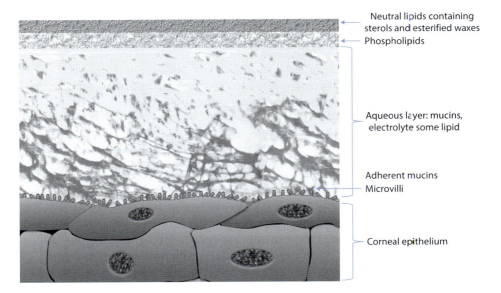

Neutral lipids containing sterols and esterified waxes
Phospholipids

Aqueous layer: mucins, electrolyte some lipid

Adherent mucins
Microvilli

Corneal epithelium

FIGURE 13.3   The structure of the tear film.

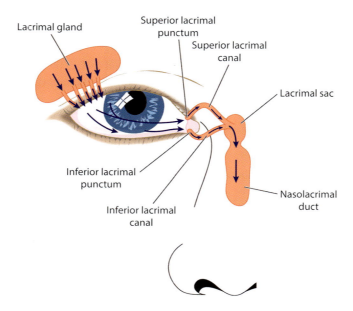

FIGURE 13.4 Lacrimal flow and drainage. The lacrimal gland produces tears, which are distributed by blinking. At normal rates of flow, tears are cleared by nasolacrimal drainage. Sudden increases produced by instillation, or reflex tearing, overspill the lower fornix and roll over the cheek. (Modified from Blamb/Shutterstock.com.)

of a solution can be incorporated in the precorneal film without causing it to destabilize. Eyedrop administration results in a sudden increase in tear volume, causing rapid reflex blinking. The volume of a drop is typically 33–38 μL and consequently some of the eyedrop can be spilled on the cheeks and eyelashes, resulting in drug wastage. Further, a large proportion of the retained instilled volume is pumped into the nasolacrimal duct, resulting in drug loss and the risk of systemic side effects.

Any irritation of the eye surface will induce a reflex lacrimation and blinking, as a protective mechanism. Again, this has implications for topical drug delivery, in particular for the formulation of topical products. The exposure of the eye surface to an acidic fluid may cause protein denaturation, forming insoluble complexes. Alkalization of the tear film tends to cause a saponification of lipids due to an interaction of the hydroxyl ions with cell membranes. At a high pH, alkalis behave as hydrophobic species and are more damaging to the cornea than acids (since precipitation of protein limits further ingress). Mildly acidic solutions are therefore better tolerated than slightly alkaline solutions, and reflex tears neutralize the irritant solution by dilution but will also wash away the instilled drug.

13.2.2.3   Epithelial Permeability Barriers

The eye presents sequential epithelial barriers, which reduce flux of nonnutrients into the eye, as a further natural defense mechanism against the entry of hazardous agents. As described in detail in Chapter 4 (Section 4.3), drug permeation across epithelial barriers is generally restricted to two main routes: the paracellular route (between adjacent epithelial cells) and the transcellular route (across the epithelial cells). The presence of tight junctions with resistances in excess of 120 Ωcm^2 between the epithelial cells of the eye severely restricts paracellular transport, confining the route to very small, hydrophilic molecules. The alternative, transcellular transport, is a viable transport pathway for drugs that possess the requisite physicochemical properties to allow for passive diffusion across many layers of epithelial cells. The physicochemical properties that favor transcellular diffusion are broadly the same as those affecting transepithelial absorption at any site and have been discussed in Chapter 4. In summary, favorable transport properties for transcellular passive diffusion include high lipid solubility and partition coefficient, low molecular weight and compact molecular volume, absence of charge, aqueous solubility, and chemical stability.

In medicines development through the turn of the century, massive jumps were made in the production of materials from cell sources rather than by the chemical synthetic route. As described in other chapters, these new "biologics" include peptide and protein drugs, as well as DNA-based therapeutics. For the eye, an early interest was in agents that restricted the uncontrolled growth of immature vessels following retinal ischemia—notably compounds that regulated expression of vascular endothelial growth factor (VEGF) and gene vectors. Such compounds had already appeared in cancer compound selection, since anti-VEGF molecules restrict the growth of solid cancers. All the drugs in this group are large and polar and therefore tend to have very low flux across membranes. Usually, they must be introduced by intravitreal injection, as the epithelial barriers restrict their penetration to the BOTE (see Section 13.7.1.1).

The principal epithelial barrier to drug transport in the eye was thought to be the cornea, but since it makes up only 20% of the outer eye area in contact with an instilled drop, the conjunctiva must also play a significant role in the absorption process. The conjunctiva is a thin, transparent mucous membrane composed of highly vascularized, nonkeratinized, stratified, columnar epithelium, which is approximately 5–15 cell layers thick. It also contains goblet cells that help lubricate the eye by producing the mucous component of the tear film. The conjunctiva lines the inside of the eyelids (tarsal conjunctiva) and the anterior one-third of the eyeball, apart from the cornea (bulbar conjunctiva). The tight junctions and multilayered nature of the conjunctiva make it an important barrier against infection but also limit the permeation of topically applied drugs. Since the conjunctiva is highly vascularized, significant drug loss to the systemic circulation can occur, resulting in drug wastage and the risk of systemic adverse effects.

The adult cornea measures 11–12 mm horizontally and 9–11 mm vertically. Corneal thickness varies, being thinnest centrally (550 μm) and gradually increases to the periphery (700 μm). It is an avascular five-layered structure comprising (anteriorly to posteriorly) the epithelium, Bowman's membrane, stroma, Descemet's membrane, and endothelium (Figure 13.5).

The outer corneal epithelium of the cornea constitutes the most important barrier for corneal drug flux. It comprises five to six layers of nonkeratinized, stratified, squamous epithelium, with an abundance of tight junctions. As introduced earlier, the tight junctions form a highly resistant barrier to paracellular transport, so that transport across the cornea is usually limited to the transcellular route. The inner stroma layer of the cornea (about 90% of the thickness of the cornea in most mammals) is composed of a modified connective tissue, rich in collagen, other proteins, and mucopolysaccharides. About 70%–80% of the wet weight is water, which presents a barrier to highly lipophilic drugs. The innermost layer of the cornea is a monolayer of endothelium, but this has leaky endothelial junctions that do not impede drug transit.

13.2.2.4 Other Physical Barriers of the Eye

Deep to the outer corneal and conjunctival epithelial barriers, the eye presents further obstacles to drug transport.

The sclera: The sclera is a layer of dense connective tissue, consisting primarily of collagen and elastin fibers, as well as proteoglycans, embedded in an extracellular matrix. It is thinnest at the site of extraocular muscle attachment (0.3 mm) and thickest at the posterior pole (1 mm). The essentially hydrophilic proteoglycan extracellular matrix can be expected to impede the diffusion of hydrophobic drugs through its tissue. Positively charged drugs may also interact with negatively charged proteoglycans and thus show reduced flux.

The retina: The retinal pigment epithelium (RPE) comprises a monolayer of cuboidal epithelial cells, lining the back of the retina (Figure 13.2). Tight junctions between the RPE cells form a blood/retina barrier and limit the access of drug molecules from the blood circulation

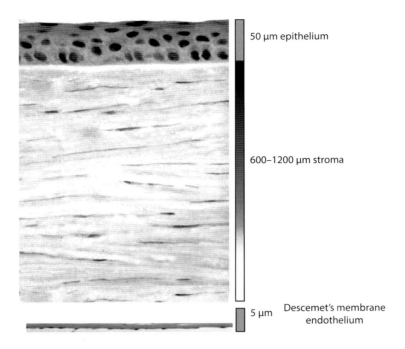

FIGURE 13.5   The corneal epithelial barrier.

to the retina. The RPE also contains several efflux pumps including P-glycoprotein and multidrug resistance–associated protein, which further limit the passage of compounds into the vitreous. The inner retina receives a blood supply from the central retinal artery, which passes with the optic nerve into the eye. These blood vessels are lined with tightly packed endothelial cells thus forming the inner limiting membrane (ILM). The barrier in the inner globe is the meshwork of ILM and neuronal cells.

Vitreous humor: The vitreous is a viscoelastic gel composed of Type II collagen fibers, which trap coiled hyaluronic acid and water molecules. As the volume of adult human vitreous is around 4 mL, the vitreous humor presents a large diluting sink. Additionally, the network of collagen and hyaluronic acid presents a diffusional barrier, limiting the passage of cells and macromolecules, particularly cationic molecules and cationic carriers, due to charge interactions.

13.3   DISEASES OF THE EYE

The eye is prone to a wide variety of diseases, which may be of a systemic origin, such as diabetes or hypertension, or peculiar to the eye, such as glaucoma, cataract, and macular degeneration. Many of these conditions, such as macular degeneration, glaucoma, and diabetic retinopathy are chronic conditions that require continuous therapy for treatment. Furthermore, since the eye is located on the surface of the body, it is also easily injured and infected. According to the location of the disease, ocular disorders are grouped as disease of the anterior cavity or vitreous chamber.

13.3.1   DISEASES OF THE ANTERIOR CAVITY

Dry eye is a common problem, estimated to affect over 4 million Americans (Schein et al. 1997) and its incidence increases with age. Dry eye will result if the composition of tears is changed, or an inadequate volume of tears is produced. Dry eye conditions are not just a cause for ocular discomfort, but prolonged dry eye can lead to corneal damage, keratitis, and compromised vision.

A variety of different inflammatory eye conditions are described, according to the region of the eye affected. For example, blepharitis is an inflammation of the eyelid, most commonly caused by a bacterial infection, usually *Staphylococcus aureus*. Conjunctivitis is an inflammation of the conjunctiva, resulting in redness to the visible white of the eye. Allergic conjunctivitis (AC) is defined as inflammation of the conjunctiva in response to an allergen. AC may be seasonal or perennial. Primary symptoms are itching, burning, and then a watery discharge. Bacterial conjunctivitis can result in pain and a purulent discharge, while viral conjunctivitis results in a clear discharge. Trachoma is an infectious cause of conjunctivitis caused by the organism *Chlamydia trachomatis*. The resulting chronic granular inflammation causes a fibrosis of the conjunctival lid surface, resulting in inversion of the lid margin. Subsequent abrasion from eye lashes results in keratitis and can lead to blindness unless treated; it is the most common cause of blindness in North Africa and the Middle East (Burton and Mabey 2009). Keratitis is an inflammation or infection (bacterial, viral, or fungal) of the cornea. Patients with keratitis present with decreased vision, ocular pain, foreign body sensation, red eye, and often a cloudy/opaque cornea.

Glaucoma refers to a group of different eye conditions, in which damage to the optic nerve leads to progressive, irreversible, vision loss. It is sometimes categorized as a stand-alone disease, separate from the posterior segment diseases, since it is treated by local topical therapy to the anterior eye. It is considered one of the major global ophthalmic clinical problems; it is the second leading cause of blindness, after cataracts, worldwide. In the United States, primary open-angle glaucoma (POAG) is the most common form of the disease, accounting for 90% of all cases. It occurs mainly in the over 50-year-olds and is characterized by an increase in the intraocular pressure (IOP) caused by a restriction of outflow and decreasing oxygenation. The associated nerve damage involves loss of retinal ganglion cells. Visual loss develops first in the peripheral visual field and then progresses to the central visual field. POAG is a painless, insidious disease in which injury develops over years. Symptoms are absent until extensive optic nerve damage has been produced. If the condition is left untreated, blindness will result.

13.3.2 Diseases of the Vitreous Chamber and Retina

Intraocular infections are those infections of the inner eye, including the aqueous humor, iris, vitreous humor, and retina. They may occur after ocular surgery, trauma, or may arise from systemic infections. Intraocular infections carry a high risk for loss of vision and may result in the spread of infection from the eye into the brain.

Uveitis refers to an inflammation of the anterior uvea (that portion of the eye including the anterior chamber, iris, and ciliary body) and surrounding tissues. In addition, a posterior segment component (retinochoroidal) may be involved. Uveitis may arise due to an autoimmune disorder, following an infection, or other causes. If untreated, inflammation in the anterior segment may lead to glaucoma and cataracts, while inflammation in the posterior segment may lead to macular edema.

Cytomegalovirus retinitis (*CMVR*) is an inflammation of the retina of the eye, which may lead to blindness. It is caused by an infection of human CMV (HCMV), a herpes-type virus. HCMV opportunistically infects patients with a weakened immune system; it came to the attention of the public with the appearance of HIV in the 1990s and may be masked by the concomitant use of corticosteroids. Much of the early research on ocular implants was stimulated by the urgent need to treat this disease.

*AMD** is a degenerative disorder of the retina, typically in persons aged 50 years and older. Central vision becomes increasingly blurred, thereby limiting perception of fine detail, which is required for activities like reading, driving, recognizing faces, and seeing the world in color. The disease begins as the more common dry AMD (85% of patients) but can later progress to the second,

---

* Age-related macular degeneration is referred to as AMD in the United States and ARMD in much of Europe.

much more severe form, known as wet AMD (neovascular AMD; 15% of patients), which is associated with acutely progressive vision loss.

In dry AMD, debris associated with the turnover of photoreceptors surrounds the most light-sensitive region of the retina, the macula lutea. The pigmented layer atrophies and degenerates, resulting in a gradual deterioration of central vision. In wet AMD, degeneration of the macula is due to the growth of new subretinal blood vessels, which are fragile and leaky; fluid leakage causes permanent macula injury.

The emergence of CMVR (as a consequence of HIV infection) in the late 1970s was a significant driver at the time for the development of new ocular delivery technologies such as ocular inserts. In a similar manner, a large section of current ophthalmic drug delivery research is driven by the need to achieve successful drug delivery to the BOTE in order to treat AMD. AMD remains a leading cause of irreversible blindness and visual impairment in the world (Agarwal et al. 2015). As many as 11 million people in the United States have some form of macular degeneration. Although previous estimates suggested that age is a prominent risk factor, with 2% risk of contracting the disease for those ages 50–59, to nearly 30% for those over the age of 75, the incidence of AMD in the United States is declining in each successive generation and adults born after World War II have a significantly lower risk of early disease progression (Cruickshanks et al. 2014).

Diabetic retinopathy is one of the major complications of diabetes and results from hyperglycemia-induced changes in the retinal microvasculature. Initially, the damage is limited to microaneurysms in the retinal capillary walls, which can leak fluid. It then progresses to scarring, followed by the proliferation of a new leaky vasculature, stimulated by VEGF. The overgrowth of new, unstable, leaky retinal capillaries reduces visual acuity. Diabetic macular edema (DME) is caused by leaking macular capillaries, which cause fluid and protein deposits to collect on, or under, the macula of the eye, causing it to thicken and swell, and distorting central vision. Retinopathy is accelerated by hyperglycemia, hypertension, and smoking.

13.4 OVERVIEW OF DRUG DELIVERY TO THE EYE

In the search for an efficient route for ocular drug delivery to the eye, various complex interrelated factors must be considered. These considerations include the limitations generated by the anatomy and physiology of the eye, the pathology and location of the ocular disease, and the accessibility of a site for drug delivery. Thus, a number of different approaches are possible, which are summarized in Figure 13.6 and described in detail in the following texts.

13.4.1 TOPICAL DELIVERY

Topical delivery is useful in the treatment of local conditions of the surface of the eye, for example, conjunctivitis, but it is limited by the short residence time of conventional eyedrops. Mechanisms to enhance residence time include the incorporation of mucoadhesive polymers into eyedrop formulations, as well as the use of micro- and nano-DDS, contact lenses, and intraocular inserts.

Applying a drug topically to the eye can also, in principle at least, be used to treat BOTE targets such as the retina and macula; in animals such approaches seem to be successful due to the size of the globe. In humans, topically applied drugs may fail to achieve sufficient conjunctival and corneal penetration to access the required BOTE targets. A successful drug would have to be extremely potent and have an exemplary safety profile. Strategies to increase corneal penetration after topical administration include the use of micro- and nanoparticulate DDS and prodrugs.

13.4.2 INTRAVITREAL DELIVERY

For the treatment of BOTE conditions such as AMD and diabetic retinopathy, an intravitreal injection provides therapeutic concentrations of the drug adjacent to the intended site of activity; a much

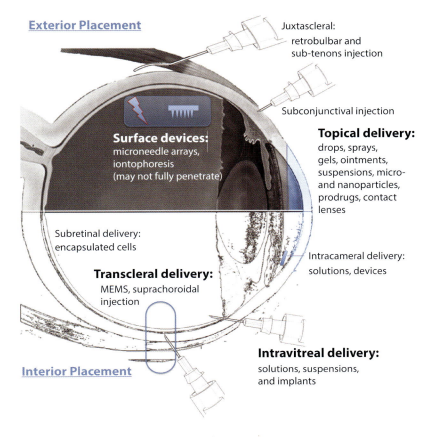

FIGURE 13.6   Different routes of drug delivery to the eye.

smaller dose is required than via the topical route. Intravitreal injections are invasive, and repeated treatment carries the small risk of retinal damage and infection. Consequently, current research is focused on the use of long-acting depots and implants, in order to minimize injection frequency.

13.4.3   TRANSSCLERAL ROUTE

In recent years, attention has turned to forming a scleral depot for achieving sustained drug delivery to the BOTE, as an alternative to topical and intravitreal delivery. Mechanisms to achieve delivery to the retina include transscleral and suprachoroidal injections. Newly emerging technologies to achieve transscleral delivery include iontophoresis and implantable diffusion pumps (Figure 13.6). Other approaches investigate the placement of a device *within* the sclera, for electroporation and iontophoresis. The suprachoroidal space (SCS) can also be targeted using short needles and microneedle arrays. Some of the methods require highly advanced drug delivery technologies and are still at an early stage of exploration.

13.4.4   SYSTEMIC ROUTE

The systemic route is another potential means by which medications can gain access to the BOTE. A drug administered orally or intravenously (i.v.) that is circulating in the blood can distribute into the eye, although distribution is poor as the eye only receives a relatively small proportion of the total blood flow. Drug entry into the posterior segment is also often limited by two significant barriers (Figure 13.2): (1) the outer barrier of the retinal pigment epithelium (RPE) and (2) the inner blood–retinal barrier, comprising the endothelial cells of the retinal blood vessels.

Assuming that the drug can achieve access to the eye via the bloodstream, the route still suffers from the disadvantage that only a very small volume of tissue in the eye requires treatment, yet all the organs of the body are subjected to the action of the drug. Interactions of the drug with other body systems may result in adverse, or even toxic, effects, particularly if the drug has a narrow therapeutic index. High doses of a drug are also required (resulting in increased drug costs and the risk of adverse effects), due to drug dilution in the bloodstream and possible premetabolism of the drug, prior to reaching the eye. For these reasons, there was a move away from oral carbonic anhydrase inhibitors such as Diamox® (acetazolamide) at the end of the 1990s, and currently the systemic route is not generally used for the delivery of pharmacologically active materials to the eye.

There is, however, an important exception: ocular photodynamic therapy (PDT). In this treatment, a nontoxic light-sensitive compound is administered i.v. and selectively accumulates in the retina. It is then exposed to laser light, whereupon it becomes toxic to targeted malignant (or other disease) cells (Kim and Morley 2006). On shining the light, free radicals and super oxide ions generate singlet oxygen, a primary mediator of tissue damage. Singlet oxygen is highly reactive and therefore can only exert effects within 10–20 nm of its generation, which confines the cytotoxic effects to a very close locus near to the target.

PDT is currently used for the treatment of wet AMD, whereby a light-sensitive agent, verteporfin (Visudyne®) is administered i.v. and preferentially accumulates in the abnormal submacular blood vessels of patients with AMD. Laser treatment then selectively generates free oxygen radicals that cause cytotoxic damage and the occlusion of new vessels. A recent update on PDT and the applications of the range of therapeutic approaches in managing AMD is published in an excellent review by Agarwal et al. (2015).

13.5 TRADITIONAL TOPICAL DRUG DELIVERY

Traditional FOTE drug delivery involves instilling drops of a solution (or suspension) of the drug from an eyedrop bottle onto the surface of the eye. 95% of all ophthalmic drugs are delivered using a traditional eyedrop bottle. Drugs for topical delivery include beta-blockers and anticholinergics for the management of glaucoma, NSAIDs and corticosteroids for pain and inflammation, antibacterials and antivirals for eye infections, and antihistamines for allergies. Topical drug delivery localizes the drug effects, facilitates drug entry that is otherwise hard to achieve with systemic delivery, and avoids first-pass metabolism. Furthermore, ophthalmic solutions are easy to prepare, filter, and sterilize.

Pragmatically, the delivery method is imprecise, inaccurate, and inefficient. In practice, topical application frequently fails to establish a therapeutic drug level for a desired length of time within the target ocular tissues and fluids. This is because the eye is extremely efficient at eliminating topically instilled medications. As described in Section 13.2.2.2, the eye has a built-in drainage system designed to protect it from damage and irritation. Eyedrops are subjected to the physiological processes of reflex lacrimation, blinking, and nasolacrimal drainage. Due to the nasolacrimal clearance system, most of an administered dose is lost within seconds of instillation. This means that to produce a therapeutic effect, conventional eyedrops must be administered several times a day, which is cumbersome for the patient and decreases patient compliance. It also leads to issues of poor efficacy and safety problems, with the danger of systemic adverse effects if the drug is absorbed.

In order to optimize drug delivery via conventional eyedrops, a number of key parameters must be controlled. Proper instillation is essential to ensure efficacious treatment. A recent study showed that 90% of glaucoma patients were not administering their drops correctly (Gupta et al. 2012). An eyedrop should be placed in the inferior fornix by tilting the head back and gently pulling the lower lid away from the globe and creating a pouch to receive the drop. The lid is then gently returned to the globe, entrapping a small amount of liquid in the inferior conjunctival sac. Gentle pressure should be applied at the inner corner of the eye to prevent tears from diluting the eyedrops and also to reduce unwanted systemic absorption of the drug. Administration in this way means the drop is retained in the eye up to twice as long as if it is simply dropped over the superior sclera.

A reduced instilled volume can also contribute to improved efficacy in ocular delivery. Volumes of 30–60 µL are typically dispensed from eyedrop bottles, yet an introduced volume of more than 10 µL will result in nasolacrimal drainage. A larger volume will induce reflex blinking and increased lacrimation, which dilute the drug and may cause it to be washed out of the eye, resulting in drug wastage and possibly systemic side effects. The size of the drop is affected by multiple parameters, including the diameter of the opening of the bottle, the handling angle of the bottle, the concentration of polymer in the formulation (higher concentrations giving rise to larger drops), and the viscosity and surface tension of the formulation. Tips capable of delivering a drop of 8–10 µL have been designed by varying the relationship between the inner and outer diameters of the end of the tip. The use of smaller eyedroppers in commercial containers has not been popular and they may pose a safety risk due to the acute angle of the bottle tip. Although a smaller drop may be retained longer in the conjunctival sac, an instilled volume less than 8 µL is not recommended due to the difficulty in making up a suitable drug concentration in such a small volume.

The normal osmolality of tears is almost equivalent to that of normal saline solution. Variations in osmotic pressure between 100–340 mOsm/kg appear to be well tolerated by the eye. Beyond these values irritation takes place, again eliciting reflex tears and reflex blinking. Accordingly, the ophthalmic preparation should be formulated with optimal pH, surface tension, and osmolality values (see USP guidance in Aldridge et al. 2013). The pH should be between 7.0 and 7.7. Some drugs are unstable in this pH range, and therefore need to be formulated at other pH values, but it is preferred that little or no buffering is employed. The surface tension of tear fluid at the eye temperature has been measured as 43.6–46.6 mN/m for normal eyes and 49.6 mN/m for patients with dry eye. The instillation of a solution that lowers the surface tension may disrupt the outermost lipid of the tear film into numerous oily droplets, which become solubilized. The protective effect of the oily film against evaporation of the tear film aqueous layer disappears and dry spots will be formed. The dry spots are painful, irritant, and elicit reflex blinking. The symptoms in sufferers are worsened by activities in which the rate of blinking is reduced, including watching television. In severe dry eye, punctal plugs may be employed to increase the resident tear volume.

Self-administration of eyedrops is often problematic, with one study reporting at least 50% of patients admitting to difficulty in instilling their own eyedrops (Connor and Severn 2011). The most common reported difficulties include problems with aiming the bottle accurately (with patients missing the eye entirely, or dropping the dose onto the eyelids, eyelashes, or cheeks) and being unable to squeeze it sufficiently to expel the dose. Elderly patients and those suffering from rheumatoid arthritis, osteoarthritis, carpal tunnel disease, and stroke have lower grip strengths, which make squeezing the bottle particularly difficult. A patient's hand may wobble because they have to squeeze the bottle at the limit of their capability, and this may compromise their ability to aim accurately.

Eyedrop containers have to be flexible to allow squeezing, yet rigid enough to prevent flooding of the ocular surface. Factors that influence the force requirements include the rigidity of the bottle, the ratio of bottle height to width, the viscosity of the medication, and the length of the dispensing tip. Small, single-dose units tend to have higher force requirements. The force exerted on the bottle determines, in part, the drop size and number of drops actually expelled: wide variability in both these parameters has been described, leading to imprecise drug dosing. There are also psychological factors at play: the eye is such a delicate sensory organ that many patients have a natural aversion to using eyedroppers, being instinctively nervous that they will damage their eyes. Again, this is especially true of the elderly, with their weak eyesight, limited strength, and poor balance.

In order to improve FOTE delivery and reproducibility, some bottles have flexible areas, or a pump action, to facilitate the action of dispensing a drop accurately. A number of eyedrop delivery aids have been developed, including, for example, the Xal-Ease® plastic eyedrop dispenser.

This fits over the existing eye bottle and helps with the removal of the bottle cap and the positioning of the drug over the eye; it then dispenses only a single drop of the medication into the eye. The VersiDoser™ is capable of delivering a precise dose of drug into the eye (ranging from 10 to 50 µL) in virtually any hand/head orientation. Other eyedrop delivery aids include a system of mirrors to help the patient see the dropper tip and devices such as the Autosqueeze™ Eye Drop Dispenser, a tongs-like device to assist with bottle squeezing.

Suspensions are also used in ophthalmic formulations, particularly for anti-inflammatory steroids, which typically exhibit poor aqueous solubility. They are commonly formulated by dispersing micronized drug powder (<10 µm in diameter) in a suitable aqueous vehicle. Suspensions provide a longer duration of action than solutions, as the particles persist in the conjunctival sac, giving rise to a sustained-release effect; however, the presence of particles may result in a foreign body sensation in the eye causing reflex lacrimation and blinking.

13.6   ADVANCED TOPICAL DELIVERY

A wide variety of technologies have been developed in recent years to improve the efficacy of topical delivery to the eye. These have been reviewed extensively, although it must be noted that many of the texts cover the same material from similar viewpoints. A good review of the field is found in Nakhlband and Barar (2011), and other sources are cited later in this chapter. It is established that simple aqueous formulations have relatively low efficacy with regard to drug delivery, which gives scope for innovation. The principal technologies focus on mechanisms to (1) prolong residence time at the ocular surface, to prevent premature clearance, and (2) enhance transcorneal penetration, to deliver topically applied drugs to the interior ocular tissues.

13.6.1   POLYMERIC EXCIPIENTS

Viscosity enhancing polymers provide a thickened solution that reduces lacrimal drainage and increases residence time at the ocular surface. This results in improved efficacy, less frequent dosing, and improved patient compliance. Hydrophilic polymers hold water by weak hydrogen bonding and if they wet surfaces, resist dehydration. A further application of polymeric solutions as functional ingredients is therefore in the supplementation of mucin-deficient tears in postmenopausal dry eye.

The function of a polymer is to interpenetrate tear and surface mucins: chain length, polymer flexibility, and chain segment mobility are key properties of nonionic polymers. The most well-known classes of polymers in this group include cellulosic polymers, such as methylcellulose (MC), hydroxyethyl cellulose, hydroxypropyl MC, and hydroxypropyl cellulose (HPC). They provide a wide range of viscosities (400–15,000 cps) and are compatible with many topically applied drugs. Polyvinyl alcohol (PVA) is also widely used as a drug delivery vehicle and a component of artificial tear preparations. This polymer can reduce interfacial tension at the oil/water interface, enhance tear film stability, be easily sterilized, is compatible with a range of ophthalmic drugs, and is nontoxic.

When extensive hydrogen bonding occurs between a polymer and surface or a solute macromolecule, the polymer may be classified as a bioadhesive. Bioadhesion is an interfacial phenomenon in which a synthetic or natural polymer becomes attached to a biological substrate by means of interfacial forces. If it involves mucin or mucous-covered membrane, the narrower term "mucoadhesion" is employed. Mucoadhesion has been used to enhance bioavailability of drugs via various other routes, including oral/gastrointestinal, oral cavity, nasal, and vaginal routes; further information can be found in the relevant chapters describing these routes. The presence of mucin in the tear film means that this phenomenon can also be exploited for ophthalmic drug delivery and charged anionic polymers can be utilized. As with the nonionic polymers, it seems that the length of the polymer tails must be long enough and mobile to facilitate molecular entanglement. The threshold has been defined as around 100,000 Da in flexible chain motif polymers. When anionic polymers are used, the maximum interaction occurs at an acid pH, suggesting that the polymer must be in its protonated form for viscoelastic synergy with surface mucins.

Sodium hyaluronate is a high-molecular-weight polymer extracted by a patented process from sources including animal sources, bacterial fermentation, and most recently directly using hyaluronate synthetase acting on UDP-sugar monomers. It consists of a linear, unbranched, nonsulfated, polyanionic glycosaminoglycan, composed of one repeating disaccharide unit of D-sodium glucuronate and N-acetyl-D-glucosamine. The polymer is mucoadhesive: the carboxyl groups of hyaluronate form hydrogen bonds with sugar hydroxyl groups of mucin, producing an intimate contact with the cornea. Furthermore, hyaluronate solutions exhibit pseudoplastic behavior (where viscosity is higher at the resting phase), which provides a thickened tear film, slows drainage, and ensures an improved distribution on the cornea during blinking. Sodium hyaluronate is also mixed with xanthan gum, the objective being to produce a tear mucin mimetic with similar viscoelastic and rewetting properties. Addition of an antibiotic, such as netilmicin, is used in the treatment of corneal abrasions (Faraldi et al. 2012).

In situ gelling systems are also used for ophthalmic delivery. A sol–gel transition can be triggered by a change in pH, temperature, or ionic strength of the formulation, upon instillation in the eye. One of the early materials investigated was cellulose acetate phthalate, which provided a pH-triggered system. The polymer has a very low viscosity up to pH 5 but coacervates in contact with the tear fluid at pH 7.4, forming a gel in few seconds and releasing the active ingredient in a sustained manner. Unfortunately, high polymer concentrations (25%) are required, and the instilled solution has a low pH, both of which cause discomfort to the patient.

Poloxamer F127 is a solution at room temperature, but when it is instilled onto the eye surface, the elevated temperature (34°C) causes the solution to become a gel, thereby prolonging its contact with the ocular surface. Again, the system suffers from the disadvantage that it requires a high polymer concentration (25% poloxamer); also, the surfactant properties of poloxamer may be detrimental to ocular tolerability.

Timoptic XE® is an in situ gel-forming solution of the beta-blocker timolol maleate and Gelrite® gellan gum. Gellan gum is an anionic polysaccharide formulated in aqueous solution, which forms clear gels under the influence of an increase in ionic strength. The gelation increases proportionally to the amount of either monovalent or divalent cations. The concentration of sodium in human tears (≈2.6 µg/µL) is particularly suitable to induce gelation of gellan gum following topical instillation into the conjunctival sac. The reflex tearing further enhances the viscosity of the gellan gum by increasing the tear volume and thus increasing the cation concentration. Scintigraphic studies showed that Gelrite (0.6% w/v) significantly prolongs ocular retention in man by forming a gelled depot on the scleral margin (Greaves et al. 1990).

Carbomers comprise poly(acrylic acid) polymers that undergo both temperature- and pH-dependent changes in structure. They are acidic, low viscosity, aqueous dispersions that transform into stiff gels when instilled into the conjunctival sac upon instillation. Carbomers offer several advantages for ophthalmic delivery, including high viscosities at low concentrations, strong adhesion to mucosa, thickening properties, compatibility with many active ingredients, and low-toxicity profiles. DuraSite® is a synthetic polymer of cross-linked poly(acrylic acid) that stabilizes drug molecules in an aqueous matrix, maintaining therapeutic doses of a drug on the eye surface for up to 6 hours. The technology is used in Azasite® (azithromycin ophthalmic solution) for the ocular delivery of the antibiotic azithromycin in the treatment of bacterial conjunctivitis: only once a day dosing is required (Friedlaender and Protzko 2007).

Combinations of different phase-transition polymers are also being investigated, in order to improve the gelling properties while also reducing the total polymer payload in the system, thereby improving tolerability and reducing discomfort. Simple mixing sometimes produces synergistic effects on thickening but may result in reduced mucoadhesion.

13.6.2 Contact Lenses and Cul-de-Sac Inserts

Drug-soaked contact lenses can be placed on the corneal surface, where they remain for up to 12 hours, thereby providing a reservoir of drug that desorbs following the Arrhenius equation and sustains drug release over a short period of time (Bengani and Chauhan 2013). Many methods of drug loading have been explored; the most common is presoaking the lens in a solution of the drug. Drug loading by soaking of lenses in ophthalmic formulations not designed for the task may cause toxicity to the corneal epithelium because preservatives such as benzalkonium chloride have a great affinity for the hydrophilic contact lens material and can become concentrated in the lens. Alternative drug-loading approaches include (1) using molecularly imprinted polymeric hydrogels; (2) chemical conjugation of the drug, or drug-loaded microparticles, to the lens surface; and (3) using liposome-loaded contact lenses.

There can be difficulty in adequately controlling drug release from the lens over the desired time frame. The drug on the inner surface has a different microenvironment to the outer surface, much of which will be lost to the external ocular tissue. There are also problems associated with patient comfort, acceptability, and compliance. The lenses must not interfere with optical clarity. Many people, particularly the elderly, find wearing lenses uncomfortable and inconvenient. They also cause foreign-body sensations and blurring and may decrease oxygen tension on the corneal surface. Dry eye syndrome and infections due to poor hygiene have been associated with prolonged lens use, which is a further concern. Furthermore, the wearing of contact lenses is contraindicated in many inflammatory conditions, which thus limits their applicability in a wide variety of ocular disorders.

Cul-de-sac inserts are designed to be left in the conjunctival sac and release their drug load over time. One of the first systems utilizing this principle was the wafer-like insoluble implant, Ocusert® from Alza, commercialized back in 1974 (see also Chapter 1). The system was designed to release pilocarpine at a constant rate for a week, to treat chronic glaucoma. It consisted of an inner layer containing pilocarpine in an alginate gel with di-(ethylhexyl) phthalate as a release enhancer, which was sandwiched between two outer, rate-controlling layers of poy(ethylene-co-vinylacetate) (EVA). Although the system represented very sophisticated drug delivery technology for the time, it nevertheless proved unpopular with patients, who complained of foreign body sensation, as well as difficulties in handling and insertion of the inserts. Approximately, 20% of all patients accidentally removed the device without being aware of the loss.

In spite of the poor patient acceptance of Ocusert, a number of other ocular inserts are almost always being studied. One of the much-copied originators, soluble ocular drug insert (SODI), consists of a small oval wafer of polyacrylamide used to deliver a variety of drugs, including ciprofloxacin and acyclovir. Porcine collagen shields have been designed to promote corneal healing but the use has not been accepted. Hybrid systems have also been developed; for example, liposomes containing cyclosporin A have been incorporated into collagen shields and shown enhanced drug delivery in comparison to both the free drug and liposomally associated drug.

Although not used as a drug delivery carrier, Lacrisert® is a rod-shaped ocular insert, which is used for treatment of dry eye syndromes. Made of HPC, the insert begins to dissolve within minutes of being inserted in the conjunctival sac. As it slowly dissolves, the inserts soften, stabilizing the tear film.

13.6.3 Use of Micro- and Nanoparticulate Drug Delivery Systems

Micro- and nanoparticulate DDS have been described in detail in Chapter 5 for parenteral delivery. They comprise an extensive array of drug carriers in the micro- and nanometer size range, including microparticles, nanoparticles, liposomes, niosomes, and dendrimers. Many of these systems have also been investigated for ophthalmic delivery, both for topical delivery as described here, as well as for intravitreal injection to the BOTE (described in Section 13.7).

Micro- and nanoparticulate DDS can offer a number of advantages for ophthalmic drug delivery, including the potential to provide (1) enhanced retention time at the corneal surface, which can facilitate drug diffusion across the corneal barrier and access to the posterior compartment; (2) high drug loading, thereby ensuring a high concentration gradient that will drive transcorneal drug permeation via passive diffusion; (3) the incorporation of mucoadhesive polymers, which also enhance corneal retention; (4) the incorporation of targeting moieties to target specific areas of the eye; (5) the protection of labile drug molecules within the carrier construct; and (6) sustained release of the drug from the carrier, thereby providing a prolonged release profile and minimizing the frequency of dosing.

Against these advantages, there are limitations associated with their use. For example, they can cause clouding of the corneal surface and interfere with the visual field. Further problems include the limited drug-loading capacity that is possible in many cases, as well as difficulties in sterilizing the formulations and the limited shelf life of the products.

Piloplex is a nanoparticle formulation comprising poly(methyl methacrylate) (PMMA)—acrylic acid copolymers, loaded with the antiglaucoma drug pilocarpine, and was one of the first DDS formulations to be studied for ophthalmic drug delivery (Mazor et al. 1979). Although the formulation necessitated fewer applications than conventional pilocarpine eyedrops, it was discontinued due to various formulation-related problems, including its nonbiodegradability, local toxicity, and the difficulty of preparing a sterile formulation.

Since then, many other nanoparticle systems have been investigated, with particular emphasis on biodegradable systems. For example, poly-ε-caprolactone nanoparticles were shown to improve the efficacy of betaxolol in the treatment of glaucoma, compared to betaxolol delivered in commercial eyedrops. The enhancement was ascribed to two factors: (1) the nanoparticles increased the precorneal retention of the drug by agglomeration and (2) the entrapped drug was in the nonionized form in the oily core of the carrier and thus could diffuse at a faster rate into the cornea. Pilocarpine-loaded poly(DL-lactic-co-glycolic acid) nanoparticles have also shown potential for controlled drug delivery, with an enhanced ocular pharmacological response than that of the free drug (Bourges et al. 2003).

The first application of liposomes in ocular drug delivery involved the use of a liposomal suspension of idoxuridine for the treatment of herpes simplex keratitis in rabbits: the liposomal formulation was found to give more efficient results compared to the aqueous solution (Smolin et al. 1981). Chitosan-coated liposomes (chitosomes) improved precorneal retention due to mucoadhesion and reduced drug metabolism at the precorneal epithelial surface. Liposomal formulations have shown improved efficacy over the free drug in the ophthalmic delivery of ganciclovir and fluconazole solution, in the candidiasis-associated keratitis model in rabbits (Habib et al. 2010). In the rabbits treated with fluconazole solution, 50% healing was observed in 3 weeks, whereas 86.4% healing was observed in rabbits treated with fluconazole-encapsulated liposomes.

13.6.4 PRODRUGS

Prodrugs are pharmacologically inactive derivatives of drug molecules that require a chemical or enzymatic transformation, in order to release the active drug within the body. In most cases, prodrugs are simply chemical derivatives that are one or two steps away from the parent drugs. In ophthalmic delivery, research has focused on producing more lipophilic derivatives of the parent drug, which display enhanced transcorneal permeability characteristics. Many ocular drugs contain hydroxyl or carboxyl groups that can be esterified to lipophilic ester prodrugs, which are subsequently acted on by esterases in the eye, to produce the parent drug.

An example is dipivefrine, a dipivalyl ester prodrug of epinephrine, which has now taken the place of epinephrine in the treatment of glaucoma. The prodrug is 600 times more lipophilic at pH 7.2 than epinephrine, and the penetration rate across the cornea is about 20 times higher. As such, a much smaller dose, with far less side effects, can be administered. Other prodrugs conferring increased corneal permeability have been developed for a wide variety of drugs including (1) the acetyl, propionyl, butyryl, and pivalyl ester prodrugs of timolol, (2) aliphatic acyl ester prodrugs of acyclovir, and (3) a range of dexamethasone esters (Ye et al. 2013).

Ocular prodrugs are also associated with a number of disadvantages, including poor aqueous stability and solubility, and an increased incidence of eye irritation. Also, from the standpoint of regulatory agencies, chemical derivatization of the drug results in the formation of a new chemical entity, which puts a further regulatory burden on development.

13.7 INTRAVITREAL DRUG DELIVERY

Intravitreal drug delivery allows a direct application of the drug near to the retina, thus eliminating the access barriers encountered when using topical administration. As a result, a much higher dose of administered drug can actually reach the target site and yield a more efficacious treatment of posterior eye diseases. Adverse drug effects are also substantially reduced. This route is currently the most acceptable and effective method to treat vitreoretinal disease. Intravitreal drug delivery can be achieved by (1) direct injection and (2) using implantable systems (Wilson et al. 2011). A review of approaches from a joint working party of Association for Research in Vision and Ophthalmology illustrates many of these approaches (Edelhauser et al. 2010).

13.7.1 INTRAVITREAL INJECTIONS

Intravitreal injection uses a small hypodermic needle to penetrate across the sclera, choroid, and retina into the vitreous body (Figure 13.6). Following administration of a small molecule, the drug diffuses from the center of vitreous to the edge of the retina in around 1.5 hours. For small molecules, the diffusion coefficient through the vitreous humor approaches that for water, whereas for larger antibodies and particulates, evidence of steric restriction is noted, particularly if the molecules or carriers are cationic. This reveals the structural characteristics of the vitreous, which consists of a network of hyaluronate holding collagen nanofibrils apart. Drug loss from the center of the vitreous takes place via two routes: (1) an anterior route involving diffusion forward into the anterior chamber, followed by drainage with the aqueous humor and removal to the systemic circulation and (2) a posterior route involving passive permeability and/or active secretion across the retina. All molecules can access the anterior route, whereas only some materials engage with the transport processes. Small drugs, such as the floxacillins, are lost primarily by anterior chamber diffusion and have a short half-life in the vitreous, which can be as little as 2–3 hours. In contrast, we have measured the half-life of 150 kDa fluorescein dextran as 30 days in the rabbit vitreous (Tan et al. 2011). With large molecules and particulate carriers, the influence of synchisis becomes important. In the ageing eye, the vitreous liquefies as the hyaluronans separate from the collagen fibers that collapse toward the back surface of the lens, allowing greater convective clearance forward. In this population, sustained delivery devices may fail to be effective.

Repeated intravitreal injection carries small but significant risks, such as infection, clouding of the vitreous humor, nonclearing vitreous hemorrhage, injury to delicate ocular structures (such as retinal detachment), and endophthalmitis. The procedure also requires specialist administration and may be painful, and the necessity of repeated injections can be highly unpleasant for patients. Drugs given by intravitreal injection include many anti-inflammatory corticosteroids, as well as antibiotics and antivirals. New advances involve the delivery of anti-VGEF drugs and gene therapies; these specific applications are described next.

13.7.1.1 Intravitreal Injections for Anti-VEGF Drugs

VGEF is an endogenous compound that induces angiogenesis, increases vascular permeability, and promotes inflammation. Recently, four new anti-VEGF drugs have been introduced for the management of wet AMD and DME (Table 13.1). The first three drugs listed in Table 13.1 are either

TABLE 13.1

Antivascular Endothelial Growth Factor Drugs for Wet Age-Related Macular Degeneration

Angiogenesis Inhibitor	Type of Molecule	Dosage
Bevacizumab (Avastin®)	mAb against VEGF	Once a month
Aflibercept (Eylea®)	Fab/VEGF receptor hybrid	Once a month
Ranibizumab (Lucentis®)	Fab	Once a month
Pegaptanib (Macugen®)	Oligonucleotide aptamer	Every 6 weeks

Abbreviation: VEGF, vascular endothelial growth factor.

monoclonal antibodies (mAb) or antibody fragments (Fab). On injection, these drugs penetrate the RPE, then bind and neutralize VGEF. The fourth drug (pegaptanib) represents a gene therapy approach, described in the next section. Anti-VGEF injections are needed every month/6 weeks, which is a significant health-care burden for this chronic condition.

13.7.1.2 Intravitreal Injections for Gene Therapy

The use of gene therapy in ophthalmic drug delivery is based on the retinal delivery of specific nucleotide sequences, which include sequences of DNA, RNA, and their modifications. The nucleotide sequences may work by a number of molecular transcription mechanisms, such as (1) induction of gene expression (gene therapy), (2) suppression of translation of target mRNA (antisense oligonucleotides), and (3) binding to specific protein targets (aptamers). In this way, the therapeutic problem is corrected at the level of molecular expression, in contrast to alleviation of symptoms as in conventional therapy. A significant problem concerns the difficulty in the administration of gene therapies, which are generally hydrophilic, large molecules, which exhibit suboptimal cellular loading when administered as simple solutions in saline. Progress is relatively slow, and although antisense oligonucleotides in particular have moved as far as proof of concept in small rodent models, access to the retina is easier in rodents than in man, which tends to overstate expectations (Short 2008).

As described in Chapter 16, gene delivery systems can be classified into viral and nonviral vectors. To augment the delivery of both vector types, techniques are used that span the full range of ocular administration technologies and include topical administration, gene gun, electroporation, sonoporation, intrastromal injection, and iontophoresis. Potential viral vectors for ocular delivery include adenovirus, adeno-associated virus (AAV), retrovirus, and lentivirus vectors. The AAV vector has been used to deliver the gene RPE65 to the retina and has restored vision in human patients suffering from Leber's congenital amaurosis, an autosomal recessive blinding disease (Simonelli et al. 2010). A recent review of the output from current clinical trials has concluded that although AAV approaches are useful, there remains a need for combinatorial approaches, including cotreatment with neuroprotective factors, antiapoptotic agents, and antioxidants (Dalkara and Sahel 2014).

Although viral vectors are more efficient in the delivery of genes, nonviral systems provide a complementary approach, offering easier production, unlimited gene size, and minimized immune reactions. The principal negative aspect of nonviral vectors is their poor transduction efficiency compared to viral vectors. Nonviral vectors include plasmid DNA, dendrimers, lipids, polymers, and nanoparticles (see also Chapter 16).

At the time of writing, there are two ocular gene-based drugs in clinical use, both delivered via intravitreal injection. The first one is Vitravene® (fomivirsen sodium), a phosphorothioate oligonucleotide for the treatment of CMV infection in acquired immunodeficiency syndrome (AIDS) patients. The second is Macugen® (pegaptanib sodium), an anti-VEGF aptamer for the treatment of wet AMD (Table 13.1). In addition, two siRNA molecules (bevasiranib and Sirna-027), which modify the activity of VEGF and its receptor (VEGFR-1), are also given by intravitreal injection and have recently entered clinical trials. For a current review on gene therapy in ocular applications, see the article by Solinís and colleagues (2014).

13.7.1.3 Long-Acting Intravitreal Injections

Risks, inconvenience, and other problems associated with intravitreal injections have been outlined earlier. To improve therapy, long-acting intravitreal injections have been developed. These are usually formulated by restricting the solubility of the active material. More advanced systems are described here.

The Verisome™ delivery technology is dependent on a phase change in the formulation to effect sustained release. Verisome™-based products can be injected into the vitreous as a liquid, using a standard 30-gauge needle. On injection, the liquid coalesces into a sphere that settles to the bottom of the globe, where it can be directly observed by the physician. The depot is biodegradable and studies have demonstrated, for example, the sustained release of triamcinolone acetate over a 1-year period.

Cortiject is a preservative-free emulsion formulation, composed of an oily carrier with phospholipid as a surfactant, which encapsulates a dexamethasone prodrug. The prodrug is de-esterified at the target site by a retina-specific esterase and activated to dexamethasone. A single intravitreal injection has been shown in clinical studies to provide sustained release over 6–9 months (see the excellent review by Haghjou et al. 2011).

Dispersed systems including liposomes and microparticulates can release their drug payloads gradually and over an extended period of time, with improved safety profiles. For example, liposome-encapsulated amphotericin B produced less toxicity than the commercial amphotericin B solution injected intravitreally. Microspheres formulated from poly(lactide-*co*-glycolide) (PLGA) loaded with 1 mg triamcinolone acetonide (TA) were injected via the intravitreous route into patients with diffuse DME. The efficacy measured by resolution of the macular edema was superior to conventional TA 4 mg injections. Intravitreal injection of liposomal tacrolimus was highly effective in suppressing the process of experimental autoimmune uveoretinitis in a rat model, without any side effects on retinal function or systemic cellular immunity (Zhang et al. 2010). Genentech and SurModics have collaborated on an investigation of biodegradable microspheres loaded with the anti-VGEF agent, ranibizumab. Preliminary data show that ranibizumab-loaded microparticles can deliver the agent over a period of approximately 4–6 months.

Larger particles can provide a longer period of treatment, but progress in this area will carry a risk of poor patient acceptance if the DDS interferes with vision, for example, appearing as "floaters" in the visual field. Using smaller particles may still cause vision hazing, as well as lessening the period of treatment.

13.7.2 Intravitreal Implants

Another approach to providing a prolonged intravitreal action is to use an intravitreal implant (Kuno and Fujii 2011). As described in Chapter 6, implants usually have a matrix or reservoir architecture and are largely diffusion-driven devices, although erosion is a further important release mechanism for biodegradable systems. Using implants, predictable and reproducible delivery rates are possible for prolonged periods of time.

One of the earliest systems, the Vitrasert® implant, is a sustained-release intravitreal implant of the antiviral ganciclovir, which is indicated for the treatment of CMVR, in individuals with AIDS. The implant comprises ganciclovir pellets dispersed in a PVA matrix, coated with a discontinuous film of the hydrophobic polymer poly(ethylene-co-vinylacetate) (EVA). The entire assembly is further coated with PVA, to which a suture tab made of PVA is attached, so that the implant can be sutured into place after insertion into the vitreous cavity. The drug is released gradually over a 5- to 8-month period (Dhillon et al. 1998).

Many intravitreal implants have been developed for the sustained release of anti-inflammatory corticosteroids. For example, Retisert® delivers fluocinolone acetonide (FA) over a 30-month period for the treatment uveitis and other retinal diseases. The implant, about the size of a grain of rice, comprises an FA core, encased in a silicone elastomer cup. A single- or double-release orifice is covered by a PVA membrane to control drug release. The entire device is attached to a PVA suture tab that again anchors the implant inside the eye.

FA is also delivered via intravitreal implant using the Iluvien® implant, which has recently been granted FDA approval for the treatment of DME. The implant, a tiny cylindrical polyimide tube, is so small that it can be injected directly into the vitreous cavity with a proprietary 25-gauge needle, in an office-based procedure. This is in contrast to the larger-sized Retisert®, which requires surgical implantation in the operating room. Controlled release of FA from Iluvien® is possible for up to 36 months. As the device is nonbiodegradable, it remains in the vitreous cavity even after the drug payload has expended. With chronic dosing, it is thus possible that many empty devices may accumulate in the vitreous of a patient over time.

The *I-vation* intravitreal implant was designed to release TA for up to 36 months. Made of titanium, the device has a corkscrew design and is coated with TA and the rate-limiting, nonbiodegradable polymers, PMMA, and EVA copolymer. The helical coil shape is designed to both maximize the surface area for drug release and also facilitate tissue attachment: once screwed into the eye, the implant is anchored to the sclera, which keeps it out of the visual field. Due to major complications, including raised IOP and cataract development, the Phase 2b trials were terminated.

Nonbiodegradable implants require either surgical removal once the drug is depleted, or in some cases the empty implant remains in the eye. As an alternative, biodegradable implants can be used. A recent FDA-approved example is the Ozurdex® intravitreal implant, which delivers the corticosteroid dexamethasone using the Novadur™ solid polymer delivery technology, based on biodegradable PLGA. Ozurdex® is preloaded into a single-use, specially designed applicator to facilitate injection of the rod-shaped implant directly into the vitreous cavity. The PLGA matrix slowly degrades in vivo to lactic acid and glycolic acid, releasing dexamethasone over a 6-month period. PLGA is also used for the manufacture of a rod-shaped biodegradable implant for the delivery of brimonidine, an alpha agonist. Brimonidine causes the release of various neurotrophins in vivo, which have the potential to prevent apoptosis of photoreceptors and/or RPE. A brimonidine-loaded PLGA intravitreal implant, dose 400 µg, is now in clinical trials.

The PRINT™ (Particle Replication in Non-Wetting Templates) technology under development by Envisia is an interesting idea using microtemplates that are filled with polymer–drug mixture to produce a range of shapes, including rods, disks, and doughnut shaped particles. The technology appears to be suitable for encapsulating large biologics, and the company has reported preliminary data for the ophthalmic delivery of prostaglandin analogs in animals.

13.7.2.1 Implanted Cells

Encapsulated cells, or stem cells, implanted subretinally, are under investigation at several centers (Tao 2006). Neurotech Pharmaceuticals has described the delivery of proteins by genetically engineered human retinal pigment epithelial cells entrapped in polymeric microcapsules, or hollow fibers, and implanted directly to the back of the eye. The implanted cells subsequently produce therapeutic proteins for the posterior segment. Renexus® (NT-501) consists of a semipermeable hollow fiber membrane capsule, surrounding a scaffold of strands of poly(ethylene terephthalate) yarn, which can be loaded with cells. The cells (NTC-200) are a human cell line of RPE, genetically modified to secrete recombinant human ciliary neurotrophic factor (CNTF). The device is surgically implanted in the vitreous through a scleral incision and is anchored by a suture at one end of the device. Once in place, the NTC-200 RPE cells continuously produces CNTF at the site of implantation, allowing for its controlled, continuous delivery. The concept exploits the function of the semipermeable membrane, which facilitates oxygen and nutrient influx into the cells and the efflux of CNTF, while simultaneously protecting the implanted cells from host cellular immunologic attack.

The safety and tolerability of subretinally implanted RPE (derived from human embryonic stem cells), as a cell-replacement therapy for patients with AMD and Stargardt's disease, was ongoing in early 2015 (Schwartz et al. 2015), but not recruiting patients at the time of writing. In support of this approach, several studies in Yucatan mini-pigs have been reported over the last 2 years, which indicate minimal changes to the retina (e.g., Stefanini et al. 2014).

13.8 TRANSSCLERAL DRUG DELIVERY

Targeting the sclera has received recent attention in the field of ophthalmic drug delivery, particularly in the application of protein and gene delivery. A number of approaches are being investigated.

13.8.1 TRANSSCLERAL INJECTION

Transscleral injections are typically used to deliver a local anesthetic for ophthalmic surgery, or any procedure requiring globe anesthesia and akinesia (Figure 13.6). A retrobulbar injection of anesthetic (retrobulbar block) is an injection of local anesthetic between the inferior and lateral rectus muscles. Subtenons injection (i.e., injection into a fascial sheath of connective tissue between the conjunctiva and episcleral plexus) is increasingly being used as an alternative to retrobulbar in anesthesia, as it is thought to be associated with less risk because a blunt needle is used.

A subconjunctival injection involves an injection of a drug beneath the conjunctiva, bypassing this barrier (Figure 13.6). The drug must then diffuse through the sclera and choroid. This method provides a localized and minimally invasive means of delivery to the posterior eye. The route is being researched for the delivery of bioactive proteins, prostaglandins, and dexamethasone. Recently, a liposomal formulation of the prostaglandin latanoprost (Lipolat) produced a sustained reduction in IOP over 3 months when given by subconjunctival injection in a pilot study in humans (Natarajan et al. 2012).

13.8.2 IONTOPHORESIS AND ELECTROPORATION

Current-assisted drug delivery, which includes iontophoresis and electroporation, has been widely studied in transdermal applications (see Chapter 9, Section 9.3.2) and therefore is of obvious interest in ocular delivery. Iontophoresis is a local, noninvasive drug delivery method that involves the use of low voltages to neutralize the charge on molecules, thereby assisting the movement of weak electrolytes and ions across membranes. By varying the intensity and duration of the electric field, precise amounts of a drug can be delivered, relatively specifically, to local targets. The mechanism of enhancement is thought to proceed through three mechanisms: (1) electric field interaction driving ions through tissue, (2) increase in tissue permeability caused by current flow, and (3) bulk movement of solvent or "electroosmosis" (Behar-Cohen 2012).

Electroporation, in contrast, involves the application of ultrashort high-voltage pulses, which cause a transient opening of the membrane allowing genes and large molecules to travel though the membrane, which then reseals. Plasmid delivery of genes directly to the ciliary muscle of rats has been reported, which then functions as a protein factory for long periods, perhaps several months (Sanharawi et al. 2013).

In ophthalmology, both transscleral and transcorneal iontophoretic drug delivery are possible, but the transscleral route is preferred, as it offers a number of advantages, including (1) a larger surface for transport, (2) enhanced transfer of drugs to the posterior segment, (3) less chance of systemic absorption, and (4) enhanced safety: resistance may cause scarring, so it is more prudent to put the electrode away from the cornea.

The EyeGate® II Delivery System is an ocular iontophoresis device that uses cathodic iontophoresis with an inert electrode for the ocular delivery of various drugs. The outcomes of clinical trials carried out using the device for the delivery of EGP-437 (a customized dexamethasone phosphate, formulated specifically for iontophoresis), for the treatment of noninfectious uveitis, were reported in 2012 (Behar-Cohen 2012).

13.8.3 MICROELECTROMECHANICAL SYSTEMS

Microelectromechanical systems (MEMs) technology utilizes microfabrication techniques used in the semiconductor industry to produce miniaturized structures, sensors, actuators, and systems. The MEMs can be utilized with miniaturized infusion pumps, to offer very precise control of the volume of drug delivered and the timing of that delivery, even offering the potential for feedback control.

MEMs-enabled implantable drug infusion pumps are refillable, making them amenable to long-term use, in contrast to the sustained-release intravitreal implants described in Section 13.7.2, which suffer from short device lifetime due to the limited drug payloads possible. The pumps being investigated for ophthalmology are driven by electrolysis, a low power-requiring process. The electrochemically induced phase change of water to hydrogen and oxygen gas generates pressure in a drug reservoir, thereby forcing the drug out through a transscleral cannula. The reservoir is implanted in the subconjunctival space and the cannula can be inserted into either the anterior or posterior segment, depending on the delivery requirements.

Such devices are currently at a very early stage of development; a number of prototypes have been described (Saati et al. 2010). For example, Replenish, Inc.™ has developed a small, refillable, implantable ocular drug pump. The device is programmable and can dispense nanoliter-sized doses (a volume sensor gives closed feedback) of drugs every hour, day, or month as needed, before refills. The tiny MicroPump™ can be "replenished" (hence the name of the technology) using a disposable, proprietary 31-gauge needle tubing kit. Although the device must initially be implanted into the eye, the Replenish device can subsequently last more than 5 years before needing replacement, i.e., much longer than current treatments. A 1-year feasibility trial on the safety profile in beagle dogs was published in 2014 and no inflammatory reactions reported (Gutiérrez-Hernández et al. 2014).

13.8.4 MICRONEEDLE DELIVERY

Microneedles, i.e., individual needles or arrays of needles, in the micrometer size range, have also been widely investigated for transdermal delivery (see also Chapter 9, Section 9.3.3.4), and the technology is now being applied to the ophthalmic route (Patel et al. 2011).

Microneedles can be fabricated out of numerous materials, including metal, polymer, glass, and ceramics, and they can be constructed in a variety of shapes, sizes, and drug-loading modalities. Solid microneedles (or microneedle arrays) can be coated with a drug: once inserted into the eye, the drug coating dissolves from the microneedle support and diffuses into the surrounding tissue, after which the microneedle can be removed. Alternatively, a hollow microneedle can be used to administer liquid formulations, including solutions and suspensions, as well as nanoparticles and microparticles, which also offer the potential of controlled release from the carrier. Biodegradable polymer microneedles can also be used, which do not have to be removed from the eye and can be designed to degrade slowly, thereby providing sustained drug release effects.

The microneedle insertion depth can be adjusted so that they penetrate just a few hundred microns into the sclera, without fully crossing the tissue, making the procedure minimally invasive. Drug released into the sclera then diffuses through the sclera tissue, to access the underlying tissues of the choroid, although retinal exposure is low. Solid microneedles release the drug in the immediate region around the placement site: the sclera then functions as a drug reservoir, for localized drug delivery to the underlying tissues. A hollow microneedle, with a liquid system, means that the drug can potentially flow over a more expanded area.

Increasingly, microneedles are being used to deliver drugs into the suprachoroidal space (SCS), i.e., the space between the sclera and choroid, which goes circumferentially around the eye. Normally, the sclera and the choroid are close together. Pushing fluid into the junction of the tissues creates the SCS, displacing the choroid from its normal position. This route has many attractive features for drug delivery: (1) it can expand and thus accommodate larger volumes of a drug formulation; (2) its proximity to the choroid and retina may provide higher drug concentrations to the chorioretinal tissues, without interfering with the optical pathways; and (3) targeting the SCS decreases exposure of nontarget tissues to the drug. Previous approaches to deliver to the SCS have included using scleral incisions or the use of long cannulas or hypodermic needles. These methods are highly invasive and cannot be performed as a simple office procedure. In contrast, microneedles offer the potential to access the SCS route in a minimally invasive manner. Preliminary results are encouraging. Microneedles, with lengths of 800–1000 µm, have successfully injected micro- and nanoparticles into the SCS, without surgery, in rabbit, pig, and human ex vivo eyes (Chen et al. 2015).

13.9 CONCLUSIONS

In conclusion, the eye continues to present significant challenges to successful drug delivery. Developments are focusing on the need to provide prolonged release of disease modulators, with less risk and easier access than are currently available. In the last few years, the field has been characterized by many exciting new advances, featuring highly sophisticated drug delivery technologies. Such technologies include the use of MEMS, iontophoretic devices, microneedle arrays, nanoparticulate DDS, gene therapies, long-lasting implants, and encapsulated cells. Such innovation and technological expertise offers considerable hope for the future.

REFERENCES

Agarwal, A., W.R. Rhoades, M. Hanout et al. 2015. Management of neovascular age-related macular degeneration: Current state-of-the-art care for optimizing visual outcomes and therapies in development. *Clin Ophthalmol* 9:1001–1015.

Aldrich, D.S., C.M. Bach, W. Brown et al. 2013. Ophthalmic preparations. *U.S. Pharmacopeia* 39(5):1–21.

Behar-Cohen, F. 2012. Current-mediated ocular drug delivery: Iontophoresis and electroporation as drug-delivery systems. *Retin Physician* 9:52–56.

Belmonte, C. and F. Cervero. 1996. *Neurobiology of Nociceptors*. Oxford, UK: Oxford University Press.

Bengani, L.C. and A. Chauhan. 2013. Extended delivery of an anionic drug by contact lens loaded with a cationic surfactant. *Biomaterials* 34(11):2814–2821.

Bourges, J.L., S.E. Gautier, F. Delie et al. 2003. Ocular drug delivery targeting the retina and retinal pigment epithelium using polylactide nanoparticles. *Invest Ophthalmol Vis Sci* 44(8):3562–3569.

Burton, M.J. and D.C.W. Mabey. 2009. The global burden of trachoma: A review. *PLOS Negl Trop Dis* 3(10):e460.

Chen, M., X. Li, J. Liu et al. 2015. Safety and pharmacodynamics of suprachoroidal injection of triamcinolone acetonide as a controlled ocular drug release model. *J Contr Release* 203:109–117.

Civan, M.M. and A.D. Macknight. 2004. The ins and outs of aqueous humour secretion. *Exp Eye Res* 78:625–631.

Connor, A.J. and P.S. Severn. 2011. Force requirements in topical medicine use—The squeezability factor. *Eye (Lond)* 25(4):466–469.

Cruickshanks, K.J., D.S. Dalton, R. Klein et al. 2014. Better eye health for aging baby-boomers: Generational differences in the 5-yr incidence of AMD. *Invest Ophthalmol Vis Sci* 55(13):6006.

Dalkara, D. and J.A. Sahel. 2014. Gene therapy for inherited retinal degenerations. *C R Biol* 337(3):185–192.

Dhillon, B., A. Kamal, and C. Leen. 1998. Intravitreal sustained-release ganciclovir implantation to control cytomegalovirus retinitis in AIDS. *Int J STD AIDS* 9(4):227–230.

Edelhauser, H.F., C.L. Rowe-Rendleman, M.R. Robinson et al. 2010. Ophthalmic drug delivery systems for the treatment of retinal diseases: Basic research to clinical applications. *Invest Ophthalmol Vis Sci* 51(11):5403–5420.

Elkington, A.R. and H.J. Frank. 1999. *Clinical Optics*, 3rd ed. Oxford, U.K.: Blackwell Publishing.

Faraldi, F., V. Papa, D. Santoro et al. 2012. A new eye gel containing sodium hyaluronate and xanthan gum for the management of post-traumatic corneal abrasions. *Clin Ophthalmol* 6:727–731.

Friedlaender, M.H. and E. Protzko. 2007. Clinical development of 1% azithromycin in DuraSite®, a topical azalide anti-infective for ocular surface therapy. *Clin Ophthalmol* 1(1):3–10.

Greaves, J.L., C.G. Wilson, A. Rozier et al. 1990. Scintigraphic assessment of an ophthalmic gelling vehicle in man and rabbit. *Curr Eye Res* 9(5):415–420.

Gupta, R., B. Patil, B.M. Shah et al. 2012. Evaluating eye drop instillation technique in glaucoma patients. *J Glaucoma* 21(3):189–192.

Gutiérrez-Hernández, J.C., S. Caffey, W. Abdallah et al. 2014. One-year feasibility study of replenish MicroPump for intravitreal drug delivery: A pilot study. *Transl Vis Sci Technol* 3(4):1–13.

Habib, F.S. E.A. Fouad, M.S. Abdel-Rhaman et al. 2010. Liposomes as an ocular delivery system of fluconazole: In-vitro studies. *Acta Ophthalmol* 88(8):901–904.

Haghjou, N., M. Soheilian, and J.M. Abdekhodaie. 2011. Sustained release intraocular drug delivery devices for treatment of uveitis. *J Ophthalmic Vis Res* 6(4):317–329.

Kim, I.K. and J.W. Morley. 2006. Photodynamic therapy. In *Intraocular Drug Delivery*, eds. G.J. Jaffe, P. Ashton, and P.A. Pearson, pp. 129–141. New York: Taylor & Francis.

Kuno, N. and S. Fujii. 2011. Recent advances in ocular drug delivery systems. *Polymers* 3(1):193–221.

Mains, J. and C.G. Wilson. 2013. The vitreous humour as a barrier to nanoparticle distribution. *J Ocul Pharmacol Ther* 29(2):143–150.

Mazor, Z., U. Ticho, U. Rehany et al. 1979. Piloplex, a new long-acting pilocarpine polymer salt. B: Comparative study of the visual effects of pilocarpine and Piloplex eye drops. *Br J Ophthalmol* 63(1):48–51.

Nakhlband, A. and J. Barar. 2011. Impacts of nanomedicines in ocular pharmacotherap. *Bioimpacts* 1:7–22.

Natarajan, J.V., M. Ang, A. Darwitan et al. 2012. Nanomedicine for glaucoma: Liposomes provide sustained delivery of latanoprost in the eye. *Int J Nanomed* 7:123–131.

Patel, S.R., A.S.P. Lin, H.F. Edelhause et al. 2011. Suprachoroidal drug delivery to the back of the eye using hollow microneedles. *Pharm Res* 28:166–176.

Saati, S., R. Lo, P.-Y. Li et al. 2010. Mini drug pump for ophthalmic use. *Curr Eye Res* 35(3):192–201.

Sanharawi, M.E., E. Touchard, R. Benard et al. 2013. Long-term efficacy of ciliary muscle gene transfer of three sFlt-1 variants in a rat model of laser-induced choroidal neovascularization. *Gene Ther* 20:1093–1103.

Schein, O.D., B. Munoz, and J.M. Tielsch. 1997. Prevalence of dry eye among the elderly. *Am J Ophthalmol* 124:723–728.

Schwartz, S.D., C.D. Regillo, and B.L. Lam. 2015. Human embryonic stem cell-derived retinal pigment epithelium in patients with age-related macular degeneration and Stargardt's macular dystrophy: Follow-up of two open-label phase 1/2 studies. *Lancet* 385:509–516.

Short, B.G. 2008. Safety evaluation of ocular drug delivery formulations: Techniques and practical considerations. *Toxicol Pathol* 36(1):49–62.

Simonelli, F., A.M. Maguire, F. Testa et al. 2010. Gene therapy for Leber's congenital amaurosis is safe and effective through 1.5 years after vector administration. *Mol Ther* 18(3):643–650.

Smolin, G., M. Okumoto, S. Feiler et al. 1981. Idoxuridine-liposome therapy for herpes simplex keratitis. *Am J Ophthalmol* 91(2):220–225.

Stefanini, F.R., M.J. Koss, P. Falabella et al. 2014. Comparative analysis of retinal layers after subretinal stem cell implantation in Yucatan mini-pigs. *Invest Ophthalmol Vis Sci* 55:3999.

Tan, L.E., C.G. Wilson, J. Burke et al. 2011. Effects of vitreous liquefaction on the intravitreal distribution of sodium fluorescein, fluorescein dextran and fluorescent microparticles. *Invest Ophthalmol Vis Sci* 52(2):1111–1118.

Tao, W. 2006. Application of encapsulated cell technology for retinal degenerative diseases. *Expert Opin Biol Ther* 6(6):717–726.

Wilson, C.G. and L.E. Tan. 2012. Nanostructures overcoming the ocular barrier: Physiological considerations and mechanistic issues. In *Nanostructured Biomaterials for Overcoming Biological Barriers*, eds. M.J. Alonso, N.S. Csaba, D. Thurston et al., pp. 173–189. Cambridge, U.K.: RSC Publishing.

Wilson, C.G., L.E. Tan, and J. Mains. 2011. Principles of retinal drug delivery from within the vitreous. In *Drug Product Development for the Back of the Eye*, eds. U.B. Kompella and H.F. Edelhauser, pp. 125–158. New York: Springer.

Ye, T., K. Yuan, W. Zhang et al. 2013. Prodrugs incorporated into nanotechnology-based drug delivery systems for possible improvement in bioavailability of ocular drugs delivery. *Asian J Pharm Sci* 8(4):207–217.

Zhang, R., R. He, J. Qian et al. 2010. Treatment of experimental autoimmune uveoretinitis with intra-vitreal injection of tacrolimus (FK506) encapsulated in liposomes. *Invest Ophthalmol Vis Sci* 51(7):3575–3582.

Section IV

Emerging Technologies

14 Hydrogels

Ronald A. Siegel and Carmen Alvarez-Lorenzo

CONTENTS

14.1 INTRODUCTION

Hydrogels are polymer networks that contain a substantial amount of water. Dry polymer networks can absorb tens, hundreds, or even thousands of times their weight in water without dissolving, as illustrated in Figure 14.1. As a result, hydrogels have properties that are very similar to those of soft biological tissues such as cornea, cartilage, and intercellular matrix, and because of these characteristics, they are of great utility in pharmacy and in other biomedical fields. First, they are soft and slippery and have a very low interfacial tension when they are in contact with tissues. As a result, they often cause little irritation. Second, hydrogels can behave both like solids and like liquids. Like solids, they are elastic and return to their original shape after they are momentarily deformed. Like liquids, they permit diffusion of small molecules such as salts, metabolites, and drugs. Under proper circumstances, they also allow larger molecules such as proteins and nucleic acids to diffuse.

A hydrogel can be envisioned as a bulk of water through which polymer strands are suspended. As illustrated in Figure 14.2, these strands are connected together by *cross-links*, which provide elasticity and prevent the strands from coming apart. The distance between strands and the frequency of

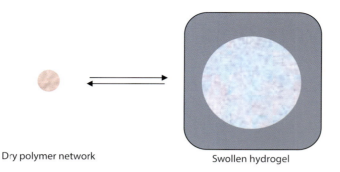

Dry polymer network   Swollen hydrogel

FIGURE 14.1   A hydrogel is a dry polymer network that is swollen with water but does not dissolve.

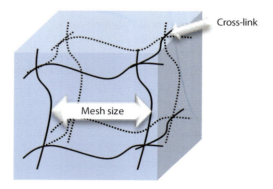

Cross-link

Mesh size

FIGURE 14.2   At the microscopic level, a hydrogel consists of cross-linked polymer chains immersed in water. The distance between chains and the density of cross-links determine the mesh size of the hydrogel.

cross-linking, which determine the *mesh size*, can be controlled; by this means, the partitioning of molecules into the hydrogel and their rates of diffusion can be regulated. For this reason, hydrogels are often used in controlled-release drug delivery.

As elastic materials, hydrogels can exert forces against confining or attached structures. As will be described later, the degree of swelling and shrinking of hydrogels can be affected by physical and chemical stimuli. Thus, the hydrogels can act as "artificial muscles," which direct the motion of such structures. This property has also been utilized in controlled-release applications.

Table 14.1 provides a list of present and potential future applications of hydrogels. Soft contact lenses were initially developed in the early 1960s as alternatives to hard contacts, which are irritating and can damage the cornea. While soft contact lenses may contain only about 40% water, this is enough to confer flexibility, slip, and compatibility with the underlying cornea. Because optical properties are of foremost importance, hydrogels in contact lenses must remain constant in their swelling during use. Other applications involving hydrogels with constant swelling are listed in bold in Table 14.1.

A tantalizing property of hydrogels is their ability to alter their swelling depending on an applied stimulus, as depicted in Figure 14.3. Commonly studied stimuli include changes in temperature, pH, and ionic strength, but other possible stimuli include the concentration of specific ions such as Ca^{2+} or glucose and other sugars, and the presence of specific antigens or DNA sequences, electric or magnetic fields, and even light. Applications involving "stimulus-sensitive" hydrogels are listed in italics in Table 14.1. (*Note*: some investigators refer to stimulus-sensitive hydrogels as "smart" or "intelligent." We eschew such anthropomorphic terminology.)

TABLE 14.1

Present and Future Applications of Hydrogels

Present	Future[a]
Soft contact lenses	**Burn, wound dressings**
Gel permeation chromatography	**Catheter, suture coatings**
Ion exchangers	**Fire retardants**
Gel electrophoresis	*Stimulus-sensitive drug delivery*
Drug delivery (e.g., capsules, matrices)	*(Bio)sensors*
Cosmetics	*Microparticulate drug and gene carriers*
Food additives	*Tissue culture, tissue engineering, and bioartificial organ supports*
Fragrances	*"Active" separations*
Dessicants	*Artificial muscle*
Toys	*Switchable chromatography*

[a] Applications in which size and shape are expected to be constant in time are in bold, while applications in which the hydrogel changes size or shape in response to various stimuli are in italics.

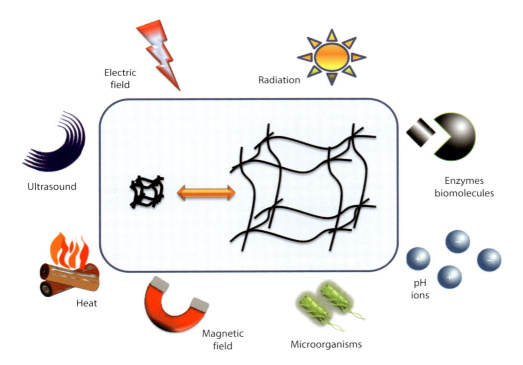

FIGURE 14.3 Stimuli that may affect hydrogel swelling.

This chapter serves as an introduction to hydrogels. More details can be found in the references listed at the end of the chapter. We first describe methods of synthesis. We then discuss factors that affect their swelling, elasticity, solute partitioning, and permeability properties. Finally, examples of present and future applications in the pharmaceutical arena are presented.

14.2 FORMATION OF HYDROGELS

14.2.1 Hydrogels Based on Biopolymers

"Natural" hydrogels are produced from biological polymers. Because they are the result of bio-logical synthesis, the structure of biopolymers comprising the hydrogels is often well controlled. Their building blocks, by nature, are often biocompatible, although care must be taken to purify the biopolymers and remove antigens and pyrogens. Examples of biopolymers used to form natural hydrogels include fibrous proteins such as collagen, gelatin (denatured collagen), and fibrin and polysaccharides such as calcium alginate, agarose, hyaluronic acid, cellulose and its derivatives, dextrans, chitosan, carrageenans, pectins, and pullulans.

Long-chain biopolymers often form physical cross-links by winding two or more chains to form helical domains. Such cross-links can be loosened by melting or by applying chemical denatur-ants. Calcium alginate hydrogels are produced when alginic acid chains form coordination com-plexes that are wrapped around calcium ions, leading to a so-called "egg case" structure. When calcium is replaced with sodium ions, these cross-linking structures are dissolved and the alginate becomes fluid. Physical cross-linking can also occur by formation of microcrystalline domains, usually driven by hydrogen-bonding interactions, as in cellulose, or by aggregation of hydrophobic domains as in pullulans.

Physical cross-links are advantageous since they can often be formed and reversed by mild processing techniques. On the other hand, physical cross-links may not be permanent, and with persistent stress, physically cross-linked hydrogels may deform permanently. Hydrogels formed from fibrous proteins can form more permanent, stress-resistant structures if they are chemically cross-linked. Here, a linker molecule such as glutaraldehyde forms covalent bonds between side chains on separate protein molecules. Companies specializing in biochemicals sell a variety of such cross-linkers, which are also used for protein conjugation. Chemical cross-linking of polysac-charides is often accomplished by chemistries that have been developed for synthetic polymers, as will be described in the following texts. Alternatively, biopolymers can be cross-linked by exposure to radiation, which works by forming free radicals along different polymer chains, which find each other and recombine, forming covalent cross-links.

14.2.2 Synthetic Polymerization

Synthetic polymers and their hydrogels are usually formed from petroleum-based organic chemicals. Except for some unusual cases, these systems exhibit less intrinsic structural reg-ularity than biopolymers. Also, only a limited set of synthetic hydrogels are biocompatible. Nevertheless, synthetic hydrogels can be produced cheaply, and their chemical compositions and network structures are easily studied and controlled. In recent decades, there have been major advances in the synthesis of organic polymers, resulting in improved quality of hydrogels. Here, we will describe several methods for preparing hydrogels, beginning with a brief review of polymer synthesis.

14.2.2.1 Step-Growth Polymerization

In *step-growth polymerization*, polymer-building blocks or monomer units contain end groups that react with each other to form larger molecules having the same end groups. Imagine a mono-mer with two end groups, say A and B, that react with each other. Two monomers of structure A–B will form a dimer of structure A–B′–A′–B. The prime symbols in the middle indicate that the joined A and B units might be modified as a result of the reaction. For example, if A is an acid group and B is a hydroxyl group, then B′–A′ will be an ester linkage. If A is an acid group and B is an amide group, then B′–A′ will be an amide linkage. In these two examples, a water molecule is liberated in a *condensation reaction*. However, the dimer still retains unmodified A and B end groups, which can enter into reaction with other A–B monomers. Similarly, already formed chains with A and B end groups can react with each other to form even longer chains. Thus, reactions such as

$$A-[B'-A']_x-B + A-[B'-A']_y-B \rightarrow A-[B'-A']_{x+y}-B$$

are possible, for any chain lengths x and y. If the reactions are condensations, then the step-growth polymerization can also be called a *condensation polymerization*. The two reactions mentioned earlier lead, respectively, to polyesters and polyamides. It should be noted that A and B need not be distinct, provided that an end group can react with an identical end group. For example, two organic acids, –COOH, can combine to form an anhydride linkage, with a water molecule liberated. It is also possible to perform step-growth polymerizations from the initial monomers, A–A and B–B, since these can react to form A–A'–B'–B, which can then go on to form longer polymer chains.

Since condensation reactions are reversible, water removal is usually required for sustained growth. Also, step-grown reactions are particularly sensitive to impurities containing only one reactive end, which will *cap* the polymer and prevent further growth.

The step-growth mechanism just described will only form linear chains. As with biopolymers, such chains can be cross-linked through their side groups. However, an even simpler method exists. Along with the A-B monomers, a fraction of monomers with structure R-A$_f$ can be included, where R is a central group that is surrounded by f A groups, where $f > 2$. Similarly, other monomers of form R-Bf can be included. The R groups in these two monomers can be the same, but they need not be; nor do the fs, so long as they are both greater than 2. The fs are called *functionalities*. With these *multifunctional* monomers, the ends of growing linear chains can be linked together into networks, as diagrammed in Figure 14.4.

One characteristic of step-growth polymerization is that the individual polymer chains, or the chains connecting cross-links, are of variable length. The growth sequences of different chains vary, and the result is a distribution of chain lengths. Therefore, while the average chain length can be specified or determined experimentally, this variability must be regarded as a fundamental property.

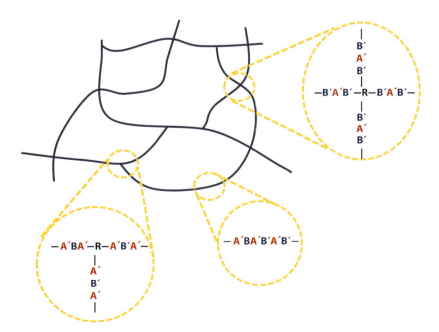

FIGURE 14.4 Schematic of a cross-linked hydrogel with chains formed by AB units, cross-linked by R–A$_3$ and R–B$_4$ units.

14.2.2.2 Chain-Growth Polymerization

In *chain-growth* or *addition* polymerization, individual chains form by *initiation* at one end, where a *reaction center* is formed. Monomers are then added, one by one, during *propagation* steps, with the reactive center shifting to the last monomer added. Propagation proceeds until a *termination* event occurs, in which the growing chain reacts with a species that permanently blocks further monomer addition.

Free radical polymerization, the most common version of addition polymerization, is illustrated in Figure 14.5. In this class of reactions, the monomer units contain *vinyl* groups, denoted by ⊣, which signifies a carbon double bond along with a (here unspecified) side group. When attacked by a *free radical* on one side, the double bond is converted to a single bond, and the free radical is shifted to the other side, where it is now available to attack a second monomer. This second monomer then attacks a third, and so on. Initiation of this chain reaction is brought about by thermal, photo, radiation, or redox decomposition of an *initiator* species, I, converting it to a free radical, which we denote by I·. The result is a long chain with the initiator at one end and a still reactive free radical on the other.

In most free radical polymerizations, propagation is ultimately terminated when two free radicals meet and either form a covalent bond joining the corresponding chains (combination) or having one chain transfer its unpaired electron to the other (disproportionation). Two other possibilities are as follows: (1) polymerization is terminated by stray free radicals that are present in the feed. Reactive oxygen species (ROS) are of particular concern, and care must be taken to minimize the presence of molecular oxygen, and (2) the propagating free radical extracts a hydrogen atom from another molecule, terminating the growing chain. However, a new free radical is formed on the "victim" molecule, initiates a new polymerization. This mechanism is called *chain transfer* and is sometimes used as a means to control molecular weight by addition of *chain transfer agents*. When the hydrogen atom is extracted from another polymer chain, a secondary *branch* may grow off the latter, creating a treelike or *branched* architecture.

Following initiation, which is often rate limiting, the propagation and termination steps are fast. Thus, individual chains are formed rather quickly and at different times within a polymerization run. This is in contrast to step-growth polymerization, in which all chains tend to grow at the same time.

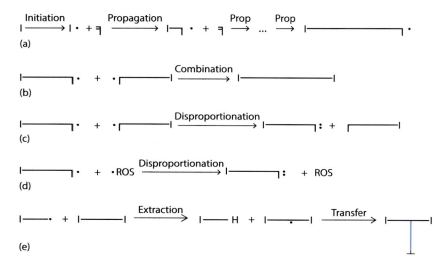

FIGURE 14.5 Steps in vinyl polymerizations: (a) initiation and propagation; (b) termination by recombination; (c) termination by disproportionation; (d) termination by stray free radicals, e.g., reactive oxygen species (ROS); and (e) chain transfer.

Free radical polymerization, along with other addition polymerizations, possesses several advantages. It is relatively easy to carry out, and there exists a wide variety of vinylic monomers containing different side chains, which endow the polymer with desired properties. More than one monomer species can be included in a polymerization in order to fine-tune these properties or to confer two or more properties at once. Polymers containing more than one monomeric species are called *copolymers*. The structure of a copolymer depends on the rate that each monomer reacts with its own kind rather than with other types of monomers. When there are two types of monomers, chain propagation proceeds according to the following mechanism:

$$-M_1 \cdot + M_1 \xrightarrow{k_{11}} -M_1 - M_1 \cdot$$

$$-M_1 \cdot + M_2 \xrightarrow{k_{12}} -M_1 - M_2 \cdot$$

$$-M_2 \cdot + M_1 \xrightarrow{k_{21}} -M_2 - M_1 \cdot$$

$$-M_2 \cdot + M_2 \xrightarrow{k_{22}} -M_2 - M_2 \cdot$$

where k_{11}, k_{12}, k_{21}, and k_{22} are the propagation rate constants. The *reactivity ratios* $r_1 = k_{11}/k_{12}$ and $r_2 = k_{22}/k_{21}$ provide information regarding the preference of monomers to react with their own kind and, hence, the tendency of monomers to be arranged (1) in blocks or (2) in alternating sequences. If r_1 and r_2 are much larger than unity, the likelihood of forming blocks of similar monomers increases, while if both are less than unity, the monomers tend to be arranged alternately. When the product $r_1 r_2$ is close to unity, then the monomers are randomly incorporated in the copolymer chains, so long as the supply of both monomers remains adequate. It should be noted that unless special precautions are made, the monomer feed ratio will drift as the polymerization progresses, and there will be changes in monomer content.

Like step-growth polymerization, free radical polymerization produces chains whose lengths present a rather broad distribution. The reason for this is that the termination step can occur at any time, independent of the extent of propagation. Other kinds of addition polymerization can minimize such variability in chain length by suppressing the termination step. In *anionic* and *cationic* polymerizations, the growing chains have charged reaction centers at their growing ends, so termination by encounter between two chains does not occur. In *ring-opening polymerization*, the terminal monomer in a chain reacts with a cyclic-free monomer, which opens as it joins the polymer. Normally, there is no basis for interchain termination reactions.

In the past two decades, novel free radical polymerization schemes, including atom transfer radical polymerization (ATRP) and reversible addition-fragmentation chain transfer (RAFT) polymerization, have received much attention. In these schemes, most of the growing chain ends are "masked" by added reagents, slowing both the propagation and termination steps. Since termination is bimolecular, the likelihood of two unmasked chains finding each other and recombining or disproportionating is severely reduced. At the cost of slower polymerization, much more uniform polymer chain lengths are obtained.

Anionic, cationic, ring-opening, ATRP, and RAFT polymerizations are often called *living*, because the chain ends remain active even when the monomer supply is exhausted. It is therefore possible to produces *block polymers*, in which chains consisting of one kind of monomer serve as *macroinitiators* for another kind of chain. While typical applications of block polymers lie in the adhesive area, particular systems have also been shown to self-assemble into micellar and vesicular (polymersome) systems (see also Chapter 5, Section 5.5.3). As will be discussed in Section 14.2.5, properly constructed block polymers can also form physically cross-linked hydrogels.

14.2.3 STRUCTURAL CHARACTERISTICS OF LINEAR POLYMERS

We have already emphasized that the molecular weight of synthetic polymers is subject to random variation. Further structural variation can be introduced by copolymerization. Besides molecular weight and composition, another characteristic of a polymer is its *tacticity*. During an addition polymerization, the reactive center may be chiral with respect to the side chain, and sequential pairs of side chains may take on *trans* or *cis* configurations. Usually, this sequence of chiral additions is random, and the polymer is *atactic*. Chains in which all the side groups are added on the same side are called "isotactic," whereas if the side groups alternate consistently, then the chain is considered *syndiotactic*. The latter two situations may require special polymerization conditions, e.g., catalysts, but there is a potential benefit insofar as the resulting polymers can form crystalline domains, or *crystallites*, by aligning with neighboring chains. Such crystalline domains form physical cross-links, which may be stable even in the presence of solvent. For example, properly processed polyvinyl alcohol (PVA) can form stable hydrogels that absorb a substantial amount of water, even though there is no chemical cross-linking. Atactic polymers cannot crystallize. Similarly, copolymers must either consist of long homomonomeric blocks (blocks containing the same monomer) or long alternating sequences, in order to crystallize.

14.2.4 CHEMICALLY CROSS-LINKED NETWORKS

In Section 14.2.2.1, it was shown how step-growth polymers could be cross-linked by including multifunctional monomers. Irradiation followed by radical recombination was also mentioned as a general method for forming cross-links between polymer chains. In vinyl polymerization, the most common method of cross-linking is to include a small amount of a *divinyl* monomer into the polymerization reaction. As illustrated in Figure 14.6, this monomer can be incorporated into two independent growing chains. If several cross-linkers are incorporated into each chain, then the chains will have multiple connections to each other, forming a macroscopic network. Cross-linkers are incorporated randomly so that the distance between cross-links is random, the same as was pointed out for hydrogels formed by step-growth polymerization. Because of this randomness, important properties such as mesh size (Figure 14.2), which determines in part the diffusivity of molecules through the hydrogel, can only be characterized as averages.

A nonobvious characteristic of many hydrogels is their nonuniformity, which goes beyond the randomness seen in chemical structure distances between cross-links. Usually, hydrogels are synthesized in the presence of a solvent, such as water, in order to prevent overheating, which results from the exothermic nature of polymerization reactions. As monomers join to form polymers, they exhibit a tendency to phase separate from the solvent, forming small domains that are rich in polymer, alternating with regions that are rich in solvent. These variations in concentration are frozen in upon cross-linking. This phenomenon called "microsyneresis" may, if not controlled, produce a cloudy appearance. It can also lead to substantial weakening of the hydrogel structure and deviation in properties that are predicted assuming uniformity.

In recent years, there have been efforts to form "regular" hydrogels with better defined distances between cross-links and fewer problems with microsyneresis. One popular technique favored by those interested in preparing hydrogels for embedding of cells, e.g., for tissue engineering, is to

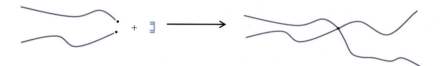

FIGURE 14.6 Cross-linking of two growing chains by propagating their free radicals through a divinyl monomer.

use polyethylene glycol (PEG) chains that are terminated on both ends by vinyl (usually acrylate) groups. These *macromers*, which are reasonably monodisperse (nearly uniform in length), are "zipped" together by free radical polymerization of the end groups. Because the PEG chains are of nearly uniform length, are already in a polymerized state before cross-linking, and are hydrophilic, microsyneresis is reduced. Similar approaches involving modification of biopolymers such as dextrans with acrylate groups, followed by "zippering," have been tried. Copolymers with a small fraction of side chains that can be cross-linked in response to light (photocross-linkers) can also be used to form uniform hydrogels, although the distance between cross-links in this case is more randomized. Most recently, nearly uniform length PEGs have been end-functionalized with groups that can condense with tetrafunctional RA-f ($f = 4$) cross-linkers, to form the so-called diamond hydrogels. Such hydrogels have been claimed to have "perfect" structure insofar as they are essentially free of inhomogeneities, and it has been shown that they have mechanical strengths greatly exceeding those of conventionally cross-linked hydrogels.

Since cross-links provide cohesiveness to the hydrogel, it follows that by incorporating biodegradable cross-linkers, the hydrogel can be programmed to lose its strength over time. This may be important for injectable and implantable systems, since it obviates the need to retrieve the hydrogel from the body after its purpose (e.g., for drug delivery or cell engraftment) has been served.

14.2.5 PHYSICALLY CROSS-LINKED NETWORKS

Physically cross-linked hydrogels can be formed using block polymers. Of particular interest are systems that are fluid at room temperature but self-assemble into 3D network structures at body temperature (Figure 14.7). Here, we will discuss two kinds of *triblock* systems (consisting of three polymer blocks in series), one is called ABA and the other ABC. The B block in both systems is a hydrophilic polymer, while the A and C blocks are hydrophilic at room temperature, but hydrophobic at body temperature. This *thermosensitivity* is attributed to the tendency of water to form hydrogen-bonded "cage" structures around hydrophobic groups. These cages are broken up with increasing temperature, causing water to migrate away, followed by association of hydrophobic segments. (Incidentally, this mechanism of hydrophobicity is also responsible for the stability of globular proteins, in which hydrophobic amino acids are sequestered in the protein molecule's "core" and are surrounded by a "shell" of hydrophilic amino acids.)

Self-assembly of ABA and ABC triblocks is depicted in Figure 14.7. The hydrophobic A and C blocks form physical cross-links by associating with their own kind. If the A and C blocks are incompatible, then they will form separate cross-linking domains. At low concentrations, ABA triblocks may form "flowers" due to A blocks associating with each other within the same chain, while this should not occur for ABC triblocks. Thus, ABC triblocks with incompatible A and C blocks are disposed to form hydrogels at lower overall polymer concentrations than ABA triblocks. Figure 14.8 shows how an increase in temperature converts a free-flowing polymer solution to a rigid hydrogel. This transformation is reversible—the fluid state is recovered upon cooling.

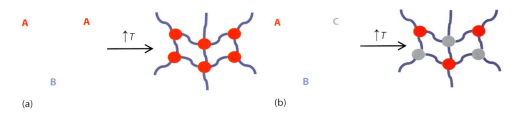

FIGURE 14.7 Self-assembly of (a) ABA and (b) ABC triblock polymers into networks, with coagulated A and C blocks forming physical cross-links.

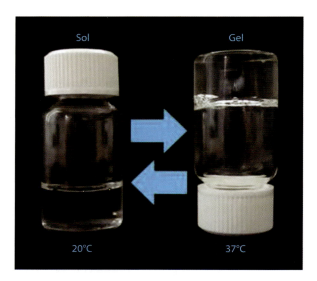

FIGURE 14.8   Reversible conversion of a triblock polymer solution with thermosensitive outer blocks, from a free-flowing solution to a rigid gel (suspended on top of the inverted vial), upon heating.

"Thermogelling" systems are of considerable interest because they can be prepared and stored at room temperature and then injected through a syringe needle into a body space. Once in the body where the temperature is 37°C, the polymers form gels, which act as depots for drug release. By incorporating biodegradable elements, these systems resorb after they have carried out their task.

14.3   FUNDAMENTALS OF PARTITIONING, PERMEABILITY, AND SOLUTE-RELEASE PROPERTIES OF HYDROGELS

A primary role of hydrogels in pharmaceutical applications is to modulate the rate of drug release. Delivery can be from a reservoir, with the hydrogel serving as the rate-determining membrane. In some cases, the hydrogel should simply allow free movement of substances from one side of a hydrogel membrane to the other. For example, soft contact lenses enable corneal access of oxygen from air and nutrients from lachrymal (tear) fluid. In other cases, the hydrogel is designed to retard or control movement.

Alternatively, the hydrogel can be loaded with drug, either in the dissolved or in a solid particulate form, and then release it at the target site. The hydrogel is then called a *monolith*, and it acts both as a storage medium and as a matrix that controls the rate of drug release. A simple example of a monolith is a contact lens soaked in a drug solution. When the drug concentration inside the lens reaches its equilibrium level, the lens is placed on the cornea, where it releases the drug into the lachrymal fluid. The drug then diffuses through the cornea to the aqueous humor, providing therapeutic levels over several hours. Alternatively, drug can be incorporated into a hydrogel during polymerization, or the hydrogel can self-assemble around drug molecules or particles in response to an environmental change such as increase in temperature, as occurs in injectable triblock systems.

The difference between a membrane and a monolith structure is depicted in Figure 14.9. Roughly speaking, one may say that drug diffuses *through* a membrane and *out of* a monolith.

For a hydrogel membrane of uniform structure, a simple equation relates the *steady-state flux*, J_{ss}, of drug to its *concentration difference* ΔC, across the membrane:

$$J_{ss} = \left(\frac{KD}{h} \right) \Delta C$$

(14.1)

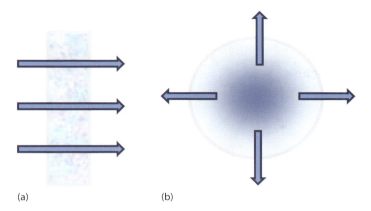

(a) (b)

FIGURE 14.9 (a) Membrane versus (b) monolith. Arrows depict drug flux.

That is, the rate of transport of molecules across the membrane is determined by (1) ΔC, which is the driving force for diffusion; (2) the molecule's ability to enter the membrane, as determined by the *partition coefficient, K*; (3) the speed at which the molecules can move inside the membrane, as determined by the *diffusion coefficient, D*; and (4) the distance that the molecules must travel, as determined by the thickness, h, of the membrane. The combination KD/h is called the *permeability* of the membrane. Since the partition and diffusion coefficients are fundamental characteristics of the hydrogel and the drug molecules, we discuss them in detail here.

The partition coefficient is the ratio of solute concentrations inside and outside of the hydrogel at equilibrium. If the drug molecule is hydrophilic, then it will prefer the aqueous space inside the hydrogel, and an important factor affecting partitioning is the water content or degree of hydration of the hydrogel. When both the drug and polymer are more hydrophobic, the drug may be attracted to the polymer chains, and decreased water content may actually lead to increased partitioning. Partitioning is also favored when the molecule matches a template receptor, as discussed later.

Polymer chains must rearrange themselves to accommodate the drug molecule's volume, and in doing so, they lose some of their freedom to move. This effect is stronger with increasing drug molecule size, leading to reduced partitioning. The presence of cross-links further limits the degree that chains can accommodate large drug molecules. At sufficiently high cross-link densities or above a critical drug molecule size, it will not be possible for a drug molecule to fit between cross-links, and no partitioning or permeation will occur. Figure 14.10 depicts how the diameter of various protein molecules affects their ability to partition into microgels of methacrylic acid, which decreases as the proportion of cross-linking monomer (N,N'-methylenbis(acrylamide)) increases and the length of polymer chains between cross-links decreases.

Similar factors affect the diffusion coefficient. Very small, hydrophilic molecules will diffuse freely through the aqueous space in a highly swollen hydrogel, hindered only by occasional obstruction and by added fluid drag due to the presence of the polymer chains. With increasing size of a hydrophilic molecule, the same factors that limit partitioning come into play. Motion of the drug molecule requires displacement of polymer chains in its path, which again will be constrained by cross-links. Hydrophobic molecules that are attracted to the polymer chains may exhibit slower transport, as they may diffuse by "inching" along the chains.

Ionized drugs may exhibit various permeability behaviors in ionized hydrogels. When the drug and hydrogel are both positively or both negatively charged, they will repel each other, leading to reduced partitioning. When the drug and the hydrogel are oppositely charged, partitioning increases due to electrostatic attraction. This is the *Donnan effect*, which is discussed in Section 14.4.1 with regard to swelling of polyelectrolyte hydrogels. Monovalent drug molecules that partition into the hydrogel will diffuse relatively freely, especially in the presence of physiological electrolytes.

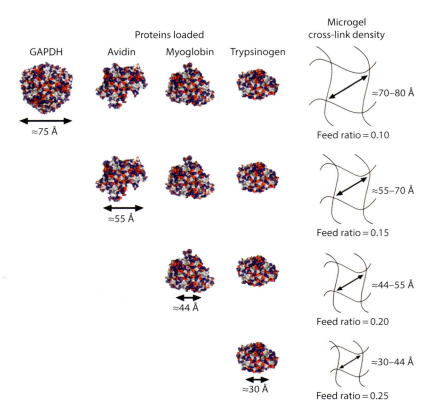

FIGURE 14.10 Effect of the proportion of cross-linker agent (feed ratio 0.10–0.25) on the hydrogel mesh size, which, in turn, determine the molecular size of proteins that can enter in the hydrogel (GAPDH, glyceraldehyde-3-phosphate dehydrogenase). (Reprinted with permission from Eichenbaum, G.M., Kiser, P.F., Dobrynin, A.V. et al., Investigation of the swelling response and loading of ionic microgels with drugs and proteins: The dependence on cross-link density, *Macromolecules*, 32(15), 4867–4878, 1999. Copyright 1999 American Chemical Society.)

Divalent or more highly charged ions may, however, be strongly attracted to oppositely charged ionic groups and linger in their vicinity, slowing down diffusion. Hydrogels bearing ionizable acid groups are particularly useful in pharmacy since more than 75% of drugs have base groups. The formation of ionic complexes depends on the degree of ionization of both species, which, in turn, depends on the pH of the medium.

When the drug and the polymer attract each other, then high partitioning and reduced diffusivity effects may sometimes cancel each other out. In addition to hydrophobicity and strong ionic attraction, there may be specific interactions. For example, hydrogels have been produced with sites that strongly bind heparin, and heparin partitions strongly but diffuses very slowly through these hydrogels. Templated hydrogels (described in Section 14.4.2.5) may also have this feature.

In monolithic systems, partitioning is not an issue per se, but drug solubility may limit the amount of freely diffusible drug inside the hydrogel. It may be convenient to have a dispersion of undissolved drug particles in the hydrogel, which act as miniature drug reservoirs. Solubility and, hence, dissolution rate of drug from these particles into the hydrogel matrix will be affected by the same factors that affect partitioning, such as hydrophilicity/hydrophobicity, volume fraction of water, and cross-link density. Also, diffusivity and, hence, release rate will be affected by these structural factors, as before.

While a hydrogel is formally defined as a water-swollen network, it can be prepared and administered in the dry state, and it is possible to control the rate of water uptake into an initially dry

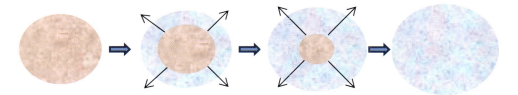

FIGURE 14.11 Drug release from a glassy polymer network. The network swells to form a rubbery hydrogel periphery surrounding a glassy core. Drug is released through the outer periphery to the external fluid as the core recedes. Eventually, the core vanishes and drug release ceases.

polymer and, hence, its rate of swelling and release properties. A useful control parameter is the polymer's glass transition temperature, or T_g, which is the temperature above which the polymer is soft and rubbery, but below which it is hard and glassy. Drug diffuses through the polymer extremely slowly when it is in its glassy state. When water enters the polymer, it plasticizes it, converting it to the wet, rubbery state that permits more rapid diffusion. The rate of plasticization and hence swelling decreases with increasing T_g, in general. The polymer must also be intrinsically hydrophilic for water to enter the polymer and swell it. For example, poly(methyl methacrylate) (PMMA) and poly(hydroxyethyl methacrylate) (PHEMA) both are glassy at room temperature, but only the latter, which has pendant hydroxyl groups, will absorb sufficient water to become rubbery. By combining HEMA and MMA at specified ratios into cross-linked networks, a series of monoliths with different swelling and drug-release rates can be formed. Typically, a *moving front* separating a glassy core from a rubbery periphery is observed, and drug release is determined by the rate of advancement of this front, as depicted in Figure 14.11 (see also Chapter 7, Figure 7.2).

Biodegradability is another property that, when incorporated into a hydrogel, can influence its drug-release properties. As chains and cross-links are degraded either by water or by enzymatic reaction, the hydrogel loosens and the diffusion of drug molecules can proceed more quickly. Alternatively, degradation can be designed to occur after the release process is finished, with the degradation products resorbed or cleared from the body.

14.4 STIMULUS-SENSITIVE HYDROGELS

14.4.1 A BALANCE OF FORCES DETERMINES THE DEGREE OF SWELLING

We have already noted that polymer synthesis often leads to considerable structural variability in chain length, tacticity, sequence (for copolymers), and distances between cross-link points. Almost all synthetic and many natural polymer chains also manifest randomness in their individual *geometrical configurations*. Naively, one might assume that polymer chains will have a linear, rod-like configuration in space. Although polymer chains that are crystallized may contain linear runs that are aligned with each other, noncrystalline polymer chains take on a "random walk" configuration. An analogy is a metal-linked chain. When tossed on the ground, the chain will take on a randomized shape, and each time it is tossed, a different shape results. (A linked chain would also adopt a randomized configuration, now in 3D, in a satellite or space station.) On the microscopic scale, thermal energy is very important, and individual chain segments are constantly being bombarded by each other and by solvent molecules, leading to randomized configurations. Because of incessant thermal bombardment, individual chain configurations are forever changing with time, and they can be characterized only statistically.

When chains are not cross-linked, this constant thermal motion leads to diffusional migration, and when sufficiently hydrophilic, the chains will dissolve in water. When cross-linked, the polymer chains cannot migrate, but they nevertheless are in a constant state of undulation. As will be explained later, these motions are important for permeation of drugs or proteins, since chains need to remove themselves from the diffusing molecule's path. The larger the molecule, or the

Mixing   Elasticity

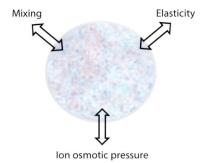

Ion osmotic pressure

FIGURE 14.12   "Forces" acting on a hydrogel. Swelling equilibrium occurs when these forces are in balance. For nonionic hydrogels, only the mixing and elasticity forces are present.

shorter the distance between cross-links, the less likely that a whole polymer chain can "get out of the way" by random motion.

We are now in a position to describe, in general, the forces that determine the degree of swelling of a hydrogel. These forces are depicted in Figure 14.12. We consider a cross-linked polymer that is immersed in a uniform aqueous medium at a fixed temperature. Generally, water will attempt to enter any material due to its thermal motion, i.e., its kinetic energy. However, the degree to which it can do so is determined by the hydrophilicity or hydrophobicity of the polymer, which are governed by the polymer's molecular structure. Polymers containing hydrogen bond donor (e.g., —OH and —COOH) and acceptor groups (—NH$_2$, —O—, and >C=O) or ionized groups are generally hydrophilic, while those lacking such groups (hydrocarbons, silicones) tend to be hydrophobic. The combination of water's thermal motion and its degree of affinity with the polymer chains corresponds to the "mixing" force contribution to swelling.

As water enters the hydrogel, it stretches the chains, pushing the cross-links farther apart. As this happens, the randomly configured chains straighten. However, as we already noted, thermal energy counteracts straightening, and this leads to a retractive force. Similarly, if too much water is withdrawn from the hydrogel, the chains crumple below their most randomized configuration, and now the thermal forces resist this tendency. Thus, the polymer chains act as miniature "springs" that resist deformation. This force is called "polymer elasticity" since it is identical to the force underlying the elastic action of rubber. The strength of the elastic force in a hydrogel is proportional to the *cross-link density*, and hence it is inversely related to the polymer chain length between cross-links.

For many hydrogels, the balance between the mixing and polymer elasticity forces determines the degree of swelling. This balance contains both enthalpic and entropic components, so swelling can be temperature dependent. Molecules that adsorb to the polymer and alter its hydrophilic/hydrophobic balance (e.g., proteins, lipids) can also affect swelling.

When the hydrogel contains ionic groups, usually ionized acids or bases residing on the polymer backbone or on the side chains, a third swelling force is present. These so-called fixed charges on the polymer network are balanced electrically by mobile *counterions*, and the total concentration of mobile ions inside the hydrogel is greater than the concentration of mobile ions outside. This distribution is analogous to the *Donnan equilibrium* that results when charged colloidal solutions are separated from colloid-free solutions by a semipermeable membrane. Due to the relative abundance of mobile ions inside the hydrogel, an *ion osmotic* force drives water into the network. Swelling equilibrium is now determined by the balance of the mixing, elastic, and ion osmotic forces. The ion osmotic force increases with the fixed charge density inside the hydrogel and decreases with increasing ionic strength of the external solution. It may also depend on the concentrations of specific ions in the external solution, which partition into the hydrogel according to their charge and their affinity for water. For example, calcium ions (valence +2) are more effective at reducing swelling of negatively charged hydrogels than sodium ions (valence +1).

At a fixed valence, ions that are considered "structure breaking" tend to reduce swelling compared to "structure-making" ions.

14.4.2   STIMULUS-SENSITIVE HYDROGELS FOR DRUG RELEASE

Each of the three forces governing hydrogel swelling is susceptible to external stimuli, provided the right chemistry is built into the hydrogel. By taking advantage of this fact, hydrogels can be designed to release their contents in an environmentally sensitive manner, either by altering the drug's diffusion coefficient in the hydrogel network or by changing the drug's affinity to the hydrogel's chains.

Hydrogels that modulate drug release as a function of specific stimuli can work in open or closed loop. *Open-loop* systems control drug release by responding to an externally applied stimulus such as ultrasound, electric or magnetic fields, or light irradiation, as illustrated in Figure 14.3. Release rate is not critically determined by conditions of the biological environment. In contrast, *closed-loop* or *self-regulated* systems detect changes in the physiological environment and release drug in order to restore the body's function to a desired *set point*. Thus, closed-loop systems can be viewed as artificial means for enhancing *homeostasis* in the body. Perhaps the most familiar example of a closed-loop system is an insulin delivery device that senses increases in blood glucose levels and responds by releasing insulin, leading to a recovery of normal glucose levels.

Imprinted hydrogels, formed by coagulating monomers around the template when the polymer is in the unswollen state, provide another example of stimulus sensitivity. If the imprinted hydrogel swells due to a stimulus, the monomers forming part of the receptors separate from each other and the attraction for the drug is lost. Drug is then released.

Although sustained release of the drug often leads to adequate therapeutic responses, discontinuous release as a function of specific signals may be desirable in many situations. Regarding systemic administration, the efficacy of a number of treatments has been shown to improve if the release occurs in phase with certain biorhythms; such drugs include reproductive hormones, growth hormone, gastric acid inhibitors, β-blockers, antiarrhythmics, and antiasthmatics. Since many of these molecules are associated with conditions that require repeated and often chronic administration, implantable hydrogel networks that can switch release on and off at the adequate time of the day would be beneficial, provided that the drugs are sufficiently potent that only small amounts need to be implanted.

In the following subsections, we discuss how the principles that were introduced regarding control of hydrogel swelling can be applied to modulate drug release through and out of hydrogels, in either open or closed loop. We pay particular attention to means by which release can be modified according to external conditions, such as temperature, and the chemical or biochemical environment.

14.4.2.1   Temperature-Sensitive Systems

In recent decades, there has been considerable interest in strongly thermosensitive hydrogels, which alter their swelling properties over a narrow range of temperatures. Thermosensitive-swelling controlled-release systems can be activated when exposed to body temperature (37°C), which is higher than typical ambient temperatures (20°C–25°C). As will be discussed later, temperature can also be controlled locally by directed delivery of different forms of energy to the hydrogel, such as ultrasound or electromagnetic fields.

To date, the most widely investigated thermosensitive hydrogels are those that contain the monomer *N*-isopropyl acrylamide (NIPAm). These hydrogels, whose structure is illustrated in Figure 14.13, typically shrink between room and body temperature. As discussed previously, shrinkage is due to the release of bound "clathrate" water surrounding the NIPAm side chains, followed by attraction of NIPAm groups to each other. The "critical" temperature, T_c, associated with shrinkage of pure poly(NIPAm) networks, is about 33°C. This temperature can be raised by coincorporating more hydrophilic monomers such as acrylamide or lowered by adding more hydrophobic monomers such as butyl methacrylate.

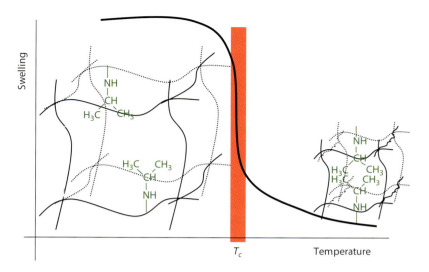

FIGURE 14.13 Structure- and temperature-dependent swelling of poly(*N*-isopropyl acrylamide) hydrogels. A sharp transition in swelling occurs at a "critical" temperature, T_c.

Recently, hydrogels containing oligo(ethylene glycol) (OEG) side chains have also been shown to have favorable thermoshrinking properties. Because they are closely related to PEG, a biocompatible polymeric material, some researchers believe that these OEG-based hydrogels will be more readily accepted than NIPAm-based hydrogels.

While shrinkage of a hydrogel might normally be expected to slow down drug release due to tightening of the hydrogel mesh (Figure 14.13), the opposite trend is sometimes observed. Increased drug release upon shrinkage has been attributed to a squeezing effect, in which drug is carried out of the hydrogel along with the water that is squeezed out, i.e., by convection.

14.4.2.2 pH-Sensitive Systems

When the ionizable groups in the hydrogel are weak acids or bases, then swelling is controlled by external pH, as depicted in Figure 14.14. In *polyacid* hydrogels, an increase in pH over a range near the pKa of the pendant acid groups leads to an increase in fixed charge density and, hence, swelling. Common polyacid hydrogels contain acrylic acid and its alkyl derivatives, or monomers containing sulfonated side chains. In *polybase* hydrogels, fixed charge density and swelling increase with decreasing pH near the pKa of the pendant base groups. Such hydrogels are usually based on imines or tertiary amines. (Quaternary amines have a positive charge regardless of pH.) It is also possible to form mixed polyacid/polybase hydrogels that exhibit increased swelling at the extremes of pH but are relatively shrunken at intermediate pH values. Shrinking is enhanced by electrostatic attractions between the oppositely charged acid and base groups; this is the so-called polyampholyte effect. Polyampholyte hydrogels can also be formed using monomers containing zwitterionic side chains such as sulfobetaines. Interestingly, the shrinking behavior of polyampholyte hydrogels is weakened and the hydrogels swell when ionic strength increases, the opposite of what occurs when the hydrogel is negatively charged.

Thus, pH can affect hydrogel's hydrophilicity by ionizing acid or base groups. Considering these factors, initially glassy systems have been studied for oral delivery of drugs in which controlled release is desired to a specific part of the GI tract. Depending on whether the polymer contains basic or acidic side chains, it can be programmed to swell and release its contents when it hits either the acidic environment of the stomach or the alkaline environment of the small intestine (see also Chapter 7, Figure 7.7). Hydrogels that swell and release their contents in the alkaline gut region are

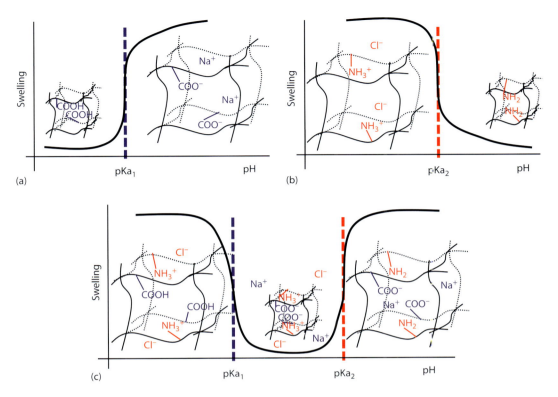

FIGURE 14.14 Structure- and pH-dependent swelling of (a) polyacid, (b) polybase, and (c) polyampholyte hydrogels. Swelling/shrinking transitions occur near the pKa values for the acidic (—COOH: pKa_1) and basic (—NH₂: pKa_2) groups.

reminiscent of enteric coatings, which remain intact in the stomach but dissolve in the small intestine, thereby preventing drug release in the acidic and enzyme-rich gastric fluids.

14.4.2.3 Glucose-Sensitive Systems

While temperature and pH provide relatively simple means to control hydrogel, there are many examples in which the decision regarding release rate should be determined by the level of a more specific biomarker, providing closed-loop control. Here we discuss glucose-sensitive swelling of hydrogels, which have been thoroughly studied due to the interest in using them to treat diabetes. We will then turn to more general biomarker-responsive systems that are under investigation.

Hydrogels containing phenylboronic acid (PBA) side chains alter their ionization and swelling in the presence of molecules containing *cis*-diol groups, most notably sugars. Attempts have been made to use such hydrogels to provide glucose-triggered release of insulin. The mechanism of sensitivity of PBAs to glucose is illustrated in Figure 14.15a. In the absence of glucose, PBA acts as a *Lewis acid*. At low pH values (low OH⁻ concentrations), the boron atom in PBA is coordinated in an uncharged, trigonal configuration, with two pendant hydroxyl groups and an empty orbital. As pH increases, the increasingly abundant OH⁻ ions fill the empty orbital, and the boron atom is now tetragonally coordinated, with three pendant hydroxyls and a net negative charge. The pKa of this Lewis acid–base reaction is controlled by the electron density distribution of the phenyl group attached to the boron atom.

For a typical monomer, methacrylamidophenylboronic acid (MPBA), pKa ≈ 8.5, the boron group is mostly uncharged. Remarkably, however, the tetragonal boronate is able to undergo a *bidentate condensation reaction* with *cis*-diol groups on the glucose molecule. This reaction stabilizes

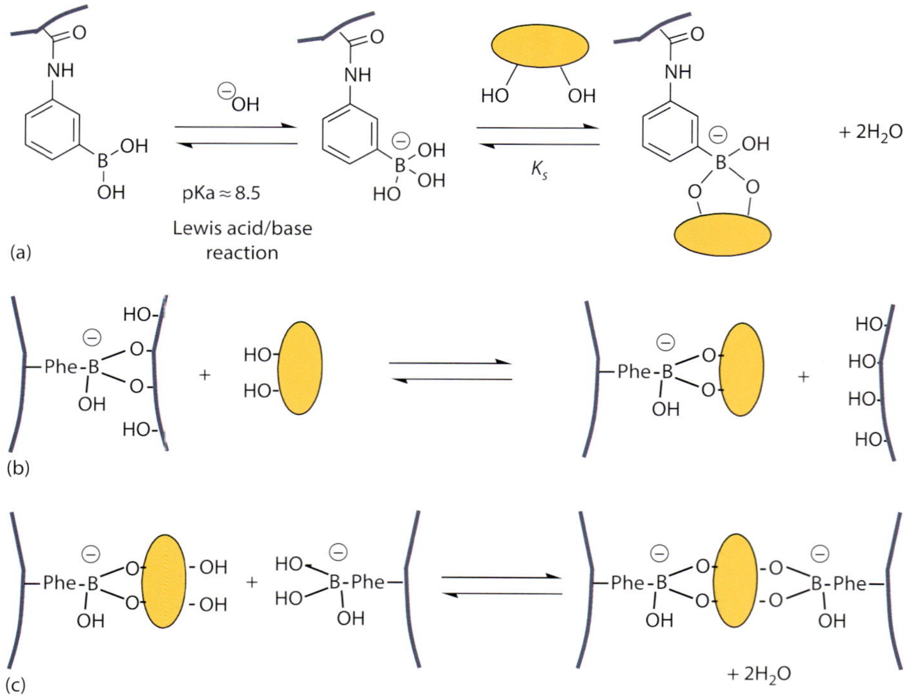

FIGURE 14.15   (a) Mechanism by which sugars such as glucose can affect the ionization of phenylboronic acids (PBA). The uncharged trigonal form (1) reacts with OH⁻ to form a charged tetragonal form and (2) is further stabilized by a *cis*-diol on the sugar molecule, through reversible bidentate condensation. (b) A complex formed between chains containing –PBA and polyvinyl alcohol forms cross-links, which break up in the presence of free sugar molecules, causing the hydrogel to swell. (c) When the pKa of the –PBA moiety is well below physiological pH, then almost all –PBA groups are charged and will bind sugars. Glucose, having two *cis*-diols, will bind to two chains containing –PBA groups, cross-linking them. At high glucose concentrations, cross-linking is reduced, however, due to occupation of each –PBA group by a glucose molecule (not shown). (Reprinted from *J. Control. Rel.*, 190, Siegel, R.A., Stimuli sensitive polymers and self regulated drug delivery systems: A very partial review, 337–351, copyright 2014, with permission from Elsevier.)

the charged boron atom. Hence, increasing glucose concentrations effectively lowers the pKa of MPBA, leading to a higher fixed charge density and greater swelling in hydrogels containing PBA side chains. Both the Lewis acid–base and the di-condensation reactions, while forming covalent bonds, are completely reversible, and when glucose is withdrawn, the hydrogel shrinks back to its original size.

There are alternative ways that glucose can affect the swelling of PBA-containing hydrogels. For example, hydrogels have been synthesized with both PBA-containing chains and PVA chains. Closely spaced –OH side groups on PVA can arrange to form bidentate condensation complexes with PBA, cross-linking the two chains together, as depicted in Figure 14.15b. Glucose introduced into such hydrogels competes with the PVA for tetragonal PBA, breaking some of these cross-links and reducing the elastic retractive force, leading to increased swelling. Again, these reactions are reversible, and swelling in the presence of glucose is reversed upon withdrawal of glucose.

When screening hydrogels for sensitivity to molecules such as glucose, specificity against other similar molecules or conditions that might interfere or produce unwanted responses must be checked. For example, the Lewis acid–base reaction that is involved in the swelling of PBA hydrogels (Figure 14.15a) is affected by changes in blood pH from its normal physiological value, pH 7.4. Acidosis, a common symptom of diabetes, will lead to a reduced hydrogel swelling at a given glucose concentration. Also, other sugars such as fructose and nonsugar diols such as lactate bind to PBAs. Fortunately, these interfering species are at relatively low concentration compared to glucose. Nevertheless, efforts have been made using molecular design of PBA derivatives to minimize both pH sensitivity and interference from molecules other than glucose.

When the PBA side chain is derivatized such that its pKa lies below physiologic pH, then binding by OH^- to –PBA is essentially complete and independent of pH. The binding propensity of *cis*-diols is maximal in this case. Interestingly, glucose has two diols that can bind and cross-link PBAs residing on different polymer chains, as illustrated in Figure 14.15c. At low glucose concentrations, this leads to shrinkage of hydrogels with increasing glucose concentration. However, hydrogel volume eventually reaches a minimum and starts to increase again as glucose concentration continues to increase, due to "flooding" of PBAs by glucose. Interestingly, this cross-linking mechanism seems to be specific to glucose.

An alternative method to confer glucose sensitivity is to immobilize the enzymes *glucose oxidase*, *catalase*, and *gluconolactonase* inside a pH-sensitive hydrogel. Together, these enzymes catalyze the net reaction: glucose + O_2 → gluconic acid + $\frac{1}{2}O_2$. The gluconic acid dissociates readily, releasing H^+ and lowering pH inside the hydrogel, thus affecting the hydrogel's fixed charge density and swelling and, hence, its permeability to insulin. Since *glucose oxidase* is specific to glucose, interference from other sugars is minimal.

14.4.2.4 Other Biomarkers

Are there methods to provide high binding specificity for other kinds of molecules, with possible control of swelling? Recall that the body generates antibodies to specific molecules or antigens. Suppose now that one constructs a hydrogel with some side chains containing an antigen and others containing the corresponding antibody. Then, we expect that the tethered antigens and antibodies will form reversible cross-linking complexes (Figure 14.16). Suppose now that the level of free antigen fluctuates in the blood. When free antigen enters the hydrogel, it will compete with the tethered antigen for the tethered antibody and break the cross-linking complex, causing the hydrogel to swell. Again, this swelling will be reversed when antigen is withdrawn from the hydrogel as its blood level falls, since the intra-hydrogel complexes will reform. This strategy has been pursued—in fact, it is not necessary to include the whole antibody—only the Fab chain containing the antigen recognition site is needed.

Thus, it is possible to incorporate into the hydrogel recognition elements (e.g., antibodies or Fab fragments) in such a way that the hydrogel undergoes changes in volume in response to a

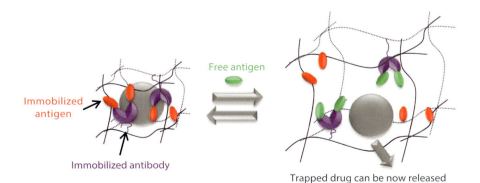

FIGURE 14.16 Antigen-responsive reversible swelling of hydrogels prepared by copolymerization of complementary antigen and antibody monomers that can act as reversible cross-links. Interaction of free antigen with the immobilized antibodies leads to the rupture of the initial cross-links, which cause the hydrogel to swell and thus trigger drug release. If the free antigen vanishes, the cross-links can be reformed and the release stops.

target biomolecule, because their cross-linking density changes by formation or dissociation of the biomolecular complexes. Nowadays, it is possible to prepare antibodies for many biomarkers, and reversible antigen-sensitive hydrogels are promising biomaterials for constructing self-regulated drug-release systems. In the absence of the antigen of interest (i.e., the biomarker), the hydrogel does not release the drug because of the high cross-linking density (thus, low mesh size) of the network due to the internal antigen–antibody interactions. Drug release starts when the antigen biomarker appears in the surrounding medium.

Other hydrogels modeled on biological recognition have been studied. Monomers such as acrylamide can be functionalized with single-strand DNA (ssDNA) oligonucleotides and their complements. Upon polymerization, hybrid pairs are formed, which act as cross-links. These hybrids are broken, reversibly, upon exposure to a free ssDNA containing one of the complementary sequences. *DNA aptamers* to a specific target molecule can also be generated, for example, using the SELEX process, starting with a library of ssDNAs to many target molecules. Connecting the target and the aptamer to the hydrogel, cross-links will form that will be broken, again reversibly, when exposed to the free target molecule.

14.4.2.5 Templated Hydrogels

Since antibodies and oligonucleotides have limited physical and chemical stability and may lead to allergenic reactions, the design of artificial receptors is gaining increasing interest. In *molecular-imprinting technology*, the spatial distribution of the monomers is optimized to achieve the maximum efficiency of interaction between the substance of interest and the polymer network. This technology was born in the context of analytical chemistry, with the goal of preparing synthetic materials with receptors able to specifically bind and separate analytes. Thus, tailor-made receptors are created by synthesizing the polymer network in the presence of the target substance, which acts as a *template*, as depicted in Figure 14.17. Incorporation of the target in the monomer solution, which may contain several comonomers, is expected to cause the spatial arrangement of the monomers as a function of the strength of their interactions. Monomers with high affinity are drawn close to the template molecules, while those with less affinity tend to be excluded from contact. Different monomers can also "cooperate" due to their affinity to different parts of the target molecule. The arrangement of the monomers is "frozen in" during polymerization and cross-linking.

Once the molecularly imprinted network has been formed, the template molecules are removed, revealing imprinted cavities, sometimes called *receptors*, in the polymer network. These cavities are complementary in size, shape, and functional groups to those of the target molecules. If the polymer network again enters into contact with a solution containing the target molecules, then the latter will be absorbed and bind more strongly to the network than other, indifferent molecules.

The success of molecular imprinting relies on the strength of interactions between the template molecules and the monomers responsible for creating the imprinted cavities, which are called *functional monomers*. These interactions should occur favorably throughout the polymerization

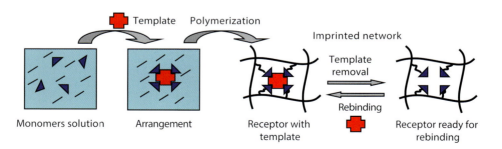

FIGURE 14.17 Steps of the preparation of a molecularly imprinted network. Drug molecules can act as templates to drive the arrangement of functional monomers.

process and during the rebinding step. Ideally, all receptors would be uniform and optimally configured around the template molecule. However, this can only occur if the template is covalently bound to the functional monomers during polymerization, with the covalent bonds broken after polymerization. Unfortunately, this approach requires difficult or sometimes impossible chemical derivatizations.

Hence, the most common approach to molecular imprinting is noncovalent self-assembly, depicted in Figure 14.17. Multipoint interactions between a template molecule and various functional monomers are required to form strong complexes, and the strengths of the monomer–monomer, template–template, solvent–monomer, and solvent–template interactions during polymerization determine the extent and the quality of the imprinted cavities, which are not identical to each other. If the molar ratios between components in the complex are not appropriate or if the assemblies dissociate to some extent during polymerization, then the functional monomers will not correctly arrange around the template molecule, resulting in a small difference between imprinted and non-imprinted (conventional) networks. Thus, the simple presence of target molecules in the monomer solution does not ensure an imprinting effect if the interactions are not favorable. The interactions will determine the number of receptors that can be formed and their heterogeneity.

Traditionally, imprinted networks have been prepared with a high proportion of cross-linking agent (>50% mol/mol) in order to obtain rigid structures with very limited swelling. This rigidity confers physical stability to the receptors. For a variety of applications, such as isolation or analysis of target substances, imprinted systems are prepared as nondeformable particles or rigid plastics or surfaces, using a solvent that acts as porogen to facilitate the removal and rebinding of the target molecules. Such an imprinted receptor resembles a permanently engraved hole. The adaptation of the molecular-imprinting principles to loosely cross-linked networks (as those that are responsive to stimuli) is quite challenging since the receptor structure must be "memorized" in order to maintain the recognition ability after several swelling/collapse cycles. The undulations of polymer chains tend to reduce the probability that monomers attached to them will remain in the optimal receptor configuration. At particular degrees of swelling, the receptors should maintain some memory of their structure and ability to bind the target molecule, while, at other degrees of swelling, this memory should be reduced, as will be the affinity for the target molecule.

Thus, when the imprinted cavities are located in a stimulus-sensitive hydrogel, the conformation of the receptors can be deformed and reconstituted as a function of an external or a physiological signal. Stimulus-responsive imprinted hydrogels are particularly suitable for developing advanced intelligent drug delivery systems able to (1) selectively and effectively load a certain drug, (2) release the drug at a rate modulated by a stimulus, and (3) uptake again the released drug from the environment if the drug remains around the hydrogel when the stimulus stops or diminishes its intensity, at which point the cavities are reformed.

14.4.2.6 External Stimulus-Sensitive Systems

For some pathophysiological processes, there is no biomarker that can be readily used to alter the drug-release properties of a hydrogel. To overcome this limitation, hydrogels can be endowed with responsiveness to external stimuli, such as light, electric or magnetic fields, and ultrasound, which can be applied to the target area when the hydrogel is either physical placed there, or accumulates there following injection as a nanocarrier. Externally triggered hydrogels switch the release on/off as a function of the intensity and the time that the external stimulus is applied. These latter systems have the advantage of providing on-demand (open loop) regulation of the release process, which can be adapted to the progress of the illness and the patient conditions.

Ultraviolet (UV), *visible*, and *near-infrared* (NIR) *radiation sources* can be used to trigger drug release at a specific site of the body, offering a very precise control of the release site. Light responsiveness is most commonly provided by photoactive groups such as azobenzene, cynnamonyl, spirobenzopyran, or triphenylmethane, which undergo reversible structural changes under UV–visible light. However, UV light can only trigger release from hydrogels placed on the skin or against

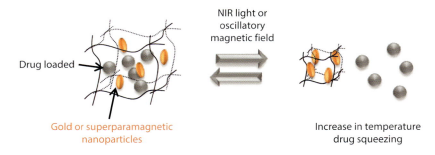

FIGURE 14.18   By incorporating metal nanoparticles into microgels or nanogels that are temperature sensitive, delivery localized in time and space can be achieved by application of electromagnetic radiation.

mucosa, since radiation of wavelengths below 600 nm does not penetrate more than 1 cm into the body. NIR light (650–900 nm) can penetrate deeper because hemoglobin (the principal absorber of visible light), and water and lipids (the main absorbers of infrared light), have their lowest absorption coefficient in the NIR region.

Metals (e.g., gold nanoparticles) can absorb *NIR light* and efficiently transform the radiant energy into local heating, increasing the temperature at the surroundings several degrees above body temperature. This phenomenon can be exploited to trigger drug release from temperature-sensitive networks, and to improve the efficiency of tumor treatments by means of thermal ablation, as depicted in Figure 14.18 (see also Chapter 18, Section 18.4.1). Tumor cells are less resilient than normal cells when exposed to elevated temperatures. Heating alone at 43°C helps to kill tumor cells, but this temperature might be reduced if anticancer drugs are simultaneously released.

It is also possible to incorporate superparamagnetic iron oxide nanoparticles (SPIONs) into nanohydrogels. *Magnetic drug carriers* containing temperature-responsive polymers possess three unique features: (1) SPIONs can be visualized by means of magnetic resonance imaging, (2) tissue distribution can be increased by placing an external magnet near the target site, and (3) drug release can be triggered when an alternating magnetic field is applied, locally, either due to squeezing of the hydrogel in a constant magnetic field or due to local heating in a radio frequency magnetic field. In a similar manner, localized heating of hydrogels followed by release of their contents can be triggered by *focused ultrasound*.

Electrical stimuli can be generated using commercially available equipment for transdermal delivery, which enables precise control of the intensity and amount of current, the duration of the pulses, and the intervals between successive pulses. Electrically sensitive networks are commonly made of polymers with a high density of ionizable groups (polyelectrolytes), similar to those used for preparing pH-responsive hydrogels. Electrically responsive networks can be used in the form of injectable drug-loaded microparticles or as subcutaneous implants. An electric field can be applied through an electroconducting patch placed on the skin over the polyelectrolyte network. The changes in pH that occur near the electrodes due to the movement of the protons to the cathode cause shrinking of the network and squeezing of the drug out of the hydrogel. The intensity of the electrical field and the time of application regulate drug-release rate and duration. When the electrical field is switched off, the hydrogel swells again. Therefore, alternating shrinking (release on) and swelling (release off) can be achieved by applying repetitive pulses of electricity.

An alternative to the polyelectrolytes is the use of intrinsically conducting polymers (ICPs), such as polypyrrole (PPy) or polyaniline, which possess the electrical, electronic, magnetic, and optical properties of a metal. Electrical conductivity is due to an uninterrupted and ordered π-conjugated backbone. ICPs are electrochemically formed as a continuous film on the surface of a working electrode. When a current is applied, the ICP undergoes reversible redox reactions that cause swelling of the polymer network and, as a consequence, can be useful to trigger drug release. ICPs have been tested in vivo as coatings of cochlear implants, which consist of an electrode array implanted into

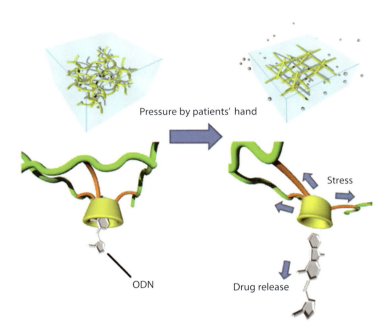

FIGURE 14.19 Pressure-sensitive hydrogels can provide on-demand drug release through a mechanical stimulus generated intentionally by the patient. In the scheme, a hydrogel composed of a β-cyclodextrin derivative and alginate hosts the antiemetic drug ondansetron (ODN). The drug is released in response to mechanical compressions by changing the inclusion ability of the cyclodextrin moieties. (From Izawa, H., Kawakami, K., Sumita, M. et al., β-Cyclodextrin-crosslinked alginate gel for patient-controlled drug delivery systems: Regulation of host–guest interactions with mechanical stimuli, *J. Mater. Chem. B*, 1, 2155–2161, 2013. Reproduced by permission of The Royal Society of Chemistry.)

the scala tympani of the cochlea to electrically stimulate spiral ganglion neurons (SGNs), providing auditory signals to individuals with hearing loss. However, the continued electrical discharges may cause apoptosis of SGNs. To overcome this problem, the electrode array is coated with a layer of PPy containing therapeutic neurotrophins, which are able to prevent the loss of SGNs. The electrode array provides electrical stimulation and thus regulates the release of the neurotrophins to the SGNs.

Hydrogels can be also designed to release drug in response to *physical pressure*. As explained before, cross-linked polymer chains behave as small springs. One can design the hydrogel with drug receptors (template hydrogels) or binding moieties (for example, cyclodextrins, which can form inclusion complexes with the drug) that exhibit binding affinity for the drug when the network is at rest, but when a mechanical pressure deforms the hydrogel, the receptors distort and the drug is released. This principle is being implemented to produce transdermal hydrogel–based formulations with drug release controlled by mild mechanical compression exerted by the patient's hand (Figure 14.19).

14.4.2.7 Multiresponsive Systems

Stimulus-responsive systems described previously rely on polymers that undergo reversible phase transitions when activated by the stimulus. An alternative approach is the utilization of dynamic covalent bonds, i.e., covalent chemical bonds that can be formed and broken under equilibrium control. These events can be made sensitive to changes in physiological parameters, such as pH or redox conditions. For example, dual pH- and redox-responsive nanogels have been prepared by cross-linking polymer chains through both imine and disulfide bonds. If only one stimulus (low pH or the presence of reducing agent) acts on the hydrogel, only one type of bond is broken and a minor increase in drug-release rate occurs because there is still adequate cross-linking. When both stimuli are present, the hydrogel, which is in a particle form (see the next section), disintegrates.

14.5   RECENT DEVELOPMENTS

14.5.1   Microgels and Nanogels

Thus far, there has been little discussion of the size of hydrogel-based drug delivery systems. While millimeter- or centimeter-sized systems can function as membranes or monoliths, hydrogels with smaller dimensions have their advantages. In the following discussion, we include both already swollen hydrogels, and networks that swell into hydrogels upon exposure to body fluids.

Microgels, of size between one and several hundreds of microns, are typically formed by suspension polymerization, wherein small droplets of water containing the monomers are suspended in a continuous oil phase, the suspension being stabilized by stirring and adding surfactants. Particle size is controlled by stirring conditions, with increasing stirring rate generally leading to a decreased particle size. Microgels created by this method tend to be polydisperse, and the particle size range may need to be trimmed by filtration. Microgels are small enough to be injected subcutaneously or intramuscularly through hypodermic needles, and they can establish a drug depot at the site of injection. They may be subject to a localized foreign body response, which can sometimes be attenuated by surface modification of the particles. Microgel depots should be programmed to disintegrate after their release function has been fulfilled.

Nanogels, which typically are of diameter less than 300 nm, can either be synthesized by polymerization and coagulation in solvents with low monomer concentrations or in microemulsions, in which water droplets containing the monomers are smaller than 1 micron. Nanogels are appropriate as circulating depots, as they can access the full circulatory system, including capillary beds. Even smaller nanogels may show enhanced extravasation into tumors, due to the fenestrated, leaky anatomy of tumor vasculature. This means localized release into tumors does not require direct injection, and it might be useful for targeting otherwise difficult-to-access primary solid tumors and their metastases. By incorporating pH sensitivity, nanogels may be able to withhold drug release until they are localized in the acidic environment that is typical of many solid tumors. By functionalizing their surfaces with tumor cell–specific ligands, nanogels may also be taken up specifically by those cells. By proper chemistry, they might be programmed to release their cargo (drug) selectively in the low-pH, high-glutathione environment of tumor cells. Uptake of nanogels by cells prior to drug release may be especially beneficial in tumors that are drug resistant due to the presence of membrane efflux pumps. Similar strategies may apply in using environmentally sensitive nanogels to deliver drug selectively to infected or inflamed tissues that have altered high temperature, acidic or alkaline pH, high ROS levels, or high glutathione levels.

Nanogels of poly(*N*-isopropylacrylamide-co-acrylic acid) bearing doxorubicin conjugated through a pH-labile bond exhibit dual temperature/pH-dependent cellular uptake and cytotoxicity. These nanogels shrink when temperature rises from 37°C to 43°C favoring cellular internalization, while the pH responsiveness enables a fine control of drug-release rate. Thus, localized thermal stimulation of a tumor may improve drug delivery in a site-specific manner.

Dually pH- and temperature-responsive hydrogels that combine NIPAm and ionizable components can adjust the release of thrombolytic agents such as heparin or streptokinase to the small changes of pH and temperature that accompany the formation of the thrombi. They can also be prepared to coat the vaginal tissue and to release an anti-HIV microbicide in response to semen-induced pH (see also Chapter 12, Section 12.7.6).

Nanogels are susceptible to opsonization and removal by the reticuloendothelial system, and surface modifications may be needed to minimize these effects. As with microgels, attention must be paid to their clearance and potential toxicities.

14.5.2 Microfluidic Chips

We conclude this section by discussing alternative ways by which the swelling and shrinking of hydrogels can be used to modulate drug delivery, focusing on microfabricated systems. In these systems, methods that were originally developed for electronic microchips and integrated circuits are used to create small solid structures that respond to their environment. Because of their origin in the electronics industry, these systems are often called "microelectromechanical systems" or MEMS. While initial work in this field was carried out with silicon, it has been expanded to include many other hard and soft materials, such as metals, glass, plastics, and silicone rubber, the latter of which is also referred to as polydimethylsiloxane or PDMS. In fact, *micromachined silicon masters* can be used to form microstructures in PDMS by molding. *Microfluidic systems*, related to MEMS, include small channels through which liquids, including drug solutions or the blood, can flow. Integrating microfluidics with microsensors, one obtains microscopic *total analytical systems* or *lab-on-a-chip*.

How can hydrogels be integrated into such microsystems? One example is to totally confine a thin hydrogel film between a solid but porous membrane, which allows small solutes to pass through it, and a slightly deformable diaphragm. When the hydrogel is exposed to external conditions (pH, glucose concentration, etc.) that would normally cause it to swell, it will exert a *swelling pressure* on the diaphragm, causing the diaphragm to distort slightly. By coupling this distortion to either a piezoresistive or capacitive microsensor, the concentration of the target analyte can be sensed. Alternatively, if the hydrogel film is bonded on one side of a microinductor but is otherwise free to swell and if SPIONs are incorporated inside the film, then swelling and, hence, analyte concentration is reported by a change in inductance, which can be monitored continuously at radio frequencies. These techniques can be used to aid in decisions regarding drug dosing, e.g., for insulin.

In an even more advanced scenario, the hydrogel controls the flow of a drug solution in a microfluidic valve. Suppose the hydrogel is pinned in a microchannel. In its swollen state, the hydrogel blocks fluid flow, but when it shrinks, fluid flow is permitted. In one proposed system, a thermoshrinking hydrogel containing gold nanoshells is synthesized in a microchannel. When exposed to infrared radiation, the nanoshell heats up, causing the hydrogel to shrink and permit flow of drug solution. Attempts have also been made to stop and start the flow of insulin solutions in microchannels using the swelling and shrinking of the glucose-sensitive hydrogels. However, the latter system may be problematic in practice, since its response is "hard wired," while flexibility is needed in insulin dosing between patients and within patients, due to circadian and other variabilities in response to insulin.

14.6 CONCLUSIONS

Hydrogels are soft materials that can function either as membranes controlling diffusion of drugs or as monoliths that both store drug and control its rate of release. They can be loaded with a variety of therapeutic substances, which makes them appealing platforms for drug delivery. As controlled-release systems that provide a variety of release rates, they can be produced in various sizes, depending on the desired application.

For hydrogels that do not suffer significant degradation or changes in volume after administration, drug release occurs via diffusion through the polymer network. If the network experiences swelling due to hydration or the action of certain stimuli, modifies its affinity for the drug, or undergoes hydrolytic or enzymatic degradation, then these processes also influence the rate of drug release. Figure 14.20 summarizes these processes.

For a given drug, the diffusion time decreases as the water content of the hydrogel increases, and for a given hydrogel, the lower the molecular weight of the drug, the shorter the release time. Thus, drug release can be regulated by tuning the swelling degree of the hydrogel, which, in turn, regulates the mesh size and the volume of water available for movement of the drug. A shrunken or

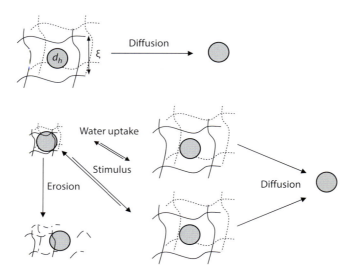

FIGURE 14.20 Release mechanisms of a drug, with hydrodynamic diameter d_h, from a hydrogel with mesh size ξ that can remain constant or that can be modified as a function of the water uptake, the action of a stimulus, or an erosion process due to hydrolytic or enzymatic degradation.

a glassy network can act as a cage and is able to completely prevent drug release. When water enters the network and swells it, the chains loosen up, enabling drug release.

This chapter has provided a review of some, but not all, of the possibilities involved in combining drugs and hydrogels. While we have cast a rather wide net, it is certain that new ideas and applications will arise as hydrogel chemistry and materials science advance. It should be recognized that, however, like any biomaterial, hydrogels must pass rigorous efficacy, biocompatibility, and toxicity tests in their development and approval. These conditions become more rigorous as the complexity of the systems increases, which may include a hydrogel, a drug, and auxiliary materials (e.g., gold on iron nanoparticles). Currently, there are already a number of stimulus-responsive products under clinical evaluation, and some of them are based on hydrogels, mainly as monoliths for controlled release of active substances on the skin during wound healing or as nanogels suitable for diabetes and cancer treatment. The information generated in the design and evaluation of these pioneering products should pave the way to the development and approval of more stimulus-responsive hydrogels.

FURTHER READING

Alvarez-Lorenzo, C. and A. Concheiro. 2014. Smart drug delivery systems: From fundamentals to the clinic. *Chemical Communications* 50:7743–7765.

Annabi, N., A. Tamayol, J.A. Uquillas et al. 2014. 25th anniversary article: Rational design and applications of hydrogels in regenerative medicine. *Advanced Materials* 26:85–124.

Davis, K.A. and K.S. Anseth. 2002. Controlled release from crosslinked degradable networks. *Critical Reviews in Therapeutic Drug Carrier Systems* 19:385–423.

Eichenbaum, G.M., P.F. Kiser, A.V. Dobrynin et al. 1999. Investigation of the swelling response and loading of ionic microgels with drugs and proteins: The dependence on cross-link density. *Macromolecules* 32(15):4867–4878.

Izawa, H., K. Kawakami, M. Sumita et al. 2013. β-Cyclodextrin-crosslinked alginate gel for patient-controlled drug delivery systems: Regulation of host–guest interactions with mechanical stimuli. *Journal of Materials Chemistry B* 1:2155–2161.

Kabanov, A.V. and S.V. Vinogradov. 2009. Nanogels as pharmaceutical carriers: Finite networks of infinite capabilities. *Angewandte Chemie–International Edition* 48:5418–5429.

Kopecek, J. 2009. Hydrogels: From soft contact lenses and implants to self-assembled nanomaterials. *Journal of Polymer Science Part A–Polymer Chemistry* 47:5929–5946.

Omdian, H. and K. Park. 2012. Hydrogels. In *Fundamentals and Applications of Controlled Release Drug Delivery*, eds. J. Siepmann, R.A. Siegel, and M.J. Rathbone, pp. 75–105. New York: Springer.

Peppas, N.A. ed. 1986. *Hydrogels in Medicine and Pharmacy*, Vols. 1–3. Boca Raton, FL: CRC Press.

Peppas, N.A., J.Z. Hilt, A. Khademhosseini et al. 2006. Hydrogels in biology and medicine: From molecular principles to bionanotechnology. *Advanced Materials* 18:1345–1360.

Qiu, Y. and K. Park. 2012. Environment-sensitive hydrogels for drug delivery. *Advanced Drug Delivery Reviews* 64:49–60.

Siegel, R.A. 2014. Stimuli sensitive polymers and self regulated drug delivery systems: A very partial review. *Journal of Controlled Release* 190:337–351.

Siegel, R.A., Y.D. Gu, M. Lei et al. 2010. Hard and soft micro- and nanofabrication: An integrated approach to hydrogel-based biosensing and drug delivery. *Journal of Controlled Release* 141:303–313.

15 Drug Delivery to the Central Nervous System

Anya M. Hillery

CONTENTS

15.1 INTRODUCTION

Treatments for disorders of the central nervous system (CNS) are currently limited because of the difficulty in successfully delivering CNS therapeutics to the brain. The CNS is protected by a barrier known as the blood–brain barrier (BBB), which regulates brain homeostasis. The BBB restricts access of blood-borne compounds to brain cells, only allowing the entry of molecules that are essential for adequate brain function. This comprehensive gatekeeper action not only ensures the maintenance of a precisely regulated microenvironment for reliable neuronal signaling, but also presents a substantial barrier to the access of therapeutic drug molecules to brain tissue.

Successful delivery of therapeutics to the CNS would be beneficial in the treatment of a wide range of conditions (Table 15.1). Furthermore, an ageing population has meant the incidence of CNS diseases of "old age," such as dementia, Alzheimer's disease, stroke, and Parkinson's disease, has increased significantly.

TABLE 15.1

Disorders of the Central Nervous System

CNS Condition	Example
Vascular disorders	Stroke and transient ischemic attack
Infections	Meningitis and encephalitis
Brain cancer	Primary and metastatic
Functional disorders	Epilepsy and neuralgia
Degenerative disorders	Alzheimer's disease
	Parkinson's disease
	Amyotrophic lateral sclerosis
	Multiple sclerosis
Psychiatric disorders	Depression, mania, anxiety

Although considerable research effort has been directed toward developing new CNS drugs for these conditions, the associated delivery problems has meant that promising new drug candidates have failed to realize their therapeutic potential. Successful CNS drug delivery and targeting strategies could therefore considerably expand the treatment possibilities currently available for a wide range of CNS pathologies. Improved drug delivery to the brain would also improve the therapeutic profile of existing CNS drugs, allowing them to be administered at lower doses, thereby reducing adverse and toxic effects.

This chapter begins with a description of the anatomical and physiological constraints afforded by the BBB and the implications therein for the successful delivery of CNS therapeutics. This is followed by a discussion of the most promising strategies currently under investigation to achieve effective CNS transport.

15.1.1 ANATOMY AND PHYSIOLOGY OF THE BBB: IMPLICATIONS FOR CNS DRUG DELIVERY

Brain capillaries are lined with specialized endothelial cells, which serve to limit the entry of xenobiotics into the brain microenvironment. These endothelial cells constitute the major physical barrier of the BBB (Brightman 1977; Patel et al. 2012). Tight junction complexes between brain capillary endothelial cells form a high-resistance barrier, which inhibits paracellular transport (i.e., drug transport between the cells; Figure 15.1b). This is in marked contrast to endothelial capillaries of the peripheral circulation that lack tight junctions between the endothelial cells and thus allow the relatively free exchange of substances between cells and tissues (Figure 15.1b). Furthermore, brain endothelial cells possess no fenestrae (openings) for the passage of molecules, again in contrast to the peripheral endothelium. Phagocytic cells known as pericytes, located on the brain side of the endothelium, share a common basement membrane with the endothelial cells and further complement the physical barrier (Figure 15.1b). Further components of the neurovascular unit are star-shaped glial cells called astrocytes. Astrocytic foot processes approach the endothelial cells to a distance of about 20 nm and secrete chemicals that maintain the unique selective permeability characteristics of the capillary endothelial cells.

This anatomical barrier is compounded by several physiological mechanisms, which further limit the transport of molecules from the blood to the brain. Brain endothelial cells

- Demonstrate minimal pinocytotic transport
- Have a high level of intracellular metabolic enzyme activity, which can degrade an active pharmaceutical ingredient (API) as it moves through the endothelial cell in transit to the brain parenchyma

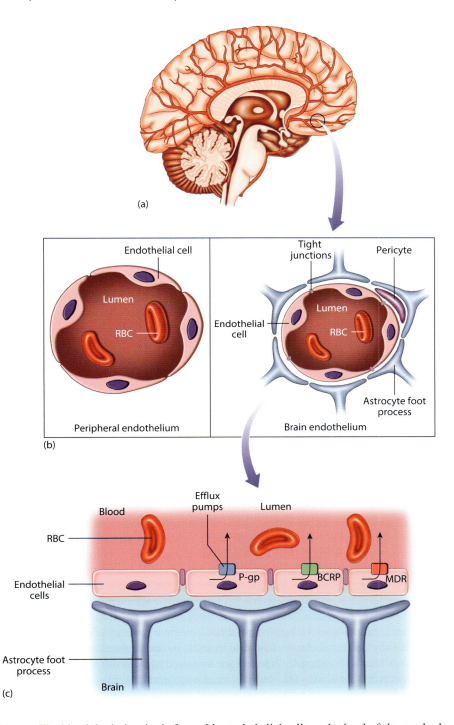

(a)

(b)

(c)

FIGURE 15.1   The blood–brain barrier is formed by endothelial cells at the level of the cerebral capillaries. (a) The blood supply of the brain. (b) The structure of peripheral endothelium and brain endothelium. Brain endothelial cells are linked via tight junctional complexes and are closely associated with pericytes and astrocytic foot processes. (c) A high level of efflux pumps is expressed on the apical surface of the endothelial cells. RBC, red blood cell; P-gp, P-glycoprotein; BCRP, breast cancer resistance protein; MDR, multidrug resistance–associated proteins.

- Express high levels of active efflux transport proteins, including P-glycoprotein (P-gp), multidrug resistance–associated proteins (MRPs), and breast cancer resistance protein (BCRP, ABCG2). The pumps are expressed at the luminal membrane of the brain capillary, so that drugs that have gained entry into the cell can be expelled back into the circulation (Figure 15.1c).

15.1.2 Transport Mechanisms across the BBB

Transport routes and mechanisms across epithelial barriers are described extensively in Chapter 4 (Section 4.3) and are equally applicable here for endothelial transport. To summarize here, the paracellular route is not normally possible across brain capillaries, due to the presence of tight junctions between the cells that form a tight physical barrier, as described earlier. Transcellular transport (i.e., through the cells) is possible via a variety of mechanisms, considered in turn.

15.1.2.1 Simple Diffusion

The majority of CNS therapeutics already on the market or in clinical trials comprise small (<400 Da), lipophilic molecules, which passively diffuse across the BBB down their concentration gradient. The rate of transcellular diffusion is governed by Fick's first law (Chapter 4, Equation 4.1). Many drugs are too large, ionized, and hydrophilic to meet these physicochemical requirements and so cannot gain access to the CNS. Furthermore, many drugs that are able to enter the cell by transcellular passive are also substrates for efflux pumps. For example, many HIV-protease inhibitors and chemotherapeutics exhibit a high degree of lipophilicity but do not access the brain, which has been attributed to efflux pumps actively exporting them back into the systemic circulation after their initial entry. A further limitation to simple diffusional transport is that drugs that diffuse into endothelial cells may subsequently be vulnerable to the degradative enzymes localized within the cells.

15.1.2.2 Carrier-Mediated Transport

Certain small, hydrophilic molecules, vital for brain function, such as amino acids, glucose, and nucleosides, use specific transporters expressed on the luminal and basolateral sides of the endothelial cells for brain entry. This carrier-mediated transport (CMT) involves stereospecific transporter proteins that mediate solute transport in the order of milliseconds. Some of these carriers have been widely studied. For example, glucose, necessary for cellular energy, enters the brain via CMT on the BBB GLUT1 glucose transporter. Large, neutral amino acids, required for brain protein and neurotransmitter synthesis, enter the brain via the BBB L-type amino acid transporter 1 (BBB LAT1).

15.1.2.3 Transcytosis

Receptor-mediated endocytosis (RME) is a highly selective type of endocytosis, by which the brain can specifically take up large, hydrophilic ligands, including transferrin (Tf), albumin, insulin, insulin growth factor, low-density lipoprotein, and ceruloplasmin (see also Chapter 4, Section 4.3.2.3). The receptors for RME are integral membrane proteins that bind specific substrates. Ligand binding triggers the membrane to invaginate, forming a clathrin-coated vesicle that contains the receptor–ligand complex. Fusion with an endosome, and then lysosomal degradation, may follow. Alternatively, in receptor-mediated transcytosis (RMT), the endocytic vesicles move across the cell and undergo exocytosis on the opposite side, discharging their cargo into the brain. In contrast to RMT, which is triggered by a specific interaction with receptors expressed on cerebral endothelial cells, adsorptive-mediated transcytosis (AMT) facilitates vesicular delivery across the BBB via a nonspecific mechanism of electrostatic attraction. Peptides and proteins with a basic isoelectric point ("cationic" proteins), bind initially to the luminal plasma membrane, mediated by electrostatic interactions with anionic sites on the membrane. Adsorption triggers endocytosis/transcytosis, again facilitated by the clathrin-mediated system.

15.1.3   OVERVIEW OF APPROACHES TO CNS DELIVERY

CNS drug delivery research is focused on a number of key areas, summarized here and discussed in detail in the following sections:

1. *Physiological approaches*: Exploiting natural transport mechanisms. Many of the physiological transport systems that the brain uses to take up nutrients and other essential molecules can be exploited for enhancing CNS drug delivery.
2. *Pharmaceutical approaches*: Using nanotechnologies, such as liposomes, nanoparticles (NPs), and dendrimers, as drug carriers for CNS delivery.
3. *Surgical approaches*: Drugs can be introduced directly into the brain via surgical techniques.
4. *Nose-to-brain delivery*: A direct transport route exists from the nose to the brain, via the olfactory nerve. Therefore, a drug given intranasally can gain access to the brain without having to surmount the BBB.
5. *Disruption of the BBB*: The integrity of the BBB can be transiently disrupted via osmotic, ultrasonic, and chemical means, thereby temporarily allowing drugs to breach the barrier.

15.2   EXPLOITING NATURAL TRANSPORT MECHANISMS

Physiological approaches exploit the natural transport mechanisms of brain endothelium for the uptake of nutrients and other essential molecules. These mechanisms have been described in Section 15.1.2 and include CMT and RMT pathways. A drug molecule that is structurally similar to an endogenous substrate can be taken up the corresponding transporter. For example, the drug L-dopa is an effective therapeutic in the treatment of Parkinson's disease because it can enter the brain via CMT, using the BBB LAT1 system described earlier (Wade and Katzman 1975). Any drug molecule can, in principle at least, be chemically modified so that it resembles the structure of an endogenous substrate, for uptake via the appropriate carrier. However, there is obviously a limit to the degree of chemical modification that can be carried out on a drug before its therapeutic activity is comprised.

Rather than focusing on the drug per se, an alternative approach is to conjugate a drug with a natural substrate of a transporter system, so that the drug–substrate conjugate can be taken up via the transporter. Monoclonal antibodies (mAbs) can also be raised against a specific transporter receptor, to form a drug–mAb conjugate for subsequent uptake. This conjugation strategy has been developed extensively by Pardridge and coworkers (Pardridge 2002, 2003, 2006), who coined the eloquent term "molecular Trojan horse" to evoke how a drug may surreptitiously gain entry to the CNS, under the cover of a natural substrate/mAb. Similarly, a nontransportable peptide can be conjugated to a BBB transport vector to form a so-called "chimeric peptide," capable of BBB transit (Coloma et al. 2000; Pardridge 1989).

CMT pathways, as described in Section 15.1.2.2, are used physiologically for the uptake of small, hydrophilic molecules (such as glucose and amino acids), so this pathway has quite limited capacity for the uptake of large drug molecules. In contrast, RMT pathways exist for the CNS uptake of much larger molecules, such as insulin, therefore exploiting RMT pathways has much wider applicability for CNS drug delivery. Research in this area has focused in particular on targeting three different RMT receptors (Chen and Lihong 2012), described here in turn.

15.2.1   INSULIN RECEPTOR

The insulin receptor (IR) is a tetrameric structure comprising two α chains that project into the plasma compartment of the brain capillaries and contain the binding site for insulin, and two β chains that contain tyrosine kinase activity and project into the intracellular compartment. Insulin per se cannot be used as a delivery vector, as the resulting drug–insulin conjugate could cause hypoglycemia. Research has instead focused on using a mAb to the IR as a targeting vector. A mouse mAb has been developed, designated mAb 83-14, which binds to an epitope contained within amino acids 469–592 of the α chain of the IR. The ability of mAb 83-14 to undergo transcytosis has been demonstrated in rhesus monkeys. Following a single i.v. dose, 4% of the Ab was delivered to the brain, whereas no brain uptake was observed for controls (Pardridge et al. 1995). Genetically-engineered forms of the human IR mAb have been now developed, and research in this area is ongoing.

15.2.2 Transferrin Receptor

The transferrin (Tf) receptor (TfR) facilitates the entry of iron-bound Tf, as a means of providing iron to brain cells (Moos and Morgan 2001). The high expression of this receptor on brain capillary endothelial surfaces has made it an important target for CNS delivery strategies (Fenart and Cecchelli 2003). The use of Tf per se as a drug delivery vector is again unsuitable, due to the high levels of competing endogenous Tf in the blood. Research has instead focused on developing antibodies against TfR. Early work focused on the OX26 murine mAb to the rat TfR (Gosk et al. 2004). This mAb has been extensively studied by Pardridge and colleagues, as a molecular Trojan horse for a wide variety of macromolecules, including glial-derived neurotrophic factor, erythropoietin, tumor necrosis factor receptor II, and various enzymes (Figure 15.2a): all have shown transport across the BBB in preclinical models (Pardridge 2012).

Recent work has focused on targeting the TfR for the CNS delivery of therapeutics for the treatment of Alzheimer's disease (Bell and Ehlers 2014). The enzyme β-secretase (BACE1) is a membrane-associated aspartyl protease that mediates the initial cleavage of amyloid precursor protein, leading to the generation of amyloid-β (Aβ) peptides, which subsequently aggregate to form the amyloid plaques characteristic of the disease. Blocking the activity of BACE1 thus represents a viable approach to reducing amyloid plaque formation. Small-molecule inhibitors of BACE1 have

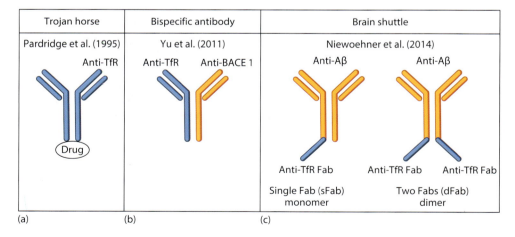

FIGURE 15.2 Exploiting receptor-mediated transcytosis via the transferrin receptor (TfR) for CNS delivery. (a) "Molecular Trojan horse": a therapeutic drug is conjugated to a monoclonal antibody against TfR (anti-TfR). (b) "Bispecific antibody": one arm comprises the anti-TfR antibody (transport vector) and the other arm comprises an antibody against BACE1 (to neutralize this plaque-forming enzyme). (c) "Brain shuttle": an antibody against aggregate-forming Aβ peptide (anti-Aβ) is conjugated with the transport vector. The vector is either a single antibody fragment against TfR (anti-TfR Fab monomer) or two antibody fragments against TfR (anti-TfR Fabs dimer).

been developed that can readily cross the BBB due to their small size; however, their lack of specificity has raised toxicity concerns. As an alternative, a highly specific anti-BACE1 antibody was developed, capable of binding to the enzyme and blocking its activity. In order to deliver this anti-BACE1 antibody across the BBB, a "bispecific" mouse antibody (Figure 15.2b) has been developed (Yu et al. 2011): one arm comprises an anti-TfR antibody (to function as a BBB transport vector) and the other arm comprises the high-affinity anti-BACE1 antibody (to neutralize the plaque-forming enzyme).

This Anti-TfR/BACE1 bispecific antibody accumulated in the mouse brain and led to a significant reduction in brain Aβ levels, even after a single systemic dose. Interestingly, the research also revealed an unexpected relationship between antibody affinity for the TfR and subsequent levels of brain uptake, which has yielded important information on the fundamental physiology of the RMT process. It was found that anti-TfR antibodies that bound with a *high* affinity to TfR could not subsequently detach from the receptor and so remained associated with the BBB, without ever reaching the brain tissue. In contrast, *lower*-affinity anti-TfR antibody variants (with more than 25 times less affinity for the receptor) were released from the receptor and showed a broad brain distribution, 24 hours after dosing.

In a different approach, a mAb against the toxic aggregate-forming Aβ peptide (i.e., anti-Aβ) was developed and coupled with a mAb fragment (Fab) against the TfR (Figure 15.2c). This "brain shuttle" demonstrated significant brain penetration and amyloid plaque reduction in a transgenic mouse model of Alzheimer's disease (Niewoehner et al. 2014). Greater than 50-fold uptake and activity was shown for the anti-Aβ antibody when delivered using the Fab "brain shuttle," than without the carrier. A construct was also prepared comprising a dimer Fab (dFab) (i.e., two Fab fragments against TfR), coupled with the anti-Aβ antibody (Figure 15.2c). Whereas the monovalent (sFab) vector mediated effective uptake and transcytosis of the conjugate, the divalent derivative (dFab) resulted in trafficking to lysosomes and subsequent destruction of the construct. This differential fate following RMT uptake needs to be studied further in order to optimize future design strategies.

15.2.3 Low-Density Lipoprotein Receptor-Related Protein 1

Low-density lipoprotein receptor-related protein 1 (LRP1), also known as apolipoprotein E receptor, is a receptor protein involved in the RMT-mediated brain uptake of cholesterol (Hertz and Strickland 2001). This receptor offers many advantages as a target receptor: (1) it is one of the most expressed receptors on the BBB, thereby offering the potential for large amounts of drug to enter the brain; (2) it is a multiligand lipoprotein receptor that mediates the internalization of a large number of different ligands in vivo; thereby offering flexibility in construct design; and (3) it demonstrates fast transport and recycling time, suggesting that it is difficult to saturate, which should minimize interference in the RMT uptake of endogenous cholesterol.

A number of biotechnology companies are actively investigating the LRP1 receptor as a means of enhancing the CNS delivery of therapeutics. Angiochem Inc. has developed a proprietary library of more than 100 peptide structures ranging from 8 to 34 amino acids, called Angiopeps™, which demonstrate a high transcytosis rate mediated by LRP1 (Demeule et al. 2008; Gabathuler 2010). The company has also developed a patented mechanism to attach drug molecules to the Angiopep™ vectors. Angiopeps™ are capable of facilitating the CNS delivery of molecules ranging from 500 Da to 150 kDa (i.e., including antibodies). The most promising drug–vector candidate to date is ANG1005, formed by the chemical conjugation of Angiopep2 with three molecules of the cytotoxic drug, paclitaxel. The Angiopep–paclitaxel conjugate is well tolerated and has shown therapeutic activity in patients with advanced solid tumors, including those with brain metastases and/or failed prior taxane therapy (Regina et al. 2008). ANG1005 is now in clinical trials for recurrent malignant gliomas and for advanced cancer and brain metastases.

The Transcend™ platform from biOasis is a further group of proprietary technologies, based on the naturally occurring transport protein melanotransferrin (human melanoma–associated antigen p97 [MTf]). Although MTf is structurally very similar to Tf, uptake appears to be mediated via the LRP1 receptor, rather than the TfR. Transcend™ demonstrates many advantages for the CNS delivery of therapeutics, including a high level of brain accumulation and transcytosis, with a very low level in the blood (Demeule et al. 2002). It has demonstrated success as a transport vector for anticancer agents such as doxorubicin and paclitaxel, as well as biologics such as mAb and lysosomal enzymes. A smaller peptide portion within MTf has recently been identified as the fragment actually responsible for crossing the BBB; this fragment is the focus of future targeting efforts.

15.2.4 ADSORPTIVE-MEDIATED TRANSCYTOSIS

AMT can enhance the ability of cationic macromolecules, such as peptides, proteins, and Abs, to enter the CNS (Hervé et al. 2008). Early studies showed that cationized albumin (pI > 8, in comparison to native albumin, with a pI = 4) was rapidly taken up by brain capillaries; it also acted as a vector for the brain uptake of beta-endorphin (Kumagai et al. 1987). Noncationic proteins can be converted to their cationized form via the amidation of their carboxylic acid groups with positively charged amines; hexmethylenediamine is widely used for this purpose.

15.2.4.1 Cell-Penetrating Peptides

Cell-penetrating peptides (CPPs) are short (<30 amino acid) peptides, derived from protein-transduction domains, which can penetrate cellular membranes and gain access to the cell interior. The precise mechanism by which CPPs gain entry to cells has not yet been fully elucidated and is a matter of ongoing research and debate, although it appears that AMT is involved in the majority of cases. CPPs can function as CNS delivery vectors when conjugated with an API. The most studied CPP vectors for CNS delivery are those derived from natural proteins, including the transactivating transcriptional activator (TAT) from human immunodeficiency virus 1 (HIV-1) and the SynB vectors, which are derived from a mammalian antimicrobial peptide that has a high affinity for biological membranes. Synthetic peptides, in particular the poly(arginines), are also showing promise (Wagstaff and Jans 2006). The potential of the approach has been demonstrated in a number of studies. For example, the enzyme β-galactosidase, fused with TAT, can be transported into almost all tissues, including the brain, when injected intraperitoneally into mice (Schwarze et al. 1999). Coupling of the cytotoxic doxorubicin to SynB vectors significantly enhanced brain uptake and also bypassed the efflux pump MDR1. Vectorization of the opioid dalargin with SynB vectors enhanced brain uptake several hundredfold after conjugation (Rousselle et al. 2003).

However, using AMT as a strategy for CNS delivery is limited because of the lack of specificity for exclusively brain uptake. Unlike RMT, there is no specific interaction with a receptor in this case: cationized ligands will adsorb to all anionic cell surfaces, not just brain endothelial cells. This can potentially cause side effects and toxicity, as drugs interact with nontargeted organs. The potential immunogenicity of these ligands is a further concern, as is the lack of information on the precise mechanism of uptake by CPPs. Further information on the physiology underpinning this transport mechanism is required, in order to optimize this delivery approach.

15.3 NANOTECHNOLOGIES FOR CNS DELIVERY

Pharmaceutical approaches to enhancing BBB transport are directed toward the use of drug delivery systems (DDS) to facilitate the transport of therapeutics to the CNS. Nanotechnology-based DDS are described in detail in Chapter 5 and comprise both soluble and particulate carriers, in the nanometer to micrometer size range. In fact, the various Drug-Antibody conjugates described in Section 15.2 are also examples of *soluble* nanocarriers and thus could also be included in the discussion here. In this section, *particulate* nanocarriers for CNS delivery are described.

Many different nanoparticulate DDS have been studied, including liposomes, NPs, polymeric micelles, and dendrimers (Patel et al. 2012; Saltzman 2001; Shah et al. 2013; Wong et al. 2012). These various types of nanocarriers are described in detail in Chapter 5. A nano- or microparticulate drug carrier is associated with many advantages. Rather than conjugating each individual drug molecule to a targeting vector in a 1:1 stoichiometry, a drug carrier such as a liposome permits the loading of thousands of drug molecules into the carrier interior. Also, there is no risk of masking the active site on the drug through covalent conjugation with a BBB vector. Targeting vectors can be coupled to the nanocarrier surface, thereby enhancing CNS uptake. Many of the targeting vectors for nanoparticulate DDS are the same targeting vectors described earlier for drug–vector conjugates (Section 15.2). These include mAb against IR, TfR, and LRP1, which can be conjugated to the nanocarrier surface. Thus, natural transport mechanisms (particularly RMT) can be exploited to enhance uptake, although in this case, uptake is for a targeted, drug-loaded DDS, rather than a targeted drug. Association with a drug carrier may also afford the API protection from enzymatic degradation, increase API solubility, and offer controlled release of the API from the carrier system. Nanocarriers can be sterically stabilized by the covalent attachment of poly(ethylene glycol) (PEG) chains or by the adsorption of surfactants, in order to avoid opsonization and capture by the reticuloendothelial systems (RES). This can serve to prolong carrier circulation times, thereby allowing the DDS more time to avail of CNS transport mechanisms.

15.3.1　Liposomes

Liposomes per se do not accumulate in the brain and require targeting using brain-specific targeting vectors. Pardridge and colleagues have extended their work using molecular Trojan horses to develop Trojan horse immunoliposomes (THILs). These comprise small, unilamellar vesicle (SUV) liposomes as the nanocarrier, with a molecular Trojan horse molecule (e.g., OX26, or mAb 83-14) to provide the targeting technology. PEG chains are covalently linked to the liposomes to provide steric stabilization and help avoid uptake by the RES (Zhang et al. 2003). Although the targeting vector can be attached directly to the liposomes, the PEG chains may then interfere with vector binding to the receptor. To avoid this, a targeting ligand can be covalently attached to the tips of 1%–2% of the PEG strands (see also Chapter 5, Figure 5.9).

Various preclinical animal studies have shown the potential of this approach. THILs bearing a mAb for targeting to the human IR have successfully delivered a DNA plasmid encoding β-galactosidase to the CNS. Widespread neuronal expression of the β-galactosidase gene in primate brain was demonstrated by both histochemistry and confocal microscopy following i.v. injection (Zhang et al. 2004). THILs loaded with an RNAi against human epidermal growth factor (EGFR, which plays an oncogenic role in 90% of primary brain cancers such as glioblastoma multiforme) caused a marked decrease in brain tumor expression of the EGFR protein, and were associated with a 90% increase in survival time, when administered i.v. on a weekly basis to adult mice with intracranial human brain cancer (Zhang et al. 2004). BBB Therapeutics have developed G-Technology®, comprising PEGylated SUVs with glutathione as the targeting vector. Glutathione, an endogenous tripeptide that possesses antioxidant-like properties, is actively transported across the BBB, although the precise molecular mechanism by which it is transported is unknown. The liposomes have demonstrated enhanced CNS delivery for a variety of drugs; trials are moving forward with this system for the CNS delivery of doxorubicin and methylprednisolone.

15.3.2　Nanoparticles

The first polymeric NP system investigated for CNS delivery comprised poly(butylcyanoacrylate) (PBCA) NPs, sterically stabilized with an adsorbed coating of Tween 80 surfactant and loaded with the opioid peptide, dilargin. Analgesic effects were demonstrated in vivo after i.v. administration of the NPs to mice, demonstrating successful arrival of the drug in the CNS. PBCA NPs have further been shown to enhance CNS delivery for a number of drugs, including methotrexate, doxorubicin, and temozolomide (Kreuter 2004; Kreuter et al. 2003). It is thought that the Tween 80 on the NPs can promote brain uptake via a number of mechanisms: (1) by providing steric stabilization, which enhances circulation times and decreases RES clearance; (2) by adsorbing apolipoprotein E and B from the bloodstream, which may facilitate RMT uptake via the LDL receptor; (3) in some cases, the surfactant action may transiently open tight junctions, thereby facilitating paracellular transport, and (4) by reducing the activity of drug efflux transporters (Kabanov et al. 2003).

Biodegradable NPs, based on the aliphatic poly(esters), poly(lactic acid) (PLA), poly(glycolic acid), and their copolymers poly(lactide-co-glycolide) (PLGA), have been widely studied for CNS delivery. For example, severalfold increases in the brain level of dexamethasone and vasoactive intestinal peptides have been demonstrated when these therapeutic agents were delivered by NPs of PLGA and PLA, respectively. Further studies have investigated PLGA NPs using TAT peptide and glutathione as CNS vectors; also, polysorbate 80 and poloaxmer 188 have been used to provide sterically stabilized systems.

Further nanoparticulate DDS showing promise include human serum albumin (HSA) NPs, loaded with loperamide as a model drug, which have been targeted to the CNS using either Tf or TfR mAb (OX26 or R17217) (Ulbrich et al. 2009). Chitosan NPs have shown promise for the CNS delivery of peptides, including caspase inhibitor and fibroblast growth factor (Caban et al. 2012). Polymer core/shell NPs, self-assembled from TAT–PEG cholesterol, have been successfully synthesized for antibiotic delivery across the BBB (Liu et al. 2008). The presence of TAT on the NP surface promoted CNS uptake.

15.3.3 DENDRIMERS FOR CNS DELIVERY

A variety of compositionally differentiated dendrimers have been studied for CNS delivery, including poly(amidoamine) (PAMAM), poly(etherhydroxylamine), and poly(propyleneimine) constructs. Again, dendrimers are amenable to the attachment of brain specific ligands and also to steric stabilization strategies. PEGylated PAMAM dendrimers containing the brain-targeting ligand Tf were investigated for the CNS delivery of DNA complexes. The transfection efficiency of vector-DNA complexes in brain capillary endothelial cells was found to be much higher for the PEGylated, targeted, systems (i.e., PAMAM-PEG-Tf-DNA complexes) than either the PEGylated, or the unPEGylated, dendrimers (Huang et al. 2007).

15.4 INVASIVE APPROACHES TO CNS DELIVERY

Invasive (neurosurgical based) approaches bypass the BBB entirely and the drug is introduced directly into the brain (Yang 2006). By introducing the drug directly into the brain, much lower levels of a cytotoxic are required, thereby lowering toxic effects. However, these strategies are very traumatic for a patient, requiring complex brain surgery and extended hospital stays, as well as involving considerable risk and high costs. Thus, the approaches described here are typically confined to the treatment of life-threatening disorders.

15.4.1 INTRACEREBROVENTRICULAR DRUG INFUSION

Intracerebroventricular infusion involves infusion of the drug into the cerebral ventricles: the interconnected cavities of the brain that are filled with cerebrospinal fluid (CSF). A major drawback of this approach is that minimal drug penetration from the ventricles into the brain tissue is possible. Drug diffusion decreases exponentially with distance, so that drug distribution into brain tissue is very limited (Jain 1990). A further obstacle is that CSF is continually replaced: CSF produced in the ventricles circulates throughout the brain and spinal cord within the subarachnoid space, prior to being reabsorbed back into the systemic circulation through the arachnoid villi. A drug infused into the ventricles is thus carried by bulk CSF flow through the CSF pathway. In about 1 hour, the infused drug is thus removed from the brain and enters the blood circulation, leaving very little time for any significant drug diffusion to occur within brain tissue.

15.4.2   INTRACEREBRAL IMPLANTS

In this approach, a drug-loaded polymeric support is implanted locally into the brain and releases the drug over time. A commercial example is the Gliadel® wafer, which comprises a biodegradable poly(anhydride) polymer (poly[bis(p-carboxyphenoxy) propane: sebacic acid]) loaded with the cytotoxic bischloroethylnitrosourea (BCNU). After surgical removal of a brain tumor, up to eight Gliadel® wafers are placed in the resection cavity. BCNU diffuses out of the polymeric support, to exert its cytotoxic action on the tumor cells. As the polymer support is biodegradable, the wafers dissolve over a period of 2–3 weeks, so a further operation to retrieve the implants is unnecessary (Brem et al. 1991).

However, the approach is again limited by the minimal drug diffusional distances which are possible within the brain parenchyma. The drug does not travel beyond the immediate region of the disk and so is ineffective in the case of a large and infiltrative tumor or when multiple tumors are located in the brain.

15.4.3   CONVECTION-ENHANCED DELIVERY

In convection-enhanced delivery (CED), a small catheter is inserted into the target site within the brain, and the drug is pumped through the catheter via an infusion pump. The continuous positive pressure generated by the infusion pump forces the drug to penetrate into the interstitial space, thereby enhancing drug diffusion into brain tissue. CED has been used in the clinical treatment of neurological diseases such as malignant brain tumors, neurodegenerative disorders, epilepsy, and stroke (Vandergrift et al. 2006). However, even when a pump is used to push the drug through brain tissue, the degree of drug distrubtion attained remains low. Furthermore, under positive pressure, the drug follows the path(s) of least resistance, which is not necessarily to the target site. Issues of catheter backflow further complicate outcomes.

There is emerging evidence that CED can be improved when used in conjunction with nanotechnology DDS. For example, drug-loaded NPs given via CED have been shown to prolong drug retention times in the brain in comparison to free drug. Drug distribution in the brain is also enhanced and controlled drug release at the site of action is possible (Patel et al. 2012).

15.5   INTRANASAL ADMINISTRATION FOR BRAIN DELIVERY

Intranasal (IN) delivery is being studied for noninvasive drug delivery to the brain. The IN route can bypass the BBB to allow therapeutic substances direct access to the CNS. This topic is discussed in detail in Chapter 10 (Section 10.6). Although the precise mechanism is unclear, it appears that drugs administered via the IN route may travel up the olfactory nerve, along the ensheathed channels surrounding the axons (Chapter 10, Figure 10.4). The olfactory nerve is localized in specialized olfactory epithelial tissue, found in the roof of the nasal cavity. This area of the nasal cavity is very difficult to access—the intricate complexity of the nasal anatomy, in combination with many physiological clearance mechanisms, makes targeting the olfactory nerve a considerable challenge. Considerable research effort is currently dedicated to optimizing IN delivery to the CNS, by optimizing nasal drug delivery devices, as well as using formulation approaches. A number of animal studies have shown a variety of biologics, including peptides, proteins, and siRNA, can directly reach the brain following IN administration. Studies in humans corroborate these findings (Lochhead and Thorne 2012).

15.6 DISRUPTION OF THE BBB

As the main component of the BBB consists of the physical barrier formed by the tightly joined endothelial cells of the brain capillaries, a transient mechanical, or chemical, disruption of this barrier can temporarily increase the permeability of the brain, thereby facilitating the entry of therapeutics. Transient disruption of the integrity of the BBB has primarily focused on the use of osmotic disruption, using a high concentration of mannitol, which is introduced into the carotid artery (Doolittle et al. 2000). The osmotic pressure exerted by the mannitol draws water from the endothelial cells, resulting in their shrinkage. This leaves an expanded space between the cells, which facilitates paracellular entry of the drug.

Chemical mediators may also be used to disrupt the BBB. Certain vasoactive molecules, including histamine, prostaglandins, and bradykinin, can interfere with endothelial tight junctions, rendering them leakier than normal. In experimental and clinical applications, the synthetic nonapeptide bradykinin agonist, Cereport®, was found to selectively increase drug delivery into brain tumors (Borlongan and Emerich 2003). There is evidence that the opening of the tight junctions occurs by activation of bradykinin B2 receptors, through a calcium-mediated mechanism. However, this approach has recently been tested in the clinical setting without success. A more targeted approach is based on the use of focused ultrasound, in conjunction with microbubbles (Kinoshita et al. 2006). Circulating microbubbles (small, air-filled, lipid-shell, or protein-shell bubbles) are activated by focused ultrasound waves, to exert a mechanical force on a targeted area of the BBB, thus causing temporary disruption for drug entry.

However, these disruption methods carry the risk of either temporarily, or even permanently, damaging the BBB. There is also a lack of specificity with this method: toxic consequences may arise if unwanted blood components enter the brain during the period of BBB disruption. Although focused ultrasound is a more specific approach, there is still the risk that proteins, pathogens, etc., enter the brain, along with the drug molecules. The entry of serum albumin to the brain, for example, is associated with seizures.

15.7 CONCLUSIONS

Current research into CNS delivery includes physiological, pharmaceutical, and invasive approaches. Although progress is being made, a number of key aspects need to be addressed. Crucially, a better understanding of the barrier nature of the BBB per se is needed—both in the physiological and pathological states. Aspects of various transport processes (e.g., CMT, AMT, and RMT) are still poorly understood but must be elucidated in order to maximize delivery strategies. New research, outlined here, is clarifying important characteristics associated with BBB transport, such as affinity binding and postuptake trafficking of RMT pathways.

More specific targeting ligands for brain endothelial cells are required. Several current targets, such as the LDL, insulin, and TfRs, are also highly expressed in other tissues. Brain-specific ligands are needed, to optimize brain traffic and minimize uptake by non-CNS sites, which results in drug wastage, as well as adverse and even toxic effects.

The use of nanocarriers to enhance systemic CNS delivery is an emerging field that has built up a solid foundation of preclinical data at this stage. However, none of the carrier approaches described here is known to be efficacious in treating human CNS disorders. Research must continue into improving both carriers targeting specificity, as well as increasing the capability of the carrier to deliver its cargo into brain tissue. Furthermore, more research is needed to examine the potential toxicity and immunogenicity of these carriers. Drug delivery to the CNS is a challenging and difficult science, which needs a multifaceted response, underpinned by a thorough understanding of the relevant physiological and pathological processes involved. Current research is making significant progress toward addressing these challenges.

REFERENCES

Bell, R.D. and M.D. Ehlers. 2014. Breaching the blood–brain barrier for drug delivery. *Neuron* 81(1):1–3.

Borlongan, C.V. and D.F. Emerich. 2003. Facilitation of drug entry into the CNS via transient permeation of blood–brain barrier: Laboratory and preliminary clinical evidence from bradykinin receptor agonist, Cereport. *Brain Res Bull* 60(3):297–306.

Brem, H., M.S. Mahaley, Jr., N.A. Vick et al. 1991. Interstitial chemotherapy with drug polymer implants for the treatment of recurrent gliomas. *J Neurosurg* 74:441–446.

Brightman, M.W. 1977. Morphology of blood–brain interfaces. *Exp Eye Res* 25(Suppl):1–25.

Caban, S., Y. Capan, P. Couvreur et al. 2012. Preparation and characterization of biocompatible chitosan nanoparticles for targeted brain delivery of peptides. *Methods Mol Biol* 846:321–332.

Chen, Y. and L. Lihong. 2012. Modern methods for delivery of drugs across the blood–brain barrier. *Adv Drug Deliv Rev* 64(7):686–700.

Coloma, M.J., H.J. Lee, A. Kurihara et al. 2000. Transport across the primate blood–brain barrier of a genetically engineered chimeric monoclonal antibody to the human insulin receptor. *Pharm Res* 17:266–274.

Demeule, M., J. Poirier, J. Jodoin et al. 2002. High transcytosis of melanotransferrin (⊃97) across the blood-brain barrier. *J Neurochem* 83:924–933.

Demeule, M., A. Regina, C. Che et al. 2008. Identification and design of peptides as a new drug delivery system for the brain. *J Pharmacol Exp Ther* 324:1064–1072.

Doolittle, N.D., M.E. Miner, W.A. Hall et al. 2000. Safety and efficacy of a multicenter study using intra-arterial chemotherapy in conjunction with osmotic opening of the blood–brain barrier for the treatment of patients with malignant brain tumors. *Cancer* 88:637–647.

Fenart, L. and R. Cecchelli. 2003. Protein transport in cerebral endothelium. *In vitro* transcytosis of transferrin. *Methods Mol Med* 89:277–290.

Gabathuler, R. 2010. Approaches to transport therapeutic drugs across the blood–brain barrier to treat brain disease. *Neurobiol Dis* 37:48–57.

Gosk, S., C. Vermehren, G. Storm et al. 2004. Targeting anti-transferrin receptor antibody (OX26) and OX26-conjugated liposomes to brain capillary endothelial cells using in situ perfusion. *J Cereb Blood Flow Metab* 24:1193–1204.

Hertz, J. and D.K. Strickland. 2001. LRP: A multifunctional scavenger and signalling receptor. *J Clin Invest* 108:779–784.

Hervé, F., N. Ghinea, and J.M. Scherrmann. 2008. CNS delivery via adsorptive transcytosis. *AAPS J* 10(3):455–472.

Huang, R.Q., Y.H. Qu, W.L. Ke et al. 2007. Efficient gene delivery targeted to the brain using a transferrin-conjugated polyethyleneglycol-modified polyamidoamine dendrimer. *FASEB J* 21(4):1117–1125.

Jain, R.K. 1990. Tumor physiology and antibody delivery. *Front Radiat Ther Oncol* 24:32–46.

Kabanov, A.V., E.V. Batrakova, and D.W. Miller. 2003. Pluronic block copolymers as modulators of drug efflux transporter activity in the blood–brain barrier. *Adv Drug Deliv Rev* 55:151–164.

Kinoshita, M., N. McDannold, F.A. Jolesz et al. 2006. Targeted delivery of antibodies through the blood–brain barrier by mRI-guided focused ultrasound. *Biochem Biophys Res Commun* 340:1085–1090.

Kreuter, J. 2004. Influence of the surface properties on nanoparticle-mediated transport of drugs to the brain. *J Nanosci Nanotechnol* 4:484–488.

Kreuter, J., P. Ramge, V. Petrov et al. 2003. Direct evidence that polysorbate-80-coated poly(butylcyanoacrylate) nanoparticles deliver drugs to the CNS via specific mechanisms requiring prior binding of drug to the nanoparticles. *Pharm Res* 20:409–416.

Kumagai, A.K., J.B. Eisenberg, and W.M. Pardridge. 1987. Absorptive-mediated endocytosis of cationized albumin and a beta-endorphin-cationized albumin chimeric peptide by isolated brain capillaries. Model system of blood–brain barrier transport. *J Biol Chem* 262:15214–15219.

Liu, L., K. Guo, J. Lu et al. 2008. Biologically active core/shell nanoparticles self-assembled from cholesterol-terminated PEG-TAT for drug delivery across the blood–brain barrier. *Biomaterials* 29:1509–1517.

Lochhead, J. and R. Thorne. 2012. Intranasal delivery of biologics to the central nervous system. *Adv Drug Deliv Rev* 64(7):614–628.

Moos, T. and E.H. Morgan. 2001. Transferrin and transferrin receptor function in brain barrier systems. *Cell Mol Neurobiol* 20:77–95.

Niewoehner, J., B. Bohrmann, L. Collin et al. 2014. Increased brain penetration and potency of a therapeutic antibody using a monovalent molecular shuttle. *Neuron* 81(1):49–60.

Pardridge, W.M. 1989. Chimeric peptides for neuropeptide delivery through the blood–brain barrier. US Patent Number 4801575 A.

Pardridge, W.M. 2002. Drug and gene targeting to the brain with molecular Trojan horses. *Nat Rev Drug Discov* 1:131–139.

Pardridge, W.M. 2003. BBB drug targeting: The future of brain drug development. *Mol Interv* 3(2):90–105.

Pardridge, W.M. 2006. Molecular Trojan horses for blood–brain barrier drug delivery. *Curr Opin Pharmacol* 6:494–500.

Pardridge, W.M. 2012. Drug transport across the blood–brain barrier. *J Cereb Blood Flow Metab* 32:1959–1972.

Pardridge, W.M., R.J. Boado, and Y.S. Kang. 1995. Vector-mediated delivery of a polyamide ("peptide") nucleic acid analogue through the blood–brain barrier in vivo. *Proc Natl Acad Sci USA* 92:5592–5596.

Patel, T., J. Zhou, J.M. Piepmeier et al. 2012. Polymeric nanoparticles for drug delivery to the central nervous system. *Adv Drug Deliv Rev* 64(7):701–705.

Regina, A., M. Demeule, C. Che et al. 2008. Antitumor activity of ANG1005, a conjugate between paclitaxel and the new brain delivery vector Angiopep-2. *Br J Pharmacol* 155:185–197.

Rousselle, C., P. Clair, M. Smirnova et al. 2003. Improved brain uptake and pharmacological activity of dalargin using a peptide-vector-mediated strategy. *J Pharmacol Exp Ther* 306(1):371–376.

Saltzman, W.M. 2001. *Drug Delivery: Engineering Principles for Drug Therapy.* New York: Oxford University Press.

Schwarze, S.R., A. Ho, A. Vocero-Akbari et al. 1999. In vivo protein transduction: Delivery of a biologically active protein into the mouse. *Science* 285:1569–1572.

Shah, L., S. Yadav, and M. Amiji. 2013. Nanotechnology for CNS delivery of bio-therapeutic agents. *Drug Deliv Transl Res* 3(4):336–351.

Ulbrich, K., E. Herbert, and J. Kreuter. 2009. Transferrin- and transferrin-receptor monoclonal antibody-modified nanoparticles enable drug delivery across the blood–brain barrier (BBB). *Eur J Pharm Biopharm* 71(2):251–256.

Vandergrift, W.A., S.J. Patel, J.S. Nicholas et al. 2006. Convection-enhanced delivery of immunotoxins and radioisotopes for treatment of malignant gliomas. *Neurosurg Focus* 20(4):E10.

Wade, L.A. and R. Katzman. 1975. Rat brain regional uptake and decarboxylation of L-dopa following carotid injection. *Am J Physiol* 228:352–359.

Wagstaff, K.M. and D.A. Jans. 2006. Protein transduction: Cell penetrating peptides and their therapeutic applications. *Curr Med Chem* 13(12):1371–1387.

Wong, H.L., X.Y. Wu, and R. Bendayan. 2012. Nanotechnological advances for the delivery of CNS therapeutics. *Adv Drug Deliv Rev* 64(7):686–700.

Yang, A.V. 2006. Brain cancer therapy and surgical interventions. In *Horizons in Cancer Research*, A.V. Yang, ed., Vol. 27. Hauppauge, NY: Nova Science Publishers.

Yu, Y.J., Y. Zhang, M. Kenrick et al. 2011. Boosting brain uptake of a therapeutic antibody by reducing its affinity for a transcytosis target. *Sci Transl Med* 3(84):84ra44.

Zhang, Y., J. Bryant, Y.F. Zhang et al. 2004. Intravenous RNAi gene therapy targeting the human EGF receptor prolongs survival in intracranial brain cancer. *Clin Cancer Res* 10(11):3667–3677.

Zhang, Y., F. Schlachetzki, and W.M. Pardridge. 2003. Global non-viral gene transfer to the primate brain following intravenous administration. *Mol Ther* 7:11–18.

16 Gene Delivery Systems

Kwang Suk Lim and Sung Wan Kim

CONTENTS

16.1 INTRODUCTION

Since it was first reported in 1972, genetic engineering has been rapidly progressing for treating many genetic and acquired diseases including cancer, diabetes, and tissue damage. Genetic engineering represented by recombinant DNA technology is applied to various fields including medicine, agriculture, and stockbreeding. Recently, recombinant DNA technology has been widely used for developing various protein drugs and for gene therapy. Nucleotides, including DNA, small interfering RNA (siRNA), and micro RNA, provide information needed by the cells and control the protein production. Protein not only works in the cell but also controls the organ and whole body functions. The absence or overexpression of specific proteins in the body can lead to various clinical symptoms due to their structural or functional changes. Gene therapy allows delivery of specific nucleotides to control the absence or overexpression of certain proteins responsible for diseases.

Protein drugs produced by recombinant DNA technology, such as granulocyte macrophage colony–stimulating factor (GM-CSF), erythropoietin (EPO), interleukins (ILs), and insulin-like growth factor-1 (IGF-1), represent one of the fastest-growing sectors in the pharmaceutical industry. Although protein drugs have exquisite therapeutic efficacy, their clinical applications are limited by poor oral bioavailability, low pharmacokinetic and pharmacodynamic profiles in the blood, high manufacturing cost, and difficult quality control of bioactivity. These require frequent parenteral

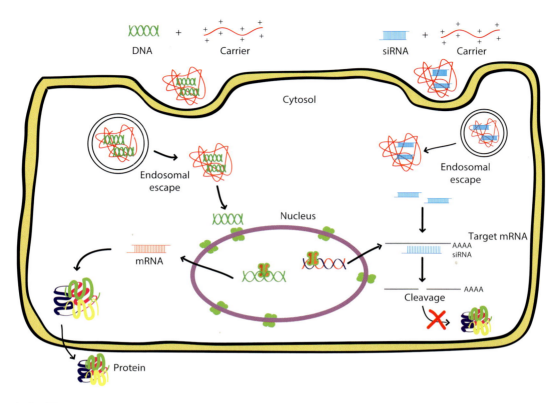

FIGURE 16.1   Scheme of gene delivery systems.

administration, which, in turn, reduced patient compliance with increases in treatment cost. These limitations of treating diseases by protein drug can be overcome by gene therapy.

Gene therapy can be used to overexpress or reduce target proteins for treatment or prevention of diseases. DNA-based gene therapy delivers exogenous plasmid DNA, which encodes a specific gene to enhance the expression of therapeutic proteins, and thus, it must be delivered to the nucleus. On the other hand, siRNA reduces protein expression by silencing target mRNA in the cytoplasm (Figure 16.1). Successful clinical application of gene therapy delivering nucleotides to the right target requires overcoming several barriers, such as penetrating the cell membrane, stability in serum, and safety concerns such as uncontrolled gene delivery. DNA and RNA require appropriate delivery vehicles to overcome these barriers. The commonly used delivery vehicles include cationic polymers, peptides, liposomes, and nanoparticles.

16.1.1   RECOMBINANT DNA TECHNOLOGY

Recombinant DNA technology refers to modification, manipulation, and synthesis of DNA expression vectors. A DNA sequence extracted from cells or synthesized in the laboratory is inserted into DNA expression systems, such as plasmid and virus, using restriction enzymes. Preparation of the recombinant DNA expression vector needs to consider the following components (Figure 16.2):

- *Template DNA*: Target DNAs are prepared by extraction from cell or by artificial synthesis.
- *Expression vector*: A plasmid or virus is used as an expression vector.
- *Polymerase chain reaction (PCR)*: The PCR is a biochemical technology that is used to amplify a single or a few copies of a piece of DNA. PCR is widely used for DNA cloning

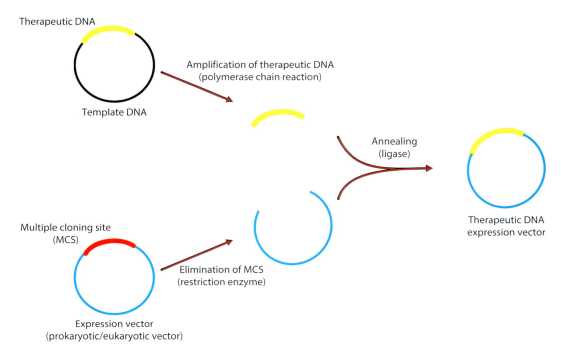

Therapeutic DNA

Amplification of therapeutic DNA
(polymerase chain reaction)

Template DNA

Annealing
(ligase)

Multiple cloning site
(MCS)

Therapeutic DNA
expression vector

Elimination of MCS
(restriction enzyme)

Expression vector
(prokaryotic/eukaryotic vector)

FIGURE 16.2 Recombinant DNA technology.

for sequencing, DNA-based phylogeny, the diagnosis of hereditary disease, and the detection and diagnosis of infectious disease.

- *Restriction enzymes*: Restriction enzymes (restriction endonuclease) cut DNA at or near specific recognition nucleotide sequences known as restriction sites.

16.1.2 PLASMID (EXPRESSION VECTOR)

Gene expression can be regulated in the cell at transcriptional, translational, and posttranslational levels, whereas expression vector can be regulated by transcription factors, promoters, and stimuli response. Plasmid vector expressing a target protein in the cell consists of the following components (Figure 16.3).

- *Promoter*: It is a region of DNA that initiates transcription of the target gene and located at the 5′ region of the antisense strand. The promoter is recognized by RNA polymerase and binds to transcription factors. Examples are cytomegalovirus, simian virus 40 (SV40), Moloney murine leukemia virus, and Rous sarcoma virus.
- *5′-Untranslated region*: All mRNA have a cap structure located between the cap site and initiation codon. The cap structure consists of a 7-methylguanosine linked to the first nucleotide via 5′-5′ triphosphate bridge. It is known to influence mRNA translation efficiency.
- *3′-Untranslated region (3′-UTR)*: The 3′-UTR is located in mRNA following the termination codon. It has a potential role in mRNA stability and AU-rich motifs.
- *Poly(A) tail*: The poly(A) consists of adenosine residues and is located at the 3′ end of most eukaryotic mRNA.
- *Multiple cloning site (MCS)*: An MCS is a short segment of DNA, which contains many restriction sites. Target DNA is inserted into the region of the MCS.
- *Restriction site*: The restriction site contains specific sequences of nucleotides, which are recognized by restriction enzymes.

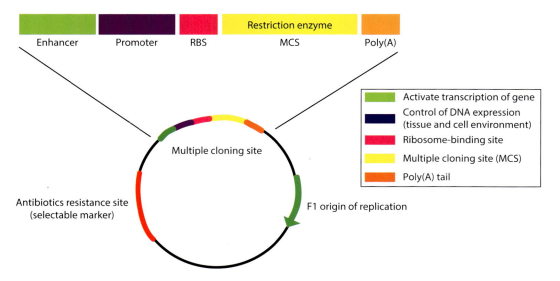

FIGURE 16.3   Structure of plasmid vector.

To treat diseases, a gene delivery system must overexpress their therapeutic gene in their target cells. Various gene carriers have been engineered to be nontoxic and have a higher transfection efficiency, but some therapeutic genes have harmful side effects when expressed in nontarget or normal tissue (Figure 16.1). For example, a growth factor gene expressed in normal tissue cells may enhance tumor formation. Therefore, careful regulation of gene expression is required to avoid deleterious side effects.

16.1.3   POLYMERASE CHAIN REACTION

Since the development by Kary Mullis in 1983, PCR has become an essential tool in a variety of applications including DNA cloning, functional analysis of genes, diagnosis of hereditary disease, and identification of genetic fingerprints. Target DNAs between 0.1 and 10 kb pairs can be amplified by PCR. These components include the following:

- *Template DNA*: A DNA vector is contained in the therapeutic DNA region.
- *Primers*: Complementary primer to the 3′ end of each of the sense and antisense strand of the template DNA. It recognizes the template DNA and has a restriction enzyme site (20–30 mer).
- *Taq polymerase*: Taq polymerase recognizes and synthesizes new DNA segments complementary to the DNA template using deoxynucleotide triphosphate (dNTP) under high temperature around 75°C.
- *dNTP*: It supplies nucleotides for synthesis of DNA segments. It consists of adenine (A), guanine (G), thymine (T), and cytosine (C).

To prepare DNA segments, PCR typically undergoes 20–40 repeated temperature changes, called cycles. The cycle consists of three steps, denaturation, annealing, and extension (Figure 16.4).

- *Denaturation step*: Template DNA is denatured under 94°C–98°C for 20–30 seconds. Sense and antisense are divided due to disrupting the hydrogen bonding between complementary bases.

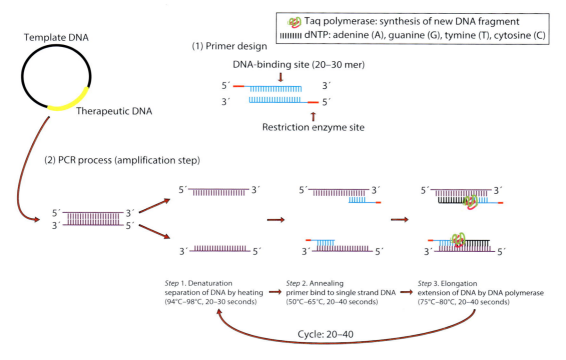

FIGURE 16.4 Polymerase chain reaction.

- *Annealing step*: This step progresses under 50°C–65°C for 20–40 seconds allowing annealing of the primers to the single-stranded DNA template.
- *Extension step*: It is called the elongation step under specific temperature depending on the DNA polymerase. Polymerase works under their optimum temperature at 75°C–80°C. The extension time depends on the DNA polymerase and on the length of the DNA fragment to be amplified.

PCR procedures have been developed for optimizing PCR conditions case by case, because PCR products are affected by various conditions, such as contamination of template DNA, size of template DNA, and primer design. Amplified DNA segments are purified and cloned into the expression vector using restriction enzymes. Constructs are verified by DNA sequence analysis.

16.1.4 GENE KNOCKDOWN

Since its discovery, RNA interference (RNAi) has been widely used for functional analysis and for therapeutic purposes. Overexpression of specific proteins may induce diseases such as Parkinson's disease (PD) and cancer. The siRNA recognizes target mRNA and leads its degradation via the intracellular mechanism of RNAi (Figure 16.1). This process is mediated by specific cytosolic proteins, dicer, RNA-induced silencing complex, and Argonaute. The siRNA molecule is localized in cytoplasm and, through complementary base pairing with the mRNA of the target gene, forms a double-stranded RNA molecule (dsRNA) molecule. The mediator proteins induce RNA silencing by cleaving the dsRNA molecule. For transportation of RNA molecules, a wide variety of nonviral vectors have been developed that induce internalization by endocytosis and subsequent escape from the endosome. These carriers are tightly bound to nucleic acids by electrostatic interactions, but this interaction may diminish the bioactivities of genes because not all genes may be released from the carrier to localize at their target locations. Various carriers showed effective endosomal escape and rapid dissociation in the cytoplasm by the proton-buffering effect and the cytoplasm-sensitive reducible polymers.

16.1.5 VIRAL VECTOR

Several types of viruses such as retrovirus, adenovirus, adeno-associated virus, and gene poxvirus have been developed for gene delivery. A modified genome is included in the virion structure of viral vectors to deliver DNA. It consists of essential viral sequences and the required transcription unit of the exogenous gene. Although engineered viruses serve as efficient delivery systems for overcoming intracellular barriers, they are more immunogenic than synthetic carriers. Some viral types are difficult to apply as a gene delivery system because of their natural tropism. Recently, various delivery systems have been developed for reducing immunogenicity using capsid engineering or regulated expression cassettes. Viruses were modified for retargeting to new cell types by vector engineering approaches such as pseudotyping or genetically engineered with peptides.

16.2 POLYMER-BASED DELIVERY SYSTEMS

A variety of synthetic polymer–based gene delivery systems have been developed to improve therapeutic effectiveness using DNA and RNA. The polymeric carriers are positively charged to overcome the negative charge of the cell membrane that blocks gene transfer into cells. After escaping from endosomes, nucleic acids have to be dissociated from the carriers. Cationic polymers have been designed to utilize the proton-buffering effect, and cytoplasm-sensitive reducible polymers were developed for effective endosomal escape followed by rapid dissociation in the cytoplasm. Bioreducible polymers, or peptides containing internal disulfide bonds in the main chain, at the side chain or in the cross-linker, have been applied to RNAi and DNA delivery. These polymers have high stabilities in extracellular spaces and are rapidly reduced in the cytosol (Figure 16.5). Internal disulfide bonds are reduced by high concentration of intracellular glutathione (GSH). Polymer reduction can lower the cytotoxicity of high-molecular-weight polycations by converting the polymers

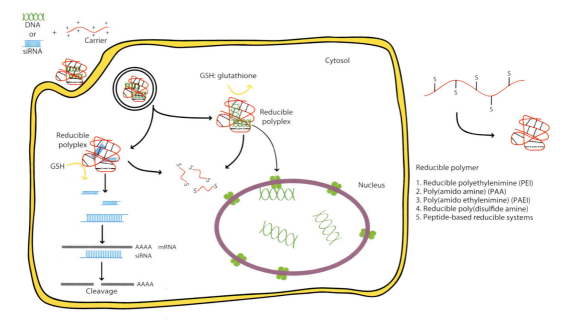

FIGURE 16.5 Reducible polymer-based gene delivery system.

back into the smaller constitutive subunits and also allow for the release of nucleic acids into the cytoplasm. Because of their potential for enhanced transfection efficiency and cytoplasm-sensitive gene delivery, the use of bioreducible polymers in gene therapy is increasing. Various reduction-sensitive cationic polymers including reducible polyethylenimine (PEI), poly(amido amine) (PAA), poly(disulfide amine), and cationic peptides have been developed for use as siRNA and DNA carriers. These polymers are highly stable in physiological conditions and rapidly reduced in the cytosol.

16.2.1 REDUCIBLE POLYETHYLENIMINE

PEI with a repeating unit composed of the amine group and two carbon aliphatic spacers is a strong polymer for transfection of DNA and siRNA (Figure 16.6a). Although PEI shows high transfection efficiency in vitro, it has serious cytotoxicity. To alleviate the toxicity, a bioreducible PEI (bPEI)-based delivery system was developed. The bPEI is dissociated in the cytoplasm having high concentrations of GSH via reduction of internal disulfide linkages in the polymers for increased transgene expression. The various thiolated PEI components (800 Da) were joined by oxidation to form disulfide cross-linked PEI (PEI-SS). The low-molecular-weight bPEI (800 Da) cross-linked with a homobifunctional reducible linker had transduction efficiency in Chinese hamster ovary cells compared with that of bPEI (25 kDa). DNA/PEI-SS was also coated by thiol-reactive poly[N-(2-hydroxypropyl) methacrylamide] through a reducible disulfide linkage or a stable thioether bond. These polymers showed lower cytotoxicities and higher transgene expressions compared with that of the high-molecular-weight bPEI (25 kDa). Dithiobis(succinimidyl propionate) was used for cross-linking PEI for preparation of reducible polymers. The cross-linked DNA/PEI improved pharmacokinetic profiles compared with those of the DNA/PEI complexes after i.v. injection in mice because of high stability during circulation and retention of transfection efficiency of cross-linked polyplexes in vivo. Recently, reducible PEI-SS was synthesized by cross-linking low-molecular-weight PEI through cystamine bisacrylamide (CBA) followed by conjugation with an anionic biocompatible polysaccharide. It significantly decreased cytotoxicity and targeted mRNA level in vitro. In addition, this polyplex suppressed tumor growth after intratumoral injection of the polyplex and decreased vascular endothelial growth factor (VEGF) mRNA and VEGF protein levels in the tumors.

16.2.2 POLY(AMIDO AMINE) AND POLY(AMIDO ETHYLENIMINE)

The linear SS-PAA is a bioreducible poly(amido amine) containing multiple disulfide linkages and synthesized using Michael-type polyaddition of various primary amines to the disulfide-containing CBA (Figure 16.6b). The reducible polymers protect genes against enzymes and serum because they are stable in physiological conditions, but they release genes from polyplex in the presence of a reducing agent. In addition, most PAAs have highly endosome escapability due to their high buffering capacity and show high transfection efficiency and are nontoxic in in vitro experiments. Various water-soluble and bioreducible SS-PAAs were synthesized using the Michael addition of N,N'-dimethylcysteamine with bisacrylamides. The effects of variations in charged density and hydrophobicity on the transfection efficiency were confirmed using the polymers with varying degrees of acetylation and benzoylation. It was demonstrated that transfection efficiency increased depending on the amount of hydrophobic benzoyl groups. In another study, reducible cationic copolymer poly(cystamine bisacrylamide-diaminohexane) (p(CBA-DAH)) was synthesized using Michael-type polyadditions of 1,6-diaminohexane (DAH) and CBA (Figure 16.6c). To deliver Fas siRNA into cardiomyocytes for the inhibition of ischemia-induced apoptosis, prostaglandin E$_2$ (PGE$_2$), a cardiomyocyte-specific ligand, was chemically conjugated with the terminal end of sense siRNA (PGE$_2$-siRNA). It showed high cellular uptake of the polypolex into cardiomyocytes via PGE$_2$ receptor–mediated endocytosis and significant silencing of Fas mRNA resulting in inhibition of cardiomyocyte apoptosis. The primary cardiomyocyte–specific peptide selected by phage display was conjugated to bioreducible p(CBA-DAH) and enhanced the cellular uptake and

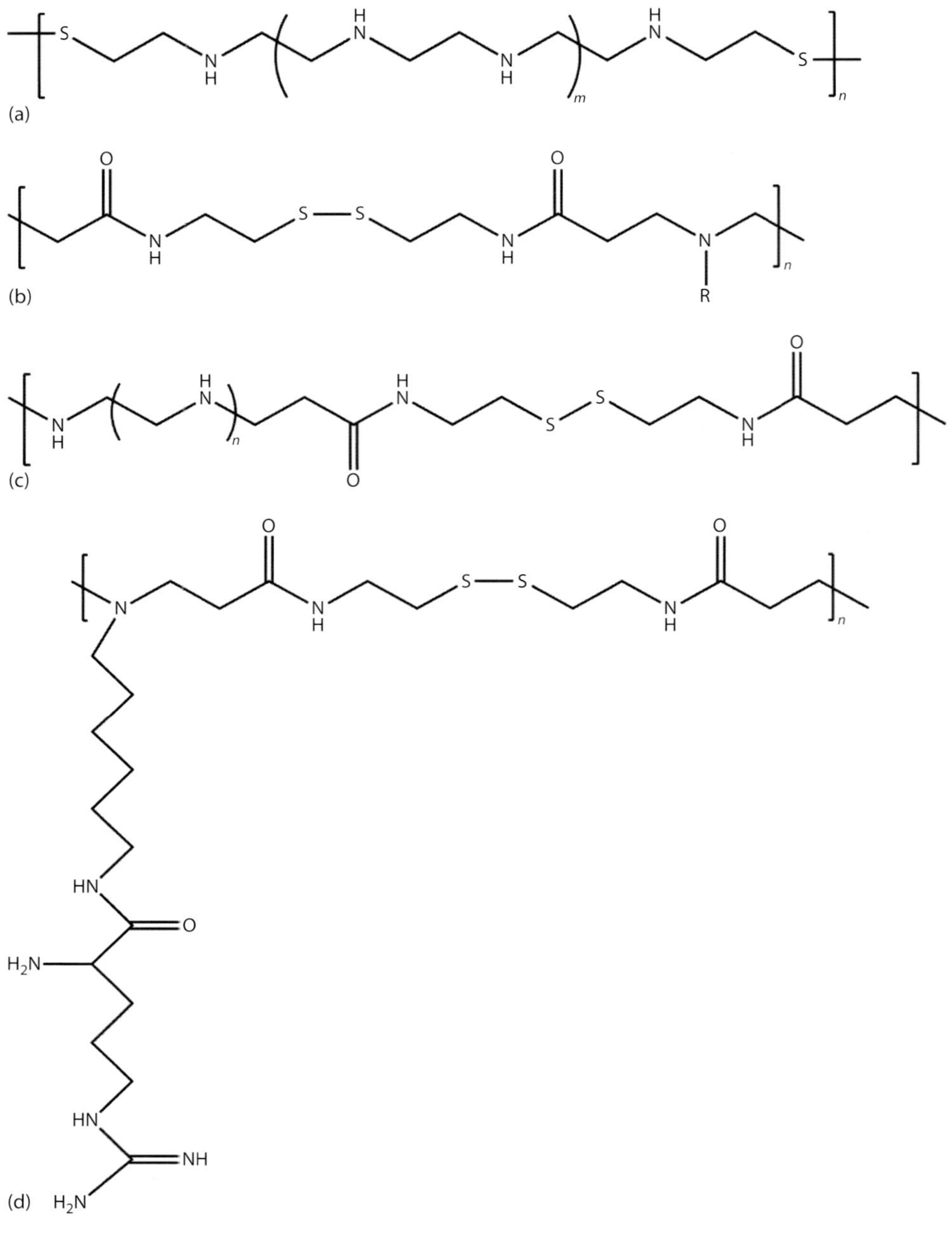

FIGURE 16.6 Structure of gene delivery polymers. (a) Polyethylenimine. (b) Poly(amido amine). (c) Poly(amido ethylenimine). (d) Arginine-grafted bioreducible poly(disulfide amine).

transfection efficiency. It significantly reduced Fas gene expression, leading to inhibition of cardio-myocyte apoptosis. Recently, repetitive Michael addition and amidation were used for introduction of dendritic poly(amido amine) into the side chains of SS-PAAs. It showed high buffer capacity, a strong DNA-condensing ability, and greatly improved transfection efficiency with low cytotoxicity compared with those of the SS-PAA.

16.2.3　Reducible Poly(Disulfide Amine)

Arginine-grafted bioreducible poly(disulfide amine) (ABP) has been synthesized as a nonviral vector (Figure 16.6d). The polymer can successfully condense plasmid DNA at the optimum ratio, while polyplex is not observed in the presence of a reducing agent. ABP was shown to enhance transfection efficiency compared with that of unmodified poly(CBA-DAH) and 25 kDa bPEI. The poly(ethylene glycol) and ABP (PEG-ABP) enhanced the expression efficiency and therapeutic effectiveness of this EPO gene delivery system in vitro. It showed efficient transfection and long-term production of EPO in various cell types. An ABP polymer was also applied to deliver human erythropoietin plasmid DNA (phEPO) to produce long-term, therapeutic erythropoiesis. The phEPO/ABP polypolex improved erythropoietic effects compared to those of traditional EPO therapies because of higher hematocrit levels over a 60-day period accompanied with reticulocytosis and high hEPO protein expression. Arginine-modified polydisulfide poly(CBA-DAH-R) was developed for siRNA-mediated cancer gene therapy. The siRNA/poly(CBA-DAH-R) polyplex was rapidly dissociated in the reductive cytoplasmic environment because of cleavage of internal disulfide bonds by intracellular GSH. Improved siRNA activity was observed due to the cytoplasm-sensitive dissociation property of this polyplex. Additionally, it inhibited VEGF expression to a greater degree than the siVEGF/bPEI polyplex did. Various ABP polymers with different polymethylene spacer lengths in the main chain and the side chain were prepared to evaluate the effects of chain length on transfection efficiency. Polymers with long side chains showed higher transfection efficiencies compared with those of their short ethylene side spacer counterparts. These polymers promise nonviral environment–sensitive vectors for their higher transgene expression level compared with that of 25 kDa bPEI.

16.2.4　Peptide-Based Gene Delivery Systems

A reductively degradable polycation (RPC) based on the Cys-(Lys)$_{10}$-Cys peptide was prepared via the oxidation of cysteinyl sulfhydryl groups into disulfide bonds as the intracellular degradable polypeptide. Sixteen lysine-based linear RPCs were also designed as a combination with an acid-dependent, anionic fusogenic peptide for DNA delivery to dividing and postmitotic cells. The RPC/DNA/fusogenic peptide polyplex showed high transfection efficiency compared with the nonreducible PLL/DNA/fusogenic peptide polyplex even in corneal endothelial cells due to reductive depolymerization of the RPC in the cytoplasm. Cytoplasm-sensitive reducible poly(oligo-D-arginine) (rPOA) was developed using Cys-(D-Arg)-Cys peptide for DNA and siRNA delivery. The in vitro and in vivo reporter gene expression of rPOA/DNA polyplex was observed with various cell lines and mouse lung after intratracheal injection. The rPOA/DNA polyplex showed comparable or similar transgene expression as that of 25 kDa bPEI in all cell lines and maintained gene expression for 1 week after a single injection. In addition, this delivery system could effectively deliver siRNA into the cytoplasm. The siRNA/rPOA polyplex was rapidly dissociated in the cytoplasm in response to intracellular GSH level and rapid internalization into the cytoplasm, which was confirmed using confocal microscopy. This system is a very promising cytoplasm-sensitive smart carrier for DNA and siRNA delivery. His-based RPC (His$_6$ RPC) polymerized using monomer of Cys-His$_6$-Lys$_3$-His$_6$-Cys by oxidation was developed for release of siRNA in the cytoplasm following effective internalization. His$_6$ RPC released cytoplasm-sensitive release of siRNA and maintained good bioactivity of RNAi. *Plasmodium falciparum* circumsporozoite (CS) was conjugated with His$_6$ RPC for targeting of siRNA to hepatocytes, and the CS-His$_6$ RPC successfully silenced target mRNA as well as reduced protein expression. Recently, stearoyl (STR) peptides with Cys (C), Arg (R), and His (H) residues were synthesized and their abilities evaluated for siRNA delivery. The STR peptides spontaneously formed disulfide linkages between Cys residues and their potential evaluated for siRNA delivery in vitro and in vivo.

16.2.5 VIRUS-BASED DELIVERY SYSTEMS

Virus particles consist of a wide array of protein molecules that form a hollow scaffold package as viral nucleic acid. Recently, viruses have been modified at the genetic level for applications as reagents, catalysts, and scaffolds for chemical reactions. Synthetic polymers, including hydrophilic polymers such as PEG and poly(N-(2-hydroxypropyl) methacrylamide) (pHPMA), were conjugated to the surface of viruses to reduce recognition by neutralizing antibodies. Bifunctional polymers have also been used to shield the surface and display new targeting ligands for changing cell tropism. Nanocarrier platforms and viruslike particles have advantages of morphological uniformity, biocompatibility, and easy functionalization. They have a variety of distinctive shapes such as icosahedrons, spheres, and tubes and vary in size from 10 nm to over a micron. Various virus-based materials have been reported in the field of nanotechnology for gene therapy. Viral RNA or DNA is packaged with functional cargo. They are more stable than empty capsids in storage due to charge neutralization.

Over the past 15 years, oncolytic viruses have been applied to cancer cell therapy. They can replicate preferentially in cancer cells and kill apoptosis-resistant tumor cells. Many different varieties of oncolytic adenoviruses have been constructed, including inherently tumor-selective virus species, viral gene–deleted mutants, promoter-engineered mutants, and pseudotyped viruses. Oncolytic adenoviruses are replicated selectively inside of the tumor cells and kill the cells by lysis. These oncolytic adenoviruses recognize the coxsackievirus–adenovirus receptor on the surface of tumor cells by Ad5-based viruses. Oncolytic adenoviruses expressing the relaxin gene and containing an Ad5/Ad35 chimeric fiber showed significantly enhanced transduction and increased viral spread throughout the tumor.

The Ad surface was modified by polymers, liposomes, or nanoparticles to extend circulation time, reduce immunogenicity, and result in increased antitumor effects due to lower accumulation and toxicity in the liver. Specific-targeting delivery systems were developed for tumor-selective oncolytic therapies against primary and metastatic cancers. The oncolytic Ad plasmid DNA with two bioreducible polymers such as bioreducible poly(disulfide amine) polymer (ABP) and PEG-5k-conjugated ABP (ABP5k) was developed and replicated only in human cancer cells. This polyplex was a reduced immune response to evade the innate and adaptive immune response. The oncolytic Ad was physically coated with PEG for systemic delivery. The tumor-to-liver ratio of this system significantly elevated 1229-fold greater than that of a naked Ad.

16.3 INTRACELLULAR ORGANELLE–TARGETING GENE DELIVERY SYSTEMS

16.3.1 NUCLEAR-TARGETED GENE DELIVERY SYSTEMS

The transport of DNA into the nucleus after entrance into the cell is essential for expression of the therapeutic gene. The DNA is transported across the nuclear membrane via the nuclear pore complex (NPC). Ions and small molecules (<40 kDa) show passive diffusion via the NPC (≈9 mm), while large molecules require active transport. Large molecules that have nucleus-targeting signals can be moved to the nucleus from cytosol. DNA carriers are modified with karyophilic molecules referred to as karyopherins. These karyopherins are divided into two classes for large molecules: importins and exportins. Beta-karyopherins, members of the importin beta superfamily, can be bound with cargo proteins, and this system is mediated by importin alpha as an adaptor molecule. The cargo protein from the mixture of cytosolic proteins recognizes importin in the first step for the nuclear import. In this step, peptide-based nuclear localization signals (NLSs) are distinguished by the importin. The classical NLSs in nuclear transport consist of one (monopartite) or two (bipartite) clusters with four or more basic amino acids. The SV40 large T-antigen NLS ([126]PKKKRRV[132]) and nucleoplasm NLS ([155]KRPAATKKAGQAKKKK[170]) are the most studied monopartite and bipartite clusters, respectively (Figure 16.7).

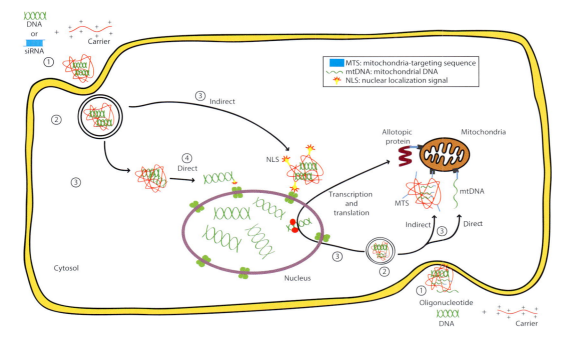

FIGURE 16.7 Schematic representation of the intracellular organelle–targeting strategies.

Since the first report, a wide variety of NLS-modified DNA delivery systems have been developed for enhancing the nuclear import, offering improved transfection efficiency. NLS directly conjugated to DNA improved the nuclear import of DNA (plasmid, linearized plasmids, and oligonucleotides). Multiple NLS peptides were bound with linearized or circular DNA using nonspecific covalent methods. Although NLS peptides improved nuclear import, the bioactivity of DNA to serve as a template for transcription was reduced because of changes in the biological behaviors and interactions with other cellular factors. Therefore, site-specific conjugation methods are required for maintenance of the bioactivity of DNA. Acridine is a heteroaromatic polycyclic molecule that binds tightly and reversibly with dsDNA through a combination of hydrophobic, electrostatic, and hydrogen bonds, along with dipolar forces. NLS-conjugated acridine allows noncovalent binding to DNA and induces the nuclear import of DNA. It showed significantly higher transfection efficiency compared to that of PEI.

Another approach is protein transduction domains (PTDs) that deliver bioactive cargos such as proteins and genes through the cell membrane. PTDs such as the TAT peptide from HIV-1 TAT protein, VP22 protein of herpes simplex virus (HSV), antennapedia (Antp) protein of drosophila, and oligo-arginine have the ability to condense DNA or RNA due to their cationic properties. TAT peptide and Antp are well-known, nonclassical NLSs and have the ability to actively transport DNA into the nucleus. Oligo-arginine showed rapid internalization and nuclear transport of DNA. However, the internalization mechanism and nuclear import of arginine-based PTDs is not conclusively known.

Cationic polymers such as peptides and polysaccharides can bind to negatively charged nucleic acids through electrostatic interaction. This condensation between polymer and nucleic acids protects nucleic acids against serum and facilitates internalization into the cells. Chitosan has been widely used for gene delivery because of its biocompatibility, biodegradability, nonimmunogenicity, and nontoxicity. In acidic aqueous solution, chitosan forms positively charged single-helicoidal stiff chains, allowing chitosan to interact with negatively charged molecules including DNA and the cell membrane. The NLS peptides were directly conjugated to the chitosan/DNA polyplex without covalent conjugation to either DNA or chitosan. Chitosan/DNA polyplexes significantly improved transfection efficiency with negligible cytotoxicity by adding NLS. Dexamethasone (Dexa), a potent glucocorticoid, is another approach to the targeted gene delivery. It is a receptor-recognized glucocorticoid, which induces the translocation of the receptor–ligand complex into the nucleus. Dexa can be used in gene delivery systems to nuclear transport of DNA. A polymer/DNA polyplex formed by PEI-Dexa conjugation showed efficient translocation of the DNA into the nucleus. This delivery system enhanced transgene expression compared to that of PEI and localized into the nucleus with low cytotoxicity, suggesting a promising nonviral gene carrier.

16.3.2 Mitochondrial Targeting

Mitochondria produce energy in eukaryotic cells through oxidative respiration and metabolize nutrients that are involved in cellular processes such as signaling, cellular differentiation, and cell death. Mitochondria also make reactive oxygen species (ROS) as a by-product of their energy production processes. The ROS induce substitution or large-scale deletions of mitochondrial DNA, which are linked to various diseases including cancer, diabetes, Alzheimer's disease (AD), and other neurodegenerative disorders. Although mitochondria have their own DNA (mtDNA) containing mitochondrial protein genes, they also require a large amount of protein from the cytosol because the mtDNA contains only a small portion (<1%) of the mitochondrial proteins. Precursor proteins, which are imported from the cytosol, are recognized and sorted by receptors in the mitochondrial outer membrane and protein translocases in the inner membrane. Mitochondrial precursors include a targeting signal (Figure 16.3). These targeting signals are diverse but have common characteristics of N-terminal-specific positioning and positive charge. Mitochondrial leader peptides and mitochondria-targeting sequences are used to deliver DNA into the mitochondria for repairing the mitochondrial gene deficiency. Use of allotropic expression is another approach to treat the mitochondria-mediated diseases for gene therapy.

16.4 DISEASE-SPECIFIC EXPRESSION SYSTEMS

16.4.1 Cancer

Various innovative gene delivery systems for the treatment of cancer have been developed. The three main gene therapy approaches for cancer therapy are immunotherapy, oncolytic virotherapy, and gene transfer. Genetically modified cells and viral particles can be used to direct an immune response toward cancer for immunotherapy. Recently, second- and third-generation vaccines in clinical trials have shown promise for the prevention of various cancers including lung cancer, pancreatic cancer, prostate cancer, and malignant melanoma. In oncolytic virotherapy, a rapidly developing field for cancer therapy, viral particles replicate within the cancer cell and induce cell death. The ONYX-015 viral therapy, currently in clinical trials for the treatment of squamous cell carcinoma of the head and neck, utilizes an adenovirus lacking the viral E1B protein that is essential for RNA export during viral replication. However, the tumor cells provide these necessary RNA export functions, allowing the virus to selectively replicate in tumor cells. Gene transfer introduces therapeutic genes into the target cancer cell for the induction of cancer cell death and reduction of cancer cell growth. Nonviral IL-12 was locally delivered using water-soluble lipopolymer (WSLP) to decrease tumor progression and increase immunogenicity with paclitaxel. Multiblock copolymers, PEG-PLL-g-16% His, increase stability of complexes in vivo via dysopsonization to mask the complexes from elements of the reticuloendothelial system. The siRNAs are a tool to reduce gene expression in cancer cells. rPOA, a peptide-based gene delivery system, delivered siRNA to the cytoplasm in cancer cells and maintained bioactivity both in vitro and in vivo. It was sufficient to condense and stabilize siVEGF and suppress tumor growth in an animal model.

16.4.2 MYOCARDIAL INFARCTION

Therapeutic genes with various delivery systems have been developed for the treatment of coronary heart disease (CHD), a leading cause of death in the United States. Genes that induce angiogenesis, such as VEGF, have been investigated for the treatment of CHD. VEGF-A induces angiogenesis in the ischemic heart. VEGF complexed with WSLP, a polymer consisting of a low-molecular-weight branched PEI (1800) and cholesterol, was expressed in cardiovascular cells in ischemic rabbit myocardium and improved capillary growth due to reduction of apoptosis. It increased myocardial blood flow and wall motion in porcine models of myocardial infarction (MI). A hypoxia-inducible plasmid expressing both heme oxygenase-1 (HO-1) and the Src gemology domain-2 containing tyrosine phosphatase-1 mircoRNA (miSHP-1) demonstrated an enhanced expression of HO-1, downregulation of SHP-1, and inhibition of cardiomyocyte apoptosis. Recently, various gene transfection techniques and the ability to regulate gene expression in the heart and circulation have improved clinical outcomes in ischemic heart disease patients.

16.4.3 DIABETES

Several polymers were applied to deliver therapeutic genes for the prevention and treatment of diabetes. Early in 2000, biodegradable poly[alpha-(4-aminobutyl)-L-glycolic acid] condense plasmid DNA pCAGGS mIL-10 plasmid and reduce the severity of insulitis in nonobese diabetic (NOD) mice to induce autoimmune diabetes mellitus. The soluble RAE-1 plasmid is another approach for the prevention of type 1 diabetes that uses an islet-specific gene delivery system. The bioreducible cationic polymer p(CBA-DAH) was modified with targeting peptide CHVLWSTRC to target the EphA2 and EphA4 receptors. This polyplex was administered to NOD mice via i.v. injection and maintained blood glucose levels in animal models for 17 weeks. Glutamic acid decarboxylase (GAD) was silenced in transgenic NOD mice to protect islet beta-cells against the development of diabetes. To deliver an antisense GAD mRNA expression plasmid (pRIP-AS-GAD), PEG-grafted poly-L-lysine was developed and showed high transfection efficiency in the MIN6 cell line and in vivo. Glucagon-like peptide-1 (GLP-1) is well known for its bioactive insulinotropic peptides derived from preproglucagon. GLP-1 reduced blood glucose levels and retained it in type 2 diabetic patients. GLP-1 gene delivery using polyethyleneimine increased insulin secretion by twofold in isolated rat islet cells and decreased blood glucose levels that was maintained for 2 weeks.

16.4.4 NEURONAL DISEASES

Neuronal diseases such as AD, PD, and neuropathy are characterized by neuron death, exitotoxicity, aging, and amyloid toxicity. Neuron growth factor (NGF), specifically targeting basal forebrain cholinergic neurons, is of particular importance in AD. Autologous fibroblasts were genetically modified to express human NGF and implanted into the forebrain as a phase 1 clinical trial. The study showed no long-term adverse effects of NGF and significantly increased cortical 18-fluorodeoxyglucose. PD, the second most common age-related neurodegenerative disorder, is caused by the progressive loss of substantial nigra pars compacta dopamine neurons and the consequent decrease in the neurotransmitter dopamine. A tricistronic lentiviral vector encoding the critical genes for dopamine synthesis were injected into the striatum and safely restored extracellular concentrations of dopamine and corrected the motor deficits for 12 months without associated dyskinesias. Chronic neuropathic pain results from a wide variety of insults to the peripheral or central nervous system including infection, autoimmune disease, and mechanical trauma. HSV-based vectors engineered to produce the inhibitory neurotransmitter IL-4 reduced mechanical allodynia in a L5 spinal nerve ligation model and reversed thermal hyperalgesia resulting from the nerve injury. IL-10 expressing HSV vector also reduced pain-related behaviors in the formalin model of inflammatory pain concomitant with a reduction in pP38 and decreased expression of the full-length membrane-spanning precursor of tumor necrosis factor alpha in spinal microglia.

16.5   CONCLUSION

Gene therapy is a rapidly emerging approach for the treatment of various diseases. DNA and siRNA are the main classes of the genes used in gene therapy. Specific protein expression and target-specific gene silencing were mediated by DNA and small RNA molecules, respectively. Various types of viral and nonviral carriers have been developed to overcome several barriers in the extracellular and intracellular spaces and to enhance transfection efficiency, minimize cytotoxicity, and improve the bioactivities of nucleic acids. These delivery systems have properties that allowed for cellular uptake through cell membranes, endosomal escape, rapid dissociation from carriers in the cytoplasm, nuclear localization of DNA, and elimination of the carrier. To improve cellular uptake and bioactivity of DNA and siRNA, charge–charge interactions or the use of binding motifs have been studied. Intracellular organelle–targeted delivery systems to target the cytoplasm, nucleus, and mitochondria have been developed using targeting peptides, sequences, and environment-responsive materials. Endosomal escape was induced by the proton-buffering effect. Recently, reduction-sensitive carriers have been reported to have improved physiological properties in extracellular spaces and facilitate reduction within cells. DNA-based gene delivery systems using classical or nonclassical NLSs can be used to achieve nuclear localization because DNA is transcribed to mRNA within the nucleus. Although various delivery systems have been developed for enhancing transfection of DNA and siRNA, the amount of carrier should be reduced because the fates and elimination routes of carrier molecules within the cell are unknown. The intracellular delivery and tracking of internalized carrier molecules and nucleic acids is important for the development of biocompatible gene delivery systems.

FURTHER READING

Brown, T.A. 1990. *Gene Cloning: An Introduction*. London, U.K.: Chapman & Hall.

Feigner, P.L., M.J. Heller, P. Lehr et al. 1996. *Artificial Self-Assembling Systems for Gene Delivery*. Washington, DC: ACS Books.

Gupta, B., T.S. Levchenko, and V.P. Torchilin. 2005. Intracellular delivery of large molecules and small particles by cell-penetrating proteins and peptides. *Adv Drug Deliv Rev* 57(4):637–651.

Lee, M. and R.I. Mahato. 2009. Gene regulation for effective gene therapy. Preface. *Adv Drug Deliv Rev* 61(7–8):487–488.

Ma, Y., R.J. Nolte, and J.J. Cornelissen. 2012. Virus-based nanocarriers for drug delivery. *Adv Drug Deliv Rev* 64(9):811–825.

Mahato, R.I., Y. Takakura, and M. Hashida. 1997. Nonviral vectors for *in vivo* gene therapy: Physicochemical and pharmacokinetic considerations. *Crit Rev Ther Drug Carrier Syst* 14:133–172.

Park, T.G., J.H. Jeong, and S.W. Kim. 2006. Current status of polymeric gene delivery systems. *Adv Drug Deliv Rev* 58(4):467–486.

Ratner, B.D., A.S. Hoffman, F.J. Schoen et al. 2013. *Biomaterials Science*, 3rd ed. Oxford, England: Academic Press.

Rolland, A. 1999. *Advanced Gene Delivery: From Concepts to Pharmaceutical Products*. Amsterdam, the Netherlands: Harwood Press.

Smith, A.E. 1995. Viral vectors in gene therapy. *Annu Rev Microbiol* 49:807–838.

Tanaka, K., T. Kanazawa, T. Ogawa et al. 2010. Disulfide crosslinked stearoyl carrier peptides containing arginine and histidine enhance siRNA uptake and gene silencing. *Int J Pharm* 398(1–2):219–224.

Won, Y.W., K.S. Lim, and Y.H. Kim. 2011. Intracellular organelle-targeted non-viral gene delivery systems. *J Control Release* 152(1):99–109.

Zanthopoulos, K. 1998. *Gene Therapy*. New York: Springer-Verlag.

17 Vaccine Delivery

Terry L. Bowersock and Suman M. Mahan

CONTENTS

17.1 INTRODUCTION

17.1.1 VACCINES BACKGROUND

The development and use of vaccines in vaccination programs has been evolving over centuries (College Physicians of Philadelphia 2015). The term vaccination was coined when Edward Jenner (1796) used cowpox virus lesions from a cow to protect milkmaids against smallpox. Following this observation, the word *vacca* (Latin for cow, hence, vaccination) was coined. Today, vaccination is considered to be the most efficient method of preventing and eradicating infectious and contagious diseases of humans, livestock, and domesticated animals. Vaccines are also being targeted as an intervention against cancers, inflammatory diseases, and allergies. Given the current drive to reduce use of antibiotics, vaccine development and usage is gaining momentum as a "green" preventative option. However, vaccines need to be safe, efficacious, and affordable. In addition, vaccines must rapidly induce a protective immune response in the host that mirrors protective immune responses induced following recovery from natural exposure. Vaccine efficacy can be influenced by the characteristics and quality of the antigen, formulation, dosing regimen, site of vaccination, and method of delivery. Vaccines broadly fall into three categories: modified live (which are attenuated and replicating), inactivated (nonreplicating whole virus or bacteria), or recombinant, which express subunits of pathogens based on protective immunity data.

Typically, modified live vaccines (MLVs) rapidly induce protective immunity because they replicate, whereas in general inactivated vaccines have a slower onset of immunity, as they require a priming and booster dose and are usually formulated with an adjuvant. Whereas MLV are developed by attenuation by classical methods (passaging in vitro or by chemical mutagenesis), the inactivated vaccines until recently have been largely developed by empirical methods whereby whole pathogens are grown up, purified or semipurified, chemically inactivated and formulated with adjuvants. Adjuvants are chemicals or a combination of chemicals added to inactivated antigens to augment the immune response. Adjuvants in their own right are considered immunomodulators as they engage the innate immune response. More recently, however, molecular immunomodulators (that are ligands to receptors of the innate immune cell system) have been added to core adjuvants to deliberately direct and regulate the pathway of an adaptive immune response. With the immense advances in immunology and in the fields of proteomics, genomics, and adjuvant research, the field of vaccinomics is advancing rapidly. Therefore, opportunities exist for the development of custom-designed vaccines partnered with strategic vaccine delivery systems, to influence the immune response for induction of a specific protective immunosignature (antibody or cell-mediated immunity [CMI]).

17.1.2 Pathogen–Host Interactions and the Immune Response

A detailed understanding of the molecular interaction of the host and the pathogen, which culminates in induction of a protective immune response, is essential for vaccine development, design, and delivery. The immune response can be functionally divided into innate and adaptive arms, which work in synergy but are executed by different cell types and their subtypes, via sophisticated interactions involving receptors (toll-like receptors [TLRs]), costimulatory ligands, and soluble mediators, e.g., cytokines. An antigen-/pathogen-specific adaptive immune response is initiated by cells of the innate immune system, via engagement of antigen-presenting cells (APCs) such as dendritic cells (see Table 17.1 for types of dendritic cells and their distribution).

APCs are the "epicenter" of the innate immune response and their activation influences downstream development of effector cells, which clear infection. Dendritic cells interact with components of pathogens, namely, the pathogen-associated molecular patterns (PAMPs), via their unique germ line–coded pattern recognition receptors such as TLRs. There are 11 different TLRs described to date. A list of the TLRs and their respective PAMPs is provided in Table 17.2 and Figure 17.1.

These interactions activate the inflammasome, i.e., an inflammatory response, resulting in the release of IL-12, IFN-γ, IFN-α, IL-1, IL-6, and IL-10 cytokines. This response is

TABLE 17.1

Dendritic Cells: Location, Toll-Like Receptors, and Cytokine Signatures

Types of Dendritic Cells	Location	TLRs Expressed	Cytokine Signatures
Myeloid (similar to monocytes)	Blood	TLR 1, TLR 2, TLR 3, TLR 4, TLR 5, TLR 7, TLR 8, and TLR 9	IL-12, IFN-α, and IFN-β
Plasmacytoid	Bone marrow	TLR 7 and TLR 9	IFN-α, TNF-α, IL-1, IL-8, and/or IL-6
Langerhans cells	Skin, nose, lungs, intestines, and stomach	TLR 1, TLR 2, TLR 3, TLR 5, TLR 6, and TLR 10	IL-1β, IL-6, IL-12, IL-18, and TNF-α

Source: Data from McKenna, K. et al., *J. Virol.*, 79, 17, 2005.
Abbreviations: TLR, toll-like receptors; IFN, interferon.

TABLE 17.2

Toll-Like Receptors and Their Agonists and Pathogen-Associated Molecular Patterns (PAMPs)

TLRs	Location	PAMP
1, 2, and 6	Surface	Lipoproteins/lipopeptides
		Peptidoglycans
		Zymosan
		LPS
		Neisserial proteins
4	Surface	LPS (Gram-negative bacteria)
		Viral envelope proteins (RSV)
5	Surface	Flagellin
3	Intracellular	dsRNA, poly(I:C)
8	Intracellular	ssRNA, imidazoquinolines
7	Intracellular	ssRNA, imidazoquinolines
9	Intracellular	CpG DNA
11	Liver, bladder, kidney (weak expression in spleen)	Uropathogenic *Escherichia coli*
		LPS, poly(I:C), and peptidoglycan are not agonists

Abbreviations: LPS, lipopolysaccharides; poly(I:C), polyinosinic–polycytidylic acid; RSV, respiratory syncytial virus.

accompanied simultaneously by processing of antigens and presentation of short peptides to T and B lymphocytes, culminating in the development of an adaptive immune response. Whereas B cells fall into one lineage, T lymphocytes are subdivided into CD4+ (further divided into Th0, Th1, and Th2 subtypes; Figure 17.2) and CD8+ T lymphocyte subpopulations and T regulatory cells. The cytokines produced by APCs influence the interaction of naïve CD4+ T lymphocytes (Th0) and antigen toward the development of either a specific Th1-mediated (these activated cells are defined by their secretion of IL-12 and IFN-γ and IFN-α) or Th2-mediated (these activated cells are defined by secretion of IL-10, IL-4) immune response (Figure 17.2). Activation of Th1 responses leads to the development of a cell-mediated immune (CMI) response mediated by CD4+ and cytotoxic CD8+ T lymphocytes and release of granzymes, perforins, and IFN-γ. These cells clear infection by killing cells infected with intracellular pathogens such as viruses, most protozoa, and some bacteria, e.g., rickettsia and *Brucella* spp. In contrast, activation of Th2 responses results in the development of a humoral immune response and production of antibody by mature B cells (plasma cells). Typically, extracellular pathogens (most bacteria and helminths) trigger humoral immunity mediated by specific antibody responses (IgM, IgA, IgG antibody subclasses). In many situations, antibody-mediated responses in concert with CMI are advantageous in the prevention/reduction of infection. Ultimately, induction of a balanced humoral and CMI response would be a goal of a good vaccine.

Whereas the innate immune response is nonspecific and is initiated within a few hours of pathogen exposure, it does not develop any immunological memory. In contrast, adaptive immune responses take a few days to mature and are specific, inducing memory, which is recalled upon repeated exposure to the same and related pathogens. Figure 17.2 represents the currently accepted activation pathways for adaptive immunity. In addition to the dendritic cells, antigen can be processed and presented to T lymphocytes by other cells, namely, the NK cells and the γδ T cells, which are considered components of the innate immune system. These cells are abundant in neonates. Cells that form the structure of the physical barriers throughout the body are also equipped with TLR receptors and receptors (damage-associated molecular patterns, DAMPs) that respond to damaged integrity of the barriers. Vaccine efficacy relies on the induced immunological memory for the persistence, and renewal, of immunity.

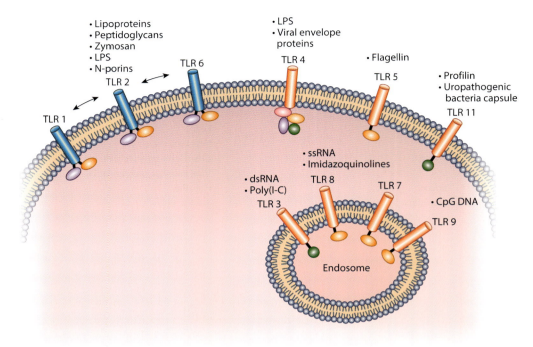

Cytokine and cellular pathways activated by TLR:
TLR 1,2,6: Th0/Th2/Treg: IL-10 (mainly), IL-12p70 (weak)
TLR 4,5,3,8,7,9: Th1 (and CTL), IL-12p70, and IFN-α

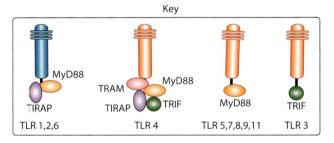

FIGURE 17.1   Toll-like receptors and ligands. TLRs, toll-like receptors.

17.2 VACCINATION-INDUCED IMMUNE RESPONSES

Vaccines are most commonly delivered to the parenteral (intramuscular [i.m.], subcutaneous [s.c.], intradermal [i.d.] sites) or the mucosal (e.g., oral and intranasal [IN]) sites. Vaccines delivered via the parenteral routes attract APCs for antigen processing at the site of injection, which then migrate to the regional lymphoid organs (lymph nodes and or spleen) for interaction with T and B lymphocytes and initiation of the adaptive immune responses.

Vaccines delivered at the mucosal sites interface with the mucosal immune system, which represents 70% of the mammalian immune system; the lymph nodes and spleen form the rest. In the mucosal immune system, antigens at mucosal sites are processed by the mucosal-associated lymphoid tissue (MALT). In the gastrointestinal (GI) tract, MALT is also called GALT, i.e., gut-associated lymphoid tissue, which comprises the Peyer's patches, appendix, and isolated lymphoid follicles. A schematic organization of the GALT is shown in Figure 17.3. Other MALTs include the nasopharynx

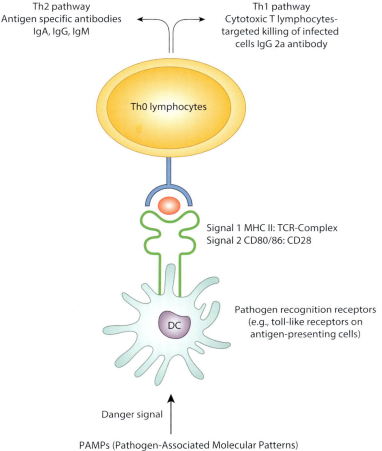

Th2 pathway
Antigen specific antibodies
IgA, IgG, IgM

Th1 pathway
Cytotoxic T lymphocytes-
targeted killing of infected
cells IgG 2a antibody

Th0 lymphocytes

Signal 1 MHC II: TCR-Complex
Signal 2 CD80/86: CD28

DC

Pathogen recognition receptors
(e.g., toll-like receptors on
antigen-presenting cells)

Danger signal

PAMPs (Pathogen-Associated Molecular Patterns)
DAMPs (Damage-Associated Molecular Patterns)

Cytokine and cellular pathways activated by TLR:
TLR 1,2,6: Th0/Th2/Treg: IL-10 (mainly), IL-12p70 (weak)
TLR 4,5,3,8,7,9: Th1 (and CTL), IL-12p70, IFN-α, IFN-γ, and TNF

FIGURE 17.2 Activation pathways for adaptive immunity. (Modified from *Trends in Immunology*, 30(1), Reed, S., Bertholet, S., Coler R.N. et al., New horizons in adjuvants for vaccine development, 23–32, Copyright 2009, with permission from Elsevier.)

(or nose)-associated lymphoid tissue (NALT) and bronchus-associated lymphoid tissue (BALT), having similar organizations as GALT. MALTs are unique in that they have inductive sites where the antigen is captured by M cells (mucosal APCs or antigen-sampling cells) and transported to underlying follicles where APCs process and present to T and B lymphocytes. These activated lymphocytes filter out of local lymph nodes to peripheral blood. They return via the blood to mucosal effector sites (Figure 17.3). The result of these inductive responses is secretion of antibodies of the IgA, IgM, or IgG subclasses and homing of mature effector T cells (CD4+ and cytotoxic CD8+ T lymphocytes).

There are significant advantages to delivering vaccines at mucosal sites because of their extensive size and coverage of the mammalian body (nose, mouth, respiratory tract, GI tract and the reproductive tract). Furthermore, the mucosal surfaces of the intestinal, respiratory, and urogenital tracts are the most common sites of pathogen entry, and over 90% of all infections are acquired by mucosal routes. Mucosal immunization also offers the potential to immunize in the face of maternal

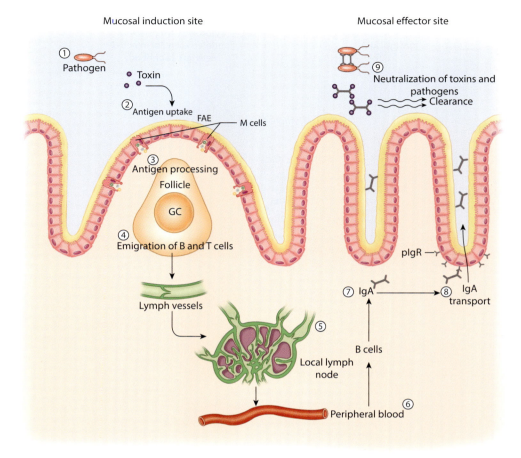

FIGURE 17.3 Interaction of antigens with the mucosal-associated lymphoid tissue. The gut-associated lymphoid tissue is shown here. (1) Antigens (pathogens, toxins, etc.) present in the gastrointestinal tract lumen are (2) taken up by the M cells of the follicle-associated epithelium (FAE). (3) Antigens interact with antigen-presenting cells that engage B and T lymphocytes in the follicle and, (4) once activated, migrate via lymph vessels to the lymph nodes (5) and then via peripheral blood (6) back to the immune response induction site. (7) Mature B cells/plasma cells secrete IgA or other immunoglobulins for (8) release to the mucosal surface via polymeric immunoglobulin receptors (pIgR). (9) Antigens are then neutralized or destroyed. GC, germinal centre.

antibody, thus avoiding potential interference observed with parenteral route of immunization. The noninvasiveness of mucosal vaccines offers the advantages of increased compliance in comparison to parenteral delivery.

In addition, mucosal immunization has the potential to induce mucosal and systemic immune responses that span both humoral and CMI. The local immune response can also, in some cases, be supplemented by humoral immune responses at distant mucosal sites. Of the humoral response, the secretory IgA antibody response is the signature response at mucosal sites. It contributes to the prevention of infection at these sites via immune exclusion, by neutralization of toxins produced by pathogens, and control of infections caused by viruses, bacteria, protozoa, and helminths that gain access via the mucosal system (Brandtzaeg and Pabst 2004).

Recognizing the huge potential of mucosal sites for vaccination, one also has to keep in mind the innate barriers that need to be overcome to achieve successful immunization, namely, the mucus layer that coats epithelial surfaces, degradation by proteases, the mucociliary apparatus, low pH of the stomach, and limited entry into the mucosal lymphoid tissues below the epithelial layers (Woodrow et al. 2012). Mucosal vaccination via various mucosal sites is discussed in detail in Section 17.4.3.

17.3 TYPES OF VACCINES

17.3.1 MODIFIED LIVE VACCINES

MLVs are extremely effective in controlling and eradicating diseases in the human and animal populations. MLVs are usually unadjuvanted, self-replicating, and attenuated, to induce immunity rapidly and provide a robust long-lived immunity mimicking natural infection without causing disease. The advantage of MLV vaccines is that they can be a single-dose vaccination. The oral MLV polio is a good example. This vaccine induces an IgA protective response in the intestinal mucosa. Other examples of MLVs are for smallpox and measles and these vaccines have been extremely effective for the control and eradication of these diseases globally. Similar to use in humans, MLV vaccines are used routinely in veterinary disease control programs and provide robust protection against respiratory and reproductive diseases. The disadvantages of MLVs are that they have the potential of going latent and shed inadvertently, and there is a risk, though small, of reverting to virulence. Typically, MLVs consist of viruses that are relatively easy to attenuate by passaging in vitro. MLVs are aqueous in nature and administered via the parenteral, oral, or IN routes.

17.3.2 RECOMBINANT VACCINES

Development of vaccines using recombinant DNA approaches promises to be the vaccine technology of the future, since this provides the potential to custom design and deliver vaccines. This is being facilitated by the availability of whole genome sequences of many bacteria and viruses, in addition to whole genome sequences of various mammalian species (humans, bovine, swine, poultry, etc.). This approach is enhanced because whole genome sequencing of bacteria and viruses can be completed rapidly using high-throughput methods and alignments that allow identification of conserved immunoprotective antigens. As of spring 2015, there are approaching 7000 bacterial genome sequences in GenBank and many more are stored at the Sanger Institute in the United Kingdom. Similarly, genomes of many viruses have been sequenced (http://www.ncbi.nlm.nih.gov/genome/viruses/).

Using the genome information and knowledge of immune responses to pathogens and their specific immunogenic or protective proteins/antigens, development of genetically engineered vaccines is gaining momentum. Reverse vaccinology uses the whole genome sequence data to predict protective antigens *in silico* (Rappuoli and Covacci 2003). These antigens can then be expressed and tested in vitro for reactivity with convalescent serum or immune cells, before testing in vivo as DNA, virus vectored, or subunit vaccines, to determine efficacy. Many mouse experimental models are used initially to obtain proof of concept before testing in target species. Reverse vaccinology was applied successfully to develop vaccines to pathogens that are difficult to grow such as hepatitis B virus. Today, immunization of children against hepatitis B is a testimony of this technology. Another success of this approach is the Group B meningococcal vaccine, which had been unsuccessfully researched for several decades using conventional vaccine development approaches.

17.3.3 DNA VACCINES

DNA vaccines contain DNA that codes antigens from a pathogen. These plasmids have a promoter, which permits them to be expressed in eukaryotic cells. After injection into a host, they are taken up by APCs; the antigenic proteins are expressed and presented to B and T lymphocytes in the context of MHC class II and I molecules, thereby priming an immune response and generating long-term memory and protection against disease. DNA vaccines or "naked DNA vaccines" can be categorized as a type of MLV, as they express their genetic load, but without propagation in the mammalian host. They have a distinct advantage over traditional vaccines (adjuvanted) because they can induce both antibody and CMI (cytotoxic T cells, T helper 1 cells). Additionally, plasmid DNA per se appears to have inherent immunomodulatory properties; the backbone possesses CpG motifs that bind to the TLR 9 receptors in APCs, stimulating the innate immune responses and adaptive CMI responses. DNA vaccines can be easily produced using bacterial fermentation methods and purification, and they do not need the cold chain for storage and shipping, as DNA is a highly stable molecule.

DNA vaccines have been a focus of attention since 1990 and to date more than 1900 patents have been filed, and more than 700 clinical trials have been approved by the National Institutes of Health (Ghanem et al. 2013). These trials are for infectious diseases such as malaria, HIV, Ebola, and hepatitis B. DNA vaccines are being investigated as treatments for cancers (breast, liver, kidney, prostate, etc.). In this case, tumor-specific antigens are cloned in the vectors, to induce a tumor-specific immune response.

There are currently four DNA vaccines licensed for control of veterinary diseases (Findik and Cifti 2012) and one for human use. Although significant effort has been invested globally, progress has thus far been disappointing due to their low immunogenicity. Although efficacious in mouse models, this efficacy has been poor to reproduce when tested in the target larger animal/mammalian hosts. Improvement of DNA vaccine efficacy has been attempted by enhancing gene expression by codon optimization and by tandem expression of cytokine and chemokine genes (Saade and Petrovsky 2012). Various methods of delivery and routes of administration are also being explored, intradermal being one because of the abundance of dendritic cells in this location. Intramuscular and gene gun (pressurized) delivery are also under investigation. Typically, gene gun delivery requires coating gold beads with the DNA vaccine. DNA vaccines work better in a prime–boost regimen (i.e., prime with DNA vaccine and boost with subunit antigen, with or without adjuvant, or with a virus vector as a heterologous boost) to optimize initialization and maturation of the immune response. Saade and Petrovsky (2012) present an excellent review of this technology.

17.3.4 Virus-Vectored Vaccines

Virus-vectored vaccines can be either replicating or nonreplicating. The most highly studied vectors are the double-stranded DNA (dsDNA) viruses, i.e., vaccinia (or poxvirus) and adenovirus (Ad) vectors (Robert-Guroff 2007). In addition, other vectors have been tested, namely, herpes virus (dsDNA), alphaviruses (ssRNA), and vesicular stomatitis virus. The advantage of virus vectors is that they are able to infect epithelial cells and transfer their genetic load for the induction of systemic and mucosal immunity. Similar to DNA vaccines, they stimulate the CMI responses in addition to antibody responses. The differences between the vectors are how they infect, and whether they alter the DNA of the host cell transiently or permanently. The review article by Robert-Guroff (2007) summarizes the advantages and disadvantages of the different virus vectors so far tested as vaccine delivery systems (see also Chapter 16, Section 16.2.5 for a further discussion on virus-based delivery systems).

Ad vectors have the special ability of infecting dendritic cells, inducing costimulatory molecules and CMI. They have a capacity to express their genetic load, develop to high titers in vitro, do not integrate into host genetic material, and are stable. The nonreplicating Ad5 vector has been tested for delivery of HIV antigens in heterologous prime–boost trials, priming with Ad5 vaccine and boosting with a plasmid DNA vaccine. This combination of immunization has generated enthusiasm for clinical trials in humans. A disadvantage of the Ad5 vector is that the host develops immunity to the vector, rendering it ineffective after repeated dosing. In addition, there is a high rate of seroprevalence in the population. Therefore, additional Ad vectors are being developed. These include Ad11, Ad26, Ad35, Ad49, Ad viruses from chimpanzee, and Ad vectors that target the gut, for induction of mucosal immunity.

Poxvirus vectors have been extensively evaluated as vaccine vectors because they also express their genetic load well. The two most researched poxvirus vectors are the NYVAC and Ankara strains, which have been rendered safe by passaging or by gene deletion. These vectors have been tested in HIV research. The fowlpox and canarypox vectors are attractive alternatives since their propagation is restricted in mammalian hosts. Recombinant vaccines using canarypox vector (Alvac™) have been licensed for use in cats against feline leukemia virus (FeLV) and rabies (Purevax® FeLV and Purevax® Feline Rabies, respectively). This nonadjuvant canarypox vaccine expresses the gag and env antigens of FeLV and is delivered orally. It expresses its genes without propagation. Merial has licensed recombinant canarypox vaccines for control of canine distemper and equine West Nile virus infections, for dogs and horses, respectively (Recombitek® vaccines). Viral-vectored vaccines are also being considered for cancer treatment. As per DNA vaccines for cancer, tumor-specific antigens are cloned in these vectors, to induce tumor-specific immune responses.

17.3.5 BACTERIAL-VECTORED VACCINES

As an alternative to viral vectors, several bacterial vectors are being developed to deliver genes into mammalian cells. These include *Escherichia coli*, *Salmonella typhimurium*, *Shigella flexneri*, *Listeria monocytogenes*, and *Lactococcus lactis*, which have been tested for treatment and control of cancers, infectious diseases, and inflammatory conditions. These nonpathogenic, but cell-invasive, vectors are unique because they deliver their genetic information to the mammalian host nucleus by the process of bactofection. In this process, the bacterial vector commits "suicide" after entering the host cell, by releasing its recombinant plasmid that subsequently associates with the host cell nucleus for expression. Induction of both humoral and cellular immune responses has been reported by the use of these bacterial vectors as delivery vehicles of antigen. The use of these vectors also provides the opportunity to deliver the gene of interest in the GI tract and induce mucosal immunity.

Although these vectors have potential for use in various therapies, the safety issues they pose are a challenge as they have virulence factors, such as LPS, which could have detrimental effects on health, and these risks have to be mitigated. Bacterial vectors can also be used for cancer treatment, by the same principle of bactofection of tumor cells. This approach, though viable, must ensure that nontumor healthy cells do not also get destroyed by the tumorlytic genes encoded in these bacterial vectors (Grillot-Courvalin et al. 2011).

17.3.6 VIRUS-LIKE-PARTICLES

Virus-like particles (VLPs) resemble viruses and have the potential to become a platform of new generation vaccines (Roldão et al. 2010). VLPs can express the full complement of the structural and target proteins of a virus, but they are noninfectious because they lack viral genetic material. As they cannot replicate in the host, they are safe; they are also able to induce a robust immune response. Expression of envelope or capsid genes in this platform can generate VLPs by self-assembly. VLPs are in the size range of 20–220 nm, which is dependent on the number of proteins to be expressed.

The baculovirus/insect cell system is currently used as an expression system for the generation of VLPs. Recombinant baculovirus with desired gene expression is transfected into insect cells; these cultures can potentially generate large amounts of VLPs expressing the target proteins. However, a drawback of the insect cell line is that sufficient glycosylation of proteins may not be achievable, especially if it is needed for full protein immunogenicity. Other mammalianized cell systems are being explored. The other challenge for this platform is to develop a manufacturing system that is scalable and profitable for commercialization.

A few notable examples of commercialized human vaccines using the VLP technology exist, namely, Gardasil™ (human papillomavirus), Engerix™ (hepatitis B virus), Cervarix™ (human papillomavirus), and Recombivax HB™ (hepatitis B virus). Several additional human clinical trials are underway for other important infectious, contagious diseases (Roldão et al. 2010).

17.4 VACCINE DELIVERY SYSTEMS AND ROUTES OF ADMINISTRATION

As summarized in Table 17.3, a wide range of vaccine delivery technologies, using various routes of administration, are currently under evaluation. Here, the discussion opens with a description of parenteral delivery, as this is the most commonly used route. Current adjuvant technology is described, as well as the emerging use of implants in order to avoid repeated injections. Transdermal vaccine delivery is then described, with particular reference to new delivery devices that have been developed for this route. As outlined in Section 17.2, vaccine delivery to mucosal sites is associated with a number of key advantages. Mucosal vaccine delivery is described here for a range of mucosal routes, including IN, pulmonary, oral, and oral cavity delivery.

17.4.1 PARENTERAL VACCINE DELIVERY SYSTEMS

17.4.1.1 Injections

Parenteral delivery, via i.m. or s.c. injection, is currently the most common method to deliver a vaccine. Systemic vaccination involves needle and syringes, which is not pleasant for the patient, requires administration by specialized personnel, and may require cold storage. Most inactivated vaccines and subunit vaccines are less potent antigens and thus require adjuvants and delivery systems to stimulate a good immune response. Micro- and nanoparticles act as both adjuvants (in part due to characteristics of the materials they are made from) and delivery systems. These include antigens associated with emulsions, microparticles, nanoparticles, liposomes, immune stimulating complexes (ISCOMs), virosomes, and VLPs (see Chapter 5 for a full discussion of these different types of delivery systems). There is considerable evidence, mostly in laboratory animal studies, that micro and nanoparticulate antigens stimulate better immune responses than soluble antigens, because of their improved antigen delivery to APCs in comparison to soluble antigens. Not only do particle-based vaccines induce better immune responses, but in some cases, they also tend to be safer than injectable adjuvants.

Polymeric microparticles (see also Chapter 5, Section 5.5.5) are synthetic or natural materials that can be formulated with antigens to make nano- or microparticles that can be injected or administered to mucosal surfaces. Synthetic polymers have the advantage of good batch-to-batch consistency,

---

TABLE 17.3

Examples of Vaccine Delivery Systems and Routes of Administration

Route of Administration	Technology	Antigen	Status of Product	References
Injectable	Implant: bioneedle	Influenza	Research	Soema et al. (2014)
	Implant: gel	Ova	Research	Nishikawa et al. (2014)
	Transdermal MN	Influenza	Research	Norman et al. (2014)
	Transdermal gun	Various	Review	Logomasini et al. (2013)
Intranasal	Microparticles	Cancer	Research	Yao et al. (2013)
Pulmonary	Nanoparticles	Multiple	Review	Kunda (2013)
Oral	Microparticles	Multiple	Review	Mane and Muro (2012)
Sublingual/buccal	Fast-release tablets	Ova	Research	Borde et al. (2012)
Conjunctival	Mannosylated nanoparticles	*Brucella ovis*	Research	Da Costa Martins et al. (2012)
Intravaginal	Multiple approaches	HIV 1	Review	Pavot et al. (2012)
Rectal/colonic	Microparticles	Colorectal target	Position paper	Zhu and Berzofsky (2013)

Abbreviation: MN, microneedle.

well-developed formulation methodology, established safety (e.g., poly(lactide-*co*-glycolide) [PLGA] microparticles), and, in some cases, demonstrate an inherent adjuvanticity. A disadvantage is that they often require organic solvents in their formulation, which can adversely affect antigens either during encapsulation or when they degrade to release antigen. The biodegradation of some polymers in vivo can produce a low pH, which also can adversely affect the antigen load.

Microparticles can be modified or inherently contain properties that enhance their utility for delivery of vaccines to APCs. In general, uncharged hydrophobic microparticles are most readily taken up by APCs. Cationic microparticles are taken up more readily than negatively charged particles. Cationic charges are used to encapsulate DNA or bind antigens to the surface of microparticles, especially in cases where degradation of the microparticles may adversely affect the antigen if it is within the matrix itself (Foged 2011; Wilkhu et al. 2011). Incorporation of adjuvants and immunostimulatory molecules within or on the surface of microparticles can further enhance and direct the immune response, via enhanced Th1 responses, as well due to synergy between polymer and adjuvant.

17.4.1.2 Implants

As mentioned earlier, systemic vaccination involves needle and syringes. Most subunit vaccines need adjuvants and immunomodulators to assure an effective immune response. Usually, there is a need for multiple injections to maintain adequate immunity. There is also an ongoing interest to develop thermostable vaccines that would not require refrigeration, which adds tremendous cost and logistical barriers to vaccine programs in developing economies, especially in warm environments. Several strategies have been developed to address these issues. One approach is the administration of implantable vaccines. As described in Chapter 6, a variety of delivery options are available for implants, including (1) in situ forming systems, i.e., formulations that are liquids at ambient conditions but that transform into a semisolid when injected into a target tissue site (e.g., Atrigel™), (2) poly(lactide-co-glycolide) (PLGA) microparticles of different sizes, and (3) solid implants.

A novel injectable, biodegradable, self-gelling, immunomodulatory, and immunostimulatory hydrogel formulation has recently shown promise in mice. The formulation comprises chemically synthesized short DNA strands which form polypod-like structured DNA (polypodna), as well as water and salts. Under physiological conditions, properly designed polypodna preparations form a hydrogel, which can be delivered using a small (29 gauge) needle (Nishikawa et al. 2014).

Solid implant technologies include a spring-loaded device that is used in place of a needle and syringe, with wider acceptance by patients. Solid formulations also have the advantage of being more likely to be thermostable. A vaccine using the recombinant protective antigen from *Bacillus anthracis* is currently in development and delivered as an implant. This allows long-term stockpiling of a vaccine that could be used with short notice in the event of an emergency situation. The vaccine is incorporated into a biodegradable pellet with a hypodermic-like sharp point on one end. A special spring-loaded delivery device is used to inject the pellet into the patient. Patients report that they prefer this method of injection to a standard needle and syringe.

Injectable implants, in solid formulations, are also available for veterinary medicine. Single-dose vaccines, as implants are intended, are particularly sought after in veterinary medicine, as it is difficult to vaccinate animals multiple times. Solidose® dual-dose vaccine implant technology is based on hydroxypropylcellulose (HPC)-coated, polyethylene glycol (PEG)-cross-linked, poly(glycolide) or poly(lactide) oligomeric pellets. Two types of pellets facilitate either immediate release of antigen, or delayed release, designed to release its antigen several weeks later. Both types of pellet are implanted simultaneously (hence "dual dose"), to provide both primary and booster doses, in a single handling. Lipid implants comprising pellets constructed with microparticles made of phospholipids, cholesterol, and Quil-A (a potent saponic immunostimulant) have been tested as a single-dose administration, with the microparticles released over time. Sustained antibody responses were induced, but apparently long-standing CMI was not (Myschik et al. 2008). Further work is necessary to develop this delivery method.

17.4.2 TRANSDERMAL VACCINE DELIVERY

The skin is an immunologically active tissue and a good target for vaccine administration. Professional antigen presenting dendritic cells and Langerhans cells, mast cells, and other cells located just beneath the skin surface are adept at taking up and processing microbes and antigens that cross the skin and transferring these via lymphatic drainage to initiate antigen-specific adaptive immune responses. Delivery of vaccines to or across the skin (transdermal administration) offers several advantages over traditional injection using needles and syringes. Transdermal delivery requires much smaller doses of vaccine, requiring as low as one-tenth of the dose of typical injectable vaccines. Clinical trials with an intradermal trivalent influenza vaccine showed equivalent immune responses in adult subjects 18–60 years of age, but with 40% less dose than the i.m. injected vaccine. Transdermally vaccinated adult subjects over 60 years old had higher seroconversion rates than subjects injected i.m. with a similar vaccine dose. There is evidence that good CMI (increase in gamma interferon, for example) can be induced by transdermal delivery and has been associated with protection. The very short needles used for transdermal delivery, typically ≤900 μm long, are virtually pain free, are well accepted by patients, eliminate the risk of needlestick injuries in personnel administering vaccines, and dramatically reduce the mass of biohazard waste of needles and syringes.

The simplest technology for transdermal vaccination is the microinjector, a short, small gauge needle attached to a syringe that delivers a small liquid volume of vaccine, typically no more than 0.5 mL. The Soluvia™ technology from Becton–Dickinson is used for Fluzone® and Intanza® intradermal seasonal influenza vaccines and delivers accurate doses in a consistent manner. Transdermal seasonal influenza vaccine using a microinjector was approved in 2009. The ballistic method of transdermal administration delivers liquid vaccine through the skin using a high-speed, pressurized liquid jet. Tension-loaded springs, pressurized air, or batteries, can provide the force needed for the ballistic delivery method.

Single to arrays of microneedles 65–700 μm long can be used to penetrate the stratum corneum (the impervious outer layer of the skin) or the epidermis (see also Chapter 9, Section 9.3.3.4). This delivery technology offers convenient application and the potential for patient self-administration. Another potential advantage of microneedle technology is that the vaccine can be formulated to be stable for an extended period of time at room temperatures. This has major implications for vaccine usage in developing nations where cold chain is limited or unavailable.

Different microneedle systems are under study. Hollow microprojectiles can deliver liquid vaccine from a reservoir by way of a pressurize-driven mechanism, such as in the hollow Microstructured Transdermal System (hMTS™) by 3M and MicronJet™ microneedle array technologies by Nanopass. Dried vaccine formulations can also be coated onto metallic or polymeric microprojections, for example, the ZP Patch™ and the solid MTS (sMTS™). Another example of this technology is the Nanopatch™, which is the subject of a case study in Chapter 9 (Section 9.3.3.5). An alternative delivery approach for dried vaccine formulations is the dissolvable microneedle technologies, such as MicroCor® and TheraJect™, which use a molding process with vaccines incorporated within the microneedles. Once applied, the microneedles penetrate the skin and vaccine is released as the microneedles dissolve (Chen et al. 2015).

Currently, four vaccines have been used in clinical applications using intradermal administration: the Bacillus Calmette–Guerin (BCG) for *Mycobacterium tuberculosis*, smallpox, rabies, and seasonal influenza vaccines. BCG and smallpox vaccines have a long history of intradermal administration dating to the early twentieth century (Roldão et al. 2010). Pre-exposure prophylaxis against rabies by intradermal vaccination was recommended by the World Health Organization, starting in 1984, in places where the vaccine is in short supply. These earlier vaccines were administered by needle and syringe using the Mantoux method to inject a bleb of vaccine into the skin—this is technically difficult and can be painful. Advances in skin immunology, combined with new delivery devices that ensure consistent delivery to the skin, are helping to pave the way for further development of transdermal vaccines.

17.4.3 Mucosal Vaccines

Although in some cases mucosal vaccination is still at an early developmental stage, this field represents an exciting area for vaccine research, due to the many advantages associated with the approach, as outlined in Section 17.2. Here, various routes for mucosal vaccination are described.

17.4.3.1 Intranasal Vaccine Delivery

There are examples of licensed vaccines delivered by the IN route that provide protection against disease in human and animal health. The FluMist® vaccine is a live attenuated influenza vaccine that induces protection against a wide range of influenza strains by inducing cytotoxic CD8+ T lymphocytes. There are many more examples of licensed IN vaccines in animal health. Commonly, IN vaccines have been in use for control of respiratory infections that impact production in cattle (TSV 2®, Inforce 3®, Nasalgen®) and swine populations (Maxi/guard® Nasal Vac). IN MLV vaccines are aqueous in nature and are administered as a spray or via a needle-free device. Delivery of inactivated vaccines by the IN route is also being researched extensively.

Many respiratory pathogens invade the host via the upper respiratory tract where they replicate in epithelial cells, releasing microbes that are then inhaled to the lower respiratory tract causing bronchitis, bronchiolitis, pneumonia, and potentially death in infants, children, or immunosuppressed or elderly adults. Local immunity in the upper respiratory tract is key to reducing virus or bacterial invasion and replication, thereby limiting access of pathogens to the lower, as well as upper, respiratory tract. Local immunity can be induced by modified live viruses or pathogens. Although effective, these can induce disease in immunosuppressed patients. Therefore, alternative methods of IN immunization are sought that can induce effective immunity with a higher safety profile. Strategies include the use of live vectors to express key proteins of pathogens, inactivated pathogens or subunit antigens coadministered with potent adjuvants, or inactivated pathogens or subunit vaccines encapsulated within polymeric micro- or nanoparticles.

Alum, the most common adjuvant used in injectable vaccines, is a poor mucosal adjuvant when used alone. Mucosal adjuvants investigated for IN administration of antigens include bacteria toxins (including LT [heat-labile enterotoxin], AB toxins from *Vibrio cholera* or *E. coli*, and mutated versions of the same), cyclic dinucleotides, cytokines, as well as TLR agonists and TLR combinations containing alum, saponin, and/or monophosphoryl lipid A (MPL) (Lawson and Clements 2011). Adjuvants can be mixed with antigens with or without encapsulation to enhance immune responses in a manner similar to injectable adjuvants, but with additional traits to enhance mucosal immunization. These include enhancing interaction with the mucosal surface to prolong contact time and facilitate uptake of the antigen by M cells (the primary APC in the NALT) and modulation of the immune response once the antigen is introduced to the lymphoid follicles where the immune response is shaped. Mucosal adjuvants are most useful in inducing both an effective local IgA and CMI response to respiratory pathogens but also can help stimulate an effective systemic immune response and, in some cases, immune responses to nonrespiratory pathogens that are associated with distant mucosal sites, such as the GI and reproductive tracts, in ways not fully understood (Pavot et al. 2012).

Polymeric micro- and nanoparticles have been widely investigated for IN vaccination. Micro- and nanoparticles offer many advantages for IN vaccination, including (1) a reduction in antigen exposure to mucosal enzymes; (2) the provision of a slow release of antigen, to induce a longer-lasting immunity; (3) an ability to bind to mucosa (in particular for mucoadhesive nanoparticles) and more reliably access efferent immune cells (M cells, dendritic cells, lymphocytes, etc.); (4) the incorporation of cell-specific targeting ligands; and (5) the evidence that using microparticles (or microparticles with mucosal adjuvants) induces a more reliable mucosal immunity (Lawson and Clements 2011).

Since M cells are the entry into the immune induction site of the mucosa, M cell targeting of nanoparticles should greatly increase the chance of success of mucosal vaccines, while potentially reducing the dose of antigen/vaccine needed to induce such a response. To this end, research is directed toward the surface modification the nanoparticles with targeting vectors such as lectins or sugars such as mannans, to improve specific binding to M cells.

PLGA polymers are approved for use in humans and can be formed into micro- or nanoparticles with good biocompatibility. Biodegradable poly-ε-caprolactone has been used to generate nanoparticles as well. However, these materials are hydrophobic in nature, limiting their usefulness for IN administration. Coating these particles with PEG improves the nanoparticle hydrophilicity and facilitates mucoadhesion, making them more effective for IN delivery. Other polymers used in the synthesis of micro- and nanoparticles for IN delivery include poly(anhydrides), poly(carbonates), poly(methyl methacrylate), and poly(phosphazene) cross-linked with albumin and gelatin.

Hydrophilic polymers such as alginate (a linear copolymer of mannuronate and guluronate residues, derived from brown seaweeds) and chitosan (derived by deacetylation or trimethylation of chitin) are inherently mucoadhesive and can prolong contact time of particles to mucosal surfaces. Chemically treating chitosan with N-trimethylation, or cross-linking with tripolyphosphates, further enhances the efficacy of particles for IN vaccine delivery. Chitosan, as well as being strongly mucoadhesive, can further enhance effective delivery as it loosens tight junctions between cells, permitting paracellular delivery of microparticles across the mucosal epithelium and thereby enhancing access to APC. There have been extensive investigations of chitosan micro- and nanoparticles for IN delivery of subunit vaccines.

Liposomes, described in Chapter 5 (Section 5.5.1), have also been investigated for the IN administration of vaccines. Due to their bilayer amphiphilic nature, liposomes can deliver both hydrophilic and hydrophobic antigens; the delivery of multiple numbers of a particular antigen is also possible, thus increasing immunogenicity of the system. Cationic or fusogenic liposomes can be generated for optimal delivery of DNA vaccines. Fusogenic liposomes have hydrophilic polymers bound to their surface that are released in a way to allow liposomes to fuse with and deliver a load within a targeted cell. Liposomal surfaces can be surface modified, i.e., coated with mucoadhesive materials like chitosan, to enhance delivery to mucosal surfaces; antibodies or other ligands can be coupled to liposomes to target them to surface receptors in mucosal tissues. Lipoplexes can be made by combining liposomes with DNA. Lipoplexes can be further optimized by formulation with protamine to prevent aggregation, thereby increasing stability during storage. Liposomes also have associated disadvantages, in that they are susceptible to host enzymes, stability is limited, and manufacturing is associated with high costs (Riese et al. 2014).

Other lipid-based microparticles include virosomes, niosomes, cochleates, proteasomes, and proteoliposomes. Each of these has the advantage of being more stable than liposomes, but concerns about toxicity in humans have limited their usage to some degree. Further development is needed before they can be considered seriously for human vaccines.

ISCOMs are immune-stimulating complexes of hexagonal-shaped cage-like structures, typically of 30–80 nm in diameter, made of cholesterol, phospholipids, and Quil-A, along with the antigen to be delivered. ISCOMs act both as an adjuvant (because of their particulate structure) and a delivery system. They demonstrate greater stability than liposomes and target a strong CMI response as well as a Th2 antibody response. ISCOMs have been studied as carriers for an equine influenza vaccine. An induced mucosal immune response to RSV envelope proteins was observed in mice for 22 weeks (Pavot et al. 2012).

17.4.3.2 Intrapulmonary Vaccine Delivery

The lungs are a good target for vaccine administration due to their large surface area and the presence of large numbers of antigen-processing alveolar macrophages and dendritic cells (see also Chapter 4, Figure 4.10). There are numerous respiratory pathogens that provide ample targets for a needle-free intrapulmonary (IP) route of vaccination.

The phenomenon of vaccination at one mucosal site inducing an immune response at distant mucosal sites has also been demonstrated for IP vaccination. For example, IP administration of antigens and two TLR agonists within multilamellar nanoparticles induced potent cross-protective T cells with effector memory immune responses, not only in the lung but also in other mucosal sites such as the intestine and reproductive tract, as well as inducing systemic immunity. Trials have also shown that IP administration can be as safe as parenteral administration of vaccines.

Despite these advantages, there are also limitations associated with the development of this route for vaccination purposes. As discussed in Chapter 11, the respiratory tract presents a number of anatomical and physiological obstacles to efficient drug/vaccine delivery, including a complex branching architecture, which makes it difficult for vaccines to access and penetrate deep into the lungs. Furthermore, mucociliary clearance mechanisms can remove deposited particles. There are a variety of delivery devices for achieving drug delivery to the lungs, including inhalers and nebulizers (see also Chapter 11) that could be used for vaccination purposes, although in many cases these devices are difficult to use and do not deliver precise dosages. Pulmonary vaccination will require highly specialized delivery devices to deliver a precise dose of vaccine, especially for the very young and elderly patients. Proper technique for using the devices will also be required, to ensure correct delivery of the full vaccine dose to the lung. Patient acceptance and delivery challenges could be greatly aided if delivery is not required to the lower pulmonary tract, but instead acceptable efficacy is attained by delivery to the upper airways (Tonnis et al. 2013).

17.4.3.3 Oral Vaccine Delivery

Oral vaccination is the ultimate goal of vaccine administration: it is the preferred route for patients, it is easy to administer to all ages of patients (and to most host species in veterinary medicine), it allows for self-administration, it is pain free, and it eliminates the use of sharps and potentially the need for cold storage. Oral vaccination is much easier to arrange for large-scale vaccination programs. There are examples of oral administration of MLVs including oral polio, which has made huge inroads into dramatically reducing disease and raised hopes for eventual eradication. Oral typhoid and cholera vaccines are also effective in human medicine. In veterinary medicine, there are several MLVs administered to poultry, swine, and calves for salmonella, rota, and coronaviruses.

Oral administration can be used to target the GALT and specifically, the M cells of the Peyer's patches (Figure 17.3; see also Chapter 7, Figure 7.8). This is especially potentially effective for immune responses directed to intestinal pathogens including helminthes and protozoa. M cells (the primary APC in the GALT) may be targeted using specific vectors (see later). Methods are also being developed to deliver antigens across intestinal epithelial cells using transcytotic receptors, especially the neonatal Fc receptor, which could potentially greatly enhance the delivery of oral vaccines targeted to the intestinal tract (Devriendt et al. 2012). However, too large or too low a dose of antigen could induce systemic tolerance, whereby the immune system fails to respond to an antigen.

Orally administered MLVs have had much more success than inactivated or subunit oral vaccines. In fact, although there are many reports of success with inactivated vaccines in laboratory animals, there are few reports experimentally in larger host species and no such vaccines are used in human medicine. A major hurdle to development of inactivated oral vaccines is the formulation and delivery of these vaccines, given the hostile environment of the GI tract. As described in Chapter 7 (Section 7.2), the GI tract exposes antigens to many challenges, including digestive enzymes, bile, extremes of pH, intestinal microflora, intestinal motility, and a mucus layer coating epithelial cells that interferes with vaccine uptake. These conditions can either destroy the antigens, or luminal contents may affect intestinal integrity and prevent antigens from being taken up by M cells.

Microencapsulated vaccines often also include other adjuvants (for example, lymphotactin, LT or CT toxins, or derivatives) and immunomodulators (e.g., CpG). Receptor ligands may also be attached to the surface of particles to facilitate uptake by the M cells of the Peyer's patches. As described in detail in Chapter 7 (Section 7.7), M cell–targeting ligands that have been studied include lectins (such as *Ulex europaeus* agglutinin 1), as well as specific pathogen ligands directed to M cells such as *Salmonella flagellin*, cholera or heat labile enterotoxin from *E. coli*, influenza hemagglutinins, and ricin (Devriendt 2012).

Particle uptake can be enhanced by reducing particle size; altering surface charge, pH, hydrophobicity; or inclusion of bioadhesive materials (see also Chapter 7, Section 7.7). Mucoadhesive compounds such as chitins and alginates have also been evaluated, as a means to enhance binding to the intestinal mucosal surface. Evidence suggests that it is important that the antigens and adjuvants be contained within the same microparticle to be delivered to the APC, to stimulate the best immune response. Formulation, delivery method, and dose are key factors that must be addressed for optimal immune stimulation. Studies in mice have shown promise, but evidence in higher animal species, as well as the balance of manufacturability and cost-effectiveness of these approaches, remains to be clarified.

As described earlier for IN immunization, a wide number of materials can be used to make micro- and nanoparticles. The most studied material is PLGA, but nanoparticles synthesized using poly(esters), poly(anhydrides), poly(methyl methacrylate), starches, albumin, chitosan, alginate, cellulose acetate, and dextrin have all been reported. There are also a range of lipid-based microparticles for oral vaccination, including liposomes, niosomes, bilosomes, ISCOMs, cochleates, and virosomes, as well as hybrid polymer lipid-based microparticles. The size of the microparticles is important for adequate delivery to the APCs. In general, a particle size of less than 10 μm is needed for uptake by dendritic cells or APCs and induction of a good immune response. There is contradictory information on the effectiveness of nanoparticles for vaccine delivery compared to microparticles, although there are good data suggesting nanoparticles can be very effective in DNA vaccine delivery. Nanoparticles offer advantages compared to microparticles: they can be made with higher percentage of antigen incorporation due in part to higher surface area, and they can be manufactured under milder conditions, which are less likely to damage antigens.

Realistically, even at the best efficiency of delivery, a tiny percentage of particles actually reach the target. What is not known at the current time is what target number/dose of antigen is required to be effective (Wilkhu et al. 2011). Unfortunately, micro- and nanoparticle technology does not appear applicable as a platform technology at this point in time. Encapsulation efficiency and success tends to vary from one antigen to another. Stability of antigens within microparticles has been an issue, due in part to the process of making the microparticles and exposure to solvents and shear forces. Methods to address this have been developed including binding of antigens to the surface of the micro or nanoparticles. Evidence suggests that antigens bound to surface of particles induce stronger immune responses than antigens encapsulated within particles.

A potentially revolutionary oral vaccine delivery system would be the expression of antigens in plants, which could be either orally ingested as part of the diet or extracted from the plant. The first proof-of-plant expression of an antigen was in a 1997 patent, describing the expression of a surface protein of the gingivitis-causing bacterium *Streptococcus mutans* in tobacco leaves, and since then other antigens have been expressed, including subunits from norovirus, hepatitis B, and rabies and measles viruses. Antigens have been expressed in multiple plants including tobacco, tomatoes, potatoes, bananas, spinach, lettuce, tomatillos, corn, and rice. Ideally, plants are used that do not require cooking to be ingested.

Vaccines based on plant-expressed antigens have issues with antigen stability, consistency of dose within food (depending on how much is ingested), long-term storage, and the risk of induction of oral tolerance. Seeds can be stored at ambient temperatures for years with apparent little loss of protein activity. Oral delivery of crude or minimally processed antigens expressed in seeds, such as rice or corn, is possible. Animal-specific vaccines for foot and mouth disease, transmissible gastroenteritis, infectious bursal disease virus, *Actinobacillus pleuropneumoniae*, and *Mannheimia haemolytica* have been expressed in a variety of plants, including tobacco, *Arabidopsis thaliana*, quinoa and cowpea leaves, corn, clover, and alfalfa. However, due to the variability in protein structure and biochemistry, it is difficult to predict how well a protein antigen will be expressed, or whether it will be folded correctly to display desired epitopes when expressed in plants.

17.4.3.4   Oral Cavity Vaccine Delivery

One simple approach to mucosal vaccination that would avoid the harsh conditions of the GI tract is to target delivery to under the tongue. Sublingual (SL) vaccines also lead to systemic uptake of antigens into the bloodstream, avoiding first-pass metabolism by the liver and facilitating a more direct stimulation of immunity. This method of immunization also suppresses IgE immune responses and has been used to treat Type I allergies. Tablets or mucoadhesive films create thermostable vaccines that dissolve under the tongue. As described in Chapter 8, the SL epithelium is thinner and less keratinized than most of the oral cavity, making the absorption of vaccines possible. Moreover, there are also dendritic cells under the tongue to process antigens, so lower antigen doses are needed than for oral vaccines. Experimentally, SL administration of vaccines, including the influenza virus and the gingivitis caries-causing bacterium *Porphyromonas gingivalis*, induces mucosal as well as systemic antibody responses (Fukuyama et al. 2012). Tablets with fast protein release appear to function better than extended-release tablets in this scenario. LT or mutations of LT toxin are toxic when administered by IN or oral administration. However, when administered sublingually, they can stimulate systemic IgG, local IgA, and cytotoxic T cell responses without the toxicity seen with IN or oral administration. SL administration can also stimulate immune responses in the stomach, intestine, and nasal passages. It is possible that mucoadhesive compounds, combined with antigens and nanoparticles, may positively affect SL administration.

An example of a mass vaccination project using oral pharyngeal immunostimulation is in veterinary medicine, using oral rabies bait vaccines, which have been a major management tool in controlling rabies in wild animal vectors. In this approach, animal bait is designed to entice an animal to chew it; the bait also contains MLV or a vaccinia vector expressing the rabies glycoprotein (Freuling et al. 2013). There is no cure for rabies infection in animals or humans other than postexposure treatment with antisera, and the use of these vaccines has significantly reduced the incidence of both animal and human rabies. The approach has been used to eradicate rabies from wildlife in Europe, as well as in North and South America.

17.5   FUTURE VACCINE DEVELOPMENT

Many studies evaluating vaccine antigens, adjuvants, and delivery systems are performed in rodent models. Unfortunately, although these models can differentiate immunogenic vs nonimmunogenic immunogens, they often do not predict efficacy in nonhuman primates or humans. Differences in anatomy, physiology, genetics, and immunity and even differences in gut flora or local flora at mucosal sites can affect how one species responds to a vaccine. Humanized mice have been developed and offer some advantages. Ideally, more predictive models need to be carefully developed, but eventually the vaccines need to be tested in humans or in host target species (dog, cat, swine, etc.).

Vaccine sales have been growing at a tremendous rate, from sales of $5 billion in 2000 to $24 billion in 2013. The WHO expects this growth to continue to a value of U.S. $48 billion in 2017 and hit $100 billion by 2025. There has been a resurgence in the vaccine industry for several reasons. There have been changes in strategies of the pharmaceutical industry—from a previous focus on blockbuster pharmaceutical drugs to a focus where vaccines are considered as blockbusters or significant contributors to a portfolio. There have been changes in laws to limit liability for adverse responses to vaccines, as well as changes in research and manufacturing processes, to improve the return on investment on vaccines. Companies are working on ways to reduce costs and improve profitability of vaccines in worldwide markets (Kadder 2014).

Vaccine products and sales are also increasing steadily in developing nations, due again to changes in pharmaceutical strategies, as well as the actions of charitable foundations and efforts of world health organizations. Foundations are funding research and initiatives to improve supply and distribution of vaccines to developing countries. In some cases, foundations team with pharmaceutical companies, as well as other research institutions, to address specific issues such as providing rotavirus vaccines, oral polio vaccines, the development of malaria vaccines, and stimulating development of novel injection free and thermostable formulations, which would greatly improve vaccine usage in remote areas in developing countries. There are also initiatives to fund global mucosal vaccines to bring together experts in immunology, nanobiotechnology, and vaccine development, to develop oral and IN vaccines, with the goal of increasing patient compliance by decreasing injections as well as reducing the incidence of blood-borne diseases associated with the reuse of needles.

REFERENCES

Borde, A., A. Ekman, J. Holmgren et al. 2012. Effect of protein release from tablet formulations on immune response after sublingual admin. *Eur J Pharm Sci* 47:695–700.

Brandtzaeg, P. and R. Pabst. 2004. Let's go mucosal: Communication on slippery ground. *Trends Immunol* 25:570–577.

Chen, D., T. Bowersock, R. Weeratna et al. 2015. Current opportunities and challenges in intradermal vaccination. *Ther Deliv* 6(9):1101–1108.

College Physicians of Philadelphia. 2015. The history of vaccines. http://www.historyofvaccines.org/content/timelines/all, last accessed March 5, 2016.

Da Costa Martins, R., C. Gamazo, M. Sánchez-Martínez et al. 2012. Conjunctival vaccination against *Brucella ovis* in mice with mannosylated nanoparticles. *J Control Release* 162(3):553–560.

Devriendt, B., B.G. De Geest, B.M. Goddeeris et al. 2012. Crossing the barrier: Targeting epithelial receptors for enhanced oral vaccine delivery. *J Control Release* 160(3):431–439.

Findik, A. and A. Cifti. 2012. Bacterial DNA vaccines in veterinary medicine: A review. *J Vet Adv* 2(4):139–148.

Foged, C. 2011. Subunit vaccines of the future: The need for safe, customized and optimized particulate delivery systems. *Ther Deliv* 2(8):1057–1077.

Freuling, C.M., K. Hampson, T. Selhorst et al. 2013. The elimination of fox rabies from Europe: Determinants of success and lessons for the future. *Philos Trans R Soc B* 368:2012–2142.

Fukuyama, Y., D. Tokuara, K. Kataoka et al. 2012. Novel Vaccine delveopment statregies for inducing mucosal immunity. *Expert Rev Vaccines* 11(3):367–379.

Ghanem, A., R. Healey, and F.G. Adly. 2013. Current trends in separation of plasmid DNA vaccines: A review. *Anal Chim Acta* 760:1–15.

Grillot-Courvalin, C., S. Goussard, and P. Courvalin. 2011. Bacterial vectors for delivering gene and anticancer therapies. *Microbe Magazine*. http://www.microbemagazine.org/index.php?option=com_content&view=article&id=3195:bacterial-vectors-for-deliviering-gene-and-anticancer-therapies&catid=719&Itemid=953, last accessed March 5, 2016.

Kadder, K. 2014. Global vaccine market features and trends. http://who.int/influenza_vaccines_plan/resources/session_10_kaddar.pdf, last accessed March 5, 2016.

Kunda, N.K. 2013. Nanocarriers targeting dendritic cells for pulmonary vaccine delivery. *Pharm Res* 30:325.

Lawson, L.B., E.B. Norton, and J.D. Clements. 2011. Defending the mucosa: Adjuvant and carrier formulations for mucosal immunity. *Curr Opin Immunol* 23:414–420.

Logomasini, M.A., R.R. Stout, and R. Marcinkoski. 2013. Jet injection devices for the needle-free administration of compounds, vaccines, and other agents. *Int J Pharm Compd* 17(4):270–280.

Mane, V. and S. Muro. 2012. Biodistribution and endocytosis of ICAM-1 targeting antibodies vs nanocarriers in the GI in mice. *Int J Nanomed* 7:4223–4237.

McKenna, K., A. Beignon, and N. Bhardwaj. 2005. Plasmacytoid dendritic cells: Linking innate and adaptive immunity. *J Virol* 79:17–27.

Myschik, J., W.T. McBurney, T. Rades et al. 2008. Immunostimulatory lipid implants containing Quil-A and DC-cholesterol. *Int J Pharm* 363(1–2):91–98.

Nishikawa, M., K. Ogawa, Y. Umeki et al. 2014. Injectable self-gelling biodegradable and immunomodulatory DNA hydrogel for antigen delivery. *J Control Release* 180:25–32.

Norman, J.J., J.M. Arya, M.A. McClain et al. 2014. Microneedle patches: Usability and acceptability for self-vaccination against influenza. *Vaccine* 32(16):1856–1862.

Pavot, V., N. Rochereau, C. Genin et al. 2012. New insights in mucosal vaccine development. *Vaccine* 30:142–154.

Rappuoli, R. and A. Covacci. 2003. Reverse vaccinology and genomics. *Science* 302:602.

Reed, S., S. Bertholet, R.N. Coler et al. 2009. New horizons in adjuvants for vaccine development. *Trends Immunol* 30(1):23–32.

Riese, P., P. Sackthiverl, S. Trittel et al. 2014. Intranasal formulations: Promising strategy to delivery vaccines. *Expert Opin Drug Deliv* 11(10):1619–1634.

Robert-Guroff, M. 2007. Replicating and non-replicating viral vectors for vaccine development. *Curr Opin Biotechnol* 18:546–556.

Roldão, A., M.C. Mellado, L.R. Castilho et al. 2010. Virus-like particles in vaccine development. *Expert Rev Vaccines* 9(10):1149–1176.

Saade, F. and N. Petrovsky. 2012. Technologies for enhanced efficacy of DNA vaccines. *Expert Rev Vaccines* 11:189–209.

Soema, P.C., G.J. Willems, K. van Twillert et al. 2014. Solid bioneedle-delivered influenza vaccines are highly thermostable and induce both humoral and cellular immune responses. *PLOS ONE* 9(3):e92806.

Tonnis, W.F., A.J. Lexmond, H.W. Firjlink et al. 2013. Devices and formulations for pulmonary vaccination. *Expert Opin Drug Deliv* 10(10):1383–1397.

Wilkhu, J., S.E. McNeil, D.J. Kirby et al. 2011. Formulation design considerations for oral vaccines. *Ther Deliv* 2(9):1141–1164.

Woodrow, K.A., K.M. Bennett, and D.D. Lo. 2012. Mucosal vaccine design and delivery. *Annu Rev Biomed Eng* 14:17–46.

Yao, W., Y. Peng, M. Du et al. 2013. Preventative vaccine-loaded mannosylated chitosan nanoparticles intended for nasal mucosal delivery enhance immune responses and potent tumor immunity. *Mol Pharm* 10(8):2904–2914.

Zhu, Q. and J.A. Berzofsky. 2013. Directed safe passage to the front line of defense. *Gut Microbes* 4(3):246.

18 Theranostic Nanoagents

Anthony S. Malamas and Zheng-Rong Lu

CONTENTS

18.1 INTRODUCTION

The field of theranostics (a portmanteau of "therapeutics" and "diagnostics") aims to integrate therapeutics with diagnosis, in order to develop more individualized therapies. Life-threatening diseases, particularly high-risk cancers, are highly heterogeneous, so that treatments are typically effective for only limited patient populations and at certain stages of disease development. Merging of the paradigms of diagnosis and therapy will provide timely assessment of therapeutic response, allowing optimization of treatments and tailoring personalized medicine based on individual needs, to improve therapeutic outcomes. A current clinical example is the use of Herceptin® (a monoclonal antibody used to treat patients with breast cancer), which is used in conjunction with the diagnostic tool, HercepTest®. Herceptin® targets the HER2 protein, which is overexpressed in approximately one-third of breast cancers. HercepTest™ specifically demonstrates overexpression of the HER2 protein in breast cancer tissues and so is used to identify those patients who are most likely to benefit from Herceptin® treatment.

Currently, research activities on theranostics are predominantly focused on cancer diagnosis and treatment. Cancer is the third most common cause of death in the world, following cardiovascular and infectious diseases. There are significant challenges in the successful delivery of cytotoxic drugs specifically to tumor cells while avoiding normal healthy tissues and minimizing side effects. In addition, it is essential to carry out diagnostic imaging to understand the cellular phenotypes, biological activity, and heterogeneity of each tumor. In response to these challenges, theranostics offers the potential to allow physicians to monitor the drugs given to each patient while assessing drug pharmacokinetics and biodistribution, as well as tumor response, all in a noninvasive and

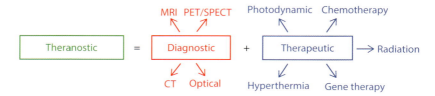

FIGURE 18.1 Diagram outlining the basic principles of a theranostic agent. Theranostic agents possess both diagnostic and therapeutic elements in a single-nanoparticle construct.

real-time manner. As a result, physicians will be able to avoid the prolonged use of nonresponsive therapies to treat cancer; instead, they will be able to alter or design innovative treatment regimens, tailored to each individual case, to improve overall survival.

Due to the continued advances in nanotechnology, including design and synthesis of new nano-carriers, the field of theranostics has evolved so that now it is directed more toward using multifunctional nanosystems, which simultaneously deliver both therapeutic and imaging moieties. Nanoparticles (NPs) can accommodate imaging probes, therapeutic agents, and/or targeting agents. This "all-in-one" approach allows assessment of both the pharmacokinetics and pharmacodynamics of the therapies. Figure 18.1 depicts the concept of theranostics based on NP platforms combining the therapeutic and diagnostic functions into one package.

The chapter will focus on the use of theranostics in cancer imaging and therapy. Medical imaging modalities are first introduced, followed by an overview of commonly used nanocarriers for drug delivery. A selection of multifunctional nanoplatforms for theranostic applications are then presented in detail.

18.2 MEDICAL IMAGING MODALITIES

Various types of medical imaging procedures allow the evaluation of organ functions and assist in the diagnosis of disease. Many of these imaging procedures are now being investigated for their application to theranostics.

Magnetic resonance imaging (MRI) produces images by measuring the radio-frequency signals that are given off by magnetized protons in living tissue. Image contrast of different tissues is created because protons in each tissue possess their own characteristic magnetic properties. Image contrast also can be enhanced with the aid of contrast agents that are delivered into the tumor tissue. Contrast agents are paramagnetic materials that can alter the magnetic properties of water protons to enhance the signal in the tissue of interest and help delineate it from that of the normal surrounding tissue. MRI is a beneficial imaging technique because it offers high spatial resolution without any tissue-penetrating limitations. Currently, highly stable gadolinium (Gd) complexes are commonly used to produce bright signal enhancement, while paramagnetic iron oxide nanoparticles (IONPs) generate dark signal enhancement.

Computed tomography (CT) is one of the most commonly used modalities in diagnostic imaging–based x-ray attenuation. CT provides superior visualization of structures with high densities but is limited when used to distinguish soft tissues that have similar densities. CT contrast agents are introduced in order to improve image contrast of soft tissue structures with similar or identical densities. CT contrast agents are composed of biocompatible materials containing elements of high atomic numbers. Clinical CT contrast agents are generally iodinated compounds. NPs containing high-atomic-number elements, such as gold, are now being investigated as CT contrast agents.

Positron emission tomography (PET) and single-photon emission computed tomography (SPECT) are nuclear imaging techniques used to map physiological and biological processes in the body following the administration of radiolabeled tracers. In PET, positrons (positively charged electrons) emitted from the nucleus during decay travel a few millimeters in the tissue, where they undergo annihilation by colliding with electrons. Each annihilation event releases two gamma-ray photons of equal energy (511 keV) in opposite trajectories (180° apart). PET scanners utilize the simultaneous detection of these two photons to precisely locate the source of the annihilation event. Positron-emitting isotopes ^{15}O, ^{13}N, ^{11}C, ^{18}F, ^{64}Cu, ^{62}Cu, and ^{68}Ga are often incorporated into biorelevant materials to follow their distribution and concentration, and to detect malignant diseases. SPECT employs gamma-emitting isotopes, such as ^{99m}Tc, ^{111}In, and ^{123}I, as probes to detect biomarkers and to label NPs. SPECT imaging can quantitatively determine disease-related biomarkers for diagnostic imaging, to track biodistribution of NPs and to assess therapeutic efficacy.

Optical imaging utilizes photons of a characteristic wavelength to excite fluorescence materials, to visualize biomarkers and to study the localization of materials. Optical imaging offers a high sensitivity and the capability of multicolor imaging. A major limitation of optical imaging is the small depth of light penetration, ranging from hundreds of microns to several centimeters. Near-infrared (NIR) imaging has proven to be more desirable owing to better tissue penetration and less interference from tissue autofluorescence.

18.3   NANOCARRIERS FOR THERANOSTICS

Nanocarriers for theranostics may be constructed from a wide range of organic and inorganic materials, including lipid-based systems (including liposomes, polymersomes, polymeric micelles), polymeric NPs, gold NPs, iron oxide NPs, quantum dots (QDs), silica NPs, calcium phosphate NPs, and various types of carbon nanomaterials (including carbon nanotubes [CNTs]). These carriers have been widely investigated as carriers for drug targeting and delivery and, as such, are described in detail in Chapter 5. The drug carriers described in this chapter can be seen as an extension of this research, whereby the drug delivery system incorporates not only a drug but also imaging agents (contrast agents, fluorescent probes, and radiolabels), for use in diagnostic imaging. Some of these nanocarriers also are cytotoxic via magnetic hyperthermia and photothermal ablation, as well as photodynamic therapy (PDT).

A particular advantage that NPs offer for theranostic purposes is that they can be targeted to specific disease sites, e.g., to the tumor/tumor cells, thereby reducing unwanted side effects. Nanocarrier targeting is described extensively in Chapter 5; to summarize here, targeting can be achieved by either (1) passive or (2) active means. Passive targeting exploits the natural distribution profile of an NP in vivo. Due to their size, shape, charge, and other physicochemical characteristics, NPs tend to be recognized as "foreign" by the immune system and are taken up by the reticuloendothelial system (RES). Their accumulation in RES organs such as the liver and spleen is desirable if these are the target organs. Alternatively, NPs can be sterically stabilized, by rendering their outer surfaces more hydrophilic. Typically, this involves covalently attaching polyethylene glycol (PEG) chains to the NP surface. The highly hydrated PEG layer reduces opsonization and RES uptake, thereby prolonging NP circulation time in the blood. This "stealth" technology can then allow sufficient time for the nanocarriers to accumulate at a tumor site, via what is termed the "enhanced permeability and retention" (EPR) effect of tumor vasculature (Maeda et al. 1984). Tumor blood supply is associated with irregularly dilated and leaky blood vessels, with large pore sizes. NPs that are smaller than 100 nm (although it should be noted that the precise particle size is not fully clear and is a subject of debate—see, for example, Chapter 5, Section 5.2.1) can thus extravasate into the tumor tissue, by passing through the pores and preferentially accumulating there.

Active targeting is the use of a specific-targeting ligand, in order to improve uptake and sequestration into the tumor. The introduction of targeting ligands can help increase the target-to-background contrast in medical imaging while also improving the local concentration of the therapeutics at the tumor site and reducing its systemic toxicity. A targeting moiety is often covalently attached to the NP surface. Ideally, a ligand is used which targets a protein that is predominately overexpressed on tumor cells, in order to prevent the delivery vehicle from accumulating elsewhere in the body.

For example, folic acid is often used as a targeting motif, as the folic acid receptor is known to be overexpressed on cancer cells.

Due to their very high surface-area-to-volume ratios, NPs have high loading capacities. Drugs and imaging agents can be loaded into nanocarriers. Encapsulated within the nanocarriers, these agents are protected from in vivo degradation, allowing controlled drug release. The NP surface can be modified with targeting ligands, drugs, imaging agents, stealth coatings, etc. Furthermore, the use of external stimuli, including temperature, light, and magnetism, can create activatable NPs, which allow for the selective trigger of drug release and image-based mechanisms. Some NPs demonstrate a pH-dependent solubility profile, which can also be used to control drug release.

18.4   INORGANIC-NANOSIZED THERANOSTICS

18.4.1   GOLD NANOPARTICLE–BASED THERANOSTICS

Gold nanoparticles (AuNPs) have emerged as one of the most extensively investigated theranostic platforms for the diagnosis, imaging, monitoring, and treatment of malignant and other diseases. AuNPs possess a variety of advantageous properties that make them very useful as multifunctional theranostic vehicles. AuNPs offer inherent biological compatibility, an important advantage over many other synthetic delivery systems described later. Furthermore, their large surface area means they can serve as highly efficient carriers, capable of high drug and diagnostic loading. Using simple wet-laboratory techniques, gold nanocarriers have been constructed in a variety of shapes and sizes, including spheres, cubes, rods, cages, and wires; all of which can now be prepared with accurate quality control and in large quantity. They can also be used as either the core or the shell, for polymer–metal and metal–metal hybrid NPs.

AuNPs have high atomic number and induce strong x-ray attenuation, making them effective contrast agents for CT imaging. Because AuNPs are visible by CT, they have potential to be noninvasively assessed in pharmacokinetics, biodistribution, and tumor targeting. AuNP-based theranostics are often developed by surface modification with therapeutic agents and other imaging agents. Photosensitizers (PS), dyes, drugs, and targeting ligands can all be attached to the surface of AuNPs either directly, via amine or thiol groups, or indirectly, using a linker molecule such as bovine serum albumin. Other imaging agents, such as Gd chelates for use in MRI, and radioisotopes for nuclear medicine, are attached to AuNPs for multimodal imaging and image-guided therapy.

Encouraging preclinical data is accumulating on the use of AuNPs as theranostic agents. Chen et al. have recently synthesized and characterized folic acid–functionalized, dendrimer-entrapped AuNPs containing the MRI agent Gd, thereby developing a nanoprobe suitable for both CT and MRI methodologies (Chen et al. 2013). The AuNPs were entrapped within the dendrimer interior and the Gd was modified on the dendrimer surfaces. This multifunctional construct specifically targeted folic acid receptor–expressing cancer cells via a receptor-mediated pathway. Both in vitro cell imaging and in vivo tumor imaging demonstrated significantly improved contrast-to-noise ratios, indicating the potential of this system to detect cancer cells in the body using dual-mode CT/MRI.

AuNPs can undergo surface plasmon resonance. This is an optical phenomenon that arises from the interaction between an electromagnetic wave and the conduction of electrons in a metal. When excited with light at, or near, the absorption maximum (typically in the visible, or NIR, range), the electrons in AuNPs collectively absorb the incoming irradiation and are excited from the ground state to a higher energy level, where they oscillate at a particular resonance frequency. The nonradiative energy relaxation of the electrons that back down to their ground state results in an increase in kinetic energy and ultimately leads to the generation of intense heat into the local environment. Thus, AuNPs can act as energy transducers, converting the absorbed light into heat (see also Chapter 14, Figure 14.18). This photothermal effect can be harnessed for cancer therapy: AuNPs accumulate at a tumor site and are subsequently illuminated; the heat spike can then selectively damage the tumor tissue, while normal tissue is unharmed. In this application, the AuNPs are not functioning as carriers for a cytotoxic drug; rather the AuNPs themselves are causing tumor cell death, via hyperthermia.

Spherical AuNPs, with a characteristic absorption at 500–600 nm, are not appropriate materials for such an application. In contrast, if the morphology is changed to a hollow AuNP construct, such as a nanocage or nanoshell, this can shift the absorption to the NIR region, between 600 and 1000 nm. This is the therapeutic window whereby the interaction of light with biological tissues is low, keeping attenuation and scattering effects to a minimum, reducing unwanted interactions with the surrounding healthy tissue.

Reports on the success of hollow gold nanospheres (HAuNS) for photothermal ablation are increasing. For example, Melancon et al. conjugated the targeting ligand C225 (an antibody directed at epidermal growth factor receptor) to the surface of HAuNS and then evaluated their distribution in mice (Melancon et al. 2008). It was shown that C225-HAuNS had a significantly higher uptake in the tumors, compared with IgG-conjugated HAuNS controls. The efficacy of photothermal ablation was also assessed. Magnetic resonance thermal imaging revealed that for mice injected with C225-HAuNS, the exposure to low doses of NIR light resulted in average maximum temperatures of 65°C, whereas the saline control mice resulted in an average maximum temperature of only 47° after laser treatment. Furthermore, histological analysis showed that tumors treated with the C225-coated HAuNS developed significantly larger necrotic areas than the control tumors following exposure to the NIR laser.

Gold nanorods (GNRs) have also demonstrated potential for photothermal ablation, due to their high light absorption coefficient in the NIR region, good photothermal stability during laser illumination, and high heat generation. GNRs bearing a folate ligand for cancer cell targeting were also labeled with radioactive iodine, in order to monitor NP distribution in vivo during the treatment period (Jang et al. 2012).

18.4.2   Iron Oxide Nanoparticle–Based Theranostics

IONPs are made from magnetite (Fe_3O_4). IONPs less than 20 nm are superparamagnetic (i.e., the particles show zero magnetism in the absence of an external magnetic field but can become magnetized in the presence of one) and can be used as contrast agents in MRI. Unlike Gd-based contrast agents, superparamagnetic iron oxide nanoparticles (SPIONs) decrease MR signals of surrounding water protons, resulting in dark signal enhancement. SPIONs are extensively investigated for theranostic purposes. SPIONs need to be surface-engineered to improve their solubility, stability, and performance in vivo. SPIONs are typically coated with hydrophilic materials. A wide variety of coating materials have been used for SPION modification, including dextrans, dendrimers, and polyvinylpyrrolidone. Therapeutics and targeting agents have been covalently conjugated to SPIONs. Biodegradable spacers, e.g., peptides, have been used to facilitate drug release in response to the environment, e.g., within lysosomes. This ensures release of the drug into the cytosol of cells, upon reaching the acidic endosome/lysosome compartments after cellular uptake (Kohler et al. 2005). Small-sized SPIONs are often incorporated into other larger drug carriers, including liposomes, polymersome, polymeric NPs, and silica NPs, to enable visualization of the larger carriers noninvasively with MRI.

Similar to the photothermal ablation described in Section 18.4.1 for AuNPs, magnetic NPs also can be used as therapeutic entities per se, rather than serving as carriers for a therapeutic drug. In this case, the cytotoxic effect is due to magnetic hyperthermia, which uses a combination of both alternating magnetic fields and magnetic NPs as heating agents, to induce a localized and specific heat around a tumor region. SPIONs possess significant energy absorption properties due to the Neel relaxation phenomenon. When placed in alternating magnetic fields, SPION dipole moments are quickly reoriented, depending on the frequency, magnetic field strength, NP size, and environmental temperature. The electromagnetic energy is dissipated as heat, and the resulting temperature spike within the surrounding tissue can be exploited for hyperthermia-induced destruction of cancer cells.

SPION clusters have been prepared that are surface derivatized with folic acid for tumor cell targeting and also contain an outer PEG shield, to promote long circulation times and tumor accumulation. After i.v. injection, the SPION clusters accumulated locally in cancer tissues within the tumor and enhanced the MRI contrast (Hayashi et al. 2013). On the application of an AC magnetic field, it was found that the temperature of the tumor was approximately 6°C higher than the surrounding tissues, 20 minutes after treatment. Thirty-five days after treatment, the tumor volume of treated mice was one-tenth that of the control mice. Furthermore, the treated mice were alive after 12 weeks, whereas control mice died up to 8 weeks after treatment.

Localized magnetic hyperthermia can also be used to trigger controlled drug release from a nanocarrier. Three functionally different polymers were used to surround SPIONs and generate a thermosensitive nanocarrier (Rastogi et al. 2011): (1) the temperature-sensitive polymer, N-isopropylacrylamide; (2) poly(acrylic acid), to tune the critical temperature point and enhance conjugation to the NP surface; and (3) PEG-methacrylate, to provide a stealth coating to the NP, increasing the circulation time and introducing reactive hydroxyl groups for the coupling of folic acid. The folic acid was incorporated as a targeting ligand for cancer cells. The polymeric nanostructures, loaded with doxorubicin (Dox), were approximately 200 nm in size and functioned as MRI contrast agents in phantom gels. The thermo-responsive property of the carrier resulted in controlled release of Dox following SPION-induced hyperthermia, contributing to a release rate that was nearly 2.5-fold higher than that for normal physiological conditions, during the first 48 hours.

18.4.3 QUANTUM DOTS THERANOSTICS

QDs are luminescent semiconductor nanocrystals, typically with cadmium–selenium (CdSe) or cadmium–tellurium cores. For imaging purposes, QDs are proving particularly valuable for multicolor fluorescent applications. Some of the major advantages of using QDs over traditional fluorescent dyes include their high absorbance, narrow and symmetric emission bands, stability, and resistance to photobleaching (Resch-Genger et al. 2008). It is also possible to control their optical properties by simply tuning the particle size (the larger the dot, the redder its fluorescence spectrum). In addition, QDs are able to form fluorescence resonance energy transfer (FRET) pairs with fluorescent dyes, allowing their use for monitoring the intracellular trafficking of particles.

As well as their use as imaging agents, the high surface-area-to-volume ratio of QDs enables the construction of a multifunctional nanoplatform, where the QDs serve not only as an imaging agent but also as a nanoscaffold for both therapeutic and diagnostic modalities. To this end, there is now a considerable body of literature describing how QDs have been functionalized with a variety of drugs, including cytotoxics, as well as targeting moieties, including aptamers, antibodies, oligonucleotides, peptides, and folates (Ho and Leong 2010).

A promising example of the potential of QDs as theranostic carries is the development of a QD–RNA aptamer–doxorubicin conjugate (QD–Apt(Dox)) for synchronous cancer imaging and traceable drug delivery (Bagalkot et al. 2007). In this system, the RNA aptamer serves as a targeting ligand for the prostate-specific membrane antigen, expressed in LNCaP cells. Dox was intercalated into the aptamer, resulting in a multifunctional QD platform for imaging and therapy. QD–Apt(Dox) has demonstrated enhanced therapeutic specificity in vitro against LNCaP cells, compared to nonspecific PC3 cells.

One elegant approach has been the in situ immobilization of CdSe QDs in the interior of a pH- and temperature-dual responsive hydroxypropylcellulose–poly(acrylic acid) (HPC-PAA) nanogel (Wu et al. 2010). In this multifunctional platform, the fluorescent QDs act as an optical identification code for sensing and imaging (Figure 18.2). The hydroxypropylcellulose (HPC) chains provide rich –OH groups for sequestering Cd^{2+} ions into the gel network, thereby stabilizing the QDs embedded in the gel network and minimizing their toxicity. The presence of the hydrophilic HPC chains on the surface of QDs also provides steric stabilization, to reduce opsonization and subsequent uptake by the RES system in vivo. High drug-loading capacity of the anticancer drug temozolomide (TMZ)

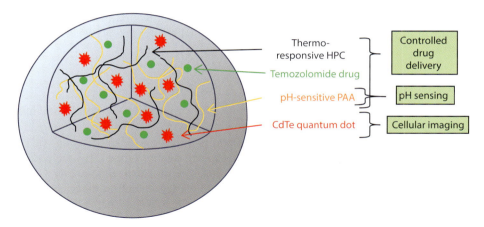

FIGURE 18.2 Schematic representation of HPC–polyacrylamide–quantum dot hybrid nanogels, for multifunctional application in drug delivery and fluorescence quantum dot imaging (HPC, hydroxypropylcellulose; PAA, polyacrylamide; CdTe, cadmium-tellurium). (Reprinted from *Biomaterials*, 31(11), Wu, W., Aiello, M., Zhou, T. et al., In-situ immobilization of quantum dots in polysaccharide-based nanogels for integration of optical pH-sensing, tumor cell imaging, and drug delivery, 3023–3031, Copyright 2010, with permission from Elsevier.)

was attributed to hydrogen bond interactions between TMZ and (1) the carboxyl groups in polyacrylamide (PAA) network chains and (2) the hydroxyl groups in HPC chains. The pH-sensitive PAA network chains were designed to induce a pH-responsive volume phase transition of the nanogel, in order to facilitate stimulus-responsive drug release. At higher pH levels, the carboxylic acid groups of the nanogels become deprotonated, which enhances nanogel swelling, thereby facilitating pH-sensitive drug release. Higher pH levels also destroy PAA–drug hydrogen bonds, further promoting drug release from the system. In vitro cytotoxicity tests indicate that the empty hybrid nanogels have very low cytotoxicity, whereas the TMZ-loaded hybrid nanogels have high anticancer activity.

Despite their impressive potential as future theranostic agents, a major disadvantage of the use of QDs in biomedical applications is their inherent cytotoxicity. Surface oxidation, and the leaching out of heavy metal ions from the core, are serious concerns. However, new strategies to minimize cytotoxicity are being developed. QDs can be associated with other theranostic carriers, to form many other types of multifunctional platforms for imaging and therapy. The association of QDs with liposomes, micelles, and CNTs is described later in this chapter. The combination of the QD with another carrier can significantly reduce the cytotoxicity of the QDs in vivo.

18.4.4 Mesoporous Silica Nanoparticles

A mesoporous material is one that contains pores with diameters between 2 and 50 nm. Mesoporous silica nanoparticles (SiNPs) are typically of size 100–200 nm and are showing considerable promise as theranostic carriers. Silica is generally regarded as a biosafe material and is used clinically in implants. The ultrahigh surface area of mesoporous SiNPs allows for extensive surface modification with targeting ligands and solubilizers; their porous interior enables them to serve as ample reservoirs, both for therapeutic drugs and imaging agents. Surface attachment with hydrophilic groups increases the solubility and also the stability of the NP dispersion in aqueous solutions.

Mesoporous SiNPs, of size 100–200 nm, were doped with SPIONs for magnetic manipulation and MRI, as well as the hydrophobic anticancer drugs camptothecin and paclitaxel (Liong et al. 2008). The surfaces of these particles were functionalized with hydrophilic phosphonate groups, in order to achieve high stability and solubility in aqueous environments. Additionally, the fluorescent dye, fluorescein isothiocyanate, was conjugated onto the surface for optical imaging capabilities, and folic acid groups were attached to the surface to promote uptake by tumor cells.

Cellular uptake studies in two types of pancreatic cancer cell lines indicated that the NPs were internalized within 30 minutes of transfection and exhibited greater toxicity levels than drug-free SiNP controls.

18.4.5 CALCIUM PHOSPHATE NANOPARTICLES

Calcium phosphate nanoparticles (CPNPs) offer a number of advantages as theranostic nanocarriers. A particular advantage over many other synthetic theranostic systems is their nontoxicity: both Ca^{2+} and PO_4^{3-} are found in relatively high concentrations in the body. The formation of CPNPs is a relatively straightforward precipitation reaction, so that encapsulation of the cargo molecules merely necessitates that the molecule of interest is present during NP formation. Additionally, CP is an easily substituted matrix, permitting the inclusion of a broad variety of substitutions, such as organic fluorophores. Fluorescent dyes can be used as a tracking device to follow the fate of the NPs in vivo and give an observable indication of cargo delivery. Encapsulation in the CP matrix can protect the drug cargo in vivo, and the incorporation of tumor-specific targeting ligands can facilitate uptake by cancer cells.

Furthermore, calcium phosphates, regardless of calcium/phosphate ratio, crystallinity, or phase, demonstrate a pH-dependent solubility, being relatively insoluble at physiological pH 7.4, but becoming increasingly soluble below pH 6.5. This pH-tunable solubility has been a major impetus in the development of CPNPs for targeting and controlled-release purposes. CPNPs can dissolve to release their cargo in the acidic environment that commonly prevails in the vicinity of solid tumors, thereby facilitating drug release in the vicinity of the tumor. CPNPs that have been taken into cells and become localized within the acidic endosomal compartment will also dissolve, releasing drug intracellularly. Thus, drugs normally with little or no solubility in physiological liquids can be delivered intracellularly, using CPNPs.

A double reverse-micelle strategy has been developed to prepare stable, nonaggregating, 20 nm CPNPs embedded with fluoroprobes and the small amphiphilic neoplastic drug, ceramide, for simultaneous bioimaging and drug delivery to a range of cell types, including melanoma and breast adenocarcinoma cell lines (Kester et al. 2008). The lifetimes and quantum properties of the fluorescent dye were shown to improve when encapsulated within the CPNPs. Furthermore, ceramide encapsulation was able to induce apoptosis, as measured by an MTS assay, in vitro. In further studies, the CPNPs were functionalized with PEG groups and targeting antibodies (Barth et al. 2011). In vivo imaging using NIR microscopy of the encapsulated indocyanine green fluorophore indicated greater tumor accumulation of targeted NPs than their nontargeted counterparts, as well as significantly better penetration through the blood–brain barrier.

18.4.6 CARBON NANOMATERIALS

Based on their structures, carbon nanomaterials are classified into a variety of categories that include fullerenes, carbon dots, nanodiamonds, carbon nanotubes (CNTs), and graphene. All of these systems have been investigated for applications in theranostics. CNTs are extensively investigated in theranostics and discussed here as an example. Structurally, single-walled carbon nanotubes (SWCNTs) are a single sheet of graphene, rolled seamlessly into a cylinder, while multiwalled carbon nanotubes (MWCNTs) comprise concentrically layered SWCNTs of increasing diameter. CNTs possess a high aspect ratio, with diameters typically in the nanometer range (0.5–3 nm [SWCNTs] and 2–100 nm [MWCNTs]), but lengths that can extend to several micrometers.

The graphite-like structure of CNTs is inert and thus not suitable for most conjugation chemistry. Researchers have applied extreme oxidative conditions to functionalize the surface, which can subsequently be utilized as mounting sites. Various molecules, including peptides and antibiotics, have been conjugated to the CNT surface in this way. CNT surfaces are also modified by both covalent (chemical conjugation of hydrophilic polymers) and noncovalent (via the physical

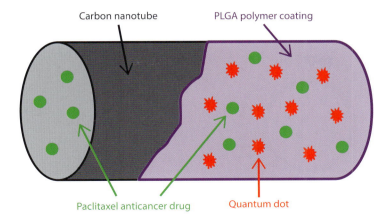

Carbon nanotube   PLGA polymer coating

Paclitaxel anticancer drug   Quantum dot

FIGURE 18.3 Schematic of a carbon nanotube loaded with the anticancer drug paclitaxel and functionalized with a poly(lactide-*co*-glycolide) (PLGA) polymer coating and luminescent quantum dots. (From Guo, Y., Shi, D., Cho, H. et al.: In vivo imaging and drug storage by quantum-dot-conjugated carbon nanotubes. *Adv. Funct. Mater.* 2008. 18. 2489–2497. Copyright Wiley-VCH Verlag GmbH & Co. KGaA. Reproduced with permission.)

adsorption of surfactants, such as sodium dodecylbenzene sulfonate) means, in order to improve their solubility and biocompatibility.

CNTs containing an ultrathin poly(lactide-*co*-glycolide) (PLGA) coating were prepared by a novel plasma polymerization method (Guo et al. 2008). The PLGA coating facilitated the conjugation of amine-containing QDs onto the surface of the CNTs, for in vivo imaging. The CNT-QD nanoconstructs were loaded with the anticancer drug paclitaxel (Figure 18.3) and demonstrated in vitro antitumor efficacy against human PC-3MM2 prostate cancer cells. They were also successfully injected in mice for in vivo optical imaging. Studies indicated predominant CNT-QD uptake in the liver, kidney, stomach, and intestine.

CNTs also demonstrate strong optical absorbance in the NIR region. The emitted photoluminescence can be used for in vivo tumor imaging while also acting as an efficient NIR absorber and heater for photothermal ablation of tumors with a low injection dose. SWNTs are able to increase the local temperature of a tumor to as high as 60°C within less than 5 minutes of 808 nm laser irradiation. Compared to other thermal ablation platforms, such as AuNPs, SWNTs can effectively eliminate tumors at 10-fold lower doses and at lower irradiation powers (Moon et al. 2009).

The primary hindrance to employing CNTs for in vivo biomedical applications arises from potential toxicity issues, which include oxidative stress and inflammatory pathways (Rothen-Rutishauser 2010). A considerable research effort is being directed toward methods to reduce toxicity, including the use of polymeric coatings on the CNT surface.

18.5 ORGANIC-NANOSIZED THERANOSTICS

18.5.1 LIPOSOMES

Liposomes (described extensively in Chapter 5, Section 5.5.1) consist of closed, spherical vesicles, composed of one or more lipid bilayers, surrounding an aqueous core (Figure 18.4a; see also Figure 5.9). With a typical size range of approximately 80–200 nm, they provide ample cargo room for the incorporation of both drug molecules and diagnostic agents. Hydrophilic molecules can be accommodated within the aqueous liposomal core, while hydrophobic species can be associated with the lipid bilayers. Liposomes are widely used and highly successful drug carriers; commercially available liposomal drug delivery systems include AmBisome®, DaunoXome®, and Doxil®. Liposomes can also be loaded with a variety of imaging agents for MR, nuclear, and fluorescence imaging applications.

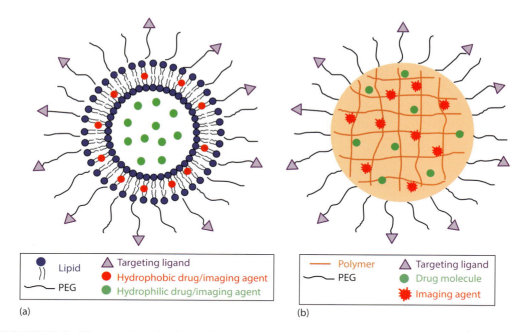

(a)

(b)

FIGURE 18.4 Therapeutic molecules and imaging agents can be simultaneously encapsulated into (a) liposomes or (b) polymeric nanoparticles. Depending on their amphiphilicity, the drug and imaging agents can be encapsulated within the aqueous core or the hydrophobic bilayer of liposomes. The particle surface of both lipid and polymeric particles can be conjugated with ligands, for targeted delivery, and with PEG, for increased circulation time in the bloodstream (PEG, polyethylene glycol).

Liposomes offer the ability to protect the encapsulated agents from harsh external environments, to prolong their systemic circulation lifetime, and to be functionalized with various targeting ligands for cell- or tissue-specific delivery. Long-circulating "stealth" constructs can be prepared by covalently attaching PEG chains to the liposomal surface.

Although still at an early stage of development, some examples serve to demonstrate the potential of liposomes as theranostic agents. A PEGylated cholesterol/DOPE/DSPC liposome was used as a theranostic carrier for Dox and IONPs as MRI contrast agents (Erten et al. 2010). In this formulation, Dox was loaded into a dextran hydrogel placed between the iron oxide core and the lipid shell. The liposomes were studied in mice bearing lung and pancreatic carcinoma xenografts and demonstrated enhanced MRI contrast. Furthermore, the IONPs also served as a heating source upon the exposure of alternative magnetic field, providing a platform for localized hyperthermia-induced drug release, in addition to their role as an MRI contrast agent (Tai et al. 2009).

Liposome-based theranostic carriers can also be used for radionuclide imaging. A PEGylated liposome formulation was loaded with a chemotherapeutic agent and a copper radionuclide (^{64}Cu) suitable for PET imaging (Petersen et al. 2011). The pharmacokinetics of this agent was studied in flank HT29 colon tumors. The accumulation of ^{64}Cu liposomes in the tumors was several orders of magnitude higher than that of the radionuclide bound to a chelating agent (^{64}Cu-DOTA). In addition, the ^{64}Cu liposomes were able to remain in the blood for over a 24-hour period of time, indicating the long-circulating nature of the PEGylated NPs, in comparison to their nonliposomal counterparts.

QDs have been associated with liposomal carriers, for use in fluorescence imaging. The fabrication and characterization of a liposomally encapsulated QD-FRET probe for monitoring the enzymatic activity of phospholipase A2 has been described (Kethineedi et al. 2013). To fabricate the probes, luminescent QDs capped with trioctylphosphine oxide (TOPO) ligands were incorporated into the lipid bilayer of unilamellar liposomes (average diameter approximately 100 nm). Incorporating TOPO-capped QDs in liposomes enabled their use in aqueous solution while maintaining their hydrophobicity and excellent photophysical properties. The probes were able to detect very low amounts of phospholipase A2 and to monitor enzyme activity in real time.

18.5.2 POLYMERSOMES

Liposomes are metastable assemblies, in which the lipid molecules can migrate between the inner and outer layers of the lipid membrane. This may result in leakiness of their core contents, in addition to liposomal fusion and fission, which are all sources of liposomal polydispersity and instability. Polymeric liposomes, or polymersomes, are a more recent variation of liposome carriers, which use amphiphilic synthetic block copolymers to form the vesicle membrane. The polymer chains in polymersomes can be tailored to enable cross-linking within the construct, in order to eliminate the "flip-flop" migration between bilayers that can occur in liposomes. Desirable functional groups can also be incorporated into the polymer chains, for the subsequent conjugation of drug molecules and imaging agents.

Spherical and wormlike polymersomes have been studied for the combined delivery of an anti-cancer drug and an MRI contrast agent (Yang et al. 2010a,b). Heterobifunctional asymmetric tri-block copolymers were synthesized, comprising folate-PEG5000-b-poly(glutamate hydrazone doxorubicin)-b-PEG2000-acrylate and folic acid-PEG114-b-PLA-b-PEG46-acrylate. These formed a self-assembling vesicular polymersome structure, with inner and outer hydrophilic PEG layers. The long PEG segments were mostly segregated in the outer hydrophilic leaflet of the membrane, whereas the shorter PEG block was mostly present in the inner hydrophilic layer. The polyglutamate segment formed the hydrophobic membrane of the vesicles, to which the chemotherapeutic agent Dox was conjugated, via a pH-sensitive hydrazone bond, in order to achieve pH-responsive drug release. Targeting was achieved by conjugating folic acid ligands onto the outer PEG segment. The polymersomes were stabilized by cross-linking the polymeric methacrylate block in the interior of the membrane bilayer. Hydrophilic SPIONs were encapsulated into the aqueous core, to allow for MRI detection. The polymersomes showed higher cellular uptake by HeLa cells than folate-free polymersomes, due to the folate receptor–mediated endocytosis process. Accordingly, Dox-loaded polymersomes exhibited higher cytotoxicity than the folate-free controls and higher contrast for MRI.

18.5.3 POLYMERIC NANOPARTICLES

A wide range of polymeric NPs have been investigated as theranostic carriers (Figure 18.4b). Polymeric NPs have been described extensively in Chapter 5 (Section 5.5.5). Many different polymers can be used in the preparation of NPs. Both the polymer cores and the NP surface can be loaded with a variety of therapeutic or imaging agents. To ensure stability while minimizing immunogenicity, polymeric NPs can be shielded by stealth materials, much like their lipid counterparts. Targeting moieties can be conjugated to the constructs, to enhance tumor uptake. Sustained and controlled release of these agents can be achieved by surface or bulk erosion, diffusion through the polymer matrix, swelling followed by diffusion, or stimulation by the local environment.

Chitosan and cyclodextrin polymers are derived from natural resources and have received extensive attention due to their biocompatibility. Theranostic chitosan NPs have been prepared, carrying both the NIR dye Cy5.5 for live imaging, and the hydrophobic cytotoxic drug paclitaxel, for cancer treatment (Na et al. 2011). The hydrophobicity of water-soluble glycol chitosan was increased by chemically modifying the polymer with 5β-cholanic acid so that the polymer would form NPs in the water and encapsulate the hydrophobic drug.

A significant amount of research is also ongoing in the design of polymeric NPs using synthetic polymers that are biocompatible and biodegradable. The most studied polymers are those based on PLGA. Ling et al. (2011) prepared theranostic multifunctional PLGA NPs, loaded with Dox and IONPs for MRI contrast. Drug-release experiments demonstrated the efficacy of the PLGA NPs to release the drug in a controlled manner. PC3 cells treated with this system, bearing a targeting moiety, were characterized by higher intracellular iron concentration and stronger MRI contrast effects, in comparison to cells that were treated with nontargeted analogs.

18.5.4 MICELLES

Polymeric micelles are prepared using block copolymers and consist of an inner hydrophobic core and an outer hydrophilic, biocompatible, shell (see also Chapter 5, Section 5.5.6). Various types of hydrophobic therapeutic and imaging functionalities can be introduced into the core of the structure, while the hydrophilic shell confers biocompatibility, stability, and extended circulation times. Multifunctional polymeric micelles have been developed as theranostic agents, composed of PEG-poly(glutamic acid), which incorporate Gd-based MRI contrast agents and platinum anticancer drugs in their core. These micelles were found to localize in the interior of an orthotopic pancreatic lesion in a mouse model (Kaida et al. 2010).

A targeted polymeric micellar platform was developed, composed of amphiphilic maleimide- and methoxy-terminated PEG-poly(lactide) block copolymers (MAL-PEG-PLA and MPEG-PLA) as carriers for the cytotoxic Dox, and SPIONs as contrast agents for MRI. The arginine-glycine-aspartic acid (RGD) tripeptide, which targets an integrin receptor, was used as a targeting ligand. A thiol–maleimide reaction scheme was used to functionalize the micelles with the $\alpha_v\beta_3$ integrin receptor–targeting cRGD peptide. Different levels of cRGD loading were achieved by controlling the amount of MPEG-PLA introduced into the system. Conjugating the targeting peptide onto the particle was able to increase cellular uptake, MRI contrast, and Dox-induced cytotoxicity in vitro (Nasongkla et al. 2006).

QD-loaded micelles have been developed to increase the biocompatibility and stability of these optical agents. In one study, a hydrophobic lipid (10,12-pentacosadiynoic acid [PCDA]) was incorporated into the micellar structure in order to facilitate drug encapsulation (Nurunnabi et al. 2010). In addition, this lipid was able to create a strong outer shell, due to its polymerizable capabilities upon UV irradiation. Both PEG-PCDA and PEG-herceptin conjugates were incorporated into the final cross-linked micellar carrier. By enabling better uptake and retention in the cancerous lesion, the QD-loaded micelles exhibited enhanced tumor activity and selective toxicity, yielding a significant reduction in tumor volume.

18.6 MULTIFUNCTIONAL-NANOSIZED THERANOSTICS FOR PHOTODYNAMIC THERAPY

Various references have been made previously to the use of magnetic hyperthermia (via SPIONs) and photothermal ablation (via AuNPs and CNTs), as methods to induce localized and specific cell death. A further approach, known as photodynamic therapy, involves the delivery of a photosensitizer (PS) to tumor tissues, followed by their irradiation with a laser of appropriate wavelength. Upon irradiation, activated PS convert molecular oxygen to toxic singlet oxygen and free radicals (reactive oxygen species), which induce apoptosis, cell death, and tissue destruction. Nontarget toxicity can be prevented with this technique because the activation of cytotoxic species only occurs at the site of illumination, which allows the therapy to be localized. A number of different theranostic carriers have been developed utilizing this approach, in which multifunctional platforms are prepared, containing a PS (as the therapeutic agent), an imaging agent, and, optionally, other components, such as targeting ligands, and stealth coatings.

Porphyrin derivatives are the most commonly used PS in PDT. A PDT theranostic agent for the imaging and treatment of brain tumors has been developed, using the PS Photofrin®, a complex mixture of porphyrin oligomers (Kopelman et al. 2005). The PS was incorporated into a PAA core, along with MRI contrast materials. The NP was further functionalized with PEG groups and RGD-targeting peptides. The major advantages of encapsulating Photofrin® with the NPs are (1) protection of the PS from degradation in vivo, (2) reduced cutaneous photosensitivity posttreatment, and (3) the waiting time between i.v. injection of the PS and subsequent laser irradiation is greatly reduced. In vivo pharmacokinetic behavior, studied in a gliosarcoma rat model by diffusional MRI, revealed a 50-fold increase in NP circulation half-life after PEGylation of the NPs. In addition, a significant increase in the diffusion coefficient of the water surrounding the tumor cells was observed, indicating a decrease in tumor growth following treatment. The PDT-Photofrin-PAA NPs were able to effectively kill engrafted brain tumors in rats within 5 minutes of light exposure.

Multifunctional NPs have been described consisting of a gold–silver nanocage core, surrounded by a silica shell containing the NIR PS, Yb-2,4-dimethoxyhematoporphyrin (Yb-HP). This PS allowed monitoring of tumor growth, as well as simultaneously administering therapy by PDT and plasmonic heating (Khlebtsov et al. 2011). Significant death of HeLa cervical cancer cells occurred in vitro when they were incubated with the NPs and irradiated with light. This was due to both the plasmonic photothermal heating effects of the gold–silver nanocages, in addition to the photodynamic effects of the Yb-HP.

PEGylated poly-(L-glutamic acid) conjugates containing a PS, as well as a Gd-based MRI contrast agent, have been developed (Vaidya et al. 2008). The MRI contrast agent provided both image guidance for the precise application of laser irradiation at the target site and noninvasive assessment of the therapeutic efficacy of PDT. MRI images revealed that PEGylated constructs achieved longer blood circulation times, lower liver uptake, and greater tumor accumulation than their non-PEGylated counterparts. Furthermore, PDT-treated animals that had been administered the PEGylated conjugates showed greater tumor growth inhibition.

A targeted PTD agent with a built-in apoptosis sensor (TaBIAS) has been designed, which both triggers and images apoptosis in cancer cells (Stefflova et al. 2006). The nanostructure consists of four components (Figure 18.5): (1) a PS, pyropheophorbide a, which localizes near mitochondria; (2) a fluorescence quencher Black Hole Quencher-3, which quenches the PS's fluorescence; (3) an

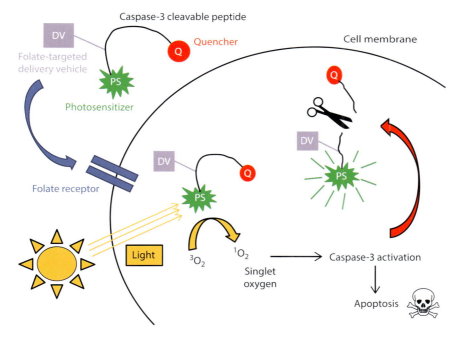

FIGURE 18.5 Schematic of the structure and function of a targeted photodynamic therapy theranostic agent with a built-in apoptosis sensor. This construct consists of a photosensitizer, a caspase-3 cleavable peptide sequence, a fluorescence quencher, and a folate-targeting vehicle. The construct accumulates preferentially in cells overexpressing folate receptors. Once activated by light, the PS produces singlet oxygen to trigger apoptosis. In the process, this leads to activation of caspase-3, which cleaves the peptide linker between the PS and the quencher, restoring the fluorescence of the PS and thus identifying those cells undergoing apoptosis using NIR fluorescence imaging. (Reprinted with permission from Stefflova, K., Chen, J., Marotta, D. et al., Photodynamic therapy agent with a built-in apoptosis sensor for evaluating its own therapeutic outcome in situ, *J. Med. Chem.*, 49(13), 3850–3856. Copyright 2006 American Chemical Society.)

enzyme-cleavable peptide linker (susceptible to the enzyme caspase-3), with the PS and quencher attached at opposing ends; and (4) a folate delivery vehicle, for targeting cancer cells that over-express the folate receptor. This nanocarrier enters cells via a folate delivery pathway, and when activated by light, the PS produces singlet oxygen molecules that destroys the mitochondrial membrane and triggers apoptosis through the activation of caspase-3. Activated caspase-3 also cleaves the peptide linker between the PS and the quencher, restoring the PS's intrinsic fluorescence and thus indicating those cells dying by apoptosis. Thus, the TaBIAS both induces apoptosis and also visualizes the event, using its own NIR fluorescence.

Mice bearing both negative and positive folate receptor tumors on contralateral sides were injected intravenously with the TaBIAS. A significantly greater post-PDT increase in fluorescence was observed in the folate receptor positive side of the tumor, confirming the targeting and apoptotic-reporting function of the photo-triggered theranostic agent.

18.7   CONCLUSIONS AND FUTURE OUTLOOK

Nanotechnology offers a variety of new materials for theranostics. As described in this chapter, various nanosized theranostic agents are being investigated, to combat life-threatening human diseases. However, despite this encouraging progress, there is currently no commercially available NP-based theranostic agent. Although many NP platforms have shown promise and present a number of advantages, extensive preclinical studies are needed to demonstrate their safety and efficacy. Some of the nanosized materials also suffer from a number of specific problems that are still without proper solutions. A particular issue is the biocompatibility of many of these synthetic constructs, as well as their limited biodegradation and clearance in vivo; this will remain a significant challenge, particularly for clinical translation and FDA regulatory approval. For example, most inorganic nanosized materials are nonbiodegradable and have a long retention half-life in the body. Although preclinical studies have shown these materials have low cytotoxicity and an absence of acute toxic side effects, prolonged retention in the body may induce toxic side effects many years later.

Current research activities in theranostics have placed a considerable emphasis on the design of ever-increasingly sophisticated and complex structures. However, the field would also benefit considerably from addressing toxicological and biocompatibility issues, as well as the relevant pharmacokinetic hurdles (absorption, distribution, metabolism, and elimination) of these agents. In particular, their degradation and clearance in vivo need to be resolved, in order to move forward and translate these systems into commercial products. Furthermore, the development of cost-effective synthesis of theranostics would accelerate the clinical translation of this new therapeutic paradigm.

Many nanosized theranostics are designed to target malignant tumors. A significant challenge in cancer treatment is tumor heterogeneity. A goal of theranostics is to help physicians identify the populations responding to the available therapies and find new therapies for the nonresponding populations. The design of safe and effective new theranostic modalities that could detect and differentiate tumor aggressiveness earlier, identify tumor responsiveness to therapies, and guide and inform therapeutic optimization, would bring us closer to the goal of curing life-threatening human diseases.

REFERENCES

Bagalkot, V., L. Zhang, E. Levy-Nissenbaum et al. 2007. Quantum dot-aptamer conjugates for synchronous cancer imaging, therapy, and sensing of drug delivery based on bi-fluorescence resonance energy transfer. *Nano Letters* 7:3065–3070.

Barth, B.M., E.I. Altinoğlu, and S.S. Shanmugavelandy. 2011. *In vivo* photodynamic therapy of leukemia. *ACS Nano* 5(7):5325–5337.

Chen, Q., K. Li, and S. Wen. 2013. Targeted CT/MR dual mode imaging of tumors using multifunctional dendrimer-entrapped gold nanoparticles. *Biomaterials* 34(21):5200–5209.

Erten, A., W. Wrasidlo, M. Scadeng et al. 2010. Magnetic resonance and fluorescence imaging of doxorubicin-loaded nanoparticles using a novel *in vivo* model. *Nanomedicine* 6:797–807.

Guo, Y., D. Shi, H. Cho et al. 2008. *In vivo* imaging and drug storage by quantum-dot-conjugated carbon nanotubes. *Adv Funct Mater* 18:2489–2497.

Hayashi, K., M. Nakamura, W. Sakamoto et al. 2013. Superparamagnetic nanoparticle clusters for cancer theranostics combining magnetic resonance imaging and hyperthermia treatment. *Theranostics* 3(6):366–376.

Ho, Y.-P. and K.W. Leong. 2010. Quantum dot-based theranostics. *Nanoscale* 2(1):60–68.

Jang, B., S. Park, S.H. Kang et al. 2012. Gold nanorods for target selective SPECT/CT imaging and photothermal therapy *in vivo*. *Quant Imaging Med Surg* 2(1):1–11.

Kaida, S., H. Cabral, M. Kumagai et al. 2010. Visible drug delivery by supramolecular nanocarriers directing to single-platformed diagnosis and therapy of pancreatic tumor model. *Cancer Res* 70(18): 7031–7041.

Kester, M., Y. Heakal, A. Sharma et al. 2008. Calcium phosphate nanocomposite particles for *in vitro* imaging and encapsulated chemotherapeutic drug delivery to cancer cells. *Nano Letters* 8(12):4116–4121.

Kethineedi, V.R., G. Crivat, M.A. Tarr et al. 2013. Quantum dot-NBD-liposome luminescent probes for monitoring phospholipase A2 activity. *Anal Bioanal Chem* 405(30):9729–9737.

Khlebtsov, B., E. Panfilova, V. Khanadeev et al. 2011. Nanocomposites containing silica-coated gold-silver nanocages and Yb-2,4-dimethoxyhematoporphyrin: Multifunctional capability of IR-luminescence detection, photosensitization, and photothermolysis. *ACS Nano* 5(9):7077–7089.

Kohler, N., C. Sun, J. Wang et al. 2005. Methotrexate-modified superparamagnetic nanoparticles and their intracellular uptake into human cancer cells. *Langmuir* 21:8858–8864.

Kopelman, R., Y.-E.L. Koo, M. Philbert et al. 2005. Multifunctional nanoparticle platforms for *in vivo* MRI enhancement and photodynamic therapy of a rat brain cancer. *J Magn Magn Mater* 293(1):404–410.

Ling, Y., K. Wei, Y. Luo et al. 2011. Dual docetaxel/superparamagnetic iron oxide loaded nanoparticles for both targeting magnetic resonance imaging and cancer therapy. *Biomaterials* 32:7139–7150.

Liong, M., J. Lu, M. Kovochich et al. 2008. Multifunctional inorganic nanoparticles for imaging, targeting, and drug delivery. *ACS Nano* 2:889–896.

Maeda, H., Y. Matsumoto, T. Konno et al. 1984. Tailor-making of protein drugs by polymer conjugation for tumor targeting: A brief review on Smancs. *J Protein Chem* 3:181–193.

Melancon, M.P., W. Lu, Z. Yang et al. 2008. *In vitro* and *in vivo* targeting of hollow gold nanoshells directed at epidermal growth factor receptor for photothermal ablation therapy. *Mol Cancer Ther* 7:1730–1739.

Moon, H.K., S.H. Lee, and H.C. Choi. 2009. *In vivo* near-infrared mediated tumor destruction by photothermal effect of carbon nanotubes. *ACS Nano* 3(11):3707–3713.

Na, J.H., H. Koo, S. Lee et al. 2011. Real-time and non-invasive optical imaging of tumor-targeting glycol chitosan nanoparticles in various tumor models. *Biomaterials* 32(22):5252–5261.

Nasongkla, N., E. Bey, J. Ren et al. 2006. Multifunctional polymeric micelles as cancer-targeted, MRI-ultrasensitive drug delivery systems. *Nano Letters* 6(11):2427–2430.

Nurunnabi, M., K.J. Cho, J.S. Choi et al. 2010. Targeted near-IR QDs-loaded micelles for cancer therapy and imaging. *Biomaterials* 31(20):5436–5444.

Petersen, A.L., T. Binderup, P. Rasmussen et al. 2011. (64)Cu loaded liposomes as position emission tomography imaging agents. *Biomaterials* 32:2334–2341.

Rastogi, R., N. Gulati, R.K. Kotnala et al. 2011. Evaluation of folate conjugated pegylated thermosensitive magnetic nanocomposites for tumor imaging and therapy. *Colloids Surf B Biointerfaces* 82(1):160–167.

Resch-Genger, U., M. Grabolle, S. Cavaliere-Jaricot et al. 2008. Quantum dots versus organic dyes as fluorescent labels. *Nat Methods* 5:763–775.

Rothen-Rutishauser, B., D.M. Brown, M. Piallier-Boyles et al. 2010. Relating the physicochemical characteristics and dispersion of multiwalled carbon nanotubes in different suspension media to their oxidative reactivity in vitro and inflammation in vivo. *Nanotoxicology*. 4:331–342.

Stefflova, K., J. Chen, D. Marotta et al. 2006. Photodynamic therapy agent with a built-in apoptosis sensor for evaluating its own therapeutic outcome *in situ*. *J Med Chem* 49(13):3850–3856.

Tai, L.A., P.J. Tsai, and Y.C. Wang. 2009. Thermosensitive liposomes entrapping iron oxide nanoparticles for controllable drug release. *Nanotechnology* 20:135101.

Vaidya, A., Y. Sun, Y. Feng et al. 2008. Contrast-enhanced MRI-guided photodynamic cancer therapy with a pegylated bifunctional polymer conjugate. *Pharm Res* 25(9):2002–2011.

Wu, W., M. Aiello, T. Zhou et al. 2010. *In-situ* immobilization of quantum dots in polysaccharide-based nanogels for integration of optical pH-sensing, tumor cell imaging, and drug delivery. *Biomaterials* 31(11):3023–3031.

Yang, X., J.J. Grailer, I.J. Rowland et al. 2010a. Multifunctional stable and pH-responsive polymer vesicles formed by heterofunctional triblock copolymer for targeted anticancer drug delivery and ultrasensitive MR imaging. *ACS Nano* 4:6805–6817.

Yang, X., J.J. Grailer, I.J. Rowland et al. 2010b. Multifunctional SPIO/DOX-loaded wormlike polymer vesicles for cancer therapy and MR imaging. *Biomaterials* 31:9065–9073.

FURTHER READING

Ahmed, N., H. Fessi, and A. Elaissari. 2012. Theranostic applications of nanoparticles in cancer. *Drug Discov Today* 17(17/18):928–934.

Choi, H.S. and J.V. Frangioni. 2010. Nanoparticles for biomedical imaging: Fundamentals of clinical translation. *Mol Imaging* 9:291–310.

Kim, K., J.H. Kim, H. Park et al. 2010. Tumor-homing multifunctional nanoparticles for cancer theragnosis: Simultaneous diagnosis, drug delivery, and therapeutic monitoring. *J Control Release* 146:219–227.

Krasia-Christoforou, T. and T.K. Georgiou. 2013. Polymeric theranostics: Using polymer-based systems for simultaneous imaging and therapy. *J Mater Chem B* 1:3002–3025.

Namiki, Y., T. Fuchigami, N. Tada et al. 2011. Nanomedicine for cancer: Lipid-based nanostructures for drug delivery and monitoring. *Acc Chem Res* 44(10):1080–1093.

Rai, P., S. Mallidi, X. Zheng et al. 2010. Development and applications of photo-triggered theranostic agents. *Adv Drug Deliv Rev* 62:1094–1124.

Rizzo, L.Y., B. Theek, and G. Storm. 2013. Recent progress in nanomedicine: Therapeutic, diagnostic and theranostic applications. *Curr Opin Biotechnol* 24(6):1159–1166.

Svenson, S. 2013. Theranostics: Are we there yet? *Mol Pharm* 10:848–856.

Wang, L.-S., M.-C. Chuang, and A. Ho. 2012. Nanotheranostics: A review of recent publications. *Int J Nanomed* 7:4679–4695.

Yu, M.K., J. Park, and S. Jon. 2012. Targeting strategies for multifunctional nanoparticles in cancer imaging and therapy. *Theranostics* 2(1):3–44.

19 Nanofabrication Techniques and Their Applications in Drug Delivery

Erica Schlesinger, Cade Fox, and Tejal Desai

CONTENTS

19.1 INTRODUCTION

The National Nanotechnology Initiative defines nanotechnology as any technology conducted or containing features at 1–100 nm. Nanodevices are systems designed on the micro- or nanoscale, with nanoscale features. When attempting to design features in the nanometer range, not only is fabrication a challenge, but verification and visualization of the results is a hurdle in itself. As this chapter will discuss, both micro- and nanoscale features offer a range of possibilities for improving medical technologies and drug delivery, but it is only with advances in fabrication technologies and analytical techniques that we are able to design, control, and characterize these features.

Nanoscale features confer unique and tunable properties to materials that can be leveraged by scientists and engineers to design materials and devices with novel properties and functions. For example, relative to their mass, materials with nanoscale features have much higher surface areas and a greater potential for surface interactions and for significant intermolecular forces than standard materials. When considering bulk material properties, physical and chemical properties are independent of size; aluminum will have the same material properties whether it is 1 in.^{2} or 1 ft^2.

However, nanosized particles exhibit properties such as electrical conductivity, chemical reactivity, and melting point, among others, which can depend on the particle size. Scientists can leverage these size dependencies to fine-tune material properties by controlling particle size.

In drug delivery and targeting, a multitude of barriers must be overcome to successfully deliver a therapeutic to its site of action. While the physicochemical properties and intended target of a drug often dictate these challenges, drug delivery systems (DDS) mitigate these obstacles to improve efficacy and bioavailability. Through their unique properties, micro- and nanofeatures bolster drug delivery systems by increasing membrane permeation, improving mucoadhesion, controlling drug release, and minimizing immune response. Because molecular and cellular biology occurs primarily at the nanoscale, utilizing nanoscale features allows scientists to interact precisely and directly with key components in biological systems. For example, nanopillar arrays can interact with individual cells to change cell morphology and increase drug permeability through a cell layer.

In this chapter, we will describe top-down micro- and nanofabrication techniques, outline some challenges in drug delivery that are being addressed through micro- and nanofabrication, and give examples of some specific applications of nanotechnology to drug delivery.

19.2   TOP-DOWN MICRO- AND NANOFABRICATION TECHNIQUES

Here we describe the fabrication and analytical technologies that have made manipulation and measurement at the nanoscale possible. Micro- and nanofabrication leverage materials and techniques originating in the microelectronics field to offer new approaches for overcoming traditional challenges in drug delivery. As a "top-down" process, micro- and nanofabrication provides a high level of control over size, shape, and surface features. Modifications of basic techniques such as lithography, molding, and extrusion allow for micro- and nanoscale manipulation of a range of materials, including polymers, as well as metals and metal oxides. The development of new fabrication techniques and materials with novel properties continues to expand the application of micro- and nanofeatures in drug delivery systems, but here we will cover only the basic fabrication techniques that are most relevant to creating nanotechnologies for drug delivery.

19.2.1   SOLVENT CASTING

Solvent casting is a technique to form uniform layers of a material, most often a polymer, on a surface. The general idea behind solvent casting is to dissolve the material of interest in an organic solvent, to cast the solution into the desired shape, and to allow the solvent to evaporate, leaving behind the material of interest in its solid form. This technique typically works best with materials that in their solid state have structural integrity and using volatile solvents. A key component is finding a compatible solvent for the material of interest. Solvent casting is utilized in micro- and nanolithography in a number of ways. It is often used in conjunction with spin-coating systems to create uniform thin films. Spin-coating systems use centrifugal force to coat a substrate with a solution, by spinning the substrate. Solvent casting can also be used to fill molds to form materials into specific shapes. Templates and molds, especially on the micro- and nanoscale, can be combined with spin-coating systems, to create thin films with topographies defined by the template or mold over which the solution is cast.

19.2.2   ETCHING

Etching is a process by which a material is removed or degraded. When controlled, etching is a powerful technique in fabrication processes to selectively remove materials, leaving protected areas of the same material, or compatible materials, behind. There are a number of different types of etching that are important in micro- and nanofabrication, including wet etching, photoetching, and reactive-ion etching. The type of etching is defined by how the material is being removed. In wet etching, a material is in contact with a solution that either dissolves the material or chemically attacks it, causing it to degrade. In photoetching, material is removed by exposure to UV light, and in reactive-ion etching, chemically reactive plasma is generated by an electromagnetic field that when directed at the substrate surface, removes the material that it contacts. Etching is most useful when combined with lithographic techniques that protect regions and patterns of a material, leaving only the exposed areas for etching. Etching can also be useful in fabrication processes to remove sacrificial templates or scaffolds. The etching process can be controlled by the duration and strength of exposure to the etchant.

19.2.3 Lithography

Lithography was first developed as a printing technique in the late 1700s, that involved creating a template stamp (by creating a grease-based image on lithographic limestone and subsequently etching the stone that was not protected by the grease-based image), which could then be used to transfer ink onto paper. In micro- and nanofabrication, lithography typically refers to photolithography, one of the most commonly used techniques.

Photolithography uses a similar approach to print lithography, but with different materials and more precision (Figure 19.1). Photolithography is made possible by the development of photopatternable materials called photoresists. Photoresists are polymeric materials that cross-link, polymerize, or cleave, when exposed to UV light. Photoresists that polymerize or cross-link upon UV exposure are called negative photoresists because the areas exposed to UV light are the areas that are retained in the final template, leaving the inverse of the UV-blocking photomask. Positive photoresists are degraded with UV exposure, leaving only the areas of photoresist that were blocked from exposure by the photomask pattern.

To create patterns using photoresists, exposure to UV radiation is limited by a micropatterned photomask with the desired features. Photomasks can be designed for either positive or negative photoresists to achieve the desired patterns. For example, when using a negative photoresist, the desired pattern will be clear with all other areas blocked out, and for a positive photoresist, the desired pattern will be blocked out, with all other areas transparent.

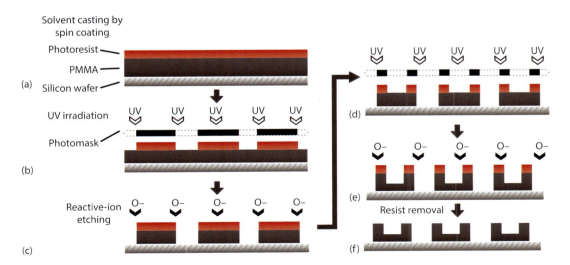

FIGURE 19.1 The photolithography process. (a) A polymer substrate layer (poly(methyl methacrylate) [PMMA]) and a positive photoresist layer are solvent cast through spin coating onto a support substrate. (b) Photoresist is exposed to ultraviolet irradiation with a photomask in place to block PMMA degradation. (c) Reactive-ion etching is used to etch away exposed PMMA substrate. (d, e) A second photomask and etching step are used to get more advanced features. (f) Finally, the photoresist is removed leaving just the polymer substrate with the desired pattern.

The photoresist pattern can be the final template, or the photoresist can be used to transfer the pattern to an underlying substrate in a subsequent etching step. In the latter case, the photoresist is first coated over the substrate of interest before UV exposure. The negative pattern of the photoresist (i.e., the areas where the photoresist is absent) allows the underlying substrate to be exposed for etching, resulting in a final pattern on the substrate that reflects the positive photoresist pattern from the areas blocked to etching by the photoresist. Lithography is often applied in a layer-by-layer approach, leveraging solvent casting and spin coating to deposit each layer and create complex patterns of multiple materials. For example, after depositing and lithographically patterning one material onto a substrate, a second material may be deposited through solvent casting on top of the first layer and lithographically patterned with a different pattern. In such a way, one is able to produce hierarchical features with multiple materials.

Photolithography is the basis for many microfabrication procedures and leverages additional techniques such as solvent casting and etching in the process. Depending on the materials being used and the precision of the masks, photolithography can be applied in various approaches to create a breadth of designs, structures, and devices, limited only by the creativity of the designer. The resolution of photolithography is limited by the wavelength of the light used to treat the photoresist. While standard photolithography can achieve at best 1 μm resolution, special light sources such as deep ultraviolet light can improve resolution to as small as 50 nm.

While photolithography allows the designer precise control over shapes and sizes, it can only be applied to a flat substrate and is not flexible enough to create geometric shapes that are not flat on the ends, such as spheres or arches. A further limitation of photolithography is the need for a cleanroom environment, free of particles and dust.

While photolithography is the most commonly used lithographic technique in micro- and nanofabrication, there are a number of other lithographic approaches, and modifications of existing approaches, that are also being used. Techniques such as multibeam interference lithography, probe lithography, electron-beam lithography, and nanoimprint lithography complement photolithography, to increase patterning capabilities by expanding the range of materials that can be patterned, as well as the sizes and shapes of patterns that can be achieved.

19.2.4 MOLDING

Molding is a technique that uses a stamp or a mold to transfer a pattern onto a surface. There are a number of different molding techniques utilized in micro- and nanofabrication. Hard-pattern molding uses a micro- or nanopatterned template as a stamp or mold for repeated patterning of polymer surfaces. Micro- and nanofeature hard mold templates are created by using other techniques such as reactive-ion etching or electron-beam lithography to transfer a pattern into a hard substrate. A monomer, polymer, or prepolymer is then applied to the mold and cross-linked or cured. There are several techniques such as nanoimprint lithography and step–flash imprint lithography that use hard-pattern molds to transfer patterns in the lithographic process.

Soft-pattern molding, also known as soft lithography, uses a micropatterned elastomeric material as the stamp or mold. Replica molding transfers a patterned master to polydimethylsiloxane (PDMS) and then solidifies a photocurable or thermally curable prepolymer against the PDMS mold to produce a replica of the original master. Solvent-assisted micromolding uses a solvent to swell or dissolve a polymer against a PDMS mold. Upon evaporation of the solvent, the polymer substrate solidifies around the PDMS mold and is thus inversely patterned with the mold features.

Hard-pattern and soft-pattern moldings are the broadest categories of molding techniques, but in recent years a number of more specialized molding techniques have been developed to overcome challenges in specific applications. For example, in order to prevent thermal degradation of drugs incorporated directly into polymer devices, a low-temperature vacuum-molding technique was developed to cure the molded polymer without the need for elevated temperatures.

The approaches introduced here are not an exhaustive representation of fabrication techniques used in micro- and nanofabrication, but rather an overview of the basic techniques that are critical to fabrication processes. Many specialized and novel techniques are emerging based upon these basic approaches. Most advanced fabrication procedures do not simply utilize a single technique but rather combine approaches in multistep processes to create complex systems. In Section 19.4, we will discuss specific examples of micro- and nanotechnologies in drug delivery that utilize these approaches in their fabrication processes.

19.3   NANOSCALE CHARACTERIZATION

A systematic approach to developing nanodevices requires not only the ability to create and control structures at the nanoscale but also the power to measure and characterize these features. Whether the aim is to demonstrate improved drug delivery or to gain a mechanistic understanding of the impact of nanofeatures, we must be able to verify the nanofeatures that we have created, their dimensions, and their properties. While standard microscopy is not powerful enough for nanoscale resolution, techniques such as scanning electron microscopy (SEM) or atomic force microscopy (AFM) provide means for directly observing and measuring our nanofeatures. Because nanodevices are often composed of micro or even macro components in combination with nanofeatures, other analytical techniques such as profilometry for microscale measurements are also critical in nanodevice technology.

19.3.1   SCANNING ELECTRON MICROSCOPY

Because of the wavelike nature of light, a light microscope can magnify an image up to, at best, 1500×. While this may be adequate to make out micron-sized features, it falls short of the resolution needed to visualize nanofeatures. The SEM, however, is able to magnify up to 1 million times, which is sufficient to resolve features of the order of a single nanometer. It scans the sample using a beam of electrons, which excite atoms in the sample, producing secondary electrons that are emitted by the excited atoms. The quantity of secondary electrons produced depends on the angle between the surface of the sample and the scanning electron beam. By detecting and analyzing the secondary electrons produced, relative to the position of the scanning beam, an image is produced.

In order to ensure consistent and adequate electron excitation, nonconductive surfaces are coated with a layer of conductive metal such as iridium or gold prior to imaging. This layer is thin enough so as to not obstruct any surface features of the underlying sample but also provides consistent and adequate electron density for excitation. However, once coated and imaged by SEM, samples cannot be recovered for use. SEM is a powerful tool for visualizing micro- and nanofeatures as well as for quantifying dimensions through comparison to scale bars that accompany the magnified images (Figure 19.2).

19.3.2   ATOMIC FORCE MICROSCOPY

Not commercially available until the end of the 1980s, AFM is capable of subnanometer resolution. In AFM, an image is collected from a probe that passes across the sample surface. The AFM probe has a nanometer-sized cantilever that does not actually touch the surface, but when brought in close proximity to the sample surface, forces between the sample and the cantilever deflect the cantilever. The force of deflection is measured and translated into an image of the surface. The AFM is able to measure different forces depending on the situation; some of these forces include mechanical contact, electrostatic, magnetic, and van der Waals forces.

19.3.3   PROFILOMETRY

Profilometry is another technique that uses a probe to scan the surface of a sample (Figure 19.3). However, profilometry does not measure the forces between the probe and the surface, but rather the

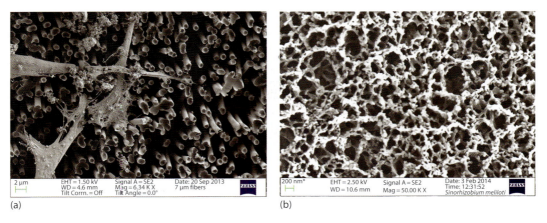

FIGURE 19.2 Scanning electron microscope images. (a) Cells attached to micron-sized structures on a poly(propylene) film. (b) Nanosized pores in a poly(caprolactone) film.

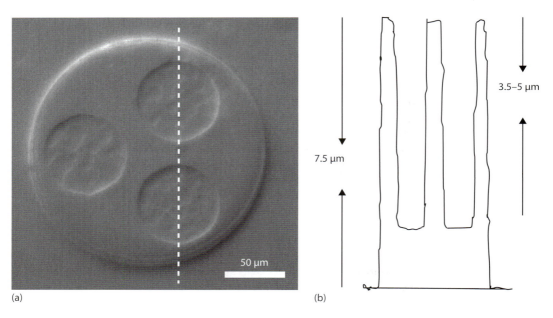

FIGURE 19.3 Profilometer used to measure changes in height across a surface. The scanning electron microscopy image (a) shows a microdevice with three wells on the surface. The dashed line indicates the path of the profilometry measurement, which measures changes in height across the surface. The profilometry data (b) are represented with height on the y-axis and distance on the x-axis. (From Chirra, H. and Desai, T.A.: Multi-reservoir bioadhesive microdevices for independent rate-controlled delivery of multiple drugs. *Small.* 2012. 8(24). 3839–3846. Copyright Wiley-VCH Verlag GmbH & Co. KGaA. Reproduced with permission.)

displacement of the probe due to changes in surface features. Profilometry is not sensitive enough to detect submicron features, but it is still an important tool in the characterization of nanodevices, which often involve both micro- and nanofeatures. Profilometry is also useful in determining the thickness of thin films, surface roughness, and microscale topographical features. Compared to SEM or AFM, profilometry is much less expensive and time intensive.

SEM, AFM, and profilometry are examples of advanced instrumentation that allow direct visualization and measurement of key features at the micro- and nanoscale. The following sections provide specific examples of how nanotechnologies are being used in drug delivery.

19.4   APPLICATIONS OF NANOTECHNOLOGY IN DRUG DELIVERY

19.4.1   MICRO- AND NANOTOPOGRAPHY TO OVERCOME EPITHELIAL BARRIERS

In drug delivery, we face the general challenge of passing molecules through biological membranes in practically all routes of administration. As described in detail in Chapter 4, these membranes limit passage in and out of the tissues, organs, and individual cells, conferring protection against invasion of pathogens and exposure to toxins. At the outermost surface, the skin presents a stratified epithelial barrier to drug entry into the body, while the plasma membrane of individual cells prevents the free diffusion of substances in and out of the cell interior. Additionally, mucus membranes line various tracts (gastrointestinal, respiratory, reproductive, etc.) that are regularly exposed to pathogens and foreign substances. In all cases, these membranes protect the body against external agents, but these barriers also limit the delivery of drugs via these sites.

As described in Chapter 4, the tight junctions between adjacent cells and epithelial cells restrict the diffusion of large molecules between cells via the paracellular route. Transcellular transport is typically limited to drugs that are low molecular weight and lipophilic. With an increased focus on delivery of macromolecules and particles, new approaches that do not require hypodermic needle injections are being explored to improve delivery of large molecules through epithelial layers.

Micro- and nanotopography, in the form of wires, needles, pegs, and grates, can be leveraged in drug delivery systems to overcome challenges associated with passage through epithelial barriers. Recent developments in transdermal drug delivery utilize microneedles to penetrate the skin with "needle-free" systems that minimize pain and offer the potential for sustained local, or systemic, delivery. Nanofeatures have also been shown to enhance mucoadhesion, improving transport through, and reducing clearance from, mucus layers that overlay epithelial cells. These techniques are described further here.

19.4.1.1   Microneedles

Some of the most successful advances in the micro- and nanotechnology for drug delivery have been in the development of microneedle systems for transdermal drug delivery. This technology is also described in Chapter 9 (Section 9.3.3.4). Transdermal delivery has long been a focus in drug delivery as the skin is an easily accessible site, with high potential for controlled or sustained release for local or systemic delivery. For successful transdermal delivery, a drug must permeate through the multiple layers comprising skin, starting with the toughest barrier to pass, the stratum corneum. Microneedles are designed to penetrate through the stratum corneum and form channels into the dermis through which drug can diffuse (Chapter 9, Figures 9.3 and 9.5). With microscale length and width, these needles are long enough to penetrate the toughest layer of the skin but short and narrow enough that they do not contact nerve endings in the dermis and thus cause no pain (Tuan-Mahmood et al. 2013). The microneedles alone can be enough to achieve sufficient transdermal delivery, or the technology can be combined with additional chemical, enzymatic, or mechanical components to enhance drug permeation.

There are a variety of different microneedle approaches, including the following systems:

- Using a microneedle patch to introduce pores into the skin before applying a topical drug
- Incorporating the drug directly into the microneedle structure
- Using microneedles that dissolve or break off within the dermis
- Using hollow microneedles that allow drug solutions to be actively delivered through the microneedle array from an external source

Microneedles are most commonly fabricated using molding and etching techniques and often utilize molds made from lithographic procedures. With these fabrication tools, the geometry, length, and array density of the microneedles can be precisely designed and produced. There are many recent reports of how microneedle design affects performance, but studies tend to focus on microneedle size and shape effects.

For example, needle length and array density have been shown to affect drug flux in a solid silicon microneedle array applied to the skin prior to application of acyclovir to the treated area (Yan et al. 2010). The results showed that needles longer than 600 μm increase drug flux and that with these longer needles, a needle density less than 2000 needles/cm^2 further increases flux. Another study demonstrated that for dissolving microneedles made from polysaccharides, a lower aspect ratio (height to width), with pyramidal rather than conical tips, increases mechanical strength of the microneedles thereby improving insertion into the skin (Lee et al. 2008). These reports highlight the importance of microneedle design, but additional work is still needed to better understand the complexities of both shape and size on the effect of microneedles made from different materials for the delivery of different therapeutic agents.

While work still continues on optimizing, characterizing, and designing microneedle systems, several commercial products and clinical studies have already proven the potential of microneedles for commercial application. Among the commercially available microneedle products are the Mi-Roll® Derma Rolling System and the MicronJet®. The Mi-Roll® is an FDA-approved skin treatment system composed of microneedles that, when applied to the skin, are intended to increase collagen production in addition to creating microchannels to enhance the effect of topical treatments and creams. The MicronJet® is an attachment with microneedles that fits standard syringes for intradermal injection. In clinical studies, low-dose influenza vaccines delivered with the hollow microneedle MicronJet® system were compared to standard intramuscular influenza vaccines (Van Damme et al. 2009). These results showed a comparable immune response obtained using the MicronJet® delivery system, as obtained with standard vaccination.

Varieties of other microneedle systems are currently under development for vaccination via the transdermal route and are described in detail in Chapter 17. These technologies include the ZP™ Patch; the hollow microstructured transdermal system (hMTS™); and the Nanopatch™ (which is further described in a case study in Chapter 9, Section 9.3.3.5). Dried vaccine formulations are being investigated in dissolvable microneedle technologies, such as the MicroCor® and Vaxmat®.

While not yet tested in humans, several groups have demonstrated the use of microneedle systems for transdermal delivery of insulin. One example is a study in rats with induced diabetes, which compared the subcutaneous infusion of insulin through a hypodermic needle with the intradermal infusion of insulin through hollow silicon microneedles. The results showed successful delivery of insulin through the microneedle patch, as indicated by a reduction in glucose levels after administration (Nordquist et al. 2007). Even though the comparison to subcutaneous infusion showed a shorter duration in suppression of glucose levels, this early work suggests that a microneedle method for insulin delivery is possible, which is not painful, does not require specialist training, and will thus improve patient compliance.

19.4.1.2 Nanostructures for Enhancing Paracellular Transport

Whereas microneedles operate by creating temporary pores in the layers of the skin to improve transdermal drug delivery, a different approach to improving permeability is by directly disrupting the structures between epithelial cells, without damaging the surrounding cells. Epithelial interfaces are associated with tight junctions between adjacent epithelial cells, which prevent drug molecules from diffusing through the paracellular space (see also Chapter 4, Figures 4.1 through 4.5 inclusive). Precisely engineered nanostructures, of the order hundreds of nanometers, could probe the space between adjacent cells, thereby facilitating paracellular transport. Several research studies have carried out preliminary investigations into the use of nanowires and nanopillars for improving paracellular permeation. Figure 19.4 shows nanowire-coated particles prepared by growing nanoscopic silicon wires from the surface of narrowly dispersed, microsized, silica beads (Uskoković et al. 2012).

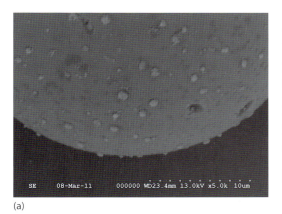

(a)

(b)

FIGURE 19.4   Silica beads without (a) and with (b) silicon nanowire coating. Nanowires enhance epithelial drug delivery. (Reprinted with permission from Uskoković, V., Lee, K., Lee, P.P. et al., Shape effect in the design of nanowire-coated microparticles as transepithelial drug delivery devices, *ACS Nano*, 6(9), 7832–7841. Copyright 2012 American Chemical Society.)

Nanowire-coated particles showed a twofold increase in permeability of small molecules when compared to smooth particles (Fischer et al. 2011). Two mechanisms have been proposed to explain this observed disruption in the tight junctions. One hypothesis is that the nanowires interact with the cell membrane and, through mechanotransduction, increase the transepithelial transport of the drug possibly by increasing the surface area of the cell membrane. Another hypothesis is that when epithelial cells contact the nanowires, the cells stretch out in order to adhere to the structures, thus widening the paracellular space leading to increased diffusion of drug through this space (Uskoković et al. 2012).

Permeability-enhancing effects caused by a disruption of cell–cell junctions were also observed using arrays of nanowires and nanopillars through monolayers of epithelial cells (Kam et al. 2013). Figure 19.5 shows clearly that without the nanostructures, epithelial cells formed a cobblestone pattern with smooth cell–cell interfaces. After exposure to nanostructured surfaces, the interface between cell membranes appeared jagged or rippled, suggesting a disruption in the tight junctions between adjacent cells. This morphological change in the cells was reversible on removing the nanostructured surface. Interestingly, the disruption to the tight junctions, as well as enhanced permeability, was observed for nanopillars with an aspect ratio of 1.5 (300 nm high, 200 nm wide) but not for pillars with a larger aspect ratio of 20 (16 µm high, 800 nm wide) (Kam et al. 2013). Additional work is still needed to further investigate the effect of size, geometry, and density of nanofeatures on transepithelial transport.

19.4.1.3   Nanodevice Interactions with the Mucus Layer

In addition to penetrating through epithelial cells, many routes for drug administration (e.g., oral, nasal, pulmonary, and vaginal) first require permeation through a mucus layer. Mucus is a viscous colloid composed of glycoproteins, water, and enzymes, which overlays the epithelial cells. The viscous nature of mucus allows it to entrap particulates, and the hydrophobic nature of its lipid constituents can prevent the passage of polar drugs. In some cases, such as the gastrointestinal tract, mucous consists of two layers: a static layer adhered to the epithelium and a motile layer that actively clears captured particulates.

There are two basic approaches to design delivery systems that overcome the mucosal barrier. In the first, devices or particles are engineered for improved mucoadhesion and in the second, for enhanced mucus penetration. The primary goal of these systems is to achieve both a higher concentration and quantity of drug delivered to the underlying epithelial cells, in order to improve absorption.

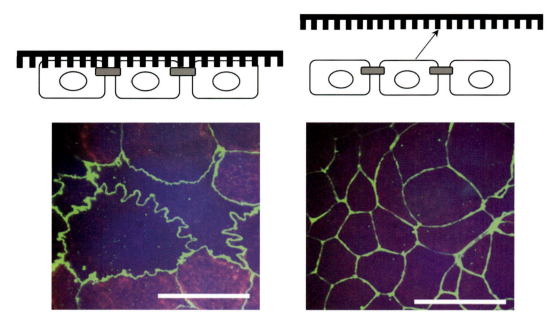

FIGURE 19.5 Effect of nanostructured surfaces on epithelial cell–cell junctions. Contact with nanostructures causes a disruption to the tight junctions between epithelial cells, as visualized through immunofluorescent staining of zonula occluden (ZO-1), a tight junction protein. The nanostructures caused a ruffling of the cell–cell boundary (left), compared to the normal smooth cell–cell interface (right). (Reprinted with permission from Kam, K., Walsh, A.L., Bock, S.M. et al., Nanostructure-mediated transport of biologics across epithelial tissue: Enhancing permeability via nanotopography, *Nano Letters*, 13, 164–171. Copyright 2013 American Chemical Society.)

When tethered to the surfaces of particles or microdevices, nanowires were shown to improve mucoadhesion in vitro. Improved mucoadhesion is thought to be because the nanowire features are able to entangle with the proteins and fibers in the mucosal layer, thereby anchoring the attached particle or device (Fischer et al. 2009). The resulting increased contact time of the device with the mucosal surface improves bioavailability.

These systems are being designed for sustained release of the drug during the time that the device is adhered to the mucus. To further increase the percentage of drug available for absorption through the epithelium, devices are being developed to provide unidirectional release. In these planar devices, one surface is modified with nanotopography to improve mucoadhesion, and asymmetric layering allows for drug release from only this side of the device (Figure 19.6). For oral delivery, this is intended to focus drug release in the vicinity of the epithelial cells for absorption and to prevent release of drug into the intestinal lumen. Additionally, the uptake of many drugs is limited by efflux transporter proteins and metabolizing enzymes, and increased localized drug concentrations may result in the saturation of efflux transporters and enzymes, in turn improving the bioavailability of the released drug (International Transporter Consortium 2010). Unidirectional drug release is typically achieved by using an outer drug-impermeable layer, followed by a drug-permeable layer, such as a hydrogel, for drug loading. The drug-permeable layer is often contained within a reservoir in the drug-impermeable layer to prevent escape of drug from the sides of the device (Figure 19.6).

Most of the work discussed in this section is still in the early stages of development and has not yet been proven in vivo. As knowledge and understanding of the interaction between nanostructures and cellular systems advances, we can hope to see similar features integrated into more devices in development as well as to see in vivo and clinical results demonstrating the efficacy of such systems. These systems are of particular interest as delivery systems for macromolecules, such as protein

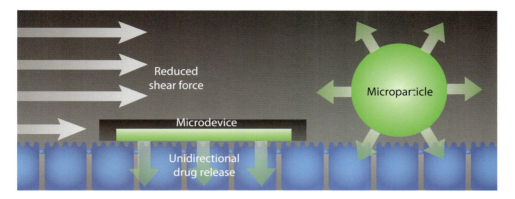

FIGURE 19.6 Advantages of a planar, asymmetric microdevice design for oral drug delivery. A planar device geometry minimizes the amount of shear force per mass on microdevices, improving device adhesion to mucosal and epithelial surfaces. A drug reservoir on one side of the device allows for unidirectional drug release, and asymmetric surface modifications allow for selective binding of the side of the device releasing the drug.

and peptide drugs. Due to their size and complexity, these therapeutics have low permeability and can be sensitive to the harsh conditions of mucosal environments. Microdevices have potential to simultaneously protect therapeutics from degradation and provide sustained, unidirectional release directed toward the epithelium, for improved absorption.

19.4.1.4 Gecko-Inspired Nanotopography for Improved Adhesion

There is a need and interest in the medical field for improving adhesion in a variety of applications such as wound dressings, drug delivery patches, or mucus-targeted delivery vehicles, among others. A novel approach inspired by a naturally occurring phenomenon is a gecko-inspired surface. One of the unique properties of the nanoscale is the drastic increase in surface area that accompanies the presence of nanostructures on a surface. Taking advantage of this, gecko-inspired surfaces are modeled after the microscale angled fibers and nanotopography found on gecko toes (Figure 19.7). By creating a large number of nanoscale surface contact points, gecko toes and gecko-inspired surfaces increase adhesion through an increase in intermolecular forces. For geckos, this allows them to cling to a vertical wall or even upside down from a ceiling. In medical applications, gecko-inspired surfaces show promise in both wet and dry environments. Similar to the microneedles and nanopillars previously discussed, the adhesion characteristics of gecko-inspired nanostructures depend on geometry, density, and material properties.

Nanostructured poly(glycerol-sebacate-acrylate) (PGSA) polymer surfaces were tested both in vitro using porcine intestine tissue and in vivo in a rat model to look at the use of these gecko-inspired materials as a biodegradable adhesive. With PGSA polymer structures, nanopillars with larger tip-to-base diameter ratios show higher adhesion; independently, larger tip diameters to pillar length ratios were also found to produce stronger adhesion (Mahdavi et al. 2008). While these and many other early studies on gecko-inspired materials do not focus on a drug delivery applications, there is potential for translating these results to improve adhesion of microneedle arrays or muco-adhesion of drug delivery devices.

19.4.2 NANOTECHNOLOGY IN CONTROLLED-RELEASE DRUG DELIVERY

Nanofabrication techniques are also used to create nanopores and nanochannels for controlled drug release. Demand is growing for the development and commercialization of controlled- and sustained-release drug delivery systems. Controlled-release systems aim to eliminate frequent dosing, provide more consistent blood levels, improve the efficacy and efficiency of therapeutics, and

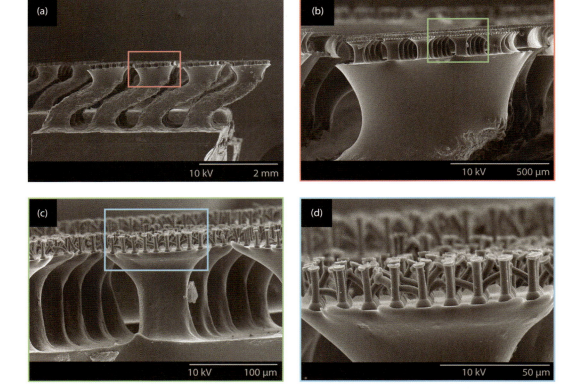

FIGURE 19.7 Hierarchical gecko-inspired topography: three-level hierarchical polyurethane fibers. (a) Curved base-level fibers. (b, c) Midlevel fibers on top surface of base fibers. (d) Third-level fibers on top surface of midlevel fibers. (Reprinted with permission from Murphy, M.P., Kim, S., Sitti, M., Enhanced adhesion by gecko-inspired hierarchical fibrillar adhesives, *ACS Appl. Mater. Interfaces*, 1(4), 849–855. Copyright 2009 American Chemical Society.)

reduce adverse reactions. As the biopharmaceutical industry continues to grow, there is a strong emphasis on developing controlled-release delivery systems for peptides and proteins in particular.

19.4.2.1 Nanochannels in Membrane-Controlled Drug Delivery Devices

Porosity has been utilized as a tool in drug delivery for a long time, but it is traditionally through the use of naturally porous materials, leaching techniques, or self-assembly, that porosity has been introduced into a system. Micro- and nanofabrication approaches to create porosity allow for greater precision in design and control of pore size, distribution, and material in delivery systems. These fabrication approaches not only allow for more precise control of pore features but also make it possible to create monodisperse pores throughout a substrate. There are a number of examples of continuous-release implantable devices in development and on the market that use micro- and nanochannels or pores in a membrane in order to control diffusion of a therapeutic out of a reservoir (see also Chapter 14, Section 14.5.2).

Release rates and profiles for small- and large-molecule therapeutics are controlled in these systems through the size of the pore relative to the drug molecule, pore or channel length, and properties of the membrane material or surface coating such as surface charge, polarity, and hydrophobicity. When the pore size is large compared with the molecule, release of drug from a reservoir through the porous membrane is described by Fickian diffusion (see also Chapter 2).

However, when pore size is on the order of the hydrodynamic radius of the molecule, diffusion of the drug molecule through the pore is constrained, resulting in "single-file" diffusion and a linear release rate. Being able to fabricate devices with nanoscale channels and pores facilitates this single-file diffusion for macromolecules such as peptides and proteins. For example, Nanopore™ technology comprises a small subcutaneously implantable reservoir, fitted at each end with membranes that are microfabricated to contain pores or channels that are 1–5 times the hydrodynamic diameter of the selected drug molecules. Pore diameter is empirically "tuned" with molecular size, to enable sustained release of drug molecules held within the reservoir, via a constrained passive diffusion mechanism.

These systems can be made from silicon, metals, polymers, and biopolymers. For silicon and metal membranes, lithographic techniques such as electron-beam lithography are powerful tools for precise fabrication of pores. To create nanochannels in a polymeric material, a series of techniques including solvent casting and template molding can be leveraged, as shown in Figure 19.8 (Bernards et al. 2012).

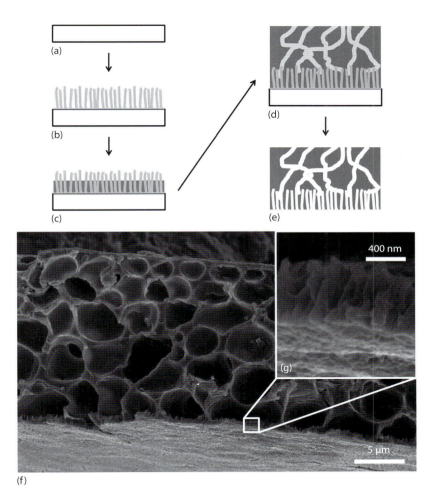

FIGURE 19.8 Fabrication of nanoporous polymer films. (a) A clean silicon substrate. (b) Zinc oxide nanorods are grown on the substrate from a zinc-oxide seed layer deposited using spin-coating techniques. (c) Using solvent-casting and spin-coating, the template is coated with a layer of polycaprolactone thin enough to not cover the nanorods. (d) An additional support layer of a mixture of polycaprolactone and PEG is added. (e) Deionized water dissolves the PEG-phase from the supporting layer and sulfuric acid etches the zinc-oxide template to generate a supported, nanostructured, polycaprolactone thin film. (f) SEM image of a typical nanostructured polycaptrolactone film. (g) The supporting nanochannel layer. (Reproduced with permission from Bernards, D.A., Lance, K.D., Ciaccio, N.A. et al., Nanostructured thin film polymer devices for constant-rate protein delivery, *Nano Letters*, 12, 5355–5361. Copyright 2012 American Chemical Society.)

19.4.2.2 Nanochannels in Cell Encapsulation

Another important feature of nanoporous reservoir devices is their capacity for immunoisolation. Nanoporous membranes can allow for the diffusion of small-molecule therapeutics and/or nutrients, while limiting the flux of large molecules and cells. This feature is particularly important in the development of biocapsules in cell or tissue transplants, which aim to prevent immune rejection (Desai et al. 1999). Much ongoing research focuses on the use of these technologies and approaches to deliver pancreatic islet cells for the treatment of diabetes. A further application is in ophthalmic drug delivery: Chapter 13 (Section 13.7.2.1) describes how genetically engineered cells loaded into polymeric microcapsules can be implanted directly to the back of the eye and subsequently produce therapeutic proteins. The microcapsules provide immunoprotection for the implanted cells while simultaneously allowing efflux of the therapeutic proteins.

19.5 CONCLUSIONS

Micro- and nanofabrication techniques allow for precise control and patterning at micro- and nanometer scales for an increasing variety of materials. These techniques are being used for the top-down fabrication of nanoparticles, nanopillars, and devices with specific features including size, geometry, surface topography, and porosity. Micro- and nanofabrication techniques are being applied through numerous avenues to improve drug delivery and targeting. In particular, nanosystems are being leveraged to improve permeability and transport across biological membranes, to control drug release, to improve pharmacokinetics, to target therapeutics, and to achieve stimuli-responsive drug delivery.

   As the complexity of these systems and the breadth of their features increases, there is a growing need for additional studies investigating the mechanistic and functional impact of these features on biological processes. Many of the systems referenced in this chapter, as well as others not mentioned, combine multiple features and approaches to achieve their drug delivery goal. A better understanding of how each individual component and the combined effects of material properties and micro- and nanofeatures affect the biology and the drug delivery process is needed. Additionally, only a few of the systems under study have been tested in vivo or commercialized. Additional studies focusing on in vivo considerations are needed to ascertain the impact of system features on biodistribution, toxicity, and immunogenicity. This rapidly expanding field of nanomedicine offers exciting opportunities to optimize drug delivery and targeting.

REFERENCES

Bernards, D.A., K.D. Lance, N.A. Ciaccio et al. 2012. Nanostructured thin film polymer devices for constant-rate protein delivery. *Nano Letters* 12:5355–5361.

Chirra, H. and T.A. Desai. 2012. Multi-reservoir bioadhesive microdevices for independent rate-controlled delivery of multiple drugs. *Small* 8(24):3839–3846.

Desai, T.A., D.J. Hansford, L. Kulinsky et al. 1999. Nanopore technology for biomedical applications. *Biomed Microdev* 2:11–40.

Fischer, K.E., B.J. Alemán, S.L. Tao et al. 2009. Biomimetic nanowire coatings for next generation adhesive drug delivery systems. *Nano Letters* 9:716–720.

Fischer, K.E., A. Jayagopal, G. Nagaraj et al. 2011. Nanoengineered surfaces enhance drug loading and adhesion. *Nano Letters* 11:1076–1081.

International Transporter Consortium. 2010. Membrane transporters in drug development. *Nat Rev Drug Discov* 9:215–236.

Kam, K., A.L. Walsh, S.M. Bock et al. 2013. Nanostructure-mediated transport of biologics across epithelial tissue: Enhancing permeability via nanotopography. *Nano Letters* 13:164–171.

Lee, J.W., J.H. Park, and M.R. Prausnitz. 2008. Dissolving microneedles for transdermal drug delivery. *Biomaterials* 29(13):2113–2124.

Mahdavi, A., L. Ferreira, C. Sundback et al. 2008. A biodegradable and biocompatible gecko-inspired tissue adhesive. *Proc Natl Acad Sci USA* 105(7):2307–2312.

Murphy, M.P., S. Kim, and M. Sitti. 2009. Enhanced adhesion by gecko-inspired hierarchical fibrillar adhesives. *ACS Appl Mater Interfaces* 1(4):849–855.

Nordquist, L., N. Roxhed, and P. Griss. 2007. Novel microneedle patches for active insulin delivery are efficient in maintaining glycaemic control: An initial comparison with subcutaneous administration. *Pharm Res* 24 (7):1381–1388.

Tuan-Mahmood, T.-M., M.T.C. McCrudden, B.M. Torrisi et al. 2013. Review: Microneedles for intradermal and transdermal drug delivery. *Eur J Pharm Sci* 50:623–637.

Uskoković, V., K. Lee, P.P. Lee et al. 2012. Shape effect in the design of nanowire-coated microparticles as transepithelial drug delivery devices. *ACS Nano* 6(9):7832–7841.

Van Damme, P., F. Oosterhuis-Kafeja, and M. Van der Wielen. 2009. Safety and efficacy of a novel microneedle device for dose sparing intradermal influenza vaccination in healthy adults. *Vaccine* 27(3):454–459.

Yan, G., K.S. Warner, J. Zhang et al. 2010. Evaluation of needle length and density of microneedle arrays in the pretreatment of skin for transdermal drug delivery. *Int J Pharm* 391:7–12.

Section V

Toward Commercialization

20 Rational Drug Discovery, Design, and Development

Haizhen A. Zhong, Osman F. Güner, and J. Phillip Bowen

CONTENTS

20.1 INTRODUCTION

The widespread use of computer-based methods has had a dramatic impact on research in the physical, biological, and medical sciences. Applications of computational chemistry to drug design have clearly advanced the discovery of bioactive compounds. Computer speeds have allowed scientists in all fields to tackle problems that were not possible 20 years ago. Using highly specialized software, drug discovery scientists can apply energy-based calculations to help understand drug–receptor interactions at the molecular level. The molecular structure of drug candidates can be inserted into binding sites in silico to determine optimal interactions. The information can be used to design novel compounds with improved fits for the binding sites under examination.

When the macromolecular drug targets are not known, homology-modeling methods may be used to construct reasonable 3D models of the putative binding site based on similar protein structures. Where there are scant experimental structural data regarding the shape of macromolecular drug targets, medicinal chemists can examine the common structural, electronic, and conformational features of known biologically active compounds to construct pharmacophore models for database searching. Complex correlations between molecular structure and predicted physical properties may be used to suggest the best potential drug leads for animal model studies and clinical trials, providing investigators with candidates that are more likely to be successful. Over the last decade, there has been increasing focus on the prediction of toxicity.

With advances in technology, computers and laboratory devices have become faster and the information obtained is more accurate, but despite these advances, the difficulty of trying to design and develop a drug with all the required physical and biological properties necessary for FDA approval remains an incredibly expensive and time-consuming challenge. The latest estimates on the cost of getting a drug approved now exceed $2 billion (DiMasi and Grabowski 2012). Nevertheless, in the last 50 years, there have been great intellectual strides made in the development and use of in silico methods. This chapter will provide a brief history of the major advances in the field, as well as a description of some of the more robust and successful methods currently used in drug design and discovery, with particular emphasis on drug delivery considerations such as optimizing the solubility of lead compounds and studying drug–transporter interactions.

20.2 COMPUTATIONAL CHEMISTRY

Computational chemistry is a generic term that describes a broadly based set of theoretical methodologies that can trace their roots to the development of mathematical physics. One of the common themes found in computational chemistry is the extensive use of computers to solve complex problems that range from polymer chemistry and nanotechnology to biochemistry and pharmacology. The methods of classical physics routinely used today in molecular modeling have foundations based on Newton's equations and/or later formulations developed by Hamilton, Lagrange, and others. They include molecular mechanics and molecular dynamics calculations. These methods work reasonably well for large molecular systems but only when the mathematical models (equations and equation parameters) have been carefully developed. For many areas of molecular modeling, the use of quantum physics is critical, particularly when the explicit treatment of electrons is essential. In 1998, the Nobel Prize in Chemistry was awarded equally to John Pople and Walter Kohn for their independent development of methods in computational quantum chemistry and the development of density functional theory (DFT), respectively. For enzyme–substrate interactions, a combination of classical and quantum physics is necessary if the goal is to examine the bond scission and/or bond formation. A general strategy is to treat the active site with quantum mechanics to account for the shifts in electron density involved in enzyme–substrate interactions (binding, bond formation, bond breakage, etc.), while the rest of the molecule is subjected to classical methods. The marriage of molecular mechanics and quantum mechanics is called QM/MM, and the importance of this approach was underscored with the 2013 Nobel Prize in Chemistry for Martin Karplus, Michael Levitt, and Arieh Warshel. The term quantum pharmacology has been coined and is applicable when computational quantum chemistry is used to calculate the molecular structures of pharmacologic interest (Richards 1984). Many medicinal chemists have used classical and/or quantum mechanical calculations to determine preferred conformations, molecular shapes, electron distributions, enzyme–substrate reactions, and drug–receptor interactions. A brief survey of some computational methods is presented in the following texts.

20.2.1 MOLECULAR MECHANICS

Molecular mechanics is widely used in computational schemes (Bowen 2004). In this method, the atoms within a molecule are treated as soft spheres that may be viewed as being held together by springlike forces. Earlier terms include the Westheimer method and the force field method. Molecular mechanics emerged from spectroscopy, and the basic approach was outlined by D.H. Andrews in the early twentieth century. Three key papers appeared in 1946 that applied classical physics to chemical problems. One was by Westheimer and Mayer, the other by Hill, and the third by Dostrovsky, Hughes, and Ingold. These papers represent the first examples of what today is recognized as molecular mechanics. Originally, this mathematical approach was called the force field method or the Westheimer method because of his demonstration of the use of the method to understand a specific chemical problem.

Molecular mechanics divides the total potential energy (U_{total}) among various component potential energy terms with which chemists can readily identify, including but not limited to stretching and compression ($U_{stretch}$), bending (U_{bend}), torsions ($U_{torsion}$), and nonbonded potentials ($U_{non\text{-}bonded}$). These nonbonded potentials account for the electrostatic and van der Waals repulsive and attractive (London dispersion forces) interactions (Bowen and Zhong 2013):

$$U_{total} = U_{stretch} + U_{bend} + U_{torsion} + U_{non\text{-}bonded} \tag{20.1}$$

Simple mathematical expressions are commonly used (e.g., Hooke's law for the stretching and bending potential energies and Coulomb's law for electrostatic interactions). The advantage in using less rigorous equations is translated into faster computational time, but the big disadvantage is the loss of accuracy and predictability when simple models are used. Is it better to get less accurate information faster or the right answer in due course?

One of the major difficulties initially encountered with molecular mechanics was the simultaneous minimization of the potential energy functions with respect to the coordinates. Westheimer was able to demonstrate the utility of the force field method by doing an energy calculation by hand. Hendrickson is credited with doing the first computer-based molecular mechanics calculation in 1961. In 1965, Snyder and Schachtschneider demonstrated that force constants were essentially transferable from molecule to molecule if key off-diagonal terms among neighboring atom pairs were included. There was no general energy minimization algorithm until one was developed in the labs of Kenneth Wiberg. In the late 1970s, the MM2 method was introduced and widely used. Clark Still and others modified and improved the original MM2 method and incorporated the computational scheme into molecular modeling graphical programs such as MacroModel. Subsequent versions of MM2 were not popular with organic and medicinal chemists because of the limitations of functional group parameterizations. Today, the standard molecular mechanics method found in most software programs is the Merck molecular force field (MMFF) developed by Thomas Halgren. Today, MMFF has replaced MM2 and subsequent versions (MM3 and MM4), which are no longer being actively developed. The Jorgensen optimized potential for liquid simulations (OPLS) molecular mechanics method is also used by many scientists. OPLS is distinguished from other methods because it is parameterized to reproduce solution phase data; other molecular mechanics schemes are fit to either gas phase experimental data and/or high-level quantum mechanics calculations.

Understanding drug–receptor, substrate–enzyme, and inhibitor–enzyme interactions is critical for many of the drug design approaches. Molecular mechanics methods designed for the accurate calculations of small molecules are not necessarily used to calculate macromolecule structures. The assisted model building and energy refinement (AMBER) and Chemistry at HARvard Molecular Mechanics (CHARMM) programs reign supreme. They use simple potential energy functions, which work remarkably well for large, unstrained macromolecules.

20.2.2 MOLECULAR DYNAMICS SIMULATIONS

Molecular dynamics (MD) simulations are popular ways to understand macromolecular behavior as a function of time (Bowen 2004; Bowen and Zhong 2013). The trajectories of the atoms of a macromolecule can be determined using classical physics (equations of motion and molecular-kinetic theory). Molecular dynamics is based on molecular mechanics energy equations and the fact that the force is equal to the negative of the potential energy (Equation 20.2) where U is the potential energy as a function of the generalized coordinates r. Based on the molecular mechanics potential energy equation, knowing the configuration of the atoms and utilizing Equation 20.2, allows the calculation of the force and, in turn, using Newton's equations of motion, the calculation of velocities and trajectories. This approach is useful for treating large molecular systems with water solvation models or the explicit use of water:

$$F = -\frac{\partial U(r)}{\partial r} \tag{20.2}$$

Pharmaceutical companies rarely use molecular dynamics simulations due to lengthy time requirements for MD simulation studies, but they remain of interest to many academic scientists. This approach has yielded important insights.

20.2.3 COMPUTATIONAL QUANTUM MECHANICS

Without question, computational quantum mechanics has emerged as an extremely useful method for examining and determining the predicted physical properties of molecular structures in silico. Unlike molecular mechanics, which can be viewed as an *ad hoc* collection of potential functions, quantum mechanics is a rigorously based theory that emerged in the mid-1920s with further developments in later years. Although several of the early pioneers of quantum mechanics did not receive Nobel Prizes for their contributions, numerous scientists (e.g., Bohr, Planck, de Broglie, Einstein, Heisenberg, Schrödinger, Pauli, Dirac, Born, and others) were recognized. In 1998, the importance of this field and two of its major contributors, John Pople and Walter Kohn, were recognized as outlined earlier. (It should be noted that independent work by Roald Hoffmann and Kenichi Fukui led to the 1981 Nobel Prize in Chemistry, and there are many seminal advances in chemistry and physics based on quantum mechanics that are not discussed in this chapter.)

Quantum mechanics has fundamentally changed the way physicists and chemists view the subatomic nature of the universe. For the most part, medicinal chemists are more involved in applying quantum mechanics to problems of pharmacological interest, rather than trying to understand the consequences of Bell's inequality and the fundamental debate of subatomic reality. Nevertheless, the field of drug design benefits directly from the latest achievements. One interesting example is the growing recognition of the importance of what are termed halogen bonds. In a halogen bond,

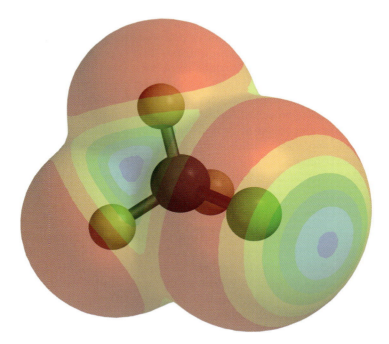

FIGURE 20.1   The molecular structure of chlorotrifluoromethane with its electrostatic potential energy surfaces superimposed. The structure was calculated with the Spartan molecular–modeling software program at the Hartree–Fock level of theory using the 6-31(d,p) basis set. The colors blue and red indicate the most electropositive and electronegative regions, respectively. The sigma hole is clearly visible. (Courtesy of the Bowen Group, Center for Drug Design, Mercer University, College of Pharmacy, Atlanta, GA. Graphics rendered using Spartan software by Pamela Ohsan, Wavefunction, Inc., Irvine, CA.)

electron-rich donor groups form stable interactions with the electron-deficient region on the surface of the halogen. This electron-deficient area is aligned with the carbon–halogen bond and is known as the sigma hole. Figure 20.1 shows the sigma hole for a very simple molecule, chlorotrifluoromethane, calculated at the HF/6-31G(d,p) level of theory.

It should be noted that the vast majority of molecular problems of particular interest in quantum pharmacology are systems with potential functions that are not changing with time. These systems may be described with the time-independent Schrödinger equation (Equation 20.3), which is an eigenvalue equation where $\hat{H}$ is the Hamiltonian operator and $\Psi(r)$ is the molecular wave function that depends on the position vectors, $\vec{r}_i$ (Hehre et al. 1986). The following equation is written in a compact form, where the symbolism masks the complexity of the equation for each atom:

$$\hat{H}\Psi(r) = E\Psi(r) \tag{20.3}$$

The Hamiltonian operator, named in honor of the accomplished nineteenth-century Irish physicist Sir William Rowan Hamilton, is the quantum mechanical equivalent of the summation of the kinetic and potential energy of a system in classical physics. The first two terms of the Hamiltonian operator $\hat{H}$ (Equation 20.4) are the kinetic energy operators for the electrons and nuclei, respectively, with the summation over all electrons i and all protons A. The Laplacian ∇_i^2 is defined in the following texts. The symbol h is Planck's constant, and m and M are the masses of the electrons and the protons, respectively. The last three terms of Equation 20.4 are based on Coulomb's law and represent the proton–proton repulsion, electron–proton attraction, and electron–electron repulsion electrostatic terms, respectively. The symbol e is the absolute value of the charge of the electron and proton; the negative sign in front of the summation signs indicates an attractive potential energy, whereas the positive sign indicates a repulsive potential energy. The signs are derived based on the charges of an electron and protons. The charge of an electron is $(-e)$, and the charge of a proton is $(+e)$:

$$\hat{H} = -\frac{h^2}{8\pi^2 m}\sum_i^{\text{Electrons}} \nabla_i^2 - \frac{h^2}{8\pi^2 m}\sum_A^{\text{Nuclei}} \frac{1}{M_A}\nabla_A^2 + \frac{e^2}{4\pi\varepsilon_0}\sum_A^{\text{Nuclei}}\sum_{B<A}^{\text{Nuclei}} \frac{Z_A Z_B}{|R_{AB}|}$$

$$-\frac{e^2}{4\pi\varepsilon_0}\sum_i^{\text{Electrons}}\sum_A^{\text{Nuclei}} \frac{Z_A}{|r_{iA}|} + \frac{e^2}{4\pi\varepsilon_0}\sum_i^{\text{Electrons}}\sum_{j<i}^{\text{Electrons}} \frac{1}{|r_{ij}|} \tag{20.4}$$

Equation 20.4 can be simplified by recognizing that on the timescale of the electronic motion, the nuclei can be considered at rest (Born–Oppenheimer approximation) on a relative basis. Thus, the second term in Equation 20.4 vanishes, as shown in Equation 20.5. The third term in Equation 20.4 is a constant for each fixed nuclear configuration and may be removed (Equation 20.6). This term may be added later to the final electronic total energy, E_{el}:

$$-\frac{h^2}{8\pi^2 m}\sum_A^{\text{Nuclei}} \frac{1}{M_A}\nabla_A^2 = 0 \tag{20.5}$$

$$\frac{e^2}{4\pi\varepsilon_0}\sum_A^{\text{Nuclei}}\sum_{B<A}^{\text{Nuclei}} \frac{Z_A Z_B}{|R_{AB}|} = \text{Constant} \tag{20.6}$$

Therefore, by making the appropriate adjustments according to Equations 20.5 and 20.6, Equation 20.4 reduces to the following formulation:

$$\hat{H}_{electronic} = -\frac{h^2}{8\pi^2 m}\sum_i^{\text{Electrons}} \nabla_i^2 - \frac{e^2}{4\pi\varepsilon_0}\sum_i^{\text{Electrons}}\sum_A^{\text{Nuclei}} \frac{Z_A}{|r_{iA}|} + \frac{e^2}{4\pi\varepsilon_0}\sum_i^{\text{Electrons}}\sum_{j<i}^{\text{Electrons}} \frac{1}{|r_{ij}|} \tag{20.7}$$

The Laplacian operator, commonly found in mathematical physics, is defined in Equation 20.8. It is the dot product of the gradient, ∇, which yields the scalar ∇^2 (Bowen 2004):

$$\nabla^2 = \left(\hat{i} \frac{\partial}{\partial x} + \hat{j} \frac{\partial}{\partial y} + \hat{k} \frac{\partial}{\partial z} \right) \cdot \left(\hat{i} \frac{\partial}{\partial x} + \hat{j} \frac{\partial}{\partial y} + \hat{k} \frac{\partial}{\partial z} \right) = \frac{\partial^2}{\partial x^2} + \frac{\partial^2}{\partial y^2} + \frac{\partial^2}{\partial z^2} \tag{20.8}$$

For most molecular problems, it is easier to identify the position of a particle with a vector in spherical polar coordinates (r, θ, φ) rather than the Cartesian coordinate system (x, y, z). The Laplacian (Equation 20.8) can be transformed into spherical polar coordinates (Equation 20.9).

$$\nabla^2 = \frac{1}{r^2} \frac{\partial}{\partial r} \left(r^2 \frac{\partial}{\partial r} \right) + \frac{1}{r^2 \sin \theta} \frac{\partial}{\partial \theta} \left(\sin \theta \frac{\partial}{\partial \theta} \right) + \frac{1}{r^2 \sin^2 \theta} \frac{\partial^2}{\partial \varphi^2} \tag{20.9}$$

Removing the nuclear terms, as described earlier, produces the electronic Schrödinger equation. The Schrödinger equation can be solved exactly for only a few simple and special cases, which provide firm examples of dealing with the math. We have not discussed spin, which is a purely quantum mechanical property. For molecular systems, the approximations are too numerous and involved to be reviewed here, but the interested reader is encouraged to investigate the References at the end of the chapter. Two important approximations that must be mentioned are the following: (1) the Hartree–Fock method, where the electrons are moving independently of one another, leads to higher energies, and (2) the so-called linear combination of atomic orbitals–molecular orbitals, where molecular orbitals are viewed as a combination of atomic orbitals. In turn, the atomic orbitals can be represented by a judicious selection of a summation of Gaussian functions. The approach described earlier represents what is termed as computational ab initio theory (Hehre et al. 1986). Electron–correlation methods have been devised for greater accuracy. In this method, higher energy levels are included with ground-state energy levels and are referred to as post-Hartree–Fock calculations. Some of the problems associated with Hartree–Fock theory have been overcome with density functional theory (DFT).

The DFT method involves the direct use of electron densities (Parr and Yang 1989). It competes with and may surpass post-self-consistent field (SCF) calculations in terms of accuracy because the basic formulation, the Kohn–Sham equations, explicitly include electron correlation. DFT, compared to ab initio methods, avoids working with the many-electron wave function. This reduces the computer time of a calculation. Computer time is an important consideration when the systems are moderate to large. It has been estimated that ab initio calculations are proportional to z^4, where z represents the number of electrons. DFT calculations are also a function of the number of electrons but are proportional to z^3. The accuracy and reduction in time have made DFT a popular option (Bowen and Zhang 2013).

Computational quantum mechanics calculations can easily be carried out with software that can run readily on laptops. Modern quantum mechanics–based software allows one to calculate molecular structure, energy, thermodynamic values, and physical properties of small and large molecular systems. For example, the anticancer drug, ixabepilone, can readily be calculated on inexpensive laptops at the 3-21G level (or higher) and displayed with its electrostatic energy surface, Figure 20.2. While the more that is known about the underlying theory the better off one is when doing calculations, it is not necessary to be a theoretician. For example, most experimental organic and medicinal chemists use NMR spectroscopy on a regular basis, but not many consider all of the underlying quantum physics when interpreting spectra.

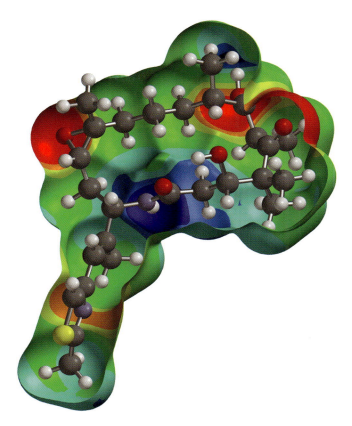

FIGURE 20.2 The molecular structure of the anticancer drug ixabepilone with its electrostatic potential energy surfaces superimposed. The structure was calculated with the Spartan molecular–modeling software program at the Hartree–Fock level of theory using the 3-21G basis set. The colors blue and red indicate the most electropositive and electronegative regions, respectively. (Courtesy of the Bowen Group, Center for Drug Design, Mercer University, College of Pharmacy, Atlanta, GA. Graphics rendered using Spartan software by Pamela Ohsan, Wavefunction, Inc., Irvine, CA.)

20.3 PHARMACOPHORE DEVELOPMENT AND DATABASE SEARCHING

A pharmacophore represents the pattern of features of a molecule that is responsible for a biological effect. The official definition proposed by the International Union of Pure and Applied Chemistry (IUPAC) states: "A pharmacophore is the ensemble of steric and electronic features that is necessary to ensure the optimal supramolecular interactions with a specific biological target structure and to trigger (or to block) its biological response" (Wermuth et al. 1998).

The pharmacophore concept dates back to 1898 when Paul Ehrlich proposed the name *toxophore* to define the features of a molecule that are responsible for the injurious biological effects on bacteria, viruses, or parasites (Ehrlich 1898). Even though Ehrlich never used the word pharmacophore and referred to these effects as toxophore, his contemporaries were referring to the same features as pharmacophore as early as 1904 (Marshall 1904). In those early days, the definition of pharmacophore involved "chemical groups" in a molecule responsible for a biological effect. The definition has evolved over time to mean the pattern of "abstract features" in 1960 by Schueler (Schueler 1960), which constitutes the earliest version of the modern definition preferred by IUPAC (Güner and Bowen 2014).

The evolved version of the pharmacophore concept is due to practical considerations. Accurate representation of the nature and 3D arrangement of features that are involved in receptor–ligand interactions is critical for the successful use of pharmacophore modeling (Leach et al. 2010). Once pharmacophore models are evaluated from the perspective of the receptor, it has become important to categorize the features in more abstract definitions, such as hydrogen-bond donors or acceptors, positive or negative centers, and lipophilic regions.

Pharmacophore modeling as a computer-aided drug design tool became broadly available in the 1980s. The pioneering pharmacophore-based database screening software, MOLPAT, was developed in 1973 (Gund et al. 1973), while the first automated pharmacophore pattern–recognition software (active analog approach) was introduced in 1979 (Marshall et al. 1979). The first software tools became commercially available in 1989 and the early 1990s (Christie et al. 1990; Güner et al. 1991, Van Drie et al. 1989; Murrall and Davies 1990; Hurst 1994; Sprague 1995). Since that time, more software tools have become available (Labute et al. 2001; Wolber and Langer 2005; Dixon et al. 2006). Pharmacophore modeling is a widely used and successful technique in drug discovery, specifically in lead discovery and optimization.

20.3.1 IMPORTANCE OF PHARMACOPHORE MODELING

Medicinal chemists have embraced pharmacophore modeling more readily than most other computer-aided design tools. This may be attributed to the similarity between pharmacophore modeling and the concept of bioisosterism that medicinal chemists employ in traditional structure–activity relationship (SAR) analyses. Furthermore, pharmacophore models are typically used in screening large databases of compounds as an alternative to high-throughput screening (HTS) that is commonly employed in pharmaceutical research. This "virtual" or in silico HTS approach eliminates some of the problems of experimental HTS, primarily expensive setup and low percentage of hits that include many false negatives (active compounds missed) and false positives (inactive compounds picked) (Langer and Wolber 2004). Hence, the earlier expectation of identifying all of the active compounds through HTS has been proven overly optimistic. Meanwhile, hypothesis-driven, pharmacophore-based screening is more narrowly focused in chemical space, speeding up discovery significantly. Pharmacophore screening also allows one to explore compounds that are not necessarily in the stockroom but are commercially available. All of these factors make pharmacophore-based screening a very effective alternative to HTS with significant cost and time savings.

The discovery of a potent (27 nM) small molecule inhibitor of transforming growth factor β type 1 by Eli Lilly through HTS provides a good comparison of pharmacophore screening and HTS. At the same time, the same compound was discovered by scientists from Biogen Idec Inc., through screening a database with a receptor-based pharmacophore model (Singh et al. 2003). The near simultaneous discovery of the same compound by two different companies employing two different approaches demonstrates the cost- and time-saving benefits of pharmacophore modeling (Güner 2005).

Caporuscio and Tafi summarized the benefits of pharmacophore modeling in their 2011 review article (Caporuscio and Tafi 2011):

> In fact, pharmacophores may be integrated in a drug discovery pipeline to rationalize structure-activity and structure-selectivity relationship data; identify new hits; guide lead optimization; profile ligand activity in silico; predict sites of metabolism, drug–drug interactions, side-effects, and toxicity; and, finally, design de novo inhibitors, in a computationally efficient way. Pharmacophores are a sufficiently general and abstract representation of ligand structures to allow the discovery of structurally diverse compounds, while providing an easy and intuitive interpretation of structure–activity relationships. Therefore, pharmacophore models are an exceptionally useful tool at every stage of the drug discovery and development process. In fact, they may suggest new chemical scaffolds potentially exerting the same biological activity of a lead compound (scaffold hopping), while showing a better ADME/Tox profile or escaping patenting restrictions. Pharmacophores may also lead to the identification of a hit for an orphan target.

Medicinal chemists who are already familiar with two-dimensional (2D) substructure and similarity-based searches of their corporate databases enjoy the chemical diversity that pharmacophore modeling provides. Because pharmacophore models involve abstract features involved in binding, the approach tends to retrieve compounds with different chemotypes (i.e., compounds with a different scaffold or chemical framework) (Leach et al. 2010). This makes pharmacophore modeling an important research tool if, for example, an institution has a number of active leads with safety issues. A new active lead of a different chemotype presents an opportunity to identify a compound series with better pharmacokinetic properties.

20.3.2 Pharmacophore for Modeling Drug Transporters

Drug transporters are expressed in many tissues, such as the intestine, liver, kidneys, and brain. As such, they play a major role in the pharmacokinetic processes of absorption, distribution, metabolism, and elimination (ADME). Absorption transporters were introduced in Chapter 4 (Section 4.3.2) as specific carriers located in the apical plasma membrane of enterocytes, which transport specific nutrients (e.g., sugars, vitamins, di- and tripeptides) across the gastrointestinal epithelium. Rational drug design offers the potential to develop drugs with substrate-like properties, for subsequent uptake by absorption transporters, thereby optimizing oral bioavailability. Furthermore, targeting vectors can be developed to specific transporters, to facilitate the delivery of drugs to target organs (e.g., the central nervous system [CNS]). Much like the use of pharmacophore modeling to identify new leads for drugs, the same general approach has been applied for drug transporters, to try and optimize drug design, delivery, targeting, pharmacokinetics, and safety. The following sections give specific examples of how rational drug design is providing strategies for the use of transporters to optimize drug delivery and targeting.

Structure-based pharmacophore modeling is employed if the three-dimensional (3D) structure of the transporting protein is available; unfortunately, not many structures are available, and homology modeling has not been very useful for these transporting proteins. Therefore, we are left with ligand-based pharmacophore modeling for drug transporters, which use the known active substrates or inhibitors to develop a model that represents a pattern of common features in space.

There are several examples of the successful use of pharmacophore modeling for identifying or predicting drug transporters. An example application is the human intestinal peptide transporter (hPEPT1), which broadly transports oligopeptides and peptidomimetics, including β-lactam antibiotics and angiotensin-converting enzyme inhibitors. A pharmacophore model for hPEPT1 was developed by Ekins et al. (2005), and the model was tested on 500 commonly prescribed drugs. Of the 65 compounds retrieved, 27 were known substrates or inhibitors, validating the notion that this approach is very efficient for identifying peptide transport potential of new chemical entities prior to in vitro testing.

In their 2006 review article, Chang and coworkers detailed a list of transporters for which pharmacophore models were developed (Chang et al. 2006), including a dopamine active transporter (DAT), i.e., a monoamine transporter associated with the reinforcing effects of cocaine. Pharmacophore-based identification of novel DAT inhibitors with a different structural framework is of interest for the treatment of cocaine abuse. They also listed a number of different efflux transporters:

1. *Permeability glycoprotein (P-gp)*: A membrane-bound protein that results in reduced oral drug absorption. Its expression at the blood–brain barrier (BBB) is critical in preventing CNS entry of drugs.
2. *Multidrug resistance protein 1 (MRP1)*: An efflux transporter of a broad range of compounds, including anticancer drugs and organic anionic substrates.
3. *Breast cancer resistance protein*: Consumes one ATP molecule per substrate transported, and its expression induces resistance to cancer cells.

In addition to the transporters listed earlier for which pharmacophore models have already been developed, Chang and coworkers also list a number of other transporters that are amenable for pharmacophore-based screening (Chang et al. 2006): (1) organic cation transporters (OCTs), (2) organic anion transporters, (3) vitamin transporters (SVCT1), (4) nucleoside transporters (NTs), (5) sodium taurocholate transporting polypeptide, and (6) apical sodium-dependent bile acid transporter (ASBT). Finally, in another 2006 review article, Chang and Swaan detail computational techniques used for modeling transporters including pharmacophore modeling and 2D and 3D quantitative SAR (3D-QSAR) approaches (Chang and Swaan 2006).

20.4   CHEMINFORMATICS

The past decade has witnessed tremendous expansion in publicly available database development for drug discovery purposes. Many of these databases have been curated by painstakingly obtaining information from the literature, e.g., analysis of physicochemical parameters that led to the development of Lipinski's rule of five (RO5) (see also Chapter 4, Section 4.3.4.5). Based on the observation that most drugs are relatively small and lipophilic, Lipinski and coworkers proposed the following rules for compounds with poor absorption or permeation: (1) the molecular mass is more than 500 Da; (2) the lipophilicity, determined by caculated Log P (Clog P) is greater than 5; (3) the number of hydrogen-bond donors is greater than 5; and (4) the sum of nitrogen and oxygen atoms is greater than 10 (Lipinski et al. 1997). Following the Lipinski's Rule of Five (RO5), variations on using different physicochemical parameters as filters have been proposed. Ghose et al. found that 80% of calculated compounds in the comprehensive medicinal chemistry database have log P between −0.4 and 5.6, with a median of 2.5, a molecular weight (MW) between 160 and 480 with median of 357, and the molar refractivity between 40 and 130 with median of 97 (Ghose et al. 1999).

Both Lipinski and Ghose predicted an increased failure in compounds with MW greater than 500 Da (with exceptions for certain classes of drugs). Oral bioavailability, however, is not determined by the MW alone. It has been proposed that molecular flexibility (measured by the number of rotatable bonds) can be used to predict oral bioavailability. A GlaxoSmithKline team observed that oral bioavailable compounds tend to have 10 or fewer rotatable bonds, with polar surface area (PSA) less than 140 Å^2 (Veber et al. 2002). Many current drug discovery programs generally incorporate one or more of the aforementioned rules as filters to screen drug-like molecules, with Lipinski's RO5 as the most popular.

Many biologically relevant databases have been developed and have been made available publicly. A brief description of these databases are given herein. Readers are recommended to go to literature to find out more details (Nicola et al. 2012).

ChEMBL (http://www.ebi.ac.uk/chembl) is an open bioactivity database containing more than 1.3 million unique compound structures and 12 million bioactivity data points with more than 2800 human proteins as drug targets. This database currently includes annotations such as the United States Adopted Name (USAN) or International Nonproprietary Name (INN). A compound's USAN or INN is typically assigned in the development stage, and thus a compound with a USAN or INN name is of biological significance. A study showed that within 4 years following the assignment of a USAN, a candidate drug can be approved (Bento et al. 2014). The interface of current ChEMBL website contains the information of Lipinski's RO5, compound chirality, binding affinity, toxicity, and ADME information. ChEMBL also provides information on Kinase (Kinase SARfari) and G protein-coupled receptors (GPCR) (GPCR SARfari). It contains protein structures and compounds that are reported to interact with related proteins.

Studies on 500,000 compound activity data reported in the ChEMBL database showed that not many oral drugs have nM potency and that many oral drugs yield off-target activities. Other research recorded that 62% of nearly 800 oral drugs have a minimum dose of greater than 10 mg, and approximately 40% have a dose of 50 mg or above (Gleeson et al. 2011). This indicates that using potency of nM or better as a filter would have screened out some drugs on the market.

BindingDB (www.bindingdb.org) is a public, web-accessible database with quantitative information such as K_i, IC_{50}, EC_{50}, and K_D provided wherever a well-defined target can be found. Currently, BindingDB contains over 1 million binding data with close to 7,000 protein targets and 450,000 small molecules. It provides downloadable databases that have been curated from literature and other database sources. It allows structure-based or simplified molecular-input line-entry system (SMILES) string–based searches for user-defined compounds. BindingDB also enables one to perform virtual screening by allowing users to upload a list of compounds and returning the ranking based on Tanimoto similarity method or the binary kernel discrimination method or the support vector machine, a machine-learning approach; the latter two methods use a training set and a test set to validate the prediction.

The PubChem database (pubchem.ncbi.nlm.nih.gov), released in 2004, provides information on small molecules. Users can search a given compound or search a type of compounds with defined physicochemical properties. Users can define downloadable compounds based on the range of MW, log P, H-bond donor count, H-bond acceptor count, topological polar surface area (TPSA), heavy atom number, and even the presence or absence of chiral centers. Compounds available in PubChem generally are reported with target proteins if any. The metabolism information of compounds on PubChem, however, is not as comprehensive as those found in the DrugBank.

Compared to BindingDB, ChEMBL, and PubChem database, DrugBank (www.drugbank.ca) is a much smaller yet information-rich database. It hosts 7700 drug entries, which include approximately 1600 FDA-approved small molecule drugs and over 6000 experimental drugs. It contains very useful information regarding many pharmacokinetic data such as bioavailability, clearance, half-life, dosage, solubility, drug–drug interactions, and drug metabolism, as well as available metabolite information, information on drug targets, and physicochemical parameters like log P, pKa, and MW.

Nuclear Receptor Ligands and Structures Benchmarking DataBase (NRList BDB; http://nrlist.drugdesign.fr) is a special database curated for nuclear receptors and their interacting ligands. Nuclear receptors are proteins that bind to DNA. After binding to a small molecule, hormones can switch on or off transcription of DNAs. Hence, nuclear receptors are transcription factors. Androgen receptors, estrogen receptors, glucocorticoid receptors, and progesterone receptors are all nuclear receptors. NRList BDB has 339 structures and 9900 compounds (agonists and antagonists) (Lagarde et al. 2014).

Guide to Pharmacology (www.guidetopharmacology.org) is a new open-access resource maintained by the International Union of Basic and Clinical Pharmacology/British Pharmacological Society that provides an in-depth, expert-curated overview of ligands and targets. The web resource provides an "authoritative" and "complete" landscape of current and research drug targets with "accurate" basic science on drug action. It offers guidance for selecting appropriate compounds for in vitro and in vivo experiments. Readers can browse the ligand list to choose the drugs under study or can use the search function to locate a query. For each recorded ligand, computed physicochemical data, such as H-bond acceptors, H-bond donors, number of rotatable bonds, TPSA, MW, and log P, as well as biological activities (pK_i) and clinical indication, are provided. For target proteins, allele variations with mutations and associated references are provided, which is a very good resource for polymorphism study (Pawson et al. 2014).

Isosteres and bioisosteres have long been considered in the design and synthesis of novel therapeutic compounds. By curating ChEMBL compounds with annotated IC_{50}, EC_{50}, K_i, or K_D, a pair of molecules was identified by the Hussain and Rea algorithm to identify the difference that lies in a small substructural component. Such pairs of molecules differing only by small substructural exchanges are called matched molecular pairs. Algorithms were developed to classify the structural exchanges into three categories: single cut (substituent replacements), double cut (linker replacement), and triple cut (scaffold replacements). A scoring scheme has been implemented to aid ranking of different replacements. This program is SwissBioisostere (http://www.swissbioisostere.ch) (Wirth et al. 2013).

With so many databases available, each with tens of thousands of compounds, if not more (for some databases there are more than a million compounds), how can one be certain how to evaluate the chemical space with biological relevance, or how can one measure the diversity of the dataset?

To evaluate chemical similarity, dissimilarity, and diversity among a set of compounds, one needs to calculate the distance between chemical compounds positioned in certain multidimensional space, i.e., the "chemistry space," as proposed by Pearlman and Smith (1998). The dimensionality of a chemistry space can be chosen using "molecular fingerprint" to describe each compound in a population. The fingerprint is represented by bit strings to indicate whether certain substructural features are present or not. In a recent paper, Reymond and coworkers used 34 fingerprints (or 34D chemical space) to calculate the chemistry space for compounds deposited in ChEMBL, DrugBank, and PubChem and compared such chemistry space to those derived from decoys and from ZINC database. Compounds from these database were analyzed using SMILES fingerprint (SMIfp) method, and it was found out that the chemical space of compounds in DrugBank is much smaller than those observed in ChEMBL, PubChem, and ZINC (Schwartz et al. 2013). Examples of SMIfp descriptor are C, representing the presence of a nonaromatic carbon, and the c for an aromatic carbon. Other than fingerprint-based descriptors (ECFP4 and FCFP4), pharmacophore-based descriptors (TAT, TAD, TGT, TGD, and so on), shape-based descriptors, connectivity-matrix-based descriptor (BCUT), physicochemical property-based descriptors (2D descriptors, such as log P, MW, log S, and the like), and the recent so-called Bayes affinity fingerprints (in silico predicted bioactivity spectra) have been proposed to use to select compounds from libraries of diverse dataset (Koutsoukas et al. 2014).

20.5   QSAR

Structural and biological activity predictions based on mechanical molecular models were accomplished in the mid- to late twentieth century. There are two classical examples: (1) Using paper models and a remarkable command of structural chemistry, Linus Pauling predicted two conformations of proteins (the alpha helix and beta sheet). (2) The same molecular modeling approach was taken by Watson and Crick, which, coupled with x-ray data from Rosalind Franklin, resulted in the prediction of the famous B form of DNA, the so-called Franklin–Watson–Crick structure. Using simple regression statistics, Corwin Hansch demonstrated excellent correlations with physical properties (hydrophobicity [log P], electronic [σ], and steric [E_T] effects) with biological activity (Hansch 1969). The essential idea was that the biological activity of a molecule could be expressed as a mathematical function of the physical properties. This approach works because the physical properties of a molecule are a reflection of its structure. It could be argued that the success of the Hansch method was the beginning of a new age in computer-assisted drug design. It was during this time that many pharmaceutical companies turned their serious attention to the use of molecular modeling. There have been many contributors to this field who are not mentioned in this book chapter.

20.5.1   3D-QSAR

Rather than focusing on the physical properties of a compound, the natural question was how the molecular structure itself could be correlated with biological activity. In the 1980s, Richard Cramer developed a method, comparative molecular field analysis (CoMFA), to answer this question (Cramer et al. 1988). The essential strategy behind this approach was to explore the steric and electrostatic interactions between a positively charged probe atom and the molecular structure directly. Using partial least squares fitting, the information can be used to create the so-called 3D-QSAR equation as a function of the calculated molecular descriptors. Structural alignment of molecular structures *in silico* is a critical question, and it has a major impact on the effectiveness of the CoMFA-based QSAR analysis. The steric and electrostatic fields generated can be displayed graphically within the Sybyl software so that at a glance medicinal chemists know where to make molecular modifications.

The limitations to this method are many; mostly, there is concern about the ability to extrapolate to diverse compounds not used in the development of the original CoMFA equation. In 1994, Klebe, Abraham, and Mietzner proposed an improved comparative molecular similarity indices analysis (CoMSIA), which is now widely used and easier to interpret (Klebe et al. 1994). During this time, one of us (JPB) was involved in the development of a different 3D-QSAR approach, molecular skeleton analysis, with improved results (Liang et al. 1996). These developments and subsequent work by others ushered in a new age of molecular modeling.

20.5.2 QSAR for Modeling Drug Transporters

Looking at drug delivery applications, both 2D- and 3D-QSAR techniques have been extensively used for modeling drug transporters. These models are detailed in the Chang and coworkers review article, which we recommend the reader to consult, for the actual references for individual models (Chang et al. 2006). P-gp is the most extensively modeled system through QSAR methods. Apart from the several pharmacophore-based models developed by Ekins (Catalyst) and Yates (DISCO), 3D-QSAR approaches through CoMFA and CoMSIA are published by Cianchetta and coworkers. Even though traditional 2D-QSAR models were developed for the human OCTs (hOCT1 and hOCT2), 3D-QSAR approaches have been more extensively used for a broader range of transport proteins including MR-associated protein 2 (MRP2), hOCT2, human concentrative NTs (hCNT1 and hCNT2), human equilibrative NT1 (hENT1), ASBT, DAT, serotonin transporter (SERT), BBB choline transporter, organic anion transporting polypeptide 1a5, and PEPT1.

20.6 STRUCTURE-BASED DRUG DESIGN

The combination of molecular modeling methods with protein or DNA x-ray data serves as the foundation for structure-based drug design. Extensive work in the 1980s with dihydrofolate reductase demonstrated the usefulness of the method (Kuyper et al. 1985). Although there have been many highly visible achievements utilizing this approach, the major difficulty is the uncertainty of how a target macromolecule will adjust itself to accommodate the small molecule. How much conformational flexibility does the macromolecule ligand have? Specifically, using molecular modeling software, the deletion of a ligand bound to a ligand-macromolecule complex followed by the insertion of a new ligand is complex, and it does not guarantee an accurate depiction of what happens in nature.

The development of new HIV protease inhibitors represents a major advance in a therapeutic area of international concern (Bowen 2004). The design of the HIV protease inhibitors and subsequent approval by the FDA represents state-of-the-art molecular modeling. All of these compounds involved structure-based drug design or related methods. These drugs include saquinavir (Invirase®, Roche, 1995; Fortovase®, Roche, 1997), indinavir (Crixivan®, Merck, 1996), ritonavir (Norvir®, Abbott, 1996), nelfinavir (Viracept®, Pfizer, 1997), amprenavir (Agenerase®, GlaxoSmithKline, 1999), lopinavir (Kaletra®, Abbott, 2000), atazanavir (Reyataz®, BMS, 2003), tipranavir (Aptivus®, Boehringer-Ingelheim, 2005), and darunavir (Prezista®, Tibotec, 2006). HIV protease is a symmetrical dimer. Each monomer is composed of 99 amino acids. Figure 20.3 shows saquinavir bound in the active site of HIV protease (3OXC, 1.16 Å).

20.6.1 Structure-Based Approaches for Modeling Drug Transporters

Due to lack of availability of 3D structures of the transporter proteins, substrate-based models via pharmacophore and QSAR approaches have been more extensively used and summarized earlier. There are, however, several attempts to perform structure-based approaches through homology modeling of these proteins. Homology models are constructed using the structural information of a template protein possessing a large sequence similarity with the target protein. Once a homology model is constructed, docking and structure-based pharmacophore models are typically utilized (Chang and Swaan 2006).

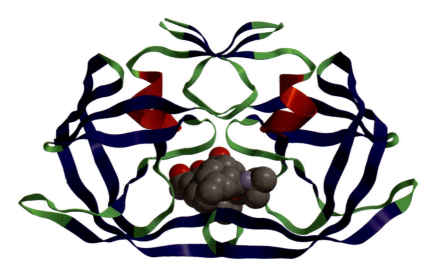

FIGURE 20.3   The molecular structure of saquinavir bound in the active site of HIV protease (3OXC, 1.16 Å). (Courtesy of the Bowen Group, Center for Drug Design, Mercer University, College of Pharmacy, Atlanta, GA. Graphics rendered using Spartan software by Pamela Ohsan, Wavefunction, Inc., Irvine, CA.)

The membrane transporter homology modeling studies include (1) ASBT using bacteriorhodopsin crystal structure, (2) glucose transporter using lactose permease (LacY), and (3) glycerol-3-phosphate transporter (GlpT) crystal structures. GlpT and LacY structures are the most utilized structures for homology modeling of transporters. The models that use these structures also include (1) glucose-6-phosphate transporter, (2) Tn10-encoded metal-tetracycline/H⁺ antiporter, (3) rat vesicular monoamine transporter, (4) *Escherichia coli* NT Pho84 phosphate transporter, and (5) oxalate transporter.

Another popular protein whose structure is used as template for homology modeling is ATP-binding cassette transporter (MsbA) crystal structure (either *Vibrio cholerae* or *E. coli*). The constructed homology models in this manner include (1) lactococcal multidrug transporter, (2) P-gp, and (3) MRP1. Other homology models are the sulfate transporter, using SpoIIAA crystal structure, and the OCT2, using GlpT crystal structure.

Ravna and coworkers creatively chained one homology model to construct another and then another. They developed Na⁺/H⁺ antiporter (NhaA) using NhaA election density projection map and chained this homology model to develop the model for DAT (Ravna 2003). Similarly, they used the DAT homology model to construct SERT and subsequently constructed noradrenaline transporter using their SERT homology model.

Clearly, many different computational techniques are being used to model transporter proteins and help us understand the interaction between these proteins and their substrates.

20.7   PROTEOMICS AND DRUG TARGET IDENTIFICATION

The past decade has witnessed explosive growth in proteomics research. Proteomics studies the entire set of proteins, which can be produced or modified by cellular and tissue-/organ-level systems under either physiological or pathological conditions. Examples of posttranslational protein modifications include phosphorylation and ubiquitination, two crucial events regulating cell signaling, particularly in cancer cells.

Currently, the most common method used in proteomics is the untargeted approach—shotgun liquid chromatography coupled with tandem mass spectrometry (LC–MS/MS). This approach starts with trypsin digestion of proteins, which degrades proteins into small peptides, separated by

LC, followed by the identification of small peptides using MS. Each individual piece of peptide, corresponding to the MS peptide peak, can be further fragmented giving rise to smaller peptide fragments, or even amino acid residues, whose information is contained in the secondary MS/MS spectrum. Therefore, the LC–MS/MS method allows for identifying and quantifying proteins. This method can identify many proteins; however, the complexity in determining the peptide mixtures involved in complex protein networks complicates the identification process.

An alternative to the untargeted LS–MS/MS method is LC coupled with selected reaction monitoring mass spectrometry (LC–SRM–MS). As the name indicates, the SRM method is used to identify and quantify target proteins using preselected stable isotope-labeled internal references. However, the disadvantage of this SRM method is its dependence on reference proteins/peptides. The problem is that labeled reference proteins sometimes cannot be obtained, and it would be very costly to have all the labeled reference proteins for identification of proteomics in a cell. Another challenge of proteomics is the vast amount of information generated from proteomics experiments. How to utilize such a tremendous amount of data to interpret the biological processes or to identify target proteins for a given drug molecule remains a challenge.

One advance in computational proteomics that is helping to tackle the problem of labeled reference proteins in the SRM method is the development of an open-source software SparseQuant (http://www.stat.purdue.edu/~ovitek) (Chang et al. 2014). This software uses a statistical approach, also called the label-sparse quantification method. In this method, a statistical model is first used to describe the possible sources of variation for each protein with observed log intensities (a measure of conditions, biological and technical replicates, and transition peptides/proteins). The log intensity is then normalized to the log intensity of reference proteins. The statistical inference for predicting the log intensities of the unobserved reference transitions comes from the estimation of the overall deviations of the observed reference transitions between two sets of experiments. This prediction model includes a probability distribution of plausible values. This label-sparse quantification method has successfully identified proteins GLGB, Q8CE68, SRBP1, and SIRT2 among others in the mouse liver study whose identities might otherwise be overlooked.

In addition to identifying proteins, computational proteomics has been applied to predict a novel therapeutic indication for an approved drug, also called "repurposing" or "repositioning" a drug. The cost of bringing a new drug to the market from the original discovery phase is prohibitively high, estimating to be US $900 million (Zhong and Bowen 2013), and, as mentioned in Section 20.1, the cost has now been estimated to exceed US $2 billion. Repurposing an already approved drug has its quick reward, without risking the attrition, because drugs under study already demonstrate an established safety/side effect profile. The practice of repurposing a drug is quite successful in that approximately 30% of approved new drugs in 2009 were repurposed drugs. Finding a new target protein for a drug is nontrivial. Three approaches have been proposed to minimize the cost and to reduce the complexity of identifying appropriate target proteins: (1) pharmacophore based, (2) network based, or (3) target based. The pharmacophore-based method utilizes the essential pharmacophoric features (i.e., pharmacophores) of a target protein and maps such a pharmacophore to a database of drugs. Drugs satisfying such a pharmacophore are predicted to be active against the target protein. The network-based method requires a large-scale database to predict how a drug molecule may affect the target protein based on data mining of ontology networks. The target-based approach relies on the knowledge of known drug target and drug mechanism of action.

An example of target-based drug repurposing is the discovery of inhibitors targeting protozoan parasites, shown in Figure 20.4. The workflow started with obtaining drug molecules from DrugBank and ChEMBL (two of the popular databases for obtaining drugs and compounds; see Section 20.4). The drug information was then loaded onto MySQL. The drug target protein sequences obtained from ChEMBL were searched against the GenBank database at the National Center for Biotechnology Information (NCBI) website, using the Basic Local Alignment Search Tool for Proteins (BLASTP) approach with a certain cutoff number. The drug molecular information and the newly identified protein information were combined and calculated using the National

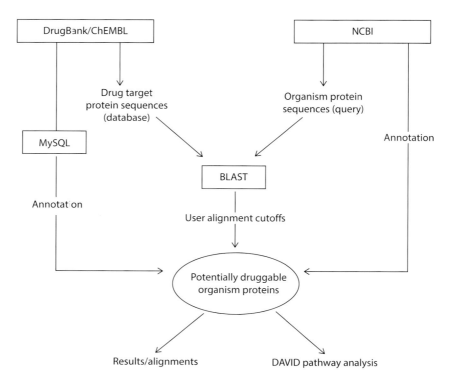

FIGURE 20.4 Overview of the workflow used for a target-based drug repurposing approach, used in the discovery of inhibitors targeting protozoan parasites. (Reprinted from Sateriale, A., Bessoff, K., Sarkar, I.N. et al., Drug repurposing: Mining protozoan proteomes for targets of known bioactive compounds, *J. Am. Med. Inform. Assoc.*, 21, 238–244, 2014, by permission of Oxford University Press.)

Institutes of Health DAVID's API services. The results were quite exciting: among 427 approved drugs in DrugBank, 45 were predicted to be efficacious in *Cryptosporidium parvum*, and 4 exhibited confirmed activity (Sateriale et al. 2014).

Network-based drug repurposing, on the other hand, depends on how well the network is constructed. In order to construct a network or systems approach that is meaningful to reflect the complex network of diseases under study, there are three concerns to be addressed: (1) What systems biology knowledge should be included in the network? Is the network comprehensive? (2) How such network is used (presumably including signature genes/proteins for certain diseases) to characterize disease states? (3) Is the information of the genomic and epigenomic variations effectively included in the network? These variations are critical for they may affect the susceptibility of individuals or different racial groups to various drug dosages.

One of the key concepts related to building a systems biology network is modularity (Hansen and Iyengar 2013). For instance, the BCR-ABL (a fused protein between Break point Cluster gene, BCR, and Abelson gene, Abl) protein tyrosine–kinase inhibitor (TKI) imatinib (IM) has been used successfully for the treatment of chronic myelogenous leukemia. Yet more than 17 mutations are observed in the ABL kinase domain (Zhong et al. 2009). To identify drugs that may be able to overcome the mutations, particularly the T315I mutant, four TKI inhibitors (IM, nilotinib [NILO], dasatinib [DASA], and danusertib [DANU]) were used in a study along with the construction of mesoscale network that started with a black box model and ended up with fully mechanistic models. These four TKIs were applied to K562, Ba/F3-210, T315I, and M351I cells. Proteins in cells were prepared by 2D gel electrophoresis (2D-PAGE), and protein identification was performed using mass spectrometry. The mesoscale models for the mechanism of action of TKIs, models to assess the efficacy of TKIs on protein expression and

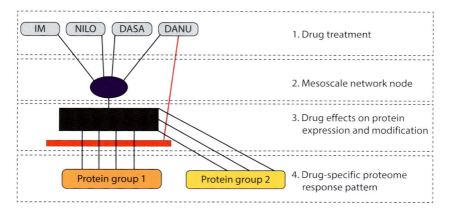

FIGURE 20.5 Structure of mesoscale pathways for induced protein expression, for prediction of the mode of action for tyrosine kinase inhibitors. The black block represents induction of the protein expression by the main pathway, whereas the red block indicates an inhibition via the main pathway (see text for details). (IM, imatinib; NILO, nilotinib; DASA, dasatinib; DANU, danusertib). (Adapted from Balabanov, S. et al., *PLOS ONE*, 8, e53668, 2013. Copyright Public Library of Science.)

apoptosis level, hence were constructed. Figure 20.5 depicts the observations in Ba/F3-p210 wild-type cells, IM, DASA, and NILO activate the same key node (oval, blue), which induces a black box of protein expression (thick rectangle, black), whereas DANU also activates a second node (thin rectangle, red). Further analysis on the mesoscale network models suggested that IM, DASA, and NILO inhibited the same BCR-ABL pathway, albeit with different inhibitory profile, and that DANU may have a distinct protein induction profile via the Aurora family of kinases pathway (Balabanov et al. 2013).

Several software programs have been developed to assist computational proteomics analyses. TARGETgene (http://bmsr.usc.edu/Software/targetgene) was developed to help identify mutation drivers and to predict therapeutic targets for a given molecule (Wu et al. 2012). The graphics user interface of TARGETgene allows user to select cancer types (e.g., breast cancer) as therapeutic targets, to load a group of candidate genes. The program then searches candidate genes and returns gene panels, along with drug panels that predict what drugs may interact with the genes, with related disease information provided. The prediction can be assessed by enrichment factors or evaluated by comparing the ratio of true positives versus false positives.

SynSysNet is a web server–based software used to identify protein–drug interactions, particularly for synapse-specific proteins (von Eichborn et al. 2013). This website covers Huntington disease, epilepsy, multiple sclerosis, Parkinson's disease, and schizophrenia, with over 1,000 genes, 6,000 protein structures, and 46,000 compounds including 750 approved drugs. Users can input the name of a drug molecule, and the program will return all potential interactions of a given drug.

Search Tool for Interactions of Chemicals is another web-based database search engine for studying protein–drug interactions (Kuhn et al. 2014). This software uses a text-mining method in which all ligands are input using SMILES strings. Upon receiving the SMILES strings for a ligand, the program then searches the compound database to identify the input compound and subsequently to identify the potential partners. The program covers interaction network of 390,000 compounds and 3.6 million proteins from over 1,100 organisms. The confidence level for protein–ligand interaction prediction is said to be 45%.

Kyoto Encyclopedia of Genes and Genomes (www.genome.jp/kegg) is a database server containing protein, genome, compounds (including drugs), and systemic pathways network information (Kanehisa et al. 2012, 2014). The disease and drug information is also provided. Users can just enter the name of a drug or a compound; the server will return all the information regarding all potential target proteins, metabolism (if any), and very useful structures of the molecules.

20.8 SOLUBILITY AND COMPUTATIONAL PREDICTION

As described in Chapter 3, poorly soluble drugs are problematic in many key areas: (1) in the drug discovery process, as potentially promising lead candidates may have to be terminated; (2) in formulation development, as the drug may be too insoluble for formulation in a suitable vehicle for in vivo administration; and (3) in in vivo performance.

A report has shown that approximately 35%–40% of immediate-release oral drugs among the top 200 drugs listed in the United States, Great Britain, Spain, and Japan are practically insoluble (solubility < 0.1 mg/mL), with approximately 15% sparingly soluble (10–33 mg/mL), 15% slightly soluble (1–10 mg/mL), and 5% very slightly soluble (0.1–1 mg/mL). In other words, only approximately 25%–30% of oral drugs show good solubility (>33 mg/mL) (Figure 20.6). Values for drug aqueous solubility can be obtained from the Merck Index, the USD DI (Drug Information). Approaches to increasing solubility of oral drugs include the ionizations of the active drug ingredients into different salts of sodium, calcium, and potassium for acidic drugs and into salts of hydrochloride, citrate, tartrate, or other acids complexes for basic drugs. In addition, cocrystallization with solvents such as ethanol or with water (becoming hydrates) also improves water solubility (Zhong et al. 2013). These and other approaches are described further in Chapter 3.

To minimize the likelihood of synthesis of water-insoluble compounds that may pose challenges for formulation, or even risks of being terminated, many computational approaches have been developed to predict aqueous solubility of organic molecules. Some 1D parameters such as MW, count of electrons, and hydrophobicity have been used to evaluate aqueous solubility. Hansch and coworkers have developed a simple equation (Equation 20.10) to use the octanol/water partition coefficient (log P) to correlate aqueous solubility of liquid organic compounds (Hansch et al. 1968), where n is the number of organic compounds used, s is the standard deviation, and r^2 is the correlation coefficient.

$$\log S = -1.339\log P + 0.978, \quad n = 156, \quad s = 0.473, \quad r^2 = 0.874 \tag{20.10}$$

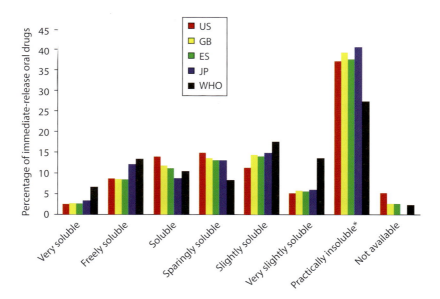

FIGURE 20.6 Comparison of the distribution of drug solubility on the United States, United Kingdom, Spain, Japan, and WHO lists. (Reprinted with permission from Takagi, T., Ramachandran, C., and Bermejo, M. et al., A provisional biopharmaceutical classification of the top 200 oral drug products in the United States, Great Britain, Spain, and Japan, *Mol. Pharm.*, 3, 631–643. Copyright 2006 American Chemical Society.)

In Equation 20.10, the log P is the measurement of hydrophobicity, and the log S is the measurement of a compound's solubility, with S being the concentration of the compound in mol/L for a saturated aqueous solution in equilibrium with the most stable form of the crystalline compound. Generally, a drug's log S ranges between −1 and −5. Equation 20.10 works well for the study set of liquid organic compounds. It is not as accurate, however, when applicable to solid compounds, due to the crystal lattice effects (Yalkowsky and Valvani 1980), drug melting points, and entropy of fusion (Jouyban 2008). Another simple parameter that can be used to measure hydrophilicity is the 3D PSA or topological PSA that measures the topological surface areas contributed by the polar atoms such as N and O. By combining log P, TPSA, MW, and other topological molecular descriptors via artificial neural network, Liu and So were able to predict the molar solubility of a diverse set of 1312 organic compounds with correlation coefficient of 0.92 between the predicted and experimental log S (Liu and So 2001). The topological indices can be as simple as the number of H-bond acceptors and aromaticity indicator, or as complicated as atom-type electrotopological state indices that classify molecular fragments into small different atoms or groups of atoms, each with different aqueous solubility contribution. The total solubility of a molecule is a summation of contribution from all these molecular descriptors (or structural fragments) (Jorgensen and Duffy 2002):

$$\log S = \sum_i a_i n_i + a_0 \qquad (20.11)$$

These topological indices, however, cannot fully account for the conformational, configurational, and other 3D effects. But the correlation factors between predicted and experimental log S appeared to be very impressive with r^2 ranging from 0.89 to 0.92 (Huuskonen 2000). Other studies have shown that the aqueous solubility log S is negatively correlated to log P, MW, and melting point (t_m), all of which suggest that increasing hydrophobicity and increasing MW will reduce aqueous solubility (Equation 20.12). The relationship between experimentally measured log S (pH = 7.4) and log P (measured log D 7.4) for ≈500 neutral AstraZeneca proprietary compounds also showed a negative relationship with log $S = -0.62$ log $P + 3.10$ (Wenlock and Barton 2013). In the following equation, the log P is the octanol/water partition coefficient, MW is the molecular weight, and f_i is the correction factor for each specific functional group (or molecular fragment):

$$\log S = a_0 - a_1 \log P - a_2 \text{MW} + \sum_i f_i \qquad (20.12)$$

The QikProp program in the Schrödinger software suite was developed by Jorgensen's group, and this program takes the following factors into account when calculating log S: solvent accessible surface area (SASA), number of H-bond donors, number of H-bond acceptors, number of amino atoms (#N), number of amide (NHCO), number of rotatable bonds, and weakly PSA. The SASA is calculated with a 3D structure as an input and measures the surface area of a molecule by using a probe radius of 1.4 Å to travel over the surface of the probed molecule. Therefore, the SASA is a 3D parameter, also called the solvent-accessible volume. The correlation coefficient of using this QikProp algorithm to predict log S was 0.90, which is very impressive (Jorgensen and Duffy 2002).

20.9 CONCLUSION

Computer-based methods are now common in drug discovery. Scientists have routinely used molecular mechanics and quantum mechanics, as well as the relatively new hybrid QM/MM methods, to design biologically active agents. The explosive growth of the power of computers has a direct impact on drug discovery. Today, energy-based methods can be applied to large drug–receptor and enzyme–substrate interactions. Twenty and thirty years ago, specialized and expensive graphics workstations were required for computer-assisted drug design, but this is no longer the case. Complex molecular interactions can be modeled, displayed, and manipulated using relatively inexpensive laptops. One of the main challenges facing medicinal chemists remains the prediction of physical properties such as solubility and other physical phenomena that influence the effectiveness of a drug, including its drug delivery and targeting capabilities. There will be continued efforts to develop predictive methods for absorption, distribution, metabolism, excretion, and toxicity. Too often, promising drug candidates have failed clinical trials. Due to lack of availability of 3D structures of the transporter proteins, substrate-based models via pharmacophore and QSAR approaches have been more extensively used.

Future advances in computer-assisted drug design will increase the likelihood of the successful of a drug candidate, but it can never be forgotten that drug discovery is primarily driven by experimental work. The goal of computer-based methods is not to replace experiments but to assist with them.

ACKNOWLEDGMENTS

This chapter makes use of Spartan molecular modeling software from Wavefunction, Inc. (www. wavefun.com). Students may purchase the Spartan Student Edition from: http://store.wavefun.com/ product_p/spstudent.htm. For instructors wishing to incorporate molecular modeling activities in conjunction with this chapter, contact Wavefunction, Inc. for details on student adoptions.

REFERENCES

Balabanov, S., T. Wilhelm, S. Venz et al. 2013. Combination of a proteomics approach and reengineering of meso scale network models for prediction of mode-of-action for tyrosine kinase inhibitors. *PLOS ONE* 8:e53668.

Bento, A.P., A. Gaulton, L.J. Hersey et al. 2014. The ChEMBL bioactivity database: An update. *Nucleic Acids Research* 42:D1083–D1090.

Bowen, J.P. 2004. Computational chemistry and computer-assisted drug design. In *Wilson and Gisvold's Textbook of Organic Medicinal and Pharmaceutical Chemistry*, 11th ed., eds. J. Block and J.M. Beale, Jr., pp. 919–947. Philadelphia, PA: Lippincott, Williams, & Wilkins.

Bowen, J.P. and H.A. Zhong. 2013. Computational chemistry. In *Encyclopedia of Pharmaceutical Science and Technology*, 4th ed., ed. J. Swarbrick, pp. 600–614. Florence, Italy: Taylor & Francis.

Caporuscio, F. and A. Tafi. 2011. Pharmacophore modeling: A forty year old approach and its modern synergies. *Current Medicinal Chemistry* 18:2543–2553.

Chang, C., S. Ekins, P. Bahadduri et al. 2006. Pharmacophore-based discovery of ligands for drug transporters. *Advanced Drug Delivery Reviews* 58:1431–1450.

Chang, C. and P.W. Swaan. 2006. Computational approaches to modeling drug transporters. *European Journal of Pharmaceutical Sciences* 27:411–424.

Chang, C.Y., E. Sabidó, R. Aebersold et al. 2014. Targeted protein quantification using sparse reference labeling. *Nature Methods* 11:301–304.

Christie, B.D., D.R. Henry, O.F. Güner et al. 1990. MACCS-3D: A tool for three-dimensional drug design. In *Proceedings of the 14th International Online Information'90 Meeting*, ed. D.I. Rait, pp. 137–161. Oxford, U.K.: Learned Information.

Cramer III, R.D., D.E. Patterson, and J.D. Bunce. 1988. Comparative molecular field analysis (CoMFA). 1. Effect of shape on binding of steroids to carrier proteins. *Journal of the American Chemical Society* 110:5959–5967.

DiMasi, J.A. and H.G. Grabowski. 2012. R&D costs and returns to new drug development: A review of the evidence. In *The Oxford Handbook of the Economics of the Biopharmaceutical Industry*, eds. P.M. Danzon and S. Nicholson, pp. 21–46. Oxford, U.K.: Oxford University Press.

Dixon, S.L., A.M. Smondyrev, and S.N. Rao. 2006. PHASE: A novel approach to pharmacophore modeling and 3D database searching. *Chemical Biology and Drug Design* 67(5):370–372.

Ehrlich, P. 1898. Über die constitution des diphtheriegiftes. *Deutsche Medizinische Wochenschrift* 24:597–600.

Ekins, S., J.S. Johnston, P. Bahadduri et al. 2005. *In vitro* and pharmacophore-based discovery of novel hPEPT1 inhibitors. *Pharmaceutical Research* 22:512–517.

Ghose, A.K., V.N. Viswanadhan, and J.J. Wendoloski. 1999. A knowledge-based approach in designing combinatorial or medicinal chemistry libraries for drug discovery. 1. A qualitative and quantitative characterization of known drug databases. *Journal of Combinatorial Chemistry* 1:55–68.

Gleeson, M.P., A. Hersey, D. Montanari et al. 2011. Probing the links between *in vitro* potency, ADMET and physicochemical parameters. *Nature Reviews Drug Discovery* 10:197–208.

Gund, P., W.T. Wipke, and R. Langridge. 1973. Computer searching for molecular structure file for pharmacophoric patterns. In *Proceedings of the International Conference on Computers in Chemical Research and Education*, Lubljana, Zagreb, July 12–17, 1973, eds. D. Hadzi and J. Zupan, Vol. 3, pp. 5–33. Amsterdam, the Netherlands: Elsevier Scientific.

Güner, O.F. 2005. The impact of pharmacophore modeling in drug design. *Investigational Drugs* 8:567–572.

Güner, O.F. and J.P. Bowen. 2014. Setting the record straight: The origin of the pharmacophore concept. *Journal of Chemical Information and Modeling* 54:1269–1283.

Güner, O.F., D.W. Hughes, and L.M. Dumont. 1991. An integrated approach to three-dimensional information management with MACCS-3D. *Journal of Chemical Information and Computer Sciences* 31:408–414.

Hansch, C. 1969. A quantitative approach to biochemical structure–activity. *Accounts of Chemical Research* 2:232–239.

Hansch, C., J.E. Quinlan, and G.L. Lawrence. 1968. The linear free energy relationship between partition coefficients and the aqueous solubility of organic liquids. *Journal of Organic Chemistry* 33:347–350.

Hansen, J. and R. Iyengar. 2013. Computation as the mechanistic bridge between precision medicine and systems therapeutics. *Clinical Pharmacology and Therapeutics* 93:117–128.

Hehre, W.J., L. Radom, P. Schleyer et al. 1986. *Ab Initio Molecular Orbital Theory*. New York: John Wiley & Sons.

Hurst, T. 1994. Flexible 3D searching: The directed tweak technique. *Journal of Chemical Information and Computer Sciences* 34:190–196.

Huuskonen, J. 2000. Estimation of aqueous solubility for a diverse set of organic compounds based on molecular topology. *Journal of Chemical Information and Computer Sciences* 40:773–777.

Jorgensen, W.L. and E.M. Duffy. 2002. Prediction of drug solubility from structure. *Advanced Drug Delivery Reviews* 54:355–366.

Jouyban, A. 2008. Review of the cosolvency models for predicting solubility of drugs in water-cosolvent mixtures. *Journal of Pharmacy and Pharmaceutical Sciences* 11:32–58.

Kanehisa, M., S. Goto, Y. Sato et al. 2012. KEGG for integration and interpretation of large-scale molecular data sets. *Nucleic Acids Research* 40:D109–D114.

Kanehisa, M., S. Goto, Y. Sato et al. 2014. Data, information, knowledge and principle: back to metabolism in KEGG. *Nucleic Acids Research* 42(Database issue):D199–D205.

Klebe, G., U. Abraham, and T. Mietzner. 1994. Molecular similarity indices in a comparative analysis (CoMSIA) of drug molecules to correlate and predict their biological activity. *Journal of Medicinal Chemistry* 37:4130–4146.

Koutsoukas, A., S. Paricharak, W.R. Galloway et al. 2014. How diverse are diversity assessment methods? A comparative analysis and benchmarking of molecular descriptor space. *Journal of Chemical Information and Modeling* 54:230–242.

Kuhn, M., D. Szklarczyk, S. Pletscher-Frankild et al. 2014. STITCH 4: Integration of protein–chemical interactions with user data. *Nucleic Acids Research* 42(D1):D401–D407.

Kuyper, L.F., B. Roth, D.P. Baccanari et al. 1985. Receptor-based design of dihydrofolate reductase inhibitors: Comparison of crystallographically determined enzyme binding with enzyme affinity in a series of carboxy-substituted trimethoprim analogs. *Journal of Medicinal Chemistry* 28:303.

Labute, P., C. Williams, M. Feher et al. 2001. Flexible alignment of small molecules. *Journal of Medicinal Chemistry* 44:1483–1490.

Lagarde, N., N. Ben Nasr, A. Jérémie et al. 2014. NRLiSt BDB, the manually curated nuclear receptors ligands and structures benchmarking database. *Journal of Medicinal Chemistry* 57:3117–3125.

Langer, T. and G. Wolber. 2004. Pharmacophore definition and 3D searches. *Drug Discovery Today: Technologies* 1(3):203–207.

Leach, A.R., V.J. Gillet, R.A. Lewis et al. 2010. Three-dimensional pharmacophore methods in drug discovery. *Journal of Medicinal Chemistry* 53:539–558.

Liang, G., S. Li, Y. Gu et al. 1996. Southeast Regional Meeting American Chemical Society (SERMACS), Greenville, SC presentation.

Lipinski, C.A., F. Lombardo, B. Dominy et al. 1997. Experimental and computational approaches to estimate solubility and permeability in drug discovery and development settings. *Advanced Drug Delivery Reviews* 23:3–25.

Liu, R. and S.S. So. 2001. Development of quantitative structure-property relationship models for early ADME evaluation in drug discovery. 1. Aqueous solubility. *Journal of Chemical Information and Computer Sciences* 41:1633–1639.

Marshall, C.R. 1904. On the physiological action of the alkaloids of jaborandi leaves. *The Journal of Physiology* 31:120–156.

Marshall, G.R., C.D. Barry, H.E. Bosshard et al. 1979. The conformational parameter in drug design: The active analog approach. In *Computer-Assisted Drug Design*, eds. E.C. Olson and R.E. Christoffersen, Vol. 112, pp. 205–226. Washington, DC: American Chemical Society.

Murrall, N.W. and E.K. Davies. 1990. Conformational freedom in 3-D databases. 1. Techniques. *Journal of Chemical Information and Computer Sciences* 30:312–316.

Nicola, G., T. Liu, and M.K. Gilson. 2012. Public domain databases for medicinal chemistry. *Journal of Medicinal Chemistry* 55:6987–7002.

Parr, R.G. and W. Yang. 1989. Density-functional theory of atoms and molecules. In *International Series of Monographs on Chemistry*, Oxford University Press, ISBN-10: 0195357736.

Pawson, A.J., J.L. Sharman, H.E. Benson et al. 2014. NC-IUPHAR. The IUPHAR/BPS Guide to PHARMACOLOGY: An expert-driven knowledgebase of drug targets and their ligands. *Nucleic Acids Research* 42:D1098–D1106.

Pearlman, R.B. and K.M. Smith. 1998. Novel software tools for chemical diversity. *Perspectives in Drug Discovery and Design* 9/10/11:339–353.

Ravna, A.W. 2003. Molecular mechanism of citalopram and cocaine interactions with neurotransmitter transporters. *Journal of Pharmacology and Experimental Therapeutics* 307(1):34–41.

Richards, W.G. 1984. *Quantum Pharmacology*, 2nd edn. London, U.K.: Butterworths.

Sateriale, A., K. Bessoff, I.N. Sarkar et al. 2014. Drug repurposing: Mining protozoan proteomes for targets of known bioactive compounds. *Journal of the American Medical Informatics Association* 21:238–244.

Schueler, F.W. 1960. *Chemobiodynamics and Drug Design*. New York: McGraw-Hill.

Schwartz, J., M. Awale, and J.L. Reymond. 2013. SMIfp (SMILES fingerprint) chemical space for virtual screening and visualization of large databases of organic molecules. *Journal of Chemical Information and Modeling* 26(53):1979–1989.

Singh, J., C.E. Chuaqui, P.A. Boriack-Sjodin et al. 2003. Successful shape-based virtual screening: The discovery of a potent inhibitor of the type I TGFβ receptor kinase (TβRI). *Bioorganic & Medicinal Chemistry Letters* 13(24):4355–4359.

Sprague, P.W. 1995. Automated chemical hypothesis generation and database searching with catalyst. In *De Novo Design*, ed. K. Müller, pp. 1–20. Leiden, the Netherlands: ESCOM Science Publishers.

Takagi, T., C. Ramachandran, M. Bermejo et al. 2006. A provisional biopharmaceutical classification of the top 200 oral drug products in the United States, Great Britain, Spain, and Japan. *Molecular Pharmaceutics* 3:631–643.

Van Drie, J.H., D. Weininger, and Y.C. Martin. 1989. ALADDIN: An integrated tool for computer assisted molecular design and pharmacophore recognition from geometric, steric, and substructure searching of three-dimensional molecular structures. *Journal of Computer-Aided Molecular Design* 3:225–251.

Veber, D.F., S.R. Johnson, H.Y. Cheng et al. 2002. Molecular properties that influence the oral bioavailability of drug candidates. *Journal of Medicinal Chemistry* 45:2615–2623.

von Eichborn, J., M. Dunkel, and B.O. Gohlke et al. 2013. SynSysNet: Integration of experimental data on synaptic protein–protein interactions with drug–target relations. *Nucleic Acids Research* 41(D1):D834–D840.

Wenlock, M.C. and P. Barton. 2013. In silico physicochemical parameter predictions. *Molecular Pharmaceutics* 10:1224–1235.

Wermuth, C.G., C.R. Ganellin, P. Lindberg et al. 1998. Glossary of terms used in medicinal chemistry (IUPAC recommendations 1998). *Pure and Applied Chemistry* 70:1129–1143.

Wirth, M., V. Zoete, O. Michielin et al. 2013. SwissBioisostere: A database of molecular replacements for ligand design. *Nucleic Acids Research* 41:D1137–D1143.

Wolber, G. and T. Langer. 2005. LigandScout: 3-D pharmacophores derived from protein-bound ligands and their use as virtual screening filters. *Journal of Chemical Information and Modeling* 45:160–169.

Wu, C.C., D. D'Argenio, S. Asgharzadeh et al. 2012. TARGETgene: A tool for identification of potential therapeutic targets in cancer. *PLOS ONE* 7:e43305.

Yalkowsky, S.H. and S.C. Valvani. 1980. Solubility and partitioning. I: Solubility of nonelectrolytes in water. *Journal of Pharmaceutical Sciences* 69:912–922.

Zhong, H., L.M. Tran, and J.L. Stang. 2009. Induced-fit docking studies of the active and inactive states of protein tyrosine kinases. *Journal of Molecular Graphics & Modeling* 28:336–346.

Zhong, H.A. and J.P. Bowen. 2013. Computer-assisted drug design. In *Encyclopedia of Pharmaceutical Science and Technology*, 4th edn., ed. J. Swarbrick, pp. 620–633. Florence, Italy: Taylor & Francis.

Zhong, H.A., V. Mashinson, T.A. Woolman et al. 2013. Understanding the molecular properties and metabolism of top prescribed drugs. *Current Topics in Medicinal Chemistry* 13:1290–1307.

21 Commercial and Regulatory Development Considerations for Nanomedicines

Donna Cabral-Lilly and Lawrence D. Mayer

CONTENTS

21.1 INTRODUCTION

Preclinical proof-of-concept studies typically provide the scientific rationale for further development of a drug candidate in a clinical setting and are usually conducted as basic research using good scientific practices with the goal of publication and enhancing basic knowledge in the field of study. Translating this promising research into a full drug pharmaceutical/clinical development program requires activities aimed at determining the safety of the drug product and its components, the efficacy, and most relevant to this review, the ability to relate these biological effects to specific chemical and physical attributes of the final product (Zamboni et al. 2012). These pharmaceutical/clinical development studies are based on the laws governing human and veterinary medicines, as well as the expectations of the regulatory agencies responsible for marketing approval.

For the most part, regulatory requirements for a drug delivery system–based nanomedicine are the same as those for a conventional drug of the same class. There are, however, significant additional required studies for this class of pharmaceuticals; some are general and applicable to different types of delivery systems, while others address a specific type such as for topicals or pulmonary delivery platforms.

This chapter discusses the regulatory development pathway for injectable human medicines that use a nanotechnology carrier, such as for liposomal/nanoparticle products and colloidal iron/gold conjugate products. For the purposes of this review, such nanomedicines are defined as drug-containing macromolecular particulate assemblies in the size range of 20 nm to <10 μm, with the majority of discussion topics being most relevant to intravenously administered products with a size range of 20–200 nm. Emphasis will be placed on considerations for chemistry, manufacturing, and controls (CMC) as well as aspects of pharmacology/toxicology requirements. A general timeline, from the Investigational New Drug (IND)/Clinical Trial Application (CTA) enabling studies, through marketing application submission, is also provided.

21.2 REGULATORY GUIDANCE AND REFLECTION PAPERS

In the United States, the development and marketing of drug products is regulated by the Food and Drug Administration (FDA) whose authority is established through the Code of Federal Regulations (CFR). The sections addressing drugs for human use can be found in 21 CFR, parts 1–100, 200s, and 300s. The Center for Drug Evaluation and Research of FDA provides a comprehensive library of guidance documents on its website (www.fda.gov/Drugs) that provide procedural and technical instructions for each stage of the product development cycle. In this context, a guidance exists specific for liposomal drug products (FDA 2015), in which many of the key elements are also applicable to other "particulate" nanomedicines such as micelles, nanoparticles, dendrimers, and colloidal particles. In 2007, FDA published a report of the findings of a Nanotechnology Task Force (FDA 2007) that highlighted the need to understand the physical and chemical properties of the delivery system and its components, which will require development of new analytical methods specific for each drug product. In 2012, FDA commissioner Margaret Hamburg summarized FDA approach to regulating nanotechnology products including nanomedicines (Hamburg 2012).

Similarly, in Europe, the European Medicines Agency (EMA) held a series of meetings on the use of nanotechnologies in medicine and published a general reflection paper in 2006 (EMEA 2006). The overall considerations mirror those of the FDA, where characterization of the drug product and justification of the appropriateness of the analytical methods used are expected. EMA has also published reflection papers on four types of intravenously administered nano-based carriers including liposomes (EMA 2013b), block copolymer micelles (EMA 2013a), surface coatings (e.g., PEGylation) (EMA 2013c), and iron-based nanocolloidal products (EMA 2015).

All of the basic product characterization regulatory criteria for conventional aqueous-based injectable pharmaceutical products (e.g., content, purity/related substances, pH, sterility, endotoxin content, particulate content, and sterility) apply to nanomedicines. However, given the macromolecular particulate assembly features of nanomedicines, significant attention must also be paid to the physical disposition of the administered drug in the milieu of the particle excipients, as well as how these physical attributes affect drug release from the carrier (bioavailability), in addition to the nascent biodistribution/PK properties of the carrier itself. These attributes provide an added level of biophysical characterization that must be considered not only for regulatory (CMC) purposes but also to ensure that drug product providing reliable exposure, efficacy, and toxicology profiles can be reproducibly manufactured in a commercial setting.

Each nanomedicine drug product will be reviewed by the regulatory agencies on a case-by-case basis, since even within the nanomedicine category of injectable drugs, different classes of drug delivery carriers exhibit unique features that must be evaluated. For example, the bioavailability and plasma half-life of paclitaxel bound to albumin particles may be very different from a taxane derivative covalently linked to PEGylated-copolymer micelles. Even within a specific class of carrier, different drug products may have unique features; for example, a liposomal product may have the active agent completely encapsulated inside the internal aqueous trapped volume (e.g., Doxil®/Caelyx®), or the active agent may be imbedded in the phospholipid membrane (e.g., Ambisome®). Each type of nanomedicine will have similar types of content and impurity quality control (QC) specifications, but each will also have a very distinct "fingerprint" of biophysical attributes such as particle size distribution and morphology, surface charge, aqueous/matrix partitioning, plasma/serum particle stability, and drug release kinetics, since these features will ultimately dictate the efficacy/safety profile in vivo.

Given the diverse physical and pharmacological properties that can be designed into nanomedicines, it is recommended that developers of such products seek regulatory advice early in the development process to gain agreement on required data and to ensure that gaps do not exist that may lead to clinical hold, or long delays, in development. Mechanisms for obtaining guidance from FDA include pre-IND meetings and guidance meetings of type A (stalled development), type B (pre-IND, end of Phase 2, pre-NDA), or type C (other general guidance), among others. Such meetings can be requested for single or multiple topics where the agency will respond to questions posed and may also remind sponsors of requirements not addressed in the questions. For EMA, advice may be obtained from the Scientific Advice Working Party and/or the Innovation Task Force. More recently, FDA and EMA have begun a process where a sponsor can obtain parallel advice from the agencies. This will save time, but does not guarantee that the agencies agree on the responses to all of the questions.

21.3 PRECLINICAL SAFETY STUDIES

As with any new drug, once a candidate formulation is established that can be produced under pharmaceutically acceptable conditions, toxicology studies are required to determine if any overt safety problems exist, as well as to establish basic pharmacokinetics and a safe starting dose before first-in-man studies may begin. Normally, these studies are done in two animal species with at least one being a non-rodent mammal, and several guidelines are available to aid in study design, based on disease indication and route of administration. If an active agent is a new molecular entity (NME), additional in vitro studies to determine carcinogenicity and genotoxicity will likely also be required. For these studies, final drug product supplies should be prepared using equipment and processes that will be utilized for production of Phase 1 clinical supplies and ideally, at the same Good Manufacturing Practices (GMP) manufacturer.

All such studies must be conducted according to Good Laboratory Practices (GLPs) (see 21CFR part 58). GLP requirements include that the personnel conducting the study are qualified by education and training, facilities are adequate and maintained, equipment is qualified and calibrated, and testing facilities and controls are adequate.

Written procedures must be in place, followed, and documented to assure that the laboratory is compliant with GLP standards. The study itself must be conducted according to a preapproved protocol that describes the objective, experimental design (e.g., number of animals, dose levels, and number of doses), samples to be collected, methods of analysis, and how results will be reported. Deviations from the approved protocol must be documented and approved, and a final report is written and approved. Analytical methods that measure drug content in plasma, urine, feces, and tissues must be validated per applicable guidelines (see FDA and International Council for Harmonisation of Technical Requirements for Pharmaceuticals for Human Use [ICH] guidelines on bioanalytical testing). Complete, accurate documentation is required, and records and samples are retained to at least the extent required in the regulations. A quality assurance unit reviews the study documents and certifies compliance with GLP standards. Full reports and summaries are submitted to FDA with the initial IND application, while only the summaries are submitted in the CTA in Canada and European countries.

For intravenous (i.v.) nanomedicine drug products, a delivery vehicle control arm is typically included in the preclinical animal studies to demonstrate the safety of the carrier itself. This is particularly important if the delivery vehicle is composed of novel excipients or exhibits physical structures that have not been tested previously in man. In such cases, more extensive toxicity studies may be warranted where chronic dosing for cumulative and/or reproductive toxicities are evaluated, as well as immunogenic toxicity effects. Dose-escalation (low, medium, high) toxicity assessments using single administration help to determine the dose to use for subsequent multiple-dose studies. Data obtained include observations of animal condition, weight, blood chemistries, plasma levels of active agent and known metabolites over time, immune reactions, tissue distribution, and full necropsy. The starting dose for the first-in-man trial (Phase 0 or Phase 1) is typically determined from the highest dose level that shows no adverse events in the most sensitive animal species.

Drug delivery carriers can often contain excipients for which no information on the safety of the compounds exists. It is likely that, at a minimum, separate carcinogenicity and genotoxicity studies will be required for each novel excipient. Drug carriers that are nonbiodegradable, as the intact carrier or components of a carrier, are of particular concern since they may remain in the body for long periods of time with the potential for adverse cumulative effects. Multiple-dose studies may be required, designed to show rates and sites of accumulation, as well as studies addressing immunogenicity, cumulative toxicity, dose dependence of toxicity, dose dependence of immunogenicity, and routes of elimination. Excipients never before administered to man may need long-term data in a suitable animal model, before human clinical studies can begin.

21.4 CHEMISTRY MANUFACTURING AND CONTROLS

The last 10–15 years have seen significant advances in technologies used to manufacture and test drug delivery–based nanomedicine products. Manufacturing equipment is now controlled by microprocessors capable of balancing temperature, pressure, shear forces, and rates of addition. Rapid mixing of multiple incoming streams of components can be achieved using readily available mixers with little or no modification. Most are also capable of clean-in-place and steam-in-place, thereby greatly reducing the risk of contamination from previous batches, other products, and microbes. On the testing side, liquid chromatography has progressed where separations of complex mixtures can be achieved in minutes and yield precise and accurate quantitation. In-line, real-time methods for characteristics such as pH and particle size are in development that will allow for more continuous processes. These advances have allowed the formulation and production of very complex drug carrier systems. Regulatory agencies are working diligently to keep up with advances as demonstrated by the nanotechnology initiatives for both FDA and EMA. Each agency relies on the innovator/sponsor to understand, explain, and justify every aspect of the new drug product and manufacturing process, and each product is evaluated individually with a high level of stringency. The required CMC information is separated into two broad categories: (1) drug substance and (2) nanomedicine drug product, which are described next.

21.4.1 DRUG SUBSTANCE

The regulatory requirements for the active pharmaceutical ingredient (API) in a medicine are the same whether the API is a totally novel compound (NME), a derivative of an existing API, or an unmodified existing API. Who is responsible for providing this information is dependent on which category applies to the API.

An active agent is considered an NME if it has never been approved by a regulatory agency. The sponsor is required to supply all information on the compound, including details on the synthesis, isolation, purification, raw materials, in-process controls, final product testing, specifications, and stability during storage. Full characterization of the molecule and its synthesis/purification is required. For example, for a small molecule, NMR, FTIR, mass spectrometry, and other such methods are used to elucidate the chemical structure. In the case of biological (protein) products, the amino acid sequence, glycosylation, and other modifications must be determined and secondary and tertiary structures provided as they are known. Beyond this characterization, each lot of API must be tested per preapproved specifications that list the attribute and its acceptance criteria. It is expected that the acceptance criteria will be wide at the beginning of development (e.g., at Phase 1) but will be refined as more experience with the manufacturing and thus data become available, with final acceptance criteria set when the marketing application is filed. Specifications should always include the following attributes: description (color, physical form), identification (often one of the characterization tests such as IR), assay, and impurities (inorganic, organic, solvents, and metal contaminates).

Physicochemical properties (e.g., melting point, refractive index), particle size, and polymorphic form should be included as appropriate. Manufacturing must be done in compliance with current GMP expectations, and the quality assurance unit must confirm, in writing, compliance with these standards. This confirmation, called a Certificate of Compliance or Certificate of Conformance, is in addition to full testing per preapproved specifications and acceptance criteria where results are tabulated in a Certificate of Analysis.

When using a drug delivery carrier, it is often necessary to modify a known drug substance in order to stably incorporate it into the carrier in the final drug product. Prodrugs are such a case where, for example, a water-soluble active agent is attached to a long-chain fatty acid through a labile linker, to make the API more hydrophobic. The prodrug would then reside in a hydrophobic core of a micellar-type nanoparticle. Alternatively, the active agent might be linked to a long-chain polymer that covers the outside of the nanoparticle. From a regulatory perspective, the prodrug is considered an NME and is subject to all of the same controls as described for a novel compound.

The active agent, however, may be purchased commercially, and information on the synthesis and stability of this now starting material may be obtained from the commercial manufacturer, who preferably will have a drug master file (DMF) established with the regulatory agencies. Characterization of the prodrug should include data to support a proposed pathway to generate the active agent. Questions to answer include as follows: What bond gets hydrolyzed first? Does the linker stay with the "pro" portion? Under what conditions does the hydrolysis take place (e.g., low pH, or is an enzyme needed)? What is the known safety profile of the "pro" portion, linker, and hydrolysis intermediates? The answers to these questions will determine what additional safety studies are required.

The most straightforward regulatory path is for a commercially available drug substance that is already approved for use in a generic drug product. A novel delivery system is usually designed in this case to improve safety or enhance efficacy, and the API can be incorporated without modification. The information on manufacturing scheme, controls, characterization, and validation may be provided in a DMF submitted by the vendor and referenced, with permission, by the drug product sponsor. The sponsor is still ultimately responsible for the quality of the API and is expected to have quality agreements in place, and conduct routine audits of the manufacturer, but does not have to duplicate the information in the DMF.

Analytical methods for the drug substance must be appropriate for the intended use and quantitative methods are preferred. If a test is described in a general chapter of a compendia (United States Pharmacopeia [USP]; European Pharmacopeia [PhEur]), it is recommended to use the procedure as described in the chapter, unless it is necessary to deviate. For example, loss on drying is a routine test performed on powders, crystals, and other solid APIs. The test can be done by placing the material in a sealed container with a desiccant, or drying a known amount in an oven or other means of drying. Both the USP, in general Chapter <731>, and the PhEur, in Chapter 2.2.32, provide procedures for each type of drying method. Using one of these compendial procedures, instead of a method developed in house, eliminates the need to fully validate the procedure. A simple verification run using a sample with a known loss-on-drying value would be sufficient.

For NME and prodrugs, novel methods for assay and impurities will need to be developed. A chromatography method is expected, where the amount of API is quantitated using a reference standard. The impurity method should be able to separate process-related impurities and degradation products. A full method development report should be written that provides details on conditions tried and showing the suitability of the final chromatography and sample preparation conditions. At the early clinical development stage (Phase 1 and 2), the chromatography methods can be used as long as they meet system suitability requirements for linearity of the standard curve and repeatability. Full method validation per ICH guidelines is expected prior to Phase 3 clinical trials.

Prior to manufacturing process validation, each lot of API should be placed on stability at the recommended storage condition, and at one or more accelerated stability conditions. Testing intervals are narrow at the beginning of the study (e.g., monthly) until some data are available and experience gained with the compound. Sensitivity to light and moisture should also be tested. Drug substances are assigned retest dates, not expiration dates, and the retest interval is based on the stability data. Once sufficient data are obtained and the manufacturing process validated, then normally one lot per year is monitored for stability through the retest date and beyond. If changes to the manufacturing process are made, a risk assessment should be done to determine if revalidation is required, although some type of comparability study will be needed. Changes made prior to commercialization of the drug product will usually require fewer of these studies.

21.4.2   Nanomedicine Drug Product

Nanomedicines are a class of drug delivery systems receiving increased attention from regulatory agencies worldwide. There are no set regulatory definitions provided by the agencies but screening tools are available: the product itself, or a component, has at least one dimension of 100 nm or less, although this size limit is not absolute. Natural products of these dimensions, including proteins, are not considered nanotechnology drug products; there must be a specifically engineered component. Consequently, although the protein albumin is not considered to be a nanotechnology product in itself, Abraxane® (paclitaxel protein–bound particles for injection) is characterized as a nanomedicine, with albumin as a carrier for the anticancer API paclitaxel.

To date, most nanomedicines, either approved or at the clinical stage of development, are carrier types, which have been described extensively in Chapter 5. Liposomes, micelles, and other nanoparticles serve as a macromolecular delivery vehicle into which the active agents are incorporated; the i.v. route of administration is predominant. Chemical and physical characterization is considered a key aspect of any regulatory submission for these drug products.

21.4.2.1   Characterization of Nanomedicine Drug Product: Chemical/Content Specifications

Table 21.1 provides an example of the specifications required for a drug product that contains a delivery carrier in the nanoscale range and is administered intravenously.

The amount of drug substance (assay) must be measured and acceptance criteria proposed and justified. Degradation studies must be done with the drug product to determine if any new related substances arise during the incorporation of the API in the nanomedicine carrier during manufacturing, and limits of any such impurities should be proposed. Further, the relative amount of the API associated with the carrier (relative to unentrapped drug) must be determined and limits proposed in the final drug product at release and during stability testing (typical acceptance range is 90%–95%). This procedure will include a step that physically separates the carrier with bound drug substance from drug substance that is free in the drug product solution. Using a liposomal product as an example, the dispersion is first diluted 10-fold in an isotonic buffer and then placed into the sample portion of a spin column with a 10,000 molecular weight cutoff and spun for 10 minutes at a temperature below the phase transition temperature of the liposome membrane. The amount of API in the filtrate is quantitated and this amount relative to the total API in the sample reported as % unencapsulated API. Uniformity of dosage per USP <905> or PhEur 2.9.40 may also be required.

The components of the drug delivery carrier are considered critical excipients and have to be controlled in the drug product much in the same way as the drug substance. The amount of each carrier component must be measured by a suitable assay and acceptance criteria listed in the drug product specifications. Uniformity of dosage should be included if it is performed for the API. An assay (or multiple assays) must be developed that can detect and separate possible degradation products for each carrier component, and which must also be able to separate and quantitate the respective degradation products for multiple components present in the final product. The method must be at least semiquantitative, with the limit of detection established and an upper limit proposed

TABLE 21.1

Example of Drug Product Specifications for a Delivery Carrier Nanomedicine

Attribute	Acceptance Criteria
Appearance	Physical state, color
Identification	Orthogonal method—corresponds to reference standard
Active substance	
Identification	HPLC retention time same as standard
% Associated with carrier	Lower limit, e.g., NLT 90%
Assay	Target ± x% (start high ±10%)
Impurities	Upper limit, e.g.,
Known 1	NMT 0.2%
Known 2	NMT 0.2%
Unknown	NMT 0.2%
Uniformity	Meets USP <905>
Carrier component 1	
Assay	Target ± x% (start high ± 15%)
Impurities	Upper limit, NMT x%
Uniformity	Meets USP <905>
Carrier component 2	
Assay	Target ± x% (start high ± 15%)
Impurities	Upper limit, NMT x%
Uniformity	Meets USP <905>
Carrier component 3	
Assay	Target ± x% (start high ± 15%)
Impurities	Upper limit, NMT x%
Uniformity	Meets USP <905>
Particle size distribution	
Mean diameter or D_{50}	Target range, 80–100 nm
D_{10}	Lower limit, NLT 50 nm
D_{90}	Upper limit, NMT 150 nm
D_{99}	Upper limit for safety, NMT 500 nm
In vitro release	
At incubation time 1	Range
At incubation time 2	Range
At incubation time 3	Range
At 24-hour incubation	Range but expected to be at least 70%
At full release condition	Lower limit, NLT 90%
Assay for buffer or bulking agent	
pH	Range
Osmolality	Range
Viscosity	Range
Other	
Residual solvents	
Process solvent 1	All as upper limit based on ICH Q3
Process solvent 2	
Etc.	

(Continued)

TABLE 21.1 (*Continued*)

Example of Drug Product Specifications for a Delivery Carrier Nanomedicine

Attribute	Acceptance Criteria
Residual metals (if not controlled completely in raw materials)	All as upper limit based on PhEur 2.4.20
Particulate matter (for injectables)	Meets USP <788> or PhEur 2.9.19
Endotoxin	Limit based on USP <85> or PhEur 2.6.14
Sterility	No growth. Test method per USP <71> or PhEur 2.6.1

Abbreviation: NMT, not more than; NLT, not less than.

for each degradation product. For example, consider a liposomal product that has a membrane composed of the phospholipid palmitoyl-oleoyl-phosphatidylcholine (POPC) and cholesterol. The oleic acid chain contains a double bond that may oxidize, so a method that measures oxidation products is required. The oleic acyl chain may hydrolyze, requiring a method to measure either the oleic acid or the lyso-PC. Cholesterol can also oxidize to form 7-ketocholesterol and this degradation product must also be measured.

Particle size distribution is of particular interest to regulatory agencies because a change here may affect the safety or performance of the nanomedicine due to altered pharmacokinetic and tissue distribution properties. Mean or median (D_{50}) diameter, as well as the size at the low end (e.g., D_{10}) and high end of the distribution (e.g., D_{90}), are reported. Acceptance criteria may be listed as a range for each measurement, but it is also acceptable to use limit specifications for the high and low ends of the distribution. The procedure and equipment used, e.g., dynamic light scattering or particle counting, must be capable of detecting particles of the full size range that exists in the product. The results will be slightly different between procedure types, and even using the same type of analysis, but with hardware and software from multiple vendors. Therefore, it is recommended to determine the most appropriate procedure while still at the preclinical stage of development, provide justification for the decision, and then use the same procedure and vendor through commercialization (with extensive measures incorporated for method/equipment calibration, to minimize method variability).

For other excipients in the final product, a direct test to quantitate them, or an indirect test where the result is dependent on the amount of excipient present, must be established. For example, if maltose is used as a stabilizing agent, osmolality could be measured instead of the actual maltose content. Tests such as appearance and pH are common, and if the product is lyophilized, the reconstitution time must be tested and acceptance limits proposed. Residual amounts of processing agents must be quantitated. These include solvents, chelation agents, and water for a lyophilized product. All drug products for i.v. administration must conform to the USP and PhEur requirements for these types of drugs, and tests for endotoxin, sterility, and particulate matter are required.

21.4.2.2 Physicochemical Characterization of Nanomedicines

In addition to the proposed product specifications, the biophysical properties of nanomedicines are considered critical to their safety and efficacy. This additional characterization is initiated during the early stages of drug development. It is necessary because differences in product performance can occur in vivo due to alterations in the physical organization of the drug and/or excipients in a nanomedicine, even if the ingredient content and purity specifications are unchanged. Biophysical characterization is particularly useful when changes to the manufacturing process, or composition of the drug product, are made. For liposomal products, the core properties include vesicle morphology and lamellarity, zeta potential, trapped volume, and membrane phase transition temperature, as well as the physical state of the API inside the liposome. Each liposomal product may have specific testing based on the formulation. For example, the intrinsic fluorescence of an anthracycline may be altered/quenched when encapsulated into a liposome. Product-specific characterization focuses on drug–excipient interaction, drug-loading mechanisms, drug–drug interactions, and physical disposition of the drug in the carrier.

A specific example is the characterization of CPX-1 (irinotecan/floxuridine) liposome injection, a clinical stage liposomal nanomedicine intended for treatment of advanced colorectal cancer (Dicko et al. 2008). The manufacturing process for CPX-1 liposomes involves the coencapsulation of the APIs irinotecan HCl and floxuridine, at a 1:1 molar ratio, into preformed liposomes. As presented in Figure 21.1, data from proton NMR spectroscopy, as well as an in vitro release (IVR) assay, were used to help characterize and determine the final liposomal drug product formulation. The aromatic region of the ^{1}H NMR spectrum of the drug product shows three peaks from irinotecan. When the buffer inside the liposomes is sodium gluconate–triethanolamine (NaGluc-TEA), the ^{1}H peaks are quite broad, suggesting the self-association of irinotecan into large aggregates (Figure 21.1). However, the presence of entrapped copper gluconate–triethanolamine (CuGluc-TEA) buffer inside the liposomes results in ^{1}H peaks that are much sharper, suggesting an interaction between irinotecan and copper, which prevents irinotecan self-aggregation (Figure 21.1). Large aggregates of irinotecan reduce membrane permeability and liposomal drug release. This was borne out by the results of the IVR assay, which showed that coordinated liposomal drug release was disrupted by removal of the copper gluconate (Figure 21.1). The liposomes with NaGluc-TEA as buffer demonstrated a slower and uncoordinated release of irinotecan relative to floxuridine and loss of the

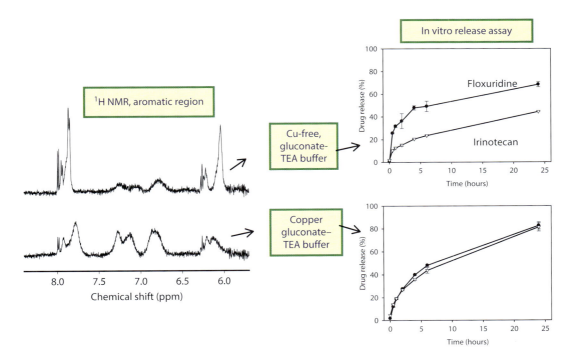

FIGURE 21.1 Identifying physicochemical features that relate to performance in CPX-1 (irinotecan/floxuridine) liposome injection. The ^{1}H NMR spectra of the aromatic region for irinotecan show broadening for the formulation without copper in the buffer inside the liposome (left top), compared to the same region in a formulation that includes copper in the entrapped buffer (left bottom). The results from an in vitro release assay (right) show that release of irinotecan is dependent on the presence/absence of copper. (With kind permission from Springer Science+Business Media: *Pharm. Res.*, Intra and inter-molecular interactions dictate the aggregation state of irinotecan co-encapsulated with floxuridine inside liposomes, 25, 2008, 1702–1713, Dicko, A., Frazier, A.A., Liboiron, B.D. et al.)

synergistic drug/drug ratio. In contrast, for liposomes with CuGluc-TEA as the entrapped buffer, the two drug substances were released from the liposomes at the same rate and at a fixed, synergistic 1:1 molar ratio. Taken together, the results strongly suggest an interaction between the API (irinotecan) and the excipient (copper), in the drug–liposome product.

An example of the full biophysical characterization for a liposomal product has been reported for CPX-351 (cytarabine/daunorubicin) liposome injection, a late clinical stage liposomal drug combination product in which two drug substances are encapsulated in a single liposome (Dicko et al. 2010). These studies demonstrate the importance of extensive biophysical study of liposomal drug products, to elucidate the key physicochemical properties that may impact their in vivo performance.

It should be noted that similar principles apply to synthetic polymer-based nanomedicines. For block copolymer micelle drug products, the biophysical characterization list in the EMA reflection paper includes morphology, zeta potential, association number, critical micelle concentration, surface properties, and physical state of the active substance. If the carrier has a surface coating, e.g., PEGylation, then (1) surface coverage heterogeneity, (2) conformational state of the coating molecules, and (3) stability of the coating under conditions of storage and use, should all be characterized. Biophysical characteristics are particularly important during scale-up from preclinical to Phase 2 and later to Phase 3/commercial where changes in equipment and manufacturing sites can introduce subtle alterations in the physical properties of a nanomedicine formulation, despite providing the desired content-related specifications. This is highlighted in FDA and EMA documents on nanomedicines where thorough biophysical characterization is advised for such manufacturing transitions to ensure reliability in correlations of preclinical/clinical results during the course of product development toward commercialization.

As alluded to earlier, a key assay for any drug delivery carrier product is a reliable and discriminatory IVR assay. Focus on this assay has significantly increased within the regulatory agencies over recent years, largely due to the fact that drug release kinetics of a nanomedicine may be one of the most important features dictating its efficacy/safety profile. This awareness has been heightened with the emergence of generic versions of liposomal anticancer agents such as Doxil®, where concerns about bioequivalence impacting in vivo performance and patient outcomes require additional characterization diligence. The IVR test is modeled after the requirements for modified-/extended-release solid oral dosage forms and is based on the premise that the API is released from the delivery carrier over time in the bloodstream. For the IVR test, the drug product is incubated in suitable media, the samples are taken at selected intervals, and the free/released API is physically separated from the drug delivery carrier. A quantitative method, usually HPLC, is used to assay the free/released API.

Method development can be extensive in order to identify conditions where the drug dissociates from the carrier, but the carrier does not completely disintegrate, in a manner that is unrepresentative of the expected in vivo situation. It is usually necessary to develop incubation media that either perturbs the carrier (e.g., an alcohol or detergent) such that the API is not stably entrapped, or allows for hydrolysis to occur (e.g., low pH, added enzyme) so that a covalently bound API is released. The method must be discriminatory, and the regulatory agencies expect to see data demonstrating that the IVR assay is capable of identifying a change in the drug product that may affect safety or efficacy in patients. Often, formulation variants are generated to demonstrate the predictive nature of the assay. For example, a liposome product can be prepared where 10% of the phospholipid degradation product, lyso-PC, is added to the normal POPC/cholesterol composition when the liposomes are made. The lyso-PC disrupts the packing of the membrane making it leaky such that an encapsulated API will diffuse out of the entrapped volume into the incubation media. A suitable IVR assay will show continuous release of the API into the incubation media where at least 70% of the API is in free/released form after 24 hours and eventually more than 90% of the drug substance becomes free in the media. An ideal IVR assay will be one able to show an in vitro–in vivo discriminatory comparability. Whereas most biophysical characterization is used for development purposes only, the IVR assay is included in the drug product specifications and routinely tested at batch release and during the stability program.

21.4.2.3 Manufacturing

The manufacture of nanomedicine drug delivery carrier products is typically a multistep process, and each step needs to be fully understood, characterized, and controlled. The expectations for how this control is demonstrated have changed with the introduction of process analytical technology initiatives and the issuance of the ICH guidance documents on Pharmaceutical Development (Q8), Quality Risk Management (Q9), and Quality Management System (Q10). A science-based and risk-based approach is used to evaluate changes and results throughout the life cycle of the drug product.

Critical quality attributes (CQA) are the features of the drug product that have an impact on the product performance—safety, efficacy, and stability. A proposed list of CQAs should be defined early (ideally at the initial IND or CTA submission) and revised as more experience with both the manufacturing process and product use is obtained. The CQAs are linked to critical process parameters (CPP): conditions during manufacturing that, if changed, would likely result in not achieving a CQA and hence resulting in a drug product that has altered characteristics and performance. A CQA specific for drug carriers is the attribute(s) that keeps the API stably associated with the carrier until it reaches the site in vivo where, by design, the API is released and becomes bioavailable. General examples include particle size distribution, surface coating density, and API-to-lipid/polymer/colloid ratio. The CQAs of the product would be assured by determining the CPPs for liposome preparation, drug substance encapsulation, lyophilization, and feasibility of reworking.

Let us use the POPC/cholesterol liposome with encapsulated API to provide a more specific example. We define a CQA to be particle size distribution, where the mean diameter of the liposome is 100 nm, D_{10} is ≥60 nm, and D_{90} is ≤200 nm, based on the following:

- Liposomes with diameters of about 100 nm have extended plasma circulation times in vivo and can be taken up by cells with a leaky vasculature, as occurs in fast growing tumors.
- The size range of 40–50 nm is near the limit at which liposomes are stable. These smaller-sized particles have different plasma circulation times, and, because of the decreased trapped volume, they encapsulate less API. Therefore, D_{10} assures that there are not too many of these very small liposomes in the drug product.
- Liposomes much above 200 nm also have altered plasma circulation times (typically more rapid elimination from the plasma) and may be too large to be taken up by a cell by passive mechanisms; consequently, less API gets delivered to the site of action. D_{90} assures limited presence of such very large liposomes in the drug product.

The next step is to determine where in the manufacturing process the liposome size becomes defined. For our example product, the liposomes form during an emulsion process, and then the size is refined by extrusion through membranes with a defined and homogeneous pore size of 100 nm. Process parameters include shear, temperature, and time. A series of experiments are designed to determine the effect of altering these three process parameters. A classical approach may be used where one parameter is changed at a time, or a design of experiments (DOE) approach may be used where multiple parameters are altered at the same time. The latter falls into the Quality by Design (QbD) described in the ICH guidelines. Both approaches are acceptable to the agencies with justification. The results of the experiments indicate that high shear can cause small liposomes to form and hence the D_{10} specification would not be met. The results also show that changes to time and temperature have little effect on particle size distribution. We then address the specific process settings that control shear, i.e., mixing rate, mixer configuration, sparge rate, and determine, for a specific batch size and defined equipment type, which two marine-type impellers fixed to a single central shaft, stirring at 20–30 rpm with nitrogen sparge set at 40–80 L/min, provide shear forces that ultimately result in the desired particle size. We now have CPPs and defined operating ranges with data to support them.

The numerous components of liposome products (multiple membrane lipids, additional excipients, possibly multiple active agents), as well as the multistep, multiday manufacturing process, pose challenges to performing development and validation studies using the QbD approach when the process is viewed as a whole; the design space would be extremely complex. It should be possible, however, to obtain a good understanding by studying each step in the manufacturing campaign separately, using enhanced science-based and process-based protocol designs.

These enhanced studies fall into the general QbD principles, which use a systematic approach to product development. Objectives are predefined and are based on a thorough understanding of the product and manufacturing process that relate raw material attributes and process controls to the CQAs. The process begins at the very early stages of product development by creating a target product profile. The formulation, manufacturing process, and controls are designed to meet this product profile, and these parameters are continually monitored throughout the product life cycle, including after commercialization. Trends are identified and the process improved as needed, to achieve product quality standards and performance.

Multivariate experiments using DOE and statistical analysis of results are conducted to understand the interactions of input variables on the overall process and product attributes. Examples of ways to present results from multivariate studies are provided in ICH guidance Q8 (R2).

Ideally, a design space is defined that describes the multidimensional interactions where product quality is assured. Changes within the design space are not considered critical and can be made without approval from the regulatory agencies. The key feature here is the use of multivariate testing; acceptable ranges that are found by testing only one parameter at a time may not provide information on the interactions between process parameters. If using only univariate studies, a design space cannot be described, and changes to the manufacturing process will have to be submitted to the appropriate regulatory agency.

21.4.2.4   Changes to Manufacturing

It is understood that the manufacturing process will undergo changes during development, with the final process not defined until validation studies are completed, prior to the filing of the marketing application. Even then, the QbD life cycle approach to development provides for continual process improvements after commercialization. The more that is understood about the manufacturing process, especially the CQAs and related CPPs, the more straightforward the pathway is to getting the regulatory agencies to agree that the change can be made.

A key aspect here is the ability to clearly identify and establish quantitative/semiquantitative assays that will detect alterations in biophysical properties of nanomedicine products that could lead to altered in vivo performance despite exhibiting identical content-related specifications. Consequently, establishing biophysical parameters and ranges that ensure reproducible blood circulation times of the delivery vehicle, and drug release rates from the vehicle after injection, are critical to the development of a nanomedicine product. The reason for this is that while content of components in the formulation typically can be readily controlled (and assayed) going from lab scale to toxicology/Phase 1 scale and ultimately to Phase 3/commercial scale, the physical disposition of the delivery vehicle and encapsulated drug(s) require more complicated and product-specific biophysical assays, which may require more than 6 months to develop and qualify. The utility of the biophysical assays is the most important during transitions in production scale, where modifications in the process, and most certainly in the equipment used, can result in altered physicochemical properties. Alterations in the physicochemical properties may occur despite generating a final drug product with no changes in content and purity of ingredients. This understanding is especially important for drug delivery carrier products, and the regulatory agencies are requesting information earlier in development. One example is the IVR assay; previously, this assay was developed and used starting in Phase 3, whereas currently FDA has suggested that the assay be used starting in Phase 1.

21.4.2.5 In Vivo Characterization of Nanomedicine Performance

While the IVR assay provides an in vitro QC-compatible method to assess drug bioavailability for a nanomedicine, ultimately this characteristic should be evaluated in patients. This requires the establishment of an assay procedure that is capable of handling plasma/serum from treated patients and separating nanomedicine carrier-associated drug from drug that has been released from the carrier. The regulatory agencies have placed increasing importance on this analysis, with the desire to differentiate the pharmacodynamics relationships for toxicity and efficacy responses as they may relate to encapsulated vs. free/released drug.

In this regard, many different assays have been developed, based on solid-phase extraction columns or ultrafiltration cartridges. The common observation arising from such analyses in patients treated with nanomedicines is that the free drug concentration is typically so small (due to the rapid tissue distribution of most free drugs) that plasma drug exposure of total drug is equivalent to encapsulated drug. Nonetheless, the development of more sensitive analysis procedures may be important in refining the elucidation of pharmacodynamics relationships for nanomedicines and warrants additional refinement efforts. For example, separation methods utilizing equilibrium conditions (e.g., ultrafiltration cartridges) appear to minimize in-process drug release and overestimations of free drug plasma concentrations that can arise with nonequilibrium methods such as solid-phase extraction columns (Mayer and St. Onge 1995).

21.5 TIMING FOR DEVELOPMENT ACTIVITIES

In the development of a drug product, the amount of information available will be limited in the early stage. The manufacturing processes, as well as analytical methods, can be revised and improved as the size of the database increases. During the clinical stage of development, CMC changes are reported in the United States to FDA as part of the annual reports for each active IND application. Substantive changes that require review and consent prior to implementation are filed as Quality Amendments to each regulatory agency where a clinical trial is taking place. As a reference, Figure 21.2 provides a general timeline for the progress of CMC development.

The identification and control of key features that dictate drug pharmacokinetics and tissue distribution for a nanomedicine drug delivery vehicle must be initiated as soon as evidence is generated that supports development as a potential product candidate. This is due to the fact that the efficacy/toxicity profile of nanomedicines can be impacted by physicochemical properties of the delivery vehicle to the same extent, or greater, than content and purity of the drugs and excipients in the final drug product. Therefore, key performance-related biophysical assays should be in place by the time that batches are being produced at the GLP toxicology/Phase 1 scale. Timing is an important consideration for assay methods such as IVR and free vs. encapsulated drug, where it can take up to 12 months to refine assay conditions to where they provide acceptable accuracy, precision, reproducibility, and relevance to product in vivo performance. Validation studies for the analytical methods, the manufacturing process, and sterility assurance are spaced throughout development. Timed well, the CMC studies will be done shortly before completion of the clinical trials and may be submitted prior to, or at the time of, the safety and efficacy results in the marketing application.

Year	1	2	3	4	5	6	7	8	9	10
Preclinical	GLP toxicology in 2 species.	Safety of novel API: genotoxicity, carcinogenicity, reproductive toxicology.		Safety studies for novel excipients.						
Clinical		Phase 1		Phase 2			Phase 3			
CMC for drug product	Analytical methods for API, drug product with system suitability. Micro methods validated.	Reference standards qualified, IVR method developed, API impurity method. Preliminary specifications.		Compendial methods verified. Methods for API and critical excipients validated. Impurity methods for carrier components developed. IVR added to specifications. Limit tests for residual solvents and metals in place.			All drug product methods validated per ICH guidelines. Impurities for API and carrier components identified.			
	Basic stability: API content, particle size, other critical quality attributes.	Stability for drug product per ICH. Key biophysical testing established.		Continue ICH stability. Add stability for in-use period. Final biophysical methods with results.			Continue stability – need full data for three batches that usually include the validation batches.			
	Manufacturing per general instructions.	Manufacturing under GMP environmental controls. Basic batch record. Compliance oversight but deviations allowed. Sterilization process validated at small scale.		Batch size increases to pilot scale. Manufacturing per batch record with minor deviations. Compliance oversight. CQAs identified. CPPs proposed and studied to define range begin.			CPPs refined and gap studies completed. Validation of manufacturing process, container closure, cleaning, repeat sterilization verification, extractables, etc. At least one batch at commercial sale.			

FIGURE 21.2 Overview of timeline for chemistry, manufacturing, and controls development studies from preclinical stage to marketing application.

REFERENCES

Dicko, A., A.A. Frazier, B.D. Liboiron et al. 2008. Intra and inter-molecular interactions dictate the aggregation state of Irinotecan co-encapsulated with floxuridine inside liposomes. *Pharm. Res.* 25:1702–1713.

Dicko, A., S. Kwak, A.A. Frazier et al. 2010. Biophysical characterization of a liposomal formulation of cytarabine and daunorubicin. *Int. J. Pharm.* 391:248–259.

EMA. January 2013a. Joint MHLW/EMA reflection paper on the development of block copolymer micelle medicinal products. EMA/CHMP/13099/2013. http://www.ema.europa.eu/docs/en_GB/document_library/Scientific_guideline/2013/02/WC500138390.pdf.

EMA. February 2013b. Reflection paper on the data requirements for intravenous liposomal products developed with reference to an innovator liposomal product. EMA/CHMP/806058/2009/Rev.02. http://www.ema.europa.eu/docs/en_GB/document_library/Scientific_guideline/2013/03/WC500140351.pdf.

EMA. June 2013c. Reflection paper on surface coatings: General issues for consideration regarding parenteral administration of coated nanomedicine products. EMA/325027/2013. http://www.ema.europa.eu/docs/en_GB/document_library/Scientific_guideline/2013/08/WC500147874.pdf.

EMA. March 2015. Reflection paper on the data requirements for intravenous iron-based nano-colloidal products developed with reference to an innovator medicinal product. EMA/CHMP/SWP/620008/2012. http://www.ema.europa.eu/docs/en_GB/document_library/Scientific_guideline/2015/03/WC500184922.pdf.

EMEA. June 2006. Reflection paper on nanotechnology-based medicinal products for human use. EMEA/CHMP/79769/2006. http://www.ema.europa.eu/docs/en_GB/document_library/Regulatory_and_procedural_guideline/2010/01/WC500069728.pdf.

FDA. October 2015. Guidance for industry. Liposome drug products. Chemistry, manufacturing and controls; human pharmacokinetics and bioavailability; and labeling documentation. www.fda.gov/ucm/groups/fdagov-public/@fdagov-drugs-gen/documents/document/ucm070570.pdf.

FDA. July 2007. Nanotechnology. A report of the U.S. Food and Drug Administration Nanotechnology Task Force. http://www.fda.gov/downloads/ScienceResearch/SpecialTopics/Nanotechnology/ucm110856.pdf.

Hamburg, M. 2012. FDA's approach to regulation of products of nanotechnology. *Science* 336:299–300.

Mayer, L.D. and G. St. Onge. 1995. Determination of free and liposome-associated doxorubicin and vincristine levels in plasma under equilibrium conditions employing ultrafiltration techniques. *Anal. Biochem.* 232:149–157.

Zamboni, W.C., V. Torchilin, A.K. Patri et al. 2012. Best practices in cancer nanotechnology: Perspective form NCI nanotechnology alliance. *Clin. Cancer Res.* 18:3229–3241.

22 Marketing Perspectives for Drug Delivery

Louise Rosenmayr-Templeton

CONTENTS

22.1   INTRODUCTION

This chapter examines drug delivery from a market and marketing perspective. It deals mainly with the end market for drug delivery prescription-only medicines (POMs). The market for over-the-counter (OTC) medicines will not be discussed in detail, but differences between it and the prescription market will be highlighted. The chapter also touches on business-to-business marketing that has been traditionally carried out by specialist drug delivery companies, which offer their technologies for licensing to other companies.

With respect to the end market for POMs, the following will be discussed:

- The characteristics of the global market for pharmaceuticals, its segmentation, and how it differs from that of consumer goods

- The drug delivery market—variations in the way it is defined, its size, and other characteristics
- The need for drug delivery products within the global pharmaceutical market and how they benefit patients, doctors, and other stakeholders
- The role of drug delivery in product development, product life-cycle management, and defending branded medicines against generic competition
- The trends within the global pharmaceutical market and how they impact on the development of drug delivery products
- The factors that make a successful product

With respect to business-to-business marketing, the following questions will be considered:

- How do companies market drug delivery technologies and drug delivery products for licensing to other companies?
- How do pharmaceutical and biotech companies evaluate these technologies and products?

For the purpose of this chapter, drug delivery products are further subdivided into two separate areas (Bossart 2005):

- *Drug delivery–enabled products*: These are products containing drugs that could not be developed by a particular route without the use of drug delivery technology, e.g., protein delivery to the lungs and formulations for silencing RNA.
- *Drug delivery–enhanced products*: These are products containing drugs that can be delivered by a particular route, but drug delivery technology has been used to improve their delivery characteristics, e.g., sustained-release formulations for the oral route.

For the convenience of the reader, a list of marketing and other terms used in this chapter has been included in Appendix 22.A. The defined terms are marked in italics the first time they are used in the text.

22.2 CHARACTERISTICS OF THE GLOBAL MARKET FOR PHARMACEUTICALS

22.2.1 Market Size and Growth

According to IMS Health data, global spending on medicines was $1069 billion in 2015 and this figure is estimated to grow to between $1.4 trillion and $1.43 trillion in 2020, with an overall *compound annual growth rate (CAGR)* in the range 4%–7% (Aitken and Kleinrock, 2015).*

As shown in Figure 22.1, the United States is currently the largest market for pharmaceuticals, followed by the combined top five European markets (Germany, Italy, United Kingdom, France, and Spain) (Atkin and Kleinrock, 2015). However, many of these established markets are stagnating or growing slowly, for reasons that will be discussed in Section 22.6. In recent years, most rapid market growth is observed in countries like China and India whose economies have been growing rapidly, with the result that both governments and individuals are able to spend more on health care.

22.2.2 Segmentation

All markets can be broken down into *segments* with the same characteristics or customer profile. The customers within these segments are therefore likely to have the same needs and preferences.

---

* Market size is calculated in general based on wholesaler invoice prices using variable exchange rates, while CAGR is based on constant exchange rates.

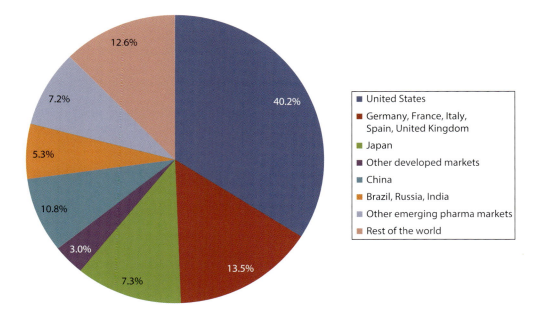

FIGURE 22.1 Global spending on medicines per country/region. Percentages calculated based on wholesaler prices. Wholesaler prices may not include discounts and rebates. Developed markets: United States, Japan, the top five European markets (Germany, France, Italy, Spain, United Kingdom), Canada, and South Korea. Emerging pharma markets countries: China, Brazil, India, Russia, Algeria, Argentina, Bangladesh, Chile, Colombia, Egypt, Indonesia, Kazakhstan, Mexico, Nigeria, Pakistan, Philippines, Poland, South Africa, Saudi Arabia, Turkey, and Vietnam. (Data from Atkin, M. and Kleinrock, M., *Global Medicines Use in 2020 Outlook and Implications.* IMS Institute for Healthcare Informatics, Parsippany, NJ, 2015, accessed February 14, 2016, ww.imshealth.com/en/thought-leadership/ims-institute/reports/global-medicines-use-in-2020.)

Companies use segmentation to characterize markets and identify gaps within them. Products/services can then be developed to meet the needs of particular segments. The products/services are then advertised, promoted, distributed, and sold through channels best suited to the needs of the target segment. For example, in the 1990s, low-cost airlines revolutionized short-haul air travel in Europe when they identified a need for cheap no-frills flights within the target geographical segment, which they sold via the Internet.

Like any other market, the global pharmaceutical market can be segmented in different ways. From a marketing perspective, there are two overarching segments: POMs and OTC products, based on different legal and regulatory requirements and the restrictions they place on marketing, sale, and supply. However, these are two very broad areas and the pharmaceutical market is typically divided into segments and subsegments based on one or more additional criteria. Examples of the way the pharmaceutical market can be segmented and subsegmented are shown in Table 22.1.

Some markets can be defined as niche. These are small, defined markets with very specific unmet needs. There are often additional barriers, e.g., technological or high investment, to entry into these markets. The size and special requirements of these markets make them unattractive to most companies, which means firms that target them face little or no competition and can charge premium prices for their products. Niche markets include those for the so-called orphan drugs for the treatment of rare diseases and conditions (see Section 22.6.5).

Segmentation is also used to identify subsegments within the "customers" of pharmaceutical companies, i.e., physicians, wholesalers, pharmacies, *payers*, and for OTC medicines and, where direct-to-consumer (DTC) advertising is allowed, patients. This segmentation is used to tailor promotional activities to these subsegments and/or develop specific deals.

TABLE 22.1
Examples of Segmentation within the Pharmaceutical Market

Segmentation	Subsegments	Possible Further Segmentation
Legal class of medicine, e.g., prescription only or OTC	Some medicines are pharmacy only, while others can be also sold in nonpharmacies.	NA
Therapeutic area, e.g., diabetes	Specific disease/condition, e.g., type 1 and type 2 diabetes. Unsurprisingly, this is the most common form of segmentation, but a medicine may have a number of indications, and these may be in different therapeutic areas. Sometimes a company will choose to focus on one indication or therapeutic area or create different brand strategies for different indications.	Mechanism of action, e.g., proton pump inhibitors and H_2 receptor antagonists. Length of action, e.g., basal and mealtime insulin. First-line therapy, second-line therapy, etc., medicines can be split into first- or second- or third-line therapy based on the conditions of their product license or based on clinical or formulary guidelines, e.g., NICE guidelines on type 2 diabetes (NICE 2012). Certain cancer drugs are only licensed when used in combination with others and/or if treatment with other products has failed.
Geography—most products developed for sale globally	Large markets, e.g., United States, Europe, Japan. Geographical regions where particular diseases/conditions are more prevalent than in others.	Individual countries with specific local needs.
Route of administration or formulation type	Parenteral, oral, pulmonary controlled-release or immediate-release solid dosage form.	Injections for self-administration or hospital use.
Location of treatment	Hospital only, out-patient clinic, medical practice in the community with self-administration at home.	Specialist hospital centers/clinics, e.g., oncology products. Some companies focus their *product portfolio* on such centers.
Age of patient, life stage, and lifestyle	Pediatric medicine, older people (easy-to-swallow formulations), pregnancy, menopause.	Vitamins and mineral supplements are very often targeted at very specific age groups and lifestyles.
Genetics/the presence or absence of biological markers/metabolic differences	This type of segmentation is not common but is set to increase due to the greater understanding of disease at a genetic and cellular level. Probably the best known example is the use of Herceptin® to treat a subset of breast cancer patients whose tumor is human epidermal growth factor receptor 2-positive.	NA

Abbreviation: NICE, National Institute for Health and Care Excellence.

22.2.3   Differences between the Pharmaceutical Market and Consumer Markets

As with consumer goods, the marketing of pharmaceutical products can be discussed with respect to product, price, place, and promotion, otherwise known as the 4Ps of marketing or *marketing mix*. However, there are a significant number of differences between marketing pharmaceutical products and consumer goods and also between the marketing of prescription-only and OTC medicines (see Table 22.2 for a comparison). These differences are discussed under the following headings:

22.2.3.1   Stakeholders

The stakeholders in the prescription pharmaceuticals market are shown in Box 22.1.

The number of stakeholders in the prescription pharmaceuticals market tends to be greater than in typical consumer markets, and their influence is differently distributed. There are a number of reasons for this. First, for prescribed medicines, a third-party payer typically covers all or part of the *retail cost of the medicine*. The patient is usually only charged a fixed amount per product

---

TABLE 22.2
Differences between Pharmaceuticals and Consumer Goods

Pharmaceuticals	Consumer Goods
Highly regulated.	Less regulation and, at most, manufacturers have only to ensure that their products are safe, comply with the law, and are "fit for purpose."
Long period of development.	Development period varies with product but typically is a lot shorter than for pharmaceuticals.
Product development affected by clinical, technological, and scientific advances.	Product development affected by technological advances and fashion.
The development of new medicines (nongenerics) results in new intellectual property.	Generation of new intellectual property is only possible for certain products.
Stakeholders—many and influential.	Number of stakeholders varies with product and supply channel but, in general, around four (the manufacturer, wholesaler, retailer, and customer).
Access to medicines is restricted in the case of prescription medicines, which require a doctor to authorize supply. In many countries, there are also restrictions on the sale/supply of OTC medicines.	Access to consumer goods is typically only limited by the customers' ability to pay. In certain cases, they are limited based on age, e.g., alcohol and cigarettes.
In most countries, the direct advertising of POMs to consumers is banned.	Goods advertised directly to customers through magazines, newspaper, TV, radio, Internet, etc.
The patient pays the retail price of OTC products.	The customer pays the retail price of the product.
However, in many countries, patients typically pay only a fixed fee or copayment per product or a percentage of the cost for prescription medicines.	This can vary depending on the supplier, e.g., Internet, independent shop, large retail chain.
Number and size of purchasers of medicines (hospitals, pharmacy chains, wholesalers, independent pharmacies) varies with the health-care system. Internet sales of prescription medicines are officially restricted to registered online pharmacies in some countries, while in others, Internet sales are forbidden.	Numerous purchasers, Internet sales are unrestricted.
Number and size of third-party payers—varies with the health-care system.	Not applicable.
Pricing and reimbursement: Pricing is typically decided by manufacturers and reimbursement by third-party payers. Pricing depends on the factors listed in Section 22.2.3.5.	Price set by the producer/seller and depends on the factors listed in Section 22.2.3.5.

BOX 22.1   STAKEHOLDERS IN THE PRESCRIPTION PHARMACEUTICAL MARKET

Governments, EU, and WHO
State insurance systems, e.g., NHS
Private insurance companies
Doctors and dentists
Pharmacists and health-care staff
Patients, carers, and advocacy groups
Pharmaceutical industry
Wholesalers
Regulatory authorities

(copayment or prescription fee) or a percentage of the price of the medicine (coinsurance). These payers are state-funded health-care bodies or programs, e.g., UK National Health Service (NHS) or the U.S. health-care program, Medicaid, and medical insurance companies dealing in cover for prescription medicines. In this market, the patient depends principally on the doctor to select the appropriate medicine based on medical knowledge, and it is only with OTC products that the patient chooses the product, with or without the advice of a pharmacist. However, with the advent of increased information on the Internet with respect to diseases and their treatment, patients and carers globally are becoming more informed and are more likely to discuss treatment options with their doctor.

Other stakeholders in the market include governments, regulatory bodies, the European Union (EU), the World Health Organization (WHO), patient advocacy groups, wholesalers, and, of course, the pharmaceutical industry itself (research-based companies and generic firms). Political stakeholders, such as governments, the EU and the WHO, are influential, as the maintenance of good health and the effective treatment, management, and prevention of disease are critical factors to both individuals and society as a whole. At a national level, the availability and affordability of effective and safe medicinal products are issues that concern politicians and voters alike.

It is therefore not surprising that the development, licensing, manufacture, analysis, distribution, sale, supply, and marketing of pharmaceuticals are the subject of strict regulatory control and that the pricing and affordability of medicines are increasingly the cause of political debate and government control through policies impacting on *reimbursement* (see Section 22.2.3.5). Of course, governments want to control spiraling public health-care spending while providing quality health services. In doing so, they must balance budget control with the need to "reward" the pharmaceutical industry for the development of new innovative medicines, and the fact that in many countries, the pharmaceutical industry is a major employer and significant contributor to gross domestic product (GDP).

22.2.3.2   Product

Pharmaceuticals differ from consumer goods in that they must receive marketing authorization prior to their launch onto a particular market. In order to receive this authorization, companies must demonstrate that their products have *efficacy*, are safe, and meet the appropriate quality standards throughout their shelf life. Importantly, the benefits of taking the medicine must outweigh the risks. Developing a new therapeutic entity and gaining market authorization takes on average 12–13 years from the first synthesis of active substance (EFPIA 2013). In 2012, the average cost of this process was estimated to be $1506 million in 2011 dollars (Mestre-Ferrandiz et al. 2012). This figure includes the cost of drugs, which failed in the development (for every 10,000 molecules at the discovery stage, only 1–2 will reach the market), and the *cost of capital*.

The length of time taken and investment required to develop pharmaceuticals is therefore much higher than typical consumer products. It also means that there is a much shorter time to recoup development costs and obtain a return on investment before patents protecting the active molecule expire and generic versions enter the market. In order to compensate partially for the time lost during regulatory review of the first application, certain countries and regions, e.g., United States and the EU, allow companies a period of additional patent protection for new molecular entities (NMEs). The rules on patent-term extension (US and EU) are listed elsewhere (35 U.S. Code 156; Europa 2009a).

22.2.3.3 Place

The sale and/or supply of POMs to patients requires physician preapproval in the form of a valid prescription, and the dispensing of medicines is mainly restricted to pharmacies (hospital and retail). In some countries, doctors are also allowed to fill prescriptions especially if the number of pharmacies is limited. Most pharmacies are retail. However, mail-order pharmacy is well established in the United States and Canada, but in some other countries, it is in its infancy or is forbidden.

For prescription-only drugs, the requirement for a prescription and the limitations on the place of sale make the pharmaceutical market very different from that of consumer goods. For example, it is not possible to buy direct from pharmaceutical wholesalers.

Restrictions on the place of sale and supply of OTC medicines vary from country to country. For example, in countries such as Germany and Austria, sales of OTC medicines are mainly restricted to registered pharmacies, while in others like the United Kingdom and Ireland, some OTC medicines can be sold in nonpharmacy outlets, e.g., supermarkets, but the purchase of others is limited to pharmacies.

22.2.3.4 Promotion

Another factor differentiating the pharmaceutical market from that of consumer goods is the regulatory and industry code of practice controls placed on the promotion of medicines (Rollins and Perri 2014).

Advertising of POMs to the general public (commonly known as direct-to-consumer [DTC] advertising) is currently banned in all countries around the world with the notable exceptions of the United States and New Zealand. Therefore, in most countries, pharmaceutical companies target their promotional efforts at physicians. This has been traditionally carried out through scientific presentations and posters at conferences, through sponsored continuing professional development for physicians, and by teams of sales representatives who "detail" the doctors on the benefits and risks of the product and provide them with free samples. In the age of the Internet, social media, and smartphone applications, new technology is being used to promote pharmaceuticals to doctors and, where permissible, also to patients. Another effective way of marketing pharmaceuticals are disease awareness campaigns, in which no specific product is mentioned, but the common symptoms, the importance of early diagnosis, and treatment are highlighted, and people experiencing these symptoms are advised to consult a physician. These campaigns are important as they promote discussion of the disease and its treatment, facilitate timely diagnosis, and, together with other promotional efforts, such as physician *detailing*, impact on sales. They also, to an extent, circumvent the ban on direct advertising of prescription medicines to patients. For example, in Austria, there have been disease awareness campaigns on chronic obstructive pulmonary disease and various vaccines on television despite the direct advertising of prescription medicines being forbidden.

The controls on the promotion of medicines, whether they require a prescription or not, are stricter than for consumer goods. TV, radio, print, or Internet advertisements for prescription (where allowed) and OTC medicines must make the patient aware of both the risks and the benefits and/or inform them where they can access this information. In addition, all claims made during detailing or any form of advertising or promotion must be backed up by data and be in line with the regulatory-approved product information, and firms cannot actively promote use of the medicine for nonlicensed indications (the so-called "off-label" use). DTC advertising in the United States is checked and monitored by the FDA's Office of Prescription Drug Promotion (FDA 2014b).

TV commercials account for most of the company spending on DTC advertising, but recently there has been an increase in the use of the Internet. In the United States, pharmaceutical company spending on DTC advertising reached a peak of over $5.4 billion in 2006 (quoted in Ventola [2011]) as a result of heavy TV promotion of small-molecule blockbuster drugs prescribed by community doctors. Since then, spending has declined due to former blockbuster drugs going off-patent and losing *market share* to generics, which are not advertised. In addition, a significant proportion of medicines approved over the last 10 years have been for diseases with small patient populations and/or otherwise fall into the Specialty Pharma segment (typically prescribed in hospital/specialist clinics) making DTC advertising less relevant. Despite this, spending has recently started to rise and in 2014, $4.5 billion was spent (excluding digital advertising) on DTC adverts, an increase of almost 21% on 2013 (Dobrow, 2015). The use of DTC advertising is still controversial. For a full discussion of the pros and cons of DTC advertising, the reader is referred to reviews by Rollins and Perri (2014), Donohue et al. (2007), and Ventola (2011).

In recent years, as a result of several scandals, the promotional practices of pharmaceutical companies have come under increased scrutiny by regulatory bodies and the media. In the United States, this has resulted in regulatory changes and a tightening of the voluntary code of practice of the Pharmaceutical Research and Manufacturers of America (PhRMA 2008), a body representing the research-based pharmaceutical industry. Despite the restrictions on the promotion of medicines, pharmaceutical companies still have to position their product within a market segment and develop a *brand* and *brand image*. A full discussion of branding and the development of *brand awareness* and *loyalty*, and how companies carry out product positioning and launch and promote *product adoption*, can be found in the reviews by Rollins and Perri (2014) and Landsman et al. (2014).

22.2.3.5 Price

For consumer goods, price is a critical component of the marketing mix and in the positioning of a product within a market. For example, higher prices are often associated with products of better quality and added value. Prices for many consumer goods can be said to be elastic in that as their price increases, demand falls. Prices are also affected by the cost of developing, manufacturing, and selling the product, including indirect fixed costs such as rent for facilities. The greater the cost of developing, manufacturing, and selling the product, the higher the price, as companies at the very least have to cover their costs. The company's business strategy with respect to a particular product is a further factor. This may change during the product's life cycle, e.g., special offers to entice consumers to buy a newly launched product and/or to gain market share, a pricing strategy that maximizes profit for well-established products facing little competition. Other factors that affect the price include the size and characteristics of the target market segment, e.g., budget or luxury goods; the relative advantages of the product compared with its competitors; the number of products competing within a target market and how differentiated the products are from each other; the type of *marketing* or *distribution channel*, e.g., independent shop versus Internet; and the relative *bargaining power of purchasers* and *suppliers*, e.g., large purchasers have the power to negotiate better prices and terms and conditions. The prices of OTC medicines sold directly to patients obey these rules to a greater extent than POMs. This is because the patient is the sole payer.

The pricing of prescribed medicines is complicated by the fact that the majority of medicines in many countries are not paid for in full by patients, but through a process of reimbursement by third-party payers. Third-party payers include governments (in the case of state-funded health-care systems/programs), private health insurance companies offering cover for prescription medicines, and *pharmacy benefit managers* (companies who manage the drug claims, and benefits aspects, of health-care insurance plans).

The pricing and reimbursement of POMs is variable and complex, as reimbursement policies and market conditions vary from country to country (Europa 2013; Rollins and Perri 2014). However, a number of general points regarding pricing policies can be made. Typically, the pharmaceutical company sets the price of the medicine, but third-party payers influence it through various cost-control mechanisms. The company's pricing strategy depends on whether the product is a patented originator product, an originator product that is off-patent or about to go off-patent or a generic. The price of an originator product will also depend on the stage of its life cycle, while those for generic products tend to be purely market driven. When pricing an originator medicine, companies need to consider development and manufacturing costs (e.g., biopharmaceuticals cost more to produce than small-molecule drugs), the innovativeness of the product, its key selling points and its advantages over existing therapies (including nondrug options), and the level of competition in the same therapeutic area. Pricing of innovative medicines is, in general, inelastic, i.e., a price increase does not influence market demand. However, the degree of price elasticity depends on the extent of competition (other molecular entities and/or generics) within the therapeutic segment, and the ability of a company to set a particular price may be limited by the pricing of other products already on the market. Geography also impacts on the prices that pharmaceutical companies can charge, due to local market conditions and differences in the rules and regulations affecting reimbursement. Price differentials between countries can result in the importing of medicines from countries where the price is low, a practice known as parallel importing. Where this exists, it creates competition and a downward pressure on prices in the importing country despite the owner of the originator product being the same in both countries. Pricing can also vary depending on the customer, with hospitals or other large customers being able to negotiate a lower price than smaller ones. Pricing decisions will be influenced by whether the medicine is for one-off, short-term, or chronic therapy and the frequency of administration. The place of administration—community, general hospital, or specialist center—can also affect price, with products administered predominately in specialist centers being typically able to command a higher price than those used in general practice. Pricing of drug therapies is increasingly becoming the subject of political and public debate. Two recent examples are the price of Sovaldi, a game-changing therapy for the treatment of hepatitis C but which costs $84,000 for the 12-week course making it unaffordable for many patients and its use in programs like Medicaid severely restricted (Palmer, 2015). Another example is Daraprim, a treatment for toxoplasmosis encephalitis, originally developed over 60 years ago whose price was increased by over 5000% after it was acquired by Turing Pharmaceuticals (Thielman, 2015).

For medicines containing NMEs, companies only have a limited time period to recoup product development costs and make a profit before patents expire and market protection is lost. After this, generic versions of the product can enter the market and impact adversely on the originator's market share. Pricing by manufacturers must balance this against the willingness of payers and patients to bear the costs of the medication, the need for access to affordable medicine, the price of competing products, and internal factors such as production capabilities and business targets for revenue and profit. Sometimes a higher price can be justified if *pharmacoeconomic* analysis shows that the medicine reduces the overall cost of therapy, e.g., shorter period of treatment, or enables community care as opposed to hospital treatment. The success of such a strategy depends on whether the payer is focused on the cost of the medicine or the total cost of treatment and the amount the patient has to contribute. If out-of-pocket expenses are too high, patients may simply decide not to have their prescriptions dispensed or demand cheaper therapy.

Product sales of patented proprietary medicines are highly dependent on the product being approved for use, and reimbursement, by third-party payers. Countries and organizations within countries have different methods of determining whether a product can be reimbursed, to which extent the costs will be refunded, and the circumstances under which the product can be used. In countries with extensive public health-care systems, e.g., United Kingdom, Germany, and Austria, reimbursement decisions are taken principally at a national level. For example, the UK's advisory body, the National Institute for Health and Care Excellence (NICE) issued guidance in 2006 advising against the use of inhalable insulin (except under special circumstances), on the grounds that the benefits of avoiding injections did not justify the higher cost of the new product; the product (Exubera) was later withdrawn by Pfizer due to poor sales. In others, e.g., United States, where there is more extensive use of private medicine, decisions on reimbursement are made at an individual third-party payer level, and these may differ between health-care insurance plans and between private and state-funded schemes such as U.S. Medicaid. Ways that the use of prescription medicines can be controlled by third-party payers include inclusion/exclusion of products in drug formularies or preferred drug lists, the placing of medicines into different tiers based on the extent of reimbursement and required patient contribution, restrictions on the use of the medicine, and/or the requirement that the prescribing of certain medicines must be preapproved by third-party payers. Such restrictions tend to affect new and expensive medicines more than cheaper and well-established products even if the latter are still protected by patents. Whenever possible, pharmaceutical companies negotiate with the responsible authorities for state-funded health care, employer, and private prescription insurance benefits, in order to agree prices/rebates and conditions, which minimize restrictions on the use of their products and maximize sales. For some state-funded systems, for example, Medicaid rebates are mandatory. However, it is important to note that prices for reimbursement purposes are often different than the actual amounts paid by the wholesaler, hospital, or pharmacy as a result of discounts and rebates.

In recent years, *health technology assessment* (*HTA*) (Sorenson et al. 2008; Arnold 2009) has been increasingly used to assess the cost and benefit of new medicines compared with existing treatments, prior to recommendations being made on their use and reimbursement. Countries like Germany have taken the pharmacoeconomic argument one step further and introduced a system of *value-based pricing*, in which newly approved products containing a new therapeutic entity or a new combination of actives can be priced higher than competing therapies based on the added benefit they bring to patients (AMNOG 2010). The price of products deemed to be without added benefit are based on comparative therapies. Both HTA and value-based pricing have resulted in pharmaceutical companies having to present the pharmacoeconomic case for their products and in some cases modify their pricing strategies through patient access schemes in order to ensure that their products will be used in a particular market. For examples of such arrangements for the NHS in England and Wales, see the NICE website (NICE 2014).

22.3   DRUG DELIVERY MARKET

The drug delivery market is a subset of the pharmaceutical market and shares almost all of the characteristics described for the main market in Section 22.2. However, in some respects, the market for drug delivery products is slightly different, and certain drivers, constraints, and trends within the main market have a specific effect on the demand for improved delivery of pharmaceuticals. These will be discussed in this section and later sections.

22.3.1   MARKET DEFINITION AND SIZE

The exact definition of a drug delivery product is open to interpretation. From a formulation viewpoint, a drug delivery product can be described as one that has been deliberately designed to modify or localize the release and/or distribution of a therapeutic entity and/or improve its bioavailability in a manner that could not be achieved if it were formulated as an immediate-release solid dosage form, a simple solution, suspension, cream, or ointment. For this reason, the term "drug delivery" has been traditionally associated with the use of formulation technologies to enhance or enable delivery of therapeutic molecules and, in particular, the use of polymers, lipids, or specialized excipients such as cyclodextrins and permeation enhancers. Therefore, the term has been associated with dosage forms, such as sustained-release tablets, implants, nanoparticles, liposomes, and technologies which enhance the solubility or permeability of drugs. Similarly specialized formulation and device combinations such as dry powder inhalers, iontophoretic patches, and ocular inserts would typically be classed as drug delivery products.

However, drug delivery can also be enabled or enhanced by chemical modification of the pharmacological entity, e.g., PEGylation of a protein or by the use of specialized devices. In particular, in

TABLE 22.3

Examples of Drug Delivery Market Size Estimates in Recently Issued Commercial Reports

Estimate of Drug Delivery Market	Time Frame	Report
Global market	2015–2020	Markets and Markets (2015)
$1048.1 billion in 2015 estimated to reach $1504.7 billion by 2020 giving a CAGR of 7.5% from 2015 to 2020		
Global market	2013–2018	BCC Research (2014)
$181.9 billion in 2013 with the market estimated to reach $212.8 billion in 2018 resulting in a CAGR of 3.2%		
U.S. market	2014–2019	Freedonia (2015)
$187 billion in 2014 increasing to $251 billion in 2019 giving a CAGR of 6.1%		

the rapidly growing area of biopharmaceuticals, around 14% CAGR is forecast between 2014–2020 to reach a global market size of $1.671 billion (Mordor Intelligence, 2015), such options are most often the only way that delivery can be improved. Differences in the type of products included in the analysis of the drug delivery market are part of the reason why estimations of this market can vary considerably, as illustrated by the data in Table 22.3.

However, differences in the type of product included in the analyses are not the only reason for varying estimates (Market Size 2014). Market research is not an exact science, especially on a global basis. Variations in market estimates can occur due to factors such as difficulties sourcing certain data, differences in market research methodology, the exact time period evaluated, and the need to make assumptions. Despite this, all of the recent data point to the global drug delivery market being worth at least $182 billion and having a growth rate similar to that of the total market for prescription pharmaceuticals.

22.3.2 DRUG DELIVERY MARKET SEGMENTATION

The drug delivery market is typically segmented by therapeutic area, geography, and/or route of administration. With respect to therapeutic area, drug delivery products are commonly used for relief of pain (sustained and breakthrough pain relief—multiple formulation types), in the central nervous system (sustained-release, orodispersible formulations, and depot injections), for type 2 diabetes (sustained-release depot injections [Bydureon®]) and fatty acid-peptide conjugate [Victoza®]), for prostate cancer (sustained-release depot injections), and in cardiovascular segments (sustained-release oral dosage forms). The treatment of asthma and other respiratory diseases is almost totally reliant on the use of inhaler technology, while drug delivery devices such as pens and autoinjectors have simplified the lives of type 1 diabetes sufferers.

With the exception of prostate cancer, oncology traditionally was a segment where the focus of innovation was on the active therapeutic and not on delivery. However, this has begun to change in recent years with the development of PEGylated biopharmaceuticals, liposomal products, antibody–drug conjugates, targeted nanoparticle systems, and subcutaneous injection formulations of biopharmaceuticals that previously could only be administered intravenously, e.g., Herceptin®. This trend is set to continue with the overall growth in the anticancer segment (see Section 22.6.5 for further details). North America (United States and Canada) is currently the largest market for drug delivery products (40%–50% of the global drug delivery market), followed by Europe (25%–35%) and then Asia/Pacific, with Asian markets showing the most growth (Freedonia 2015; Challener 2014). Table 22.4 shows how the drug delivery is typically segmented by route of administration and drug delivery technology.

TABLE 22.4

Typical Segmentation and Subsegmentation by Administration and Technology

Route	Technology Subsegment	Comments
Oral	*Controlled release*: Matrix and reservoir systems, multiparticulates (coated pellets/nonpareils in a capsule), osmotic systems, orodispersible, taste masking, and others	The oral segment is the largest of all, with around 40%–50% of the market. Unsurprisingly, controlled-release formulations represent the largest subsegment. It has been estimated that the U.S. oral drug delivery market will grow from $107 billion in 2014 to $130.4 billion in 2019 (CAGR 4%) (Freedonia, 2015). However, future growth in this segment is likely to be impacted by the loss of patent protection for some highly successful controlled-release formulations, e.g., Seroquel® XR in 2017, and the increase in biopharmaceuticals both as a percentage of the total number of approved therapeutic entities and the total pharmaceutical sales.
Injectable	Pen injectors Autoinjectors Needle-free Depot injections Encapsulated delivery systems	There is often considerable variation in how the injectable market is defined with some market analyses including monoclonal antibody products in the definition. However, in all cases, this segment is growing rapidly for the reasons given in Section 22.6.5.
Pulmonary	Metered-dose inhalers Dry powder inhalers Nebulizers	Pulmonary delivery is third largest segment in the drug delivery market. It represents typically around 10%–15% of the drug delivery market.
Transdermal	Passive transdermal Active transdermal, e.g., iontophoresis	This segment is fairly small. Expansion is limited by the low numbers of drugs that can be delivered across the skin.
Nasal	Inhalers Sprays	Currently, this route of administration is mainly limited to local action except a few notable exceptions, e.g., flu vaccine, calcitonin, and fentanyl.
Implants	Drug-eluting stents Implantable infusion pumps Intravitreal implants Contraceptive implants Brachytherapy seeds	Small segment whose growth is driven by stents and intravitreal implants.
Topical Ophthalmic Transmucosal	Semisolid formulations plus others Eyedrops Ointments Inserts Buccal and sublingual Rectal and vaginal	These are small segments concerned mainly with local therapy, with the exception of buccal and sublingual. Transmucosal delivery is typically cited as the smallest of all drug delivery segments, but there has been considerable growth in the use of the buccal route in recent years due to advances in buccal tablet and film technology.

22.4 NEED FOR DRUG DELIVERY PRODUCTS

22.4.1 TYPES OF NEED

In general, products and services are conceived, developed, and launched to satisfy identified customer needs or wants that are not met by those currently on the market. In this respect, the market for pharmaceutical products is no different from any other. Market needs can be divided into three groups: real and clearly identified, latent, or perceived, as shown in Table 22.5.

TABLE 22.5

Different Types of Marketing Needs[a]

Real and Clearly Identified	Latent	Perceived
Patients, doctors, and companies recognize there is a need for a treatment or an improvement in therapy, e.g., an effective treatment for Alzheimer's disease.	Patients/doctors are not actively conscious of the need, but when questioned about or offered the product, they immediately see its advantages. For example, buccal films that can be administered without water.	A perceived need for a pharmaceutical product is mostly associated with OTC medicines although it can arise as a result of direct-to-consumer advertising of prescription medicines. In this case, the product may not be necessary for health, e.g., the purchase of vitamin and nutrient supplements by healthy nonpregnant adults.

[a] Depending on the circumstances, drug delivery products meet one or more of these types of needs.

22.4.2 How Drug Delivery Products Meet Market and Stakeholder Needs

Drug delivery products meet market needs in a number of ways. These needs can be viewed from the standpoint of the main stakeholders involved in health care. Patients, doctors, and other health-care professionals want products that are more efficacious, safer, and more convenient to use than those already on the market. Doctors in particular desire a broad palette of therapeutic options so that they can tailor therapy to the needs of individual patents. In addition, patients want products with a pleasant taste, which are nongritty and are easy to swallow. Reductions in dosing frequency and improved ease of administration or taste are often accompanied by an improvement in adherence to the medication regimen, which, in turn, has a knock-on effect on therapeutic effectiveness. The same can be said for medicines with an improved side-effect profile either over the original product or others in the same therapeutic class. The price of the new therapy is also important for OTC and POM products in cases where patients have to make a large copayment, or a contribution, which is a percentage of the total price.

Payers want therapies with greater effectiveness and safety or that aid adherence to therapy, as this should reduce the total time of treatment and reduce the likelihood of adverse effects. Nonadherence with chronic medication is currently thought to be around 50% and this costs the U.S. health-care system alone between $100 billion and $289 billion annually (Viswanathan et al. 2012). Unsurprisingly, the issue of nonadherence is of importance to payers, and although the reasons for it are multifactorial, it has been repeatedly shown that simplifying the dosage regimen facilitates compliance with therapy (Laliberté et al. 2013; Medic et al. 2013). Payers also want medications that can be self-administered or administered in an out-patient setting, to reduce the need to treat patients in hospital, which is very expensive. Often drug delivery approaches are required to develop such formulations.

Despite the potential benefits of drug delivery–enabled and drug delivery–enhanced medicines, it should be remembered that payers are principally focused on the cost effectiveness of medicines compared with therapeutic alternatives already on the market, including generics. The cost-effectiveness of a new drug delivery product may be relatively easy to demonstrate if its use results in an overall reduction in the cost of care, e.g., a change from in-patient to out-patient treatment or a reduction in drug waste. Benefits, such as better therapeutic outcomes as a result of increased patient compliance, can be difficult to prove for a new product, as in the real world, other factors affect adherence to therapy including the level of patient out-of-pocket expenses (see Section 22.6.9 for further comments).

Companies that develop pharmaceutical products need to develop products that will sell well, capture market share, allow them to recoup their development costs, make a profit, and satisfy shareholders. Research-based pharmaceutical companies, drug delivery companies, and generic companies use drug delivery technologies to achieve these goals. In the case of drug delivery companies (firms whose principal business is the development and exploitation of in-house delivery systems), the products are mainly developed on behalf of clients. The importance of drug delivery technology to the pharmaceutical industry is discussed next.

22.5 ROLE OF DRUG DELIVERY IN PRODUCT DEVELOPMENT AND PRODUCT LIFE-CYCLE MANAGEMENT

Drug delivery technology is primarily used to develop second-generation formulations for *product life-cycle management* purposes, but it is also used increasingly during the development of new compounds. Creating products that benefit or add value to patients and other stakeholders makes good financial sense, and a significant number of drug delivery products have become blockbusters, with annual sales in excess of $1 billion (for examples, see Table 22.6).

22.5.1 IMPROVED EFFICACY AND COMPLIANCE AND REDUCED ADVERSE EFFECTS

The overriding goal of early development is to initiate animal, toxicology, and later clinical testing as soon as possible, in order to get an initial readout on a new therapeutic entity's safety and pharmacokinetics. Typically, pharmaceutical scientists employ as simple formulations as possible (solutions, suspensions) in order to achieve sufficient drug exposure during early development.

TABLE 22.6
Examples of Top-Selling Drug Delivery Products

Product	Active	Technology	2015[a] ($ Million)	2014 Sales[a] ($ Million)
Seroquel® XR	Quetiapine	Controlled-release tablet	1,025	1,224
Sandostatin®SC/LAR[b]	Octreotide	PLGA microparticle depot injection	1,630	1,650
Symbicort®	Budesonide and formoterol	Inhaler	3,394	3,801
MabThera®/Rituxan®	Rituximab	Human hyaluronidase enzyme in	5,640[b,c]	5,603[b,c]
Herceptin®[b,c]	Trastuzumab	subcutanous injections	6,538[b,c] (Swiss francs)	6,275[b,c] (Swiss francs)
Neulasta®	Pegfilgrastim	PEG conjugate On-body injector enables delivery 1 day after chemotherapy	4,715	4,596
Invega Sustenna®/ Xeplion®/Invega Trinza®	Paliperidone palmitate	Nanocrystal® Technology to improve solubility of insoluble ester	1,830	1,588
Victoza®	Liraglutide	Lipid conjugate in pen injector	18,027 (Danish kroner)	13,426 (Danish kroner)
Seretide/Advair	Fluticasone propionate and salmeterol xinafoate	Inhaler	3,681 (Sterling)	4,229 (Sterling)

[a] Sales figures from company reports. In US$ unless otherwise stated.

[b] Sales figures include both standard formulations and drug delivery product.

[c] Subcutaneous injections at the time of writing this article were available only in some markets but proving popular as they decrease the time for injection to around 5 minutes and are less invasive.

Even in later clinical stages, companies tend to develop simple immediate-release formulations (tablets/capsules or injections) in order not to add complexity to the development process. More sophisticated formulations and drug delivery approaches are often only used for these "first-generation" formulations if necessary, e.g., to achieve adequate bioavailability and/or stability of the compound.

The number of compounds in pharmaceutical companies' pipelines with suboptimal absorption, distribution, metabolism, and excretion (ADME) properties has risen in recent decades. This is, in part, due to the increased use of combinatorial chemistry and high throughput screening in drug discovery. These techniques tend to identify highly potent lead compounds, but such leads also often have poor aqueous solubility (*Biopharmaceutics Classification System* [BCS] Class 2 and 4 drugs; see also Chapter 3). Depending on the dose, simple formulations of such compounds may not result in sufficient bioavailability, and solubility-enhancing technologies are therefore required to achieve this.

Another factor increasing the number of therapeutic entities with suboptimal ADME properties is the fact that biopharmaceutical and peptide drugs now occupy a larger proportion of company pipelines than in the past. Many protein and peptide drugs have short half-lives in vivo and drug delivery technologies such as PEGylation or microparticle-based depot injections are necessary to reduce the frequency of injection, decrease administration costs, and improve convenience (see also Chapter 6).

There is continuing interest in the delivery of therapeutic entities via noninvasive routes. This is driven by the general unpopularity of injections, the rise in the incidence of diabetes, the number of peptide drugs in development, the desire to reduce or eliminate first-pass metabolism, and an improved understanding of how to overcome the epithelial barriers presented by noninvasive routes for molecules with a MW of 6000 or less. The systemic bioavailability of most compounds is poor via routes such as nasal, buccal, transdermal, or pulmonary, and drug delivery technology is required to enable or facilitate delivery.

A long-standing goal of drug delivery is to target drugs to specific cells, tissues, and organs where they can exert their action and keep them away from those where they would cause side effects. For parenteral delivery, targeting can either be (1) passive, i.e., reliant on the ability of colloidal nanosystems with the correct particle size and surface properties to accumulate in, or avoid, certain tissues, or (2) active, due to the presence of ligands, which bind to cell surface receptors on the target cells. Such targeting can be used for both therapeutic and diagnostic purposes. Due to the complexity of developing such systems, very few products have reached the market so far. Examples include liposomal formulations of daunorubicin and doxorubicin, paclitaxel albumin–bound nanoparticles, and radiolabeled monoclonal antibodies. However, this may be set to change with a number of antibody drug conjugates (ADCs) in the late clinical development. At the time of writing this chapter, only two such ADCs were on the market, Adcetris® (brentuximab vedotin) and Kadcyla® (trastuzumab emtansine).

22.5.2 PRODUCT DIFFERENTIATION AND HOW TO INCREASE MARKET SHARE

Product differentiation is important in crowded markets. As a rule of thumb, the first product on the market in a segment has the best chance of capturing and maintaining the largest market share, providing it can overcome initial resistance to its use due to physician/patient unfamiliarity. Products entering the market later and particular "me-too" members of the same therapeutic class must position themselves against the first-to-market product. Unless these products have superior efficacy, safety, and/or convenience, then this can be an uphill struggle, although the extent to which they are disadvantaged depends on a number of factors, such as the time between their launch and that of the first product, promotional efforts, and pricing (Schulze and Ringel 2013).

Analysis of the factors affecting the market success of Specialty Pharma products showed that an improved side-effect profile and/or an increase in convenience are powerful motivators for doctors to prescribe either a "me-too" drug in the same class (Gudiksen et al. 2008) or a new formulation of a previously marketed drug.

In some cases, the development of a second-generation drug delivery–enhanced or drug delivery–enabled product, which is clinically superior or provides distinct advantages, can result in product sales that outstrip that of the original formulation and promote wider use of the therapeutic compound. Examples of this include (1) the transdermal patch formulation of rivastigmine (Exelon®), which offers reduced side effects, improved convenience, and carer satisfaction, compared with oral formulations of the same drug (Darreh-Shori and Jelic 2010; Adler et al. 2014), and (2) Pegasys®, which offers increased efficacy in hepatitis C, reduced side effects, and improved convenience over the non-PEGylated version (Rasenack et al. 2003; FDA 2013).

22.5.3 FOR LIFE-CYCLE MANAGEMENT PURPOSES TO EXTEND THE TIME OVER WHICH A PARTICULAR BRAND OF MEDICINE REMAINS PROFITABLE

Companies have always developed new formulations and dosage forms as part of product life-cycle management. However, the application of drug delivery, e.g., controlled-release technologies, allows the development of products with clear competitive and clinical advantages over the original product and provides a defense against other competitors already in, or entering, the market. This defense comes in the form of new intellectual property and, in some cases, exclusivity on the clinical data generated to support the market authorization of the drug delivery product. Generic companies therefore cannot immediately make direct copies of these new formulations/dosage forms even if the drug is off-patent. They must wait until the period of exclusivity (if applicable) is over, and the patents have expired or have been proven to be invalid or not infringed by the generic company's product. For research-based companies, which develop new therapeutic options, the strategy typically involves getting as many patients transferred to the newly improved product prior to the approval of the first generic copy of the original formulation.

22.5.4 TO ENABLE THE DRUG TO BE USED FOR NEW INDICATIONS

Sometimes new formulations of already approved drugs are required in order to allow these drugs to be used for new indications. For example, the short-acting opiate drug, fentanyl, was initially marketed as a short-term painkiller for use during surgical procedures. However, the approval of a transdermal patch formulation, Duragesic®, in 1990 expanded its use to chronic pain control. More recently, numerous noninvasive fentanyl formulations indicated for breakthrough cancer pain have reached the market. The development of such products creates new markets for the drug, results in new IP and, in some cases, exclusivity for the clinical data used to support the market authorization of the new indication.

22.5.5 TO ENABLE USE IN SPECIALIST POPULATIONS OR TO FACILITATE PRESCRIPTION TO OTC SWITCHES

All patients want easy-to-swallow and pleasant-tasting medicines, inhalers that are easy to use, and injection devices that cause minimal pain and distress. However, these aspects are particularly important in pediatric populations and to a certain extent in elderly people. For example, developers of liquid, dispersible, and orodispersible formulations often employ taste-masking technologies, e.g., polymeric-coated drug particles that work together with the vehicle to disguise an API's bitter taste.

In the OTC market, competition is fierce. Branded products must offer additional advantages and perceived higher quality to cheaper generic alternatives, in order to justify their higher price. Formulation approaches that improve taste and convenience (especially those that remove the need to take the dosage form with water) and accelerate onset of action are popular with patients.

22.5.6   To Create and Exploit Intellectual Property and Other Barriers to Market Entry by Competitors (Especially Generics)

There are three main ways that pharmaceutical originator products can achieve market exclusivity: The first way is via patents and associated patent-term extensions (e.g., United States) or supplementary protection certificates (e.g., EU). The second is a result of receiving marketing authorization from a regulatory authority. The third type of barrier relates to technical and investment barriers that are associated with very innovative and complex delivery systems. These three mechanisms are now considered in turn.

22.5.6.1   Patents

The development of drug delivery–enabled or drug delivery–enhanced products leads to new patents and patent applications that competitors must challenge or work around in order to develop and commercialize a product with similar advantages. However, formulation patents do not provide as strong protection as those for new therapeutic entities, as it is often possible for competitors to create a product with similar benefits by using a different formulation approach. The strength of a product's patent position depends on a number of factors including the uniqueness of the drug delivery technology used, whether it itself is patented and has a strong patent position, the claims granted by patent offices, and the skill of the patent lawyers and scientists involved in drafting and defending the patents.

Although a number of factors affect the actual length of patent protection, the term of a patent generally runs from the date the patent is granted, to 20 years from the date on which the first application was filed. Since the development and regulatory approval process for pharmaceuticals is so lengthy, companies obtaining market authorization for originator products in many countries are eligible for extensions to patent protection, called patent term extensions in the United States and supplementary protection certificates in the EU. The rules governing these extensions vary, but in general, the United States and EU allow a maximum of 5 years additional IP protection (one patent and first authorization of drug) (Ellery and Hansen 2012). Limited additional patent protection may also be granted for approval of use of medicines in pediatric populations.

22.5.6.2   Market Exclusivity

Companies can receive certain market exclusivity rights for their products after regulatory approval. Different rules on exclusivity apply to new chemical entities and certain hormones, biologics, and those designated as orphan drugs. Typically, exclusivity is also granted, albeit for a shorter period, if a company obtains regulatory approval for new indications and/or pediatric use of a previously approved therapeutic. A detailed description of the rules and regulations pertaining to market exclusivity are found in the further reading list (FDA 1999; Small Business Assistance 2010; Ellery and Hansen 2012; European Commission 2013; Frias 2013).

In the United States, companies are rewarded with *new drug product exclusivity* for new products containing previously approved active moieties whose marketing authorization was dependent on the results of new clinical investigations (not just bioavailability/bioequivalence studies). Such products can be approved through the FDA's 505(b)(2) new drug application (NDA) route (FDA 1999; Small Business Assistance 2010) provided that the drug is not protected by patents or exclusivity. Under this regulatory route, the applicant (often a small specialist or generic company) must provide full reports of safety and effectiveness studies, but at least some of the supporting data comes from studies not conducted by or for the applicant (e.g., preclinical and clinical studies that supported the U.S. marketing authorization of the reference product). Formulations/dosage forms approved by this route differ from those submitted as an abbreviated new drug application (ANDA) in that they are not a duplicate, or essentially a duplicate, of an approved originator reference product (FDA 1999; Small Business Assistance 2010).

For successful 505(b)(2) submissions for new formulations/dosage forms, the period of new drug product exclusivity is 3 years if new clinical studies were essential for approval, although other types of exclusivity may apply, e.g., pediatric. The 3-year new product exclusivity prevents approval of ANDA or other 505(b)(2) applications for the same conditions of approval. However, it does not prevent approval of 505(b)(2) applications for other formulations or dosage forms of the same active moiety, if the conditions of approval are different, e.g., different indications or generics of the originator product. It also does not protect against a duplicate product coming onto the market if it has been the subject of a full NDA submission (505(b)(1)) (Small Business Assistance 2010). The previous discussion refers to small-molecule drugs and noncomplex biologics approved through the NDA pathway. Complex biologics approved through a biologics license application in the United States are entitled to 12 years of market exclusivity; legislation on biosimilars has only been introduced (FDA 2014a) in recent years. To date the FDA has approved only one biosimilar product, Zarxio™, a biosimilar to Amgen's Neupogen® (filgrastim) (FDA 2015). For biologics there is no additional exclusivity awarded for new indications, dosage forms, routes of administration, delivery systems etc., if the biopharmaceutical has not been structurally altered to improve purity, safety or potency (FDA 2014d).

The FDA also takes into account if there are patents in force protecting a product or its use before granting approval or in some cases allowing dossier submission. Applicants submitting NDAs (505(b)(1) or 505(b)(2)) are required to list any patents protecting their products and their use in their applications. These patents, their expiry data, and any U.S. exclusivity associated with the product are listed in the FDA publication "Approved Drug Products with Therapeutic Equivalence Evaluations" commonly known as the Orange Book.

In addition, as with ANDAs, the 505(b)(2) applicant must provide information on any unexpired exclusivity and all U.S. patents associated with the referenced approved active moiety, as given in the Orange Book. They have to make certifications, e.g., that all relevant patents have expired or, in the case of nonexpired patents, that these are not valid or not infringed (*Paragraph IV Patent Certification*). If protection still exists, approval and under certain circumstances filing will be delayed. Unsurprisingly, Paragraph IV Patent Certification is typically a trigger for patent infringement litigation by the originator company.

Therefore, in the United States, there are two mechanisms (patents and new product exclusivity) by which new formulations and dosage forms can be protected from generic competition, provided they required new clinical studies to support their approval. In Europe, the first time a therapeutic entity (drug or biological) is approved, the product benefits from an 8-year *data exclusivity* period during which competitors cannot submit a dossier referencing data used to support market authorization of the reference product. This is followed by 2 years of market protection during which competitors can submit dossiers for regulatory review and approval but cannot launch the product (European Commission 2013; Frias 2013). Market protection can be extended by a further year if marketing authorization for a new indication is obtained and certain conditions are met. This is generous in comparison with the new drug product exclusivity granted to new chemical entities entering the U.S. market (only 5 years). There is also provision for 1 year of data exclusivity for a new indication for a well-established product and a 1 year for a change of legal status to OTC if significant preclinical or clinical studies were necessary to support authorization (applies to study data only) (European Commission 2013; Frias 2013). Different data exclusivity rules apply to orphan drugs.

However, in the EU, the start of market protection begins with the first authorization of a therapeutic entity, and subsequent approval of different dosage forms, strengths, etc., does not affect it except in limited cases, e.g., where it is required for a new indication (European Commission 2013; Frias 2013). However, new formulations, strengths, and dosage forms are developed and licensed within the EU, some as so-called *hybrid products* (an EU-authorized medicinal product is used as a reference product but new clinical data are required to obtain approval). The development of these typically results in additional patent protection, which, in turn, facilitates market protection.

22.5.6.3 Technical and Investment Barriers

Other barriers are created when the delivery system is complex and difficult to formulate. Examples of such delivery products include microparticle-based implants, liposomes, and nanoparticles for parenteral use. The use of these technologies is restricted to drugs with specific delivery or compliance issues, which treat serious, and very often chronic, conditions. The products arising from such development typically provide distinct advantages and solve serious delivery issues. As a result, they often dominate markets with high profit margins. For competitors, such complex formulations represent a technological and investment barrier as their development requires specialist expertise and equipment and additional analytical capabilities compared with traditional dosage forms such as tablets. In addition, there is often very little regulatory guidance or experience with these types of products, particularly those involving intravenous colloidal systems (see also Chapter 21).

22.5.7 To Bolster Pipelines and Decrease Risk

New products containing already approved actives can be developed faster than those containing new therapeutic moieties and for considerably less investment (time and investment depends on the need for, and extent of, new clinical data required). They also carry less risk than new therapeutic moieties, as the clinical effectiveness of the active is known and, if developed by an originator company, will benefit from existing brand recognition.

A significant proportion of line extensions incorporate drug delivery technology to add advantages to the second-generation product. However, originator companies are not the only ones who exploit drug delivery technology to help fill their pipeline and balance risk. A significant number of companies focus their activity on developing drug delivery–enabled or drug delivery–enhanced products that can be approved through the FDA's 505(b)(2) pipeline. These are often small companies, but also generic giants such as Teva employ such tactics (see Section 22.8.1 for further details).

22.6 TRENDS WITHIN THE PHARMA MARKET AND THEIR IMPACT ON THE DEVELOPMENT OF DRUG DELIVERY PRODUCTS

22.6.1 Increasing Delivery Challenges

The number of poorly soluble compounds in drug development pipelines has increased over the years for the reasons given in Section 22.5.1 (see also Chapter 3). This, of course, impacts on the need for technologies to enhance the solubility of these compounds. Several technologies for improving aqueous solubility via the oral route exist, and the state of technology development could be described as fairly mature with several commercially viable technologies being available. However, there is a technology gap for poor solubility via the parenteral route as restrictions on the excipients considered safe for injection and issues with reprecipitation of drug following administration limit the current formulation options.

Silencing RNA and gene therapy development are currently hampered by delivery issues. Delivery technologies for these types of therapeutics exist, but they are far from ideal, and the area is one of active research (see also Chapter 16). Delivery system development in this area is set to continue and will increase dramatically if siRNA candidates currently in development fulfill their therapeutic promise.

Other trends within the marketplace include a sharp rise in the percentage of new product approvals for biopharmaceuticals and peptide drugs. According to IMS data, biopharmaceuticals were responsible for 18% of the global pharmaceutical prescription market in 2012 with insulin and monoclonal antibodies being responsible for the majority of sales (Rickwood et al. 2013) and this figure is set to grow.

Formulation-type approaches for improving delivery are often not suitable for very large molecules such as recombinant proteins and monoclonal antibodies due to instability, release, and/or cost of goods issues. In these cases, chemical conjugation approaches (e.g., PEGylation) are used to reduce dosing frequency, and delivery devices (pens and autoinjectors) are employed to make administration more convenient. One exception to this is the use of recombinant human hyaluronidase (Halozyme Therapeutics) to break down hyaluronan in the extracellular space and improve the delivery of large molecules via the subcutaneous route. The use of this technology has allowed the development of subcutaneous injections of antibodies that previously could only be administered intravenously (see Table 22.6). Other exceptions include a once-weekly injection of somatropin (Somatropin Biopartners), which has been approved in Europe, and several formulations of insulin and growth hormone for delivery via noninvasive routes, which are in clinical development.

Peptide drugs are unstable in gastrointestinal fluids and plasma and have poor membrane permeability. They therefore have short half-lives in body fluids and poor bioavailability when delivered by noninvasive routes. However, compared to proteins, they represent a more reasonable challenge for commercial drug delivery product development in terms of stability within microparticle systems, percentage absorbed in the presence of permeation enhancers, and lower drug substance manufacturing costs. A number of highly successful peptide drug delivery products have already been marketed for example, Victoza®, Zoladex®, and Sandostatin®. The growing pipeline for peptide drugs (mainly for the treatment of cancer and neuropeptides) therefore represents an opportunity for the development of drug delivery–enabled and drug delivery–enhanced products.

22.6.2 USE OF DEVICES TO IMPROVE CONVENIENCE AND SAFETY

The rise in the number of large biopharmaceuticals on the market and increasing levels of diabetes have had a significant impact on the size of the current and potential market for injection devices with increased ease of use, reduced pain, and more controlled delivery. As discussed in Section 22.6.1, large and fragile molecules such as proteins and monoclonal antibodies do not lend themselves to commercial development as polymer-based sustained-release depots or noninvasive dosage forms. Until technical challenges are solved, pharmaceutical developers will continue to focus on the use of prefilled syringes, pens, and autoinjectors to provide patients and physicians with greater convenience and reduced risk of contamination while device companies continue to develop and commercialize more sophisticated products. There is also an increase in the number and sophistication of complex drug–device combinations with ongoing improvements in the design of pulmonary inhalers and nasal delivery devices. Iontophoretic patches reentered the U.S. market with the launch of Zecuity® (sumatriptan iontophoretic transdermal system) in 2015, however postmarketing reports of burning and scarring in patients have led to a recent voluntary product withdrawal.

22.6.3 CHANGING DEMOGRAPHICS

According to WHO data, the number of people aged 60 years and older has doubled since 1980 and is forecast to reach 2 billion by 2050. This coupled with falling birth rates, especially in developed countries, means that by 2050 it is projected that globally the percentage of children less than 15 years of age will equal that of adults over 65. An increase in the elderly population will not only affect wealthy countries but also low-to-middle income ones. For example, it is estimated that Chile, China, and Iran will have a greater proportion of older people than the United States.

Changing demographics will have a dramatic impact on health-care needs and costs. Diseases associated with age such as cancer, osteoporosis, and Parkinson's and Alzheimer's disease will increase, as will the need for effective and affordable treatment of pain and rheumatic conditions. At the same time, there will be proportionally fewer people of working age paying taxes to cover the higher public expenditure on health care. Both will put enormous pressure on state-funded health care and pricing and reimbursement policies. Medicines that are easy to swallow and administer will become more important, because aging, and some of the diseases associated with it, increases the incidence of dysphagia and results in a reduction in motor and sensory skills.

In addition, many older people have multiple health issues resulting in complex medication regimens and a higher risk of adverse reactions.

Changing demographics represents an opportunity and a threat to drug delivery. The opportunities arise from the new products that can be developed to meet the needs of an aging population, e.g., oral films, taste-masked liquids, effective but noninvasive delivery to the eye, targeted diagnostics, and treatments for cancer. In developing such systems for use in older populations, companies need to consider changes in physiology that may impact on drug absorption and pharmacokinetics and the fact "old" people are a diverse group both in terms of age and extent of fitness. The threat comes from control of pricing and reimbursement and the willingness of payers to recognize the value of drug delivery–enabled and drug delivery–enhanced products and reimburse them appropriately.

22.6.4   INCREASING LEVELS OF OBESITY AND SEDENTARY LIFESTYLES

Rising levels of obesity and sedentary lifestyles in both developed and emerging markets have been accompanied by an increased incidence of metabolic syndrome, type 2 diabetes, and heart disease. Since these conditions often need treatment with multiple drugs, there are opportunities to simplify therapy and improve compliance through application of drug delivery. For example, Amylin used Alkermes' PLGA microparticle technology to develop Bydureon®, a sustained-release subcutaneous depot injection containing the glucagon-like peptide-1 receptor agonist, exenatide, which enables an injection frequency of 7 days, compared with twice daily for the conventional injection.

22.6.5   FOCUS ON CANCER THERAPIES, SPECIALIST AREAS, AND NICHE MARKETS

Oncology is by far the largest therapeutic segment in the pharmaceutical market (see Table 22.7), followed by treatments for antidiabetics. In recent years, there has been a focus on the development of cancer treatments and other medicines for specialist areas in which a premium price can be charged. At the same time, there has been a move away from the search for potential blockbuster drugs (prescribed in general practice to large patient populations).

The drivers for this change can be divided into three groups: commercial, technical, and regulatory. Commercial and technical drivers include the steadily rising cost of drug development, the high rate of failure in drug development pipelines, and the dramatic erosion of market share through generic competition when drugs lose their market exclusivity. Focusing on smaller specialized markets with significant unmet needs means that companies can concentrate their development efforts on conditions for which there are no, or few, satisfactory therapies. In such segments, successful products can make a clinical and commercial impact and there is less competition. In addition, new therapies for areas such as oncology are more likely to be given priority for review by regulatory authorities and, since they are administered by specialists, require a smaller sales force for their promotion than that required for a medication prescribed by community doctors.

Often the drugs used in specialized markets are biological products. To date, they have not suffered from generic competition to the same extent as small-molecule drugs, for the reasons described in Section 22.6.7.

Regulatory incentives promoting the development of specialist treatment include the orphan drug regulations and regulatory fast tracking and assistance for clinically superior new therapies for serious conditions (FDA 2014c). The former exist in a number of countries, and, although the definition of an orphan drug varies, all offer incentives to the development and commercialization of medicines for rare and mostly chronic diseases and conditions.

The increased focus on specialist markets opens up opportunities for drug delivery products within markets, in which effective products can command a premium price. For example, there is an urgent need for the improved targeting of many cancer drugs so that they reach the target cancer cells without causing nontarget side effects. However, tumor structure, composition, and biochemistry are complex and, together with the ability of cancer cells to become resistant to therapy, leads

TABLE 22.7
Global Sales Based on Therapeutic Classes

	Ranking	2014 Sales ($ Million)	2014 Growth (% $ LC)	2013 Sales ($ Million)
Oncology	1	74,449	12.2	67,486
Antidiabetics	2	63,573	18.0	54,850
Pain	3	59,786	6.5	57,625
Antihypertensives, single agent and combination	4	47,537	−1.2	49,648
Antibacterials	5	40,272	0.8	40,823
Respiratory agents	6	39,570	5.6	37,985
Mental health	7	39,134	0.6	39,533
Autoimmune diseases	8	35,906	17.5	30,952
Lipid regulators	9	28,412	0.2	28,947
Dermatologics	10	28,223	9.5	26,561
Anticoagulants	11	26,619	12.5	24,198
GI products	12	25,135	9.9	23,667
Antiulcerants	13	24,811	−1.1	25,650
HIV antivirals	14	22,678	10.9	20,615
Other cardiovascular	15	22,625	9.3	21,277
Nervous system disorders	16	22,106	11.7	20,191
Other CNS	17	19,652	5.5	19,036
Viral hepatitis	18	18,079	212.6	5,941
Kanpo, Chinese medicines	19	16,054	9.5	14,662
Vaccines (pure, combined, and others)	20	15,116	8.4	14,265

Sources: IMS Health Midas Data 2014; IMS Health Topline Data, http://www.imshealth.com/en/about-us/news/top-line-market-data, accessed March 20, 2016.

Notes: Sales and rank are in U.S. dollars with quarterly exchange rates; $LC, growth is in constant $ to normalize for exchange rate fluctuations. Sales cover direct and indirect pharmaceutical channel wholesaler and manufacturers. The figures above include prescription and certain over-the-counter data and represent manufacturer prices.

to huge technical challenges in trying to achieve targeting. Liposomes, nanoparticles, and drug–antibody conjugates are approaches currently being pursued (see also Chapter 5). In addition, drug delivery technology is used extensively to improve chronic and breakthrough pain therapy and most recently to prevent the abuse of prescription narcotics.

22.6.6 EMERGENCE OF PERSONALIZED MEDICINE

Personalized medicine is a potential game changer in pharmaceutical development and drug delivery. The segmentation of markets based on diagnostic results will enable patients to receive optimal therapy, but it further increases specialization and therefore will raise medication costs. At present, its use is mainly limited to certain monoclonal antibodies, which were developed in tandem with a diagnostic test. With the emergence of personalized medicine, the opportunities for advanced drug delivery approaches increase, in order to enable the delivery of therapeutics and diagnostics to the target cells. However, the development of personalized medicine, and the use of drug delivery to enable it, may be limited by pressure to control burgeoning health-care bills. For example, in April 2014, the NICE deemed the high cost of the antibody–drug conjugate, trastuzumab emtansine, "unaffordable" and did not recommend its funding by NHS budgets (NICE 2015).

22.6.7 INCREASING GENERIC COMPETITION AND WEAK DEVELOPMENT PIPELINES

In the last few years, a number of blockbuster small-molecule drugs have lost patent protection in key markets and started to face competition from generics. In the United States, typically the first company to file an ANDA containing a Paragraph IV certification for a generic copy of a medicine is entitled to 180-day market exclusivity (FDA, 2016). In Europe, the first-to-file generic does not receive any form of market protection. However, in both cases, price erosion occurs rapidly especially once there is more than one generic on the market (Simoens 2012; Rollins and Perri 2014). With payers becoming ever more cost-conscious, physicians are encouraged to prescribe generics and in many cases pharmacists are obliged or permitted to substitute a generic for a prescribed prescription medicine. It is estimated that in the United States, generics account for 90% of the sales volume of off-patent medicines with originator products losing about 80% of their sales within the first year. For example, in 2012, Lipitor, Pfizer's blockbuster statin, lost 59% of its worldwide sales and 81% of its U.S. sales due to generic competition (Pfizer 2012; Sheppard 2014).

To date, generic competition has had little impact on sales of biopharmaceuticals despite biosimilar legislation being in place in the EU since 2004. Biosimilars currently only account for 1% of the biologics market (Rickwood et al. 2013). This is due to a number of factors. First, until the passing of the Patient Protection and Affordable Care Act in 2010, there was no biosimilar legislation in the United States meaning that the world's largest pharmaceutical market was closed to such products. Second, a biosimilar can never be an exact copy of the originator product due to the molecular complexity of proteins and monoclonal antibodies. Biosimilar regulations therefore demand that certain preclinical, clinical, and immunogenicity studies are carried out to prove the biosimilarity of the new product to the comparator. Thus, the development of biosimilars is more costly and time-consuming than that of generic copies of small-molecule drugs, which can reference all preclinical and clinical data carried out on the originator product.

At the same time, development pipelines of many pharmaceutical companies have been drying up leading to a flurry of merger, acquisition, and in-licensing activity. This is due to the technical difficulties in discovering new drugs for many diseases that offer real therapeutic advantages compared with current treatments, and also the high attrition rates within development pipelines. The combination of the loss or imminent loss of patent protection, with fewer potential blockbusters in development to replace them, has led many companies to focus on new formulations and drug delivery–enhanced products as a defense against the generics and to manage product life cycle. In addition, drug delivery–enhanced products can be developed more cheaply and rapidly as there is no discovery stage and fewer clinical trials are required for them to gain marketing authorization.

22.6.8 INCREASED REGULATIONS AND REGULATORY SCRUTINY

The demands placed on pharmaceutical companies by regulatory authorities continue to increase. However, in recent years, a number of incentives for the development of particular medicines have been introduced. These include orphan drug regulations, increased regulatory fast tracking for innovative therapies, and the pediatric medicine regulations. The orphan drug and pediatric drug regulations give companies increased market protection for their products. The orphan drug regulations pertain to drugs that have been officially designated as orphan according to the regulations of individual regulatory authorities due to the low number of patients suffering from a serious disease or condition. These drugs are often biologics, but drug delivery products also benefit from them. For example, Abraxane® (paclitaxel albumin–bound particles for injectable suspension) has currently an orphan drug status in the United States for the treatment of stages IIb–IV melanoma and pancreatic cancer.

Pediatric regulations, which have been introduced in a number of countries, are designed to promote the clinical testing of medicines in children and therefore increase the need for appropriate formulations for younger patients. This, in turn, creates a demand for taste-masking technologies and easily swallowed dosage forms.

22.6.9 HEALTH-CARE COST CONTROL

The cost of health care is rising dramatically due to increased longevity and the greater prevalence of diseases associated with old age. This is coupled with soaring rates of type 2 diabetes and the rise in the number of medicines produced by biotechnology, the latter being significantly more expensive than small-molecule medicines.

In order to contain costs, many payers have adopted different strategies for controlling pharmaceutical budgets. This is mainly carried out by control of reimbursement. These include gating mechanisms to new drug access, reference pricing, value-based pricing, formularies and preferred product lists, and the use of HTAs to determine if a drug is a cost-effective use of resources and should be publically funded. To date, such strategies have been more prevalent in state-funded systems; however, they are also used by private medical insurance. This trend is set to continue as the impact of the U.S. Patient Protection and Affordable Care Act 2010 takes full effect. This law is designed to increase the number of people with health insurance, make health care more affordable for individuals, and provide the government with better value for money.

Controls on reimbursement mean that expensive therapies or those priced higher than competitors may not be reimbursed, or only under restricted circumstances, unless they have very clear effectiveness and/or safety advantages. For example, reference has already been made earlier to the inhaled insulin product Exubera®, which was only allowed to be prescribed on the UK NHS under very limited circumstances, as it was significantly more expensive than treatment with injected insulin but offered no effectiveness and/or safety benefits.

The key selling point of many drug delivery products is increased convenience and this can benefit patients in a number of ways, e.g., improved adherence. However, although a number of studies point to reduced dosing frequency and pill burden facilitating compliance with therapy, it is often difficult to demonstrate this for a particular product within clinical studies. The lack of robust data on the benefits of the improved convenience makes it difficult to justify a price premium for a product. The pharmacoeconomic case for a higher price is easier to argue for a novel delivery product that enables self-administration of the medication or reduces the number of physician visits, or time spent in hospital, and therefore reduces the overall cost of treatment. However, in many cases the new enhanced product will be priced based on the first generation product so that it is not higher for payers with the benefit of the new product to the company being market share expansion/maintenance through increased differentiation from competitor products (pricing of second generation life cycle products discussed in Ellery and Hansen [2012]).

Despite the comments earlier, the "add-on value" placed on increased convenience by a specific HTA organization may depend on the scope of the HTA (broad or narrow), the data/evidence and pharmacoeconomic justification presented by the pharmaceutical company, and if the opinion of outside parties, such as patient advocacy groups and independent experts, are taken into account. For example, one area of increasing importance to HTA is the impact of therapy on carers.

Among patients, cost (in the form of the amount they have to copay for the medicine) has a major effect on their willingness to pay for reduced dosing. For example, in one survey of U.S. patients with type 2 diabetes, patients with a high pill burden (≥5 tablets/day or more than once/day), i.e., high medication costs, were less willing to pay more for improvements in dosing convenience, than those with a lower pill burden. However, both groups of patients were willing to pay higher copayments for more efficacious medicines (Hauber et al. 2013).

As payers, due to financial constraints, become even more influential in prescribing decisions and physicians more aware of the cost of medication, product price together with effectiveness and safety data will determine prescribing and reimbursement decisions, and the use of HTA will continue to grow. This will impact on the prices that pharmaceutical companies can charge for their products, as they will have to prove new formulations and dosage forms are just as, or more, cost-effective than the competition. This is already the case in a number of countries, for example, the UK and Germany.

22.6.10 Emergence of New Pharma Markets

The so-called pharmerging markets like China, Turkey, India, Mexico, and Indonesia are growing economically, and spending on pharmaceuticals will increase from $26 billion in 2012 to $30–$50 billion in 2017 according to IMS predictions (Rickwood et al. 2013). With more money to spend on health care and the emergence of a wealthy professional class who are likely to place greater value on convenient dosage forms, pharmaceutical markets in these countries will continue to grow in the future. For example, it is estimated that global usage of medicine will increase by one third by 2020 over 2005 figures (Atkin and Kleinrock, 2015). However, in these markets, nonoriginal brands, generic, traditional and OTC medicines occupy a greater market share than in so-called developed markets and there is overall less money to pay for innovative medicines. This is likely to limit the use of drug delivery products in these countries.

22.7 WHAT MAKES A SUCCESSFUL DRUG DELIVERY PRODUCT?

Table 22.6 shows examples of successful drug delivery–enabled or drug delivery–enhanced products that had worldwide annual sales in excess of $1 billion in 2015. Other examples of successful drug delivery products in the past and present include the following:

- *Oral route*: Procardia XL® (once-daily nifedipine tablet), Effexor® XR (once-daily venla-faxine HCl), OxyContin® (twice-daily oxycodone), Wellbutrin® XL (once-daily bupropion)
- *Injectables*: Lupron Depot® (leuprolide depot injection)
- *Transdermal route*: Duragesic® (fentanyl patch)
- *Inhalation*: Spirva® (tiotropium bromide)

Product success can never be guaranteed and is influenced by a number of factors. However, based on previous successful and less than successful products, a number of key attributes can be identified and are described here.

22.7.1 Previous Success of the Active Therapeutic Moiety

Drug delivery technology rarely impacts on the efficacy of a therapeutic compound. So unless the technology enables a product that can be used for a new indication, or dramatically alters the dose that can be administered, e.g., as in the case of paclitaxel when formulated as Abraxane® (paclitaxel albumin–bound particles), the efficacy of the original and drug delivery product remains the same. In addition, the active is associated with a brand name and this may influence physician and patient attitudes to any second-generation formulation of the same therapeutic. This means that companies typically invest in improved formulations of actives whose first-generation products were commercially successful.

The exception is when drug delivery technology can address some of the previous barriers to success e.g. reduce side effects, result in new indications, or otherwise give a product a clear competitive advantage. An example of a drug whose sales were boosted by a second-generation dosage form is rivastigmine (Exelon®). It was first developed for oral administration and then as a transdermal patch for the treatment of Alzheimer's disease. This was because the controlled transdermal delivery of the drug reduced the incidence of side effects compared to the oral dosage forms, and enabled carers to check visually for compliance with therapy (Darreh-Shori and Jelic 2010; Adler et al. 2014).

22.7.2 EQUIVALENT OR IMPROVED EFFICACY AND SAFETY TO THE IMMEDIATE-RELEASE FORMULATION

Efficacy and side effects are more important to patients and doctors than increased convenience. If a drug delivery product is less efficacious than the original product or if it results in increased side effects, then even if it is approved, it is unlikely to be commercially successful if it reaches the market. Nutropin® Depot was a PLGA microparticle depot injection containing the growth hormone somatotropin. It enabled once or bimonthly dosing of this hormone, which is normally administered daily (FDA 2004). However, children using this product did not grow as fast or to the same extent as those dosed with the standard injection. It also resulted in an increased level of injection-site reactions. This, and other factors, resulted in its withdrawal from the market for commercial reasons.

The delivery system also has to be pharmaceutically stable and drug release consistent and reliable. Ionsys®, a fentanyl iontophoretic transdermal system for patient-controlled analgesia, was approved in both Europe and United States in 2006. It was withdrawn from the European market due to device stability problems and never launched on the U.S. market (Europa 2009c).

Neupro® (rotigotine transdermal system) was approved by the European Medicines Agency (EMA) in 2006 and the FDA in 2007 for the treatment of Parkinson's disease. Reports of rotigotine precipitation on the outside of the patches appeared shortly afterward. Based on stability data at $5°C$ and dissolution data, the EMA allowed the product to remain on the market provided storage conditions were changed to refrigerated and other conditions were met (Europa 2009b). The FDA, however, requested that the product be reformulated. Neupro® reentered the U.S. market in 2012 and had global sales of €182 million in 2013. However, the loss of the sales from the United States for around 4 years was a major setback for the brand.

22.7.3 NO ADDITIONAL CONTRAINDICATIONS OR RESTRICTIONS COMPARED WITH COMPETING PRODUCTS

Drug delivery product should be suitable for administration to as wide a patient population as possible, unless it has been specifically designed for a target subsegment. This was an issue for Exubera®. It was designed to compete with rapid-acting insulin, but its use was contraindicated in patients who smoked or had lung disease. In addition, its use was only allowed if patients had satisfactory lung function test results, and this test had to be repeated annually.

22.7.4 PRODUCT MEETS MARKET NEEDS

Successful products meet market needs by solving clinical problems. Microparticle-based depot injections of gonadotropin-releasing hormone agonists such as goserelin (Zoladex®) and leuprolide (Lupron®) and antipsychotic drugs such as risperidone (Risperdal® Consta®) are highly successful products. In the case of leuprolide and goserelin, the depots remove the need for daily injections on a chronic basis, while in the case of Risperdal® Consta®, which is indicated for schizophrenia and bipolar disease, it ensures compliance in a patient population whose adherence to therapy is often poor. Controlled-release tablets such as Procardia® XL (nifedipine) and Toprol® XL (metoprolol) enable once-daily dosing of chronic medication indicated for hypertension and angina. In the case of hypertension, patients may have no symptoms and therefore may be less likely to be compliant with therapy. Once-daily dosing compared with three times (in the case of nifedipine immediate release) therefore simplifies therapy and aids adherence.

On the other hand, Exubera®, the first inhaled insulin product to reach the market, only partially addressed patient needs. Injections are never popular and patients always state that they would prefer to take medications orally or via other noninvasive routes. Exubera® removed the need for diabetics to inject prandial insulin. It also removed the requirement to store the insulin refrigerated as the spray-dried insulin powder was stable at room temperature. However, patients still had to inject basal insulin; they had to learn to use the device; dose conversion was complicated by the fact that the insulin dose was in milligrams and not in the usual international units; and the inhaler was large and could not be used discreetly (something that is important to diabetics) (Heinemann 2008). Exubera® was withdrawn from the market after sales of only $12 million.

22.7.5 ALIGNMENT OF PRODUCT PRICE AND BENEFITS

The benefits of a drug delivery product in terms of effectiveness, safety, and convenience of administration have to be in line with its price. Products based on sophisticated delivery systems are usually only developed for diseases and conditions whose treatment can demand a premium price and when there are significant issues with delivery. As health-care cost containment increases, there will be further focus on the cost of medication in terms of total cost and relative cost compared with other therapies and a need to demonstrate value for money. In reimbursement systems that are influenced by HTA, demonstrating the cost-effectiveness of a product is key to obtaining a recommendation that the product be reimbursed.

HTA typically focuses on effectiveness, safety, and cost of treatment and not on convenience per se although it is taken into account to a certain extent. The ability of a company to justify a price for their drug delivery–enabled or drug delivery–enhanced product may depend on a number of factors:

- The difference between the cost of therapy with the drug delivery product and other competitors, including other formulations developed by the originator. For example, in 2006, Clarosip®, a novel formulation of taste-masked clarithromycin granules in a drinking straw, was not recommended for use in NHS Scotland; the HTA assessment found that it was more expensive than other clarithromycin products, but it had no proven effect on compliance (Scottish Medicines 2014). Similarly, use of Exubera® was not recommended for public funding (except in a very limited group of patients) because of its premium price compared to injectable soluble insulin.
- The delivery problem solved by the product and if this improves effectiveness and/or safety.
- If an increase in compliance/adherence can be adequately demonstrated.
- The characteristics of the disease or condition and the patient population, e.g., specific formulations for pediatric patients or orphan diseases.
- If the drug delivery product removes the need for treatment in hospital or otherwise reduces the total cost of therapy. Subcutaneous Herceptin® was recently approved for use in NHS England because it reduces the time patients need to be in hospital and is less invasive than the formulation administered by intravenous infusion. Hence, it enables service redesign to take account of the reduced pharmacy and clinic time (NHS England 2013).

22.7.6 COST OF GOODS

Cost of goods manufactured is a term used to describe the cost of manufacturing a product. It includes raw material costs and direct labor costs. It therefore is affected by the cost of the drug, the excipients, the length and complexity of the production process, and the batch size. Sophisticated delivery systems such as parenteral microparticle or nanoparticle products naturally have a higher cost of goods than a controlled-release tablet or capsule and therefore are only suitable for markets in which a premium price can be demanded.

For peptides, proteins, and monoclonal antibodies, the cost of the active therapeutic moiety is considerable. This is an issue for drug delivery products being developed for biological products. The delivery of peptides and proteins by noninvasive routes results in relatively low bioavailability even after the system has been optimized, e.g., the bioavailability of Exubera® was around 10%.

This means that a considerably higher dose is required compared with injection with an associated increase in cost of goods (Heinemann 2008). Other cost of goods issues potentially affecting delivery systems include loss of drug during loading of colloidal delivery systems, the use of expensive excipients, complex production methodology, a greater degree of analytical testing and characterization than standard formulations, and small batch size.

22.7.7   PATENT PROTECTION AND UNIQUENESS OF DELIVERY TECHNOLOGY

Patent protection of the drug delivery product is a vital barrier to competitors entering the market. Ideally, the product should be protected by product-associated patents plus those covering the underlying technology. Once the drug delivery product is off-patent, then direct generic copies can be made. However, such products can face competition in advance of patent expiry if the drug itself is no longer patented and a competitor circumvents the patent using a different formulation or delivery approach. For example, two patented prolonged-release versions of a highly successful drug may coexist on the market, if they were developed using different controlled-release technology and the patents on both do not infringe each other. There are currently various different formulations of fentanyl on the market for the treatment of breakthrough cancer pain via noninvasive routes that all compete with each other.

A greater degree of protection for the drug delivery product is achieved if the technology is unique or complex, e.g., Seretide/Advair Diskus®, the delivery issue is technically challenging to solve, and it requires high investment and specific expertise to develop and manufacture the product. Even then if the market is large enough, there will be financial motivation for competitors to try to overcome the barriers. For example, several depot injections of gonadotropin-releasing hormone agonists have been launched on the U.S. market. However, the Lupron® Depot was still the leading hormone therapy for the palliative treatment of advanced prostate cancer in the United States and achieved global sales of $800 million in 2012 (Abbvie company report). This is despite being first approved in 1989.

One particular successful life-cycle management strategy was adopted by Abbott Laboratories in defense of its fenofibrate franchise in the United States. It employed a combination of improved formulations of this poorly soluble drug, obtaining additional indications for the drug, clever drafting of patent claims, and legal action against generic companies for patent infringement (this delays ANDA approval by the FDA for up to 30 months). By using solubility-enhancing technology, Abbott was able to alter the dose required on two separate occasions and remove the need to take the drug with food. In addition, the third reformulation contained fenofibric acid and not the ester. Generic companies then had difficulty copying the originator product as the dose and formulation kept changing. Those that reached the market could not be directly substituted for the branded product due to dose differences. Abbott was so successful that it itself was the subject of various court actions for abusing its market position to hinder competition. As a result, the company had to pay over $300 million in court settlements (Downing et al. 2012).

22.7.8   SUCCESSFUL PRODUCT LAUNCH AND MARKETING OF DRUG DELIVERY PRODUCTS

Product launch and the months preceding and following it are critical for the future success of any product including new medicines (Rollins and Perri 2014). Prelaunch in the market has to be thoroughly researched, the marketing strategy and tactics agreed, the product correctly positioned, and all four elements of the marketing mix decided upon; the advertising material must be prepared including medicine-specific websites and TV and radio commercials (where allowed), the sales force trained, and doctors, wholesalers, pharmacists, and the press briefed; launch stocks need to be available. If sales are not according to plan, then steps must be rapidly taken to correct this. The power of marketing cannot be overemphasized; however, the growing influence of payers in prescribing decisions means that a convincing pharmacoeconomic justification for a new medicine is critical.

22.8 BUSINESS-TO-BUSINESS MARKETING OF DRUG DELIVERY TECHNOLOGIES

22.8.1 DEVELOPERS, USERS, AND SELLERS OF DRUG DELIVERY TECHNOLOGY

Back in the 1980s and early 1990s, the term *drug delivery* was mainly associated with specialist companies, e.g., Alza, Elan, Eurand, Jago (later acquired by SkyePharma), and Biovail. These companies concentrated mainly on developing delivery technologies and using these technologies to create controlled-release dosage forms of drugs (typically for oral or transdermal administration) on behalf of clients. Research-orientated pharma formulated products to achieve adequate bioavailability and some of these products were drug delivery enabled or enhanced; however, its focus was not specifically on drug delivery technology development. Research-based pharma was the client of the drug delivery specialist, and in general, drug delivery companies did not develop and market their own products. Today, the situation is far more complex with a variety of players developing and exploiting delivery technology. As discussed in Section 22.3.1, the rise in the number of biopharmaceuticals has broadened the term *drug delivery* to include chemical conjugation approaches, delivery devices, and other approaches. The list below shows the type of firms developing drug delivery technologies and products to support their own product pipeline and/or to offer them to other firms through licensing, development, (sometimes manufacturing), and commercialization agreements or service contracts.

Big Pharma/Big Biotech. They develop, or acquire through merger and acquisition activity, drug delivery technologies that they use in the development of their own products. The development of devices is not usually part of their expertise with a few exceptions, e.g., Baxter. They also collaborate with drug delivery companies if in-house expertise is lacking and, if the product development is seen as strategic to future success, buy shares in these firms. Such collaborations are common for complex delivery systems such as microparticle-based depot injections or transdermal patches. Some companies, e.g., GlaxoSmithKline, AstraZeneca, and Boehringer Ingelheim, are experts in the field of pulmonary delivery.

Small- to medium-sized pharma. They develop their own products but are focused on a few therapeutic areas (examples include Grünenthal and UCB Pharma). They may have drug delivery technologies developed in-house or have acquired them to assist in the development of their own compounds. They may also collaborate with drug delivery companies and Big Pharma. Grünethal, for example, focuses on pain therapeutics and has developed its own tamper-resistant abuse-deterrent technology and controlled-release and transdermal patch products.

Specialty Pharma, i.e., companies focusing principally on products that are prescribed by clinical specialists (as opposed to community doctors) for particular diseases and conditions. Since many of these treatments are administered by injection, there is a focus on the use of devices to ease administration. These devices are typically not developed in-house. However, Specialty Pharma also encompasses postsurgical and cancer pain therapy, anti-tumor medicines, and siRNA. It therefore includes microparticle-based depot injections, liposomal-entrapped anticancers, Abraxane®, antibody–drug conjugates, and silencing RNA delivery systems

Drug delivery companies. These are companies whose main business is developing drug delivery technologies, which they can license to other companies for use with their drugs. The traditional drug delivery business model of developing drug delivery technologies, and then using them to develop products for third parties in return for up-front payments, development and manufacturing fees, milestone payments for pre-agreed goals, and royalties on commercial sales, has changed over the years. Most companies now also develop their own products (e.g., Alkermes, SkyePharma) and either launch them onto the market or more commonly out-license them in late clinical development when higher royalty levels can be demanded. They may also concentrate their efforts on a particular therapeutic area.

Generic companies, especially those specializing in difficult-to-formulate generics, e.g., Teva and Sandoz (the generic division of Novartis). Teva, in particular, has a product pipeline filled with drug delivery–enabled or drug delivery–enhanced products partially as a result of its "new therapeutic entity" strategy (http://www.tevapharm.com). This strategy involves developing new products containing known drugs and getting them approved in the United States through the 505(b)(2) pathway.

Contract development and manufacturing organizations. A number of these offer drug delivery technology as part of their services, e.g., Catalent and Patheon. Some of these technologies are offered on a fee-for-service basis.

22.8.2 Trends with in the Drug Delivery Business-To-Business Market

Various trends can be observed within the drug delivery business-to-business market, which are described here.

22.8.2.1 Continuing Decline of the Traditional Drug Delivery Company Business Model

The "pure" drug delivery business model (i.e., the focus is on proprietary technology development and developing products for third parties) has a number of inherent disadvantages:

1. First, the company can only access drug through clients or if it is commercially available. This limits its activities to client projects and off-patent compounds that can be sourced at a suitable grade (cGMP for clinical studies).
2. Clients are only willing to pay license fees and royalties for technologies that offer significant technical and market advantages over others available commercially and that have significant patent protection and remaining patent term. The expertise of the drug delivery company is therefore in competition with in-house and service providers' capabilities. This is particularly the case since formulation patents are, in general, easier to circumvent than those for active therapeutic moieties.
3. Drug delivery companies, in common with all service providers, are not involved in strategic decisions about the products they develop. Such decisions include client project priorities and the level of the marketing effort put into promoting a commercialized project. These decisions affect if and when a product reaches the market and the extent of sales and, hence, royalties.

The drawbacks of the drug delivery company business model were recognized early on with Alza being one of the first to develop its own products. Two early successful products developed by drug delivery companies and then licensed later on to Big Pharma include (1) Concerta® (methylphenidate extended release), which was developed by Alza and licensed to Johnson & Johnson, and (2) Wellbutrin® XL (bupropion), which was developed by Biovail and licensed to GlaxoSmithKline. Other examples include Alkermes development of Vivitrol® (naltrexone for extended-release injectable suspension), which it initially licensed to Cephalon and then took the product itself to market.

Advancing a product to, and through, Phase 3 trials requires considerable resource both in terms of finance and internal expertise, which is a heavy burden for drug delivery companies. Taking a product to the market takes a whole other set of skills and resources and usually requires partnering with other companies. Obtaining the level of investment required is often a stumbling block for many companies. The reward is being able to maximize value from the product, e.g., in terms of royalty or profit levels. The focus on own product development has led to companies previously associated with drug delivery now to describe themselves as biopharmaceutical firms (e.g., Alkermes, Halozyme) or Specialty Pharma (Flamel) in order to underline their business strategy.

22.8.2.2 Increasing Number of Biopharmaceuticals in Company Pipelines as Opposed to Small Molecules

As explained in Section 22.6.1, formulation drug delivery technologies are mainly suited to improving the delivery of small-molecule drugs and some peptides. Strategies for improving delivery of biopharmaceuticals are largely through modification of the active substance itself or via devices. This in effect reduces the potential market for formulation technologies but increases the one for novel devices or drug–device combinations. It also creates interest in companies that have technologies capable of delivering peptides and proteins by noninvasive routes or enable the slow release and/or subcutaneous delivery of larger biopharmaceuticals.

22.8.2.3 Increase in the Number of Other Service Providers Providing Drug Delivery as Part of Their Offering

There has been a rise in the number and extent to which other service providers offer drug delivery and specialized formulation expertise as part of their service offering. This has been due to an increase in demand overall for outsourcing and the increase in the number of small-molecule compounds with poor solubility issues. Some contract manufacturers such as Catalent have long been associated with drug delivery, e.g., as a result of their softgel and orodispersible expertise, while others such as Patheon have been actively promoting their know-how in this area, e.g., their SoluPath Flex™ package. Capsugel, originally a manufacturer of hard gelatin capsules, has been steadily broadening its delivery options and acquired Encap (liquid- and semisolid-filled hard gelatin capsules) and Bend Research (spray-dried powder solutions for poorly soluble compounds).

Competition is fierce in the world of outsourcing. Contract manufacturers use drug delivery capabilities to differentiate their early service offering, capture clients for their other services, and encourage long-standing business relationships. Excipient and capsule manufacturers boost sales of their main products by acquiring/developing drug delivery expertise or specific products for that particular market, e.g., the previously mentioned Capsugel acquisitions, BASF's SoluPlus® polymers, and Evonik's purchase of Surmodics pharmaceutical assets (drug delivery and PLGA polymers).

22.8.2.4 Use of Drug Delivery by Generic Companies

Some of the biggest users of drug delivery technology are generic companies that use it not only to directly copy existing formulations but also create new ones with competitive advantages that can be registered in the United States via the 505(b)(2) route. As previously mentioned, this can result in a 3-year new drug product exclusivity if clinical trials were essential to product approval. Such formulations are also likely to result in new intellectual property. In addition to Teva, Sandoz, Dr. Reddy's Laboratories (acquired OctoPlus in 2013), and Mylan also have considerable drug delivery expertise in-house. In addition, large generic companies often have API manufacturing facilities giving them easy access to supplies of drug substance at an early stage, something that most drug delivery companies do not have.

22.8.2.5 Once Novel Proprietary Technologies Become Part of the Standard Formulation Toolbox

A common delivery issue drives research in both Big Pharma and drug delivery companies, e.g., poor aqueous solubility of pipeline compounds. One company develops, patents, and commercializes a breakthrough technology, e.g., Elan's NanoCrystal® technology. Other approaches to solving the same problem are close behind. The success of these depends on the advantages/disadvantages they present, compared with the breakthrough technology, or if they serve a particular niche in the market. Gradually, the market becomes more and more crowded with competing technologies, and large companies acquire the expertise either through acquisition or in-house development.

TABLE 22.8
Drug Delivery Technology Maturity Continuum

Emerging technologies	Early	Crossing the BBB barrier
		RNAi delivery
		Nose-2-brain delivery
		Cell encapsulation technology
	Late	Active transdermal
		Peptide conformation stability
		Permeability enhancement
		Protein delivery via nonparenteral routes
		Intravitreal drug delivery
Products, but still evolving	Early	Drug–antibody conjugates
		PEGylation
		Liposomes/nanoparticles
	Late	Buccal delivery
		Nasal delivery
		Solubility enhancement
		Implants/depot injections
		Powders for inhalation
		Bilayer tablets
		CR liquids
Mature technology		Passive transdermal
		Sustained-release tablets and capsules
		Taste masking

Abbreviation: BBB, blood–brain barrier.

Later, the patents on the novel technology start running out and they are no longer so commercially valuable to their owners, although products manufactured using them may continue to be blockbusters. Eventually, the technology becomes part and parcel of the standard formulation toolbox. An example of this process is the history of the use of enteric coating. The exception is for highly specialist dosage forms, e.g., there are only a handful of companies with passive transdermal capabilities. Table 22.8 shows where certain formulation and chemical solutions to drug delivery issues sit on the technology maturity continuum (in the author's opinion).

22.8.2.6 Mergers, Acquisitions, Joint Ventures, and Spin-Outs

Like the rest of the pharmaceutical industry, drug delivery specialty companies have been the subject of merger, takeover, and spin-out activity. Of the six companies mentioned at the start of Section 22.8.1, only SkyePharma is still a stand-alone company. Alza was independent, then part owned by CibaGeigy, then independent once more, and finally taken over by Johnson & Johnson in 2001. Drug delivery products developed by Alza are still key assets within the J&J portfolio. However, much of the delivery technology assets were spun off and formed the basis of companies like Zosano Pharma.

The Elan Corporation merged with Athena Neurosciences in 1996. In the years that followed, the drug delivery business concentrated on oral administration and used joint ventures (JVs) with other delivery specialists to codevelop potential next-generation technologies and spread risk. Some of these JV partners such as NanoSystems and the Liposome Company were later bought out by Elan. However, the JV strategy was to be the company's undoing when questions were raised about its accounting practices. The drug delivery business later became a separate division of the company and was bought by Alkermes in 2010. Another example is Biovail, which was acquired by Valeant Pharmaceuticals in 2010 and Eurand became the drug delivery division of Aptalis Pharma in 2011, which was itself taken over by Forest Laboratories in 2014. In 2015, the drug delivery business of Aptalis was divested following Actavis plc's takeover of Forest and renamed Adare Pharmaceuticals.

Merger, acquisition, and spin-out activity in the field of drug delivery continues as larger companies buy out the smaller specialists and either run them as separate service divisions or integrate their assets within the larger organization and divest those that do not fit with the business strategy. JV formation is much less popular than previously although specialist companies with complementary skills still collaborate. New delivery specialists arrive on the market all the time, but they are more likely to use services to fund their own product development and call themselves biopharmaceutical or Specialty Pharma firms.

22.8.3 MARKET FOR DRUG DELIVERY TECHNOLOGIES

22.8.3.1 Current Status

When reading through Section 22.8.2, it could be construed that the drug delivery technology market is not in a healthy state, but this could not be further from the truth. Big Pharma always played a leading role in the creation of drug delivery products either through in-house development or investment in specialist firms. In contrast to earlier times, a number of service providers and generic companies see drug delivery technology as key to their commercial strategy.

It is true that certain oral formulation technologies have matured, but a significant number of unmet needs exist including improved solubility solutions for drugs administered intravenously, noninvasive options for peptides and proteins, efficient targeting, and intracellular uptake of molecules while avoiding their lysosomal breakdown. In addition, there are a number of evolving technologies in need of optimization, e.g., polymer conjugation to peptides and proteins. With the return of iontophoresis products onto the market, and the continuing interest in painless injections and microneedle technology, the area of drug–device combinations looks like an area of significant technological progress and commercial success in the future. Table 22.9 shows examples of companies which exploit drug delivery technology and expertise (research-based Big Pharma examples have been excluded).

22.8.3.2 Marketing a Technology

Companies offering drug delivery technology to third parties need to market it to their potential clients. They do this in a number of ways. For example, they conduct the necessary development, animal, and Phase 1 clinical studies to prove that their technology is robust and stable and works in humans. These studies are conducted using readily available model drugs for the type of delivery challenges they are hoping to overcome, e.g., fenofibrate for certain solubility-enhancing technologies. They later carry out such studies with client compounds in the course of limited-term feasibility studies. They also protect their technology through patents on different aspects of the technology itself and its use with as wide a variety of types of molecule as possible, as well as specific patents on products developed using it. Companies advertise their technology through scientific publications, talks, and posters at conferences, articles in industry magazines, their website, and press releases. They can also present their technology and its benefits to technology scouts and/or the relevant business development personnel within companies that could be potential clients.

22.8.3.3 How Pharmaceutical and Biotech Companies Evaluate Technologies and Products

Figure 22.2 shows the reasons why potential clients (e.g., small and big pharma) in-license drug delivery technology.

TABLE 22.9

Companies Using Drug Delivery to Drive Their Business

Company Name	Commercial Type/Strategy[a]	Examples of Technologies
3M Drug Delivery Systems	Drug delivery division of a global conglomerate with considerable device expertise/contract development and manufacturer of specialized dosage forms.	Inhalation, nasal, transdermal, oral, and topical dosage forms including microneedles.
Alkermes	A global biopharmaceutical company focused on CNS disorders/contract services.	Long-acting depot injection, Medifusion™, extends circulation time of biologics, solubility enhancement, oral controlled release.
Catalent	Global contract developer and manufacturer.	Softgel, oral fast-dissolving dosage forms, inhalation, oral controlled release, injections.
Capsugel	Originally a hard capsule manufacturer with dosage-form support. It has widened its delivery offering through acquisitions.	Abuse-deterrent, colonic dosage form design, spray drying for solubility enhancement, and liquid-filled and semisolid-filled capsules.
Flamel Technologies	A specialty pharmaceutical company whose roots are in drug delivery.	Oral controlled release—solid and liquid, anti-abuse formulations, hydrogel depot injections.
Hovione	API manufacturer, specialist in the field of inhalation and particle design.	Proprietary inhalers, particle engineering.
LTS Lohmann Therapie-Systeme	Contract developer and manufacturer of transdermal and oral thin-film dosage forms. Part owned by Novartis.	Transdermal, oral-cavity thin films.
Nektar Therapeutics	A clinical-stage biopharmaceutical company that exploits its PEGylation and polymer conjugate technology platforms.	Small molecule, prodrug, large molecule, and antibody fragment conjugates.
OctoPlus	Now a subsidiary of Dr. Reddy's Laboratories. It focuses on the formulation of injectables.	Microparticle technologies, liposomes and lipids, complex injectable formulations.
SkyePharma	Drug delivery company.	Oral controlled release and inhalation expertise.
Teva Pharmaceutical Industries	Global generics company with OTC and Specialty Pharma products.	Oral controlled release, inhalation, buccal delivery, abuse deterrent opioids, and other delivery systems used to produce products for its strategy to fill its pipeline with improved formulations of already approved products.

[a] Based on description of company or strategy on website.

As previously stated, companies do not wish to pay for drug delivery technology licenses or share their revenue in the form of royalties if they have suitable in-house technology or can access it as part of a contract manufacturer's services. The technology must therefore be patented (to prevent it being copied) and be unique or offer significant advantages over other similar technologies. Companies are also often interested in licensing in a specific drug delivery product developed by a drug delivery provider that it is in late clinical development. Products licensed in after Phase 2 or Phase 3 are associated with fewer development and regulatory risks, and the time to reach the market is shorter.

In evaluating these technologies or drug delivery products, the potential client will consider a number of different points including, for example, the delivery challenge solved by the technology and the type of molecules for which it is suited. This will include the likelihood that the technology will work for their compound. The strengths, weaknesses, opportunities, and threats associated with the technology or drug delivery product will be evaluated and compared with competing technologies/products. The technical, preclinical, toxicology, clinical, and stability data package will also be considered, for the technology or specific drug delivery product, and what clinical stage it has reached. Other considerations include which other products have been or are being developed using the technology and

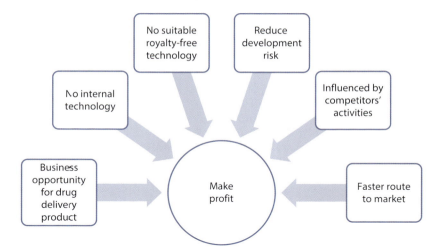

FIGURE 22.2   Why do third parties in-license drug delivery technology?

the current status of development, the strength of the patent protection and length of patent life (drug delivery technology and/or drug delivery product), current scale and ease of scale-up, and the regulatory pathway for the product under development. Regulatory considerations increase if the technology contains one or more novel (new) excipients. The FDA defines a new excipient as one that is not fully qualified by existing safety data with respect to the currently proposed level of exposure, duration of exposure, or route of administration. In such cases, additional toxicology data are required by the regulatory authorities (FDA 2005) and this increases both development costs and risks.

A further issue is the potential cost of goods, including arrangements for clinical and commercial manufacturing. For example, does the drug delivery technology provider have suitable manufacturing facilities or do they use contract manufacturers? Is the technology so specialized that it requires the drug delivery technology provider's manufacturing expertise, or can it be transferred to the client site? The reputation, financial stability, expertise, and capabilities of the drug delivery provider are important considerations, including their analytical, regulatory, and project management skills. In the case of innovative technologies being developed by small companies, Big Pharma/Biotech may even invest in the company. Other considerations include the fit between the needs of the potential partner and what the drug delivery technology provider is offering. This may also include an assessment if the client and provider can work together successfully, of the financial expectations of the drug delivery provider, and of how they fit with those of the potential client.

Companies and drug delivery technology providers typically sign a short development agreement, which allows the company's compound to be assessed in combination with the technology. If this is successful, the client may decide to sign further agreements including licensing of the technology.

22.8.3.4   Drug Delivery Deals

According to the Global Formulation Report (Vitaro and Kararli, 2015) there were 325 deals involving drug delivery and formulation in 2014, 58 of which were for technologies and 168 were for products (deals and acquisitions). The number of technology deals was down from 113 in 2013 and 128 in 2012, while the number of product deals was similar in 2014 and 2013 at 146 and 158 respectively, a sharp rise from the figure of 43 in 2012. These figures may point to less risk-taking on the part of potential licensees who are less willing to invest in developing products from scratch with delivery technologies but are willing to license in promising products based on them. Examples of drug delivery deals carried out in Q1 2016 are listed in Table 22.10. The sample selected again highlights the growing importance of devices in drug delivery and also the search for ways to improve the targeting of cancer drugs.

TABLE 22.10

Examples of Drug Delivery Deals Signed in Q1 2016

Companies	Technology Involved
Insulet Corporation/Eli Lilly and Company	Insulet's pump technology. A development agreement to develop a new version of Insulet's OmniPod tubeless insulin delivery system, specifically designed to deliver Lilly's Humalog 200 units/mL.
Grünenthal Group/Patheon	Patheon selected as Grunenthal's preferred development partner for its products made using Grünenthal's abuse deterrent formulation technology INTAC.
Bind/Synergy Pharmaceuticals	Research collaboration to develop a product based on ACCURINS®, Bind's targeted nanomedicine platform, incorporating Synergy's proprietary uroguanylin analogs for GI cancer.
Unilife/Amgen	Wearable device for the delivery of large molecules at home that are normally administered intravenously.
Cosmo/Ferring	CORTIMENT® MXX, budesonide in an oral tablet containing Cosmo's MMX® multi-matrix technology which is designed to result in the prolonged release and better distribution of drugs in the colon.
Capsugel/Pulmatrix	The companies will collaborate to develop novel inhaled medicines using Pulmatrix's iSPERSE dry powder technology and Capsugel's spray-drying technology.
Flamel Technologies/FSC Pediatrics	Flamel acquired FSC Pediatrics, a company with two drug delivery products and a medical device in the pediatrics market.

22.9 CONCLUDING REMARKS

This chapter reviewed the key characteristics and trends in the market for drug delivery products and discussed the factors associated with product commercial success. It also touched on the business-to-business marketing of drug delivery technologies and products. It is clear that drug delivery–enabled or drug delivery–enhanced products can bring clinical benefits to patients and their carers and financial rewards to the companies that develop and take them to market. However, like all other pharmaceuticals, these products will come under increasing price pressure due to cost-containment measures by third-party payers. Companies will have to increasingly demonstrate the cost-effectiveness of their therapies including those involving drug delivery technology, in order to ensure that their products will be reimbursed by third-party payers.

22.A APPENDIX: MARKETING, BUSINESS, AND OTHER TERMINOLOGY DEFINITIONS

Bargaining power of purchasers: The bargaining power of purchasers increases with their size and the quantities of product(s) they purchase from suppliers. It also is higher when there are too many suppliers compared with the number of purchasers in a particular market. In addition, it rises when individual purchasers band together to negotiate lower prices and/or better terms and conditions from suppliers.

Bargaining power of suppliers: The bargaining power of suppliers is high when their goods are unique or clearly differentiated from the competition and cannot be easily substituted. It also increases when the number of suppliers is limited and there is high demand for a product. Other factors that increase supplier-negotiating power include control of the product distribution channels or the bundling of the product sales with associated value-added services or when the cost of changing suppliers is high.

BCS: A regulatory-approved system of classifying drugs based on their aqueous solubility and membrane permeability.

Class 1 high permeability, high solubility; *Class 2* high permeability, low solubility; *Class 3* low permeability, high solubility; *Class 4* low permeability, low solubility (http:// www.fda.gov/AboutFDA/CentersOffices/OfficeofMedicalProductsandTobacco/CDER/ ucm128219.htm).

Brands, branding, and brand image: Medicines are branded using a proprietary name for the product. The company logo and packaging facilitate branding; however, the need for the latter to look professional and to protect the product means that for medicines, there is less room for distinctive design than with consumer goods. All originator medicines (patented or off-patent), many OTC and even some generic products have brand names. Branding results in a name that is linked to a set of product attributes and a market position through advertising and promotion to create a brand image. Brand name and positive image promotes product adoption and prescriptions/sales and facilitates product differentiation from competitors.

Brand awareness and loyalty: Branding awareness is the extent to which a brand is associated correctly with a particular product by potential customers. Raising awareness is the goal of promotional efforts during and after product launch. Brand loyalty in the form of repeat prescriptions/sales is the desired result and for OTC products is often exploited to facilitate introduction of new products within the "brand family"

CAGR: The year-over-year growth rate (in this case of a market) over a defined period of time.

Cost of capital: The opportunity cost of funds used for financing a business. The rate of return investors could have earned, if they had put their money in an alternative investment and not in drug development.

Data exclusivity: Period of protection for an originator product during which the EMA will not accept a competitor's marketing dossier referencing data used to support market authorization of the reference (originator) product. Data exclusivity can pertain to an entire dossier or to specific preclinical and clinical studies.

Detailing: This is promotion of pharmaceutical products by sales representatives through visits to physicians. The purpose of these visits is to disseminate information on the product, educate the doctor, raise product awareness, and encourage prescribing of the product.

Effectiveness versus efficacy: There is a subtle difference between a product's efficacy (its clinical effect in a randomized double-blinded clinical trial, i.e., under almost ideal conditions) and its effectiveness (its clinical effect in the real world). Payers in particular are interested in comparing product effectiveness with therapeutic alternatives, and therefore, for the purpose of this article, the term effectiveness is used when discussing the needs of payers.

GDP: GDP is the market value of all final goods and services produced within a country in a fiscal year excluding net income from abroad.

HTA: Health Technology Assessment is a way of assessing the ways science and technology are used in health care and disease prevention. Diagnostic and treatment methods, medical equipment, pharmaceuticals, rehabilitation, and prevention methods and also organizational and support systems used to deliver health care are examples of health technologies. (*Definition*: http://ec.europa.eu/health/technology_assessment/policy, accessed on January 7, 2013).

Hybrid product: An EMA term for nonoriginator (generic) products, which reference studies carried out on an EU-approved originator product but do not meet all the criteria for a generic medicinal product and require the applicant to carry out certain new clinical studies.

Marketing or distribution channel: Marketing or distribution channels are the ways that goods and services are advertised, promoted, distributed, sold, or otherwise reach consumers from the producer/supplier.

Marketing mix or four Ps of marketing: The strategic manipulation of product, price, place, and promotion to meet the needs of a target group of customers and, hence, optimize income.

Market share: The total sales of a product in a particular market or market segment expressed as a percentage of total sales in that market/segment.

New drug product exclusivity: FDA definition—A 5-year period of exclusivity is granted to new drug applications for products containing active moieties never previously approved by FDA either alone or in combination. No. 505(b)(2) application or ANDA may be submitted during the 5-year exclusivity period except that such applications may be submitted after 4 years if they contain a certification of patent invalidity or non-infringement. A 3-year period of exclusivity is granted for a drug product that contains an active moiety that has been previously approved, when the application contains reports of new clinical investigations (other than bioavailability studies) conducted or sponsored by the sponsor that were essential to approval of the application.

Paragraph IV Patent Certification: A competitor (generic) drug company submitting either an ANDA or a 505(b)(2) application can choose one of the four certifications with respect to the patents listed in the FDA's Orange Book for the reference (originator) product used in their dossier. A Paragraph IV Patent Certification states that any unexpired patents relating to the reference product are either noninfringed by the competitor's product or are invalid.

Payer: The person or third-party organization that pays for the medicine. In many countries, third-party payers cover all or part of the retail cost of the medicine. Examples are the state-funded NHS in the United Kingdom and private medical insurance companies providing insurance cover for medicines.

Pharmacoeconomics: Pharmacoeconomics is a branch of economics that uses cost–benefit, cost-effectiveness, cost-minimization, cost-of-illness, and cost–utility analyses to compare pharmaceutical products and treatment strategies (as defined in Arenas-Guzman et al. [2005]).

Pharmacy benefit manager: A company that administers prescription drug claims and benefits of health-care insurance plans, e.g., Express Scripts in the United States. Services include processing of claims from pharmacies, contracting with pharmacies, formulary development, and oversight, and negotiating with pharmaceutical manufacturers.

Product adoption: Often described as a five-stage process that starts with the customer first learning about the product and ends with them either becoming regular purchasers or deciding not to buy. In the case of POMs, adoption results in regular prescribing, and pharmaceutical companies often target doctors identified to be "early adopters" during product launch and in the period afterward.

Product differentiation: The factors (real or perceived) that make a product different from its competitors and are desirable to potential customers. For pharmaceuticals, product differentiation is often achieved by promoting the positive aspects of the product (e.g., reduced side effects or fast action) and/or strategic use of the marketing mix.

Product life-cycle management: The management of the product life cycle (development, market introduction, growth, maturity, and decline) to maximize profit by alteration of the marketing mix to suit the life-cycle stage. Pharmaceutical companies use a number of strategies to manage product life cycle including the development of line extensions, approval of the drug for new indications and, if possible, obtaining approval for the OTC use of the medicine. Drug delivery technologies facilitate life-cycle management by enabling the development of new products with therapeutic advantages.

Product portfolio: The products marketed by a company and/or in its development pipeline.

Product positioning: Deliberate placing of a product within a market through strategic use of the marketing mix so that its advantages over competitors are clearly highlighted.

Reimbursement: The money paid by state-funded health-care systems or insurance companies/pharmacy benefit managers to pharmacies to compensate them for medicines dispensed on prescription. The amount reimbursed for a medicine and the amount that the patient has to contribute to its cost varies from country to country and even between different organizations and health-care insurance plans within the same country, e.g., in the United States.

Retail cost of medicine: The retail cost is the price of the dispensed prescription medicine in the pharmacy. It includes the cost of the medicine and wholesaler and pharmacy markups and may include government taxes. Typically, only a part is paid by the patient and the rest by a third-party payer.

Segment and segmentation: Most markets can be divided into segments and subsegments based on the specific needs or particular characteristics of a subset of potential customers within that market.

Value-based pricing: The pricing of a product based on its estimated or perceived value to customers and/or stakeholders as opposed to its development and production costs.

REFERENCES

35 U.S. Code 156. 2011. Extension of patent term. USPTO website. Accessed January 21, 2014. http://www.uspto.gov/web/offices/pac/mpep/s2751.html.

Adler, G., B. Mueller, and K. Articus. 2014. The transdermal formulation of rivastigmine improves caregiver burden and treatment adherence of patients with Alzheimer's disease under daily practice conditions. *Int J Clin Pract* 68(4):465–470.

AMNOG. 2010. Evaluation of new pharmaceutical. Accessed April 28, 2014. http://www.gkv-spitzenverband.de/english/statutory_health_insurance/amnog___evaluation_of_new_pharmaceutical/amnog___evaluation_of_new_pharmaceutical_1.jsp.

Arenas-Guzman, R.A. Tosti, R. Hay, and E. Haneke. 2005. National Institue for Clinical Excellence Pharmacoeconomics – an aid to better decision-making. *J Eur Acad Dermatol Venereol* 19(Suppl 1): 34–39.

Arnold, R.J.G. 2009. *Pharmacoeconomics: From Theory to Practice*. Boca Raton, FL: CRC Press.

Atkin, M. and M. Kleinrock. 2015. *Global Medicines Use in 2020 Outlook and Implications*. IMS Institute for Healthcare Informatics, Parsippany, NJ. Accessed on February 14, 2016. ww.imshealth.com/en/thought-leadership/ims-institute/reports/global-medicines-use-in-2020.

BCC Research. 2014. Global markets and technologies for advanced drug delivery systems. Accessed April 24, 2014. http://www.bccresearch.com/market-research/pharmaceuticals/advanced-drug-delivery-markets-phm006j.html.

Bossart, J. 2005. Benchmarking drug delivery—Defining the potential of drug delivery. *Drug Deliv Technol* 5(3):22.

Challener, C. 2014. Drug delivery technologies provide growth opportunities for biopharmaceuticals. Accessed May 7, 2014. http://www.pharmtech.com/pharmtech/Drug+Delivery/Drug-Delivery-Technologies-Provide-Growth-Opportun/ArticleStandard/Article/detail/839349.

Darreh-Shori, T. and V. Jelic. 2010. Safety and tolerability of transdermal and oral rivastigmine in Alzheimer's disease and Parkinson's disease dementia. *Expert Opin Drug Saf* 9(1):167–176.

Dobrow, L. 2015. Pharma DTC spending jumps almost 21% in 2014. MM&M. Accessed March 13, 2016. http://www.mmm-online.com/agency/pharma-dtc-spending-jumps-almost-21-in-2014/article/404922/.

Donohue, J.M., M. Cevasco, and M.B. Rosenthal. 2007. A decade of direct-to-consumer advertising of prescription drugs. *N Engl J Med* 357:673–681.

Downing, N.S., J.S. Ross, C.A. Jackevicius et al. 2012. How Abbott's fenofibrate franchise avoided generic competition. *Arch Intern Med* 172(9):724–730.

Ellery, T. and N. Hansen. 2012. *Pharmaceutical Lifecycle Management: Making the Most of Each and Every Brand*. Hoboken, NJ: Wiley.

Europa. 2009a. Supplementary protection certificate for medicinal products. Accessed January 21, 2014. http://europa.eu/legislation_summaries/other/l21156_en.htm.

Europa. 2009b. EMEA recommends changes in the storage conditions for Neupro (rotigotine). Accessed May 28, 2014. http://www.ema.europa.eu/ema/index.jsp?curl=pages/news_and_events/news/2009/11/news_detail_000047.jsp&mid=WC0b01ac058004d5c1.

Europa. 2009c. Questions and answers on the recommendation to suspend the marketing authorisation of Ionsys. Accessed May 28, 2014. http://www.ema.europa.eu/docs/en_GB/document_library/Medicine_QA/2009/11/WC500014766.pdf.

Europa. 2013. Pricing and reimbursement of medicinal products and associated references. Accessed January 7, 2014. http://ec.europa.eu/enterprise/sectors/healthcare/competitiveness/pricing-reimbursement/european-initiatives/index_en.htm.

European Commission. 2013. Volume 2A—Procedures for marketing authorisation. Marketing Authorization. Accessed May 23, 2014. http://ec.europa.eu/health/files/eudralex/vol-2/a/vol2a_chap1_2013-06_en.pdf.

European Federation of Pharmaceutical Industries and Associations (EFPIA). 2013. The pharmaceutical industry in figures. Accessed January 4, 2014. http://www.efpia.eu/.

FDA. 1999. Draft guidance: Applications covered by section 505(b)(2). Accessed May 23, 2014. http://www.fda.gov/downloads/Drugs/Guidances/ucm079345.pdf.

FDA. 2004. Entry for Nutropin® depot on drug@FDA. Accessed May 28, 2014. http://www.accessdata.fda.gov/drugsatfda_docs/label/2004/21075s008lbl.pdf.

FDA. 2005. Guidance for industry. Nonclinical studies for the safety evaluation of pharmaceutical excipients. Accessed February 4, 2014. http://www.fda.gov/downloads/drugs/guidancecomplianceregulatoryinformation/guidances/ucm079250.pdf.

FDA. 2013. Entry for pegasys on Drugs@FDA. Accessed May 22, 2014. http://www.accessdata.fda.gov/drugsatfda_docs/label/2013/103964s5245lbl.pdf.

FDA. 2014a. Biosimilars. Accessed May 23, 2014. http://www.fda.gov/drugs/developmentapprovalprocess/howdrugsaredevelopedandapproved/approvalapplications/therapeuticbiologicapplications/biosimilars/default.htm.

FDA. 2014b. The office of prescription drug promotion. Accessed April 27, 2014. http://www.fda.gov/aboutfda/centersoffices/officeofmedicalproductsandtobacco/cder/ucm090142.htm.

FDA. 2014c. Fast track, breakthrough therapy, accelerated approval and priority review. Accessed January 30, 2014. http://www.fda.gov/forpatients/approvals/fast/ucm20041766.htm.

FDA. 2014d. Draft Guidance for Industry. Reference Product Exclusivity for Biological Products Filed Under Section 351(a) of the PHS Act. Accessed March 20, 2016. http://www.fda.gov/downloads/Drugs/GuidanceComplianceRegulatoryInformation/Guidances/UCM407844.pdf.

FDA. 2015. FDA approves first biosimilar product Zarxio. Accessed March 20, 2016. http://www.fda.gov/NewsEvents/Newsroom/PressAnnouncements/ucm436648.htm.

FDA. 2016. Small Business Assistance: 180-Day Generic Drug Exclusivity. Accessed March 20, 2016. http://www.fda.gov/drugs/developmentapprovalprocess/smallbusinessassistance/ucm069964.htm.

Freedonia. 2015. Drug Delivery Products. Accessed March 15, 2016. http://www.freedoniagroup.com/industry-study/3354/drug-delivery-products.htm.

Frias, Z. 2013. Data exclusivity, market protection and paediatric rewards. Accessed January 28, 2014. http://www.ema.europa.eu/docs/en_GB/document_library/Presentation/2013/05/WC500143122.pdf.

Gudiksen, M., E. Fleming, L. Furstenthal et al. 2008. What drives success for specialty pharmaceuticals? *Nat Rev Drug Discov* 7(7):563–567.

Hauber, A.B., S. Han, J.C. Yang et al. 2013. Effect of pill burden on dosing preferences, willingness to pay, and likely adherence among patients with type 2 diabetes. *Patient Prefer Adherence* 7:937–949.

Heinemann, L. 2008. The failure of exubera: Are we beating a dead horse? *J Diabetes Sci Technol* 2(3):518–529.

IMS Topline Data. 2012. Top-line market data trends. Accessed January 8, 2014. http://www.imshealth.com/deployedfiles/ims/Global/Content/Corporate/Press%20Room/Top-Line%20Market%20Data%20&%20Trends/Top_20_Therapeutic_Classes_2012.pdf.

Laliberté, F., B.K. Bookhart, W.W. Nelson et al. 2013. Impact of once-daily versus twice-daily dosing frequency on adherence to chronic medications among patients with venous thromboembolism. *Patient* 6(3):213–224.

Landsman, V., I. Verniers, and S. Stremersch. 2014. The successful launch and diffusion of new therapies. In *Innovation and Marketing in the Pharmaceutical Industry*, eds. M. Ding, J. Eliashberg, and S. Stremersch. New York: Springer.

Market Size. 2014. Accessed May 5, 2014. http://www.frost.com/prod/servlet/mcon-mktmeasures-mkt-size.pag.

Marketsandmarkets.com. 2015. Drug Delivery Technology Market by Route of Administration (Oral (Solid), Pulmonary (Nebulizer), Injectable (Device), Ocular (Liquid), Nasal (Drop), Topical (Solid), Implantable (Active), Transmucosal (Oral)), End User (Hospital, ASC, Home Care) -Forecast to 2020. Accessed March 15, 2016. http://www.marketsandmarkets.com/Market-Reports/drug-delivery-technologies-market-1085.html.

Medic, G., K. Higashi, K.J. Littlewood et al. 2013. Dosing frequency and adherence in chronic psychiatric disease: Systematic review and meta-analysis. *Neuropsychiatr Dis Treat* 9:119–131.

Mestre-Ferrandiz, J., J. Sussex, and A. Towse. 2012. *The R&D Cost of a New Medicine*. London, U.K.: Office of Health Economics.

Mordor Intelligence. 2015. *Global Biopharmaceuticals Market Growth, Trends & Forecasts (2014–2020)*. Accessed March 15, 2016. http://www.mordorintelligence.com/industry-reports/global-biopharmaceuticals-market-industry?gclid=CKiwvq7Ww8sCFdEy0wodGssJmw.

National Institute for Health and Care Excellence. 2009. Short clinical guideline 87—Type 2 diabetes: Newer agents. Accessed January 20, 2013. http://www.nice.org.uk/nicemedia/live/12165/44318/44318.pdf.

NHS England. 2013. NHS England sanctions new breast cancer treatment injection. Accessed May 28, 2014. http://www.england.nhs.uk/2013/09/24/cancer-treatment-injctn/.

NICE. 2014. Patient access schemes liaison unit. Accessed January 22, 2014. http://www.nice.org.uk/aboutnice/howwework/paslu/ListOfPatientAccessSchemesApprovedAsPartOfANICEAppraisal.jsp.

NICE. 2015. Breast cancer drug costing tens of thousands of pounds more than other treatments 'unaffordable' for NHS. Accessed May 27, 2015. http://www.nice.org.uk/newsroom/pressreleases/NICEBreastCancerDrugCostingTensOfThousandsOfPoundsMoreThanOtherTreatmentsUnaffordableForNHS.jsp.

Palmer, E. 2015. Gilead considered $115K price for Sovaldi, Senate investigation says. FiercePharma. Accessed March 14, 2016. http://www.fiercepharma.com/story/gilead-considered-115k-price-sovaldi-senate-investigation-says/2015-12-02.

PDF. 2012. Neupro® re-approved by FDA for treatment of Parkinson's disease. Accessed May 28, 2014. http://www.pdf.org/en/science_news/release/pr_1333471029.

Pfizer. 2012. Financial report. Accessed May 27, 2014. http://www.pfizer.com/files/annualreport/2012/financial/financial2012.pdf.

Pharmaceutical Research and Manufacturers of America (PhRMA). 2008. Code on interactions with healthcare professionals. Accessed January 21, 2014. http://www.phrma.org/sites/default/files/pdf/phrma_marketing_code_2008.pdf.

Rasenack, J., S. Zeuzem, S.V. Feinman et al. 2003. Peginterferon alpha-2a (40kD) [Pegasys] improves HR-QOL outcomes compared with unmodified interferon alpha-2a [Roferon-A]: In patients with chronic hepatitis C. *Pharmacoeconomics* 21(5):341–349.

Rickwood, S., M. Kleinrock, and M. Núñez-Gaviria. 2013. The global use of medicines: Outlook through 2017. The IMS Institute of Healthcare Informatics, Parsippany, NJ. Accessed January 4, 2014. http://www.imshealth.com/portal/site/imshealth/menuitem.0be132395225d98ee566e5661ad8c22a/?vgnextoid=a64de5fda6370410VgnVCM10000076192ca2RCRD&vgnextfmt=default.

Rollins, B.L. and M. Perri. 2014. *Pharmaceutical Marketing*. Burlington, MA: Jones & Bartlett Learning.

Schulze, U. and M. Ringel. 2013. What matters most in commercial success: First-in-class or best-in-class? *Nat Rev Drug Discov* 12:419–420.

Scottish Medicines. 2014. The Scottish Medicines Consortium issues advice on clarithromycin granules for oral suspension (ClaroSip), for use within NHS Scotland. Accessed February 3, 2014. http://www.scottishmedicines.org.uk/Press_Statements/Clarithromycin_granules_for_oral_suspension__ClaroSip_.

Sheppard, A. 2014. Generic medicines: Essential contributors to the long-term health of society. Accessed May 27, 2014. http://www.imshealth.com/imshealth/Global/Content/Document/Market_Measurement_TL/Generic_Medicines_GA.pdf.

Simoens, S. 2012. A review of generic medicine pricing in Europe. *GaBI J* 1(1):8–12.

Small Business Assistance. 2010. Frequently asked questions for new drug product exclusivity. Accessed January 28, 2014. http://www.fda.gov/drugs/developmentapprovalprocess/smallbusinessassistance/ucm069962.html.

Sorenson, C., M. Drummold, and P. Kanavos. 2008. Ensuring value for money in health care: The role of health technology assessment in the European Union. Accessed January 7, 2014. http://www.euro.who.int/__data/assets/pdf_file/0011/98291/E91271.pdf.

Thielman, S. 2015. Martin Shkreli walks back on pledge to lower price of HIV drug Daraprim. The Guardian, UK. Accessed March 14, 2016. http://www.theguardian.com/business/2015/nov/25/martin-shkreli-hiv-drug-daraprim-turing.

Vitaro, R. and T. Kararli. 2015. Global formulation report. *Drug Deliv and Dev* 5(6):18–50.

Ventola, C.L. 2011. Direct-to-consumer pharmaceutical advertising. *Pharm Therapeut* 36(10):669–684.

Viswanathan, M., C.E. Golin, C.D. Jones et al. 2012. Interventions to improve adherence to self-administered medications for chronic diseases in the United States: A systematic review. *Ann Intern Med* 157(11):785–795.

23 Bringing Research to Clinical Application

Lessons from ThermoDox®: A Thermal-Sensitive Liposome for Treatment of Cancer

David Needham

CONTENTS

23.1 INITIAL PERSPECTIVE AND SCOPE: WHERE ARE THE LESSONS TO BE LEARNED?

This is the story of our low-temperature-sensitive liposome (LTSL) (Needham et al. 2000), subsequently called ThermoDox® by the company Celsion Corporation which licensed it from Duke University in November 1999. It is an account of "lessons learned," as told by me, as I saw it, experienced it, and lived it, and so much of this will be written in the first person. Everyone has their own story. I, myself, Celsion, and ThermoDox® have ours. As Bernard Malamud says in his book, *The Natural*, "We have two lives, the one we learn with, and the life we live after that" (Malamud 1952). This certainly applies to me and Celsion. So I write this not for Celsion, or for Duke, but for the next generation of inventors, carers, and entrepreneurs, who have a life to learn with and perhaps have yet to make their own mistakes or achieve their own real and verifiable successes. This chapter contains some of what you might have in store. I encourage you to heed these "lessons learned." See if you can benefit from the positive results and avoid the ones that, let's say, made life more difficult. At least be aware of what could happen if you, as the inventor, trust anyone else with your precious invention, which you will have to do. I know this is a long and detailed story, but please try and make it all the way to, what will hopefully be, a happy ending.

23.1.1 IN THE LIPOSOME FIELD ITSELF

There are always "lessons to be learned," and the liposome field is no exception, especially with its event-rich and inextricably linked journey from research to commercialization. Although slightly before my own time in liposomes,* but not in research (Eley and Needham 1984; Needham 1981), the liposome field rapidly progressed in the mid- to late 1970s. The challenge for the early pioneers

---

* After graduating in the summer of 1975, I was washing dishes in a hotel in Majorca, then working at a pigment factory 1976–1977 in the backstreets of Manchester, and then doing my PhD in gas–solid catalysis 1977–1980 with Professor Dan D. Eley, FRS, and all the while, Gregoriadis et al. were developing liposomes, who knew!

was to identify where this seemingly high potential impact, biocompatible, emerging clinical technology could, in fact, find its therapeutic application. Immunology and especially anticancer drug encapsulation and targeting took the center stage. Thus, once liposomes had been "discovered" (Bangham and Horne 1964), much of this pioneering work was led by luminaries such as Dimitri Papahadjopoulos and Gregory Gregoriadis and included reports on making and characterizing liposomes in terms of structure and permeability (Papahadjopoulos and Miller 1967; Papahadjopoulos and Watkins 1967), the development of liposomes for encapsulation (Gregoriadis and Ryman 1971; Gregoriadis et al. 1974), as adjuvants (Allison and Gregoriadis 1974), and targeting (Gregoriadis and Neerunjun 1975), and even early patents (Allison and Gregoriadis 1977). Gregory was also an enthusiastic and prolific reporter of each new advance in liposomes, bringing together the relatively small community of researchers in the then nascent field, through his series of review papers (Gregoriadis 1976) and edited books (Gregoriadis 1979), including the extremely influential three-volume *Liposome Technology* (Gregoriadis 1984), a series that still continues in its third edition (Gregoriadis 2010).

In his 1975 paper on homing liposomes, the abstract reads as follows:

> The possibility of homing liposomes to target cells was investigated. Liposomes containing an antitumor drug and associated with molecular probes, which exhibit a specific affinity for the surface of a variety of normal and malignant cells, were prepared. In vitro and in vivo experiments suggested that such probes were capable of mediating selective cellular uptake of the associated liposomes and the entrapped drug. It is anticipated that liposomes designed to home may become important tools in the control of cell behavior.

This is 1975! This kind of "homing" work is still going on today in liposomes and, indeed, with a whole host of "nanoparticles." Drug encapsulation and targeting is still being researched, and attempts are still being made to test clinically and eventually commercialize these approaches. Forty years later, it's still a challenge. While I was trying to give advice from our own experiences with ThermoDox®, I suggest that these early papers still have something to teach us (especially the new generations entering the field of "nanomedicine") about both research and the decision-making process during commercialization of all nanoparticles for drug delivery and imaging. So just because you and your advisor are choosing to work on nanoparticles of gold, iron oxide, chitosan, poly(lactic-co-glycolic acid), etc., and even antibodies, does not mean you should not look up, read, study, understand, and use this literature on liposomes to plan your studies and interpret your results.

23.1.2 EXAMPLE OF AN EARLY "LESSON LEARNED" IN LIPOSOMES: IS IT BEING HEEDED TODAY IN NANOMEDICINE?

For example, these early applications were hampered by the now relatively well-understood opsonization phenomenon that labels most nanoparticles in the bloodstream as foreign for removal by the reticuloendothelial system (RES) (Roerdink et al. 1981). In their early formulations (Gregoriadis et al. 1971, 1974), Gregoriadis and colleagues had used a composition of 40 μmol of phosphatidylcholine, 11.4 μmol of cholesterol, and 5.7 μmol of the negatively charged phosphatidic acid (molar ratio 7:2:1), giving liposome that was 10 mol% negatively charged. When Alec Bangham took me out to a Chinese dinner on the eve of my giving his 80th Birthday Lecture at Gregory's 2001 5th International Conference on Liposome Advances, in London (Needham 2001a), he told me a story from those early days. Alec had explicitly advised that the initial strategies to make the liposomes "repulsive," by including negative charge, which was thought to be electrostatically stabilizing, would conversely only result in liposomes going to the RES faster than ever. This has since been proven (Allen et al. 1988) and is fairly well understood (Liu et al. 1995) in serum and serum-free media (Rothkopf et al. 2005). As we now all appreciate, this opsonization to Kupffer cell removal of the liposome in the bloodstream was solved by applying earlier work on the PEGylation of proteins (Beauchamp et al. 1983), and the invention of the so-called "stealth" liposome (Allen 1989), with its sterically stabilizing polymer (Allen and Gabizon 1990; Klibanov et al. 1990), forming the basis for Doxil® (see Barenholz's review and references therein, Barenholz 2012). There is also a lesser-known mechanism of a mechanical effect of high cholesterol content in PC lipids or sphingomyelin (Kim and Needham 2001; Needham and Nunn 1990), which gives a similar long circulation half-life (Lasic and Needham 1995), used to good effect (slow leakage and long circulation) in the vincristine liposome (Boman et al. 1993, 1994; Kanter et al. 1994; Krishna et al. 2001; Webb et al. 1995, 1996).

I mention these early studies and "lessons learned" because newbies (and "oldbies") to the field of nanomedicine would do well to read through some of these early reports and seek to inform their own current research strategies. "Nanomedicine" does seem to know all about PEG-derived steric stabilization, but is it as well appreciated in the nanomedicine literature that a tight lipid-surface is also not well opsonized and that an amount of negative charge of only 3 mol% or greater (Allen et al. 1988) can send your precious nanoparticle rapidly to its RES fate?

23.1.3 READ THE LITERATURE

As Frank Szoka (who entered the field in its "golden age"; Szoka and Papahadjopoulos 1978), famous for ranting in an editorial *Commentary: Rantosomes and Ravosomes* (Szoka 1998), reminded us, quoting the philosopher, essayist, and poet, George Santayana: "Those who cannot remember the past are condemned to repeat it." The example Frank chose to illustrate this was liposomes carrying anti-HIV agents. He points out that

> The difficulties with the approach could be anticipated based upon the early liposomal anti-cancer drug delivery literature. The failure of those working in the field is not because they repeated previously published studies. But rather because they did not translate the lessons from the early studies into a therapeutic strategy that complemented the pathophysiology of the disease, with the pharmacological aspects of the drug and the delivery properties of the liposome.

Thus, to reiterate in this editorial, Frank's amendment to the Santayana quote was

> Those who cannot *translate the lessons of* the past are condemned to repeat it.

Frank, correct me if I am wrong, but basically what I think you were saying is

> Those of you who do not read the literature are condemned to repeat it! And guess what? We have already done it, and we are tired of hearing about your 'new work' at meetings, or seeing what you think is new and exciting in your grant proposals, that we have to take our precious time to read through, review, and reject.

Back then, the (liposome) world was certainly smaller, and as Szoka also points out, many of the potential uses of liposomes in drug delivery that have since come to be were discussed in those two 1976 review articles by Gregory Gregoriadis (1976). With only a few people involved, "discussions raged over materials, mechanisms, models, methods and structures" at the handful of liposome/drug carrier meetings that everyone attended.

Many lessons were, in fact, learned back then, face-to-face, between researchers during the meetings. But today, even with (and perhaps because of) easy access to the (voluminous) literature, this face-to-face contact is easily lost, and so the repetition of research and, indeed, the mistakes, still go on today. Frank, who rants on average about every 15 years, has another one. This time, it is directed at *all nanomedicines* and the people who research and develop them. Entitled "Cancer nanomedicines: So many papers and so few drugs!" (Venditto and Szoka 2013), folks would do well to heed it, as this most comprehensive review identifies a timeline to nanomedicine anticancer drug approval using the business model of inventors, innovators, and imitators. Unfortunately for today's researchers who are still involved in "liposomes," there is a lot of literature, some 48,772 publications currently listed (February 28, 2015) on a PubMed search for "liposome." And incredibly, 339,246 patents that concern "liposome" at free patents online. "Lessons learned" from liposomes, like Chezy Barenholz's landmark paper *Doxil®—the first FDA-approved nanodrug* (Barenholz 2012), will undoubtedly apply to all nanomedicines being researched and developed today. With only 9422 papers found on PubMed for a search of "nanomedicine," it is likely that "your precious nanomedicine" shares many of the problems that were already figured out by the liposomologists, maybe before you were born or, at least, while you were still in grade school. So where would a company turn to, in order to get the most expert and up-to-date knowledge about their licensed product? Would they sift through the 48,772 papers and pay a law firm $gazillions to search through the 339,246 patents? No. Maybe try asking the *inventor*, and first see what he or she has to say (and more about that relationship later).

23.1.4 EARLY COMMERCIALIZATION AND ACCESS TO A WEALTH OF EXPERTISE: JUST ASK US

Early commercialization efforts were considerable (as well as contentious; Free Library 2014; NeXstar Pharmaceuticals 1997). From my perspective, there were four principal liposome companies: three were established in the United States: Liposome Technology Inc. (LTI) in Menlo Park, California; The Liposome Company Inc. in Princeton, New Jersey; NeXstar, based in San Dimas, California; and Inex in Vancouver, Canada, first as a spin-off from TLC and then in its own right. These were the main entities that, for me, paved the way in mainly anticancer liposomal therapeutics, took relatively similar drug delivery ideas, and brought them through the development and translation to commercialization. Two companies focused on liposomal doxorubicin, LTI (Doxil®) and The Liposome Company (Evacet™ and Myocet®), NeXstar went with daunorubicin (DaunoXome®) and also successfully developed amphotericin B (AmBisome®), and Inex developed liposomal vincristine (Marqibo®).

With academic researchers involved at varying levels of invention, innovation, and continued support, there is a whole host of very experienced individuals (still) around the conference, industrial, and grant-review world, whom students, postdocs, and young faculty new to the field can call on for advice and informal consultation. So, if you see Chezy Barenholz (Barenholz 2012), Frank Martin, or Martin Woodle et al. (Barenholz and Haran 1994; Papahadjopoulos and Skoza 1980; Woodle et al. 1989) ex-LTI-Sequus; Andy Janoff et al. (Janoff et al. 1989, 1996; Mayer et al. 1997) ex-TLC; Pieter Cullis, Lawrence Meyer, Marcel Bally, or Murray Webb et al. (Bally et al. 1997; Webb et al. 1996; Wheeler et al. 1999), ex-Inex; or Gary Fujii, Bill Ernst, Jill Adler-Moore, or Su-Min Chang et al. (Adler-Moore and Chiang 1997; Adler-Moore and Ernst 1997; Proffitt et al. 1999; Schmidt and Fujii 1998) ex-NeXstar; or very knowledgeable and outspoken independents, like Frank Szoka (Papahadjopoulos and Skoza 1980), Theresa Allen (Allen 1989; Allen and Gabizon 1990), or Valdimir Torchilin (Torchilin et al. 2006), or me and Dewhirst, at a meeting or seminar, we are all inventors of some of the earliest or later patents and a host of peer-reviewed papers that have contributed to those 48,772 publications. You might ask us one or more of the questions I am hoping to address in this chapter: What was the original observation that motivated the product? How did you or your colleagues have it? What was the question you tried to answer? What was the scientific hypothesis that you and your colleagues addressed? What funds were available for initial development and how did you get them? How did you develop it, including preclinical animal studies, and what were the results? What collaborators did you work with? What was the initial licensing deal and how was it structured and with whom? How did you manage the scale-up of manufacturing? What particular disease did you decide to treat and why? What did the Food and Drug Administration (FDA) require you to write in order to obtain an Investigational New Drug (IND) application? What funds did you have to raise, in order to carry out the IND application and how did you get them? How did you manage the Phase 1 trials, and what was the result? How did you recruit the clinicians and the sites to carry out the trials? Were there any unexpected problems implementing the protocols? Was that next stage a Phase 2 or did you go straight to Phase 3 human clinical trials? What was the next documentation you had to provide to the FDA and how long did it take to move to this next stage? What was the protocol? How many patients and sites did this Phase 3 involve, and how was it all managed? How did it turn out? Are you still in business? If so, what are your future plans for your drug delivery technology? In your view, what's next? What's the next big thing in advanced drug delivery and why? The answers are all available if you would just ask.

23.1.5   THIS CHAPTER

Invented in 1996 (Needham 2004), the LTSL is now in Phase 3 human clinical trials for liver cancer. This chapter will describe the engineering design of the LTSL (Needham 2013) and the licensing and clinical testing that is moving the invention toward commercialization. LTSLs, in conjunction with local mild hyperthermia (HT), can release drug within seconds of entering the warmed tumor vasculature. The released, free drug diffuses into the tumor interstitium, reaching its nucleus target with greater penetration distance, and to much higher concentrations, than those achievable by either free drug or the more traditional long-circulating liposome formulations (Manzoor et al. 2012). Intravascular drug release provides a mechanism to increase both the time that tumor cells are exposed to maximum drug levels and the penetration distance achievable by drug diffusion. This establishes a new paradigm in drug delivery: *rapidly triggered drug release in the tumor bloodstream*, saturating neoplastic cells, as well as endothelia, pericytes, and stroma, with the anticancer drug.

This chapter will attempt to describe the process, the pitfalls, the good, the bad, and the ugly of translating what might be a good research idea through initial funding, development, and preclinical and clinical trials. Whether unique or not (and we suspect the issues highlighted in our particular case are probably quite ubiquitous), I will use, as the main example, the invention, development, clinical testing, and intended commercialization of the thermal-sensitive liposome, subsequently named ThermoDox®. Additionally, an update of the recent progress in human clinical trials will be given, including Celsion's Phase 3 for liver cancer, Phase 2 recurrent chest wall (RCW) cancer, and their new trials in metastatic liver cancer, ovarian cancer, pancreatic cancer, breast cancer, and glioblastoma, including the adaptation of high-frequency ultrasound (HiFu) as the source of targeted mild HT. I will say up front that the current Celsion administration should be commended for their pioneering work in taking on this invention, initiating this very impressive series of trials, with multiple sites, in different countries, and in particular for creating and carrying out what actually is the largest randomized double-blind trial ever with radio-frequency ablation (RFA). For example, following on from what was learned in the Phase 3 HEAT study on primary liver cancer, Celsion's new OPTIMA trial is currently enrolling, with the first patient enrolled Q3-2014. It will include approximately 550 patients in up to 100 sites in North America, Europe, China, and the Asia-Pacific region.

Kudos to Celsion and in particular Mike Tardugno (CEO) and Nick Borys (CMO); it's not an easy task visiting all these clinical sites and spending multiple days on airplanes.

Following the licensing of my (Oops, sorry, no, it's not mine!) Duke's invention by Celsion Corporation in November 1999, ThermoDox® has now, just over 15 years later, completed a Phase 3 trial in 701 patients with primary liver cancer, where ThermoDox® + RFA was trialed against RFA alone (Poon and Borys 2009; Wood et al. 2012). There were certainly *lessons learned* in the science (and all that is published in several reviews; Landon et al. 2011; Needham 2013; Needham and Dewhirst 2001, 2012) and the seminal paper on drug accumulation and penetration by Manzoor et al. (2012). But why we think there are real *lessons to be learned* in the development and commercialization from all this in particular, is the fact that when the 701-patient Phase 3 primary liver cancer trial results were unwrapped by Mike Tardugno and colleagues, on January 31, 2012 (Tardugno 2013), the headline read as follows:

ThermoDox® failed to meet its primary endpoint of better progression free survival than the heating modality (RFA) alone.

As described later, data are still being analyzed and some encouraging results are being revealed, and further trials are being conducted including new heating modalities like HiFu. A lot of lessons have been learned (especially) by the inventor and also by his collaborators, the university licensing and ventures office, the faculty patent committee, and the company and the clinicians, which will still bring this particular treatment option to oncologists, their patients, and the clinic. Taken largely in chronological order, this chapter will detail the events and lessons learned and address many of the questions listed in Section 23.1.4 so that others, with potentially good ideas, compelling in vitro, and in vivo data, can perhaps have a smoother path to testing, and, if successful, commercialization of their own idea.

The dedicated Section 23.7 on "Lessons Learned" will present six main topics:

- Your university administration
- Your Office of Licensing and Ventures (OLV)
- Your license agreement
- Meeting the milestones
- Your (inventor's) relationship with the company
- The distribution company

I hope this will form a basis for discussion, if not, identify pitfalls that others might avoid, or at least be aware of, and some of the reasons they could occur. There will also be a few *lessons as we go*, in dedicated boxes, just for good measure.

The chapter will end with a brief description of an approach we are currently pursuing to formulate especially hydrophobic anticancer drugs specifically to metastatic disease. This approach has again benefited from lessons learned in the liposome and other nanomedicine fields. It is an endogenous one, which is based on the way nature delivers her own hydrophobic molecules in the body, what we call "Put the Drug in the Cancer's Food." And it gives me one last chance to make a difference. With my group here in Odense, Denmark, I want to do it in such a way that it is not encumbered by the kinds of events that have hampered ThermoDox® and other products. I therefore introduce what might be a surprising concept, *translating drug delivery without profit* and *open-source pharmaceuticals*, especially for cancer. Whether this approach might help to eliminate the negatives that sometimes cloud these translational issues of hidden agendas, incompetence, and greed and replace them with transparency and the right expertise for the right job, and an opportunity to provide healthcare to the people, which is not encumbered by the promise of huge profits, is still to be determined. This may not be possible in some countries, but the more socially minded ones, perhaps in Europe, indeed like Denmark, the "cancer capital of the world" (WCRF International 2015), could maybe lead the way in this new thinking about translating drug delivery and making our research advances actually available to the people who suffer, at cost.

23.2 WHAT WE KNEW IN 1995

The story proper starts with the initial motivation by Dr. Mark Dewhirst, the then director of the Duke Hyperthermia Program, who, in about 1995, quite explicitly said "Liposomes aren't working, Hyperthermia isn't working, I need something I can heat and it releases drug, damn it." It was Mark then, who motivated the need to come up with a possible solution: a fast drug-releasing thermal-sensitive liposome, disclosed to Duke in 1996 (Box 23.1).

As shown in Figure 23.1, in 1995, three timelines intersected:

1. Dox was discovered and was one of the first choices of drug for encapsulation in some of the first liposomes to be made for testing against cancer.
2. HT as a treatment modality had been developed and was being used to heat tumors.
3. Lipid membrane mechanochemistry used micropipette manipulation techniques to measure the properties of single-walled lipid vesicles.

BOX 23.1   LESSON AS WE GO #1: LISTEN TO YOUR CUSTOMER

And here's the first lesson. Listen to the people on the front lines, the clinicians, the veterinarians, and the directors of centers, and see what "your customer" might actually want. As I discuss at the end of this chapter, the new formulation we are now working on was requested, or at least motivated, by a need expressed by several actual medical oncologists, asking, "can you help us to reformulate lapatinib, niclosamide, fulvestrant, abiraterone, and orlistat?" So does your formulation fulfill an actual clinical need? If not, consider not doing it (at least not just for the papers) and challenge your PhD or postdoc advisor and ask them: Does what you want me to do pass the "so what test"? Is it useful knowledge? Does it even attempt to meet and unmet need? If you are not sure, you probably know the answer, but you still might want to canvas some end users over at your local cancer center.

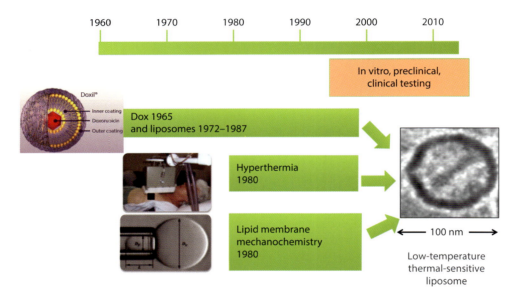

FIGURE 23.1   In 1995, three timelines intersected. Doxorubicin and its encapsulation in liposomes, hyperthermia used to heat tumors, and lipid membrane mechanochemistry, using micropipette manipulation techniques to measure the properties of single-walled lipid vesicles.

23.2.1   DOXORUBICIN AND LIPOSOMES

When we decided to develop the LTSL technology, we simply inherited the "stealth" liposome concept and the cytotoxic drug, Dox, which had already been used in liposomal delivery. And so, as a prelude to discussing actual ThermoDox®, it is worth a relatively brief (not exhaustive) review of what we knew about liposomes and Dox in 1995, highlighting a few comparisons, even a few lessons learned as we go, and exactly what it was about liposome technology and its performance at the time that motivated the LTSL invention and its development.

23.2.1.1   Doxorubicin, the Drug

Dox (also known as hydroxydaunorubicin and hydroxydaunomycin; trade name, Adriamycin®) is an anthracycline antibiotic discovered in the early 1960s by the Farmitalia Research Laboratories of Milan. Clinical trials began in the 1960s, and the drug saw success in treating acute leukemia and lymphoma. It was approved for use in the United States in 1974. However, by 1967, it was

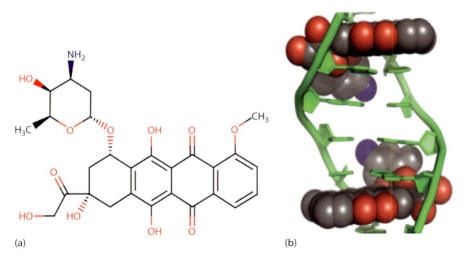

(a)   (b)

FIGURE 23.2 Doxorubicin (Adriamycin). (a) Molecular composition and structure. (b) Diagram of two doxorubicin molecules intercalating DNA.

already being recognized that Dox (and another anthracycline chemotherapeutic, daunorubicin), while active against cancer, could produce fatal cardiac toxicity.

The structure and mechanism of action of Dox is shown in Figure 23.2. It is known to intercalate between DNA base pairs (Frederick et al. 1990), resulting in DNA and DNA-dependent RNA synthesis inhibition due to template disordering and steric obstruction (Abraham et al. 2005). By virtue of this ability to intercalate with DNA, it stabilizes the topoisomerase II complex after it has broken the DNA chain for replication, preventing DNA double helix from being resealed, stopping the process of replication (Frederick et al. 1990).

Dox has one of the widest spectrums of any neoplastic agent and is commonly used to treat Hodgkin's lymphoma and some leukemias, as well as cancers of the bladder, breast, stomach, lung, ovaries, thyroid, soft tissue sarcoma, multiple myeloma, and others. Toxicities exhibited include myelosuppression, alopecia, mucositis, nausea, vomiting, and cardiomyopathy. Due to the cardiotoxicity being additive, the cumulative lifetime dose is limited to $\approx$550 mg/m^2. The cardiotoxicity is most likely a consequence of free radical generation and binding of the drug to cardiolipin in the heart muscle.

Dox also interacts with the cell's electron transport chain, to lead to the formation of superoxide anion radicals and hydrogen peroxide, which has a very damaging effect on the cellular components. Because of its anticancer activity and also because of its intense toxicity, its encapsulation in liposomes was one of the biggest early successes. Liposomal encapsulation reduced its toxicity. The flip side of reduced toxicity though could be reduced efficacy, and subsequent clinical data support this formulation conflict (O'Brien et al. 2004), where in first-line therapy for Metastatic Breast Cancer, PEGylated liposomal Dox (PLD) significantly reduced cardiotoxicity, myelosuppression, vomiting, and alopecia but only provided comparable efficacy to free Dox. Still, with the widespread prescribing of Dox, as O'Brien et al. suggest, for elderly patients and patients with specific cardiac risk factors, PLD was an important new therapeutic option.

As is well known, after intravenous (i.v.) dosing, Dox blood levels fall dramatically as the drug distributes into the tissues, followed by a slow elimination phase due to renal and biliary clearance and metabolism. Hence, worth reproducing, and as shown in Figure 23.3, Dox itself has a half-life of only $\approx$1 minute (Gustafson et al. 2002). But, as a chemical, Dox has some very interesting physicochemical properties. With a pKa of 8.3, it is a weak base cation, such that at pH 7.2, $\approx$10% of it is uncharged and 90% of it is positively charged. What this means is that the uncharged fraction of Dox can pass through the whole tissue, simply by concentration-dependent, Fickian diffusion.

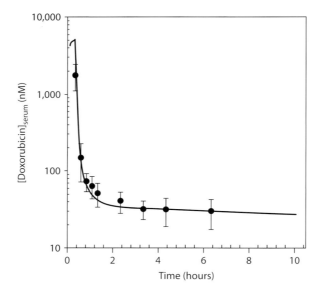

FIGURE 23.3 Plasma levels of doxorubicin in dogs, after a 20 minutes i.v. infusion at a dose of 30 mg/m². (From Gustafson, D.L., Rastatter, J.C., Colombo, T. et al.: Doxorubicin pharmacokinetics: Macromolecule binding, metabolism, and excretion in the context of a physiologic model. *J. Pharm. Sci.* 2002. 91(6). 1488–1501. Copyright Wiley-VCH Verlag GmbH & Co. KGaA. Reprinted with permission.)

Permeation (and thus, loss) of the neutral form through the membranes drives the equilibrium disassociation reaction forward, releasing more uncharged fraction in the interstitial fluid, which is then available for further transit through each cell membrane. Uncharged Dox also partitions into the cytoplasmic and organelle membranes; it passes through and may have a direct, or indirect, toxicity effect there too. In fact, following Dox treatment, cardiotoxicity develops through the preferential accumulation of iron inside the mitochondria, due to Dox becoming concentrated in the membrane-rich mitochondria and increasing both mitochondrial iron and cellular reactive oxygen species levels (Ichikawa et al. 2014). Also, because of this dual solubility afforded to such weak base cations, once partitioned in, it takes a long time to get back out. It could be lethal in so many ways, if only we could get the drug just to the tumor vasculature, and in high quantities.

23.2.1.2 Doxorubicin Encapsulated in Liposomes

Dox was already approved and formulated in a non-PEGylated liposomal form (Myocet®, formerly known as Evacet™) (Abraham et al. 2005) but, as predicted by Bangham from the outset, "traditional" liposomal formulations showed too rapid a circulatory clearance and little efficacy. This was followed by Doxil® (marketed as Caelyx® in Europe), the PEGylated "stealth"* liposome (reviewed comprehensively in the edited book *Stealth Liposome* by Lasic and Martin, 1995; also Lasic and Needham, 1995 and also described in Chapter 5, Section 5.5.1.2). Although it was later than the traditional liposomes in its development, Doxil® was the first liposomal pharmaceutical product to receive U.S. FDA approval in 1995, for the treatment of chemotherapy refractory, acquired immunodeficiency syndrome–related Kaposi's sarcoma. Thus, as we were thinking about how to solve the problem that Mark Dewhirst identified, the field was buzzing with news of FDA approval for Doxil®. Doxil® certainly played a huge part in getting this liposomal technology into mainstream use, and as reviewed recently by Chang (Chang and Yeh 2012), as of 2012, there were 12 liposome-based drugs approved for clinical use and more are in various stages of clinical trials.

---

* Sterically stabilized liposomes are often described as "stealth" carriers (but the capitalized word "Stealth" is actually a registered trade name of Johnson & Johnson).

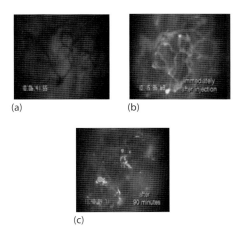

(a)   (b)

(c)

FIGURE 23.4 Intravital microscopy images of stealth liposomal injection, circulation, and extravasation into the tumor interstitium. (a) Bright-field image of tumor vasculature through window, (b) 1 minute after injection, (c) 90 minutes after injection (reduced camera gain). Larger blood vessels are about 30 μm in diameter. (Reproduced from Wu, N.Z. et al., *Cancer Res.*, 53, 3765–3770. Copyright 1993, American Association for Cancer Research. With permission.)

The success of Doxil® and its stealth, PEGylated polymer design was to escape RES uptake and create a long-circulating liposome that could stand a chance of extravasating into the tumor interstitium via the purported enhanced permeability and retention (EPR) effect. We had even already shown this ourselves in 1993 (Wu et al. 1993). Figure 23.4 shows the intravital microscopy images after fluorescently labeled stealth liposomes were injected i.v. into rats bearing Dewhirst's dorsal skin-flap window chambers, containing a vascularized mammary adenocarcinoma.

The bright-field image in Figure 23.4a shows the complex and resurgent flow of the tumor vasculature, where, for scale, the larger blood vessels are about 30 μm in diameter. Figure 23.4b is taken just 1 minute after i.v. injection of the 82 ± 24 nm diameter stealth liposomes into the rat's tail vein (Wu et al. 1993). Switching the microscope to epifluorescence illumination, the liposomes are clearly flowing in the blood vessels and, even at this early time point, have started to extravasate into the tumor interstitium. Ninety minutes later, not yet having set our optimal low-light-level camera settings, we had to turn the gain on the fluorescent camera down to get the last image in Figure 23.4c, where there was a significant, but heterogeneous, accumulation in the tumor perivascular space.

As shown in Figure 23.5, when the data were quantitatively analyzed, the longer circulation half-life correlated with there being more stealth liposome accumulation in the tumor interstitium than for the more conventional non-PEGylated (91 ± 41 nm diameter) liposomes.

And with all this actual in vivo data, we were part of helping Doxil® to get approved, by showing that the extravasation actually happened in real time within 90 minutes or less. In fact, normal vasculature showed no extravasation at all. The importance of these data then was that it seemingly supported the Doxil® data from Vaage and Mayhew (Vaage et al. 1992) that had prompted our window chamber study in the first place. The EPR effect was alive and well (in this subcutaneous animal model). The passive accumulation of long-circulating stealth liposomes was an unqualified success (again, it must be emphasized "in this subcutaneous animal model").

So, why would anyone need to improve on this?

It didn't take long for data to start accumulating in the literature that even though there was now a long enough time to allow for passive extravasation, there were other challenges:

- Stealth liposomes might not be able extravasate to the extent that we had been seeing in subcutaneous animal tumor models.

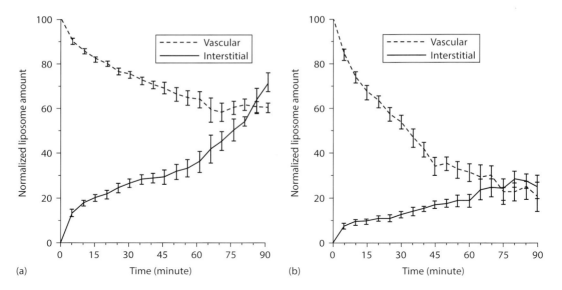

FIGURE 23.5 Time courses of normalized vascular and interstitial liposome amounts for stealth and conventional liposomes in tumor tissues. (a) Averaged vascular decay and interstitial accumulation of stealth liposomes from nine tumor preparations. (b) Averaged vascular decay and interstitial accumulation of conventional liposomes from seven tumor preparations. (Reproduced from Wu, N.Z. et al., *Cancer Res.*, 53, 3765–3770. Copyright 1993, American Association for Cancer Research. With permission.)

- If they did, "this is as far as your liposomes can go," i.e., only as far as the perivascular space.
- As Yuan et al. in 1994 had shown, using a similar window chamber technique (Yuan et al. 1994), the intramural accumulation of liposomal fluorescent spots (observed within 5 minutes after liposome injection) could be continuously observed for up to 2 weeks.
- No one knew exactly how leaky a human tumor vasculature really was to these ≈100 nm liposomes.
- No one was really sure to what extent, and at what rate, the liposomes, which had been well optimized for the best encapsulation of Dox, could even release their drug in the tumor interstitium.
- In very recent studies that have sought to provide a mechanism whereby Doxil® could potentially leak its drug, in an environment of ammonia (Silveram amd Barenholz 2015), it was found that PLD without ammonia had a "very poor cytotoxicity," demonstrating again that Doxil® does a very good job of retaining its drug, but is not designed to release it.

So, quietly, between ourselves, Mark Dewhirst and I had questions in 1993–1994, born out of our own preclinical animal studies: "could Doxil® really deliver enough drug to human tumors to achieve significant efficacy?"

Attempting to improve Doxil's efficacy by using HT to increase the vascular permeability, the Hyperthermia Center at Duke also carried out two "Doxil® + HT" clinical trials in ovarian and breast cancers (Secord et al. 2005; Vujaskovic et al. 2010). Unfortunately, the clinical trial evaluating the combination of Doxil® + HT in patients with persistent and recurrent ovarian cancer did not demonstrate increased efficacy, compared with prior reports using Doxil® alone. And in a Phase 1/2 study of paclitaxel in patients with locally advanced breast cancer in the preoperative setting, the pathological complete response (pCR) rate was only 9% for liposomal Dox (Evacet™) + local HT, although the combined (i.e., paclitaxel + liposomal Dox [Evacet™] + local HT) pCR rate was 61%.

Although we did not know it at the time, therapeutically all these concerns were, in fact, borne out in subsequent human trials, including a trial that was instrumental in gaining Doxil's approval for ovarian cancer. A trial for metastatic breast cancer showed that Doxil® really was no better than free drug alone (O'Brien et al. 2004). That is, even though Dox has a half-life of only 2 minutes and Doxil® has a half-life of 73.9 hours:

- The overall survival was comparable with both treatments.
- Median: PEGylated liposomal Dox 21 months, versus doxorubicin, 22 months; hazard ratio (HR) = 0.94 (95% CI 0.74–1.19).
- At the time of the analysis, approximately 56% of patients in each group had died. When adjusted for potential imbalances in prognostic variables using the Cox regression analysis, the HR was 0.94 (95% CI 0.75–1.19), similar to the unadjusted treatment HR.

Doxil® was approved for ovarian cancer (Gordon et al. 2001) in 2005, but again, with little therapeutic benefit, when compared to topotecan:

- Time to progression (TTP): Doxil®, 4.1 months, and topotecan, 4.2 months
- Overall median survival: Doxil®, 14.4 months, and topotecan, 13.7 months
- 18% reduction in risk of death (hazard ratio [HR] = 1.216)
- Overall tumor response rates: 19.7% (47 patients) in the Doxil® arm and 17% (40 patients) in the topotecan arm

Compared to topotecan then, improvement in TTP for Doxil® was 0.1 month, which is only a 2.4% progression-free survival (PFS), and it was approved. TTP was also a measure for ThermoDox®, and we will see later how it faired (when taken in total, it didn't fare well), in relation to the FDA-required %PFS stipulated for its approval.

23.2.1.3 New Toxicities for Doxil®

As mentioned earlier, the initial success of non-PLD (Myocet®) was to reduce the drug's cardiotoxicity, by simply encapsulating and retaining it; Doxil® certainly achieved the same retention and reduced cardiotoxicity profile. However, the long circulation half-life also introduced new toxicities: hand–foot syndrome or palmar–plantar erythrodysesthesia (PPE). PPE is a common dermatologic toxic reaction associated with certain chemotherapeutic agents, including continuous infusion Dox, cytarabine, floxuridine, high-dose interleukin-2, docetaxel, capecitabine, vinorelbine, and gemcitabine. It presents as a redness, swelling, and pain on the palms of the hands and/or the soles of the feet; sometimes blisters appear. In a study in 2007 by Lorusso et al. (2007), the incidence of PPE is increased in patients receiving PLD compared with conventional Dox. In studies that utilized the currently approved dose of PLD (50 mg/m² every 4 weeks), ≈50% of all patients receiving PLD experienced PPE and ≈20% experienced grade 3 PPE. It has been hypothesized that following local trauma associated with routine activities, PLD may extravasate from the deeper microcapillaries in the hands and feet. Interestingly then, that passive accumulation in tumors—which could help in its therapeutic effect—may be compromised because of limited extravasation, whereas a similar extravasation of PLD in active hands and feet introduces a new toxicity (Box 23.2).

Given these successes of the Doxil® formulation:

- Good encapsulation and retention of a toxic chemotherapeutic during the blood-borne delivery phase
- A reduction in the cardiotoxicity for the drug
- Long circulation half-life that at least gives the chance of EPR

but coupled with its discovered limitations of:

- Low EPR in human tumors
- Introduction of known drug-associated toxicities for the liposome formulation

> ### BOX 23.2   LESSON AS WE GO #2: DESIGN, DESIGN, DESIGN
>
> Do as much forward engineering design for your formulation as possible before and during the research and development phase. A function that has been designed for therapy can often lead to a second negative performance characteristic. In fact, I would encourage you to follow the design methodology I laid out in a review chapter for a different Kinam Park–edited book: "Reverse Engineer the Low Temperature Sensitive Liposome (LTSL)" (Needham 2013). In it, I take the LTSL as an example and outline a design process that not only describes all aspects of the liposome and how it is used, but also provides a general scheme that, when followed for any device or product or "problem that has already been solved," allows a person to formally reverse engineer that product. In doing so, the process itself becomes an invention generator for material improvement or inspiration for new designs and, when used in a forward engineering sense, is the template for your new research and development effort. As discussed further in Section 23.8, we did that for the LDL project we are pursuing now in Denmark, and this process ensures a more knowledgeable design.

- Limited penetration of the 80 nm diameter liposome into the tumor interstitium
- Lack of effective delivery to all tumor cells (due to the same good retention of the drug)

I felt that in trying to respond to Mark's need of "something I can heat and it releases drug," the main challenge was to try to keep the good parts, i.e., the retention of Dox in the liposome while in the bloodstream (although this was somewhat compromised, as we will see later), but to also get the drug out, fast, and only at the tumor. Since there was a well-established maximum tolerated dose (MTD), the challenge was actually less about reducing toxicity and more about increased bioavailability in delivery, thereby gaining actual efficacy.

23.2.2 HYPERTHERMIA

While all this liposomal development was going on, we also knew that tumors could be heated using HT, or at least Mark Dewhirst did. Mark and his colleagues had been working in HT since the early 1980s. In a monograph from 1983 (Oleson and Dewhirst 1983), Oleson and Dewhirst focused on progress at the time regarding "major aspects of the biologic effects of elevated temperatures both in vitro and in vivo, on the physical methods clinically used to produce HT, and on the results of treatment in large animals and humans." It is well recognized that HT can have at least three different effects on cells:

- Direct cytotoxicity at elevated temperatures, where the degree of cell killing is both time and temperature dependent
- Heat sensitization of radiation, where HT reduces cancer cells ability to repair sublethal and potentially lethal radiation damage
- Synergistic effects with certain drugs, including alkylating agents, nitrosoureas, cis-platinum, bleomycin, and adriamycin, which showed marked synergism above 43°C

There were also distinctions between the temperatures attained and the desired effect. *Mild HT* is a therapeutic technique in which cancerous tissue is heated above the body temperature to induce a physiological or biological effect but often not intended to directly produce substantial cell death. The goal is to obtain temperatures of 40°C–45°C for time periods up to 1 hour (Issels et al. 2010; Viglianti et al. 2010). In contrast, *ablative HT* is commonly greater than 55°C, but for shorter durations of 20 seconds to 15 minutes (Wood et al. 2002). An example would be the relatively recent,

HiFu (Kennedy 2005), and the older and more traditional, RFA, in treating lesions in primary liver cancer (Tateishi et al. 2005). RFA is used in conjunction with imaging techniques (e.g., ultrasound, computed tomography, or magnetic resonance imaging) to help guide a needle electrode into a cancerous tumor. High-frequency electrical currents are then passed through the electrode, creating heat that destroys the abnormal cells. An area of ≈30 cm^2 (≈3 cm diameter tumors) can be ablated with a single application of RFA (Goldberg et al. 1996). While HiFu is a technique that Celsion is now planning to use, RFA was the heating modality Celsion chose to use in the human clinical trials, in order to raise the temperature of the one or more liver lesions that would be exposed to ThermoDox® (more on this later in Section 23.6).

Given that there are at least 25 different sites for cancers, the ability to heat and provide HT is only limited by the engineer's ability to design and build an applicator, catheter, or focused radio-frequency (RF) array, to provide the heat. Thus, most parts of the body can be heated by HT using especially designed applicators for breast and head and neck cancer, a focused RF array for abdominal tumors, a catheter for prostate cancer, a microwave cap for brain cancers, and an RFA probe for liver cancer.

As will be discussed later, while RFA is a widely used technique and certainly has the capacity to burn away the center of the tumor, Celsion's plan was to target the micrometastases at its *edge*, thought to be responsible for disease recurrence, with ThermoDox®. However, the control over the temperatures attained at the tumor margins was not well characterized, and so this became a "heat-and-hope" strategy. Also, the continual heating that we had applied using a simple "leg-in-the-water-bath" technique in the mouse studies was not possible here. RFA is applied full-on for 6 minutes, and additional heating can only be given in a cycled fashion.

In contrast, Celsion had already fully characterized their Prolieve® Thermodilatation System (Prolieve®) technology for benign prostatic hyperplasia (BPH). This catheter-based system was approved for use in BPH in February 2004 as being capable of heating the whole prostate. Prolieve® has undertaken a Phase 4 postmarketing study to evaluate the long-term safety and effectiveness in the treatment of BPH (Weiner 2015). So with an FDA-approved treatment, Celsion started a Phase 1 study with William Gannon as the CMO, which could easily have provided good solid data on temperature and drug release and even resulted in a chance at approval for ThermoDox®. So, what went wrong here? Despite encouraging efficacy data in Phase 1, Celsion decided against finishing it for "business reasons," which are described further in Section 23.6.1.

The bottom line for HT though, in the context of the LTSL, is that now we had a drug–device combination, with several devices that could effectively heat many different parts of the body, where tumors might grow. If the engineers could heat it, LTSL would release its drug, which was the underlying tumor-targeting strategy.

23.2.2.1 Effects of Hyperthermia

As shown in Figure 23.6, HT has some effects alone (Dewhirst and Sim 1984; Oleson and Dewhirst 1983), is synergistic with drug (Storm 1989), increases blood flow and vascular permeability to macromolecules (Wu et al. 1993), and enhances liposome extravasation (Kong and Dewhirst 1999; Kong et al. 2000, 2001; Li et al. 2013; Matteucci et al. 2000). But the new concept (Needham 1999, 2001b, 2004) was HT-induced *rapid* triggered release of drug from the LTSL (Anyarambhatla and Needham 1999; Mills and Needham 2005; Needham 2001b, 2013; Needham and Dewhirst 2012; Needham et al. 2012; Wright 2006) that was only initiated in the bloodstream of the tumor (Needham and Ponce 2006), resulting in deeper penetration of Dox to all cell nuclei throughout the tumor interstitium (Landon et al. 2011; Manzoor et al. 2012), leading to an increased therapeutic effect in in vivo preclinical models (Kong et al. 2000; Needham 2001b; Yarmolenko et al. 2010) and in canine (Hauck et al. 2006) and human clinical trials (Poon and Borys 2009; Vujaskovic 2007; Zagar et al. 2014).

So what was the underlying science? What were the observations and questions that generated the invention? It was 15 years of *lipid membrane mechanochemistry*.

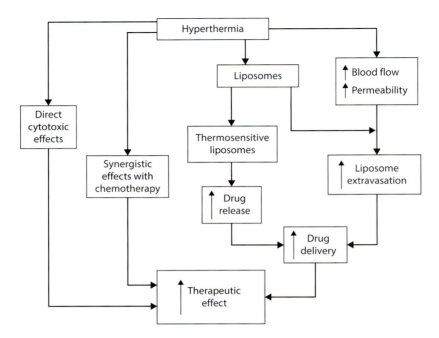

FIGURE 23.6 Hyperthermia interacts with liposomes, and at other levels, to generate a therapeutic effect, but the new concept was rapid triggered release in the bloodstream of the tumor, leading to an increased therapeutic effect. (Reproduced from Kong, G. and Dewhirst, M.W., *Int. J. Hyperthermia*, 15(5), 370. Copyright 1999, Taylor & Francis Group.)

23.2.3 Lipid Membrane Mechanochemistry

Following the pioneering characterization of the red blood cell (RBC) membrane in the 1970s (Evans and Hochmuth 1976), "lipid membrane mechanochemistry" as a field of study for pure lipid bilayers was really initiated in the early 1980s by Evan Evans, when he moved in 1980 from Duke Biomedical Engineering to the University of British Columbia, Vancouver, Canada. But he started endogenously, by checking out nature's own designs for lipid-based capsules.

23.2.3.1 Starting with Red Blood Cells

While at Duke (1973–1981), Evans had developed and used the micropipette manipulation technique to characterize the mechanical properties of RBCs. Interestingly, the biochemists (Steck 1974) were only just starting to identify the composition and organization of the spectrin membrane cytoskeleton (Bennett 1982) at the same time that Evan was measuring with the micropipette technique (Evans 1973a,b; Evans et al. 1976). Thus, sophisticated mechanical models of the RBC membrane were introduced in the early 1970s by both Skalak (Skalak et al. 1973) and Evans (Evans 1973a). Evans unified a new material concept for the RBC membrane, which provided the capability of large deformations exhibited by normal discocytes as a two-dimensional, incompressible material, and a general stress–strain law was developed for finite deformations (Evans 1973). What followed was a period of intense activity (Evans and Hochmuth 1977, 1978), culminating in the seminal book by Evans and Skalak *Mechanics and Thermodynamics of Biomembranes* (Evans and Skalak 1980). Thus, for over 50 years, these micropipette techniques have provided the unique ability to apply well-defined stresses for dilation, shear, and bending modes of membrane and cellular deformation. They laid the foundation for similar micromechanical experiments applying well-defined stresses to single-giant unilamellar vesicles (GUVs) that, in turn, eventually generated the "idea" for the LTSL. This is how that happened.

23.2.3.2 Giant Unilamellar Vesicle Experiments

The classic materials engineering approach to either understanding an existing design (like the red or white blood cell) or creating a new design for a nanoparticle drug delivery system like a liposome is the same. It involves as complete an understanding as possible for the composition–structure–property (CSP) relationships of the materials involved in the component design. To be fair, most liposomologists did not have access to the more mechanochemical techniques we were using, and so in contrast to the development of most liposome and nanomedicine formulations, we didn't just come up with an all-encompassing series of lipid–drug compositions and evaluate them for performance; we came up with one. Of course, this was based on liposomes that had already been developed and tested, especially the stealth version (Allen 1989). But with over 20,000 papers in the literature on liposomes at the time, new innovation can often require a deeper understanding of CSP relationships. And so, by the time Mark Dewhirst asked for a formulation he could "heat and it released drug," we (and Evans before me) had been studying membrane materials science for over two decades and had already evaluated many of their CSP relationships using the micropipette technique. These micromechanical methods and the resulting data on single-bilayer vesicles have allowed us to characterize the lipid bilayer membrane in both a liquid and a solid state, including the influence of cholesterol on membrane elasticity and tensile strength. It was these kinds of direct measurements of the "two-molecule-thin" material that were, literally, instrumental in helping to characterize liposomal systems and explain much of the processing and performance of the liposome.

Starting in the 1980s, the micropipette technique was adapted and developed by Evans and Kwok (Evans and Kwok 1982; Kowk and Evans 1981) and then established by (Evans and Needham 1987), to study individual GUVs of various lipid compositions. A particularly historical and newly insightful perspective on the mechanics and thermodynamics of lipid biomembranes has been given recently (Evans et al. 2013), which emphasizes "the inherent softness of fluid–lipid biomembranes and the important entropic restrictions that play major roles in the elastic properties of vesicle bilayers." In particular, the properties of importance for the LTSL were

- The membrane elastic modulus and other mechanical properties that determine drug loading and retention
- The nature of its main acyl-melting phase transition, which was the key trigger for drug release
- The behavior of the gel-phase membrane in shear and in particular its degree of yield shear and shear viscosity, as a result of in-plane shear deformations of membrane-grain structure
- Molecular exchange with water-soluble and membrane-soluble molecules like lysolipids that modified the membrane to allow it to have such a rapid release of encapsulated drug
- The adhesive and repulsive interactions that help maintain stability in blood circulation

Tables 23.1 through 23.5 show each of the most important and influential micropipette experiments that directly led to the LTSL invention. These key experiments will now be briefly described, along with the video images of the micropipette and GUV on which the experiments were carried out and their typical results. Students, please take note of the depth of inquiry, understanding, and especially mechanism, which such measurements and analyses have revealed. When I say you have to understand as much as you can about *structure–property* relationships of your nanomedicine material, in order to put *mechanism* between *composition* and *performance*, this is what I am talking about.

The first experiment and analysis that was carried out on single-walled GUVs by micropipette was to measure the elastic expansion and failure of egg lecithin bilayers (Kwok and Evans 1981). Summarized in Table 23.1, this experiment was similar to what had been used to measure the elastic modulus of RBC membranes (Evans et al. 1976)—a favorite inspiration for liposomes. It was used

TABLE 23.1
Elastic Expansion and Failure of Membranes

Video Micrographs   **Typical Result**

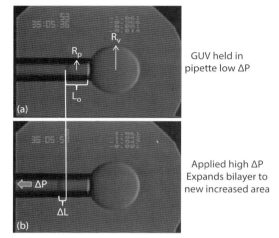

GUV held in
pipette low ΔP

Applied high ΔP
Expands bilayer to
new increased area

Video micrograph shows (a) a single vesicle (outer radius R_v) in the micropipette (pipette radius, R_p) at the starting position of a small holding suction pressure. (b) The application of increasingly higher pressures expands the membrane measured as an increase in the projection length ΔL into the pipette until the membrane eventually breaks under high tension.

The plot shows the membrane tension vs vesicle area change giving the following:

- Elastic area expansion modulus K_A
- Lysis tension T_{lys}
- Critical strain at failure α_c

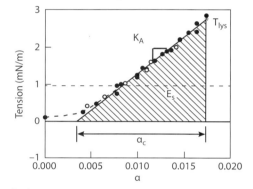

- Strain energy E_s

again by Needham and Nunn to characterize SOPC* and SOPC/cholesterol systems (Needham and Nunn 1990) and was further refined to include the contribution of membrane undulations by Evans (Evans et al. 2013), again, as a function of lipid composition including cholesterol content. For comparison, typical monounsaturated dichain lipids such as SOPC form lipid bilayers that are relatively soft (K_A = 200 mN/m), weak (T_{lys} = 6 mN/m), and somewhat permeable to water and other materials (Bloom et al. 1991). The inclusion of 50 mol% cholesterol in bilayers composed of saturated chain lipids like DSPC, or hydrogenated soy lecithin (again a favorite compositional strategy for making liposomes), have $K_A \approx$ 2000 mN/m, being 10× as stiff and 10× as strong, and relatively impermeable to water (Bloom et al. 1991). In their review of the *Stealth Liposome*, Lasic and Needham correlated such micromechanical lipid membrane data with liposome circulation half-life, showing that it is not just PEG and steric stability that can underlie extended half-life of a liposome in the bloodstream (Lasic and Needham 1995). The tighter the interface, the longer the liposomes could circulate. This explains the successful "stealth" effect of Cullis et al.'s (non-PEGylated) sphingomyelin–cholesterol–vincristine liposomes, with a measured elastic modulus of ≈2000 mN/m (Needham

---

* *Abbreviations*: Dipalmitoylphosphatidylcholine, DPPC; dimyristoylphosphatidylcholine, DMPC; distearoylphosphatidylcholine, DSPC; monopalmitoylphosphatidylcholine, MPPC; mono-oleoyl phosphatidylcholine, MOPC; monostearoylphosphatidylcholine, MSPC; stearoyl-oleoyl phosphatidylcholine, SOPC; palmitoyl-oleoylphosphatidylcholine, POPC; distearoyl-phosphatidylethanolamine-poly(ethylene glycol)-2000, DSPE-PEG[2000] (2000 molecular weight PEG).

and Nunn 1990). These data helped us understand the role of mechanics in liposome-behavior: how it retains the drug and aspects of membrane permeability to water (at least), and how it evades the body's defenses, with both steric, and mechanical, mechanisms.

By osmotically shrinking a single vesicle by just a few percent, thereby creating sufficient excess membrane area (compared to a sphere of the same volume) and supporting this membrane under low tension in the micropipette, I was able to take a single-vesicle membrane through its complete liquid-to-solid main acyl phase transition, T_m. For the SOPC/POPC-mixed system shown in Table 23.2, by measuring the area changes of the single vesicle, we could provide a phase diagram for this lipid mixture as shown in the plot. In an earlier study (actually my first ever study using the micropipette system and GUVs with Evans) reported in a very comprehensive paper on the thermomechanics of DMPC vesicle membranes (Needham et al. 1988), we showed how the liquid membrane underwent a tension-free transition into the L phase of over 30% change in lipid membrane area, consistent with the corresponding single lipid molecular area change from DMPC from 64 to 42 Å^2. We also showed how the membrane entered the known Pβ'-rippled phase with tilted acyl chains, measured the mechanics of this rippled deformation, and, from the slopes of the area vs temperate plots, obtained the thermal area expansion coefficients in each liquid (Lα), semisolid (Pβ'), and solid (Lβ) phases. It was interesting to note that the changes in relative area per molecule at T_m and the pretransition T_p of 22% and 4% were in a similar ratio to the respective excess-specific heats reported from differential scanning calorimetry (DSC) measurements (Lentz et al. 1978; Mabrey and Sturtevant 1976), i.e., in the ratio 5:l, thus showing how our single-vesicle mechanocalorimetry agreed with traditional DSC. We also studied the

TABLE 23.2
Main Acyl Solid–Liquid Phase Transition

Video Micrographs

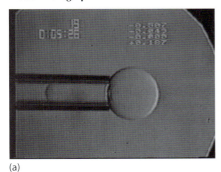

(a)

(b)

Liquid Lα
freezing

Solid Lβ
remelting

Typical Result

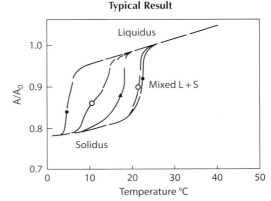

Video micrographs of (a) a single vesicle at a temperature above its melting transition (and so is a liquid-phase membrane) of 18.7°C. (b) After lowering the ambient temperature of the microchamber, the next image is a frozen vesicle at 15.3°C, initiating freezing of the bilayer. The melting is reversible although undergoes hysteresis. The white circle shows a slight kink that represents a grain boundary.

The plot shows the relative vesicle area vs temperature showing a real phase transition from liquidus to solidus lines and mixed phase region of liquid plus solid (L + S) in between these two states.

influence of cholesterol in broadening and eventually obliterating any main acyl chain–melting transition, using again DMPC GUVs (Needham et al. 1988).

During this DMPC phase transition work (Needham et al. 1988), unpublished observations showed that a single, tension-free vesicle below the phase transition was faceted in appearance. Additional evidence for the presence of grains was actually seen in this transition experiment. If you look closely at the magnified insert of the solid lipid–melting image (Table 23.2b circled for your attention), there is a slight "kink," which represents a grain boundary. Holding the vesicle with just a slight supporting suction pressure, as the temperature was raised and the lipid vesicle melted, its area per molecule expanded, the total area of the vesicle expanded, and it reversibly returned to its original melted liquid area. The (even more) interesting event was that, as it expanded, it clearly had to go through the mixed phase region of solid plus liquid domains, and we actually saw a lumpy-bumpy motion of the outer vesicle portion as "icebergs" of still frozen lipid in the melting bilayer moved past the glass pipette tip. Thus, for this SOPC/POPC lipid mixture, we showed that a vesicle became solid below its main acyl transition. But the next question was, "How solid?" In fact, the more mechanically oriented question was, "Is it a solid-like material (with elasticity), or does it have liquid-like character (and flows)?" It turns out it has both, behaving like a classic Bingham plastic with an initial elastic region, a plastic yield, and a viscous flow when deformed in in-plan shear. Next then is this experiment and the result that tested and measured this property of gel-phase bilayers, a state our LTSL would eventually be in, when injected into the bloodstream and then warmed through this transition by the local HT; whether in the blood stream or in a test tube, *a property is a property*.

By considering the gel-phase vesicle again with excess membrane area (over a sphere of the same volume), deformed into a micropipette, Evans came up with a membrane-mechanical analysis of the shear deformation of this single gel-phase vesicle that would measure the yield shear and shear viscosity of such DMPC membranes (Evans and Needham 1987). Then, as usual, it was my job to do the experiment and make it represent exactly the boundary conditions and assumptions in the theory. Table 23.3 shows this experiment for a smoothed out DMPC vesicle below its main acyl chain phase transition at a temperature of 13°C* (Needham et al. 1988).

In the experiment, a gel-phase vesicle is aspirated into the micropipette and formed into an outer spherical portion with a long projection into the pipette itself (as shown in Table 23.3d). I then release the pressure and gently blow out the vesicle, turn it around, and reaspirate it, axially with the projection. In applying a small suction pressure, the idea is to then find the point at which the elastic shear deformation of the membrane is exceeded, i.e., its elastic limit (Table 23.3a) at its plastic yield. This occurs when the projection in the pipette just reaches one pipette radius ($1 R_p$). Below this point, the material would return elastically to its original shape; beyond this plastic yield, the membrane would flow in shear. And so, a suction pressure of six times this yield pressure is applied and the vesicle duly flows into the pipette (Table 23.3b and c), with plug flow at the projections and positive and negative shear near the mouth and at the back end, respectively, until it is completely reaspirated (Table 23.3d). This behavior, of an elastic yield and viscous flow, is similar to more traditional materials that are termed "Bingham plastics" (Bingham 1922). In our highly cited 1987 review paper[†] (Evans and Needham 1987), we suggested that the shear rigidity and shear viscosity primarily reflected the density and mobility of crystal defects in the membrane bilayer surface. Thus, even a two-molecule-thick bilayer in its solid Lβ gel phase can deform with in-plane shear by grain sliding at intergrain boundaries (Kim et al. 2003). Ole Mouritsen and Martin Zuckermann (1987) had theorized that grains existed for single-phase membranes, in their analysis of the phase transitions for another similar PC lipid, DPPC, and we will see where this fits in later, when I discuss mechanisms of drug release at the phase transition of the LTSL. In fact, grain structure is clearly evident in the LTSL itself, as in the lower panel showing a transmission

---

* *Note*: You can only get these kinds of microscope-chamber temperatures on a winter day in Canada, where the humidity is so low; it would never happen in North Carolina.

[†] 595 on Web of Science today.

TABLE 23.3
Yield Shear and Shear Viscosity of Solid Membranes

Video Micrographs	Typical Results

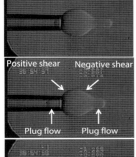

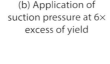
(a) Yield shear at length in pipette equal to 1 R_p

(b) Application of suction pressure at 6× excess of yield

(c) Maintaining excess suction pressure

(d) Complete re-aspiration into micropipette

Video micrographs of (a) a single solid-phase membrane vesicle at the yield shear point of the membrane. After application of suction pipette pressure that is six times the yield value, the vesicle membrane flows into the micropipette under shear viscous deformation, at t = 2 s (b), 3 s (c), and 6 s (d).

The plot shows the surface shear viscosity (left ordinate) and surface shear rigidity (right ordinate) plotted as a function of temperature derived from measurements on frozen DMPC vesicles below the main acyl chain transition (T_c) and pretransition (T_p).

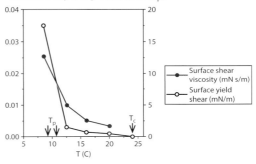

Grain structure in lipid monolayers: (a) and (b) microparticle shells, (diameters ≈ 25 μm) doped with BODIPY FLDPPE lipid dye, viewed under epifluorescence; (c) Transmission electron micrograph of an air-filled diC18:0 PC microparticle cooled at a rate on the order of 100°C/min, also showing the grain structure that is evident even at the nanoscale in (b) for the low-temperature-sensitive liposome.

electron micrograph (TEM) of a Dox-loaded LTSL of only 100 nm in diameter (Ickenstein et al. 2003). The faceted nanostructures are as evident as similar grain structures in monolayers identified by fluorescent lipids that phase separate to the grain boundaries, as studied by Dennis Kim (Kim 1999; Kim and Needham 2001; Kim et al. 2003). Importantly then, the microparticle diameter is ≈5 μm with 0.5 μm grains; in the lower panel, the liposome is 100 nm, with 20 nm grains. Thus, grain structure is commensurate with total domain material size.

Now comes the experiment that generated the concept for lysolipid modification of the phase transition, to generate such rapid drug release from the thermosensitive liposomes. At the time, we were not studying heat-triggered drug release, but we were studying molecular exchange with lysolipids (Needham et al. 1997). As shown in Table 23.4, by using three micropipettes in the microscope chamber, our very talented postdoc, Natalia Stoicheva, could hold a single-test vesicle with one pipette and then introduce one of the two flow pipettes in order to either deliver a lysolipid solution (the lysolipid was MOPC) or, after such delivery, wash the vesicle with MOPC-free solution at controlled flow rates. As shown in Table 23.4a, the initial vesicle projection length

TABLE 23.4

Molecular Exchange with Lysolipids

Video Micrographs

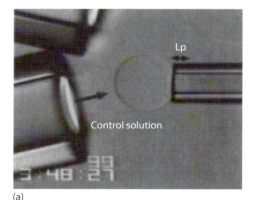

(a)

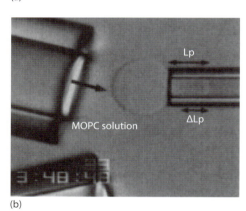

(b)

Typical Results

Video micrographs showing arrangement of the holding pipette and the two flow pipettes that deliver MOPC and MOPC-free solution at controlled flow rate to the test vesicle. (a) Initial vesicle projection length (Lp) inside the pipette is established when the lower pipette is used to flow MOPC-free bathing solution over the vesicle; (b) the lower pipette is replaced by the upper pipette and MOPC solution is made to flow over the vesicle, causing an increase of the vesicle Lp.

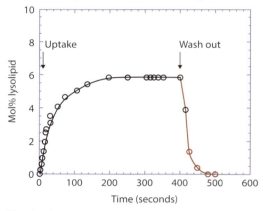

The plot shows uptake of MOPC monomer at its CMC. MOPC adsorbs into the lipid vesicle bilayer, measured by the area change of the membrane reaching 6 mol% before a change back to the bathing solution rapidly washes back out the MOPC.

(Lp) inside the pipette is established when the lower pipette is used to flow MOPC-free bathing solution over the vesicle. Then, the lower pipette is replaced by the upper pipette, and MOPC solution is made to flow over the vesicle, causing an increase of the vesicle projection length, ΔLp (Table 23.4b). This ΔLp then gives the membrane area change due to individual molecules of lysolipid entering the outer monolayer of the vesicle, and the plot shows uptake of MOPC monomer reaching 6 mol% before the MOPC-free bathing solution rapidly washes back out the MOPC.

So, what was the idea? "If lysolipid can be included in a liquid bilayer and then trapped in the bilayer in its gel state, would it desorb or form defects that would enhance the release of encapsulated membrane-impermeable contents?" The answer is yes, as described in Section 23.3.2.

Finally, the whole story about stabilizing vesicle–vesicle adhesive contact and aggregation, as well as protein binding due to a stealth effect (as opposed to mechanical stiffness), starts with fundamental measurements and theory of what it is to be adherent and what it is to be stabilized against adhesion, i.e., the accumulation of all the adhesive and repulsive potentials involving colloidal long-range forces (Evans and Needham 1987, 1988). This is what we were studying in the mid-1980s: by bringing two lipid vesicles together in salt solutions. Much of this work is described in a series of papers with Evans, where we completely stripped back the nature of the attractive and repulsive potentials at lipid bilayer interfaces and introduced them one at a time: measure one thing (adhesion energy) and change one thing (interactive potentials by changing the lipid, salt solution concentration, or polymer composition) in the system. In 1980, Evans had already modeled the vesicle adhesion experiment and trialed some preliminary systems, but again, it was my job to make an experiment that matched exactly his required boundary conditions of enough excess membrane so that—as shown in the Figures in Table 23.5—the left-hand vesicle tension could be gently lowered, allowing the vesicle membrane to spread on the tensioned, right-hand vesicle of the same material (SOPC).

TABLE 23.5

Adhesive and Repulsive Interactions Involving Colloidal Long-Range Forces

Video Micrographs   **Typical Results**

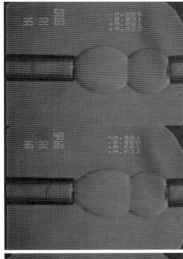

(a) Starting position

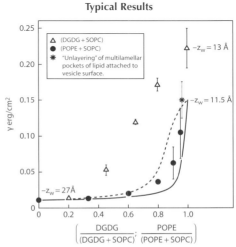

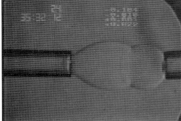

(b) Allowed to spread

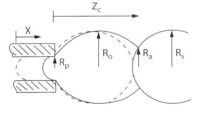

(c) Adherent vesicles

$$\left(\frac{DGDG}{(DGDG + SOPC)}; \frac{POPE}{(POPE + SOPC)} \right)$$

Video micrographs of vesicle–vesicle adhesion test.

a. Vesicles maneuvered into close proximity but not forced into contact.

b. Spontaneous adhesion allowed to progress in discrete steps controlled by pipette suction applied to vesicle on left.

The plot shows free energy potentials for adhesion by van der Waals attraction measured with neutral bilayer vesicles vs composition, either (filled circle) POPE/(POPE + SOPC) or (open triangles) DGDG/(DGDG + SOPC) fitted to theory that comprises van der Waals attraction opposed by very-short-range hydration repulsion for compositions of decreasing hydration and so interbilayer separation distance.

From Evans' 1980 paper, showing equilibrium contours for vesicle adhesion to a rigid spherical surface. The particular adhesive contact shown here was sufficient to produce a displacement of the aspirated projection equal to two pipette radii, and that's what was measured as a function of micropipette suction pressure.

(d) Evans' original analysis of spread contact

In the experiment, we first measured a fundamental force of nature, van der Waals attraction, limited at PC membrane surfaces by a very short-range hydration repulsion. We then started adding in all the other potentials, as described in Evans and Needham (1987). By adding in charged lipids, this attraction could eventually be overcome by ≈5 mol% phosphatidylserine, which fitted according to the well-known DLVO theory for colloid stability. For neutral vesicles, adding extraneous nonadsorbing polymers, like PEG or dextrans, generated large attractive stresses and highly measured adhesion energies, which again were well modeled by depletion–flocculation theory (Evans and Needham 1988). These attractive potentials were so large they could overcome even mutual negative repulsions at the interfaces of up to 30 mol% charged lipids. Finally, on this baseline, the inclusion of PEG lipids caused separations of the membranes and overcame all long-range colloidal forces and generated the "stealth" effect, as described in more detail with Dan Lasic (Lasic and Needham, 1995).

Although not part of the LTSL story, we (with Dorris Noppl, a student exchange from Germany) used the same two-vesicle experiment to observe and measure receptor-mediated adhesion (Needham and Kim 2000; Noppl-Simson and Needham 1996), an effect that could have some bearing on ligand targeting.

To summarize then, and as described more fully in another recent review (Bagatolli and Needham 2014), these experiments have characterized a two-molecule-thick membrane:

- It is a soft elastic material with a compressibility somewhere between that of a bulk liquid and a gas.
- It is stiffened considerably by the inclusion of cholesterol to levels equivalent to polyethylene.
- It is permeable to water in relation to its compliance.
- It displays a 25% change in area when taken through its main acyl chain freezing transition.
- As a gel-phase material, it shows the yield shear and shear viscosity of a Bingham plastic.
- It can exchange small amounts of other lipids and surfactants with its surrounding milieu.

When manipulated in pairs, lipid vesicles have also been found to be subject to the same range of attractive and repulsive colloidal interactions as many other colloidal particles, including

- Van der Waals attraction, limited by a very-short-range hydration repulsion and the variable-range power law of electrostatic repulsion
- Steric repulsive barriers due to the presence of bound aqueous polymers like PEG, incorporated as, for example, PEG lipid
- Depletion and flocculation by water-soluble polymers like PEG and dextrans when free in surrounding solution
- Intimate mixing (fusion) when manipulated into contacts that reduce the hydration barrier and allow membrane–membrane fusion

DMPC is not much smaller ($2\times$ (CH_2)) than DPPC, the lipid used in ThermoDox®. And so, even in 1983–1987 (over 10 years before I invented the LTSL), these thermomechanic studies of GUVs were solidifying in my mind exactly what happens when a liposome is taken through a phase transition by changing the temperature, in terms of its physical behavior, the membrane area change, the broadening by a second lipid or cholesterol component, and the shear and viscosity as a result of sliding grain boundaries in the membrane surface. All that was needed now to provide the "light bulb" of the LTSL idea was a fundamental understanding of what happens when a lipid vesicle is exposed to a soluble lipid that can partition into the membrane and can be washed out again, upon changing the bathing medium.

This section may seem arduous, in the context of a chapter on commercialization, but it is absolutely necessary, in order to create survivable patents and to understand the mechanism that the technology is based on (as opposed, again, to patenting a whole series of compositions and not having much clue at all about CSP relationships that influence performance and indeed are part of processing). This then is the level of understanding that is needed, in order to fully characterize your nanomedicine and develop one that actually works in the way it was designed to. I would, therefore, gently suggest that you might need at least some of the same kinds of insights, if you are going to invent anything of use in nanomedicine. There is another lesson to be learned here (see Box 23.3)

BOX 23.3   LESSON AS WE GO #3: FULL CHARACTERIZATION OF ANY NANOMEDICINE INVOLVES A COMPLETE UNDERSTANDING OF COMPOSITION–STRUCTURE–PROPERTY RELATIONSHIPS

Attending many of the drug delivery meetings as I do, it is often the case that a "new" nanomedicine is presented in terms of its material composition, and this is compared directly to its performance in vitro or in vivo experiments. "This is what it's made of and my graduate student carried out these studies showing that this is the IC_{50} for these particular cancer cells," or "this is the tumor growth delay compared to other systems that also do not work." While these kinds of feasibility studies are certainly advancing the field, it is left to others to gain the deeper understanding of if, and why, the design works or, most often, does not work, especially in humans, if it even gets that far. IMHO simply is not good enough for researchers to proceed like this. I appreciate that not everybody has all the expertise or techniques required for a complete understanding of structure–property relationships (or is lucky enough to be working with Evan Evans), but that does not mean you cannot collaborate with people who do, or that you "read the damn literature." At least recognize that not only it is necessary to more fully understand your system, but also the design process itself is your invention generator. If you and your advisor do not know the mechanisms, then simply carving out as big a compositional space as possible to protect your idea, and trying to persuade the patent examiner that it's okay to patent every composition that you (or your lawyer) can think of, is cheating. And do not let me catch you at a meeting, or worse still, in your PhD defense if I am an examiner, presenting something you don't understand, or have not at least tried to understand, mechanistically.

23.3   ENGINEERING DESIGN OF THE LTSL

Invention generation, to my mind, relies on an ability to make connections of fundamental characteristics of materials involved in the new design. It is then a design methodology process to generate the whole concept from functions to evaluating performance in service. Placing this kind of knowledge in the context of a design scheme is the invention generator I keep referring to. I have shown how to go through the whole process of "reverse engineering" a design that already works, using the LTSL as the prime example (Needham 2013). If you do this yourselves and adhere to the rigor of the process, who knows? You might generate an invention yourself; furthermore, this process of reverse engineering a design guarantees that you at least have a mechanism, and not just a series of compositions.

23.3.1 Lysolipid Exchange with Membranes Generates the "Idea"

The LTSL formulation comprises gel-phase DPPC membranes, containing nonbilayer-forming lipids like the lysolipids MPPC (Anyarambhatla and Needham 1999; Needham 2001b) or later, MSPC (Mills and Needham 2005; Wright 2006), as well as DSPE-PEG. Our optimized formulation consists of DPPC/MSPC/DSPE-PEG^{2000} in the molar ratio of 86.5:9.7:3.8 mol% (Mills and Needham 2005; Wright 2006). As described earlier, and in reference to the Figures in Tables 23.1 through 23.5, the idea for the lysolipid incorporation to potentially generate a more rapid and complete release of an encapsulated drug came from a series of micropipette experiments, theory, and interpretation, spanning at least 15 years. Taken together, in the context of lipid membrane CSP relationships, they explain much of the mechanisms (further described next) that provide for Inex's mantra of "load, retain, avoid, target, fuse" and our innovation of "release."

23.3.2 Mechanism of Release

The mechanism of release relies on three main features of gel-phase DPPC membranes containing nonbilayer-forming lipids like MSPC and DSPE-PEG:

1. The membrane composition can freeze with the formation of grains and grain boundaries.
2. Lipid-chain-compatible lysolipids, like MSPC, are readily incorporated and trapped in the gel-phase DPPC membranes and form porous defects as the membrane is warmed into its acyl-melting phase transition.
3. Coincorporated PEG lipids, while not generating enhanced porosity themselves, do seem to stabilize the equilibrium pores.

As first shown by Anyarambhatla (Anyarambhatla and Needham 1999; Needham 2001b), confirmed in more detail mainly by my graduate students Jeff Mills (Mills 2002; Mills and Needham 2004, 2005) and Alex Wright (2006) and supported by data from other labs (Banno et al. 2010; Li et al. 2010; Negussie et al. 2011; Tagami et al. 2011), the release of encapsulated Dox and other drugs like cisplatin (Woo et al. 2008) can occur within only a few seconds of reaching the transition temperature of the lipid bilayer (mixture) around 41°C.

The mechanism has been described in detail in a paper we wrote for the Faraday discussions (Needham et al. 2012) and, in the context of its design, in the reverse engineering chapter (Needham 2013). To summarize briefly here, using the example illustrated in Figure 23.7, which shows an LTSL formulation consisting of MSPC and DSPE-PEG^{2000} in the DPPC host bilayer, the enhanced permeability in the phase transition region is through MSPC pores.

That is, as the transition temperature is approached and the grain boundaries begin to melt, the lysolipid forms lysolipid-lined nanopores at these now liquid boundaries. These MSPC pores also seem to be stabilized by the presence of PEG lipid, since in the absence of PEG lipid, the release is slightly slower. As a consequence, the hydrogen ion gradient rapidly equalizes and DoxH$^+$ comes out in seconds (large red arrow), as does any remaining embedded Dox in the bilayer.

The presence of only a few mol% of lysolipid contained in the gel-phase DPPC bilayer of the LTSLs significantly increases both the rate and amount of drug released, allowing for a "burst" release in vivo, in only a matter of seconds, upon the application of HT directly to the tumor. Furthermore, compared with DPPC alone, the slight lowering of the bilayer transition temperature by the presence of the $\approx$10 mol% MSPC is offset by the inclusion of the disaturated acyl chains of the

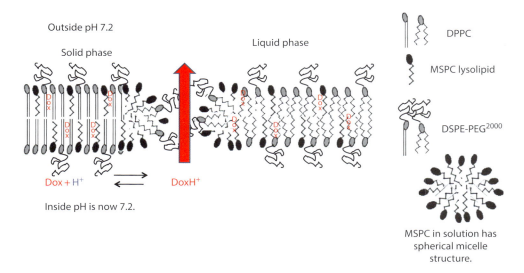

FIGURE 23.7 Proposed mechanisms for thermally triggered release at 41°C: grain boundaries melt, lysolipid forms, PEG lipid stabilizes the nanopores, and doxorubicin is released. (From Needham, D. et al., *Faraday Discuss.* (*Lipids Membr. Biophys.*), 161, 515, Copyright 2012. Reproduced by permission of Faraday Division, Royal Society of Chemistry.)

≈4 mol% DSPE-PEG, which slightly raises T_m compared to DPPC alone. The result is that the transition temperature of the optimized LTSL bilayer maintains the transition around 41.3°C, thereby maximizing the release parameters for mild HT clinical use.

To get an idea of the pore size, we tested for permeability through the LTSL membrane at its phase transition for a range of ions and molecules including dithionite, cisplatin, carboplatin, carboxyfluorescein, Dox, manganese porphyrin, and both 10,000 and 40,000 Da dextran. The results showed that only the 40,000 Da dextran could not pass through the membrane. Thus, the conclusion is that the pore is at least 5 nm in diameter and less than 10 nm (Wright 2006).

23.4 PERFORMANCE IN PRECLINICAL STUDIES

23.4.1 ALL 11/11 MICE ARE "CURED"

As presented in two papers in Cancer Research in 2000 (Kong et al. 2000; Needham 2001b), the first preclinical assessment of the Dox-containing LTSL (Dox-LTSL) formulation could not have gone any better. The first growth delay study we did showed that the Dox-LTSL could actually "cure" mice from an implanted tumor (FaDu, a squamous cell carcinoma) out to 60 days, after just a 1 hour heating to mild HT temperature of 42°C with LTSL in the bloodstream. We tested an unheated saline control against several treatments including a nonthermal-sensitive, Doxil-like liposome (NTSL), both with and without heating to 42°C for 1 hour. In the experiment, FaDu tumors implanted on the leg of the test mice grew to five times the original tumor volume (of 5–7 mm diameter) in 10 days in the unheated control. Depending on the test "treatment," the growth of the tumor was delayed by

- 3.5 days for a single dose of free Dox at normothermic (34°C) skin temperature (13.5 days)
- 10 days for a heated tumor with no drug (20 days)
- 14 days for a heated tumor with drug (24 days)
- 11 days for a normothermic tumor with NTSL (21 days)

- 1 day for a normothermic tumor with LTSL (1 day)
- 22 days for a heated tumor with NTSL (32 days)
- 40 days for a heated tumor with LTSL (40 days)

Thus, while free drug could only delay the growth by 3.5 days, HT alone had a significant effect, causing a delay in growth of 10 days. When the two approved treatments were combined, heating to 42°C with injected drug, the delay was 14 days. NTSL had some effects (11-day growth delay) at normothermic temperatures, showing how the EPR effect must have been operating for this long-circulating liposome, in this particular implanted tumor. But LTSL at normothermic was less than useless. Producing only a 1-day growth delay, its lack of effect was comparable to saline and worse than free drug. This showed that LTSL must be leaking drug and not even performing like a stealth liposome. This was later confirmed by others who measured the drug-to-lipid ratio (i.e., a measure of the encapsulated drug compared with the circulating lipid of the liposomes) showing a half-life of Dox retention in vivo of only 1 hour (Banno et al. 2010), which was confirmed also in humans (Wood et al. 2012). When the tumor was heated to 42°C for 1 hour, NTSL had an even greater effect, causing a growth delay of 22 days. Since it is not thermally sensitive, this was probably a direct result of the increased extravasation enabled by mild HT, which is known to enhance vascular permeability (Gaber et al. 1996; Kong et al. 2000) but could also have included some increase in membrane permeability to H^+ ions and the deprotonated Dox that would follow through the membrane.

The main event though came with the LTSL cohort, where the tumor was heated to 42°C, the Dox-LTSL was injected, and the tumor heating was continued for 1 hour (Dox-LTSL + HT). In the second paper in this series, all 11/11 tumors remained regressed out to the full endpoint of the study, which was 60 days, with a minimum 50-day growth delay. It was this result, published in 2000 (Needham 2001b), but obviously obtained earlier, that sealed the license agreement with Celsion in November 1999. It is time for another quick lesson (Box 23.4).

BOX 23.4   LESSON AS WE GO #4: IN VIVO, IN VIVO, IN VIVO

As an inventor, you will not see your precious idea be even remotely appreciated by anyone with investment capital unless you have very positive in vivo data, and quite rightly so.

23.4.2 NOT ALL CANCERS RESPOND THE SAME

So, did we get lucky with FaDu? Kind of. Expanding the preclinical testing to a series of other tumors, the efficacy of the commercial formulation (ThermoDox®) was reexamined in FaDu and compared with HCT116, PC3, SKOV-3, and 4T07 cancer cell lines (Yarmolenko et al. 2010). It turned out that variations in antitumor effect of Dox-LTSL + HT are primarily related to in vitro doubling time. In all five tumor types, Dox-LTSL + HT increased median tumor growth time, compared with untreated controls and HT alone (Yarmolenko et al. 2010). Compared with the Dox-LTSL without heating, Dox-LTSL + HT yielded significantly higher drug concentrations in the tumor. Thus, heating is critical for drug release; the EPR effect alone for this formulation is simply not good enough.

The study also evaluated the cell lines for sensitivity to the drug. FaDu was the most sensitive to Dox (IC_{50} = 90 nM) in vitro, compared to the other cell lines (IC_{50} = 129–168 nM). Again, of the parameters tested for correlation with efficacy, only the correlation of in vitro doubling time and in vivo median growth time was significant. Slower-growing SKOV-3 and PC-3 had the greatest numbers of complete regressions and longest tumor growth delays, which are clinically important parameters. Thus, while in this second series of preclinical model tests, we did not get the same 11/11 tumors "cured" and the results were still incredibly positive. Dox-LTSL + HT again resulted in the best antitumor effect in each of the five tumor types. Interestingly, these variations in efficacy were most correlated to in vitro cell doubling time. It was suggested in this 2010 paper that in the clinic (and perhaps Celsion should consider this), the rate of tumor progression must be considered in the design of treatment regimens involving Dox-LTSL + HT. Another study from Brad Wood's lab at National Institutes of Health (NIH) largely confirmed the ability of LTSL to release Dox in vitro and in vivo, by combining LTSL with noninvasive- and nondestructive-pulsed HiFu exposure. This study also showed enhanced delivery of Dox and, consequently, its antitumor effects (Dromi et al. 2007). More on these clinical studies will be discussed later.

23.4.3 NEW PARADIGM FOR LOCAL DRUG DELIVERY: DRUG RELEASE IN THE BLOODSTREAM PROVIDES FOR GREATER AMOUNTS OF FREE DRUG AND DEEPER PENETRATION INTO TUMOR TISSUE AND SHUTS DOWN THE TUMOR VASCULATURE

Thus, from the in vitro studies, the release time of Dox from the LTSL at only mild HT temperatures was faster than the transit time of liposomes through the warmed tumor vasculature. Drug release could therefore occur in the bloodstream of the tumor! The problematic EPR effect (for such large 80–100 nm liposomes) was not needed at all. We had previously shown the importance of dosimetry (Kong et al. 2000), achieving approximately 50 µM Dox in the tumor tissue. When compared to the IC_{50} for FaDu of 90 nM, drug concentrations by release in the bloodstream were 500 times the IC_{50}. Tumor drug levels for Dox-LTSL + HT were up to 30 times higher than those achievable with free drug administration and 3–5 times elevated compared to NTSL, which must rely on extravasation to deliver any drug at all to the tumor interstitium (Needham and Dewhirst 2001; Yarmolenko et al. 2010), and their leakage of drug would be slow at best. Moreover, bioavailability was also key; heating the tumor to 42°C and continuing to heat for 1 hour while administering Dox-LTSL resulted in half of the Dox being bound to DNA and RNA of tumor cells after only that one 1 hour of treatment (Kong et al. 2000). In stark contrast, the amount of Dox bound to DNA and RNA of tumor cells after free Dox + HT or NTSL + HT was not even detectable (Kong et al. 2000) (Box 23.5).

BOX 23.5   LESSON AS WE GO #5: STAND AND DELIVER AND RELEASE

From a drug delivery perspective, it is not enough to deliver drug to the perivascular space of the tumor interstitium: that drug must be bioavailable. It is also not enough to deliver what seems to be sufficient drug to cause tumor growth delay and abolish tumors over the first 10 days, if not enough drug is delivered to prevent them from growing back. It will be important to remember these data when discussing the inherent limitations that may have been responsible for the underperformance of the subsequent Celsion-run, Phase 3 liver cancer trial using ThermoDox® with RFA.

So, if drug was being released into the blood vessels of the tumor, what effects could it be having, perhaps on the vasculature itself? Fan Yuan, a close collaborator of ours and ex-Rakesh Jain's group, measured the RBC velocity in tumor vasculature before and after the 1 hour LTSL "treatment." Using fluorescent red cells as "tracers bullets" (Chen et al. 2004), he found that the average RBC velocity was reduced by almost 150 times, from 0.428 to 0.003 mm/s and the microvascular density was reduced from 3.93 to 0.86 mm/mm². In addition, blood flow stasis and severe hemorrhage occurred immediately after treatment and there was no blood flow in microvessels in five out of six tumors at 6 and 24 hours after the treatment. Thus, at 24 hour, after just a 1 hour treatment, tumor blood flow can actually be shut down by Dox-LTSL + HT in FaDu tumors.

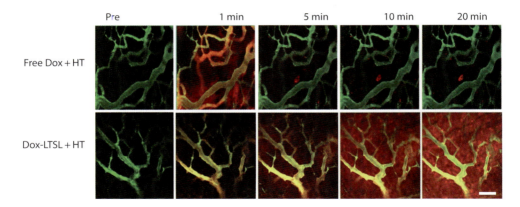

FIGURE 23.8 Tumor uptake of doxorubicin vs time. Time sequence images of blood vessels (green) and doxorubicin (red) for preinjection and at 1, 5, 10, and 20 minutes after injection. Shown are injections to a warmed tumor (42°C) of free doxorubicin (free Dox + HT) and the doxorubicin-loaded LTSL (Dox-LTSL + HT). Scale bar = 100 μm. (Reprinted from Manzoor, A.A. et al., *Cancer Res.*, 72(21), 5575. Copyright 2012, American Association for Cancer Research. With permission.)

Finally, the most compelling and dramatic evidence for not only release in the bloodstream, but deeper penetration of Dox into a tumor than has ever before been achieved and measured in vivo, was presented by Manzoor et al. (2012). As shown in Figure 23.8, adapted from that paper, real-time confocal imaging of Dox delivery to the FaDu xenograft in window chambers illustrates that compared to administering free drug alone, intravascular drug release from Dox-LTSL in the prewarmed tumor massively increased the amount of free drug in the interstitial space, after only 20 minutes of heating. Clearly, this increased both the time that tumor cells were exposed to maximum drug levels and the drug penetration distance, compared with free drug or traditional PEGylated liposomes. Maximum measureable drug penetration from tumor vasculature vs. treatment group shows that drug delivered with Dox-LTSL penetrates twice as far as Doxil® liposomes (78 vs 34 μm). The released, now free, drug diffuses into the tumor interstitium, reaching its nucleus target with greater penetration distance, and to much higher concentrations, than those achievable by either free drug administered alone or the more traditional long-circulating liposome formulations (Eley and Needham 1984). Intravascular drug release provides a *mechanism* (see properties of Dox above, Section 23.2.1.1) to increase both the time that tumor cells are exposed to maximum drug levels and the penetration distance achievable by drug diffusion. This establishes a new paradigm in drug delivery: rapidly triggered drug release in the tumor bloodstream, saturating neoplastic cells, as well as endothelia, pericytes, and stroma, with the anticancer drug.

23.4.4 ThermoDox® Phase 1 in Canine Patients

A Phase 1 canine trial of the Dox-LTSL was also carried out in spontaneous tumors. Of the 20 dogs that received 2 doses of Dox-LTSL, 12 had stable disease (SD) and 6 had a partial response (PR) to treatment (Hauck et al. 2006). The conclusion from this work was that "doxorubicin-LTSL offers a novel approach to improving drug delivery to solid tumors. It was well tolerated and resulted in favorable response profiles in these patients. Additional evaluation in human patients is warranted."

23.4.5 New Data Shows a Systemic Effect: Heat One Leg Tumor and the Other Tumor Also Shows Growth Delay

As we continued to explore the potential for the LTSL to advance local tumor drug therapy, anecdotal evidence from the chest wall recurrence trial seemed to indicate there was even some systemic effects. The MD carrying out the clinical studies, Dr. Kim Blackwell, told Mark Dewhirst about this and Mark immediately established a new study to determine whether treatment of a tumor site with systemically administered LTSL, with HT for triggered release, would have a dual antitumor effect on the primary heated tumor and an unheated secondary tumor in a distant site (Viglianti et al. 2014). They also wanted to determine the ability of noninvasive optical spectroscopy to predict treatment outcome. As discussed in their paper, mice were inoculated with SKOV3 human ovarian carcinoma in both hind legs. Only one tumor was selected for local HT and subsequent systemic treatment, and the size and characteristics of both would be measured.

Data for the four treatment groups (control, Doxil®, and two different LTSL formulations containing Dox) showed that similar to previous studies (Kong and Dewhirst 1999; Kong et al. 2000; Needham 2001b; Ponce et al. 2007), tumor growth delay was seen with both Doxil® and the thermally sensitive liposomes, in the tumors that were heated. As before, there was significant growth delay with the Doxil® and two LTSL treatment groups on the primary tumor side, since HT enhances both EPR and Doxil® uptake into the interstitium (Gaber et al. 1996), as well as releasing drug from the LTSL in the tumor vasculature. The data are also consistent with Mark Dewhirst's previous work showing that growth time correlates with intratumoral drug levels (Kong et al. 2000; Palmer et al. 2010). However, the startling result was that *tumor growth delay was also seen in the opposing tumor* in the thermally sensitive liposome–treated groups, but not with Doxil®. This mechanism of the so-called abscopal effect* of Dox-LTSL is most likely due to recirculation of intravascularly released drug. Thus, these thermally sensitive liposomes affect the primary heated tumor and also bring systemic efficacy. Learning from the past, as you will read about next, we have sent this paper to Celsion and informed them to look out for these effects in their continued clinical trials. We hope they do (listen that is) (Box 23.6).

23.5 LICENSING

LTSL was licensed by Duke University to Celsion Corporation in November 1999. Here is a brief historical synopsis about Celsion. Augustine Y. Cheung, a well-known microwave expert with a PhD in electrical engineering from the University of Maryland, established Cheung Laboratories, Inc. in the State of Maryland in 1982. It was a device company that focused on HT and developed various types of HT equipment, including a balloon catheter technology for enhanced thermotherapy of BPH, called Prolieve®. We started talking with Dr. Cheung (an old friend of Mark Dewhirst's) in the mid-1990s, as we were starting to develop our LTSL. Boston Scientific exclusively distributed Prolieve® from 2004 and eventually purchased the technology from Celsion, in 2007. They then licensed it back to Dr. Cheung in 2013. Dr. Cheung was the president and CEO of Celsion, but as Celsion transitioned from "just" a device company to a biotech company, he "stepped down" in 2005, to be replaced by Lawrence Olanoff (ex-Forest Laboratories); Dr Cheung eventually became president and CEO of Medifocus. Olanoff himself (whom I did not get a good feeling about, right from the first time we met) only lasted 14 months and quit Celsion to go back to Forest Labs in October 2006. Celsion Executive Vice President, Chief Operating Officer, and Chief Finance Officer Anthony P. Deasey, replaced Olanoff on an interim basis, until a replacement was found. Luckily, for the project, this turned out to be Mr. Michael Tardugno, who joined Celsion on January 3, 2007, as president and chief executive officer. Mike is really the person (with his team including Nick Borys, CMO), who eventually moved the technology along to where it is today. Celsion is finally under a stable and strong leadership in Tardugno and Borys. But the aforementioned is typical of the merry-go-round of small company wheelings and dealings, ins and outs, and ups and downs, which your invention and potential product might be exposed to. If it is, little will get done until senior management is capable and stable, competent, and free from hidden agendas and greed.

---

* The *abscopal effect* is a phenomenon in the treatment of metastatic cancer where localized treatment of a tumor causes not only a shrinking of the treated tumor but also a shrinking of tumors in different compartments from the treated tumor.

BOX 23.6 LESSON AS WE GO #6: HEAT TUMOR, ADMINISTER LTSL, KILL TUMOR (SIMPLE, RIGHT?)

From a drug delivery perspective, it is not enough to inject liposomes and hope to deliver drug to the perivascular space using the LTSLs, imagining that they would release Dox inside the tumor vasculature. To reiterate, the tumor has to be first heated to 42°C (Manzoor et al. 2012) and be at the release temperature before the liposomes are administered. Otherwise, the loss of drug from the liposomes and loss of liposomes themselves from the circulation can deplete the reservoir of Dox-LTSL in the blood stream, and so reduce the amount of doxorubicin that can be available and be released in the tumor vasculature. So, from all the preclinical studies that we did, both in vitro (Anyarambhatla and Needham 1999; Ickenstein et al. 2003; Matteucci et al. 2000; Mills 2002; Mills and Needham 2005; Needham 2001b; Wright 2006; Wu et al. 1993a,b) and in vivo (Chen et al. 2004, 2008; Gaber et al. 1996; Kong et al. 2000a,b, 2001; Manzoor et al. 2012; Matteucci et al. 2000; Needham 2001b; Ponce et al. 2007; Wright 2006; Wu et al. 1993a,b; Yarmolenko et al. 2010)—and these are original peer-reviewed manuscripts and PhD theses—not just reviews, which we also published (Kong and Dewhirst 1999; Landon et al. 2011; Mills and Needham 2004; Needham 2013; Needham and Dewhirst 2001, 2012; Needham and Ponce 2006) (for Celsion and others to read), here is our recommended, research-derived protocol:

- Heat tumor to 42°C.
- Administer LTSL while maintaining this mild HT level for preferably 1 hour.
- Kill tumor.
  - Release drug in the bloodstream, enough to kill a horse.
  - Reduce tumor blood flow within the first few hours.
  - Shut down the tumor vasculature in the first 24 hours.
  - Penetrate the drug deep into endothelial, pericyte, stroma, and neoplastic cells of the tumor in the first 20 minutes.
  - Half the Dox in the tumor is already bound to the DNA of every cell in the tumor within the first hour.

Simple, right? You would think so, but as we will see later, this was not what Celsion did.

The license agreement is actually available on line. As an example of a license agreement, just search *Celsion license agreement Duke* and this will pop up: Home > Sample Business Contracts > Celsion Corp. The contracts are at the Onecle web page (Onecle 1999): "Business Contracts from SEC Filings." Thus, the license agreement—Duke University and Celsion Corporation—is there for all to see (why this is public or who put it there; I have no idea).

Celsion got this "field":

```
"FIELD" shall mean the use of the HEAT TRIGGERED DELIVERY TECHNOLOGY and
PATENT RIGHTS in thermally sensitive formulations designed to release,
activate, or express pharmaceutically active agents locally, such release,
activation, or expression being initiated by local application of heat and
being made for the purpose of treating any disease or altering a physi-
ological process in animals, including, without restriction, in humans.
```

All drugs, all diseases, for the lifetime of all the patents: not a bad deal. So what did they do with it?

23.6   CLINICAL TESTING

Having licensed this "technology," Celsion, in their various guises, first as a device company and then as a full-blown biotech company, to their credit, started a series of human clinical trials. These included a Phase 1 study in prostate cancer, a Phase 2 RCW, and a Phase 1 for liver cancer that rapidly progressed to a Phase 3 trial, but without the usual intermediate stage of a Phase 2. Some of this is newly reported in this chapter (see prostate cancer trial in the following section), while some—the RCW and liver trial data—has already been published (Needham 2013) and should be consulted for the fuller, more technical story.

23.6.1   PHASE 1 PROSTATE CANCER

The first Phase 1 human trials were actually in prostate cancer—a perfect place to start testing Thermodox®. The prostate cancer trial (Celsion 2003–2009) was initiated in 2003 in order to determine the maximum tolerated dose of doxorubicin released from Thermodox® via thermal microwave therapy in patients with adenocarcinoma of the prostate. However, the trial was terminated in 2009. I asked why, and did not get a real answer for a while until someone said, "*business reasons.*" The suggestion was that there already were lots of treatments for prostate cancer, and so there was no money to be made in it. Really?

The reason I say this was a perfect place to start is that Celsion already had a tried and tested prostate heating system—their own Prolieve® Thermal Dilatation System and so the pairing of their own device with our drug seemed like a match made in heaven. A paper was actually published in *Urology* (Larson et al. 2006) using interstitial temperature mapping during Prolieve® transurethral microwave treatment to show that the temperatures we required for Thermodox® could actually be achieved in the prostate capsule. Other data on clinical efficacy obtained in the Phase 1, as far as I know, was never released. According to their current web site at MediFocus Inc. (2016), "The Prolieve® System is an in-office technology, a medical device that both heats the prostate and dilates the prostatic urethra" using a balloon catheter, as shown schematically in Figure 23.9. However, even back then when they were naturally considering also using it to heat prostate cancer, they had already done a significant amount of characterization of what this technology could do for benign prostate hyperplasia and had an FDA application to test it (FDA 2004).

The Larsen study was performed using the Prolieve® Thermodilatation System "funded by an unrestricted educational grant from Boston Scientific Corporation, Marlborough, Massachusetts" (Larson and Robertson 2006) that simultaneously compressed the prostate with a 46°C balloon circulating heated fluid and delivering microwave energy into the prostate. Results by actual interstitial temperature mapping showed that average peak temperatures of 51.8°C were attained at an average of 7 mm away from the prostatic urethra. These temperatures were greater near the bladder neck and midgland than toward the prostatic apex. Magnetic resonance imaging also revealed necrotic zones that were consistent with sustained temperatures greater than 45°C. Thus, as shown in Figures 23.9b and 23.10, Celsion were in possession of actual data showing that the Prolieve® system heated the prostate right out to its boundaries, to a temperature that was more than enough to cause ThermoDox® to release its drug, in less than 4 minutes. So, while BPH actually requires higher temperatures in the 50°C range, they could already attain temperatures that were perfect for ThermoDox® in the prostate. And, moreover, unlike the trial they did choose to invest all their resources in (the primary liver cancer trial using RFA), here was a treatment modality where they could "heat and know" they were attaining the desired temperature, in contrast to the "heat-and-hope" strategy of the subsequent primary liver cancer trial. It was a system, completely owned and developed by Celsion, that was perfect for the ThermoDox® technology they had licensed from Duke, and yet they abandoned the trial because of "business reasons." Go figure! (Box 23.7).

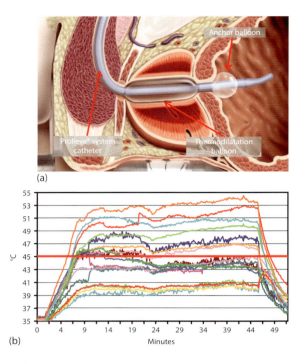

(a)

(b)

FIGURE 23.9 The Prolieve® system. (a) Schematic illustration of the Prolieve® benign prostatic hyperplasia treatment device. (b) Intraprostatic temperature plot of representative subject showing temperature and time. All recorded temperatures measured are displayed, where each line represents temperature of individual sensor. (a: Courtesy of MediFocus Inc.; b: Adapted from *Urology*, 68(6), Larson, B.T. and Robertson, D.W., Interstitial temperature mapping during Prolieve® transurethral microwave treatment: Imaging reveals thermotherapy temperatures resulting in tissue necrosis and patent prostatic urethra, 1206–1210. Copyright 2006, with permission from Elsevier.)

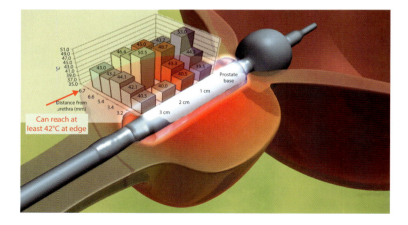

FIGURE 23.10 Schematic illustration of the Prolieve® Thermodilatation System treatment, showing temperatures attainable throughout the prostate, as measured directly by thermal probes. *Note*: Even temperatures at the edge are at least 42°C, which is enough to release doxorubicin from ThermoDox® in any prostate-encapsulated tumor. (Adapted from *Urology*, 68(6), Larson, B.T. and Robertson, D.W., Interstitial temperature mapping during Prolieve® transurethral microwave treatment: Imaging reveals thermotherapy temperatures resulting in tissue necrosis and patent prostatic urethra, 1206–1210. Copyright 2006, with permission from Elsevier.)

BOX 23.7   LESSON AS WE GO #7: BE ON THE LOOKOUT FOR DECISIONS THAT ARE BASED ON "BUSINESS REASONS"

When CEOs and accountants do "analysis," it's not the same analysis that you or I might do as scientists, clinicians, chief medical officers, or patients. This Phase 1 trial was apparently already showing some efficacy in the limited dosing range (three patients per dose, 20–50 mg/m^2 dose escalation), even before it reached the expected MTD. And it was heatable using Celsion's own Prolieve® technology, then actually being used for benign prostate HT. Hmm, "business reasons" that probably ended what could have been an approvable procedure, the first for ThermoDox®.

23.6.2   Phase 2 Recurrent Chest Wall Cancer

As an advisor to Celsion, Mark Dewhirst was also very proactive in developing ThermoDox®, within his (our) NIH-funded Hyperthermia Program. RCW was another ideal cancer for initial clinical testing on the way to FDA approval. But guess what? Despite Mark's expert efforts and advice, it didn't cut through the commercialization strategy either. This one wasn't abandoned, but its progress was slowed considerably for a series of commercial reasons, including the fact that again, this was not a huge market. However, in Phase 1 trials, ThermoDox® showed very promising results, at doses that were even 60% less than the expected MTD. This is where it gets really interesting and frustrating: interesting, in terms of ThermoDox's outstanding results, and frustrating, because these positive results were in spite of what Celsion and their clinical investigators were deciding on, writing in the trial, and implementing, which was a *protocol* that went against everything we had been telling Celsion to do. What we had learned from our research suggested that *their protocol* almost completely missed the bioavailability of ThermoDox®. It amounts to another series of "business decisions" (by definition, it's a "company") and "hospital decisions" (because they have budget rules), as opposed to "rational" decisions (because it's the right scientific thing to do, and because it could actually benefit the patient), which probably set back any potential approval of ThermoDox® by years.

This second human Phase 1/2 trial (called the "Duke study") in breast cancer recurrence (Celsion 2012) was designed to evaluate the MTD, pharmacokinetics, safety, and efficacy of approved ThermoDox® + HT, in patients with breast cancer recurrence at the chest wall. In the initial Phase 1 (which was actually started in 2001, but later became nonrecruiting (Celsion 2001), HT was administered with an eminently achievable temperature goal of 40°C–42°C, for 1 hour, using the BSD-500 PC System. However, the company supplying this HT equipment also seemed to be dragging its feet, because it was (historically) in competition with Celsion and (almost unbelievably) no one could get past the old bad blood. Surprisingly (but maybe not that surprising) given the preclinical data Manzoor et al. got later (2012) (see Figure 23.8), there were several instances of SD, PR, and two of the complete responses (CR), for a dose escalation of only 20–30 mg/m^2, compared to a final MTD of 50 mg/m^2. Several patients in this trial achieved either PR or CR (Vujaskovic 2007). As shown in Figure 23.11, for one patient, her widely disseminated chest wall tumor had completely disappeared after only four 1 hour mild HT treatment cycles with ThermoDox® in her bloodstream.

This test dose (30 mg/m^2) was thus only 60% of the expected MTD, and so there were also no side effects of the drug. A second patient had a similar CR at 30 mg/m^2 (Vujaskovic 2007). Again, here was a heating system where the temperatures required for release of Dox from ThermoDox® were known, controlled, and eminently achievable, using this BSD machine (unlike the ensuing RFA trial that went horribly wrong).

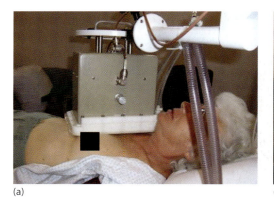
(a)

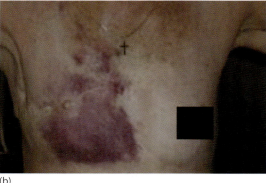
(b)

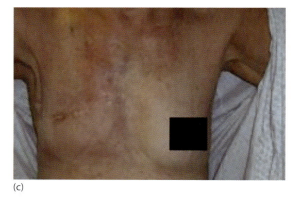
(c)

FIGURE 23.11 (a) Treatment of chest wall recurrence of breast cancer, using a BSD-500 PC System to achieve a temperature goal of 40°C–42°C. (b) Same patient before treatment. (c) Precycle 5 given 30 mg/m^2. (Adapted from Zagar, T.M. et al., *Int. J. Hyperther.*, 30(5), 285. Copyright 2014, Informa Healthcare.)

A recent update provided by Celsion (April 15, 2015) and published in an abstract and poster (Rugo et al. 2013) and peer-reviewed paper (Zagar et al. 2014), for this ongoing open-label Phase 2 DIGNITY trial of ThermoDox® in RCW breast cancer, now shows that

- Of the 16 patients enrolled and treated, 12 were eligible for evaluation of efficacy
- 67% of patients experienced a clinical benefit of their highly refractory disease with a local response rate of 58% observed in the 12 evaluable patients
- Notably, there were five CR, two PR, and one patient with SD

When taken together with the previous (stalled) Duke study, of the 29 patients treated in the two trials (comprising 11 patients in this now-termed "DIGNITY" study, with 18 patients in the Duke study), 23 were eligible for evaluation of efficacy (Zagar et al. 2014). A local response rate of over 60% was reported in 14 of the 23 evaluable patients, with 5 CR and 9 PR. The simple Phase 1 study concluded that 50 mg/m^2 ThermoDox® with mild HT is safe (the MTD of Dox itself) and produces objective responses in heavily pretreated RCW patients. It recommended that "future work should test thermally enhanced LTSL delivery in a less advanced patient population."

Given the protocol, these really were/are very promising results—so what's all about this *protocol*? Glad you asked, because here is another lesson to be learned.

23.6.2.1 RCW Protocol and ThermoDox® Pharmacokinetics

There have been several measures of the plasma clearance of ThermoDox®, its encapsulated Dox, free Dox, and even metabolites, over the years; for this discussion, it is important to review as many as we can find, some published in peer-reviewed papers, in investigator's brochures, or in presentations made by the Celsion-hired clinicians carrying out the studies at international meetings. Although the data are mainly from the Phase 1 liver trial, the graphs serve the purpose of presenting the pharmacokinetics that the protocol *should have* taken into account. Figure 23.12 shows a graph of human plasma clearance of 50 mg/m² for ThermoDox®, published in 2008 in the investigator's brochure for ThermoDox® + HT, for the treatment of solid tumors (Celsion 2008), and also shown in Poon and Borys's (chief clinician and Celsion's CMO, respectively) *Expert Opinion* paper (Poon and Borys 2009). It was for six liver cancer patients. In this study, patients were treated with a combination of RFA and ThermoDox®. A 30 minute infusion of ThermoDox® was given 15 minutes prior to ablation (Ravikumar et al. 2010). (*Note*: More details about ablation and RFA will be discussed next, but for now, let's focus on the pharmakokinetics [PK].) It is very similar to the one produced by Brad Wood's Phase 1 trial also in liver cancer using RFA (Wood et al. 2012) and in a preclinical animal study published by Banno et al. (2010). Assuming that all the analyses and assays are correct, the graph shows three main things*:

1. The peak in both ThermoDox®-encapsulated Dox and free (unentrapped) Dox occurs just after the end of the 30 minute infusion.
2. Free Dox represents 43% of the total Dox area under the curve.
3. The half-life of the ThermoDox®-entrapped "liposomal Dox" is only about 1 hour.

This means that as soon as the infusion is stopped, the formulation is already losing ground. As mentioned earlier and shown in Figure 23.3, it is well known that free Dox has a half-life itself after i.v. injection of only 2 minutes. Thus, the fact that the unencapsulated Dox tracks the encapsulated

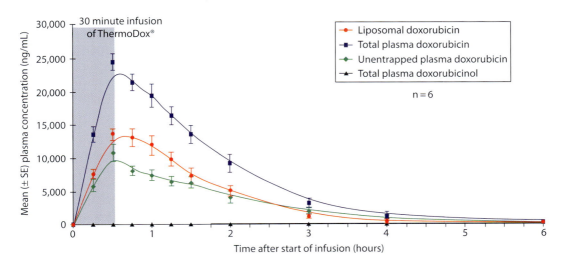

FIGURE 23.12 Human plasma clearance of 50 mg/m² ThermoDox® (Mean ± SE). (Courtesy of Celsion Corporation, Lawrenceville, NJ.)

---

* On one of my, too-infrequent, invited visits to Celsion, I noticed that this very same graph was also on a poster, I guess, proudly displayed on the wall of the CEO, who must have walked past it every day going in and out of his office.

Dox in ThermoDox® suggests that it is actually not as stable as the other more traditional liposomes and is, in fact, leaking from the formulation. So let's call what it really is, leaked Dox, from ThermoDox®. As is also well known and was, in fact, one of the major goals in its design, Doxil®, on the other hand, retains its drug very well (maybe too well), and the total Dox and encapsulated Dox (i.e., the drug-to-lipid ratio) of Doxil® tracks the liposomal half-life. Even though the stealth-like formulation of the ThermoDox® liposome (containing, as it does, PEG) should probably circulate as long as Doxil® does (half-life on the order of 24–48 hours) (Northfelt et al. 1996). ThermoDox® does not keep its drug inside for too long, as also confirmed by Ravikumar et al. (2010). Preclinical animal data from other labs (Banno et al. 2010) have also confirmed that the Dox can slowly leak out of the Dox-LTSL when administered to mice in vivo.

To summarize, while the LTSL lipid itself has a plasma half-life of approximately 8 hour, consistent with PEGylation, the encapsulated drug half-life is only 1.3 hours. But looking at the preclinical data produced later by Manzoor et al. (2012), it should be pointed out that Dox doesn't need to remain encapsulated for long periods; drug comes out in a few seconds when heated to the right 41°C–42°C mild HT temperature and fills the whole tumor. So, if this is the PK of ThermoDox®, a somewhat leaky liposome (compared to the more drug-retaining traditional ones) that nevertheless, if given the chance, can completely fill the tumor and all its cells with one of the most deadly drugs on the planet in 20 minutes, how did Celsion and Duke set up their protocol?

Here's how they describe ThermoDox® at Clincaltrials.gov, for the trial called "Phase 1/2 Study of ThermoDox® With Approved Hyperthermia in Treatment of Breast Cancer Recurrence at the Chest Wall (DIGNITY)" NIH identifier NCT00826085 (ClinicalTrials 2010):

> ThermoDox® is a 30-minute intravenous infusion followed by hyperthermia within 60 minutes of infusion completion.

WHAT?! "Within 60 minutes of infusion completion"? After all, we have published on the order of how to use our LTSL—"heat first, infuse ThermoDox®, keep heating, and kill cancer"—and told Celsion countless times, and this is the protocol they come up with? Some of you may think this next figure, Figure 23.13, is somewhat flippant. It is not; it is deadly serious. You can imagine, as in Figure 23.13, the medical oncology doctor coming in to see and sign off on the patient, who as just had her infusion of ThermoDox® in medical oncology. She is waiting for the orderly to come and wheel her down to the radiation oncology suite, where the radiation oncologists are waiting to heat the patient, saying to her and her family (because they are just adhering to the written Celsion protocol):

> Sure, have another cup o' tea, we got plenty of time

No! We *do not* have plenty of time. Look at the PK profile. If you heat the patient "within 60 minutes of infusion completion," you will miss almost all of the drug availability. It's not *just* a bottle or bag of red liquid, it's ThermoDox®, which has been designed to be given *only* when the tumor has been heated to 41°C–42°C. As to why the protocol was written this way in the first place—it seems it wasn't Celsion's fault. The anecdotal story was that the hospital administration would not let a medical oncology nurse work in a radiation oncology suite, because of "budgetary reasons." And so, ThermoDox® *had* to be administered first in medical oncology, and the patient wheeled all the way down to radiation oncology (which is always in the basement, with all their nasty radiation) to get the heating. Unbelievable. We did discuss the idea of Celsion hiring their own nurse, in order to make sure the correct protocol was strictly adhered to. And Celsion, correct me if I am wrong, but as far as I know, this never happened. In fairness, I understand that a company such as Celsion is bound by and must follow the rules of the particular hospital administration, including departmental boundaries and concomitant budgetary constraints. Furthermore, one could imagine that if Celsion were to pay for their own oncology nurse, it might compromise their impartiality with respect to their own clinical trial (Box 23.8).

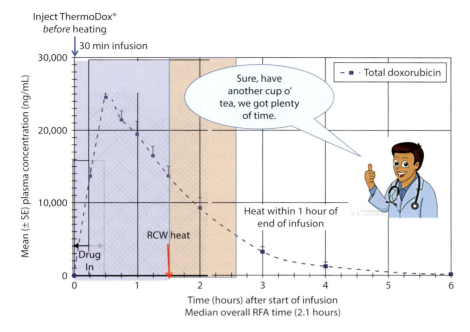

FIGURE 23.13   Plasma pharmacokinetics for the maximum-tolerated dose cohort (50 mg/m²) with superimposed median radio-frequency ablation (RFA) time, showing times that the area under the curve (AUC) was exposed to RFA. (Graph reprinted from *J. Vasc. Interv. Radiol.*, 23, Wood, B.J., Poon, R.T., and Lockin, J., A phase 1 study of heat-deployed liposomal doxorubicin during radiofrequency ablation for hepatic malignancies, 248–255. Copyright 2012, with permission from Elsevier.)

BOX 23.8   LESSON AS WE GO #8: BE ON THE LOOKOUT FOR PROTOCOL DECISIONS THAT ARE BASED ENTIRELY ON A "HOSPITAL ADMINISTRATION'S BUDGET SHEET"

When a company takes your invention and then tries to implement a protocol at one of its medical sites (like Duke), that requires two clinical departments, and therefore two different paid staff to cooperate; make sure that it is not the hospital administration's budget sheet that determines your protocol; in our case this meant, essentially, that the patient missed almost all of the ThermoDox® bioavailability.

23.6.3   PHASE 1 FOR LIVER CANCER

In 2006–2007, all this RCW was in the background when Celsion decided to focus almost exclusively on primary liver cancer (also known as hepatocellular carcinoma [HCC]). Clearly, this was a cancer that needed clinical treatment options. There is a much greater population with primary liver cancer than RCW, and so patient accrual could be achieved in a more reasonable amount of time. But also, with an eye on "business reasons," for patent lifetime, the clock is always ticking. Many patients die quickly from primary liver cancer, taking the inverse of "survival" statistics from Cancer UK for primary liver cancer in general; the rates are 70% dead at 1 year and 90% dead at 5 years. Focusing on such a high mortality rate, a pharmaceutical company can determine much more quickly if their newly developed commercial product is working, compared to say, the situation for prostate cancer, which is characterized by both a slow progression of the disease and a low mortality rate. Given these considerations, the reasoning behind the "business decisions" to abandon the prostate treatment (even though it was showing to be relatively safe and actually seemed to be working) becomes a little clearer (although obviously still not necessarily good, from the perspective of a prostate patient's health and well-being). And as scientists (who deal in the tens to hundreds of thousands of dollars that last us about 1–5 years in our small research lab budgets with lowly paid professors and even cheaper postdocs and graduate students), remember this all costs money, lots of money, in millions. Almost by definition, "business decisions" have to start and end with cost, risk, and benefit to the company. If the company goes under, no patient can benefit. And it wasn't as though Celsion had buckets of cash; they were budgeting and rebudgeting on a monthly basis, with quite a high burn rate given the cost of setting up and managing clinical trials. So while you want to keep them on their toes, do cut your licensee some, but not too much, slack.*

Now that Tardugno† was in charge (from January 3, 2007), the primary liver cancer Phase 1 trial was initiated in February 2007 and completed in good time by December 2009, with a primary completion date of October 2008 (final data collection date for primary outcome measure) (Celsion 2009). The results looked so good that it advanced to a Phase 3 without going through a Phase 2.

Primary liver cancer is one of the deadliest forms of cancer and ranks as the fifth most common solid tumor cancer. The incidence of primary liver cancer today is approximately 26,000 cases per year in the United States and approximately 40,000 cases per year in Europe. It is rapidly growing worldwide, at approximately 750,000 cases per year, 55% of which are in China, due to the high prevalence of hepatitis B and C in developing countries. The World Health Organization estimates that primary liver cancer may become the number one cancer worldwide, surpassing lung cancer, by 2020. The standard first-line treatment for liver cancer is surgical resection of the tumor; however, 90% of patients are ineligible for surgery. RFA has increasingly become the standard of care for nonresectable liver tumors, but the treatment cannot adequately ablate larger tumors. There are few nonsurgical therapeutic treatment options available, since radiation therapy and chemotherapy are largely ineffective in the treatment of primary liver cancer.

While single-agent Dox has been found to be effective, it has not become a standard treatment for primary liver cancer, due to its relatively high incidence of severe toxicity, including congestive heart failure and neutropenia. Hence, the new initiative was to attempt to increase primary liver cancer cure rate by combining two approaches: ThermoDox® + RFA. So a drug that is known to be effective can now be delivered directly to the tumor site in excess quantities. Couple this with a heating system that is widely used and can heat tumors to ablate their centers, while peripheral tumor heating could also release the drug. What could go wrong?

As shown in Figure 23.14, placement of an RFA electrode in a liver tumor can produce temperatures in the ablation zone upward of 60°C. Clinically, RFA induces in situ thermal coagulation necrosis, through the delivery of high-frequency alternating current to the tissues. However, RFA still has its limits (Choi et al. 2001). With currently available devices, the largest focus of necrosis that can be induced with a single application is approximately 4–5 cm in greatest diameter, and lesions that size have a high frequency of marginal recurrences. Thus, the diameter of suitable lesions must be less than 3–4 cm. Furthermore, tumors located near large vessels may not be effectively ablated because the heat-sink effect (i.e., blood flow cooling) of these vessels prevents ablation temperatures from being reached. Brad Wood and colleagues found that the "drug works no matter how you heat" (Locklin et al. 2008). Drug release was independent of the heat source; equivalent cytotoxicity could be obtained via heating using RFA or a warm water bath.

As we saw in studies that established its in vitro performance (Mills and Needham 2005), ThermoDox® releases drug at significant rates with a peak at 41°C–42°C. However, the ablation

---

* I can't believe I said that; I must be getting soft in my old age.
† And I really like Mike; his heart and mind are in the right place. He believes in our LTSL concept and always tries very hard to do the right thing. Thanks Mike. See http://celsion.com/docs/about_management for more details.

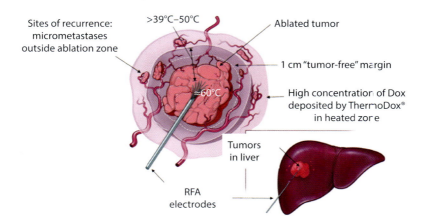

FIGURE 23.14 Radio-frequency ablation (RFA) of liver tumors + ThermoDox®. By itself, RFA does not target micrometastases outside the so-called ablation zone. However, the lower temperatures in the outer "thermal zone" can facilitate the release of doxorubicin from ThermoDox®, thereby expanding the treatment area. (Courtesy of Celsion Corporation, Lawrenceville, NJ.)

temperature at the tumor center is >55°C. As the ablation temperature drops off from the center out to the periphery, there has to be a zone (the "thermal zone") where temperatures are in the range 50°C–39°C (Figure 23.14). And so, if temperatures are in this range in the periphery, then, when ThermoDox® is infused into the bloodstream of an already warmed tumor, it will deposit high concentrations of Dox throughout this heated tumor interstitium and perhaps kill the dangerous micrometastases and the growing periphery of these deadly tumors.

Before this idea could be tested in an efficacy study, it had to be approved for safety. The ThermoDox® primary liver cancer Phase 1 study (Celsion 2009) was a multicenter, open-label, single-dose, dose-escalation study. The objective of this Phase 1 study was to determine the MTD of ThermoDox® when used in combination with RFA, in the treatment of primary and metastatic tumors of the liver (Celsion 2009). The protocol was that "patients with unresectable liver cancers underwent RFA with a 30-minute i.v. infusion of ThermoDox® starting 15 minutes before percutaneous or surgical RFA." (There's that "give-ThermoDox®-first, heat-later" problem, creeping in again.)

As reported by Poon et al. (2008), a total of 24 patients (9 with primary liver cancer and 15 with metastatic liver cancer were treated (3, 6, 6, 6, 3 patients at 20, 30, 40, 50, and 60 mg/m², respectively). Median tumor size was 3.7 cm (the range, however, was 1.7–6.5 cm). In total, 28 tumors were treated. The important toxicity findings were as follows:

- RFA + LTSL is safe and feasible.
- Neutropenia is an important toxicity.
- It has similar toxicity profile to Dox.

Even as a Phase 1, the trial did provide some interesting data on clinical efficacy. After treatment, 20 (83%) of the patients had no evidence of local tumor failure. Despite this only being a Phase 1 dose-escalation toxicity study, as shown in Figure 23.15, there was actually a dose–response relationship in terms of time to tumor progression (of 32, 53, 135, 185 days, respectively), giving ≈500% increase in PFS for the MTD (50 mg/m²) compared to the lowest starting dose.

Encouragingly, then, there appeared to be a preliminary dose–response relationship in terms of time to tumor progression as the study reached its MTD. A 2011 Future Medicine paper predicted that "Lyso-thermosensitive liposomal doxorubicin could be an adjuvant to increase the cure rate of radiofrequency ablation in liver cancer" (Poon and Borys 2011).

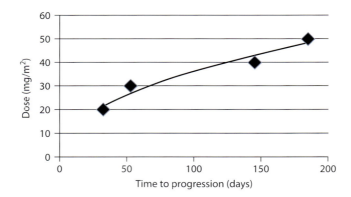

FIGURE 23.15 Results for Phase 1 dose-escalation study to determine maximally tolerated dose for ThermoDox® in conjunction with radio-frequency ablation, in the treatment of primary liver cancer. Dose and response (mg/m²) vs time for tumors to progress (days). (Courtesy of Celsion Corporation, Lawrenceville, NJ.)

23.6.4 PHASE 3 FOR LIVER CANCER HEAT STUDY (HEPATOCELLULAR CARCINOMA STUDY OF RFA AND THERMODOX®)

Uncertainty at so many levels clouded this trial.

Given the efficacy seen in the Phase 1 trial, clinical testing was moved rapidly to a Phase 3 trial, bypassing Phase 2. Phase 3 was a randomized, double-blind, placebo-controlled study of the efficacy and safety of ThermoDox® + RFA, compared to RFA alone, in the treatment of nonresectable primary liver cancer (Celsion 2012). Let's start with the results.

23.6.4.1 Phase 3 Trial Fails to Meet 33% PFS

Given all the encouraging preclinical data and the clinical responses seen in Phase 1 trials in prostate, RCW, and primary liver cancer itself, it was therefore somewhat surprising for all concerned when the top-line data, announced January 31, 2013, showed that

> ThermoDox® in combination with radiofrequency ablation (RFA) did not meet the primary endpoint of the Phase 3 HEAT Study in patients with hepatocellular carcinoma (HCC), also known as primary liver cancer. Specifically, Celsion has determined, after conferring with its independent Data Monitoring Committee (DMC) that the HEAT Study did not meet the goal of demonstrating persuasive evidence of clinical effectiveness that could form the basis for regulatory approval in the population chosen for study. The HEAT Study was designed to show a 33% improvement in PFS with 80% power and a p-value = 0.05. In the trial, ThermoDox® was well-tolerated with no unexpected serious adverse events. The HEAT Study was conducted under a Special Protocol Assessment agreed to with the U.S. Food and Drug Administration (FDA).

> **Celsion (2013)**

23.6.4.2 The Trial Itself

This Phase 3 trial engaged 75 different sites in 11 different countries and was the largest study ever conducted in the treatment of unresectable primary liver cancer (i.e., HCC) (Celsion 2014). It looked to treat the usually untreatable large 3–7 cm primary liver tumors. It was conducted under FDA Special Protocol Assessment, received FDA Fast Track Designation, and had been designated as a Priority Trial for liver cancer by the NIH. ThermoDox® was granted orphan drug designation in both the United States and Europe for this indication. The European Medicines Agency (EMA) had confirmed the HEAT Study was acceptable as a basis for submission of a marketing authorization application. In addition to meeting the U.S. FDA and European EMA enrollment objectives, the HEAT Study also enrolled a sufficient number of patients to support registration filings in China, South Korea, and Taiwan, the three other large and important markets for ThermoDox®.

The arms of the study were the following:

- Experimental 1. ThermoDox® (50 mg/m² in 5% dextrose solution). Start 30 minutes infusion about 15 minutes before RFA begins.
- Sham comparator 2. Sham (5% dextrose solution). Start 30 minutes infusion about 15 minutes before RFA begins.

The primary outcome measures were as follows:

PFS was measured from the date of randomization to the first date on which one of the following occurred: (1) local recurrence, (2) any new distant intrahepatic HCC tumor, (3) any new extrahepatic HCC tumor, and (4) death from any cause (time frame: 3 years). A secondary confirmatory endpoint was overall survival (OS).

The main inclusion criteria were as follows:

- Diagnosed HCC.
- No more than four HCC lesions, with at least one ≥3.0 cm and none >7.0 cm in maximum diameter, based on diagnosis at screening.
- If a subject has a large lesion (5.0–7.0 cm), any other lesions must be <5.0 cm.

By May 2012, the HEAT study reached its enrollment objective of 701 patients. The primary endpoint for the study was to measure just a 33% improvement in PFS with a P value of 0.05. A total of 380 events of progression were required to reach the planned final analysis of the study. In the late 2012, 380 PFS events occurred.

23.6.4.3 Phase 3 Trial Results

As mentioned and reviewed previously, expectations were high, especially when considering a comparison between the dose and response seen in the Phase 1 study and the criteria for this Phase 3. The PFS in the Phase 3 was required to show only a 33% improvement compared to RFA alone. This comparison is very favorable with the increase in PFS in the Phase 1 seen for the dose escalation (from 20 to 50 mg/m²) of almost 500% (the RFA-alone control was not done). A caveat though is that in the Phase 1 trial, median tumor size was 3.7 cm (range 1.7–6.5 cm), and so a handful of patients in this Phase 1 study could have had tumors that were critically smaller than those that made up the 701 subsequent cases for the Phase 3: "No more than four HCC lesions, with at least one ≥3.0 cm and none >7.0 cm in maximum diameter."

Here, in Figure 23.16, is the Kaplan Meier plot from the presentation made by Professor Riccardo Lencioni MD, at the European Conference on Interventional Oncology, entitled "New IO Approaches for HCC: An Update on Clinical Trials," for the PFS analysis (Lencioni 2013). From the same presentation by Lencioni, the OS analysis (secondary endpoint) showed similar results, with both plots of ThermoDox® + RFA treatment (Trt B) virtually coincident with the RFA control (Trt A).

We can all speculate as to why this trial failed at the first-line level (see the *Reverse Engineering* review paper [Needham 2013] for some suggestions). But basically, it would appear that the RFA-"recommended" protocol, as it was then, often failed to heat the margins, little or no drug was released there, and, in many cases, it was just like doing RFA with no drug. In fact, the control arm, just RFA, was 20% better than expected, because the protocol had the clinicians do a minimum of 12 minutes heating, when in usual practice they only do 6 minutes of intense RFA heating (that's all that is needed to burn out the center of the tumors). So the Celsion trial actually made RFA better. And worse still, this 20% better for the "control" arm ate into any advantage ThermoDox® + RFA might have had. But

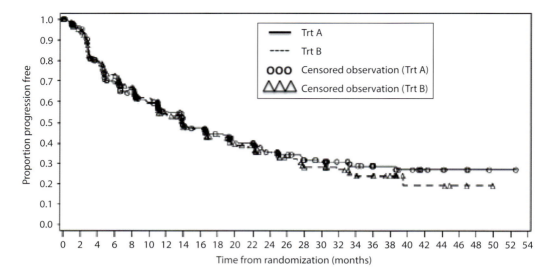

FIGURE 23.16 HEAT trial progression-free survival analysis. Trt A, Treatment A, i.e., RFA control; Trt B, Treatment B, i.e., ThermoDox® + RFA Treatment. (Courtesy of Celsion Corporation, Lawrenceville, NJ.)

there was a silver lining to this thundercloud. Some clinical sites used a different RFA device, and some sites got some of it right, at least enough to show positive results. For example, it appeared that the OS improvements correlated with studies done in South Korea where they used a Covidien cool-tip RFA probe. This contrasted markedly with little or no improvement in the Chinese sites, where they predominantly used the Angio StarBurst probe. Who knew these pieces of equipment could be so different? Figure 23.17 shows the PK plot again (i.e., Figure 23.12) but with the RFA protocol overlaid.

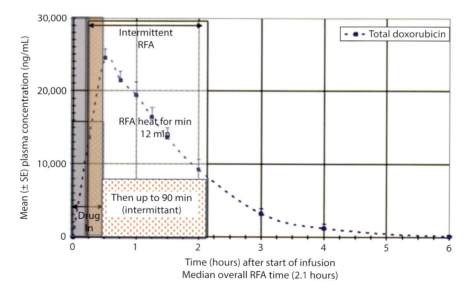

FIGURE 23.17 Radio-frequency ablation (RFA) protocol overlaid on the PK profile. Injection is started 15 minutes before the initially intense RFA heating. Following completion of the 30 minutes infusion of the clinical dose of ThermoDox® suspension, heating is then intermittent for the next 45 minutes, or up to 90 minutes in some cases. (Reprinted from *J. Vasc. Interv. Radiol.*, 23, Wood, B.J., Poon, R.T., and Lockin, J., A Phase 1 study of heat-deployed liposomal doxorubicin during radiofrequency ablation for hepatic malignancies, 248–255. Copyright 2012, with permission from Elsevier.)

With regard to dosimetry, note that a mean peak plasma concentration of Dox of 2500 ng/mL = 4.3 µM and in the five cell lines studied in the paper (Yarmolenko et al. 2010), the IC_{50} for Dox was in the range 129–168 nM. We therefore expect that as the liposomes do go through the warmed vasculature, they will be releasing all their drug and attaining even higher values in the tissue, perhaps as high as 50–100 µM, values in excess of the IC_{50} and approaching an IC_{90} or greater. Again, it is imperative to point out that ThermoDox® has a 1 hour half-life for retaining its encapsulated Dox, and so even a 15 minutes delay is missing some of the area under the curve (AUC) for the available drug.

23.6.4.4 Second-Line Data

Once the trial failed, again to their credit, and I am sure at the speedy insistence of Messrs. Tardugno and Borys, Celsion wasted no time in analyzing the second-line data, digging deeper into the actual patient-by-patient and RFA-by-RFA details. Second-line data are now showing what went wrong and how to fix it, as presented recently in the Celsion corporate presentation of April 2015 (Celsion 2015). In this presentation, Celsion is showing how expanding the treatment zone addresses RFA limitations when using ThermoDox® + RFA. This new protocol is to evaluate ThermoDox® in combination with standardized RFA (sRFA), called *ThermoDox® + sRFA 45*, after finding that heating for times greater than 45 minutes gives significant improvement. So now, although ThermoDox® is still infused i.v. ≈15 minutes prior to sRFA, heating is continued, albeit intermittently, over a 45 minute period. They are really doing their best given all the constraints of the instrumentation and infrastructure. As shown in the earlier schematic in Figure 23.14, with this expanded protocol, ThermoDox® is given a greater chance to concentrate in the "thermal zone," and so Dox is released in the "thermal zone" expanding the treatment area.

And Celsion is now publicly offering their own take on *lessons learned* (Celsion 2015), so let me simply feature that, unedited (Box 23.9).

BOX 23.9   LESSON AS WE GO #9: "REVIEW DATA FROM (THE MOST POSITIVE) 285 PATIENT SUBGROUP"

- RFA must be used within its engineered design limitations.
- 3 cm or greater lesions require multiple overlapping ablations.
- Longer RFA time (>45 minutes) results in better outcomes.
- Heating duration directly affects clinical outcome, by allowing for high local perfusion of drug at the tumor site.
- High tissue concentration of ThermoDox® prevents recurrence (supported by multivariate Cox regression analysis).
- PFS is not a reliable endpoint in HCC trials.

Evaluating all this with the FDA and their clinical partners and coming to these conclusions was, I am sure, a long and hard process. As is obvious from the previously mentioned, this post hoc analysis actually showed some very important and key aspects that are now moving the technology forward once again. If there is one thing Celsion is, it's resilient. So, let me finish off this section with some very positive results that have shaped the new OPTIMA trial (Celsion 2014).

23.6.4.5 Second-Line Data Show Very Positive Results

As reported by Celsion, again in their Celsion Corporate presentation of April 2015, second line data is showing extremely positive results, even though the protocol, from my perspective, was not optimal in terms of the tumor being heated first, and only intermittent heating after the first 12 minutes of intense RFA. It is not clear what temperatures are attained during the protracted intermittent heating period. In any event, in the second-line data from a significant number (285) of patients in the trials, it was not surprising (to us) for them to report that the duration of heat from the RFA procedure is a key factor in a successful clinical outcome when combined with ThermoDox® (Lencioni 2013). The data showed the following:

- ThermoDox® + RFA where the intermittent heating was carried out for *less than 45 minutes* (37%) was no better than the RFA control arm.
- In ThermoDox® + RFA where the intermittent heating was carried out for between *45 and 90 minutes* (40%), the *OS improved by 66%* compared to control RFA.
- In ThermoDox® + RFA where the intermittent heating was carried out for *longer than 90 minutes* (23%), the OS almost doubled compared to control RFA.
- When these results are combined, there was a 53% improvement in OS.

Thus, it is worth presenting the graph for ThermoDox® + RFA, with intermittent heating >45 minutes. As shown in Figure 23.18, the subgroup analysis of the HEAT study data in 285 patients with standardized RFA (>45 minutes) showed a much-improved response, which is now over 80 months and still ongoing.

The survival probability has not yet reached the 50% mark, and so if we interpolate from the graph at the 80% mark, we see that the improvement for the RFA + ThermoDox® is ≈13 months over RFA alone, and at 60% survival probability, it's ≈24 months. Also in these time-to-event curves (and in the one shown later for Doxil®), the HR is used as an expression of the hazard, or chance, of events occurring in the treatment arm, as a ratio of the hazard of the events occurring in the control arm. An HR of 1 means that there is no difference in survival between the two groups. Thus, the HR of 0.628 signifies that the TTP for the RFA arm is 0.628 of the average TTP in the treatment arm. Hence, the correlation with the simple interpolation of the data showing 26 vs 39 months at 80% survival (0.666) or 49 vs 73 months (0.671) for the 60% survival.

Celsion performed another comparison for these data, to evaluate the subgroup analysis (single lesion) of the HEAT study: 285 patients who received the standardized RFA for >45 minutes ± ThermoDox® vs the 167 patients who received RFA for <45 minutes ± ThermoDox®. Figure 23.19 shows the difference between the less-than-45-minute heating and the greater-than-45-minute heating, with ThermoDox® given i.v. in the bloodstream, is now ≈39 months.

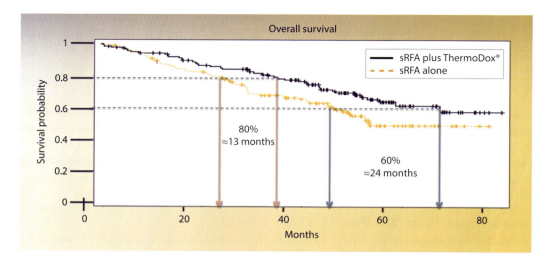

FIGURE 23.18 Subgroup analysis of HEAT study data for 285 patients with standardized radio-frequency ablation (>45 minutes), hazard ratio, 0.628 (95% CI 0.420–0.939); and P value, 0.02. (Courtesy of Celsion Corporation, Celsion Corporate Presentation, April 2015.)

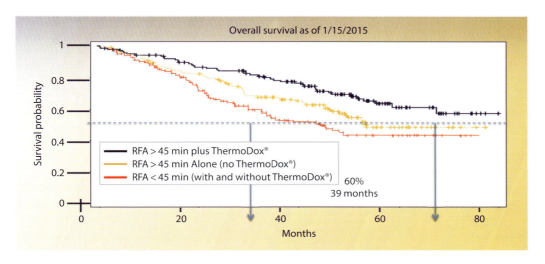

FIGURE 23.19 Subgroup analysis (single lesion) of HEAT study for 285 patients standardized RFA > 45 minutes ± ThermoDox® vs 167 patients RFA < 45 minutes. Presented are data for overall survival sRFA > 45 minutes ± ThermoDox®; hazard ratio, 0.628 (95% CI 0.420–0.939); and P value, 0.02. (Courtesy of Celsion Corporation, Celsion Corporate Presentation, April 2015.)

23.6.5 THERMODOX® VS. DOXIL®

Without Doxil®, there would be no ThermoDox®. In fact, it was the (what seemed to be) only modest successes of this formulation that motivated the invention of the LTSL: "I need something I can heat and it releases drug, damn it." Chezy, Alberto, Frank, Martin, Theresa, the late Dan Lasic, and Demitri, and the many, many more, too many to mention, are dedicated people that pioneered this whole field. In my opinion, they cannot get enough praise, thanks, and gratitude for all they achieved. And so this comparison is presented merely so that readers can put these two "nanomedicines" and later view their own nanomedicines, with some perspective.

Let's briefly compare the ThermoDox® results with those that got Doxil® its approval, i.e., the Doxil® vs topotecan study (Doxil® 2015), and published by Gordon et al. (2001). This was a Phase 3, randomized, multicenter study of Doxil® as a single agent, versus topotecan, in patients with recurrent epithelial ovarian cancer. As reported by Gordon and colleagues, "data show that PEGylated liposomal doxorubicin and topotecan, an established, efficacious agent for recurrent ovarian carcinoma, demonstrate equivalence in efficacy measures" (Gordon et al. 2001). Thus, Doxil® was no better than the then current treatment. The data, as a table and just one of the graphs from the website and in Gordon's paper, both in Figure 23.20, show overall clinical benefit assessed by the endpoints of overall response rates and OS* in the single-agent recurrent ovarian cancer study. The number of patients was 474, which represents the protocol-defined intent-to-treat population. The CR to Doxil® given to treat ovarian cancer was just 3.8%, just 9 patients out of 474, while the standard of care, topotecan, showed 4.7% CR as 11 patients of the population are treated. Compare this with the Phase 1/2 for ThermoDox® in RCW, where a local response rate of over 60% was reported in 14 of the 23 evaluable patients, with 5 CR and 9 PR, and some of these were doses that had not even reached the MTD. For example, the picture of the patient in Figure 23.11 is at only 30 mg/m², only 0.6 of the final MTD of 50 mg/m². And this was in a study where the investigators may have missed almost all the AUC.

---

* The time to progression (TTP) is also available for viewing at https://www.doxil.com/hcp/for-recurrent-ovarian-cancer/efficacy, accessed March 17, 2016. Here Doxil® median TTP was 4.1 months, range 1.3–106.9 weeks since first dose, and the hazard ratio was 0.955.

Response rate	Doxil® (n = 239)	Topotecan (n = 235)
Overall response, n (%)	47 (19.7)	40 (17.0)
Complete response, n (%)	9 (3.8)	11 (4.7)
Partial response, n (%)	38 (15.9)	29 (12.3)
Median duration of response (months)	6.9	5.9
Range (weeks)	5.0+, 93.1	7.0+, 93.9+

(a)

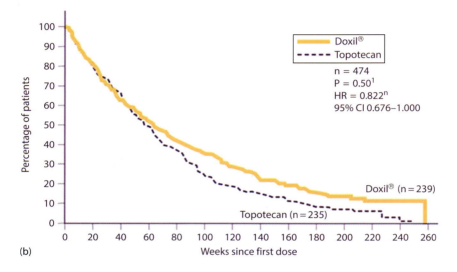

(b)

FIGURE 23.20 (a) The table shows the overall response and median duration of the response for the two arms of the study. (b) The graph shows that the Doxil® median overall survival was 14.4 months. The range was 1.7–258.3 weeks since the first dose. (Courtesy of Celsion Corporation, Celsion Corporate Presentation, April 2015.)

Comparing now to the Phase 3 liver trial, the Doxil® graph in Figure 23.20 shows a similar analysis to how ThermoDox® was analyzed in subgroup analysis, as OS.

Obviously, ovarian cancer is a very challenging disease, and the recurrent ovarian cancer population that Sequus chose to trial their Doxil® formulation in was in very dire straits. In the Doxil® arm, the median TTP was just 4.1 months (Doxil® 2015), and as listed in the table of Figure 23.20, the median duration of response was only 6.9 months, compared to 5.9 months for Doxil® and topotecan, respectively. And the HR is 0.822 for the 239 patients in the Doxil® treatment arm for the previous median duration of the response. But it was 0.955 for the Doxil® median TTP, which was 4.1 months, and so is as close to a value of 1 as makes no difference. This comparison could be even more striking when Celsion release the new data from the OPTIMA and RCW trials. Given all their lessons learned, Celsion have now stipulated "standardized" and not just "recommended" protocols and have (I am sure) done as much as they can to optimize the protocols given the constraints of a drug–device combination.

23.6.6 OUR RECOMMENDED PROTOCOL

No matter what heating system is being used, as shown in Figure 23.21, and repeated again here in the text (just for good measure), here is our recommended protocol:

* Establish steady-state temperature of 41°C–42°C *prior* to infusion of ThermoDox®.
* Maintain temperature of tumor at 41°C–42°C throughout the whole infusion.

Mean (± SE) plasma concentration (ng/mL)
30,000
20,000
10,000
0
Total doxorubicin
Establish steady-state temperature of 41°C–42°C *prior* to infusion
Maintain at 41°C–42°C throughout infusion
And if possible for at least 30–60 min after
Drug In
0
1
2
3
4
5
6
Preferred protocol
Time (h) after start of infusion
Median overall RFA time (2.1 h)

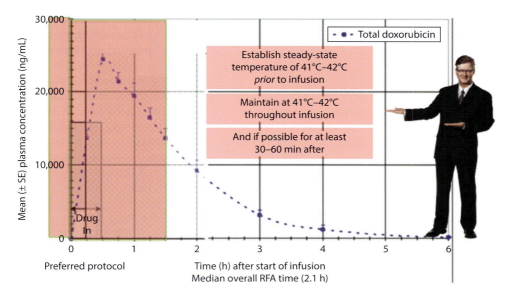

FIGURE 23.21 Our recommended protocol doxorubicin/lipid half-life is only 1 hour. *Do not miss it.* (Suit from online clothing store, my head from photo shoot by Malou Reedorf, photographer.)

- And if possible for at least 30–60 minutes after the infusion is terminated, catch the majority of the circulating ThermoDox®, and release the drug in the warmed tumor.

For short, we call it "heat first, administer ThermoDox®, keep heating for at least 1 hour, kill tumor."

23.6.7 New and Ongoing Human Clinical Trials for ThermoDox®

Celsion has made a huge commitment to a full clinical program that is underway for ThermoDox®. As shown in Figure 23.22 (from their web page Celsion "Pipeline" 2015), in addition to the trials

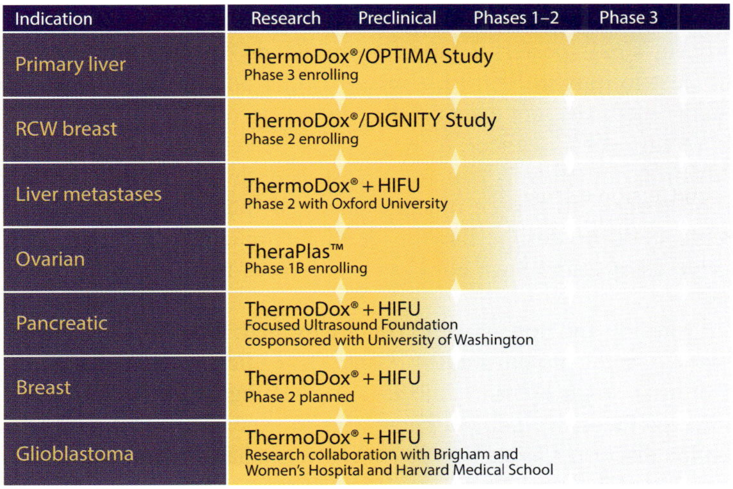

Indication
Research
Preclinical
Phases 1–2
Phase 3
Primary liver
ThermoDox®/OPTIMA Study
Phase 3 enrolling
RCW breast
ThermoDox®/DIGNITY Study
Phase 2 enrolling
Liver metastases
ThermoDox® + HIFU
Phase 2 with Oxford University
Ovarian
TheraPlas™
Phase 1B enrolling
Pancreatic
ThermoDox® + HIFU
Focused Ultrasound Foundation
cosponsored with University of Washington
Breast
ThermoDox® + HIFU
Phase 2 planned
Glioblastoma
ThermoDox® + HIFU
Research collaboration with Brigham and
Women's Hospital and Harvard Medical School

FIGURE 23.22 ThermoDox® clinical programs at Celsion Corporation (July 2015).

reviewed earlier (primary liver, OPTIMA; RCW, DIGNITY), new research and preclinical development has started in metastatic liver cancer, ovarian cancer, pancreatic cancer, breast cancer, and glioblastoma. One of the most exciting new inclusions here is the adaptation of HiFu as the source of targeted mild HT.

23.7 LESSONS FROM THERMODOX®

23.7.1 THE FASTEST WAY TO DO RESEARCH IS LARGELY INCOMPATIBLE WITH ITS DEVELOPMENT AND COMMERCIALIZATION

Before we get to specifically what I, personally, learned from ThermoDox®, let me give some general advice about the process of research and development, sometimes called big R and little D and little R and big D, depending on who is doing the R and who is doing the D.

There's an old saying that "the fastest way to herd cattle is slowly." In my view, this also applies to research. The fastest way to do research is … slooooooowly, and surely. We spend so much time making sure we get it right by reviewing our current data, identifying the key observation, posing the right questions founded in scientific depth and rigor, and composing scientific hypotheses to be tested. Based on this, you try to carve out enough concentrated time away from teaching class, your ongoing research, one- and two-day trips to give invited seminars, conducting administration duties, and not to mention the social and domestic life that you do not have, in order to write the proposed *Research Plan*.

And this is a plan with full and referenced descriptions of its overall goal, specific aims, background and significance, preliminary studies, research design, methods, timetable, the budget, budget justification, resources and biographical sketches of key personnel, responding to the first rejection and resubmitting the proposal that has a 1-in-14 chance of being funded in the top 7 percentile; with 28 grants in the study section, yours has to be one of the top two. Say it does get selected for funding, and you are then busied by advertising, interviewing, and recruiting the best students and postdocs; training them; carrying out all the experimentation, modeling, and/or simulations; obtaining data in at least triplicate; analyzing and interpreting the data, writing up the papers; and responding to reviewers on the way to eventual publication, a publication that has to last forever, to be true. In philosophy, "truth by consensus" is the process of taking statements to be true simply because people generally agree upon them. In science, truth is the process of checking everything and getting the same result; it's more like "truth by proof."

The *truth* is something that we all agree upon. Obtaining that truth is a long and detailed process. As I say, in small posters I tape up around my lab and student offices, "CHECK EVEYRTHING," just to remind them that mistakes can pop up where you least expect them, and we cannot make mistakes, not even typos. And if any of this research actually generates a useful idea, like a treatment for cancer, then we better be very sure we have a deep understanding of the laws, theories, and models that are consistent with its functions, the component design and mechanism of action, our choice of materials, and their CSP relationships, optimized to meet the requirements of the design; have made the formulation and tested its performance in in vitro and in vivo assays; and have convinced ourselves that what we think is happening is really happening.

You have done your job and done it well. You are excited; your family is proud of you, and all your hard work. Based on the animal studies, where 11/11 mice with implanted tumors were cured out to 60 days after just a single bolus i.v. injection and heating their tumor for just 1 hour, you have high expectations, even though you do not know what to expect. The University owns all this. You submit your invention disclosure as soon as you are sure you have something of utility, and the OLV logs it in with a number and date and evaluates it for potential patent application, before you submit for publication review. Okay, where is the disconnect?

It turns out that this academic, painstaking, yet very exciting journey—your life, is at almost complete odds with the development and commercialization of your invention and your collaborators' proof of concept. It turns out that for all the reasons of pressure to return investment, a stock price that is under attack from every financial blogger, and a series of revolving-door CEOs who come and go without making a positive impact on your invention, the process of development and commercialization of your invention wants it done, *quickly.*

Thus, following Frank and George from earlier, in today's commercialization, we might even say, "Those who do not listen and act on the advice of the *inventor* are condemned to fail!" So, for mistakes to still be happening in today's commercialization would seem to be avoidable, but it would take some effort to assimilate all the lessons learned from the past that is actually made so much easier (and voluminous) by efficient search engines. It would take a licensee who can strike the right balance between the enormous tasks of clinical implementation, regulatory issues, hospital administrations, and return on investment, to bravely and smartly adhere to evidenced-based practice. And since the university even managed to obtain a licensee, you might want to check the terms of the license and make sure with monthly e-mails so that you do not forget to keep track of the license until it is all too late. Having said that and having seen Celsion go through this whole process, with, I must say, considerable effort, generated financing, and some real success, researchers need to respect the entrepreneurial process that follows their research-demonstrated feasibility. To the entrepreneurs, I would say, even though you licensed the invention and feel entitled, you would do well to listen to the inventors.

23.7.2 FINALLY! LESSONS LEARNED FROM THERMODOX®

As I laid out in no uncertain terms in my presentation at the Peck Symposium (Needham 2015), here is a list of lessons I have learned and would offer, now a little more gently, as some of the area's researchers, inventors, and, yes, even university offices of licenses and ventures and the companies they deal with might consider and be highly aware of (Box 23.10).

BOX 23.10   FINAL LESSONS FROM "THERMODOX®": YOU CAN'T LEAVE EVERYTHING UP TO PEOPLE WHOSE JOB IS TO DO IT RIGHT (YES, THIS IS THE SOFTER, GENTLER VERSION)

Your university administration

- Keep track of the changing rules by your university administration
- They may change the patent rules for the inventor's share as they see fit.

Your Office of Licensing and Ventures (OLV)

- Have an ongoing and functional relationship with your OLV.
- They may not know how to create a license agreement, and if they do, they may not manage it with the expectations and enthusiasm that you would.
- They are probably very busy and have to make judgment calls as how to best use their limited resources, and you might not be top of their list

Your license agreement

- Review the license agreement.
- Make sure it is compatible with what the company is capable of actually delivering.

- If they license your technology for all drugs, all diseases and physiological conditions, for the lifetime of all the patents, for a pittance, make sure that they do due diligence on the value of competing technology (and adjust the "pittance" accordingly).
- Encourage them to actually keep track of the license agreement and all its milestones, even though it's not their highest priority.

Meeting the milestones

- If (and I am sorry to add), or when, the company fails to meet the milestones, make sure your OLV renegotiates the license in an appropriate way.
- For example, do not let them give away all your sublicenses, and let the company that could not develop it keep the technology as is, only to find out that the company did manage to generate a sublicensee for $20 M and you, as the inventor, and your OLV and the University get nothing.

Your (inventor's) relationship with the company

- Develop an ongoing and mutually respected relationship with the company that your, and the University's, technology is licensed to.
- If they will communicate with you, listen to the company, and their issues, and the problems they face.
- Do as much as you can, up front, to make sure they listen to your expert advice with regard to what the technology is and how it should be used, and also listen to them.
- For example, "heat first, administer ThermoDox®, keep heating for at least 1 hour, and kill tumor."
- No matter how good the Phase 1 trial results appear to be, encourage them (and encourage your OLV to include it in the license agreement) to carry out a Phase 2 study before a Phase 3.

The distribution company (there is no way to sugarcoat this one)

- Encourage your university development office to show integrity and manage the stock themselves.
- Your university will select a distribution company to manage your stock at arm's length, for fear of being accused of conflicts of interest. Do not trust the distribution company that your university has "selected" by way of the cheapest tender (and therefore likely to be one of the least competent outfits).
- The distribution company will invariably fail to do due diligence and will sell your (and your university's) stock at the lowest value, without warning or consulting you (because of the arm's length deal), just so that they can get their fee. You and your university could lose millions of dollars.

23.8   NEW DIRECTIONS: PUT THE DRUG IN THE CANCER'S FOOD

During the summer of 2011, I was forced to take time off to recover from a major train accident. I started to think about a problem that I might work on, which had been suggested to me earlier that year, by a colleague, Neil Spector. Neil was one of the lead clinicians who had carried out a clinical trial on a new hydrophobic, reversible inhibitor of ErbB1 and ErbB2 tyrosine kinases, *lapatinib*, invented by Stephen Frey when he was at GSK. Neil arranged a meeting with me and asked if I could help to reformulate lapatinib. I started to study the problem, and by early September, the result was a 50-page white paper I called "The Formulation of Hydrophobic Anti-Cancer Drugs Part I. Exploring Mechanisms of Endogenous Uptake of Drugs by Cancer Cells - that have the Potential to Deliver them to within Ångströms of their Target Molecule." I never sought to publish it but used it to form the basis for a part of my Niels Bohr Professorship Application, which was successful and brought me to Denmark (Box 23.11).

BOX 23.11 NEW LESSON AS WE GO #11: CARVE OUT SOME TIME TO THINK!

We are all so busy that often we are not able to carve out some time to actually *think* (deeply). Give yourself time to search the literature, build your ideas, check *eveyrthing* (yes, typo on purpose). You might not want to go to the lengths I went to and get yourself in a train wreck (not recommended), but maybe just take that vacation you are owed and use it on something useful. Take the time to get back to where we started with Frank Szoka, read some literature, understand it, sleep on it, and give yourself time to have an original idea. I'm just sayin'.

Without going into too much detail, the overall idea was to see if we could utilize the nanoparticle CSP of the low-density lipoprotein (LDL) and its receptor-based natural targeting, as an endogenous mechanism to achieve greater uptake of anticancer drugs by tumor cells. Basically, it was to see if we could simply "put the drug in the cancer's food." By reverse engineering the LDL, we hope to make pure-drug nanoparticles that will bind to, and be taken in, by cancer cells. From a nanoparticle design's point of view, rather than fill the particle up with cholesteryl ester and try to dissolve or partition a drug into that matrix, the thinking is that by making pure-drug nanoparticles, we can have 2000–3000 drug molecules per particle.

The choice of the LDL receptor (LDLR) or other receptors like folate (and even passive endocytic pinocytosis) as the cell entry point is motivated by several observations already in the literature: rapidly growing cancer cells have high numbers of these receptors; and numerous malignancies are known to overexpress LDLR, including brain, colon, prostate, adrenal, breast, lung, leukemias, and kidney tumors, where LDLR can be some four hundred times greater on cancer cells than on normal cells (Ho et al. 1978). As a result, cancers are known to take in more LDL than normal cells, and in patients with cancer, their LDL count is even known to go down. An abundance of LDLR is also a prognostic indicator of metastatic potential (Rudling et al. 1986), and a propensity to store cholesteryl ester is a sign of the aggressiveness of a patient's cancer. Thus, with 48 pages of text, 44 figures, and 221 references, this white paper sought to review the LDL—nature's own hydrophobic delivery system—and establish a basis for new research and development of endogenous-inspired advanced drug delivery. I tried to answer basic questions such as the following:

- "What makes the LDL so effective at reaching its normal targets (adrenals, muscle, liver) and also cancer cells?"
- "How are its contents processed?"
- "Could it be that lapatinib, taken orally, enters this chylomicron–LDL pathway, and is it this endogenous 'cholesteryl ester delivery system' that delivers it to the tumor cells anyway?"

It was then not such a leap to ask the following:

- Rather than lose 99.3% of the administered drug via an oral tablet, can we go i.v. and use the cholesteryl ester/cholesterol pathway to deliver similarly hydrophobic anticancer drugs to tumors?

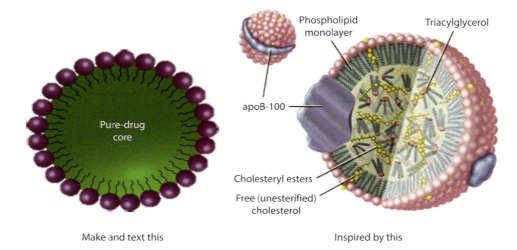

Make and test this   Inspired by this

FIGURE 23.23   Could we make and test a pure-drug, lipid-coated nanoparticle that was inspired by reverse engineering the natural low-density lipoprotein? (Courtesy of Encyclopedia Britannica, Inc., 2007.)

- Would such delivery bring the drugs into the cell in such a way that they could target growth, metabolism, and survival pathways?
- Could the same pathway also be used for diagnostics, especially PET imaging?

This led to the idea, shown schematically in Figure 23.23: "Could we make and test a pure-drug, lipid-coated nanoparticle that was inspired by reverse engineering the natural LDL?"

23.9   TRANSLATING DRUG DELIVERY WITHOUT PROFIT: OPEN-SOURCE PHARMACEUTICS?

So, if we look at what we learned, the question I have to ask is, would I do it all again? And the answer is a resounding "No." At least not in the way I just reported and certainly not within this academic–corporate–hospital complex. I cannot and will not, again, be part of a commercialization process that has to move from an academic or research setting to an eventual product for sale with profit in order to be available, requiring that it successfully crosses, what former NIH Director Elias Zerhouni has dubbed, the "valley of death." An interesting perspective of this is given by Declan Butler (Butler 2008). In it, he discusses the problem from the NIH perspective, identifying the "chasm that has opened up between biomedical researchers and the patients who need their discoveries," and considers "how the ground shifted and whether the US National Institutes of Health can bridge the gap." But he does not mention the word "profit," and so is probably missing one of the main problems in how "the ecosystems of basic and clinical research have diverged." I personally see this as a major driver—whether direct profits or the all-important "stock price." I will not put my ideas, hopes, and dreams into the hands of others, especially where profits and budgets dominate medical protocols, and "business decisions" made are not necessarily in the best interests of my invention and even the patient's treatment. So what is the alternative? I asked myself:

What would happen if we just took profit completely out of the equation?

And others are certainly asking the same question, as the concept of "Nonprofits and the Valley of Death in Drug Discovery" is gaining some traction (Moos 2010). Moos states, quite categorically, "When we get mired in the economics or politics of the pharmaceutical industry, we can lose sight of some essential facts—the development of effective drugs saves lives."

And here I am not just saying, "Pharma, open up on failed drugs," like they did in 2010, where two large pharmaceutical companies participated in the unprecedented deposition of hundreds of thousands of potential leads for new malaria drugs into an open-source database (Strauss 2010). As explained by Strauss, in an article entitled, "Pharma embraces open source models":

> The proposition is for companies to de-emphasize intellectual property rights at least on early biology and be more open about sharing negative results so that knowledge advances faster in drug discovery research

I am not sure what to make of this except, in sharing negative results. Maybe they were fishing for some much-needed help, but for perhaps less profitable drugs that are targeting third-world diseases, rather than, for example, cardiovascular or cancer—but that's just me being a little cautious in not totally trusting everything I read.

But the focus of all this is still on new drug entities. I am more concerned with effective formulations, and especially reformulations, of drugs we already have, but that have been poorly formulated in the first place.

> What if we could develop and test the feasibility of this nonprofit idea in drug delivery and nanomedicines?
>
> Could we scale up the nanoparticle production, and then offer it AT COST, NO PROFIT, and under a public license?

Thus, rather, I am saying, make the *formulations* open source so that more cancer patients actually get more of the drugs we already have (or develop in the future) to their cancers, in a much more efficient way than simply putting them in a cellulose tablet and settling for treating cancer as a chronic disease. There are strategies to help increase bioavailability of hydrophobic drugs taken orally— like a fat chaser. Just ask Burris and Spector and company (Burris et al. 2005) and their analysis of the lapatinib trial, where women who ate fatty meals seemed to have an increased response. This has since been investigated further and quantified, finding that "the AUC_{0-24} increased following Lapatinib administration 1 hour after a low-fat meal by 1.80-fold and after a high-fat meal by 2.61-fold" (Devriese et al. 2014).

Before we get there though, i.v. formulations are required that actually stand a chance of delivering the appropriate dose (as we are investigating via more endogenous-inspired formulations). But if all we nanomedicine people do is create the next "nanowidget" company, hoping to make it big and then sue each other for having similar products (like what happened in the liposome field in the 1990s), it's not going to work. There are lots of liposomologists around who have all been there, seen it, and done it, and the only people to profit are the lawyers and some of the independent expert witnesses (thank you very much).

I guess we will see, because I am determined to at least try it, and Denmark may just be the right place to do it. With 338 cancers per 100,000 people (WCRF International 2015), it is the so-called cancer capital of the world. To be fair, France is only 14 behind in second place, and the United States at 318 per 100,000 is sixth. Everyone in the Western world and Australia is in the low 300s or high 200s. They do smoke like factory chimneys here and eat a lot of red meat (and two vegetables), but is it the ubiquitous and accessible health care and available screening that makes Denmark's numbers so high? Maybe Denmark is just the "cancer *detection* capital of the world." If we can prove feasibility in my remaining 2.5 years on the job here as the Niels Bohr professor with my group and close collaborators, maybe we can make Denmark the "cancer *treatment* capital of the world." I wouldn't put it past us.

And to the young and restless researchers out there, join us. What have you got to lose? As you take your precious invention forward, with all the enthusiasm in the world, only to come up against a whole series of other agendas (some good for you, some not so good) outside your lab door, it is unlikely that you will make any significant money from your inventions anyway. So you might as well do some good. Join us.

Peace.

REFERENCES

Abraham, S.A., D.N. Waterhouse, L.D. Mayer et al. 2005. The liposomal formulation of doxorubicin. *Methods in Enzymology* 391:71–97.

Adler-Moore, J. and S.-M. Chiang. 1997. Liposomal cyclosporin formulations as agents for immuno-suppression and multiple drug resistant indications. Patent no. US5656287 A. UPTO, NeXstar Pharmaceuticals, Inc.

Adler-Moore, J.P. and W.A. Ernst. 1997. Pharmaceutical formulation and process. Patent no. US5660856 A UPTO, NeXstar Pharmaceuticals, Inc.

Allen, T.M. 1989. Liposomes with enhanced circulation time. Liposome Technology, Inc. US Patent 4837028 A. Publication Date: June 6, 1989.

Allen, T.M. and A. Gabizon. 1990. Liposomes with enhanced circulation time. Liposome Technology, Inc. US Patent 4920016 A. Publication Date: April 24, 1990.

Allen, T.M., P. Williamson, and R.A. Schlegel. 1988. Phosphatidylserine as a determinant of reticuloendothe-lial recognition of liposome models of the erythrocyte surface. *Proceedings of the National Academy of Sciences of the United States of America* 85(21):8067–8071.

Allison, A.C. and G. Gregoriadis. 1974. Liposomes as immunological adjuvants. *Nature* 252(5480):252.

Allison, A.C. and G. Gregoriadis. 1977. Immunological preparations. US Patent 4053585 A. Publication Date: October 11, 1977.

Anyarambhatla, G.R. and D. Needham. 1999. Enhancement of the phase transition permeability of DPPC liposomes by incorporation of MPPC: A new temperature-sensitive liposome for use with mild hyper-thermia. *Journal of Liposome Research* 9(4):491–506.

Bagatolli, L.A. and D. Needham. 2014. Quantitative optical microscopy and micromanipulation studies on the lipid bilayer membranes of giant unilamellar vesicles. *Chemistry and Physics of Lipids* 181:99–120.

Bally, M.B., N.L. Bowman, P.R. Cullis et al. 1997. Liposomal compositions for enhanced retention of bioac-tive agents. Inex Pharmaceuticals Corporation, University of British Columbia. US Patent 5595756 A. Publication Date: January 21, 1997.

Bangham, A.D. and R.W. Horne. 1964. Negative staining of phospholipids and their structural modifica-tion by surface-active agents as observed in the electron microscope. *Journal of Molecular Biology* 8:660–668.

Banno, B., L.M. Ickenstein, G.N.C. Chiu et al. 2010. The functional roles of poly(ethylene glycol)-lipid and lysolipid in the drug retention and release from lysolipid-containing thermosensitive liposomes in vitro and in vivo. *Journal of Pharmaceutical Sciences* 99(5):2295–2308.

Barenholz, Y. 2012. Dxoil®—The first FDA-approved nano-drug: Lessons learned. *Journal of Controlled Release* 160(2):117–134.

Barenholz, Y. and G. Haran. 1994. Method of amphiphatic drug loading in liposomes by ammonium ion gradient. Yissum Research Development Company of the Hebrew University of Jerusalem. US Patent 5316771 A. Publication Date: May 31, 1994.

Beauchamp, C.O., S.L. Gonias, D.P. Menapace et al. 1983. A new procedure for the synthesis of polyeth-ylene glycol-protein adducts; Effects on function, receptor recognition, and clearance of superoxide dismutase, lactoferrin, and α2-macroglobulin. *Analytical Biochemistry* 131(1):25–33.

Bennett, V. 1982. The molecular basis for membrane—Cytoskeleton association in human erythrocytes. *Journal of Cellular Biochemistry* 18:49–65.

Bingham, E.C. 1922. *Fluidity and Plasticity.* New York: McGraw-Hill.

Bloom, M., E. Evans, and O.G. Mouritsen. 1991. Physical properties of the fluid lipid bilayer component of cell membranes: A perspective. *Quarterly Reviews of Biophysics* 24:293–397.

Boman, N.L., D. Masin, L.D. Mayer et al. 1994. Liposomal vincristine which exhibits increased drug retention and increased circulation longevity cures mice bearing P388 tumors. *Cancer Research* 54(11):2830–2833.

Boman, N.L., L.D. Mayer, and P.R. Cullis. 1993. Optimization of the retention properties of vincristine in liposomal systems. *Biochimica et Biophysica Acta* 1152(2):253–258.

Burris, H.A., H.I. Hurwitz, E.C. Dees et al. 2005. Phase 1 safety, pharmacokinetics, and clinical activity study of Lapatinib (GW572016), a reversible dual inhibitor of epidermal growth factor receptor tyro-sine kinases, in heavily pretreated patients with metastatic carcinomas. *Journal of Clinical Oncology* 23(23):5305–5313.

Butler, D. 2008. Translational research: Crossing the valley of death. *Nature* 453:840–842.

Celsion. 2001. A randomized pivotal clinical trial in breast cancer patients of pre-operative focal microwave thermotherapy treatment for early-stage breast disease in intact Breast, NCT00036998.

Celsion. 2003–2009. *A Dose Escalation, Pharmacokinetics, and Safety Study of Doxorubicin Encapsulated in Temperature Sensitive Liposomes Released Through Microwave Therapy in the Treatment of Prostate Cancer*, NCT00061867.

Celsion. 2008. ThermoDox®, lyso-thermosensitive liposomal doxorubicin: Investigator's brochure for ThermoDox® in combination with focused heat for the treatment of solid tumors.

Celsion. 2009a. Completed. A phase I dose escalation tolerability study of ThermoDox™ (Thermally Sensitive Liposomal Doxorubicin) in combination with radiofrequency ablation (RFA) of primary and metastatic tumors of the liver (NCT00441376).

Celsion. 2009b. A dose escalation, pharmacokinetics, and safety study of doxorubicin encapsulated in temperature sensitive liposomes released through microwave therapy in the treatment of prostate cancer. NCT00061867.

Celsion. 2012. Phase 1/2 Study of ThermoDox® with approved hyperthermia in treatment of breast cancer recurrence at the chest wall (DIGNITY). NCT00826085.

Celsion. 2014a. A phase III, randomized, double-blinded, dummy-controlled study of the efficacy and safety of ThermoDox® (thermally sensitive liposomal doxorubicin) in combination with radiofrequency ablation (RFA) compared to RFA-alone in the treatment of non-resectable hepatocellular carcinoma. http://www.celsion.com, accessed March 17, 2016.

Celsion. 2014b. Study of ThermoDox® with standardized radiofrequency ablation (RFA) for treatment of hepatocellular carcinoma (HCC) (OPTIMA), FDA 2015.

Celsion. 2015. Pipeline. http://celsion.com/pages/pipeline, accessed March 17, 2016.

Celsion Corporation. 2012. A phase III, randomized, double-blinded, dummy-controlled study of the efficacy and safety of ThermoDox® (Thermally Sensitive Liposomal Doxorubicin) in combination with radiofrequency ablation (RFA) compared to RFA-Alone in the treatment of non-resectable hepatocellular carcinoma. NCT00617981.

Celsion Corporation. 2013. Celsion announces results of phase III HEAT Study of ThermoDox® in primary liver cancer. http://investor.celsion.com/releasedetail.cfm?ReleaseID=737033, accessed March 17, 2016.

Celsion Corporation. 2015. Celsion Corporate Presentation. April 2015.

Chang, H.-I. and M.-K. Yeh. 2012. Clinical development of liposome-based drugs: Formulation, characterization, and therapeutic efficacy. *International Journal of Nanomedicine* 7:49–60.

Chen, Q., A. Krol, and A. Wright. 2008. Tumor microvascular permeability is a key determinant for antivascular effects of doxorubicin encapsulated in a temperature sensitive liposome. *International Journal of Hyperthermia* 24(6):475–482.

Chen, Q., S. Tong, M.W. Dewherst et al. 2004. Targeting tumor microvessels using doxorubicin encapsulated in a novel thermosensitive liposome. *Molecular Cancer Therapeutics* 3(10):1311–1317.

Choi, H., E.M. Loyer, R.A. DuBrow et al. 2001. Radio-frequency ablation of liver tumors: Assessment of therapeutic response and complications. *RadioGraphics* 21:S41–S54.

ClinicalTrials.gov. 2010. Phase 1/2 Study of ThermoDox with approved hyperthermia in treatment of breast cancer recurrence at the chest wall (DIGNITY). clinicaltrials.gov.

Devriese, L.A., K.M. Koch, M. Mergui-Roelvink et al. 2014. Effects of low-fat and high-fat meals on steady-state pharmacokinetics of lapatinib in patients with advanced solid tumours. *Investigational New Drugs* 32(3):481–488.

Dewhirst, M.W. and D.A. Sim. 1984. The utility of thermal dose as a predictor of tumor and normal tissue responses to combined radiation and hyperthermia. *Cancer Research (Suppl)* 44:4772s–4780s.

Doxil. 2015. Doxil: For recurrent ovarian cancer. https://www.doxil.com/hcp/for-recurrent-ovarian-cancer/efficacy, accessed March 17, 2016.

Dromi, S., V. Frenkel, A. Luk et al. 2007. Pulsed-high intensity focused ultrasound and low temperature-sensitive liposomes for enhanced targeted drug delivery and antitumor effect. *Clinical Cancer Research* 13(9):2722–2727.

Eley, D.D. and D. Needham. 1984. Hybridization and Catalysis by Lanthanide Films. *Proceedings of the Royal Society A (Mathematical, Physical and Engineering Sciences)* 393:257–276.

Evans, E.A. 1973a. A new material concept for the red cell membrane. *Biophysical Journal* 13:926.

Evans, E.A. 1973b. New membrane concept applied to the analysis of fluid shear- and micropipette-deformed red blood cells. *Biophysical Journal* 13(9):941–954.

Evans, E. and D. Needham. 1987a. physical properties of surfactant bilayer-membranes—Thermal transitions, elasticity, rigidity, cohesion, and colloidal interactions. *Journal of Physical Chemistry* 91(16):4219–4228.

Evans, E. and D. Needham. 1988. Attraction between lipid bilayer membranes in concentrated solutions of nonabsorbing polymers: Comparison of mean-field theory with measurements of adhesion energy. *Macromolecules* 21:1822–1831.

Evans, E. and R. Kwok. 1982. Mechanical calorimetry of large dimyristoylphosphatidylcholine vesicles in the phase transition region. *Biochemistry* 21(20):4874–4879.

Evans, E. and R.M. Hochmuth. 1977. A solid–liquid composite model of the red blood cell membrane. *Journal of Membrane Biology* 30:351–362.

Evans, E. and R.M. Hochmuth. 1978. Mechanochemical properties of membranes. In *Current Topics in Membranes and Transport*, eds. F. Bonner and A. Kleinzeller. New York: Academic Press.

Evans, E., W. Rawicz, and B.A. Smith. 2013. Back to the future: Mechanics and thermodynamics of lipid biomembranes. *Faraday Discussion (Chemical Society)* 161:591–611.

Evans, E.A. 1973a. A new material concept for the red cell membrane. *Biophysical Journal* 13 (9):926–940.

Evans, E.A. 1973b. New membrane concept applied to the analysis of fluid shear- and micropipette-deformed red blood cells. *Biophysical Journal* 13(9):941–954.

Evans, E.A. and D. Needham. 1987b. Physical properties of surfactant bilayer membranes: Thermal transitions, elasticity, rigidity, cohesion, and colloidal interactions. *Journal of Physical Chemistry* 91(16):4219–4228.

Evans, E.A. and R. Skalak. 1980. *Mechanics and Thermodynamics of Biomembranes*. Boca Raton, FL: CRC Press.

Evans, E.A. and R.M. Hochmuth. 1976. Membrane viscoelasticity. *Biophysical Journal* 16(1):1–11.

Evans, E.A., R. Waugh, and L. Melnik. 1976. Elastic area compressibility modulus of red cell membrane. *Biophysics Journal* 16(6):585–595.

FDA. 2004. Premarket Approval (PMA) application for Prolieve™ thermodilatation system. accessdata.fda.gov.

Frederick, C.A., L.D. Williams, G. Ughetto et al. 1990. Structural comparison of anticancer drug-DNA complexes: Adriamycin and daunomycin. *Biochemistry* 29(10):2538–2549.

Gaber, M.H., N.Z. Wu, K. Hong et al. 1996. Thermosensitive liposomes: Extravasation and release of contents in tumor microvascular networks. *International Journal of Radiation Oncology, Biology, Physics* 36(5):1177–1187.

Goldberg, S.N., G.S. Gazelle, E.F. Halpern et al. 1996. Radiofrequency tissue ablation: Importance of local temperature along the electrode tip exposure in determining lesion shape and size. *Academic Radiology* 3:212–218.

Gordon, A.N., D. Guthrie, D.E. Parkin et al. 2001. Recurrent epithelial ovarian carcinoma: A randomized phase III study of pegylated liposomal doxorubicin versus topotecan. *Journal of Clinical Oncology* 19(14):3312–3322.

Gregoriadis, G. 1976. The carrier potential of liposomes in biology and medicine. *New England Journal of Medicine* 295(14):765–770.

Gregoriadis, G. 1979. *Drug Carriers in Biology and Medicine*. London, UK: Academic Press.

Gregoriadis, G. 1984. *Liposome Technology: Vol. I Preparation of Liposomes; Vol. II Incorporation of Drugs, Proteins, and Genetic Material; Vol. III Targeted Drug Delivery and Biological Interaction*, 1st ed. Boca Raton, FL: CRC Press.

Gregoriadis, G. 2010. *Liposome Technology*, 3rd ed. New York: Informa Healthcare.

Gregoriadis, G. and B.E. Ryman. 1971. Liposomes as carriers of enzymes or drugs: A new approach to the treatment of storage diseases. *Biochemical Journal* 124(5):58P.

Gregoriadis, G. and E.D. Neerunjun. 1975. Homing of liposomes to target cells. *Biochemical and Biophysical Research Communications* 65(2):537–544.

Gregoriadis, G., D. Putnam, and L. Louis. 1974. Comparative effect and fate of non-entrapped and liposome-entrapped neuraminidase injected into rats. *Biochemical Journal* 140(2):323–330.

Gregoriadis, G., P.D. Leathwood, and B.E. Ryman. 1971. Enzyme entrapment in liposomes. *FEBS Letters* 14(2):95–99.

Gustafson, D.L., J.C. Rastatter, T. Colombo et al. 2002. Doxorubicin pharmacokinetics: Macromolecule binding, metabolism, and excretion in the context of a physiologic model. *Journal of Pharmaceutical Sciences* 91(6):1488–1501.

Hauck, M.L., S.M. LaRue, W.P. Petros et al. 2006. Phase 1 trial of doxorubicin-containing low temperature sensitive liposomes in spontaneous canine tumors. *Clinical Cancer Research* 12:4004–4010.

Ho, Y.K., R.G. Smith, M.S. Brown et al. 1978. Low-density lipoprotein (LDL) receptor activity in human acute myelogenous leukemia cells. *Blood* 52:1099.

Ichikawa, Y., M. Ghanefar, M. Bayeva et al. 2014. Cardiotoxicity of doxorubicin is mediated through mitochondrial iron accumulation. *The Journal of Clinical Investigation* 124(2):617–630.

Ickenstein, L.M., M.C. Arfvidsson, D. Needham et al. 2003. Disc formation in cholesterol-free liposomes during phase transition. *Biochimica et Biophysica Acta* 1614(2):135–138.

Issels, R.D., L.H. Linder, J. Verweij et al. 2010. Neo-adjuvant chemotherapy alone or with regional hyperthermia for localised high-risk soft-tissue sarcoma: A randomised phase 3 multicentre study. *Lancet Oncology* 11:56–70.

Janoff, A.S., P.R. Cullis, M.B. Bally et al. 1989. Dehydrated liposomes. The Liposome Company, Inc. US Patent 4880635 A. Publication Date: November 14, 1989.

Janoff, A.S., P.R. Cullis, M.B. Bally et al. 1996. Method of dehydrating liposomes using protective sugars. UPTO, The Liposome Company, Inc. US Patent 5578320 A. Publication Date: November 26, 1996.

Kanter, P.M., G.M. Klaich, G.A. Bullard et al. 1994. Liposome encapsulated vincristine: Preclinical toxicologic and pharmacologic comparison with free vincristine and empty liposomes in mice, rats and dogs. *Anti-Cancer Drugs* 5(5):579–590.

Kennedy, J.E. 2005. High-intensity focused ultrasound in the treatment of solid tumours. *Nature Reviews Cancer* 5(4):321–327.

Kim, D. 1999. Mechanical properties, microstructure, and specific adhesion of phospholipid monolayer-coated microbubbles. PhD thesis, Duke University.

Kim, D.H., M.J. Costello, P.B. Dunan et al. 2003. Mechanical properties and microstructure of polycrystalline phospholipid monolayer shells—Novel solid nanoparticles. *Langmuir* 19:8455–8466.

Kim, D.H. and D. Needham. 2001. Lipids as bilayers and monolayers: Characterization using micropipette manipulation techniques. In *Encyclopedia of Surface and Colloid Science*, ed. A. Hubbard. New York: Marcel Dekker.

Klibanov, A.L., K. Maruyama, V.P. Torchilin et al. 1990. Amphipathic polyethyleneglycols effectively prolong the circulation time of liposomes. *FEBS Letters* 268(1):235–237.

Kong, G., G. Anyarambhatla, W.P. Petros et al. 2000a. Efficacy of liposomes and hyperthermia in a human tumor xenograft model: Importance of triggered drug release. *Cancer Research* 60(24):6950–6957.

Kong, G., R.D. Braun, and M.W. Dewhirst. 2000b. Hyperthermia enables tumor specific nanoparticle delivery: Effect of particle size. *Cancer Research* 60:4440–4445.

Kong, G., R.D. Braun, and M.W. Dewhirst. 2001. Characterization of the effect of hyperthermia on nanoparticle extravasation from tumor vasculature. *Cancer Research* 61:3027–3032.

Kong, G. and M.W. Dewhirst. 1999. Hyperthermia and liposomes. *International Journal of Hyperthermia* 15(5):345–370.

Krishna, R., M.S. Webb, G. St Onge et al. 2001. Liposomal and nonliposomal drug pharmacokinetics after administration of liposome-encapsulated vincristine and their contribution to drug tissue distribution properties. *Journal of Pharmacology & Experimental Therapeutics* 298(3):1206–1212.

Kwok, R. and E. Evans. 1981. Thermoelasticity of large lecithin bilayer vesicles. *Biophysical Journal* 35(3):637–652.

Lacko, A.G., M. Nair, and W.J. McConathy. 2006. Lipoprotein nanoparticles as delivery vehicles for anticancer agents. In *Nanotechnology for Cancer Therapy*, ed. M.M. Amiji, pp. 777–786. Boca Raton, FL: CRC Press.

Landon, C., D. Needham, M.W. Dewhirst et al. 2011. Nano-encapsulation technology for drug delivery. *Therapeutic Approaches and Drug Delivery: The Open Nanomedicine Journal* 3:38–64.

Larson, B.T. and D.W. Robertson. 2006. Interstitial temperature mapping during Prolieve transurethral microwave treatment: Imaging reveals thermotherapy temperatures resulting in tissue necrosis and patent prostatic urethra. *Urology* 68(6):1206–1210.

Lasic, D.D. and F.J. Martin. 1995. *Stealth Liposomes*. Boca Raton, FL: CRC Press.

Lasic, D.D. and D. Needham. 1995. The "Stealth" liposome: A prototypical biomaterial. *Chemical Reviews* 95(8):2601–2628.

Lencioni, R. 2013. New IO approaches for HCC: An update on clinical trials. In *European Conference on Interventional Oncology (ECIO)*. New York.

Lentz, B.R., E. Freire, and R.L. Biltonen. 1978. Fluorescence and calorimetric studies of phase transitions in phosphatidylcholine multilayers: Kinetics of the pretransition. *Biochemistry* 17:4475–4480.

Li, L., T.L. ten Hagen, M. Bolkestein et al. 2013. Improved intratumoral nanoparticle extravasation and penetration by mild hyperthermia. *Journal of Controlled Release* 167:130–137.

Li, L., T.L. ten Hagen, D. Schipper et al. 2010. Triggered content release from optimized stealth thermosensitive liposomes using mild hyperthermia. *Journal of Controlled Release* 143:274–279.

Liu, D., F. Liu, and Y.K. Song. 1995. Recognition and clearance of liposomes containing phosphatidylserine are mediated by serum opsonin. *Biochimica et Biophysica Acta* 1235(1):140–146.

Locklin, J.K., S. Libutti, Z. Neeman et al. 2008. Imaging features in patients undergoing liver RFA plus heat deployed nanoparticles. In *SIR 2008 Conference*. Washington, DC.

Lorusso, D., A.D. Stefano, V. Carone et al. 2007. Pegylated liposomal doxorubicin-related palmar-plantar erythrodysesthesia ('hand-foot' syndrome). *Annals of Oncology* 18(7):1159–1164.

Mabrey, S. and J.M. Sturtevant. 1976. Investigation of phase transitions of lipids and lipid mixtures by sensitivity differential scanning calorimetry. *Proceedings of the National Academy of Sciences of the United States of America* 73:3862–3866.

Malamud, B. 1952. *The Natural*. New York: Harcourt, Brace and Co.

Manzoor, A.A., L.H. Lindner, C.D. Landon et al. 2012. Overcoming limitations in nanoparticle drug delivery: Triggered, intravascular release to improve drug penetration into tumors. *Cancer Research* 72(21):5566–5575.

Matteucci, M.L., G. Anyarambhatla, G. Rosner et al. 2000. Hyperthermia increases accumulation of technetium-99m-labeled liposomes in feline sarcomas. *Clinical Cancer Research* 6:3748.

Mayer, L.D., M.B. Bally, P.R. Cullis et al. 1997. High drug:lipid formulations of liposomal antineoplastic agents. The Liposome Company, Inc. US Patent 5616341 A. Publication Date: April 1, 1997.

MediFocus Inc. 2016. Prolieve BPH treatment. http://medifocusinc.com/prolieve-bph-treatment/, accessed March 17, 2016.

Mills, J.K. 2002. Triggered release of liposome contents: Mechanism involved in membrane permeability and compromise. PhD thesis, Duke University, pp. 1–292.

Mills, J.K. and D. Needham. 2004. The materials engineering of temperature sensitive liposomes. *Methods in Enzymology* 387:82–113.

Mills, J.K. and D. Needham. 2005. Lysolipid incorporation in dipalmitoylphosphatidylcholine bilayer membranes enhances the ion permeability and drug release rates at the membrane phase transition. *Biochimica et Biophysica Acta—Biomembranes* 1716(2):77–96.

Moos, W. 2010. Nonprofits and the valley of death in drug discovery. Exome (Biotech and Health). Available from: http://www.xconomy.com/san-francisco/2010/08/19/nonprofits-and-the-valley-of-death-in-drug-discovery/, accessed March 17, 2016.

Mouritsen, O.G. and M.J. Zuckermann. 1987. Model of interfacial melting. *Physical Review Letters* 58(4):389–392.

Needham, D. 1981. Catalysis by evaporated films of rare earth metals. In *Physical Chemistry*. Nottingham, U.K.: University of Nottingham.

Needham, D. 1999. Temperature-sensitive liposomal formulation, WO99/65466, issued December 23, 1999.

Needham, D. 2001. Temperature-sensitive liposomal formulation U.S. Patent No. 6,200,598. Filed: June 18, 1998.

Needham, D. 2004. Temperature-sensitive liposomal formulation. US Patent 6726925 A. Filed: December 9, 1999.

Needham, D. 2001a. Lipid membranes and monolayers: Physical properties from micropipette studies. In *Fifth International Conference on Liposome Advances. Progress in Drug and Vaccine Delivery. A Celebration of the 80th Birthday of Alec Bangham Md, Frs*. London, U.K.: The School of Pharmacy, University of London.

Needham, D. 2001b. Temperature-sensitive liposomal formulation. US Patent 6200598.

Needham, D. 2004. Temperature-sensitive liposomal formulation. US Patent 6726925 A. Publication Date: April 27, 2004.

Needham, D. 2013. Reverse engineering the low temperature sensitive liposome (LTSL) for treating cancer (Chapter 12). In *Biomaterials for Cancer Therapeutics*, ed. K. Park. Cambridge, U.K.: Woodhead Publishing Ltd.

Needham, D. 2015. Bringing research to clinical application: Lessons from Thermodox®—A thermal sensitive liposome for treatment of cancer. In *12th Annual Garnet E. Peck Symposium. Clinical and Commercial Translation of Drug Delivery Systems*. West Lafayette, IN: Purdue University.

Needham, D., G. Anyarambhatla, G. Kong et al. 2000. A new temperature-sensitive liposome for use with mild hyperthermia: Characterization and testing in a human tumor xenograft model. *Cancer Research* 60(5):1197–1201.

Needham, D. and M.W. Dewhirst. 2001. The development and testing of a new temperature-sensitive drug delivery system for the treatment of solid tumors. *Advanced Drug Delivery Reviews* 53(3):285–305.

Needham, D. and M.W. Dewhirst. 2012. Materials science and engineering of the low temperature sensitive liposome (LTSL): Composition-structure-property relationships that underlie its design and performance. In *Smart Materials for Drug Delivery*, eds. C. Alvarez-Lorenzo and A. Concheiro. Cambridge, U.K.: Royal Society of Chemistry.

Needham, D. and R.S. Nunn. 1990. Elastic deformation and failure of lipid bilayer membranes containing cholesterol. *Biophysical Journal* 58:997–1009.

Needham, D. and A. Ponce. 2006. Nanoscale drug delivery vehicles for solid tumors: A new paradigm for localized drug delivery using temperature sensitive liposomes. In *Nanotechnology for Cancer Therapy*, ed. M.M. Amiji, pp. 677–719. Boca Raton, FL: CRC Press.

Needham, D. and D. Kim. 2000. PEG-covered lipid surfaces: Bilayers and monolayers in protein and cell repellent surfaces. *Colloids and Surfaces B: Biointerfaces* 18(3–4):183–195.

Needham, D., N. Stoicheva, and D.V. Zhelev. 1997. Exchange of monooleoylphosphatidylcholine as monomer and micelle with membranes containing poly(ethylene glycol)-lipid. *Biophysical Journal* 73(5):2615–2629.

Needham, D., T.J. McIntosh, and E.A. Evans. 1988. Thermomechanical and transition properties of dimyristoylphosphatidylcholine/cholesterol bilayers. *Biochemistry* 27:4668–4673.

Needham, D., A.M Wright, J. Tong et al. 2012. Materials characterization of the low temperature sensitive liposome (LTSL): Effects of lipid composition (Lysolipid and DSPE-PEG2000) on the thermal transition and release of doxorubicin. *Faraday Discussion* (*Lipids and Membrane Biophysics*) 161:515.

Negussie, A.H., P.S. Yarmolenko, A. Partanen et al. 2011. Formulation and characterisation of magnetic resonance imageable thermally sensitive liposomes for use with magnetic resonance-guided high intensity focused ultrasound. *International Journal of Hyperthermia* 27(2):140–155.

Noppl-Simson, D. and D. Needham. 1996. Avidin–Biotin interactions at vesicle surfaces: Surface binding, cross-bridge formation and lateral interactions. *Biophysical Journal* 70:1391–1401.

Northfelt, D.W., F.J. Martin, P. Working et al. 1996. Doxorubicin encapsulated in liposomes containing surface-bound polyethylene glycol: Pharmacokinetics, tumor localization, and safety in patients with AIDS-related Kaposi's Sarcoma. *The Journal of Clinical Pharmacology* 36(1):55–63.

O'Brien M.E., N. Wigler, M. Inbar et al. 2004. Reduced cardiotoxicity and comparable efficacy in a phase III trial of pegylated liposomal doxorubicin HCl (CAELYX/Doxil) versus conventional doxorubicin for first-line treatment of metastatic breast cancer. *Annals of Oncology* 15(3):440–449.

Oleson, J.R. and M.W. Dewhirst. 1983. Hyperthermia: An overview of current progress and problems. *Current Problems in Cancer* 8(6):1–62.

Onecle. 1999. Sample Business Contracts. License Agreement—Duke University and Celsion Corp. Available from: http://contracts.onecle.com/celsion/duke.lic.1999.11.10.shtml, accessed March 17, 2016.

Palmer, G.M., R.J. Boruta, B.L. Viglianti et al. 2010. Non-invasive monitoring of intra-tumor drug concentration and therapeutic response using optical spectroscopy. *Journal of Controlled Release* 142(3):457–464.

Papahadjopoulos, D. and N. Miller. 1967. Phospholipid model membranes. I. Structural characteristics of hydrated liquid crystals. *Biochimica et Biophysica Acta—Biomembranes* 135(4):624–38.

Papahadjopoulos, D. and J.C. Watkins. 1967. Phospholipid model membranes. II. Permeability properties of hydrated liquid crystals. *Biochimica et Biophysica Acta—Biomembranes* 135(4):639–652.

Papahadjopoulos, D.P., J. Francis, and C. Szoka. 1980. Method of encapsulating biologically active materials in lipid vesicles Patent no. US4235871 A.

Ponce, A.M., B.L. Viglianti, D. Yu et al. 2007. Magnetic resonance imaging of temperature-sensitive liposome release: Drug dose painting and antitumor effects. *Journal of the National Cancer Institute* 99(1):53–63.

Poon, R.T. and N. Borys. 2009. Lyso-thermosensitive liposomal doxorubicin: A novel approach to enhance efficacy of thermal ablation of liver cancer. *Expert Opinion on Pharmacotherapy* 10(2):333–343.

Poon, R.T.P. and N. Borys. 2011. Lyso-thermosensitive liposomal doxorubicin: An adjuvant to increase the cure rate of radiofrequency ablation in liver cancer. *Future Oncology* 7(8):937–945.

Poon, R., K. Ng, J. Yuen et al. 2008. Phase 1 study of Thermodox (thermally sensitive liposomes containing doxorubicin) given prior to radiofrequency ablation for unresectable liver cancers. In *IHPBA Conference*. Mumbai, India.

Proffitt, R.T., J. Adler-Moore, and S.-M. Chiang. 1999. Amphotericin B liposome preparation. NeXstar Pharmaceuticals, Inc.

Ravikumar, T.S., K. Fehn, C.D.V. Black et al. 2010. A phase I trial of ThermoDox® in patients undergoing radiofrequency ablation (RFA) of liver tumors. In *IHPBA*, 2010. Buenos Aires, Argentina.

Roerdink, F., J. Dijkstra, G. Hartman et al. 1981.The involvement of parenchymal, Kupffer and endothelial liver cells in the hepatic uptake of intravenously injected liposomes. Effects of lanthanum and gadolinium salts. *Biochimica et Biophysica Acta* 677(1):79–89.

Rothkopf, C., A. Fahr, G. Fricker et al. 2005. Uptake of phosphatidylserine-containing liposomes by liver sinusoidal endothelial cells in the serum-free perfused rat liver. *Biochimica et Biophysica Acta—Biomembranes* 1668(1):10–16.

Rudling, M.J., L. Stahle, and C.O. Peterson. 1986. Content of low density lipoprotein receptors in breast cancer tissue related to survival of patients. *British Medical Journal* 292:580–582.

Rugo, H., T.M. Zagar, S.C. Formenti et al. 2013. Abstract P4-15-05: Novel targeted therapy for breast cancer chest wall recurrence: Low temperature liposomal doxorubicin and mild local hyperthermia. *Cancer Research* 73(24 Suppl.):P4-15-05.

Schmidt, P.G. and G. Fujii. 1998. Method of making liposomes with improved stability during drying. NeXstar Pharmaceuticals, Inc. WO Patent 1990003795 A1.

Secord, A.A., E.L. Jones, C.A. Hahn et al. 2005. Phase 1/II trial of intravenous Doxil® and whole abdomen hyperthermia in patients with refractory ovarian cancer. *International Journal of Hyperthermia* 21(4):333–347.

Silverman, L. and Y. Barenholz. 2015. In vitro experiments showing enhanced release of doxorubicin from Doxil® in the presence of ammonia may explain drug release at tumor site. *Nanomedicine: Nanotechnology, Biology and Medicine* 11(7):1841–1850.

Skalak, R., A. Tozeren, R.P. Zarda et al. 1973. Strain energy function of red blood cell membranes. *Biophysical Journal* 13(3):245–254.

Steck, T.L. 1974. The organization of proteins in the human red blood cell membrane: A review. *The Journal of Cell Biology* 62(1):1–19.

Storm, F.K. 1989. Clinical hyperthermia and chemotherapy. *Radiologic Clinics of North America* 27:621–627.

Strauss, S. 2010. Pharma embraces open source models. *Nature Biotechnology* 28(7):631–634.

Szoka Jr., F.C. 1998. Commentary: Rantosomes and ravosomes. *Journal of Liposome Research* 8(3):7–9.

Szoka Jr., F.C. and D. Papahadjopoulos. 1978. Procedure for preparation of liposomes with large internal aqueous space and high capture by reverse-phase evaporation. *Proceedings of the National Academy of Sciences of the United States of America* 75(9):4194–4198.

Tagami, T., M.J. Ernsting, and S.-D. Li. 2011. Efficient tumor regression by a single and low dose treatment with a novel and enhanced formulation of thermosensitive liposomal doxorubicin. *Journal of Controlled Release* 152:303–309.

Tardugno, M. 2013. Celsion announces results of phase III HEAT Study of ThermoDox® in primary liver cancer. Lawrenceville, NJ.

Tateishi, R., S. Shiina, T. Teratani et al. 2005. Percutaneous radiofrequency ablation for hepatocellular carcinoma. *Cancer* 103(6):1201–1209.

The Free Library. 2014. Valentis announces settlement of PEG-liposome patent infringement. NeXstar Pharmaceuticals, The Liposome Co., San Dimas: CA. Settle Patent Litigation, 1997. Available from: http://www.thefreelibrary.com/NeXstar+Pharmaceuticals%2c+The+Liposome+Co.%2c+Settle+Patent+Litigation.-a019669609, accessed March 17, 2016.

Torchilin, V., A. Lukyanov, and Z. Gao. 2006. Micelle delivery system loaded with a pharmaceutical agent. USPTO. US Patent 20060216342 A1. Publication Date: September 28, 2006.

Vaage, J., E. Mayhew, D. Lasic et al. 1992. Therapy of primary and metastatic mouse mammary carcinomas with doxorubicin encapsulated in long circulating liposomes. *International Journal of Cancer* 51(6):942–948.

Venditto, V.J. and F.C. Szoka. 2013. Cancer nanomedicines: So many papers and so few drugs! *Advanced Drug Delivery Reviews* 65(1):80–88.

Viglianti, B.L., M.W. Dewhirst, R.J. Boruta et al. 2014. Systemic anti-tumour effects of local thermally sensitive liposome therapy. *International Journal of Hyperthermia* 30(6):385–392.

Viglianti, B., P. Stauffer, E. Repasky et al. 2010. Hyperthermia. In *Holland-Frei Cancer Medicine*, eds. B.R.J. Hong, W. Hair, D.W. Kufe et al., pp. 528–540. Shelton, CT: People's Medical Publishing House.

Vujaskovic, Z. 2007. Chest wall recurrence of breast cancer treated with hyperthermia and thermally sensitive liposomes (ThermoDox). In *Society for Thermal Medicine*. Washington, DC.

Vujaskovic, Z., D.W. Kim, E. Jones et al. 2010. A phase I/II study of neoadjuvant liposomal doxorubicin, paclitaxel, and hyperthermia in locally advanced breast cancer. *International Journal of Hyperthermia* 26(5):514–521.

WCRF International. 2015. Data for cancer frequency by country. http://www.wcrf.org/int/cancer-facts-figures/data-cancer-frequency-country, accessed March 17, 2016.

Webb, M.S., M.B. Bally, and L.D. Mayer. 1996. Sphingosomes for enhanced drug delivery. Inex Pharmaceuticals Corporation. US Patent 5543152 A. Publication Date: August 6, 1996.

Webb, M.S., T.O. Harasym, D. Masin et al. 1995. Sphingomyelin-cholesterol liposomes significantly enhance the pharmacokinetic and therapeutic properties of vincristine in murine and human tumour models. *British Journal of Cancer* 72(4):896–904.

Weiner, P. 2015. Post-marketing Study using Prolieve® for the treatment of Benign Prostatic Hyperplasia (BPH). Medifocus, Inc. Clinical Trial. Identifier: NCT02021032. ClinicalTrials.gov.

Wheeler, J.J., M. Hope, P.R. Cullis et al. 1999. Methods for encapsulating plasmids in lipid bilayers. Inex Pharmaceuticals Corp. US Patent 5981501 A. Publication Date: November 9, 1999.

Woo, J., G.N. Chiu, G. Karlsson et al. 2008. Use of a passive equilibration methodology to encapsulate cisplatin into preformed thermosensitive liposomes. *International Journal of Pharmaceutics* 349:38–46.

Wood, B.J., R.T. Poon, and J. Lockin. 2012. A phase I study of heat-deployed liposomal doxorubicin during radiofrequency ablation for hepatic malignancies. *Journal of Vascular and Interventional Radiology* 23:248–255.

Wood, B.J., J.R. Ramkaransingh, T. Fojo et al. 2002. Percutaneous tumor ablation with radiofrequency. *Cancer* 94:443–451.

Woodle, M.C., F.J. Martin, A. Yau-Young et al. 1989. Liposomes with enhanced circulation time. Liposome Technology, Inc. US Patent 5013556 A. Filing Date: October 20, 1989. Publication Date: May 7, 1991.

Wright, A.M. 2006. Drug loading and release from a thermally sensitive liposome. In PhD thesis, Duke University, pp. 1–322.

Wu, N.Z., D. Da, T.L. Rudoll et al. 1993b. Increased microvascular permeability contributes to preferential accumulation of Stealth liposomes in tumor tissue. *Cancer Research* 53:3765–3770.

Wu, N.Z., B.G. Klitzman, D. Rosner et al. 1993a. Measurement of material extravasation in microvascular networks using fluorescence video-microscopy. *Microvascular Research* 46:231–253.

Yarmolenko, P.S., Y. Zhao, C. Landon et al. 2010. Comparative effects of thermosensitive doxorubicin-containing liposomes and hyperthermia in human and murine tumours. *International Journal of Hyperthermia* 26(5):485–498.

Yuan F., S.K. Huang, D.A. Berk et al. 1994. Microvascular permeability and interstitial penetration of sterically stabilized (stealth) liposomes in a human tumor xenograft. *Cancer Research* 54:3352–3356.

Zagar, T.M., Z. Vujaskovic, S. Formenti et al. 2014. Two phase I dose escalation/pharmacokinetics studies of low temperature liposomal doxorubicin (LTLD) and mild local hyperthermia in heavily pretreated patients with local-regionally recurrent breast cancer. *International Journal of Hyperthermia* 30(5):285.

24 Conclusions

Anya M. Hillery and Kinam Park

CONTENTS

24.1 INTRODUCTION

As described in Chapter 1, the early years of drug delivery research (from about 1950 to 1980) saw the development of various mechanisms to achieve controlled drug release. The principal physical approaches to achieve controlled release were based on diffusion- and dissolution-control. Osmotic pressure and ion-exchange mechanisms were also used. Technologies at this time included sustained-release, once-a-day, formulations for oral delivery, and transdermal patches suitable for daily or weekly administration. The thrust of the research at this time was based on adjusting the physicochemical properties of the formulation, which resulted in the desired sustained-release pharmacokinetic profiles in vivo. This research was highly successful and resulted in the introduction of a large number of commercial products.

The period from the 1980s to the present time is marked by a slower pace of research, as evidenced by the lower numbers of products that have been brought to commercial fruition. During this time, more complicated challenges have emerged, which have proved much more difficult to resolve. Increasingly, formulators must work with molecules derived from drug discovery programs that have challenging physicochemical properties for satisfactory delivery, including poor water solubility, large size, and metabolic lability. Additionally, drug delivery scientists have to address a multitude of highly complex biological barriers, including epithelial and endothelial barriers, the complexity and variability of the biological environment in vivo, access and penetration barriers *en route* to the target, the complexity of the pathologies being treated (for example, cancer), and, in the case of implantable drug delivery systems (DDS), problems of biocompatibility and host-response reactions (Yun et al. 2015). The combination of formulation and biological barriers has, in many cases, limited successful efforts at delivery.

This book describes current technologies and emerging trends in drug delivery science. It comes 15 years after the publication of the first edition (Hillery et al. 2001). Significant progress, greater in some areas than in others, has been made in the last decade and a half. The preceding chapters of this book bear witness to the excellence of current research being carried out worldwide in the field, as well as to the considerable range and scope of this work. In this chapter, further discussion is confined to two important areas of drug delivery research: (1) nanotechnologies for drug delivery and (2) transport across epithelial barriers.

24.2   NANOTECHNOLOGY AND DRUG DELIVERY

Since the 1990s, the synthesis and testing of novel drug nanocarriers has constituted the most active area of R&D in the drug delivery field. A vast amount of literature is now available on the subject—for example, an Internet search for "nanotechnology and drug delivery" currently yields 1.75 million hits. Initially, nanoparticles (NPs) were defined as particles smaller than 100 nm. However, for drug delivery, NPs less than 100 nm typically do not have enough space to load drug (an exception being when NPs are made of the drug per se, e.g., DNA or RNA condensed by interaction with cationic polymers), so that NPs used in drug delivery are usually larger than 100 nm. The definition was therefore modified, to describe engineered particles that have at least one dimension in the nanoscale range (Nanotechnology Safety and Health Program 2014). Thus, current nanotechnologies for drug delivery are classified purely on the basis of their size. There is no scientific rationale for defining NPs based on their size alone—the usefulness of NPs should be due to their unique properties and functions, rather than their size. In fact, many of the current "nanotechnologies" described in this book are actually the selfsame carriers that were simply called "microparticles" and "drug-targeting systems" in the first edition of this text.

An important aspect of nanotechnologies is their targeting potential. The term "targeting" is a broad one, meaning different things depending on the context. For example, local/topical application is a type of regional targeting, e.g., antiallergy drugs can be given by nasal spray, for the topical treatment of allergic rhinitis. By delivering the drug locally, it is effectively targeted to its site of action, maximizing the therapeutic effect, while minimizing unwanted side effects. Alternatively, some drugs, such as antibiotics and hormones, possess an intrinsic targeting ability in that they distribute through the entire body but only interact with their target receptors or cells, without affecting other cells. In the context of nanotechnology, targeting is defined as either (1) "passive," taking advantage of the "enhanced permeability and retention" (EPR) effect (Maeda et al. 1985; Matsumura and Maeda 1986), or (2) "active," using a specific targeting ligand, which is attached to a drug, or drug-loaded DDS, in order to improve uptake and sequestration into the cell or tissue (see also Chapter 5). Both passive and active targeting strategies are considered further here.

24.2.1   PASSIVE TARGETING: THE EPR EFFECT

The idea that the EPR effect facilitates "passive targeting" to tumors (i.e., that nanocarriers can passively enter the leaky blood supply of a tumor and thus preferentially accumulate in the tumor tissue) is one that has gained near universal acceptance in the field. It is widely referred to throughout this book. However, in recent years there have been doubts expressed about the clinical applicability of the phenomenon, the limited evidence for the effectiveness of the mechanism, the low amount of uptake generally achieved, and the heterogeneity of the effect (Stirland et al. 2013; Nichols and Bae 2014; Bae and Park 2011). In Chapter 5, the authors further add to this note of caution (Section 5.2.1).

In spite of the doubts that are increasingly being raised, there nevertheless remains a widespread misconception in the field that, due to the EPR effect, nanocarrier DDS are taken up exclusively by tumors. It should be stressed that, even in the most favorable circumstances, this is far from the case – in contrast, most of the administered dose (i.e. $\approx95\%$), still ends up in the reticuloendothelial system (RES) organs of the liver and spleen and, to a lesser extent, in other, nontumor tissues.

There is a pressing need to resolve outstanding issues, such as the extent (currently only a few percent uptake), applicability (relevance in humans, as opposed to mouse models), and reproducibility (due to problems of intra- and interpatient tumor heterogeneity) of the EPR effect. It would be highly beneficial for future research to precisely define these parameters, as well as define the optimal nanocarrier physicochemical properties that could improve tumor uptake via the EPR effect.

Research into the EPR effect is hampered by the lack of appropriate animal models. The limitations of preclinical cancer models have been widely reviewed and are largely acknowledged by the field (Begley and Ellis 2012; Crommelin and Florence 2013). Although veterinary cancers and transgenic models would be the most physiologically relevant animal models to use, they are not usually employed in research because of their low availability, or the prolonged time necessary to generate tumors. It should be noted that a recent review concluded that most of the frequently used rodent models (such as the subcutaneous flank tumor xenograft) can lead to an overestimation of the potential usefulness of the EPR effect (Prabhakar et al. 2013). This is because animal tumors grow much faster than human ones, so that blood vessels in rodent tumors do not develop properly and are inherently much leakier, which favors NP uptake. There is obviously a pressing need to develop more predictive animal models, as discussed further in Section 24.4.6.

Crucially, the EPR effect needs to be systematically and rigorously investigated in human patients. To date, the information available in human subjects is scant and not standardized. It is also, thus far, not very convincing. Even within a single tumor, large differences with regard to vascular permeability have been shown (Lammers et al. 2012; Lammers 2013). Harrington et al. used radiolabeled PEGylated liposomes in conjunction with whole body gamma camera imaging to study liposome biodistribution in vivo (Harrington et al. 2001, 2002). The levels of tumor liposome uptake estimated from regions of interest on gamma camera images were again low— approximately 0.5%–3.5% of the injected dose at 72 h. They also showed considerable variation in the amount taken up depending on the tumor type, as well as for the same tumor type, but between different patients.

24.2.2 ACTIVE TARGETING

The use of active targeting strategies is widely referred to throughout this book, for example, using folic acid to target cancer cells, lectins and vitamins to target enterocytes, and the use of monoclonal antibodies to target specific receptors. In spite of all this research and focus, there are currently no actively targeted nanomedicines in clinical use. To understand the poor success of the strategy to date, it is important to remember the inherent limitations associated with this approach.

24.2.2.1 Nonspecificity of Current Targeting Ligands

There is currently a lack of specificity associated with the approach. Targeting ligands are often chosen for receptors that are merely overexpressed on tumor cells, rather than exclusive to tumor cells. For example, folic acid is widely chosen as a targeting motif because the folic acid receptor is known to be overexpressed on cancer cells. However, the folic acid receptor is also present on normal cells. Since normal cells far outnumber tumor cells in vivo, the overexpression differential advantage of tumor cells is lost (Bae and Park 2015). To achieve greater specificity, ligands *unique* to the target must be used, for example, in the case of cancer targeting, by using tumor-associated antigens (i.e., antigenic structures specifically occurring at the surface of tumor cells). However, ligands of such unique specificity are difficult to find.

24.2.2.2 Targeting Ligands Do Not Deliver Drugs to the Site of Action

Targeting vectors have zero ability to *deliver* their cargo to a specific cell or tissue. These ligands are not satellite navigation "homing" devices that direct a nanocarrier specifically to a tumor or any other site. In vivo, there is no greater accumulation of actively targeted NPs at a tumor site in comparison to their nontargeted counterparts. On the contrary, they are merely cell surface–recognition moieties, which are still dependent on the random encounter of a nanocarrier with the appropriate receptor, during its circulation lifetime in vivo (Crommelin and Florence 2013). There is no guarantee of this interaction. In this sense, the use of the word "active" is misleading, as it suggests an active (i.e., energy driven) process at work, to seek out the corresponding receptor, but this is absolutely not the case.

24.2.2.3 Access Barriers Limit the Effectiveness of Targeting Ligands

The likelihood of a random encounter of a targeting ligand with its corresponding receptor is severely compromised in vivo due to the difficulty in accessing target receptors (Lammers et al. 2012; Crommelin and Florence 2013). In the case of tumor targeting, the vectors are for receptors on tumor cells, which means the nanocarrier must first extravasate from the bloodstream to access these cells. If the nanocarrier does manage to extravasate from the bloodstream (and this is in no way guaranteed—see the problems associated with the EPR effect, described in Section 24.2.1), it still has to diffuse, against a high interstitial fluid pressure, through the perivascular space (which contains layers of pericytes, smooth muscle cells, and fibroblasts), in order to reach the tumor cells. If the nanocarrier does succeed in navigating through this complex tumor microenvironment, targeting vectors will then bind to the first receptors they encounter and so may not penetrate very deeply into the tumor (this phenomenon is known as the binding-site barrier).

In the limited environment of a cell line, a targeted NP may indeed show enhanced uptake by tumor cells (up to 100× greater uptake, than for nontargeted controls)—but these are in vitro studies, in which the NPs are pipetted directly on top of the test cells, so that ligand–receptor interactions are highly favored. The situation in vivo is much more complex, and nanocarriers must overcome the accompanying access and penetration barriers, before uptake can take place. Thus, increased uptake of a nanocarrier in a cell line is actually a very poor indicator of efficacy in vivo.

A number of approaches are currently being taken to address the problems of access to tumor cells, including actively targeting the nanocarrier to tumor endothelium, rather than the tumor cells (reviewed in Lammers et al. 2012). By binding and killing tumor endothelial cells, the tumor becomes deprived of oxygen and nutrients, resulting in tumor death. This approach has many advantages over targeting tumor cells: (1) it is easier to find the target in vivo, (2) it eliminates the need for the DDS to extravasate, and (3) the nanocarrier does not have to penetrate the perivascular space.

A different approach is to use triggered drug delivery, using stimuli-sensitive carrier materials. Once at the tumor, stimuli-sensitive release facilitates the release of the low-molecular-weight payload, which can diffuse more rapidly through the perivascular space than a (much larger) nanoparticulate carrier. For example, Chapter 23 describes the use of ThermoDox®, low temperature–sensitive liposomes that release their doxorubicin (Dox) payload in the vicinity of a tumor, when the tumor is subjected to localized heat treatment.

Although at a very early stage of development, a further research avenue involves the use of energy-driven mechanisms to actively propel drugs toward a target or deep into a tumor. One recent example (reminiscent of the sci-fi film Fantastic Voyage) is the use of gold–mesoporous silica nanorods that use the surrounding biological environment as a source of fuel, to self-propel themselves through the body (Wang et al. 2015). Multiple fuel sources have been explored to power these nanomotors; one model uses gastric acid, which produces bubbles of hydrogen gas for nano-motor thrust and propulsion.

24.2.3 FUTURE DIRECTIONS FOR NANOTECHNOLOGY AND DRUG DELIVERY

At the preclinical level, we are churning out nanocarriers at a prolific rate. Literally thousands of different constructs have been described in the literature, yet they all share remarkably similar baseline properties: typically, they incorporate a (nonspecific) targeting ligand, are loaded with Dox, show uptake in a cell culture study, and are variously described as "targeted," "smart," and "advanced," with the potential to "cure cancer." Although this activity keeps pharmaceutical research humming along and contributes toward the vast amount of literature now available in the field, endlessly producing new nanocarriers for the sake of newness is not enough and will not help the field to progress (Figure 24.1).

Producing ever-more complex nanoparticulate DDS will not, per se, guarantee clinical success. Increased complexity does not necessarily provide improved NP performance in vivo (Raemdonck and de Smedt 2015), also, complex DDS are expensive; they require specialized manufacturing know-how and facilities, and possibly difficult reconstitution procedures. In general, the more complex the system, the greater the scale-up, manufacturing, and regulatory challenges will be (see Section 24.4.5).

A more productive approach might be to focus on developing a more rational design strategy for nanocarriers in drug delivery, which prioritizes their ultimate purpose. The next generation of nanocarriers, developed by rational design, would consider a number of key aspects at the outset of their development, including

- The clinical need
- The scientific basis, principles, and justification for development, that underpins this research
- How this delivery system improves on all the (literally thousands of) previous nanocarriers already developed and described in the literature
- Biocompatibility and toxicity issues, which, by extension, impact on the likelihood of the DDS receiving regulatory approval (see Section 24.4.5)
- How this system will be meaningfully tested, to find relevance to clinical applications

In developing a rational design strategy, one possible way forward is to try and set more easily attainable goals. Rather than trying to find a single, omnipotent "magic bullet" that will single-handedly

FIGURE 24.1 There is a danger that drug delivery scientists may become like Sisyphus, continually toiling to produce new nanocarrier constructs, which, without the ability to overcome biological barriers, may not translate into clinical formulations.

cure cancer, the field might be better served by progressing in smaller, less ambitious, more incremental steps, which focus on carving out specific *niche* areas of expertise. Some suggestions are outlined here.

24.2.3.1 Alter the Pharmacokinetics

Many commercially available nanotechnology formulations offer improved drug pharmacokinetics, diverting drugs away from vulnerable organs (e.g., heart, brain, kidneys). For example, Dox can cause cardiomyopathy, which may lead to congestive heart failure and death. Doxil®, the PEGylated liposomal form of Dox, was found in clinical trials for ovarian cancer to have "comparable efficacy, favorable safety profile" to the control treatment, topotecan (Gordon et al. 2001). So Doxil® was actually no better than the current treatment—its added value was in the reduction of cardiotoxicity caused by Dox. Similarly, DaunoXome® improves the cardiotoxicity of Dox, rather than imparting any improved therapeutic effect, and the liposomal product AmBisome® minimizes the nephrotoxicity associated with the antifungal drug, amphotericin B. Abraxane®, an albumin-based NP formulation of paclitaxel, is—as stated by the European Medicines Agency—"a new formulation developed to overcome the water insolubility of the active component paclitaxel and prevent hypersensitivity reactions associated with solvent-containing formulations." In this case, the advantage of reformulation as NPs was to facilitate increased solubility for a practically insoluble drug. Researchers might thus consider switching emphasis away from the development of increasingly elaborate nanoparticulate DDS for tumor targeting (which, as outlined in Sections 24.2.1 and 24.2.2, is fraught with difficulties and limitations) and look instead at drug reformulation in order to improve problems such as unfavourable pharmacokinetics, or poor solubility.

24.2.3.2 Study Other (Easier) Pathologies

Nanotechnology research is currently heavily skewed toward cancer therapy. Given the inherently difficult nature of this disease (characterized as it is, by a large inter- and intrapatient tumor heterogeneity, drug toxicity, and access and penetration barriers), another way forward is to address other therapeutic needs, which afford less delivery challenges and are more inherently suited to NPs. A prime example would be the use of nanotechnology in the treatment of inflammatory disorders. The inflammatory response is characterized by increased vascular permeability and increased blood flow. Thus, the treatment of inflammatory disorders is ideally suited to passive targeting strategies using nanocarriers, with the advantage of having none of the access issues and other problems associated with tumor therapy. Focusing research here would be of use in the treatment of a wide variety of inflammatory diseases and conditions, including atherosclerosis, arthritis, Alzheimer's, hypersensitivity reactions, asthma, and infectious diseases.

Similarly, cardiovascular diseases, as well as cancers of the blood, are more amenable to DDS targeting concepts, because the nanocarrier does not have to leave the circulation for the drug to exert its effect, so the issues outlined in Section 24.2.2.3 (extravasation, penetration of the perivascular space, tumor penetration, etc.) do not arise.

A further promising research direction is the use of nanotechnologies in vaccine delivery. As described in Chapter 17, there is considerable evidence, albeit still mostly in laboratory animal studies, that antigens presented as micro- and nanoparticulate delivery systems stimulate better immune responses than soluble antigens. NPs also show considerable promise for use in imaging procedures, as described in Section 24.2.3.5.

24.2.3.3 Investigate Drug Combinations

An important feature of nanotechnologies is their ability to incorporate more than one drug as payload, even drugs that possess highly different physicochemical characteristics (see, for example, Chapter 5, Figure 5.9). However, there is a tendency in preclinical research to use a single drug (typically Dox)—with the focus primarily on increasing the complexity of the carrier, rather than paying attention to the encapsulated drug. Future research could investigate nanotechnologies as carriers for multiple drug combinations. In particular, this approach would be advantageous for cancer treatments, as multidrug chemotherapy is associated with synergistic anti-cancer effects, reduced individual cytotoxicites, and suppression of drug resistance.

24.2.3.4 Use Nature's Own Processes

There has been a recent interest in echoing nature's own processes to design novel DDS. For example, Chapter 23 describes the approach of "putting the drug in the cancers food"—i.e., utilizing the NP composition, structure and properties of the low-density lipoprotein, and its receptor-based natural-targeting properties, as an endogenous mechanism to achieve greater uptake of anticancer drugs by tumor cells. A further example is a recent study that describes the development of platelet-mimetic NPs, comprising biodegradable polymeric NPs enclosed in the plasma membrane of human platelets (Hu et al. 2015). As well as demonstrating reduced cellular uptake by macrophage-like cells, the cloaked NPs displayed platelet-mimicking properties, which resulted in an enhanced therapeutic efficacy in a model of coronary restenosis and a systemic bacterial infection.

It would seem a highly efficient approach to work in synergy with existing physiological processes in the body. Furthermore, studying natural processes increases current understanding of fundamental cell biology and cellular uptake processes. An example here is the recent research into exploiting receptor-mediated transcytosis (RMT) via the transferrin receptor, for improving CNS delivery. The research has yielded important information on the fundamental physiology of the RMT process (see Chapter 15, Section 15.2.2), which opens avenues for further research and optimization in this area. Such biologically inspired delivery technologies also offer the advantages of biocompatibility and biodegradability, which positively impact on subsequent clinical translation and FDA regulatory approval. This is in contrast to the use of synthetic nanocarriers, which can be associated with toxicity and biocompatibility issues (see Section 24.4.5). Synthetic nanocarriers are also susceptible to recognition and capture by the organs of the RES, typically resulting in high carrier uptake by the liver, which compromises targeting to other sites.

24.2.3.5 Imaging and Theranostics

A promising future direction for nanotechnology is in the field of theranostics, which is the subject of Chapter 18. The field has evolved in recent years and now is focused on the simultaneous delivery of both a therapeutic drug and a diagnostic agent, within a single multifunctional nanoparticulate platform. Although the field offers a promising future direction for nanotechnology and the development of personalized medicine, the limitations of the approach also need to be clarified, to help guide future research. The premise of using a single NP construct is based on the concept that both imaging agent and drug remain within the nanocarrier and can be followed concurrently. However, it has not been confirmed that both imaging agent and drug reach the target tumor simultaneously— both can be released before reaching the target, resulting in different biodistributions in the body (Hollis et al. 2013). The differing biodistributions of the nanocarrier, free drug, and free probes need to be resolved, before the potential of this application can be realized.

Again, this might be an area that could benefit from adopting a less complex strategy—for example, focusing on the use of nanocarriers as imaging agents *per se*, rather than as dual diagnostic and therapeutic platforms. This in itself would be a vital service—nanocarriers could play an important role in ex vivo tissue analysis, for the detection of early cancer and the profiling of molecular biomarkers. Many radionuclide-loaded nanocarriers, as well as antibody–radionuclide conjugates, are demonstrating their superiority for nuclear imaging techniques (Lammers et al. 2012; see also Chapter 5, Section 5.4.1). Development of advanced imaging nanotechnologies should better assist oncologists in the early diagnosis of cancer and in the early detection of metastases, and also provide guidance on when is the most appropriate time to stop therapy.

24.3   OVERCOMING THE EPITHELIAL BARRIER

The noninvasive delivery of biologics remains one of drug delivery's most elusive and sought-after goals, that necessitates overcoming the epithelial barrier. As described in Chapter 4, the protective barrier function of epithelial interfaces is essential to our survival but also presents a formidable challenge to the entry of drugs and DDS. An understanding of the nature of the epithelial barrier then, is crucial to the success of achieving transepithelial transport. For this reason, an entire chapter of this book (Chapter 4) is dedicated to understanding epithelial barriers in general. Additionally, each chapter of Section 3 (nonparenteral routes) begins with a detailed consideration of the relevant anatomical and physiological barriers pertaining specifically to the route in question, as well as the implications therein to successful drug delivery and targeting via this route.

In particular, researchers joining the drug delivery field from nonbiology-based backgrounds might do well to consider these issues very carefully. Sophisticated nanoengineering of DDS cannot translate into clinical success without an appreciation of the relevant anatomical and physiological challenges (Figure 24.1). It is worth briefly summarizing some of the challenges of transepithelial transport here, to fully appreciate the challenges ahead (see also Chapter 4).

1. *Access barriers:* There are considerable problems associated with gaining access to the absorbing surface of an epithelial interface. For example, for the oral route, a DDS must negotiate through chyme, intestinal fluids, pancreatic secretions, bile salts, and sloughed off intestinal cells before reaching the absorbing surface. The enterocytes themselves are coated with a mucus layer, about 500 μm thick. Access is further compromised by the relatively short, and highly variable, intestinal transport time. As described earlier, targeting moieties on the DDS surface do not *deliver* the nanocarrier to the enterocyte—as always, targeting is only relevant if the ligand system can gain proximity to the enterocyte surface, which, given all the access barriers present, is a formidable task.

2. *Large hydrophilic molecules do not passively diffuse through epithelial membranes:* The physicochemical properties of a drug molecule that favor epithelial transport are described in Chapter 4. To summarize here, in order to be absorbed via transcellular passive diffusion, a drug molecule must have a low molecular weight, be lipophilic and uncharged; it must also demonstrate some aqueous solubility and be metabolically stable. These properties are virtually the antithesis of the physicochemical properties of the new biologics that are coming through drug development programs, which means these agents demonstrate very poor epithelial transport. Although small amounts (<0.1%) of protein or peptide drugs can sometimes be absorbed orally in an intact form, this is likely to occur through the Peyer's patches, frank lesions, and senescent cell replacement events that are part of normal gut physiology.

3. *Carrier-mediated transport is restricted to small molecules:* Carrier-mediated transport is a possible alternative means of transepithelial absorption for biologics. But typically, carrier-mediated transport is confined to small molecules (see also Chapter 4, Figure 4.4). For example, in the GI tract, nutrient transporters have evolved for the absorption of the digested products of protein nutrients, i.e., amino acids and di- and tripeptides. Most peptide and protein drugs are much too large for uptake via such transporters.

4. *Increasing the contact time at the absorption surface does not automatically improve absorption:* Merely increasing the contact time of a protein drug (e.g., insulin) with the absorbing surface (for example, by incorporating a mucoadhesive or incorporating nanotopography into the DDS) does not automatically mean that the protein will be absorbed in a bioactive form. No matter how long the contact time with the absorbing surface has been extended, the fundamental characteristics of the biologic remain the same—it is still a large, charged, and labile macromolecule that demonstrates poor transepithelial transport.

Maintaining the drug at the absorption surface for prolonged periods will not change these fundamental properties and in order to improve absorption, additional strategies will also be required (e.g., via the inclusion of an absorption enhancer, enzyme inhibitor, etc.).

24.3.1   NANOTECHNOLOGY AND EPITHELIAL TRANSPORT

Micro- and nanoparticulate DDS have been intensively researched and developed for the last two decades in attempts to enhance the oral bioavailability of peptide and protein drugs, in particular, insulin. A wide variety of targeting ligands, including lectins, sugars, and vitamin B12, have been conjugated to NPs, in order to facilitate uptake by enterocytes. However, as concluded in Chapter 7, real progress remains disappointingly slow—nanoparticulate DDS are all still at an early stage of development and have not shown a significant enhancement of oral bioavailability thus far. Reasons for the slow progress include, as described earlier, difficulties in accessing the enterocyte surface, the limited targeting opportunities that are possible, and the constraints conferred by the physico-chemical properties of the drug. A further important consideration is that nanoparticulate DDS will typically be taken up across epithelia by endosomal vesicular transport (see also Chapter 4, Section 4.3.3). Endosomes are generated by a variety of receptor and nonreceptor mechanisms. At the apical surface of enterocytes, these are typically clathrin mediated and result in endosomes of ≈120–150 nm in diameter. Thus, particles that do not fit into this size of structure will not be efficiently taken into the cell. But even for those nanocarriers that do succeed in being internalized, the endosomes will then typically fuse with lysosomes, and the endosomal contents are destroyed, rather than transported across the cell and into the body.

24.3.2   SUCCESSFUL STRATEGIES FOR EPITHELIAL TRANSPORT

In the light of these challenges, it is worth highlighting strategies that are showing promise in overcoming the epithelial barrier, as a guide for future research. As described in Chapter 7, formulations that have thus far proven successful in improving oral bioavailability are typically relatively simple formulations, that include formulation excipients such as an absorption enhancer and an enzyme inhibitor, incorporated in an enteric-coated tablet or capsule.

A variety of mechanisms have been successfully developed to improve the solubility and dissolution rate of poorly soluble drugs, which can have the knock-on effect of improving transepithelial absorption (as described in Chapter 3). Strategies such as the development of NanoCrystals® and amorphous solid dispersions (ASDs) can be considered a real success story in the field. In fact, solubility enhancement technologies, which constituted a mere paragraph in the first edition of this book, have now matured sufficiently to warrant a dedicated chapter in this edition.

The emergence of alternative epithelial routes to the oral one for systemic absorption is another success story in the field. As described in the preceding chapters, the buccal, transdermal, nasal, and pulmonary routes all have licensed products for systemic delivery. Alternative epithelial sites to the GI tract offer fewer difficulties for systemic absorption, as they feature favorable conditions such as a relatively lower metabolic activity; in some cases, higher permeability; and the avoidance of hepatic first-pass metabolism. These routes also offer ease of access, ease of administration, prolonged retention, and the potential for controlled release from long-acting devices (e.g., transdermal patches, buccal films, intravaginal rings [IVRs]). Mechanisms are also being developed that actively force drug molecules across the epithelium, via the application of current or ultrasound, as for example in transdermal and ophthalmic delivery. Other mechanisms are used to physically breach the epithelial barrier, for example, using long-acting injections (LAIs) and implants, microneedles arrays, and gene guns.

The successful methods of achieving transepithelial transport described in this book all have in common that they integrate device and formulation optimization with close attention to both the physicochemical characteristics of the drug and the prevailing environmental conditions of the route.

24.4   FUTURE DIRECTIONS FOR DRUG DELIVERY

The future directions in the field are described in detail in the relevant chapters of this book. Moving forward, the following points are also worth considering.

24.4.1   TARGETED DRUG DELIVERY

The rather disappointing truth is that our ability to achieve drug "targeting" by nanocarriers, or any other type of carriers, is currently very limited. There is no still no "magic bullet," capable of delivering a drug directly to its target. Rather, as outlined earlier (Section 24.2), both passive and active targeting strategies are associated with many short-comings and weaknesses. In spite of this, the term "targeting" is widely used in the context of drug delivery, perhaps creating a somewhat misleading impression that the concept is already highly successful. There is a risk of hubris hampering meaningful progress. Moving forward, the rhetoric surrounding the issue needs to be scaled back, to more realistically reflect the reality of the situation. For this reason, we have removed the word "Targeting" from the title of this edition of the book, reverting from "Drug Delivery and Targeting," to the more representative title, "Drug Delivery."

A variety of suggestions have been made above to address targeting problems, such as researching more specific *niche* areas, rather than looking for panaceas. Local targeting has been more successful and various new technologies have emerged to improve local delivery. These technologies are described in the relevant chapters and include improved nasal delivery devices, to target the posterior nasal cavity; improved pulmonary technologies, to facilitate delivery to the alveoli for systemic absorption; the use of microneedle arrays to facilitate transdermal and transscleral delivery; and microelectronic drug delivery for programmed targeting to specific regions of the GI tract.

24.4.2   CONTROLLED RELEASE

Many parenteral depot formulations are characterized by an initial burst release, resulting in an initial peak blood concentration much larger than the therapeutically effective concentration. As technology moves toward controlled release of drugs for longer duration times (up to 1 year or longer), research, as described in Chapter 6, is focusing on the need to more precisely control the release kinetics and avoid burst effects, as well as to improve host responses.

The Section IV: Emerging Technologies part of the book describes increasingly sophisticated methods currently under development to effect highly precise control over drug release from DDS. Stimuli-sensitive systems (Chapter 14) allow drug carriers to release their contents upon exposure to external stimuli, such as heat, light, ultrasound, and magnetic fields; or internal triggers, for example, near infected or inflamed tissues that have altered high temperature, acidic or alkaline pH, high reactive oxygen species levels, or high glutathione levels. Chapter 23 describes the heat-sensitive release of Dox from Thermodox®. Many controlled-release systems are being developed by leveraging techniques from the microelectronics industry (Chapter 19), which has led to the development of, for example, microelectromechanical systems (MEMS; Chapter 14) and technologies such as the InteliSite® and IntelliCap® systems, which allow electronically programmed drug release in the GI tract (Chapter 7).

Such great strides in controlled drug release have many applications. One, of great importance in the field, is the development of a glucose-sensitive transient insulin delivery device, with on–off switching capability. There are many hurdles to be overcome in the development of such a device (Park 2014), but increasingly sophisticated technology and know-how is being applied to the problem, so that this objective is becoming a near-term reality (see also Chapter 14, Section 14.4.2).

24.4.3   Multidrug Regimens

Another future direction for the field is toward improving compliance and efficacy via the development of multidrug DDS. The potential of nanotechnologies as carriers for multiple drug combinatons in cancer therapy has been outlined in Section 24.2.3.3. For pulmonary drug delivery, combination Drug-Aerosol products (containing drugs from different therapeutic classes within the same aerosol) are the top-selling aerosol products worldwide (Chapter 11). Drug delivery to the female reproductive tract is moving toward "multipurpose prevention technologies (MPT)," i.e., biomedical interventions designed to simultaneously address multiple reproductive health needs. For example, segmented dual-reservoir design IVRs can carry both an antiviral agent and a contraceptive; many other systems are under study (Chapter 12). Three-dimensional (3D) extrusion printing has recently been used to combine complex medication regimes into a single personalized tablet, or "polypill," for oral delivery (Khaled et al. 2015). This multiactive solid dosage form contained five compartmentalized drugs, with two independently controlled-release profiles.

24.4.4   Personalized and Precision Medicines

There will be increasing emphasis placed on personalized medicine, to provide customized treatment to patients. In the treatment of cancer and other diseases, a patient's genomic data can be used to identify the best treatment mode, known as precision medicine. There are great opportunities, and challenges, for drug delivery science to provide better, more personalized, and precise treatments and services for individual patients. The emerging field of Theranostics (another entirely new subject area for this edition of the book), and the promise it holds for the development of personalized medicine, is described in Chapter 18. An interesting new development is the alignment of the field of information technology (which is being used to collect and analyze patient health data) with the pharmaceutical and health-care fields, to establish a broad foundation for personalized medicines.

24.4.5   Regulatory Issues

A new DDS must be approved by the relevant regulatory body in order to be marketed. It should be stressed that regulatory agencies are focused on ensuring the "safety and efficacy" of drug products (rather than the ingenuity and sophistication of a particular nanocarrier design). For DDS that use complex formulation designs and manufacturing processes, the task of ensuring consistency in safety and efficacy is much more challenging than for traditional simple dosage forms (Wen et al. 2015).

New DDS require extensive preclinical and clinical studies in order to gain regulatory approval. Chapter 21 describes the plethora of extra studies necessary for the regulatory approval of injectable nanomedicines such as liposomal/NP products and colloidal iron/gold conjugate products—far more than the basic product characterization criteria required for conventional parenteral products. Chapter 6 describes the issues of host response/biocompatibility and immunogenicity, and the extensive testing necessary to gain regulatory approval, for implants and long acting injections (LAIs).

Potentially promising preclinical research may have to be discontinued because of subsequent stringent regulatory demands. In order to achieve successful regulatory approval and clinical translation, preclinical drug delivery research needs to be more cognizant of potential regulatory hurdles further down the line and be more concerned with safety and toxicology issues. For example, the field has seen the recent introduction of a diverse array of nanoscale materials and processes, such as carbon nanotubes, fullerene derivatives, and quantum dots. The hazard potential and biocompatibility of many of these new synthetic nanocarriers are important concerns, as they may involve agents that are toxic, carcinogenic, have long half-lives (of the order of decades), and show widespread tissue distribution in vivo (Hardman 2006; see also Chapter 18). Again, rather than prioritizing the design of evermore sophisticated and complex nanostructures based on these materials, research should first address these inherent toxicological and biocompatibility problems.

Similarly, in considering the development of absorption enhancers for transepithelial permeation, it should be remembered that getting regulatory approval for a new absorption enhancer can be as formidable as getting approval for a new chemical entity (NCE). The required safety studies are costly and time consuming, and there is reluctance in the pharmaceutical industry to fund trials for what may have originally been intended as merely an inexpensive formulation additive.

Another hurdle: if a drug is chemically conjugated to a DDS, regulatory agencies will consider that the conjugated drug is a NCE; as such, clinical trials will again be required. Similarly, a prodrug is considered to be an NCE by regulatory bodies, even though it is bioreversible to the original drug (FDA Guidance for Industry 1999).

As drug delivery technologies increase in sophistication, the concomitant regulatory challenges become increasingly complex. For example, in the case of the "polypill" described in Section 24.4.3, each 3D-printing device that produces a personalized polypill could in principle be treated as an independent manufacturing machine, making regulatory approval highly complicated. A close collaboration between industry, academia, and regulatory agencies is therefore necessary to address these complex challenges and allow drug delivery science to progress, while also ensuring patient safety.

24.4.6   A Need for Improved In Vitro/In Vivo Testing

Given the increasing complexity of DDS formulations, as well the variability, heterogeneity, and complexity of the biological milieu that they must operate under, there is definite need to develop better in vitro and in vivo tests, to allow both the accurate assessment, and prediction, of drug activity and toxicity. A particular problem is that even though a nanocarrier may be assiduously characterized in vitro, as soon it is injected, it will be cloaked immediately with a "corona" of plasma proteins, which will fundamentally change the nature of the carrier, including its size, surface properties, stability, and other properties. For this reason, the surface of nanocarriers is frequently PEGylated for extended circulation in the blood, although the overall impact on targeted drug delivery has been modest at best.

As noted in Section 24.2.1, cell lines and the majority of animal models currently used to test nanotechnology DDS are poor predictors of in vivo performance. For tumor-targeting research efforts, it has been proposed to improve the current situation by establishing a well-defined, standardized, panel of animal models, which should include, among others, an orthotopic, metastatic, and transgenic model (Lammers 2013). Having a "gold standard" panel would allow improved testing of novel DDS formulations and provide more predictive analysis of in vivo performance. It would also allow direct head-to-head comparisons of new formulations against other DDS and against the standard therapy. A further interesting possibility is the recently described in vitro tumor model, "tumor-microenvironment-on-chip" (T-MOC) (Kwak et al. 2014). The model comprises 3D microfluidic channels in which tumor cells and endothelial cells are cultured within an extracellular matrix perfused by interstitial fluid. The T-MOC platform has been shown capable of simulating the complex microenvironment around the tumor and allows nanocarrier transport behavior to be studied while systematically and independently varying tumor microenvironmental parameters, such as cutoff pore size, interstitial fluid pressure, and tumor tissue microstructure.

The need for better in vitro and in vivo predictive models is not confined to nanotechnology for cancer research. In practically every chapter of this book, across the entire spectrum of drug delivery research, authors have expressed their concern about the limitations of current testing procedures and stressed the need for improved methodologies.

The development of more predictive in vitro and in vivo models seems an area that would be ideally suited to widespread collaboration within the pharmaceutical industry. Although pharma companies may be reluctant to collaborate on developing new drug leads for fear of loss of competitive advantage, the pooling of resources to develop improved testing procedures and improved in vitro-in vivo correlation would offer significant advantages to all.

24.4.7 A NEED FOR COLLABORATIVE, TRANSPARENT RESEARCH

A striking feature of drug delivery science is the breadth and range of the field, which encompasses pharmaceutics, polymer chemistry, biology, physiology, pathology, physics, immunology, oncology, medicine, and engineering. Overcoming the considerable challenges associated with drug delivery requires a concerted multifaceted effort by scientists from all of these disciplines.

Cross-disciplinary collaboration allows the input of fresh ideas and enables new ways of approaching old problems. Reimagining content in a new context may help drug delivery scientists to think outside the box and even facilitate the kind of creative leap necessary to "think in new boxes" (de Brabendere and Iny 2013). Recent examples of the benefits of cross-fertilization from other disciplines include the leverage of techniques from the microelectronics industry, to precisely fabricate DDS in the nanometer range (Chapter 19); the use of reverse-engineering design to generate liposomal DDS (Chapter 23); the merging of imaging technologies with DDS, to develop the field of theranostics (Chapter 18); the merging of vaccinology with DDS, to develop nanoparticulate vaccine delivery systems (Chapter 17); and the use of cell biology to understand physiological uptake processes in the BBB, to develop drug delivery strategies to the CNS (Chapter 15).

Progress in drug delivery technology, as with all true progress, occurs via trial and error. The key to moving forward in drug delivery is to try many approaches and test them out—those that work can be further improved. But there is also much to be gained by studying what did not work. At the risk of sounding like a motivational wall poster: "There is no failure, only feedback." Experiments are always an opportunity to learn, adjust, and try again. A careful consideration of why things did not turn out as expected can be the springboard to greater success. It can also save a lot of time, effort, money, and needless repetition, if negative results are shared with the scientific community.

Unfortunately, there are currently extremely limited opportunities for scientists to present negative data, thus this valuable, indeed vital, information is lost. If scientists were able to be more open and transparent about negative findings, progress would proceed far more rapidly than at the current pace. The authors strongly agree with the recent recommendations to improve the reliability of preclinical cancer studies (Begley and Ellis 2012):

> There must be more opportunities to present negative data. It should be the expectation that negative preclinical data will be presented at conferences and in publications. Preclinical investigators should be required to report all findings, regardless of the outcome. To facilitate this, funding agencies, reviewers and journal editors must agree that negative data can be just as informative as positive data.

This obviously requires a paradigm shift in current thinking on how research is conducted, at all levels of scientific research, from government funding agencies, right through pharmaceutical industry, academia, and research institutions. Such changes will not happen overnight, although there are already encouraging indicators that the tide is turning. For example, the Journal of Controlled Release (JCR) recently outlined a new editorial policy for the journal (Park 2015):

> Understanding why a certain formulation failed in animal studies and/or in clinical studies is more important than seeing another me-too data. Sometimes authors find that a widely used experimental method is misleading, and describing such information is highly valuable despite the negative data. Presenting a new insight into a presumed, and widely accepted, theory, is critical for the progress of the field, especially when the theory has not been able to explain the results in clinical studies. In short, the JCR welcomes those data that do not conform to the widely accepted dogma, proven by carefully designed experiments.

A further practical solution, easily implemented, would be to create an Internet forum for drug delivery, where unsuccessful preclinical experimental data could be freely exchanged, discussed, and analyzed. The site would need careful monitoring by moderators who are familiar with the specific topics under discussion: a lot of information is not necessarily useful information. But a carefully moderated forum, sharing negative and inconclusive results, would be of great service to the drug delivery community—and would also serve as a best practice model for scientific research in general. In particular, a collaborative Internet resource seems ideally suited to the new generation of digital-native millennials who are entering drug delivery research. Having an open forum would also serve to encourage dialog and help to forge the type of multifaceted, cross-disciplinary collaboration that is crucial for future progress in the field.

This kind of collaborative, open-forum approach is steadily gaining traction within the scientific community. An interesting recent example of an open-source model for pharmaceutical R&D is the Project Data Sphere (www.projectdatasphere.org). Developed in order to accelerate oncology drug discovery and development, it is an online resource for sharing clinical trial data. The technology platform, built and maintained by business analytics software company SAS, shows how historical cancer research data can be shared, integrated, and analyzed, bringing together diverse groups with common interests, including researchers from academia, industry, hospitals, and institutions. Additionally, Chapter 20 describes many public, web-accessible databases, for drug discovery purposes. A further possibility has been described in Chapter 23, with the idea of "translating drug delivery without profit: open source pharmaceuticals," i.e., making effective drug delivery formulations open source and freely available.

24.4.8 FINAL WORDS

In conclusion, successful progress in the field is contingent on a number of factors, including

- A full appreciation of the challenges (formulation and biological) that need to be overcome
- The use of a rational design strategy for the development of a new DDS
- An increased collaboration between complimentary disciplines
- A concerted effort to develop more appropriate in vitro and in vivo testing methodologies
- A more open and transparent reporting of preclinical research, including the publication of negative data

Considerable progress has been made in the field and, in spite of the formidable challenges, many new approaches and avenues of research have opened up. The perseverance, verve and ingenuity that has brought us from the Spansule® "tiny little time pills" of the 1950s to the sophisticated nanotechnologies of today continues to drive research forward and assures the development of safe and effective DDS for the future.

REFERENCES

Bae, Y.H. and K. Park. 2011. Targeted drug delivery to tumors: Myths, reality and possibility. *J. Control. Release* 153(3):198–205.
Barenholz, Y. 2012. Doxil®—The first FDA-approved nanodrug: Lessons learned. *J. Control. Release* 160:117–134.
Begley, C.G. and L.M. Ellis. 2012. Drug development: Raise standards for preclinical cancer research. *Nature* 483:531–533.
Crommelin, D.J. and A.T. Florence. 2013. Towards more effective advanced drug delivery systems. *Int. J. Pharm.* 454(1):496–511.
de Brabandere, L. and A. Iny. 2013. *Thinking in New Boxes: A New Paradigm for Business Creativity.* New York: Random House.

FDA Guidance for Industry. Applications covered by section 505(b)(2). 1999. Available from: http://www.fda.gov/downloads/Drugs/Guidances/ucm079345.pdf, accessed March 17, 2016.

Gordon, A.N., D. Guthrie, D.E. Parkin et al. 2001. Recurrent epithelial ovarian carcinoma: A randomized phase III study of PEGylated liposomal doxorubicin versus topotecan. *J. Clin. Oncol.* 19(14):3312–3322.

Hardman, R. 2006. A toxicologic review of quantum dots: Toxicity depends on physicochemical and environmental factors. *Environ. Health Perspect.* 114(2):165–172.

Harrington, K.J., S. Mohammadtaghi, P.S. Uster et al. 2001. Effective targeting of solid tumors in patients with locally advanced cancers by radiolabeled PEGylated liposomes. *Clin. Cancer Res.* 7(2):243–254.

Harrington, K.J., K.N. Syrigos, and R.G. Vile. 2002 December. Liposomally targeted cytotoxic drugs for the treatment of cancer. *J. Pharm. Pharmacol.* 54(12):1573–1600.

Hillery, A.M., A.W. Lloyd, and J. Swarbrick. 2001. *Drug Delivery and Targeting: For Pharmacists and Pharmaceutical Scientists.* Boca Raton, FL: CRC Press.

Hollis, C.P., H.L. Weiss, M. Leggas et al. 2013. Biodistribution and bioimaging studies of hybrid paclitaxel nanocrystals: Lessons learned of the enhanced permeability and retention (EPR) effect and image-guided drug delivery. *J. Control. Release* 172:12–21.

Hu, C.-M.J., R.H. Fang, K.-C. Wang et al. 2015. Nanoparticle biointerfacing by platelet membrane cloaking. *Nature* 526:118–121.

Khaled, S.A., J.C. Burley, M.R. Alexander et al. 2015. 3D printing of five-in-one dose combination polypill with defined immediate and sustained release profiles. *J. Control. Release* 217:308–314.

Kwak, B., A. Ozcelikkale, C.S. Shin et al. 2014. Simulation of complex transport of nanoparticles around a tumor using tumor-microenvironment-on-chip. *J. Control. Release* 194:157–167.

Lammers, T. 2013. SMART drug delivery systems: Back to the future vs. clinical reality. *Int. J. Pharm.* 454(1):527–529.

Lammers, T., F. Kiessling, W.E. Hennink et al. 2012. Drug targeting to tumors: Principles, pitfalls and (pre-) clinical progress. *J. Control. Release* 161:175–187.

Maeda, H., M. Ueda, T. Morinaga et al. 1985. Conjugation of poly(styrene-co-maleic acid) derivatives to the antitumor protein-neocarzinostatin: Pronounced improvements in pharmacological properties. *J. Med. Chem.* 28:455–461.

Matsumura, Y. and H. Maeda. 1986. A new concept for macromolecular therapeutics in cancer chemotherapy: Mechanism of tumoritropic accumulation of proteins and the antitumor agent SMANCS. *Cancer Res.* 46:6387–6392.

Nanotechnology Safety and Health Program. National Institute of Health, Technical Assistant Branch. 2014. http://www.ors.od.nih.gov/sr/dohs/Documents/Nanotechnology%20Safety%20and%20Health%20Program.pdf, accessed March 17, 2016.

Nichols, J.W. and Y.H. Bae. 2014. EPR: Evidence and fallacy. *J. Control. Release* 190:451–464.

Park, K. 2014. Controlled drug delivery systems: Past forward and future back. *J. Control. Release* 190:3–8.

Park, K. 2015. Editorial: The state of the journal. *J. Control. Release* 215:A1–A2.

Prabhakar, U., H. Maeda, R.K. Jain et al. 2013. Challenges and key considerations of the enhanced permeability and retention (EPR) effect for nanomedicine drug delivery in oncology. *Cancer Res.* 73:2412–2417.

Raemdonck, K. and S.C. de Smedt. 2015. Lessons in simplicity that should shape the future of drug delivery. *Nat. Biotechnol.* 33:1026–1027.

Stirland, D.L., J.W. Nichols, S. Miura et al. 2013. Mind the gap: A survey of how cancer drug carriers are susceptible to the gap between research and practice. *J. Control. Release* 172:1045–1064.

Wang, Y.-S., H. Xia, C. Lv et al. 2015. Self-propelled micromotors based on Au–mesoporous silica nanorods. *Nanoscale* 7:11951–11955.

Wen, H., J. Huijeong, and X. Li. 2015. Drug delivery approaches in addressing clinical pharmacology-related issues: Opportunities and challenges. *AAPS J.* (Spec. issue *Clinical and Commercial Translation of Drug Delivery Systems*) 17(6):1327–1340.

Yun, Y.H., B.K. Lee, and K. Park. 2015. Controlled drug delivery: Historical perspective for the next generation. *J. Control. Release* 219:2–7.

Index